MATHEMATICAL MODELING AND APPLIED CALCULUS

Mathematical Modeling and Applied Calculus

JOEL KILTY AND ALEX M. MCALLISTER

OXFORD
UNIVERSITY PRESS

OXFORD

UNIVERSITY PRESS

Great Clarendon Street, Oxford, OX2 6DP,
United Kingdom

Oxford University Press is a department of the University of Oxford.
It furthers the University's objective of excellence in research, scholarship,
and education by publishing worldwide. Oxford is a registered trade mark of
Oxford University Press in the UK and in certain other countries

Published in the United States of America by Oxford University Press
198 Madison Avenue, New York, NY 10016, United States of America

British Library Cataloguing in Publication Data

Data available

Library of Congress Control Number: 2018940139

ISBN 978–0–19–882472–5 (hbk.)
ISBN 978–0–19–882473–2 (pbk.)

For Alexander and Harrison
and
For Benjamin, Daniel, and Ella

Contents

Preface

Human beings have an innate desire to understand reality. From the moment we are born, we touch, we listen, we look, we taste, we smell, and we process this sensory information in an effort to make sense of our world. For millenia now, humans have recognized that mathematics serves as an impressively effective tool for understanding many aspects of reality. This book, *Mathematical Modeling and Applied Calculus* (MMAC), explores some of the most important elements of such mathematics by introducing the reader to mathematical modeling and provides a set of tools for analyzing these mathematical models by means of calculus.

Our driving motivation for writing this book is to create a learning experience that introduces the mathematical content students need for allied disciplines in the contexts they will encounter in those disciplines. Many traditional introductory courses in mathematics are oriented toward supporting students who intend to major in mathematics, engineering, physics, or chemistry. However, students who intend to major in other disciplines, such as biology, economics, psychology, or sociology, genuinely benefit from an introduction to the process of mathematical modeling and to the tools of calculus as a means to analyze models. These students often have different needs than just a simplified version of a traditional Calculus I course, and this book is designed to meet these needs. Throughout MMAC, the study of mathematical ideas takes place in real-world contexts. In the first half of this book, we study various mathematical functions and the process of mathematical modeling by creating models for diverse data sets, both large and small. This integrated approach allows us to accomplish these two goals simultaneously, and students should feel as though they are learning new ideas, rather than just reviewing the same old material from previous courses. In the second half of the book, as we study the tools of calculus, we constantly revisit the models that were created in the first half of the book, especially when we are learning the basic rules of differentiation and integration. The study of more sophisticated differentiation and integration techniques includes the introduction of more sophisticated models of real-life phenomena. By embedding the development of mathematical content in specific contexts, we integrate the study of mathematical ideas with the application of these ideas.

While the primary audience of MMAC is students enrolled in introductory collegiate courses, we do not shy away from introducing more advanced topics. For example, multivariable functions are introduced in the first chapter, and techniques for multivariable functions are introduced alongside their single-variable counterparts. We have gone to great lengths to introduce these ideas at a level that is appropriate to the audience and have field-tested our approach in multiple classes, fine-tuning our approach. Similarly, in our treatment of the method of least squares, we develop the techniques from courses in Linear Algebra and Calculus III needed to introduce the method, as well as explaining why it works. Our study of these ideas is accessible to an introductory audience and does not require the mathematical sophistication involved in the typical study of the method of least squares in Linear Algebra. Finally, we regularly provide explanations for why certain mathematical statements are true at appropriate points in the development of these ideas. We often present mathematical justifications at the end of a section or in optional sections. This approach enables students first to master the procedural aspects of these ideas and then to focus on the often more difficult process of understanding why these ideas are true.

As an example, we chose to postpone the study of limits until after the study of derivatives, which might strike some users as unusual. This choice is consistent with our desire for students to first master the procedural and conceptual aspects of the derivative, which led us to present the more theoretical underpinning of limits in an optional section at the end of Chapter 4. This section could easily be presented before derivatives and covers all of this topic's major ideas, including the definition of the limit, graphically and numerically estimating limits, algebraic techniques for computing limits, continuity, and the limit definition of the derivative. Our field tests of this approach have been quite successful with the intended audience and have shown that it improves student understanding of the idea of the derivative without side effects in subsequent courses.

On a related note, we have made choices in defining terms that are not always the more rigorous or complete and, similarly, the hypotheses of some theorems are not always the most general. These choices were made very thoughtfully and with particular care with respect to what would be the most successful approach for our intended audience, while also remaining correct within the context of the most common modeling functions. This approach works well for these students, keeps them engaged, and helps them develop informed, if still sometimes novice, understandings. As these students persist in taking a next mathematics course, they will encounter these more complete world-views of the mathematician and are well prepared to acquire more nuanced understandings.

In summary, this book enables a rich learning experience for students, accomplishing the study of sophisticated mathematical ideas and methods in real-world contexts and at an appropriate level of challenge.

Overview of the Book

The first half of this book focuses on the basics of modeling: the most commonly used functional models, the process of identifying both reasonable models for given data sets and the parameters of such models, the basics of dimensional analysis, and the method of least squares as a means for finding best possible models when an exact model does not exist. In Chapter 1, we review basic properties of the functions that are most commonly used as mathematical models of data. Many of these functions will be familiar, including linear, exponential, logarithmic, sine, and sigmoidal models. This chapter also provides an introduction to multivariable functions. In Chapter 2, we study and implement the modeling cycle. Namely, for a given data set, we learn analytic, graphical, and numerical approaches to identifying reasonable models and their parameters, as well as other tools of mathematical modeling, including dimensional analysis. Chapter 3 details the aspects of linear algebra necessary for understanding and implementing the method

of least squares, including vectors, linear combinations, matrices, matrix multiplication, projections, and residual vectors. This chapter culminates in the application of the method of least squares to linear data sets.

The second half of this book explores the calculus as a collection of tools that enable a more thorough analysis of mathematical models. In Chapter 4, we introduce the derivative as the value approached by the average rate of change of a function, when this average rate is calculated over smaller and smaller intervals. We develop the standard differentiation rules for the basic modeling functions and their various combinations, including sum, product, quotient, and chain rules. We also study partial differentiation to enable the analysis of multivariable models. In Chapter 5, we study single- and multivariable optimization, including both global and local optimization as well as multivariable and constrained multivariable optimization. In Chapter 6, we introduce the definite integral as a means for measuring the accumulation of quantities. We study left and right approximations of the definite integral, and then the first and second fundamental theorems of calculus as a means to exactly evaluate definite integrals of the common modeling functions and their combinations. Chapter 6 concludes with the two most important techniques of integration: the method of substitution and integration by parts.

Pedagogical Features

Throughout the process of writing this book, our guiding principle has been: "How can we share these ideas so as to best enable effective learning and teaching?" We were certainly informed by our own experiences as teachers and students, but also by countless conversations with colleagues and students. Motivated by our goal to enable effective teaching, we incorporate various special features into this book:

- **Real-life data and examples** are incorporated throughout the book and are drawn from multiple fields of inquiry, including economics, medicine, biology, psychology, sociology, and more.

- **Examples with justification** alongside each step of a calculation help students to follow each of the calculations presented.

- **Embedded questions** for immediate application of ideas and methods as they are introduced. When teaching from this book, we use many of these questions as in-class exercises. Alternatively, when we teach a flipped class, we assign these questions for homework.

- **Answers to embedded questions** are provided in Appendix A to provide feedback and help solidify ideas. More than just answers, these solutions provide key intermediate steps to facilitate student learning.

- **Exercise sets** are thorough and quite extensive, with 75 to 100 exercises at the end of every section. These exercises have been crafted to provide a spectrum of practice opportunities, from the straightforward to the more challenging. We also include exercises with distinctive features. These include the following:

  ○ *Your Turn* exercises, which ask for students to create their own examples and questions.

- ○ *In Your Own Words* exercises, which ask for students to explain important mathematical ideas and methods.

- ○ *RStudio* exercises, which ask for students to use RStudio, or a similar software package, to implement mathematical algorithms, model data sets, and analyze such models.

- **Section summaries** provide a focused, condensed outline of the main ideas from each section.

- **Answers to the odd-numbered exercises** are provided in Appendix B to provide immediate feedback to students about the accuracy of their solutions. Again, some of these answers are left relatively unsimplified to facilitate student learning of the mathematical ideas and processes that are being studied.

- **RStudio commands** are introduced at appropriate points throughout the book to enable the development and application of mathematical models and methods. R is the standard statistical software package used by academic and professional statisticians, and includes powerful tools for modeling and analyzing data sets. RStudio is an open source, freeware software package that provides an integrated interface for using R. In addition, the MOSAIC project team has developed an extensive library of add-ons, many of which enable the successful study of this book's focus: mathematical modeling and applied calculus. The data sets used in the book are all available in the `MMAC` package written especially for this book. Additionally, the `mosaic`, `mosaicCalc`, and `manipulate` packages are needed throughout the book. As we introduce the ideas of modeling, we detail how to use RStudio to carry out these same methods. Embedded questions and exercises with a specific technological focus are included throughout the book. Appendix C provides information about accessing and starting to use these free, open source software packages.

Course Designs

This book has been written to support the teaching of a variety of different introductory mathematics courses. For the most part, the first three chapters of the book do not depend on each other and can be studied independently. Chapter 4 on derivatives, Chapter 5 on optimization, and Chapter 6 on accumulation and integrals build upon themselves, as is typical in the study of calculus. However, Chapter 6 does not rely on Chapter 5.

Depending on course goals, audience, and pace, we envision at least four likely courses for which this book would be an excellent choice:

- The primary, intended audience is students taking a course in Mathematical Modeling and Applied Calculus. These students need an introduction to the main ideas of calculus, but may not intend to enroll in the traditional calculus sequence. This course would focus on Chapter 2 through Chapter 6, and the mostly precalculus topics in Chapter 1 can be utilized as more of a reference or as a just in time review.

- A second audience is students taking a precalculus course, which develops a careful study of functions along with some attention to modeling. This course would focus heavily on Chapters 1 and 2, and perhaps include some content from Chapter 3. The study of the method of least squares in Chapter 3 does not utilize the ideas of calculus studied later in this book and so can be readily studied in such a precalculus course.

- A third audience is students taking a one-semester course in Calculus or Applied Calculus with some review of functions. This course would consist of a thorough treatment of Chapter 4 through Chapter 6, with periodic references back to Chapter 1 as needed for a just-in-time review of precalculus topics.

- A fourth possibility is students taking a lower-level undergraduate Mathematical Modeling course. This course would use Chapter 1 through Chapter 3, and select portions of Chapter 4 through Chapter 6 depending on instructor preferences. Such a course would have a greater focus on Chapters 2 and 3 than a precalculus course, as well as probably including some of the calculus topics from Chapter 4 through Chapter 6 for analyzing the mathematical models developed in Chapters 2 and 3.

Acknowledgments

This book was only possible with the help and encouragement of all of our colleagues in the Mathematics Department at Centre College. John H. Wilson played a particularly vital role in helping us develop and design the course curriculum that led to writing this book, spending hours in conversation with us, attending every class meeting of a pilot course, and writing initial drafts of portions of this book. John's insightful wisdom and enthusiastic investment were essential for the success of this project and we are extremely grateful to him.

Our students at Centre College have proven very helpful in crafting this book. For multiple summers, we were able hire teams of students to work on various elements of this book. They helped us find many of the interesting real-world examples studied in this book and provided feedback on early drafts of the manuscript, highlighting potential points of confusion and cumulatively working every example, question, and exercise. The answers to the odd-numbered exercises in the back of the book are a direct result of their efforts. We are particularly grateful to Monica E. Fitch, Wangdong Jia, Adrienne C. Kinney, Daniel J. McAllister, Matthew D. O'Brien, Abby Quirk-Royal, Melissa Stravitz, William S. Thackery, and Anne Wilson. In addition, we used initial drafts of this book while teaching our MAT 145: Mathematical Modeling and Applied Calculus during the 2013–14, 2014–15, 2015–16, 2016–17, and 2017-18 academic years. Our students' and colleagues' feedback has made this book immeasurably better.

We thank Centre College for supporting this project through Faculty Development Funds in Summers 2014, 2015, and 2016. In addition, Centre College awarded a multi-course release to Joel through a Stodghill Research Professorship in Spring 2015 and Alex an H. W. Stodghill Jr. and Adele H. Stodghill Professorship beginning in Fall 2015, both of which enabled significant work on this project.

We thank Danny Kaplan and Benjamin Klein, who suggested that the Centre College Mathematics Department rethink how we teach our lower-level calculus course during a

2013 external review of our Mathematics Department. As leaders of the Summer 2013 MAA PREP workshop *Modeling: Early and Often in Undergraduate Calculus*, Karl-Dieter Crisman, Robyn Cruz, Danny Kaplan, and Randall J. Pruim were inspiring and encouraging in our creation of this book. Throughout this writing project, the Project MOSAIC team has been vital in enabling our effective use of RStudio as a tool for developing and analyzing mathematical models. Likewise, almost all the graphs in this book were created using Maple 16, which was provided through the Maplesoft Author Support Program. We were grateful to have access to this software package in support of our work.

Our editor Daniel Taber and the production teams at Oxford University Press have been incredibly encouraging and helpful in this past year as we have finished this project. We worked on our own from 2013 through 2016, hopeful that a publisher might take interest in this project once it had been developed sufficiently. Daniel's enthusiastic reception of our proposal and continuing support have been vital in help us carry through to the finish.

Most importantly, we thank God and our families for their unflagging support and encouragement. We dedicate this book to our children in recognition of their sacrifice of much time with us. Renee and Julie, we could not have completed this project without you, and we are forever grateful for your support in seeing this book to completion.

And whatever you do, in word or in deed, do everything
in the name of the Lord Jesus, giving thanks to
God the Father through Him.
Colossians 3:17

Joel Kilty
Alex M. McAllister

June 2018

- A second audience is students taking a precalculus course, which develops a careful study of functions along with some attention to modeling. This course would focus heavily on Chapters 1 and 2, and perhaps include some content from Chapter 3. The study of the method of least squares in Chapter 3 does not utilize the ideas of calculus studied later in this book and so can be readily studied in such a precalculus course.

- A third audience is students taking a one-semester course in Calculus or Applied Calculus with some review of functions. This course would consist of a thorough treatment of Chapter 4 through Chapter 6, with periodic references back to Chapter 1 as needed for a just-in-time review of precalculus topics.

- A fourth possibility is students taking a lower-level undergraduate Mathematical Modeling course. This course would use Chapter 1 through Chapter 3, and select portions of Chapter 4 through Chapter 6 depending on instructor preferences. Such a course would have a greater focus on Chapters 2 and 3 than a precalculus course, as well as probably including some of the calculus topics from Chapter 4 through Chapter 6 for analyzing the mathematical models developed in Chapters 2 and 3.

Acknowledgments

This book was only possible with the help and encouragement of all of our colleagues in the Mathematics Department at Centre College. John H. Wilson played a particularly vital role in helping us develop and design the course curriculum that led to writing this book, spending hours in conversation with us, attending every class meeting of a pilot course, and writing initial drafts of portions of this book. John's insightful wisdom and enthusiastic investment were essential for the success of this project and we are extremely grateful to him.

Our students at Centre College have proven very helpful in crafting this book. For multiple summers, we were able hire teams of students to work on various elements of this book. They helped us find many of the interesting real-world examples studied in this book and provided feedback on early drafts of the manuscript, highlighting potential points of confusion and cumulatively working every example, question, and exercise. The answers to the odd-numbered exercises in the back of the book are a direct result of their efforts. We are particularly grateful to Monica E. Fitch, Wangdong Jia, Adrienne C. Kinney, Daniel J. McAllister, Matthew D. O'Brien, Abby Quirk-Royal, Melissa Stravitz, William S. Thackery, and Anne Wilson. In addition, we used initial drafts of this book while teaching our MAT 145: Mathematical Modeling and Applied Calculus during the 2013–14, 2014–15, 2015–16, 2016–17, and 2017-18 academic years. Our students' and colleagues' feedback has made this book immeasurably better.

We thank Centre College for supporting this project through Faculty Development Funds in Summers 2014, 2015, and 2016. In addition, Centre College awarded a multi-course release to Joel through a Stodghill Research Professorship in Spring 2015 and Alex an H. W. Stodghill Jr. and Adele H. Stodghill Professorship beginning in Fall 2015, both of which enabled significant work on this project.

We thank Danny Kaplan and Benjamin Klein, who suggested that the Centre College Mathematics Department rethink how we teach our lower-level calculus course during a

2013 external review of our Mathematics Department. As leaders of the Summer 2013 MAA PREP workshop *Modeling: Early and Often in Undergraduate Calculus*, Karl-Dieter Crisman, Robyn Cruz, Danny Kaplan, and Randall J. Pruim were inspiring and encouraging in our creation of this book. Throughout this writing project, the Project MOSAIC team has been vital in enabling our effective use of RStudio as a tool for developing and analyzing mathematical models. Likewise, almost all the graphs in this book were created using Maple 16, which was provided through the Maplesoft Author Support Program. We were grateful to have access to this software package in support of our work.

Our editor Daniel Taber and the production teams at Oxford University Press have been incredibly encouraging and helpful in this past year as we have finished this project. We worked on our own from 2013 through 2016, hopeful that a publisher might take interest in this project once it had been developed sufficiently. Daniel's enthusiastic reception of our proposal and continuing support have been vital in help us carry through to the finish.

Most importantly, we thank God and our families for their unflagging support and encouragement. We dedicate this book to our children in recognition of their sacrifice of much time with us. Renee and Julie, we could not have completed this project without you, and we are forever grateful for your support in seeing this book to completion.

And whatever you do, in word or in deed, do everything
in the name of the Lord Jesus, giving thanks to
God the Father through Him.
Colossians 3:17

Joel Kilty
Alex M. McAllister

June 2018

Chapter 1

Functions for Modeling Data

One major goal of mathematics is to help develop a better understanding of physical and social phenomena. Humans care about many aspects of reality, from the price of textbooks to human relationships, from weather events to changing populations, from profit margins to the success of medical interventions. Researchers in diverse fields of work and study develop an understanding of such phenomena through a dynamic process known as the *modeling cycle*.

THE MODELING CYCLE. The five steps of the **modeling cycle** are as follows:

(1) Ask a question about reality.

(2) Make some observations and collect the corresponding data.

(3) Conjecture a model or modify a known model based on the data.

(4) Test the model against known data (from step (2)) and modify the model as needed.

(5) Repeat steps (2)–(4) to improve the model.

In broad strokes, the modeling cycle begins by asking a question about some phenomena of interest, gathering data relevant to this question, and conjecturing a model describing the data. The accuracy of this model is then tested against the gathered data, possibly leading to modifications in order to obtain a model that more closely matches the data set. This cycle is then repeated as we develop a better understanding of the phenomenon through the model and through additional observations and collection of data.

For these purposes, a **model** is a mathematical function whose input and output correspond to observations of some phenomenon of interest. This first chapter explores the basic idea of a function and then studies the particular functions that are used most commonly as models of data sets. Chapter 2 develops an ability to apply the modeling cycle, which will help enable better insights into the world and more informed predictions.

1.1   Functions

Many aspects of reality can be described in terms of inputs and outputs. We experience input–output relationships all the time. When purchasing textbooks, we pay the bookseller some money (an input to the bookseller), and receive a copy of the book needed for class (an output from the bookseller). In order to earn money, we work some number of hours (an input to the business) and are paid a corresponding wage in return (an output from the business). At your school, most teachers (an input to the school) teach multiple sections of various classes (an output from the school). Before reading further, take a moment to think about some additional input–output relationships from your day-to-day life.

We focus on input–output relationships that identify each input with exactly one output, known as *functions*. This focus arises from mathematicians' success in developing diverse tools for analyzing functions, which, in turn, carries over to success with the modeling cycle.

Before diving into the mathematics of functions, some graphical presentations of input–output relationships are considered. Figure 1 provides the performance of the Dow Jones Industrial Average at the end of each quarter from 2007 to 2011, which happens to be a functional input–output relationship. Namely, at each point in time (an input), the Dow Jones had exactly one stock market value (the output).

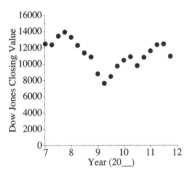

Figure 1: The Dow Jones Industrial Average from 2007 to 2011

As demonstrated in Figure 1, the inputs are typically listed along the horizontal x-axis and the outputs are listed along the vertical y-axis. In some settings, alternative points of view are explored by reversing the roles of inputs and outputs to work with the *inverse function*. In such cases, a new graph of the original input–output relationship is created, listing the objects now treated as inputs (the former outputs) along the horizontal x-axis.

Another input–output relationship is the track of Hurricane Ivan in 2004, shown in Figure 2. Following the standard practice just described, think of longitude as an input (listed along the x-axis) and latitude as an output (listed along the y-axis). From this perspective, the hurricane's track presents a different type of input–output relationship because some inputs correspond to multiple outputs. Notice that for some longitudes Ivan passed through multiple latitudes. For example, at 85°W, Ivan passed through approximately latitudes 21°N, 25°N, and 35°N, as can be seen by tracing north along the vertical line of longitude 85°W.

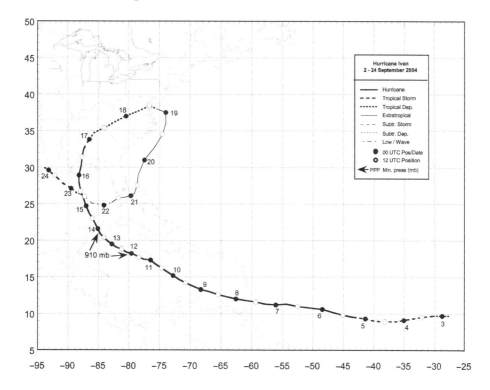

Figure 2: Path of Hurricane Ivan in 2004

Based on this observation, Ivan's track is not a function, because some inputs correspond to multiple outputs, rather than to exactly one output. In this example, there happen to be many different longitude inputs with multiple latitude outputs. Take a moment to identify a couple more longitudes for which Ivan passed through multiple latitudes. The fact that several input values have multiple outputs is interesting (and particularly important for the people who lived along the track of Hurricane Ivan), but not necessary to show that this input–output relationship is not a function. Rather, if one single input has multiple associated outputs, the input–output relationship is not a function.

For some nonfunctional input–output relationships, adopting a different perspective on the data may result in an alternative point of view that does represent a function. Continuing the discussion of the track of Hurricane Ivan, perhaps you have already identified a way to think about Ivan's track that results in a function? The key insight is to consider time as the input and the ordered pair (longitude, latitude) as the output.

At any particular time, Ivan is centered at a unique location, which corresponds to a unique output pair (longitude, latitude). From this perspective, the track of Hurricane Ivan is a function.

In addition to functions whose outputs consist of multiple components, as with the output (longitude, latitude) for Hurricane Ivan, many functions have inputs with more than one component. For example, Figure 3 presents a weather map displaying temperature and is an example of such an input–output relationship, which is known as a *multivariable function*. For this temperature map, think of the input as an ordered pair of (longitude, latitude) that outputs the unique numeric temperature at the given location on a winter's day in 2017.

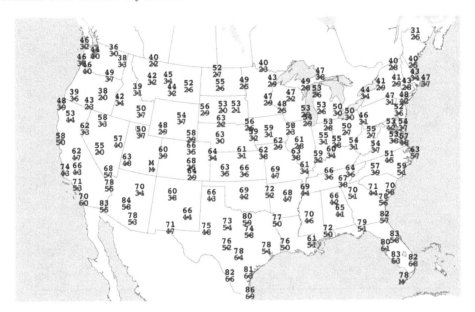

Figure 3: U.S. maximum and minimum temperatures on a winters day in 2017

Another way to interpret the map in Figure 3 is to think about the input as the name of a city and the output as the corresponding temperature. Adopting this common reading of temperature maps still requires a two-component input, because some cities have the same name. According to the U.S. Postal Service, the most common city name in the United States is "Franklin," which is used to identify 31 different cities. Therefore, an input consisting of the ordered pair (city, state) is needed in order to determine a unique output of temperature.

This section and the next extend the ideas discussed in these specific examples to more general settings. Building on this intuitive understanding of functions, the definition of a function is stated, followed by examining both examples and nonexamples of these important mathematical objects.

(listed along the x-axis) and latitude as an output (listed along the y-axis). From this perspective, the hurricane's track presents a different type of input–output relationship because some inputs correspond to multiple outputs. Notice that for some longitudes Ivan passed through multiple latitudes. For example, at 85°W, Ivan passed through approximately latitudes 21°N, 25°N, and 35°N, as can be seen by tracing north along the vertical line of longitude 85°W.

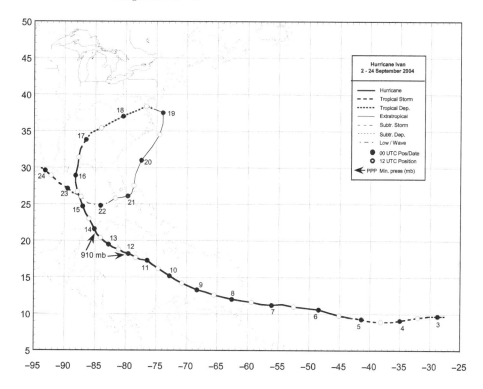

Figure 2: Path of Hurricane Ivan in 2004

Based on this observation, Ivan's track is not a function, because some inputs correspond to multiple outputs, rather than to exactly one output. In this example, there happen to be many different longitude inputs with multiple latitude outputs. Take a moment to identify a couple more longitudes for which Ivan passed through multiple latitudes. The fact that several input values have multiple outputs is interesting (and particularly important for the people who lived along the track of Hurricane Ivan), but not necessary to show that this input–output relationship is not a function. Rather, if one single input has multiple associated outputs, the input–output relationship is not a function.

For some nonfunctional input–output relationships, adopting a different perspective on the data may result in an alternative point of view that does represent a function. Continuing the discussion of the track of Hurricane Ivan, perhaps you have already identified a way to think about Ivan's track that results in a function? The key insight is to consider time as the input and the ordered pair (longitude, latitude) as the output.

At any particular time, Ivan is centered at a unique location, which corresponds to a unique output pair (longitude, latitude). From this perspective, the track of Hurricane Ivan is a function.

In addition to functions whose outputs consist of multiple components, as with the output (longitude, latitude) for Hurricane Ivan, many functions have inputs with more than one component. For example, Figure 3 presents a weather map displaying temperature and is an example of such an input–output relationship, which is known as a *multivariable function*. For this temperature map, think of the input as an ordered pair of (longitude, latitude) that outputs the unique numeric temperature at the given location on a winter's day in 2017.

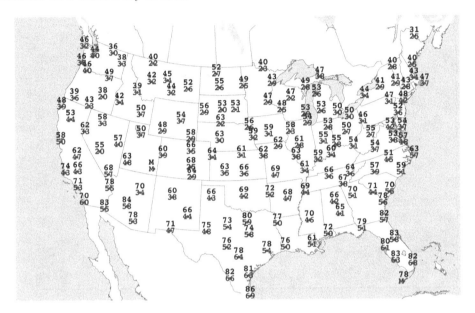

Figure 3: U.S. maximum and minimum temperatures on a winters day in 2017

Another way to interpret the map in Figure 3 is to think about the input as the name of a city and the output as the corresponding temperature. Adopting this common reading of temperature maps still requires a two-component input, because some cities have the same name. According to the U.S. Postal Service, the most common city name in the United States is "Franklin," which is used to identify 31 different cities. Therefore, an input consisting of the ordered pair (city, state) is needed in order to determine a unique output of temperature.

This section and the next extend the ideas discussed in these specific examples to more general settings. Building on this intuitive understanding of functions, the definition of a function is stated, followed by examining both examples and nonexamples of these important mathematical objects.

> **Note.** Any definition, including a mathematical definition, is meaningful only when there exist both examples and nonexamples of the object or event. For example, the word "human" is meaningful because some beings are human and other beings (such as dogs) are not human.

Intuitively, a function is an input–output relationship with the distinctive feature that each input corresponds to exactly one output, as illustrated in Figure 4(a). In this case, the input 1 maps to the output a, and the input 2 maps to the output b. The potential output c is not actually an output, but this does not impact Figure 4(a) being a function.

Figure 4: A function and a nonfunction

In contrast, a nonfunction is an input–output relationship for which some input corresponds to more than one output, as illustrated in Figure 4(b). The input 1 maps to both a and b, making this relationship a nonfunction. Note that even though the input 2 corresponds to only output c, a single repeated input is enough to make a relationship a nonfunction.

These intuitive ideas motivate the formal definition of a function as well as various adjectives for functions.

> **Definition.**
>
> - A **function** is a rule assigning every input to exactly one output.
>
> - A **single-variable** function has one input and one output.
>
> - A **multivariable** function has two or more inputs and one output.
>
> - A **vector-valued** function has one or more inputs and two or more outputs.

Single-variable functions are discussed for the remainder of this section, multivariable functions in Section 1.2, and vector-valued functions in Chapter 3. The study of single-variable functions considers tabular, graphical, and analytic presentations of input–output relationships and discusses how to determine whether each presentation is a function or not.

Tabular Functions and Nonfunctions

A horizontal tabular presentation of an input–output relationship lists the inputs in the top row of the table and the outputs in the bottom row.

◆ **EXAMPLE 1** Explain why each table does or does not define a function.

(a)
Input	1	2	3	4	5	6
Output	1	4	9	16	25	16

(b)
Input	1	2	3	4	1	6
Output	1	4	9	16	25	36

Solution.

(a) Table (a) defines a function, because each input corresponds to exactly one output. Also, notice that the two inputs 4 and 6 both map to the same output of 16. The definition of a function does *not* mandate that each output value appear just once and, in fact, many functions output the same value for multiple inputs. Instead, a function only requires that every input have exactly one output, which occurs in Table (a).

(b) Table (b) does not define a function, because the input 1 corresponds to the two distinct outputs of 1 and 25, rather than to a single output. ∎

➤ **QUESTION 1** Explain why each table does or does not define a function.

(a)
Input	a	b	c	a	b	c
Output	2	4	2	16	4	2

(b)
Input	a	b	c	d	a	e
Output	2	4	2	16	2	25

Graphical Functions and Nonfunctions

Graphically, functions and non-functions are distinguished using a result known as the *vertical line test*. When graphing a function, standard mathematical practice is to let the x-axis represent the input values and the y-axis represent the corresponding output values. This standard practice provides a useful approach to determining if a given graph presents a function.

> **VERTICAL LINE TEST**. A curve in the xy-plane is a function if and only if every vertical line intersects the curve at most once.

Applying this test requires visually deciding whether the vertical line passing through each x value intersects the graph of output values more than once. If some vertical line $x = a$ does intersect the graph more than once, then the input a has more than one output. If every vertical line intersects the graph exactly once or not at all, then the graph represents a function.

◆ **EXAMPLE 2** Using the vertical line test, explain why each curve in Figure 5 does or does not define a function.

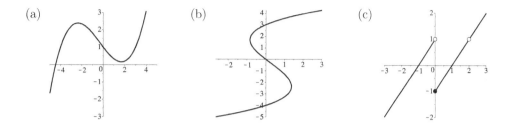

Figure 5: Curves for Example 2

Solution.

(a) Curve (a) is a function, because no matter which vertical line $x = a$ is selected, every such line intersects this curve exactly once, as illustrated for the three example vertical lines in Figure 6(a).

(b) Curve (b) is not a function, because the vertical line $x = 0$ intersects the curve more than once as shown in Figure 6(b). In fact, multiple vertical lines intersect the curve three times, such as the lines $x = -0.5$ and $x = 0.5$. Any of these vertical lines demonstrates that this curve is not a function.

(c) Curve (c) is a function. The vertical line $x = 0$ may appear to be problematic, but notice that curve (c) has only one output value, of $f(0) = -1$, as indicated by the open circle on the positive y-axis when $y = 1$ and the solid dot on the negative y-axis when $y = -1$. In addition, observe that curve (c) is not defined when $x = 2$ (as indicated graphically by an open circle), and the vertical line $x = 2$ does not intersect this curve at all. The vertical line test is still satisfied, because $x = 2$ does not intersect the curve more than once, which allows for this line to not intersect curve (c). ■

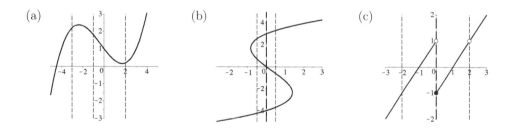

Figure 6: Vertical line test for Example 2

▶ **QUESTION 2** Using the vertical line test, explain why each curve in Figure 7 does or does not define a function.

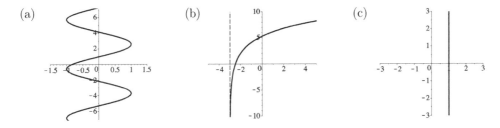

(a)   (b)   (c)

Figure 7: Curves for Question 2

Analytic Presentations of Functions

Analytic presentations of functions include such familiar expressions as $f(x) = 4x - 5$ and $g(x) = x^2 + 8$. This study of analytic presentations of functions begins with the standard mathematical vocabulary used to discuss and describe these functions.

Definition.

- Symbolically, $y = f(x)$ indicates that input x is assigned to output y and we say that $f(x)$ **maps** the input x to the output y

- The set of input values of a function $y = f(x)$ is called the **domain** of $f(x)$. Note: The domain of every function in this book is contained in the real numbers.

- The set of output values for the domain of a function $y = f(x)$ is called the **range** of $f(x)$.

Note that a function is only defined on its domain. If a potential input is not in the domain of a function, then the function is said to be **undefined** for that input. Sometimes the domain of a function is explicit, as with tabular presentations, but more often the domain is implicit, as with graphical and analytical presentations. The modeling cycle frequently requires identifying the domain and range of a function. First, consider tabular presentations of functions.

◆ **EXAMPLE 3** The table below presents how many millions of people used Twitter during each quarter of a year since 2000, where 11 represents the first quarter of 2011 (January–March 2011), 11.25 represents the second quarter of 2011 (April–June 2011), and so on. Identify the domain and range of the function described by this table.

Quarter of year	11	11.25	11.5	11.75	12.75
Millions of Twitter users	68	85	101	117	185

Solution. First, observe that this table presents a function, because every input appears only once in the top row and so maps to exactly one output. The domain of this function is the set of input values $\{11, 11.25, 11.5, 11.75, 12.75\}$ from the top row of the table. The range of this function consists of all outputs $\{68, 85, 101, 117, 185\}$ from the bottom row of the table. ∎

In Example 3, note that the domain of this tabular function is *not* the interval $[11, 12.75]$ of all real numbers between 11 and 12.75. While such an answer might seem plausible, output values are only given for $11, 11.25, 11.5, 11.75$, and 12.75, not for any intermediate values (such as 12). Because only five numbers are valid inputs, the domain is $\{11, 11.25, 11.5, 11.75, 12.75\}$.

➤ **QUESTION 3** The table below presents the U.S. annual unemployment rate during each year. Identify the domain and range of the function described by this table.

Year	2008	2009	2010	2012	2013
Annual unemployment rate	5	7.8	9.7	8.2	7.9

When a function is presented analytically, identifying its domain and range is often more complicated than for a tabular presentation. In this book, the following two main features restrict the domain of an analytic function.

DOMAIN OF A FUNCTION WITH AN ANALYTIC PRESENTATION.
Unless a domain is given, the domain of a function with an analytic presentation consists of all real numbers except those numbers that produce either

- a negative number under a square root (or any other even root) or

- a zero in the denominator of a fraction.

Similarly, identifying the range of a function presented analytically is often more subtle, especially as the presentation of the function becomes more elaborate. The ranges of common modeling functions are given as these functions are introduced throughout this chapter.

◆ **EXAMPLE 4** Identify the domain of each function:

(a) $a(x) = 3x^2 - 4x$
(b) $b(x) = \sqrt{2x - 4}$
(c) $c(x) = \dfrac{1}{x^2 - 4}$

Solution.

(a) The domain of $a(x)$ is the set of all real numbers, which is written in interval notation as $(-\infty, \infty)$, because this combination of operations can be performed on every real number.

(b) The domain of $b(x)$ does not include all real numbers, because the function involves the square root of an expression that is negative for some inputs. The square root of a negative number is undefined in the real numbers, so $2x - 4$ must be non-negative; that is, $2x - 4 \geq 0$. Solving for x in this inequality determines the domain of $b(x)$.

$$(2x - 4) + 4 \geq 0 + 4 \qquad \text{Add 4 to both sides}$$
$$2x \geq 4 \qquad \text{Simplify}$$
$$x \geq 2 \qquad \text{Divide both sides by 2}$$

Thus, the domain of $b(x)$ consists of all real numbers greater than or equal to 2, which is written in interval notation as $[2, \infty)$.

(c) The domain of $c(x)$ does not include all real numbers, because the denominator of a function cannot be zero. The real numbers that are not in the domain are found by solving for where the denominator is zero, that is, by solving $x^2 - 4 = 0$. Factoring, produces $(x + 2)(x - 2) = 0$, which has the two solutions $x = -2$ and $x = 2$. Thus, the domain of $c(x)$ consists of all real numbers except $x = -2$ and $x = 2$, which is written in interval notation as $(-\infty, -2) \cup (-2, 2) \cup (2, \infty)$.

■

➤ **QUESTION 4** Identify the domain of each function:

(a) $a(x) = x^5 - 4x^3$ (b) $b(x) = \sqrt{4 - x}$ (c) $c(x) = \dfrac{1}{\sqrt{4x - 8}}$

When a function is presented analytically, $y = f(x)$ is **evaluated** for a particular input $x = a$ by substituting the value a for every x appearing in the analytic expression that defines the function and then performing the appropriate computations. When substituting inputs into a function, be sure to confirm that they are in the domain of the function.

◆ **EXAMPLE 5** Evaluate the function $f(x) = 3\sqrt{x} - 5$ for each input x, if possible:

(a) $x = 4$ (b) $x = 0$ (c) $x = -3.2$

Solution.

(a) $f(4) = 3 \cdot \sqrt{4} - 5 = 3 \cdot 2 - 5 = 6 - 5 = 1$

(b) $f(0) = 3 \cdot \sqrt{0} - 5 = 3 \cdot 0 - 5 = 0 - 5 = -5$

(c) $f(-3.2)$ is undefined, because the square root of a negative number is undefined.

■

➤ **QUESTION 5** Evaluate the function $g(x) = x^2 + 8$ for each input x, if possible:

(a) $x = 4$ (b) $x = 0$ (c) $x = -3.2$

◆ **EXAMPLE 9** Graph the line $y = 2x - 7$ on the domain $[-10, 10]$ using RStudio.

Solution. Two possible approaches are given that produce exactly the same results. First, graph the line $y = 2x - 7$ by entering the full expression in the `plotFun` command. Second, define the function $y = 2x - 7$ in RStudio with the `makeFun` command, which is named `line1`, and then graph $y = 2x - 7$ using this name `line1`. As noted above, "`(x)`" must be included immediately after the function name `line1` in the `plotFun` command.

```
plotFun(2*x-7~x,xlim=range(-10,10))
line1=makeFun(2*x-7~x)
plotFun(line1(x)~x,xlim=range(-10,10))
```

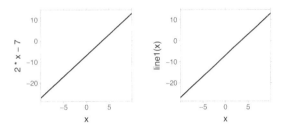

■

▶ **QUESTION 9** Graph the line $y = -3.546x - 9.128$ on the domain $[-15, 15]$ using RStudio.

Summary

- A *function* is a rule assigning every input to exactly one output. A *single-variable function* has one input and one output. A *multivariable function* has two or more inputs and only one output. A *vector-valued function* has one or more inputs and two or more outputs.

- *Vertical line test*: A curve in the xy-plane is a function if and only if every vertical line intersects the curve at most once. Alternatively, a curve is not a function if and only if some vertical line intersects the curve more than once.

- Write $y = f(x)$ to indicate that input x is assigned to output y and say that $f(x)$ *maps* the inputs x to the output y. The set of input values of a function is called the *domain* of $f(x)$; the set of output values for the domain is called the *range* of $f(x)$.

- A function is defined *piecewise* when a single expression cannot completely describe a function.

Exercises

In Exercises 1–6, give an example of a real-life input–output relationship with the specified property.

1. A single-variable function

2. A two-variable function

3. A four-variable function

4. A single-variable nonfunction

5. A two-variable nonfunction

6. A four-variable nonfunction

In Exercises 7–14, explain why the table does or does not define a function. Also, if its a function, state the domain and range.

7.
x	a	b	c	d
$f(x)$	2	3	4	5

8.
x	a	b	c	d
$f(x)$	2	3	2	5

9.
x	a	b	a	d
$f(x)$	9	8	9	6

10.
x	a	b	b	d
$f(x)$	9	8	7	6

11.
x	2.5	4.1	8.7	9
$f(x)$	π	4	π^2	16

12.
x	2.5	4.1	8.7	9
$f(x)$	π	4.1	π	9

13.
x	4.1	4.1	8.7	9
$f(x)$	3.14	π	2	3

14.
x	2.5	9	8.7	9
$f(x)$	2.6	4.1	7.3	2.6

In Exercises 15–18, explain why the table of data does or does not define a function. Also, if it is a function, state the domain and range.

15. Annual total retail sales taxes (T) collected in the United States in billions of dollars each year (Y); for example, during 2006, $141,100,000,000 in taxes was collected

Y	2006	2008	2010	2012
T	141.1	141.4	139.4	150.4

16. Average debt (D) in thousands of dollars at the end of the spring term in each year (Y) for bachelor's degree recipients attending public four-year colleges and universities who borrowed money to finance their education

Y	2001	2003	2005	2006
D	20.4	20.9	21.5	21.8

17. Price of gas (P) at different gas stations in Los Angeles, California on June 11–12, 2015 (D)

D	6/11	6/11	6/12	6/12
P	3.79	3.50	3.79	3.49

18. Floors (F) where a hotel's elevators are located at T minutes past 3 PM

T	11	11	12	15	15
F	13	22	1	5	7

Your Turn. In Exercises 19 and 20, give an example of a tabular input–output relationship with the specified property.

19. Single-variable function

20. Single-variable nonfunction

In Exercises 21–26, explain why the curve does or does not define a function. Also, if it is a function, state the apparent domain and range.

21.

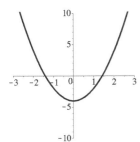

25.

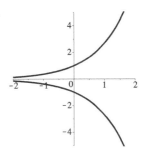

22.

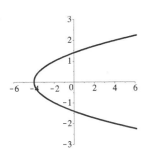

26.

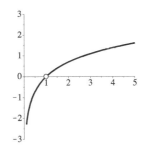

Your Turn. In Exercises 27 and 28, sketch two curves that

27. Represent functions.

28. Do not represent functions.

In Exercises 29–38, determine the domain of the function.

29. $f(x) = 3x - 7$

30. $f(x) = x^3 + \pi x$

31. $f(x) = \sqrt{x + 7}$

32. $f(x) = \sqrt{x^2 + 3}$

33. $f(x) = \sqrt{x^2 - 9}$

23.

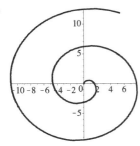

34. $f(x) = \sqrt{x} + \sqrt[4]{1 + x}$

35. $f(x) = \dfrac{3x + 1}{x + 5} + \dfrac{2}{4x}$

36. $f(x) = \dfrac{3}{x^2 - 1}$

37. $f(x) = \dfrac{\pi x}{x^3 - 9x}$

24.

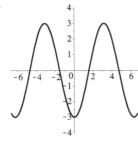

38. $f(x) = \dfrac{x}{\sqrt{4 - x^2}}$

In Exercises 39 – 44, evaluate the function $f(x) = 3x^2 + 5$ for the input x.

39. $x = 3$ 42. $x = -3$

40. $x = 2.451$ 43. $x = -5.234$

41. $x = 934.82$ 44. $x = -1000$

In Exercises 45 – 48, approximate how many million people are using Twitter in each quarter of the year using the model $f(x) = 66.8x - 666.6$, which is based on the data from Example 3.

45. $x = 12$ 47. $x = 12.5$

46. $x = 12.25$ 48. $x = 13$

In Exercises 49 – 56, evaluate the function $f(x)$ for the input x.

$$f(x) = \begin{cases} 3x + 4 & x \leq -5 \\ 2 - x^2 & -5 < x \end{cases}$$

49. $x = -10$ 53. $x = -\pi$

50. $x = 0$ 54. $x = -5$

51. $x = 0.5$ 55. $x = -4.9$

52. $x = -6.3$ 56. $x = 10$

In Exercises 57 – 64, evaluate the function $f(x)$ for the input x.

$$f(x) = \begin{cases} 2x^2 + 3 & x \leq -5 \\ 3 & -5 < x < 2 \\ 3x - 6 & x \geq 2 \end{cases}$$

57. $x = -10$ 61. $x = \pi$

58. $x = 0$ 62. $x = -5$

59. $x = 2.1$ 63. $x = -3.4$

60. $x = -6.1$ 64. $x = 10$

Your Turn. In Exercises 65 – 68, define a single-variable piecewise function and evaluate this function for the input x.

65. $x = 4$ 67. $x = 0$

66. $x = -4.92$ 68. $x = 93$

RStudio. In Exercises 69 – 76, use RStudio to evaluate the arithmetic expression.

69. $(-4) \cdot 3 - 2$ 73. $|\pi^2 - 5^3 \cdot 3^{2.1}|$

70. $5^{-3} + 6^{0.3}$ 74. $|8 \cdot 5^2 - 5 \cdot 8^2|$

71. $\dfrac{3 \cdot \pi + 4 \cdot \sqrt{2}}{2}$ 75. $\dfrac{(2.34)^3 + 7}{3.7 \cdot (4.3 - 3)}$

72. $\dfrac{4.327 - 8.269}{2.631}$ 76. $\dfrac{\pi^2 - \sqrt[3]{7}}{1.414}$

RStudio. In Exercises 77 – 80, use RStudio to define the function $f(x)$ and evaluate it for the input x, when $f(x) = 4.3(3x^5 - 2.6) + 2.5$.

77. $x = -84.658$ 79. $x = 4.341$

78. $x = -10.256$ 80. $x = 120.456$

RStudio. In Exercises 81 – 84, use RStudio to define the function $f(x)$ and evaluate it for the input x, when

$$f(x) = \frac{\pi(9.852x^3 - 10.375x)}{(9\pi)^2 + 27x}.$$

81. $x = -10$ 83. $x = 13.987$

82. $x = 0$ 84. $x = 25.491$

RStudio. In Exercises 85 – 90, use RStudio to graph the line on the given domain.

85. $y = 3x + 27$ on $[-10, 10]$

86. $y = 7x - 15$ on $[-5, 5]$

87. $y = 44.53x - 6.75$ on $[-50, 150]$

88. $y = -3x + 22$ on $[0, 15]$

89. $y = -26.79x - 145.62$ on $[-20, 5]$

90. $y = -11.55x + 16.45$ on $[-1, 3]$

In Your Own Words. In Exercises 91–100, explain the following.

91. The modeling cycle

92. Function

93. Nonfunction

94. Single-variable function

95. Multivariable function

96. Vertical line test

97. Domain

98. Range

99. Evaluate a function for a given input

100. Piecewise function

1.2   Multivariable Functions

We seek to use mathematical functions to help describe and understand reality. While some real-life applications are characterized by input–output relationships consisting of one input and one output, many settings involve several different interrelated quantities. In mathematics, such situations can often be modeled with multivariable functions, which have multiple inputs and a single output.

Section 1.1 defined a **function** as a rule assigning every input to a unique output and a **multivariable function** as a function with two or more inputs and only one output. As an example, the discussion of a temperature map of the United States highlighted the need for an input of (longitude, latitude) or (city, state) to identify the temperature at a particular location. Another example of the multivariable character of reality includes a company determining its cost for producing a particular good based on such expenses as labor, materials, equipment, and utilities. Similarly, a psychiatrist prescribing a medication must take into account the patient's symptoms, possible side effects, current medications, body mass, and ability to pay for the medicine. Before reading further, take a moment to think about an additional input–output relationship from your day-to-day life with multiple inputs and a single output.

Multivariable functions are represented analytically using essentially the same notation as for single-variable functions. Namely, write $f(x, y)$ equal to an expression with input variables x and y, write $g(x, y, z)$ equal to an expression with input variables x, y, and z, and so on. Similarly, a function $f(x, y)$ is **evaluated** for particular inputs $x = a$ and $y = b$, which is usually written as $f(a, b)$, by substituting the value a for every x in the expression and the value b for every y in the expression.

◆ **EXAMPLE 1** Evaluate $f(x, y) = 4x^2 - 5y\sqrt{x + 1}$ for each input (x, y), if possible:

(a) $(x, y) = (3, 2)$ (b) $(x, y) = (0, -2.1)$ (c) $(x, y) = (-2.1, 5)$

Solution.

(a) $f(3, 2) = 4 \cdot 3^2 - 5 \cdot 2 \cdot \sqrt{3 + 1} = 36 - 10 \cdot 2 = 16$

(b) $f(0, -2.1) = 4 \cdot 0^2 - 5 \cdot (-2.1) \cdot \sqrt{0 + 1} = 0 + 10.5 \cdot 1 = 10.5$

(c) $f(-2.1, 5)$ is undefined because $x + 1 = -2.1 + 1 = -1.1$, and there is no real number that is the square root of a negative number. ◼

➤ **QUESTION 1** Evaluate $f(x, y, z) = 3x + \dfrac{y}{z}$ for each input (x, y, z), if possible:

(a) $(1, 4, 2)$

(c) $(10, -4, -16)$

(b) $(2.2, 16.8, -4)$

(d) $(3.4, 5.6, 0)$

Multivariable functions with two inputs can also be presented in tabular form as shown in Figure 1. The numbers in the table give the output for each corresponding pair of input values indicated by the labeling of the rows and columns. For example, the output for the input $(2, 3)$ is determined by using the column labeled $x = 2$ and the row labeled $y = 3$, which gives the associated output of 44.

◆ **EXAMPLE 2** Evaluate $f(x, y)$ for each input (x, y) using the table in Figure 1:

(a) $(x, y) = (-3, 4)$ (b) $(x, y) = (5, -2)$ (c) $(x, y) = (4, 3)$

$f(x, y)$		-5	-4	-3	-2	-1	x 0	1	2	3	4	5
	4	6	12	18	24	30	36	42	48	54	60	66
	3	2	8	14	20	26	32	38	44	50	56	62
	2	-2	4	10	16	22	28	34	40	46	52	58
	1	-6	0	6	12	18	24	30	36	42	48	54
y	**0**	-10	-4	2	8	14	20	26	32	38	44	50
	-1	-14	-8	-2	4	10	16	22	28	34	40	46
	-2	-18	-12	-6	0	6	12	18	24	30	36	42
	-3	-22	-16	-10	-4	2	8	14	20	26	32	38
	-4	-26	-20	-14	-8	-2	4	10	16	22	28	34

Figure 1: A tabular presentation of a two-variable function

Solution.

(a) Using the column labeled -3 and the row labeled 4, observe that $f(-3, 4) = 18$.

(b) Using the column labeled 5 and the row labeled -2, observe that $f(5, -2) = 42$.

(c) Using the column labeled 4 and the row labeled 3, observe that $f(4, 3) = 56$.

■

➤ **QUESTION 2** Evaluate $f(x, y)$ for each input (x, y) using the table in Figure 1:

(a) $(x, y) = (-1, 4)$ (b) $(x, y) = (0, 2)$ (c) $(x, y) = (2, -2)$

◆ **EXAMPLE 3** The National Weather Service Windchill Chart in Figure 2 provides the windchill as a function of air temperature in degrees Fahrenheit (°F) and wind speed in miles per hour (mph).

(a) Find the windchill when the temperature is 30°F and the wind is blowing at 50 mph.

(b) Find the windchill when the temperature is −15°F and the wind is blowing at 10 mph.

(c) Determine the temperature and wind speed pair(s) when the wind chill is −1°F.

(d) Determine the temperature and wind speed pair(s) when the wind chill is −89°F.

Temperature (°F)

Wind (mph)	Calm 40	35	30	25	20	15	10	5	0	-5	-10	-15	-20	-25	-30	-35	-40	-45
5	36	31	25	19	13	7	1	-5	-11	-16	-22	-28	-34	-40	-46	-52	-57	-63
10	34	27	21	15	9	3	-4	-10	-16	-22	-28	-35	-41	-47	-53	-59	-66	-72
15	32	25	19	13	6	0	-7	-13	-19	-26	-32	-39	-45	-51	-58	-64	-71	-77
20	30	24	17	11	4	-2	-9	-15	-22	-29	-35	-42	-48	-55	-61	-68	-74	-81
25	29	23	16	9	3	-4	-11	-17	-24	-31	-37	-44	-51	-58	-64	-71	-78	-84
30	28	22	15	8	1	-5	-12	-19	-26	-33	-39	-46	-53	-60	-67	-73	-80	-87
35	28	21	14	7	0	-7	-14	-21	-27	-34	-41	-48	-55	-62	-69	-76	-82	-89
40	27	20	13	6	-1	-8	-15	-22	-29	-36	-43	-50	-57	-64	-71	-78	-84	-91
45	26	19	12	5	-2	-9	-16	-23	-30	-37	-44	-51	-58	-65	-72	-79	-86	-93
50	26	19	12	4	-3	-10	-17	-24	-31	-38	-45	-52	-60	-67	-74	-81	-88	-95
55	25	18	11	4	-3	-11	-18	-25	-32	-39	-46	-54	-61	-68	-75	-82	-89	-97
60	25	17	10	3	-4	-11	-19	-26	-33	-40	-48	-55	-62	-69	-76	-84	-91	-98

Frostbite Times ▢ 30 minutes ▢ 10 minutes ▢ 5 minutes

Figure 2: Windchill as a function of air temperature and wind speed for Example 3

Solution.

(a) The windchill is 12°F.

(b) The windchill is −35°F.

(c) The temperature is 20°F with a wind speed of 40 mph.

(d) Either the temperature is −45°F with a wind speed of 35 mph, or the temperature is −40°F with a wind speed of 55 mph. ∎

A tabular representation of a multivariable input–output relationship is identified as a function or nonfunction in much the same way as for single-variable functions presented in tabular form: look for a particular input with multiple associated outputs. Such nonfunctional behavior might appear in a table either as multiple entries in the same location of the table separated by a comma or as an entire column or row of repeated input values. Two-variable nonfunctions in tabular format are not common and will not be discussed further in this book.

Multivariable functions with two input values are represented graphically using either a three-dimensional *surface plot* or a two-dimensional *contour plot* as illustrated in Figure 3.

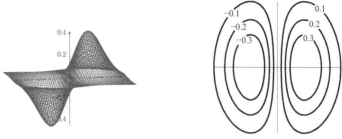

Figure 3: Surface and contour plots of $f(x,y) = xe^{-x^2-y^2}$

The advantage of a surface plot is that it provides a three-dimensional view of the relationship between the two input variables and the output variable. Most often, technology is used to draw surface plots, as described for RStudio later in the section. Contour plots are more widely used, because they provide essentially the same information but are two-dimensional and much easier to create.

Definition. Let $f(x,y)$ be a function of two variables.

- A **surface plot** of $f(x,y)$ is a three-dimensional graph where the value of each $f(x,y)$ is plotted as a height measured relative to z-axis above or below the corresponding point (x,y) on the xy-plane.

- A **contour** of $f(x,y)$ **at level** C is a curve in the xy-plane containing all points where $f(x,y) = C$. A **contour plot** or **contour map** of $f(x,y)$ is a two-dimensional graph in the xy-plane of contours of $f(x,y)$ at various levels.

The contour plot in Figure 3 shows contours at levels ± 0.1, ± 0.2, and ± 0.3. For example, consider the contour at level 0.2 on the right side of the graph. At each point on this contour, the output of the function $f(x,y)$ is 0.2. In the corresponding surface plot, all of the inputs on the contour at level 0.2 correspond to outputs at the same height of 0.2. Similarly, all inputs on the contour at level -0.3 on the left side correspond to outputs at the same height $f(x,y) = -0.3$.

◆ **EXAMPLE 4** Use the contour plot of $f(x,y)$ in Figure 4 to find the exact or an approximate value of $f(x,y)$ for each input (x,y):

(a) $(x,y) = (-1,0)$.   (b) $(x,y) = (0.5,0)$.   (c) $(x,y) = (0.75,0)$.

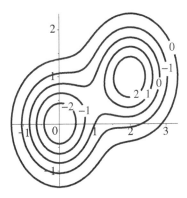

Figure 4: Contour plot of $f(x, y)$ for Examples 4, 5, and 6

Solution.

(a) For input $(-1, 0)$, the output is $f(-1, 0) = -1$.

(b) For input $(0.5, 0)$, the approximate output is $f(0, 0) \approx -2$.

(c) For input $(0.75, 0)$, the approximate output is $f(0.5, 0) \approx -1$.

∎

Contour plots allow the identification (or at least approximation) of the *maximum value* and the *minimum value*, also known as the *extreme values*, of the corresponding function. Identifying such extreme values, or *extrema*, lies at the heart of everyday life. The most efficient way to complete a task is often sought, as are investment strategies that maximize returns and directions that minimize the amount of time required to reach a particular destination. Graphically, such extreme values are located at the peaks and valleys, which can be located using a contour plot. An analytic approach to finding the exact values of extrema is studied in Chapter 5.

Definition. Let $f(x, y)$ be a function of two variables.

- The **extreme values** of $f(x, y)$ are the maximum and minimum values of the function.

- The **global maximum** of $f(x, y)$ on a domain D is the numeric value $f(a, b)$ for a point (a, b) in the domain of $f(x, y)$ such that $f(a, b) \geq f(x, y)$ for all (x, y) in D.

- The **global minimum** of $f(x, y)$ on a domain D is the numeric value $f(a, b)$ for a point (a, b) in the domain of $f(x, y)$ such that $f(a, b) \leq f(x, y)$ for all (x, y) in D.

◆ **EXAMPLE 5** Use the contour plot of $f(x, y)$ in Figure 4 to approximate the value and the coordinates of

(a) the global maximum of $f(x, y)$ on the domain provided in Figure 4, and

(b) the global minimum of $f(x, y)$ on the same domain.

Solution.

(a) The global maximum of $f(x, y)$ is approximately 2.5 at approximately $(2, 1)$. While the exact value and location of the maximum cannot be known for sure from the given contour plot, a reasonable conjecture is that it lies inside this highest contour and has a value greater than 5.

(b) The global minimum of $f(x, y)$ is approximately -2.5 at approximately $(0, 0)$. ∎

   In addition to identifying the extreme values of a function, determining whether the outputs of a function are immediately increasing, constant, or decreasing as input values increase is also of interest. For multivariable functions, such questions also require identifying a particular direction for inputs increasing. For two-variable functions, the standard choice is to focus on the particular directions of x-inputs increasing while y-inputs are constant (or moving straight right) and of y-inputs increasing while x-inputs are constant (or moving straight up).

◆ **EXAMPLE 6** Continuing the study of the contour plot of $f(x, y)$ in Figure 4, determine if $f(x, y)$ immediately increases or decreases as

(a) the x-inputs increase starting at $(1, 0)$, and

(b) the y-inputs increase starting at $(2, 0.4)$.

Solution.

(a) Starting at $(1, 0)$ and moving straight right (or letting the x-inputs increase while the y-inputs are constant), $f(x, y)$ decreases because of moving from the contour at level 0 toward the contour at level -1.

(b) Starting at $(2, 0.4)$ and moving straight up (or letting the y-inputs increase while the x-inputs are constant), $f(x, y)$ increases because of moving from the contour at level 1 toward the contour at level 2. ∎

◆ **EXAMPLE 7** The image below shows Mount St. Helens in the state of Washington after it erupted in 1980.

Figure 5 shows a countour map corresponding to this image. The units on the contour map are meters above sea level. Use this contour map to answer each question.

(a) What is the altitude for input $(4, 3)$?

(b) What is the altitude for input $(3, 4)$?

(c) What are the approximate coordinates of the global maximum?

(d) Starting at $(5, 3)$, does the altitude immediately increase or decrease as the x-inputs increase?

(e) Starting at $(5, 3)$, does the altitude immediately increase or decrease as the y-inputs increase?

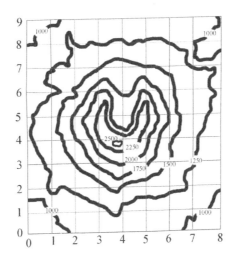

Figure 5: Contour map of Mount St. Helens for Example 7

Solution.

(a) For input $(4, 3)$, the altitude is 2000 meters.

(b) For input $(3, 4)$, the altitude is 2250 meters.

(c) The global maximum is approximately 2550 meters above sea level and the coordinates are approximately $(3.75, 3.75)$

(d) Starting at $(5, 3)$ and moving straight right (or letting the x-inputs increase while the y-inputs are constant), the altitude immediately decreases.

(e) Starting at $(5, 3)$ and moving straight up (or letting the y-inputs increase while the x-inputs are constant), the altitude immediately increases.

                                                                                                ■

➤ **QUESTION 3** Use the contour plot of $f(x, y)$ in Figure 6 to answer each question.

(a) Determine the value of $f(x, y)$ for input $(0, 1.25)$.

(b) Approximate the value of $f(x, y)$ for input $(-2.5, 2)$.

(c) Approximate the value and the coordinates of the global minimum of $f(x, y)$ on the domain shown in Figure 6.

(d) Approximate the value and the coordinates of the global maximum of $f(x, y)$ on the same domain.

(e) Starting at $(2, -2)$, determine if $f(x, y)$ immediately increases or decreases as the x-inputs increase.

(f) Starting at $(-1, 1.4)$, determine if $f(x, y)$ immediately increases or decreases as the y-inputs increase.

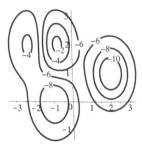

Figure 6: Contour plot for Question 3

We can often find an analytic expression for the contours of a given function $f(x, y)$ at specific levels. This fact allows us to sketch a contour plot when given an analytic representation of a multivariable function.

◆ **EXAMPLE 8** Find and sketch contours of $f(x,y) = x+y+5$ at levels $C = 2, 4, 6, 8$.

Solution. Recall that a contour of $f(x,y)$ at level C identifies all the inputs where $f(x,y) = C$. In each case, the graph plots all the inputs where $x + y + 5 = C$ or, solving this equation for y, where $y = C - 5 - x$. Thus, when $C = 2$, the function is $y = C - 5 - x = 2 - 5 - x = -3 - x$. Similarly, the contours at levels $C = 4$, 6, and 8 are $y = -1 - x$, $y = 1 - x$, and $y = 3 - x$, respectively. Therefore, all contours of $f(x,y)$ are lines with slope -1. Plotting each of these lines produces the sketch of the contour plot for $f(x,y) = x + y + 5$ given in Figure 7. ∎

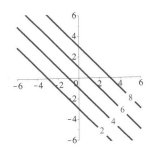

Figure 7: Contour plot of $f(x,y) = x + y + 5$ for Example 8

◆ **EXAMPLE 9** In the mid-twentieth century, Charles Cobb and Paul Douglas modeled a simplified version of the American economy between 1899 and 1922 using the multivariable function $P(K,L) = bK^\beta L^\alpha$, where P is total production, L is the total amount of labor, K is the total capital input to the production process, and b, α, β are constants (note that α is the Greek letter "alpha" and β is the Greek letter "beta"). This common model is often referred to as the **Cobb–Douglas function**. Using the method of least squares (which is studied in Chapter 3), they determined approximate values for the constants and so the model $P(K,L) = 1.01K^{0.25}L^{0.75}$. Find and sketch the contours of $P(K,L)$ at levels $C = 5, 10, 15, 20$, and 25.

Solution. All contours of the Cobb–Douglas function are of the form $C = 1.01K^{0.25}L^{0.75}$. In order to sketch the contours, solve for L as a function of K as follows:

$$C = 1.01K^{1/4}L^{3/4} \qquad \text{General form of a contour, where } 0.25 = 1/4 \text{ and } 0.75 = 3/4$$

$$\frac{C}{1.01K^{1/4}} = L^{3/4} \qquad \text{Divide by } 1.01K^{1/4}$$

$$L = \left(\frac{C}{1.01K^{1/4}}\right)^{4/3} \qquad \text{Raise both sides to the 4/3 power}$$

$$L = \left(\frac{C}{1.01}\right)^{4/3} \frac{1}{K^{1/3}} \qquad \text{Simplify using properties of exponents}$$

Next, substitute in $C = 5, 10, 15, 20,$ and 25 and simplify to obtain the following:

$$L = \frac{8.44}{K^{1/3}} \qquad L = \frac{21.26}{K^{1/3}} \qquad L = \frac{36.51}{K^{1/3}} \qquad L = \frac{53.57}{K^{1/3}} \qquad L = \frac{72.14}{K^{1/3}}$$

Graphing these curves produces the contour plot in Figure 8.   ∎

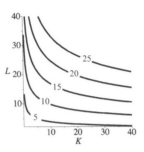

Figure 8: Contour plot of Cobb–Douglas function $P(L, K)$ for Example 9

➤ **QUESTION 4** Find and sketch the contours of $f(x, y) = y - x^2 + 1$ at levels $C = -3, -2, -1, 0, 1$.

All contour plots share two fundamentally important features.

PROPERTIES OF CONTOUR PLOTS.

- Contours at different levels cannot cross.

- The relative closeness of the contours indicates the relative steepness of the function.

The first property stating that contours at different levels cannot cross is a consequence of the fact that contour plots represent functions. If two contours at different levels crossed, there would be an input with two distinct outputs, which cannot happen because contours correspond to functions.

The second property concerning relative steepness matches our real–world experiences. For example, imagine climbing a steep mountain versus a shallow hill. When climbing a steep mountain, small changes in position (i.e., lattitude and longitude) often make a big difference in elevation. In contrast, on a relatively flat field, a small change in position usually makes a very small, if any, difference in elevation.

These physical observations tie into the second property about the closeness of contours on a contour plot. A contour indicates all of the inputs that have the same output $f(x, y)$ and so are all at the same height. Therefore, if contours of different levels are close together, then a small change in the input moving you to a higher contour indicates the relative steepness of the function. Typically, contour plots present contours with equally spaced levels so that the relative steepness of the function can be more readily ascertained.

In Figure 9, the top two graphs show a surface plot and a contour plot of a function that is relatively flat, and the bottom two graphs show a surface plot and a contour plot of a function that is relatively steep. Notice that the contours are further apart in the top contour plot because the function is comparatively flatter, while the contours are closer together in the bottom contour plot because the function is comparatively steeper.

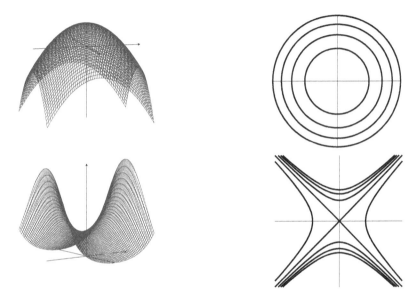

Figure 9: Closeness of contours indicating the relative steepness of a function

---

Working in RStudio

Defining and evaluating multivariable functions in RStudio is essentially the same as for single-variable functions, except that all the input variables must be explicitly identified. For example, $f(x, y) = x^2 - |y|$ is defined in RStudio by entering

```
f=makeFun(x^2-abs(y)~x&y).
```

In this command, the "~x&y" tells RStudio that both x and y are the input variables.

Multivariable functions are evaluated by substituting numbers into the name of the function. For multivariable functions, you are *strongly* encouraged to explicitly name the variables in order to ensure you are substituting the desired number for the desired variable. For example, evaluate $f(x, y) = x^2 - |y|$ for $(x, y) = (2, 8)$ by entering `f(x=2,y=8)`.

Examples of Commands:

- `f=makeFun(1.01*L^{0.75}*K^{0.25}~L&K)`

- `f(L=10,K=20)`

◆ **EXAMPLE 10** Define and evaluate the function $f(x, y, z) = x^2 + 7y - z$ for inputs $(x, y, z) = (3, 4, 5)$ and $(x, y, z) = (1.813, -4.267, 5.1)$ using RStudio.

Solution. For $f(3, 4, 5)$, the function values are entered in two different orders to illustrate that the order of stating the values does not matter, provided the value being assigned to each variable is explicitly identified:

`f=makeFun(x^2+7*y-z~x&y&z)`
`f(x=3,y=4,z=5)`

`[1] 32`

`f(z=5,y=4,x=3)`

`[1] 32`

`f(x=1.813,y=-4.267,z=5.1)`

`[1] -31.68203`

Based on this RStudio output, $f(3, 4, 5) = 32$ and $f(1.813, -4.267, 5.1) \approx -31.682$. ∎

➤ **QUESTION 5** Define and evaluate the function $g(x, y) = |x| - y^3 + \pi$ for each input (x, y) using RStudio:

(a) $(1, 2)$ (b) $(-1, 2)$ (c) $(4.65, 1.1)$ (d) $(4.65, -1.1)$

Working in RStudio

RStudio plots functions of two variables using the `plotFun` command, following the same basic form as with single-variable functions. For example, RStudio graphs the surface plot of $f(x, y) = x^2 + y^2$ by entering

    `plotFun(x^2+y^2~x&y,surface=TRUE)`.

As when defining multivariable functions, the input variables are separated with the ampersand symbol "&". Alternatively, RStudio graphs a contour plot of a multivariable function by omitting the `surface=TRUE` option in the `plotFun` command; for example, enter `plotFun(x^2+y^2~x&y)`. Without specifying additional options, the contour plot includes color shading to help indicate the height of the function. If you prefer an unshaded contour plot, then include the option `filled=FALSE`; for example, enter `plotFun(x^2+y^2~x&y,filled=FALSE)`. The levels to be drawn can be specified with the option `levels=c()`, where the desired levels are included in a comma-separated list inside of `c()`. For example, RStudio graphs a contour plot of $f(x, y) = x^2 + y^2$ with levels $C = 0, 1, 2, 3, 4$ by entering

```
plotFun(x^2+y^2~x&y,filled=FALSE,levels=c(0,1,2,3,4),
        xlim=range(-3,3),ylim=range(-3,3)).
```

The inputs are adjusted with `xlim=range(-3,3)` and `ylim=range(-3,3)` so that the desired contours can be seen in the resulting plot. With some practice, you will become adept at making such adjustments.

The `manipulate` package enables the rotation of surface plots and viewing them from different angles. This is accomplish by clicking the wheel in the top left corner of the surface plot.

Examples of Commands:

- `plotFun(x^2-y^2~x&y,surface=TRUE,xlim=range(-5,5),`
  `ylim=range(-5,5))`

- `plotFun(x^2-y^2~x&y,filled=FALSE,xlim=range(-5,5),`
  `ylim=range(-5,5), levels=c(0,1,2,3,4))`

◆ **EXAMPLE 11** Let $f(x, y) = 5 - x - y$.

(a) Graph a surface plot of $f(x, y)$ for $-5 \le x \le 5$ and $-5 \le y \le 5$ using RStudio.

(b) Graph a contour plot of $f(x, y)$ with contours at levels $C = -4, -2, 0, 2, 4$ for inputs $-5 \le x \le 5$ and $-5 \le y \le 5$ using RStudio.

Solution.

(a) `plotFun(5-x-y~x&y,surface=TRUE,xlim=range(-5,5),ylim=range(-5,5))`

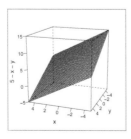

(b) `plotFun(5-x-y~x&y,filled=FALSE,xlim=range(-5,5),ylim=range(-5,5),`
  `levels=c(0,2,4,6,8,10),lwd=5)`

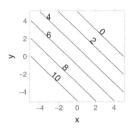

➤ **QUESTION 6** Let $f(x, y) = x^2 - y$.

(a) Graph a surface plot of $f(x, y)$ for $-3 \leq x \leq 3$ and $-3 \leq y \leq 3$ using RStudio.

(b) Graph a contour plot of $f(x, y)$ with contours at levels $C = 0, 2, 4, 6, 8, 10$ for inputs $-3 \leq x \leq 3$ and $-3 \leq y \leq 3$ using RStudio.

Summary

- A *surface plot* of $f(x, y)$ is a three-dimensional graph where the value of each $f(x, y)$ is plotted as a height measured relative to the z-axis above or below the corresponding point (x, y) on the xy-plane.

- A *contour of $f(x, y)$ at level C* is a curve in the xy-plane containing all points where $f(x, y) = C$. A *contour plot* or *contour map* of $f(x, y)$ is a two-dimensional graph in the xy-plane of contours of $f(x, y)$ at various levels C. Typically, contours at equally spaced levels are used.

- Among the properties of contour plots are that contours at different levels cannot cross and the relative closeness of the contours indicates the relative steepness of the function.

- The *extreme values* of a function $f(x, y)$ are the maximum and minimum values of the function.

- The *global maximum* of a function $f(x, y)$ on a domain D is the numeric value $f(a, b)$ for a point (a, b) in the domain D such that $f(a, b) \geq f(x, y)$ for all (x, y) in D.

- The *global minimum* of $f(x, y)$ on a domain D is the numeric value $f(a, b)$ for a point (a, b) in the domain D such that $f(a, b) \leq f(x, y)$ for all (x, y) in D.

Exercises

In Exercises 1–4, give an example of a real-life input–output relationship with the specified property.

1. A single-variable function

2. A two-variable function

3. A three-variable function

4. A four-variable function

In Exercises 5–8, evaluate the function $f(x,y) = xy - x^2$ for the point (x,y).

5. $(1,3)$ 7. $(5.14, -1.97)$

6. $(-6,7)$ 8. $(-2.1, -7.6)$

In Exercises 9–12, evaluate the function $f(x,y) = x^2y^3 - y^2 + x$ for the point (x,y).

9. $(3,-2)$ 11. $(-0.31, 0)$

10. $(-4,-1)$ 12. $(-1.1, 6.2)$

In Exercises 13–16, evaluate the function $f(x,y,z) = xy + yz + z^2$ for the point (x,y,z).

13. $(1,1,1)$

14. $(0,-3,5)$

15. $(-1.85, 3.67, 0)$

16. $(2.1, -3.2, -4.3)$

In Exercises 17–20, evaluate the function $f(w,x,y,z) = xyz - 3w + 2$ for the point (w,x,y,z).

17. $(-1,3,7,-2)$

18. $(-6.7, 0, -8.9, 5)$

19. $(1, 3.1, 6, -9.4)$

20. $(71, 9.2, -0.1, 4)$

In Exercises 21–26, determine and explain the meaning of the output value for the given input to the multivariable function in the table, which presents the Body Mass Index values for males by percentile (in the top row) and age (in the left column).

	10	25	50	75	90
20s	20.7	22.9	25.6	29.9	33.8
30s	22.4	24.9	28.1	32.0	36.2
40s	22.9	25.4	28.2	31.7	36.1
50s	22.9	25.5	28.2	32.0	37.1
60s	22.7	25.3	28.8	32.5	37.0

21. $(50, 40-49)$ 24. $(50, 20-29)$

22. $(90, 30-39)$ 25. $(10, 30-39)$

23. $(25, 50-59)$ 26. $(75, 60-69)$

In Exercises 27–32, determine and explain the meaning of the output value for the given input to the multivariable function given in the table, which presents the sea surface temperatures in degrees Fahrenheit in the North Atlantic by latitude (in the top row) and longitude (in the left column).

	−68	−66	−64	−62	−60
44	48.2	50	50	51.8	48.2
42	57.2	51.8	60.8	68	60.8
40	77	75.2	73.4	75.2	73.4
38	75.2	78.8	75.2	75.2	71.6
36	75.2	73.4	75.2	73.4	73.4

27. $(-68, 40)$ 30. $(-62, 38)$

28. $(-68, 36)$ 31. $(-60, 44)$

29. $(-64, 42)$ 32. $(-66, 40)$

In Exercises 33–40, answer the question about the following contour plot of a function $f(x, y)$:

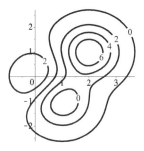

33. Find the value of $f(2.5, 0)$.

34. Find the value of $f(1, 1)$.

35. At $(2.5, 0)$, does $f(x, y)$ immediately increase or decrease as the x-inputs increase?

36. At $(1, 1)$, does $f(x, y)$ immediately increase or decrease as the x-inputs increase?

37. At $(2.5, 0)$, does $f(x, y)$ immediately increase or decrease as the y-inputs increase?

38. At $(1, 1)$, does $f(x, y)$ immediately increase or decrease as the y-inputs increase?

39. Approximate the value and the coordinates of the global minimum of $f(x, y)$.

40. Approximate the value and the coordinates of the global maximum of $f(x, y)$.

In Exercises 41–48, answer the question about the following contour plot of a function $f(x, y)$:

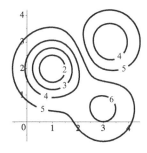

41. Find the value of $f(3, 0)$.

42. Find the value of $f(1, 3)$.

43. At $(3, 0)$, does $f(x, y)$ immediately increase or decrease as the x-inputs increase?

44. At $(1, 3)$, does $f(x, y)$ immediately increase or decrease as the x-inputs increase?

45. At $(3, 0)$, does $f(x, y)$ immediately increase or decrease as the y-inputs increase?

46. At $(1, 3)$, does $f(x, y)$ immediately increase or decrease as the y-inputs increase?

47. Approximate the value and the coordinates of the global minimum of $f(x, y)$.

48. Approximate the value and the coordinates of the global maximum of $f(x, y)$.

In Exercises 49–56, answer the question about the following contour plot of a function $f(x, y)$.

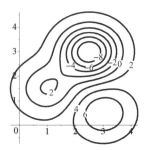

49. Find the value of $f(1, 2.5)$.

50. Find the value of $f(4, 2.5)$.

51. At $(1, 2.5)$, does $f(x, y)$ immediately increase or decrease as the x-inputs increase?

52. At $(4, 2.5)$, does $f(x, y)$ immediately increase or decrease as the x-inputs increase?

53. At $(1, 2.5)$, does $f(x, y)$ immediately increase or decrease as the y-inputs increase?

54. At $(4, 2.5)$, does $f(x, y)$ immediately increase or decrease as the y-inputs increase?

55. Approximate the value and the coordinates of the global minimum of $f(x, y)$.

56. Approximate the value and the coordinates of the global maximum of $f(x, y)$.

In Exercises 57–62, match the contour plot with its corresponding surface plot.

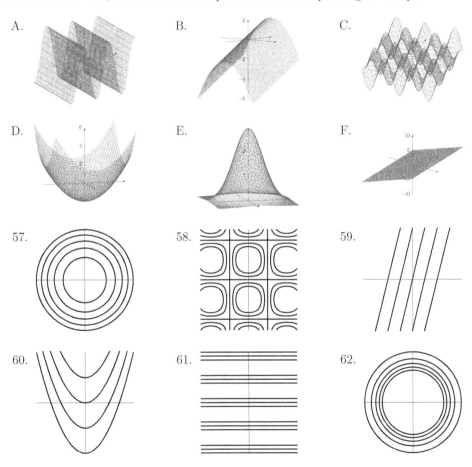

In Exercises 63–68, sketch the contour plot of $f(x, y)$ with the given levels C.

63. $f(x, y) = 2x + y$; $C = -4, 0, 4, 8$

64. $f(x, y) = 2y - 3x - 7$; $C = -1, 0, 2, 5$

65. $f(x, y) = xy$; $C = -3, -1, 1, 3, 5$

66. $f(x, y) = x^2 - y$; $C = -2, -1, 0, 1, 2$

67. $f(x, y) = x - y^2$; $C = -5, 0, 5, 10, 15$

68. $f(x, y) = x^2 + y^2$; $C = 0, 3, 6, 9, 12$

RStudio. In Exercises 69–72, use RStudio to define $f(x, y) = |x + 5y^3|$ and evaluate it for the point (x, y).

69. $(-10, 4)$

71. $(-\pi, 2\pi^3)$

70. $(-3.62, -4.81)$

72. $(42.6, 223.9)$

RStudio. In Exercises 73–76, use RStudio to define $f(w, x, y, z) = w - x \cdot z^2 + 5y$ and evaluate it for the point (w, x, y, z).

73. $(1, 2, 3, 4)$

74. $(-3.8, 2.1, 32, -4.6)$

75. $(9.623, 1.579, -4.332, 21.65)$

76. $(2434, -890345, 4574, -34917)$

RStudio. In Exercises 77–82, use RStudio to create a surface plot of the function.

77. $f(x, y) = x + y + 2$

78. $f(x, y) = xy + x$

79. $f(x, y) = x - y^2 + 2$

80. $f(x, y) = x^2 + y^2$

81. $f(x, y) = x^2 - xy + y^2$

82. $f(x, y) = xy - x^2 y^3$

RStudio. In Exercises 83–88, use RStudio to create a contour plot of the function with five distinct levels.

83. $f(x, y) = x + y + 2$

84. $f(x, y) = xy + x$

85. $f(x, y) = x - y^2 + 2$

86. $f(x, y) = x^2 + y^2$

87. $f(x, y) = x^2 - xy + y^2$

88. $f(x, y) = xy - x^2 y^3$

In Your Own Words. In Exercises 89–94, explain the following.

89. Multivariable function

90. Evaluation a multivariable function for an input

91. Surface plot

92. Contour plot

93. Contour

94. Properties of contour plots

1.3   Linear Functions

Our study of the most commonly used modeling functions begins with linear functions, which are the simplest functions and provide the most basic models. Despite being relatively simple, linear models describe many aspects of reality quite well. The word *line* often brings to mind the equation $y = mx + b$, and a *linear function* is typically written in the form $f(x) = mx + b$. Linear functions are special in many ways, including the fact that they are completely determined by just two input–output pairs (x_1, y_1) and (x_2, y_2). Even more, there exists exactly one single-variable linear function corresponding to the line through any such pair of points.

Definition. A **linear function** is a function of the form $f(x) = mx + b$.

- The **slope** m of the line through the points (x_1, y_1) and (x_2, y_2) can be calculated in a variety of ways, including the following when $x_1 \neq x_2$:

$$m = \frac{y_2 - y_1}{x_2 - x_1} = \frac{y_1 - y_2}{x_1 - x_2} = \frac{\Delta y}{\Delta x} = \frac{\text{rise}}{\text{run}}$$

  A vertical line $x = a$ has an *undefined* slope.

- The **vertical intercept** of a line is the y-coordinate b of the point $(0, b)$ where the line intersects the y-axis, if such a point exists.

- The **horizontal intercept** of a line is the x-coordinate a of the point $(a, 0)$ where the line intersects the x-axis, if such a point exists.

As indicated by the preceding definition, the slope of a line is usually found by computing the difference of the y-coordinates in the numerator and the difference of the x-coordinates in the denominator in either order, provided a consistent ordering is used. In fact, the distinguishing feature of a line is that the slopes between any two points on the line have the same numeric value. The tradition of using m to denote the slope of a line originated in France based on the word "monter," which translates into English as "to climb" or "to rise." Figure 1 provides a generic illustration of a line and the components of the slope.

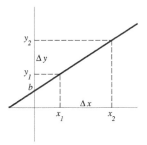

Figure 1: The graph of a line $y = mx + b$

◆ **EXAMPLE 1** Find the equation and sketch the graph of the line through $(1, 2)$ and $(6, 3)$.

Solution. The generic equation of a line has the form $y = mx + b$, so both the slope m and the vertical intercept b must be found. First, the slope is computed with the two possible orders of differences in the numerator and denominator presented as follows; observe that they produce the same numeric value:

$$m = \frac{3 - 2}{6 - 1} = \frac{1}{5} = 0.2 \qquad\qquad m = \frac{2 - 3}{1 - 6} = \frac{-1}{-5} = 0.2$$

Substituting 0.2 for m, the equation of this line is of the form $y = 0.2x + b$. The vertical intercept b is obtained by substituting either of the two given points into this intermediate linear function. This example uses $(1, 2)$ as follows:

$$y = 0.2x + b \qquad \text{Intermediate version of the linear function}$$
$$2 = (0.2) \cdot 1 + b \qquad \text{Substitute } (x, y) = (1, 2)$$
$$2 - 0.2 = b \qquad \text{Subtract 0.2 from both sides}$$
$$1.8 = b \qquad \text{Simplify}$$

Therefore, the equation of the line through $(1, 2)$ and $(6, 3)$ is $y = 0.2x + 1.8$.

The basic approach to graphing a line is to plot any two points on the line, and then draw the line through those two points, as illustrated in Figure 2. Graph (a) shows the plot of the two given points $(1, 2)$ and $(6, 3)$ with the line then drawn through them. Alternatively, Graph (b) plots the vertical intercept $(0, b) = (0, 1.8)$ on the y-axis and uses the slope to find some second point; for example, $(4, f(4)) = (4, 0.2 \cdot 4 + 1.8) = (4, 2.6)$ lies on the line. The line is then drawn through those two points.

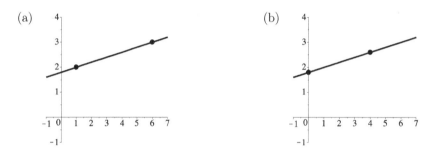

Figure 2: The graph of the line $y = 0.2x + 1.8$ for Example 1

> **QUESTION 1** Find the equation and sketch the graph of the line through $(-1, -3.5)$ and $(4.5, 5)$.

In some settings, determining the horizontal intercept of a linear model will be of interest. This process is illustrated as follows, along with showing how to graph a line using its two intercepts.

◆ **EXAMPLE 2** Find the horizontal intercept of the line $y = 0.5x + 7$, and sketch the graph of this line using its vertical and horizontal intercepts.

Solution. The horizontal intercept of a line is found by setting $y = 0$ and solving for x. This approach gives the following computations:

$$0 = 0.5x + 7 \qquad \text{Substitute } y = 0 \text{ into the linear function}$$
$$-7 = \frac{x}{2} \qquad \text{Subtract 7 from both sides}$$
$$(-7) \cdot 2 = x \qquad \text{Multiply both sides by 2}$$
$$-14 = x \qquad \text{Simplify}$$

Therefore, $x = -14$ is the horizontal intercept of $y = 0.5x + 7$. From the equation of the line, the vertical intercept is $y = 7$. The following graph plots the horizontal intercept $(-14, 0)$, the vertical intercept $(0, 7)$, and the line through these two points:

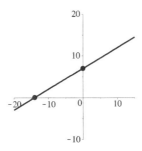

▶ **QUESTION 2** Find the horizontal intercept of the line $y = 3x - 9$, and sketch the graph of this line using its vertical and horizontal intercepts.

In addition to algebraically determining the vertical and horizontal intercepts of a function, finding where two given lines intersect will also be of interest. Sometimes two lines do not intersect, because they are parallel, but most often a pair of lines do intersect and always do so at a single point. For linear functions of the form $y = mx + b$ and $y = nx + c$ with $m \neq n$, the point of intersection is found by setting the right side of the linear equations equal to each other and solving for x as follows.

$$mx + b = nx + c \qquad \text{Equate the right sides of the linear functions}$$
$$mx - nx = c - b \qquad \text{Subtract } nx \text{ and } b \text{ from both sides}$$
$$(m - n)x = c - b \qquad \text{Factor } x \text{ from the expression on the left}$$
$$x = \frac{c - b}{m - n} \qquad \text{Divide both sides by } m - n$$

These computations produce the x-coordinate of the point of intersection. The y-coordinate is then determined by substituting this value for x into either of the original linear functions. While the above calculations do provide a formula for the x-coordinate of the point of intersection, it is generally easier to reproduce the algebraic steps for two specific functions than to memorize this formula. When $m = n$, the linear functions $y = mx + b$ and $y = nx + c$ are parallel and either they do not intersect (when $b \neq c$) or they are identical to each other (when $b = c$).

◆ **EXAMPLE 3** Find the point of intersection for each pair of lines, or explain why the lines do not intersect:

(a) $y = 3x + 4$
     $y = -x - 16$

(b) $y = 2x - 3$
     $y = 2x + 5$

(c) $y = 3$
     $x = 7$

Solution.

(a) Setting the right side of the linear functions equal to each other and solving for the x-coordinate of the point of intersection gives the following:

$$3x + 4 = -x - 16 \qquad \text{Equate the right sides of the linear functions}$$
$$3x - (-x) = -16 - 4 \qquad \text{Subtract } -x \text{ and 4 from both sides}$$
$$4x = -20 \qquad \text{Simplify}$$
$$x = \frac{-20}{4} = -5 \qquad \text{Divide both sides by 4}$$

Substituting $x = -5$ into either of the linear functions provides the y-coordinate of the point of intersection. Working with the first linear function gives $y = 3(-5) + 4 = -15 + 4 = -11$. Therefore, the lines $y = 3x + 4$ and $y = -x - 16$ intersect at $(-5, -11)$. This point of intersection is illustrated in Figure 3(a).

(b) The lines both have a slope of 2, but different vertical intercepts, so they do not intersect, as illustrated in Figure 3(b).

(c) The horizontal line $y = 3$ and the vertical line $x = 7$ intersect at the point $(7, 3)$, as shown in Figure 3(c).

■

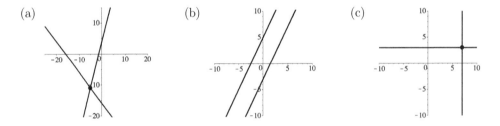

Figure 3: Points of intersection for Example 3

▶ **QUESTION 3** Find the point of intersection for each pair of lines, or explain why the lines do not intersect:

(a) $y = -2x + 7$
    $y = 5x + 9$

(b) $y = 3$
    $x = 7$

(c) $y = 6$
    $y = 2$

Parameters of Linear Functions

A key element of modeling linear data sets is thinking about the slope m and the vertical intercept b of a line as **parameters** that can take on various values. When a linear function appears to be the most reasonable model of a given data set, values for the parameters m and b are conjectured to define the equation $y = mx + b$ of such a linear model. In order to facilitate such work in conjecturing models, the effect of the parameters m and b on the graph of a line is examined in detail. Recall the following formulas for computing the slope m of a line:

$$m = \frac{y_1 - y_2}{x_1 - x_2} = \frac{y_2 - y_1}{x_2 - x_1} = \frac{\Delta y}{\Delta x} = \frac{\text{rise}}{\text{run}}$$

Usually, the input values x are thought of as increasing from some x_1 to a greater value x_2. Adopting this perspective, the denominator Δx (or the change in x) is always positive, and the sign of m is determined by the difference between the corresponding outputs y_1 and y_2 appearing in the numerator. Example 4 highlights the interconnections between the slope of a line and the graphical behavior of the line.

◆ **EXAMPLE 4** Identify each line in Figure 4 as one of $y = mx + 4$ with $m = -2$, $m = 0$, or $m = 2$. The vertical intercept of $b = 4$ for all three lines ensures that their different graphical behaviors depend only on the slope.

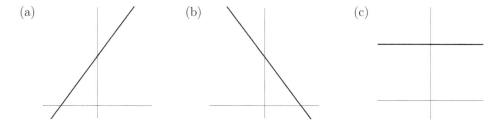

Figure 4: Slope m as a parameter for Example 4

Solution. The slope of line (a) is positive because the output values increase as the inputs increase (i.e., moving left to right across the graph). Thus, line (a) has slope $m = 2$. The slope of line (b) is negative because the output values decrease when moving left to right; consequently, the slope of line (b) is $m = -2$. Finally, the slope of line (c) is zero because the outputs do not change as the inputs increase. ∎

This example illustrates the effect of the slope parameter m on linear functions. In general, the sign of the slope of a line determines which the direction in which the line tilts. Reading inputs from left to right, lines with positive slopes have output values that are becoming larger, and such lines are tilted upward, running from the bottom left to the top right portion of the Cartesian plane. Lines with negative slopes have output values that are decreasing, and such lines are tilted downward, running from the top left to the bottom right portion of the plane. Lines with slope zero have the same output value for every possible input and are horizontal. This information can be summarized as follows:

THE EFFECT OF SLOPE ON LINES. Let $f(x) = mx + b$ be a linear function.

- If m is positive, then the outputs increase as the inputs increase.

THE EFFECT OF SLOPE ON LINES. (CONTINUED)

- If m is zero, then the outputs remain the same as the inputs increase and the line is horizontal.

- If m is negative, then the outputs decrease as the inputs increase.

All types of lines have been considered, with the exception of vertical lines $x = a$. Working graphically, the vertical line test from Section 1.1 shows that a vertical line $x = a$ does not define a function. In particular, the vertical line $x = a$ intersects itself more than once (in fact, infinitely often) because every real number on the y-axis is an output. When attempting to compute the slope of such a line, the only choices for input values are $x_1 = a$ and $x_2 = a$, which gives $\Delta x = x_2 - x_1 = a - a = 0$. The slope formula states that $m = \Delta y / \Delta x$, but dividing by zero is not allowed. For this reason, the slope of a vertical line $x = a$ is said to be *undefined*.

Definition. The slope of a line is **undefined** if and only if the line is vertical.

➤ **QUESTION 4** Determine whether the slope of each line in Figure 5 is positive, zero, negative, or undefined.

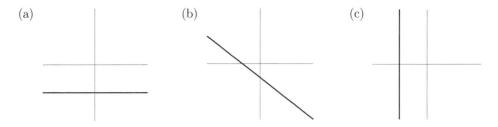

Figure 5: Slope m as a parameter for Question 4

At this point, the effect of the parameter m on the graph of a linear function $y = mx + b$ is hopefully well understood. In a similar fashion, Example 5 examines the effect of the parameter b on the graph of a linear function.

◆ **EXAMPLE 5** Identify each line in Figure 6 as one of $y = x + b$ with $b = -2$, $b = 0$, or $b = 2$. The slope of $m = 1$ for all three of these lines ensures that their different graphical behaviors depend only on the vertical intercept.

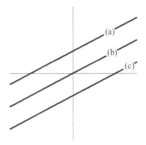

Figure 6: Vertical intercept b as a parameter for Example 5

Solution. Line (a) passes through the positive y-axis, which means that the vertical intercept is positive and line (a) is $y = x + 2$. Line (b) passes through the origin, which means that the vertical intercept is zero and line (b) is $y = x + 0 = x$. Finally, line (c) passes through the negative y-axis, which means that the vertical intercept is negative and line (c) is $y = x - 2$. ∎

This example illustrates the effect of the vertical intercept parameter b on linear functions. In general, the sign of the vertical intercept determines whether the line passes through the positive y-axis, the origin, or the negative y-axis. This information is summarized as follows:

THE EFFECT OF THE VERTICAL INTERCEPT ON LINES.
Let $f(x) = mx + b$ be a linear function. The vertical intercept b affects the graph of $f(x)$ by shifting the line up or down along the vertical axis, depending on the sign of b:

- If b is positive, then the line passes through the positive y-axis.

- If b is zero, then the line passes through the origin.

- If b is negative, then the line passes through the negative y-axis.

➤ **QUESTION 5** Determine whether the vertical intercept of each line in Figure 7 is positive, zero, or negative.

Monotonicity

An important characteristic of a function is where the function is increasing, constant, or decreasing. These behaviors are referred to collectively as the *monotonic behavior* or the *monotonicity* of a function. Monotonic behavior occurs in diverse real-life situations such as whether the value of the stock market is increasing or decreasing, whether the bloodstream concentration of a medication is increasing or decreasing, and whether the temperature is increasing or decreasing.

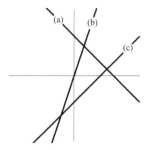

Figure 7: Vertical Intercept b as a parameter for Question 5

This section examines the monotonic behavior of linear functions from a graphical perspective based on their slope. Later, in Chapter 4, Section 4.2 will introduce the derivative as a means to investigate the monotonic behavior of functions. First consider the mathematical definitions of increasing, constant, and decreasing:

Definition. The **monotonicity** of a function $f(x)$ on an open interval (a, b) is defined as follows:

- $f(x)$ is **increasing** on (a, b) if and only if for all numbers c and d in (a, b), when $c < d$ then $f(c) < f(d)$.

- $f(x)$ is **constant** on (a, b) if and only if for all numbers c and d in (a, b), $f(c) = f(d)$.

- $f(x)$ is **decreasing** on (a, b) if and only if for all numbers c and d in (a, b), when $c < d$ then $f(c) > f(d)$.

These definitions provide a mathematically precise way to express functional behavior that is readily identifiable from a graph. When describing functions as increasing, constant, or decreasing, think about input values becoming larger, which corresponds to reading inputs from left to right on a graph.

When reading inputs from left to right, if the output values of a function become larger, then the function is increasing. Visually, the graph of an increasing function moves upwards as you move from left to right. Similarly, reading inputs from left to right, if the output values become smaller, then the function is decreasing. Visually, the graph of a decreasing function moves downwards as you move from left to right. Finally, a function is constant if every input produces the same output, which produces a horizontal line.

◆ **EXAMPLE 6** Identify the intervals on which each function in Figure 8 is increasing, constant, or decreasing, and state the sign of the slope of the line.

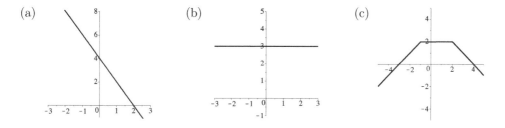

Figure 8: Monotonic behavior of linear functions for Example 6

Solution. Line (a) is decreasing on $(-\infty, \infty)$ and its slope is negative. Line (b) is constant on $(-\infty, \infty)$ and its slope is zero. Finally, the function (c) exhibits a blend of all three behaviors: increasing on $(-\infty, -1)$ where its slope is positive; constant on $(-1, 2)$ where its slope is zero; and decreasing on $(2, \infty)$ where its slope is negative. Note that a monotonic behavior for the function (c) is not identified at the "corners" corresponding to the inputs $x = -1$ and $x = 2$. ∎

▶ **QUESTION 6** Identify the intervals on which each function in Figure 9 is increasing, constant, or decreasing, and state the sign of the slope of the line.

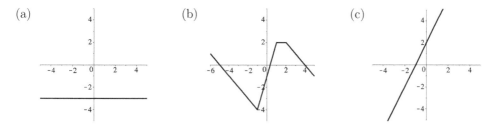

Figure 9: Monotonic behavior of linear functions for Question 6

◆ **EXAMPLE 7** Identify the intervals on which the piecewise function $f(x)$ is increasing, constant, or decreasing:

$$f(x) = \begin{cases} 7x & x < -2 \\ -3x + 5 & -2 \leq x < 5 \\ \pi & 5 \leq x \end{cases}$$

Solution. Each piece of this function is linear, which allows determination of the monotonic behavior of $f(x)$ based on the slope of each piece. On $(-\infty, -2)$, the slope of $f(x)$ is $m = 7$ and positive, which means that $f(x)$ is increasing on this interval. On $(-2, 5)$, the slope of $f(x)$ is $m = -3$ and negative, which means that $f(x)$ is decreasing on this interval. Finally, on $(5, \infty)$, the slope of $f(x)$ is zero, which means that $f(x)$ is constant on this interval. As with graphical piecewise functions, the monotonic behavior of such a function is (usually) not identified at the endpoint(s) of each piece. ∎

➤ **QUESTION 7** Identify the intervals on which $f(x)$ is increasing, constant, or decreasing:

$$f(x) = \begin{cases} -x & x < 0 \\ x & 0 \le x \end{cases}$$

Working in RStudio

Recall that RStudio graphs functions using the `plotFun` command. For example, RStudio graphs the line $y = -4x + 8$ on the interval $[-5, 5]$ by entering

        `plotFun(-4*x+8~x,xlim=range(-5,5))`

A second plot can be added to the graph using the argument **add=TRUE**. For example, the graph of $y = 3x + 1$ is added to the graph of $y = -4x + 8$ (which has already been plotted) by entering

        `plotFun(3*x+1~x,add=TRUE)`

An input interval is not included with this second `plotFun` command, because the option `xlim=range(-5,5)` in the first `plotFun` determines the plotting window for any subsequent functions added to the original plot.

Examples of Commands

- `plotFun(5*x-3~x,xlim=range(-5,5))`

- `plotFun(3*x+1~x,add=TRUE)`

◆ **EXAMPLE 8** Graph $y = 4x - 5$ and $y = -2x + 1$ on the same pair of axes using RStudio.

Solution. First, $y = 4x + 7$ is graphed and then $y = -2x + 1$ is added to the plot:

```
plotFun(4*x+7~x,xlim=range(-10,10))
plotFun(-2*x+1~x,add=TRUE)
```

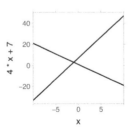

➤ **QUESTION 8** Graph $y = -x + 5$ and $y = 3x - 2$ on the same pair of axes using RStudio.

Working in RStudio

RStudio finds the horizontal intercept of a line and the point of intersection of two lines using the `findZeros` command. This command expects three inputs: the function and its input variable separated by ~, and an input interval on which the zero(s) occur. For example, the horizontal intercept of the line $y = 7x - 1$ is found (which corresponds to solving $7x - 1 = 0$) on the input interval $(0, 1)$ by entering

> `findZeros(7*x-1~x,xlim=range(0,1))`

If the input interval is not apparent, an approximate interval where a zero occurs can be obtained by graphing the line using the `plotFun` command and visually identifying the needed interval.

In order to find where $f(x) = g(x)$, first rearrange the equation by subtracting $g(x)$ from both sides to obtain $f(x) - g(x) = 0$. Now find where the function $f(x) - g(x)$ is equal to zero. If necessary, an approximate interval for where the lines intersect can be obtained by graphing the two lines on the same pair of axes and visually identifying the needed interval. For example, find where the lines $y = 2x + 3$ and $y = 4x - 5$ intersect on input interval $(0, 10)$ by entering

> `findZeros(((2*x+3)-(4*x-5))~x,xlim=range(0,10))`

Examples of Commands:

- `findZeros(4*x-3~x,xlim=range(-1,2))`

- `findZeros(((9*x-1)-(11*x))~x,xlim=range(-2,0))`

◆ **EXAMPLE 9** Find the point where $y = 4x + 7$ and $y = -2x + 1$ intersect using RStudio.

Solution. The goal is to find where $4x + 7 = -2x + 1$ or, equivalently, where $4x + 7 - (-2x) - 1 = 0$. Simplifying produces $4x + 7 + 2x - 1 = 0$. Based on the graph from Example 8, the point of intersection lies in the interval $(-5, 0)$:

```
findZeros(4*x+7+2*x-1~x,xlim=c(-5,0))
```

```
    x
1  -1
```

Thus, the two lines intersect when $x = -1$ and so at the point $(-1, 3)$, where the y-coordinate of 3 is obtained by substituting $x = -1$ into either linear function. ■

▶ **QUESTION 9** Find the point where $y = -x + 5$ and $y = 3x - 2$ intersect using RStudio.

Summary

- A *linear function* is a function of the form $y = mx + b$.

- The *slope* of a line, denoted by m, through the points (x_1, x_2) and (y_1, y_2) can be calculated using the following techniques, when $x_1 \neq x_2$:

$$m = \frac{y_2 - y_1}{x_2 - x_1} = \frac{y_1 - y_2}{x_1 - x_2} = \frac{\Delta y}{\Delta x} = \frac{\text{rise}}{\text{run}}$$

  A vertical line $x = a$ has an *undefined* slope.

- The *vertical intercept* of a line is the y-coordinate b of the point $(0, b)$ where the line intersects the y-axis, and the *horizontal intercept* of a line is the x-coordinate a of the point $(a, 0)$ where the line intersects the x-axis, if such points exist.

- The slope m and the vertical intercept b of a line can be treated as *parameters* that take on different values. The sign of the slope of a line determines the direction in which the line "tilts." The vertical intercept of a line determines where the line passes through the y-axis.

- The *monotonicity* of a function $f(x)$ on an open interval (a, b) is defined as follows:

  - $f(x)$ is *increasing* on (a, b) if and only if for all numbers c and d in (a, b), when $c < d$ then $f(c) < f(d)$.

  - $f(x)$ is *constant* on (a, b) if and only if for all numbers c and d in (a, b), $f(c) = f(d)$.

  - $f(x)$ is *decreasing* on (a, b) if and only if for all numbers c and d in (a, b), when $c < d$ then $f(c) > f(d)$.

- If $f(x) = mx + b$ is a linear function, the monotonic behavior of $f(x)$ is determined by the sign of its slope.

Exercises

In Exercises 1–8, find the equation and sketch the graph of the line through the pair of points.

1. $(1, -5)$ and $(3, 1)$

2. $(3, 4)$ and $(-6.7, 4)$

3. $(-3, -6)$ and $(3, -1)$

4. $(4, \pi)$ and $(4, -\pi)$

5. $(-7, -6)$ and $(-5, -6)$

6. $(6.2, -5.4)$ and $(6.2, 8.7)$

7. $(-2.3, 18.6)$ and $(3.7, -4.5)$

8. $(3.2, 13.1)$ and $(-32.8, -16.2)$

Your Turn. In Exercises 9–14, choose, if possible, a point in the of the given quadrants (as shown below) so that the line through the two points has the requested slope. Find the equation and sketch the graph of the corresponding line.

$$\text{II} \mid \text{I}$$
$$\overline{\phantom{\text{III}}\mid\phantom{\text{IV}}}$$
$$\text{III} \mid \text{IV}$$

9. Points in quadrants I and II for a line with positive slope

10. Points in quadrants I and II for a line with negative slope

11. Points in quadrants I and IV for a line with undefined slope

12. Points in quadrants I and IV for a line with zero slope

13. Points in quadrants II and III for a line with positive slope

14. Points in quadrants III and IV for a line with zero slope

In Exercises 15 – 24, find the equation and sketch the graph of the line with the slope passing through the point.

15. $m = 2$ through $(0, -5)$

16. $m = 1$ through $(-1.5, 5)$

17. $m = -3$ through $(-2, -5.5)$

18. $m = -2$ through $(-3.4, 6.1)$

19. $m = 2.6$ through $(0, -9)$

20. $m = 1.5$ through $(9.1, -6.2)$

21. $m = 0$ through $(-7.5, 6.4)$

22. $m = 0$ through $(\pi, -7)$

23. m undefined through $(-\pi, 2)$

24. m undefined through $(-2.5, -4.25)$

In Exercises 25 – 34, find the vertical and horizontal intercepts of the line, if possible, and use these points to sketch the graph of the line.

25. $y = 4x + 8$

26. $y = 3x - 9$

27. $y = -8x + 24$

28. $y = -5x - 21$

29. $y = -5x$

30. $y = -4.3x + 6.8$

31. $y = -23.2$

32. $x = 17.9$

33. $y = 3.6x + 19.2$

34. $y = -7.8x - 8.6$

In Exercises 35 – 44, find the point of intersection of the pair of lines, if possible. If not, explain why.

35. $y = 4x + 8$
    $y = 2x - 6$

36. $y = 3x - 9$
    $y = 6x + 2$

37. $y = 5x - 11$
    $y = 10x - 36$

38. $y = 8x + 3$
    $y = 7x + 2$

39. $y = -5x - 21$
    $y = -5x + 7$

40. $y = -5x - 3$
    $y = -5x - 3$

41. $y = -4.3x + 6.8$
    $x = 2.1$

42. $y = -23.2$
    $x = 15.7$

43. $y = 3.6$
    $y = 19.2$

44. $y = -7.8x - 8.6$
    $y = 4.2$

In Exercises 45 – 50, identify the slope and the vertical intercept of the line as positive, zero, or negative, if they exist, and approximate their values.

45.

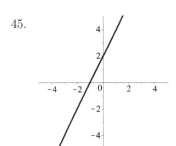

46.

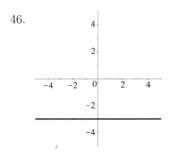

47.

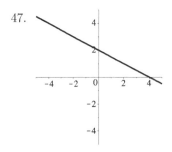

48.

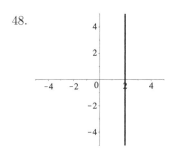

49.

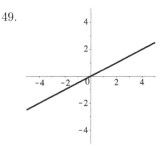

50.

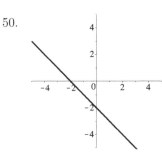

In Exercises 51 – 54, consider how many millions of people used Twitter (U) during each quarter of a year since 2000 (Q).

Q	11	11.25	11.5	11.75	12.75
U	68	85	101	117	185

The linear function $f(x) = 66.8x - 666.6$ is a reasonable model of this data set. State the value and explain the meaning of the quantity.

51. $f(12)$

52. Slope of $f(x)$

53. Vertical intercept of $f(x)$

54. Horizontal intercept of $f(x)$

In Exercises 55 – 58, consider total United States health expenditures as a percentage of gross domestic product (GDP) each year.

Year	2009	2010	2011	2012
%GDP	17.71	17.66	17.68	17.91

The linear function $f(x) = 0.062x - 106.911$ is a reasonable model of this data set. State the value and explain the meaning of the quantity.

55. $f(2013)$

56. Slope of $f(x)$

57. Vertical intercept of $f(x)$

58. Horizontal intercept of $f(x)$

In Exercises 59–62, consider the average debt load in thousands of 2012 dollars at the end of each year's spring term for bachelor's degree recipients who attended public four-year colleges and universities and borrowed money to finance their education.

Year	2001	2003	2005	2006
Debt	$20.4	$20.9	$21.5	$21.8

The linear function $f(x) = 0.28333x - 546.582$ is a reasonable model of this data set. State the value and explain the meaning of the quantity.

59. $f(2004)$

60. Slope of $f(x)$

61. vertical intercept of $f(x)$

62. Horizontal intercept of $f(x)$

In Exercises 63–66, identify the intervals on which the line is increasing, constant, or decreasing.

63. $y = 5x + 13$

64. $y = -3x + 9$

65. $y = -11$

66. $y = -7.8x - 8.6$

In Exercises 67–74, identify the intervals on which the function is increasing, constant, or decreasing.

67. $f(x) = \begin{cases} 3x + 2 & x < 5 \\ 7x & 5 \le x \end{cases}$

68. $f(x) = \begin{cases} 5x - 3 & x < 0 \\ 5x + 5 & 0 \le x \end{cases}$

69. $f(x) = \begin{cases} -x + 9 & x \le -2 \\ x - 11 & -2 < x \end{cases}$

70. $f(x) = \begin{cases} -2x - 3 & x \le 3 \\ -\pi x - 7 & 3 < x \end{cases}$

71. $f(x) = \begin{cases} 4 & x \le 8 \\ 3x + 1 & 8 < x \end{cases}$

72. $f(x) = \begin{cases} -2x & x < -3 \\ 6 & -3 \le x \end{cases}$

73. $f(x) = \begin{cases} 7x + 9 & x \le 0 \\ -3x + 1 & 0 < x < 5 \\ 17x & 5 \le x \end{cases}$

74. $f(x) = \begin{cases} x + 3 & x < -4 \\ -9 & -4 \le x \le 16 \\ 2x - 5 & 16 < x \end{cases}$

RStudio. In Exercises 75–80, use RStudio to graph the line on a domain that includes both its vertical and horizontal intercepts.

75. $y = 16x + 8$

76. $y = 3x - 27$

77. $y = -8x + 4$

78. $y = 66.7x - 666.6$

79. $y = 0.062x - 106.911$

80. $y = 0.28333x - 546.58167$

Your Turn. In Exercises 81–84, state the equation of a line with the slope and vertical intercept, and use RStudio to graph this line.

81. Positive slope and vertical intercept

82. Zero slope and negative vertical intercept

83. Negative slope and vertical intercept

84. Undefined slope

RStudio. In Exercises 85 – 90, use RStudio to find the point of intersection of the lines.

85. $y = 4x + 8$
    $y = 2x - 6$

86. $y = -3x - 9$
    $y = 6x + 2$

87. $y = -5x - 11$
    $y = 10x - 36$

88. $y = 8x + 3$
    $y = -7x + 2$

89. $y = -3.6x + 9.2$
    $y = -9.2x - 7.8$

90. $y = -7.8x - 8.6$
    $y = 4.2x$

In Your Own Words. In Exercises 91 – 100, explain the following.

91. Slope of a line

92. Vertical intercept of a line

93. Horizontal intercept of a line

94. Lines as parametrized families of functions

95. The effect of the slope parameter m on the graph of a linear function $y = mx + b$

96. The effect of the vertical intercept b on the graph of a linear function $y = mx + b$

97. Monotonic behavior of a function

98. A function increases on an interval

99. A function is constant on an interval

100. A function decreases on an interval

1.4   Exponential Functions

Linear functions are the simplest and most widely used mathematical models, followed closely by exponential functions. In popular culture, quantities "growing exponentially" identify outputs that become very large, very quickly. As an illustration of how fast exponential functions grow, Figure 1 presents the U.S. Census Bureau's estimates for world population in billions of people per year.

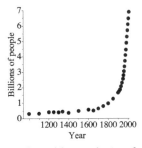

Figure 1: Estimated world population from 1000 to 2015

In fact, exponential functions increase more quickly than any of the other common modeling functions, or decrease more quickly, depending on the particular form of the exponential function. In addition to population growth, many other aspects of reality have an exponential exponent, including the bloodstream concentration of certain medications, the mass of a radioactive substance, the interest earned by an investment, and more.

The adjective **exponential** refers to the defining feature of these functions: the variable x appears in the **exponent** r of the expression a^r, where the **base** a is a real number greater than one. Examples of specific exponential functions include 2^x, 3^{-x}, and 10^{4x}. When modeling data sets, the most useful base is the number $e = 2.71828182\ldots$. In mathematics and the sciences, the letter "e" is used to represent the distinguished transcendental number $e = 2.71828182\ldots$ in exactly the same way that "π" represents the distinguished number $\pi = 3.14159265\ldots$. In fact, $y = e^x$ is often referred to as the *natural exponential function* for calculus-based reasons, which are discussed during the study of derivatives in Chapter 4. In addition to base e models, sometimes the integers 2 and 10 are used as the base for a model depending on the context.

Definition.

- An **exponential function** is a function of the form $f(x) = Ca^{kx}$, where a is a real number greater than one, $C \neq 0$ and $k \neq 0$ are constants, and x is the variable.

- Most often, the **natural exponential function** $y = Ce^{kx}$ with base $e = 2.71828182\ldots$ is used, and sometimes the exponential functions $y = C2^{kx}$ or $y = C10^{kx}$.

The effect of the parameters C and k on the graph of an exponential function are discussed in detail later in this section. For the moment, a few general observations are made based on the two specific exponential functions $y = 2e^{0.5x}$ and $y = -2e^{-0.5x}$, which are graphed in Figure 2. Note that the domain of these two exponential functions (in fact, all exponential functions) is the set of all real numbers.

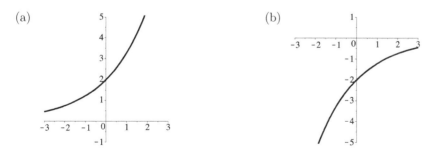

Figure 2: Graphs of the exponential functions (a) $y = 2e^{0.5x}$ and (b) $y = -2e^{-0.5x}$

As illustrated in Figure 2, the graphs of the two exponentials pass through the y-axis at the point $(0, C)$; that is, the exponential function $y = 2e^{0.5x}$ in Figure 2(a) intersects the y-axis at $(0, 2)$ and the exponential function $y = -2e^{-0.5x}$ in Figure 2(b) intersects the y-axis at $(0, -2)$. This pattern holds for all exponential functions $y = Ca^{kx}$, with the coefficient C always providing the **vertical intercept** of the function.

The graphs of all exponential functions follow one of the two characteristic bends illustrated in Figure 2, and this shape depends on the sign of C. In Figure 2(a), the exponential $y = 2e^{0.5x}$ has $C = 2$ and bends upward, which is referred to as being **concave up**. In Figure 2(b), the exponential $y = -2e^{-0.5x}$ has $C = -2$ and bends downward, which is referred to as being **concave down**. Some people remember the overall shape of such functions by identifying concave up functions as looking like a "cup" and concave down functions as looking like a "frown," (which rhymes with "down"). As with the vertical intercept, this pattern holds for all such functions: every exponential function $y = Ca^{kx}$ is exactly one of concave up or concave down.

A combination of the signs of C and k determines the monotonicity of an exponential function $y = Ca^{kx}$ as always increasing or always decreasing. In addition to effecting monotonicity, the coefficient k effects the relative flatness or steepness of an exponential function. These effects are explored in greater detail later in this section.

As illustrated in Figure 2, exponential curves become very close to the x-axis. Even more, an exponential function $y = Ca^{kx}$ is never equal to zero, and so its graph never intersects the x-axis. Based on these long-term behaviors, the x-axis (or $y = 0$) is said to be a **horizontal asymptote** of an exponential function.

As highlighted above, the vertical intercept of every exponential function $f(x) = Ca^{kx}$ always corresponds to the value of C because $a^0 = 1$ for any base $a > 1$ and $f(0) = Ca^{0k} = Ca^0 = C$. Even if the exact value of C is not given, sometimes C can still be determined. Example 1 provides one such setting: when k and another point on the graph of the exponential are known.

◆ **EXAMPLE 1** Find the vertical intercept of the exponential function $y = Ce^{x/2}$ that passes through $(3, 12)$. Also, sketch the graph of this function and discuss both its monotonicity and concavity.

Solution. The vertical intercept C is obtained by substituting the given point $(3, 12)$ on the curve into the equation $y = Ce^{x/2}$ and solving for C as follows:

$$y = Ce^{x/2} \qquad \text{Given exponential function}$$
$$12 = Ce^{3/2} \qquad \text{Substitute } (x, y) = (3, 12)$$
$$C = \frac{12}{e^{1.5}} \approx 2.678 \qquad \text{Divide both sides by } e^{3/2} = e^{1.5}$$

Therefore, the equation of the exponential function $y = Ce^{x/2}$ passing through $(3, 12)$ is

$$y = \frac{12}{e^{1.5}}e^{x/2} \approx 2.678e^{x/2}$$

with vertical intercept $C = 12/(e^{1.5}) \approx 2.678$. This exponential function is graphed in Figure 3 and, based on its graph, this function is increasing and concave up for all input values.   ∎

➤ **QUESTION 1** Find the vertical intercept of the exponential function $y = Ce^{2x}$ that passes through $(2, -60)$. Also, sketch the graph of this function and discuss both its monotonicity and concavity.

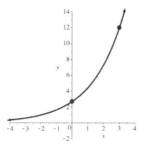

Figure 3: Graph of $y = \dfrac{12}{e^{1.5}} e^{x/2}$ for Example 1

Recognizing Exponential Functions Graphically

While studying linear functions $y = mx + b$, we learned that a constant value of the slope m is the defining feature of a line and how to compute the slope of a line using any two points on the line. In a similar fashion, a constant doubling time or a constant halving time is the defining feature of an exponential graph.

While the graph of a line can be recognized immediately, curved graphs require more care because many non-exponential graphs share a similar initial appearance to exponentials. Observe the similarity of the exponential function $y = e^x$ graphed in Figure 4(a) with the quadratic function $y = x^2 + 1$ graphed in Figure 4(b). Among other features, both curves have vertical intercept 1, are increasing, and are concave up.

Figure 4(c) identifies a graphical difference between the two functions: namely, the two functions have increasingly different output values as the inputs increase and, in fact, this difference becomes even more dramatic for inputs greater than $x = 3$. While this difference can help identify an exponential function in such a comparative setting, a stand-alone approach is needed for determining if a graph is exponential. The notions of doubling time and halving time provide the key.

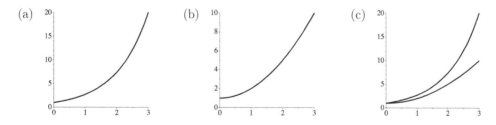

Figure 4: Graphs of (a) $y = e^x$, (b) $y = x^2 + 1$, and (c) both $y = e^x$ and $y = x^2 + 1$

First, consider the *doubling time* of an exponential function $y = Ce^{kx}$ where k is positive. The defining feature of such exponential functions is a constant width Δx for an input interval that results in the output for the right end of the interval being twice as much as the output for the left end of the interval. Figure 5 illustrates this idea for $y = e^{1.15x}$. On the input interval $[2.8, 3.4]$ with width $\Delta x = 0.6$, the outputs double from

25 to 50. Similarly, the outputs double from 50 to 100 on the input interval $[3.4, 4]$ with width $\Delta x = 0.6$ and, again, they double from 100 to 200 on the input interval $[4, 4.6]$ with width $\Delta x = 0.6$. This constant width $\Delta x = 0.6$ is characteristic of this particular exponential and is called the doubling time of $y = e^{1.15x}$.

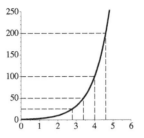

Figure 5: Doubling time for exponential function $y = e^{1.15x}$

In fact, every exponential function $y = Ce^{kx}$ with k positive has a constant doubling time Δx, and this property of exponentials can be used to graphically determine whether or not a curve is exponential. Historically, the name "doubling time" for this characteristic width of an input interval originates from various real-life settings for which the input is a time. In a similar fashion, every exponential function $y = Ce^{kx}$ with k negative has a constant *halving time*. These ideas are carefully defined as follows, and the precise numeric relationship between the parameter k and the value of the doubling or halving time of the exponential is also given:

Definition. Let $f(x) = Ce^{kx}$ be an exponential function.

- If k is positive and if $f(x_2) = 2f(x_1)$ with $x_2 > x_1$, then $\Delta x = x_2 - x_1$ is the **doubling time** of the exponential function, and $k = \dfrac{\ln(2)}{\Delta x} = \dfrac{\ln(2)}{\text{doubling time}}$.

- If k is negative and if $f(x_2) = \dfrac{1}{2}f(x_1)$ with $x_2 > x_1$, then $\Delta x = x_2 - x_1$ is the **halving time** of the exponential function, and $k = \dfrac{-\ln(2)}{\Delta x} = \dfrac{-\ln(2)}{\text{halving time}}$.

The formulas for k are both stated using the real number $\ln(2) \approx 0.6931472$. The natural logarithm function $\ln(x)$ is studied in Section 1.6. As an illustration of this formula, recall from the preceding discussion that the exponential function $y = e^{1.15x}$ has a doubling time of approximately $\Delta x = 0.6$. Applying the formula given in the definition of doubling time gives

$$k = \frac{\ln(2)}{\Delta x} = \frac{\ln(2)}{\text{doubling time}} = \frac{\ln(2)}{0.6} \approx 1.155245$$

This computed value of $k \approx 1.155$ quite closely approximates the actual value of $k = 1.15$ explicitly stated in $y = e^{1.15x}$.

◆ **EXAMPLE 2** Estimate the doubling time of the exponential function in Figure 6, and then state the corresponding exponential function $y = Ce^{kx}$.

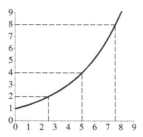

Figure 6: Estimating doubling time for Example 2

Solution. The graph given in Figure 6 identifies some specific doubling of outputs on the y-axis and input intervals on the x-axis. Namely, the outputs double from 2 to 4 to 8, and the corresponding input intervals both have the same width of $\Delta x \approx 2.5$. Therefore, the doubling time of this exponential is approximately 2.5. Applying the doubling time formula gives

$$k = \frac{\ln(2)}{\Delta x} = \frac{\ln(2)}{\text{doubling time}} = \frac{\ln(2)}{2.5} \approx 0.2772589$$

The parameter C is equal to the vertical intercept, which is 1 based on Figure 6. Therefore, the exponential function is $y = 1 \cdot e^{0.277x}$ or, more simply, $y = e^{0.277x}$. ■

Figure 6 provides specific doublings of outputs. Note that the value of the doubling time does not depend on the particular choice of outputs; rather, any doubling of outputs can be used. Example 2 considered outputs doubling from 2 to 4 to 8. Alternatively, the outputs doubling from 3 to 6 given in Figure 7 can be considered, because the same doubling time of $\Delta x \approx 2.5$ results.

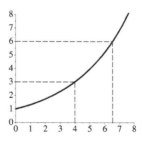

Figure 7: Independence of doubling time from output choices

The approach illustrated in Example 2 also applies to working with the halving time of an exponential function $y = Ce^{kx}$ with k negative.

◆ **EXAMPLE 3** Estimate the halving time of the exponential function in Figure 8, and then state the corresponding exponential function $y = Ce^{kx}$.

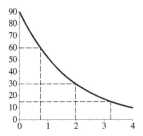

Figure 8: Estimating halving time for Example 3

Solution. The graph given in Figure 8 identifies some specific halving of outputs on the y-axis and the corresponding input intervals on the x-axis. Namely, the outputs decrease by half from 60 to 30 to 15, and the corresponding input intervals both have the same width of $\Delta x \approx 1.25$, which is the approximate halving time of this exponential. Applying the halving time formula gives

$$k = \frac{-\ln(2)}{\Delta x} = \frac{-\ln(2)}{\text{halving time}} = \frac{-\ln(2)}{1.25} \approx -0.5545177$$

The parameter C is equal to the vertical intercept, which is 90 based on Figure 8. Therefore, the exponential function graphed is $y = 90e^{-0.555x}$. ■

➤ **QUESTION 2** Estimate the doubling or halving time of the exponential functions in Figure 9 as appropriate and state the corresponding exponential function $y = Ce^{kx}$.

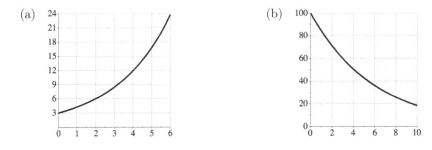

Figure 9: Estimating doubling or halving times for Question 2

Parameters of Exponential Functions

Studying linear functions $y = mx + b$ included thinking about the coefficients m and b as parameters that can take on different values and a discussion of the effect of these parameters on the graph of a line. Similarly, for exponential functions $y = Ca^{kx}$, both C and k can be thought of as parameters with distinct effects on the graph of an exponential.

We begin with the coefficient C and then consider k. The beginning of this section noted that C is always the vertical intercept of an exponential function $y = Ca^{kx}$, which means that the graph of an exponential function passes through the y-axis at the point $(0, C)$. In addition, the sign of the number C determines the concavity of $y = Ca^{kx}$. In particular, when C is positive, $y = Ca^{kx}$ is concave up and, when C is negative, $y = Ca^{kx}$ is concave down. The following example and question illustrate these graphical effects of C being positive versus negative as well as exploring the effect of C having relatively larger or smaller values.

◆ **EXAMPLE 4** Identify each graph in Figure 10 as $y = Ce^x$ for $C = -1$, -3, or -7. Describe the general effect of the parameter C on intercepts, monotonicity, concavity, and asymptotes. The other parameter of $k = 1$ for all three exponentials ensures that their graphical behavior depends only on the parameter C.

Solution. The fact that C is the vertical intercept of an exponential enables the ready identification of each curve with its corresponding function. In Figure 10, curve (a) has intercept $C = -1$ and is defined by $y = -e^x$. Similarly, curve (b) has intercept $C = -3$ and is defined by $y = -3e^x$, and curve (c) has intercept $C = -7$ and is defined by $y = -7e^x$. Exponential functions do not have horizontal intercepts, but instead have horizontal asymptotes of $y = 0$ (or the x-axis).

When the parameter C is negative, the graph of the exponential function $y = Ce^{kx}$ lies below the x-axis in quadrants III and IV. In addition, exponentials with negative coefficient C are concave down on $(-\infty, \infty)$. Finally, note that all three of these curves are decreasing on $(-\infty, \infty)$.   ■

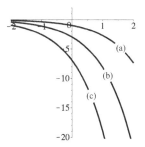

Figure 10: Studying the parameter C when it is negative for Example 4

The parameter C can also be positive. Question 3 examines the effect of this parameter when it is positive.

▶ **QUESTION 3** Graph the exponentials $y = Ce^x$ for $C = 1$, 3, and 7. Describe the general effect of the parameter C on the intercepts, monotonicity, concavity, and asymptotes.

Now consider the two effects of the exponent k on the graph of an exponential function $y = Ca^{kx}$. First, a combination of the signs of C and k determines the monotonicity of an exponential as always increasing or always decreasing. Second, the coefficient k affects the relative flatness or steepness of an exponential function. The following examples and questions illustrate these graphical effects of k being positive versus negative, as well as exploring the effect of k having relatively larger or smaller values.

◆ **EXAMPLE 5** Identify each graph in Figure 11 as $y = e^{kx}$ for $k = 1, 2$, or 3. Describe the general effect of the parameter k, when k is positive, on intercepts, monotonicity, concavity, and asymptotes. The fixed value of the other parameter $C = 1$ for all three exponentials ensures that their graphical behavior depends only on the parameter k.

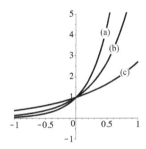

Figure 11: Studying parameter k when it is positive for Example 5

Solution. One strategy for identifying each curve with its corresponding function is to evaluate each function for the same input values of $x = 0.5$ and -0.5, and then compare the resulting outputs:

For $y = e^{kx}$	$k = 1$	$k = 2$	$k = 3$
$x = 0.5$	$y = e^{1 \cdot 0.5} \approx 1.6$	$y = e^{2 \cdot 0.5} \approx 2.7$	$y = e^{3 \cdot 0.5} \approx 4.5$
$x = -0.5$	$y = e^{1 \cdot (-0.5)} \approx 0.6$	$y = e^{2 \cdot (-0.5)} \approx 0.4$	$y = e^{3 \cdot (-0.5)} \approx 0.2$

Comparing the outputs resulting from $x = 0.5$ and the relative heights of the curves in Figure 11 shows that curve (a) is defined by $y = e^{3x}$, curve (b) $y = e^{2x}$, and curve (c) by $y = e^x$. In addition, when comparing the outputs resulting from $x = -0.5$, the exponentials switch their relative heights at their vertical intercept; so, to the left of the y-axis, curve (c) $y = e^x$ is the top function and curve (a) $y = e^{3x}$ is the bottom function. This crossing behavior happens for all such families of exponential functions.

The vertical intercept of these exponentials is $C = 1$, as graphically demonstrated by the common point of intersection $(0, 1)$, and they all have $y = 0$ as their horizontal asymptote. All three exponentials are concave up on $(-\infty, \infty)$ because $C = 1$ is positive. In addition, these exponentials are all increasing on $(-\infty, \infty)$, and, in fact, whenever the parameters k and C are both positive, the exponential function is always increasing. Finally, as k becomes larger, the corresponding exponential functions increase more rapidly from left to right.

■

The parameter k can also be negative. Question 4 examines the effect of this parameter when it is negative.

▶ **QUESTION 4** Graph the exponentials $y = e^{kx}$ for $k = -1$, -2, and -3. Describe the general effect of the parameter k, when it is negative, on intercepts, monotonicity, concavity, and asymptotes.

Finally, the relationship between the graphs of exponential functions $y = Ca^{kx}$ when k is positive versus when it is negative is examined.

◆ **EXAMPLE 6** Identify each graph in Figure 12 as $y = e^{kx}$ for either $k = -1$ or 1. Describe the general effect of the sign of the parameter k on intercepts, monotonicity, concavity, and asymptotes. The fixed value of the other parameter $C = 1$ for all three exponentials ensures that their graphical behavior depends only on the parameter k.

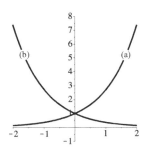

Figure 12: Studying the sign of the parameter k for Example 6

Solution. Following the strategy in Example 5, input $x = 1$ into each function and compare the outputs of the two functions as follows:

For $y = e^{kx}$	$k = -1$	$k = 1$
$x = 1$	$y = e^{(-1) \cdot 1} \approx 0.367$	$y = e^{1 \cdot 1} \approx 2.718$

Comparing these outputs and the relative heights of the curves in Figure 12 shows that curve (a) is defined by $y = e^{-x}$ and curve (b) by $y = e^{x}$. Notice that these two curves are reflections of each other across the y-axis. Also, observe the standard behaviors of exponentials: the vertical intercepts are $C = 1$, the horizontal asymptotes are $y = 0$, and the curves are concave up on $(-\infty, \infty)$ because C is positive.

The monotonic behavior of these exponentials depends on the sign of the exponent parameter k. Based on the graph $y = e^{-x}$, we conclude that when k is negative, the graph of the corresponding exponential function is decreasing on $(-\infty, \infty)$. In contrast, based on the graph $y = e^{x}$, when k is positive, the graph of the exponential is increasing on $(-\infty, \infty)$. ■

▶ **QUESTION 5** Graph the exponentials $y = -e^{kx}$ for $k = -1$ and 1. Describe the general effect of the sign of the parameter k on intercepts, monotonicity, concavity, and asymptotes.

This analysis of the effects of the parameters of exponential functions on the graphs of such functions can be summarized as follows:

EFFECTS OF PARAMETERS OF EXPONENTIAL FUNCTIONS.
For $y = Ca^{kx}$, the parameter C is the y-intercept of the graph. If C is positive, the graph is in quadrants I and II, and is concave up. If C is negative, the graph is in quadrants III and IV, and is concave down. The parameter k affects the steepness of the graph. The combination of the signs of C and k determines whether the function is increasing or decreasing as follows:

- If C is positive and k is positive, then $y = Ca^{kx}$ is increasing.

- If C is positive and k is negative, then $y = Ca^{kx}$ is decreasing.

- If C is negative and k is positive, then $y = Ca^{kx}$ is decreasing.

- If C is negative and k is negative, then $y = Ca^{kx}$ is increasing.

Concavity

While discussing exponential functions $y = Ca^{kx}$, the effect of the parameter C on how a function bends was examined. In general, the **concavity** of a function refers to its "bendiness" and whether the inside bend of the function opens upward or downward.

For exponential functions with C positive, the graph bends upward and is said to be concave up on $(-\infty, \infty)$. On the other hand, when C is negative, the graph bends downward and is said to be concave down on $(-\infty, \infty)$. Some people remember the overall shape of such functions by identifying concave up functions as looking like a "cup" and concave down functions as looking life a "frown." A more precise definition of concavity is given in Chapter 5.

Notice that linear functions are neither concave up nor concave down, because they do not bend. Therefore, the three options when discussing the concavity of a function are concave up, linear, or concave down, which parallel the three options for monotonicity: increasing, constant, or decreasing.

Linear functions and exponential functions are exactly one of concave up, linear, or concave down on $(-\infty, \infty)$, but other functions exhibit a blend of these behaviors, depending on the particular portion of the domain being used. The following examples consider such functions. .

◆ **EXAMPLE 7** Discuss the monotonicity and the concavity of each curve in Figure 13 by identifying the intervals on which the function is increasing, constant, or decreasing, and the intervals on which the function is concave up, linear, or concave down.
Solution.

(a) The curve is decreasing on $(-\infty, \infty)$. Also, the curve is concave up on $(-\infty, 2)$ and concave down on $(2, \infty)$.

(b) The curve is increasing and concave down on $(-\infty, 0)$, constant and linear on $(0, 3)$, and decreasing and concave up on $(3, \infty)$.

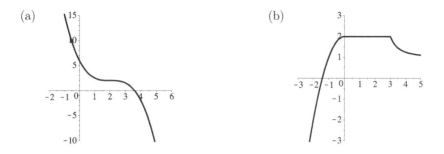

Figure 13: Curves illustrating monotonicity and concavity for Example 7

■

➤ **QUESTION 6** Discuss the monotonicity and the concavity of each curve in Figure 14 by identifying the intervals on which the function is increasing, constant, or decreasing, and the intervals on which the function is concave up, linear, or concave down.

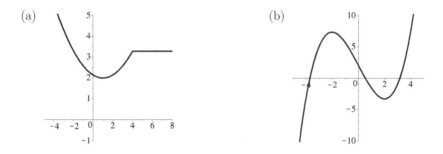

Figure 14: Curves illustrating monotonicity and concavity for Question 6

◆ **EXAMPLE 8** Discuss the monotonicity and concavity of

$$f(x) = \begin{cases} e^x & x < 2 \\ -2x + 3 & 2 \le x < 4 \\ -x^2 & x \ge 4 \end{cases}$$

Solution. On $(-\infty, 2)$, the function is increasing and concave up because $f(x)$ is an exponential with $C > 0$ and $k > 0$. On $(2, 4)$, the function is decreasing and linear because $f(x)$ is a linear function with $m = -2$. Finally, on $(4, \infty)$, the function is decreasing and concave down because $f(x)$ is a parabola of the form $y = Ax^2 + Bx + c$ with $A = -1 < 0$.

■

➤ **QUESTION 7** Discuss the monotonicity and concavity of $f(x) = \begin{cases} -3e^{-x} & x < 0 \\ 2e^x & x \geq 0 \end{cases}$

Algebra of Exponents

Studying exponential functions and other models requires the use of algebraic rules for exponents. These rules are often studied in high school algebra courses, but are recalled here because of their importance when working with mathematical models of reality.

EXPONENT RULES.
Let $a > 0$ and $b > 0$ be real numbers, and let r and s be real numbers.

(1) $a^r \cdot a^s = a^{r+s}$ (4) $\dfrac{1}{a^s} = a^{-s}$ (7) $a^r \cdot b^r = (a \cdot b)^r$

(2) $\dfrac{a^r}{a^s} = a^{r-s}$ (5) $\sqrt[s]{a} = a^{1/s}$ (8) $\dfrac{a^r}{b^r} = \left(\dfrac{a}{b}\right)^r$

(3) $(a^r)^s = a^{rs}$ (6) $\sqrt[s]{a^r} = a^{r/s}$

◆ **EXAMPLE 9** Express each quantity in the form a^r using the exponent rules:

(a) $\dfrac{1}{4^x}$ (c) $\dfrac{y^5}{y^8}$ (e) $\sqrt[4]{x}$ (g) $2^5 \cdot x^5$

(b) $(2^5)^x$ (d) $x^5 \cdot x^8$ (f) $\dfrac{2^5}{x^5}$ (h) $\sqrt[3]{5^x}$

Solution.

(a) Applying rule (4), $\dfrac{1}{4^x} = 4^{-x}$.   (e) Applying rule (5), $\sqrt[4]{x} = x^{1/4}$.

(b) Applying rule (3), $(2^5)^x = 2^{5x}$.   (f) Applying rule (8), $\dfrac{2^5}{x^5} = \left(\dfrac{2}{x}\right)^5$.

(c) Applying rule (2), $\dfrac{y^5}{y^8} = y^{5-8} = y^{-3}$.   (g) Applying rule (7), $2^5 \cdot x^5 = (2 \cdot x)^5$.

(d) Applying rule (1), $x^5 \cdot x^8 = x^{5+8} = x^{13}$.   (h) Applying rule (6), $\sqrt[3]{5^x} = 5^{x/3}$.

■

➤ **QUESTION 8** Express each quantity in the form a^r using the exponent rules:

(a) $\dfrac{11^x}{11^y}$ (c) $\dfrac{7^x}{y^x}$ (e) $11^x \cdot 11^y$ (g) $7^x \cdot y^x$

(b) $\sqrt[y]{x^2}$ (d) $(x^3)^y$ (f) $\sqrt[x]{2}$ (h) $\dfrac{1}{x^9}$

Working in RStudio

RStudio uses the command `exp(x)` for the natural exponential function e^x. Specific values of this function are computed by entering numbers into this command, such as $e = e^1 \approx 2.718282$ with `exp(1)` and $e^2 \approx 7.389056$ with `exp(2)`. Similarly, RStudio uses the command `log(2)` for the number $\ln(2) \approx 0.69314782$.

RStudio graphs exponential functions using the `plotFun` command. For example, the exponential $y = e^{0.5x}$ is graphed on the interval $[-6, 6]$ by entering

    plotFun(exp(0.5*x)~x,xlim=range(-6,6)).

Recall that the output interval can be controlled with the option `ylim=range(,)` and that multiple functions can be graphed on the same axes using the option `add=TRUE`.

Examples of Commands:

- `exp(5) + exp(-3)`

- `plotFun(exp(5*x)~x,xlim=range(-3,3))`

- `plotFun(exp(-3*x)~x,add=TRUE)`

◆ **EXAMPLE 10** Graph $y = 3x^2 + 4$ and $y = 2e^x$ on the same pair of axes using RStudio.

Solution. First make the plot $y = 3x^2 + 4$ and then add $y = 2e^x$ to the plot.

```
plotFun(3*x^2+4~x,xlim=range(-5,5),ylim=range(-10,100))
plotFun(2*exp(x)~x,add=TRUE)
```

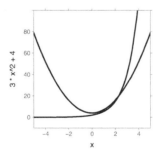

■

➤ **QUESTION 9** Graph $y = e^{-2x}$ and $y = -e^{2x}$ on the same pair of axes using RStudio.

Summary

- An *exponential function* is a function of the form $f(x) = Ca^{kx}$ or $y = Ca^{kx}$, where a is a real number greater than one, $C \neq 0$ and $k \neq 0$ are constants, and x is the variable.

- Most often, the *natural exponential function* $y = Ce^{kx}$ with base $e \approx 2.7182818$ is used, and sometimes the exponential functions $y = C2^{kx}$ or $y = C10^{kx}$.

- If $f(x) = Ca^{kx}$ with $a > 1$, if k is positive, and if $f(x_2) = 2f(x_1)$ with $x_2 > x_1$, then the difference $\Delta x = x_2 - x_1$ is the *doubling time* of the exponential function and $k = \dfrac{\ln(2)}{\Delta x} = \dfrac{\ln(2)}{\text{doubling time}}$.

- If $f(x) = Ca^{kx}$ with $a > 1$, if k is negative, and if $f(x_2) = \dfrac{1}{2}f(x_1)$ with $x_2 > x_1$, then the difference $\Delta x = x_2 - x_1$ is the *halving time* of the exponential function and $k = \dfrac{-\ln(2)}{\Delta x} = \dfrac{-\ln(2)}{\text{halving time}}$.

- For $y = Ca^{kx}$, the parameter C is the y-intercept of the graph. If C is positive, the graph is in quadrants I and II, and is concave up. If C is negative, the graph is in quadrants III and IV, and is concave down. The parameter k affects the steepness of the graph. The combination of the signs of C and k determines whether the function is increasing or decreasing as follows:

  - If C is positive and k is positive, then $y = Ca^{kx}$ is increasing.
  - If C is positive and k is negative, then $y = Ca^{kx}$ is decreasing.
  - If C is negative and k is positive, then $y = Ca^{kx}$ is decreasing.
  - If C is negative and k is negative, then $y = Ca^{kx}$ is increasing.

- *Exponent rules* Let $a > 0$ and $b > 0$ be real numbers, and let r and s be real numbers:

  (1) $a^r \cdot a^s = a^{r+s}$

  (2) $\dfrac{a^r}{a^s} = a^{r-s}$

  (3) $(a^r)^s = a^{rs}$

  (4) $\dfrac{1}{a^s} = a^{-s}$

  (5) $\sqrt[s]{a} = a^{1/s}$

  (6) $\sqrt[s]{a^r} = a^{r/s}$

  (7) $a^r \cdot b^r = (a \cdot b)^r$

  (8) $\dfrac{a^r}{b^r} = \left(\dfrac{a}{b}\right)^r$

Exercises

In Exercises 1–10, find the vertical intercept C for the exponential function $y = Ca^{kx}$ passing through the given point, and then sketch its graph.

1. $y = Ce^{2x}$ through $(1, -2)$

2. $y = Ce^{0.14x}$ through $(3, 17)$

3. $y = Ce^{-3x}$ through $(1, 2)$

4. $y = Ce^{-0.1x}$ through $(15, -4)$

5. $y = Ce^{x/7}$ through $(-2, 14)$

6. $y = Ce^{-4.1x}$ through $(-1, 10)$

7. $y = C2^{3x}$ through $(3, 5)$

8. $y = C2^{-0.2x}$ through $(-2, -7)$

9. $y = C10^{4x}$ through $(1, 14)$

10. $y = C10^{x/8}$ through $(-6, -10)$

In Exercises 11 and 12, determine the parameters C and k of the exponential function $y = Ce^{kx}$ with the doubling time and vertical intercept, and then sketch its graph.

11. Doubling time 10
    vertical intercept 3

12. Doubling time 4
    vertical intercept -2

In Exercises 13 and 14, determine the parameters C and k of the exponential function $y = Ce^{kx}$ with the halving time and vertical intercept, and then sketch its graph.

13. Halving time 3
    vertical intercept -4

14. Halving time 13
    vertical intercept 6

In Exercises 15–17, determine the parameters C and k of the exponential function $y = Ce^{kx}$ with the doubling time and through the point, and then sketch its graph.

15. Doubling time 5; $(4, 10)$

16. Doubling time 7; $(6, -5)$

17. Doubling time 0.3; $(-1, 3)$

In Exercises 18–20, determine the parameters C and k of the exponential function $y = Ce^{kx}$ with the halving time and through the point, and then sketch its graph.

18. Halving time 30; $(10, -15)$

19. Halving time 6.3; $(-15, 6)$

20. Halving time 0.7; $(-3, 21)$

Your Turn. In Exercises 21–24, state, if possible, an exponential function with points in the pair of quadrants, and then sketch its graph.

21. Points in quadrants I and II

22. Points in quadrants I and III

23. A point in quadrant IV and a positive vertical intercept

24. A point in quadrant III and a negative vertical intercept

In Exercises 25–34, give a rough sketch of the pair of exponential functions on the same axes with domain $[-2, 2]$. Label each curve.

25. e^{2x}, e^{-2x}

26. e^x, $5e^x$

27. e^x, $-e^x$

28. e^{3x}, $-2e^{-x}$

29. 2^x, -2^x

30. 2^x, $3 \cdot 2^x$

31. e^x, 10^x

32. e^{-x}, 10^{-x}

33. 2^{-x}, $-e^{-x}$

34. 2^x, e^x

38.

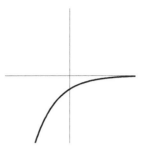

In Exercises 35–38, state the sign of C and k of the exponential function $y = Ce^{kt}$, and describe its monotonicity and concavity.

In Exercises 39–44, determine whether the curve is exponential or not by determining if the curve exhibits a constant doubling or halving time, as appropriate.

35.

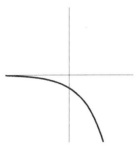

36.

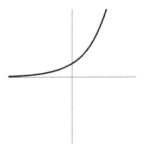

37.

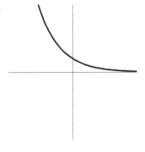

39.

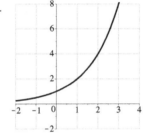

40.

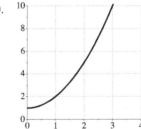

41.

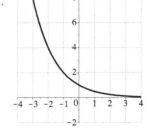

42.

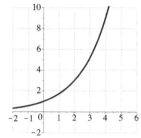

43.

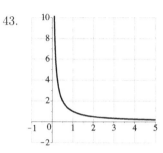

44.

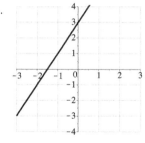

In Exercises 45 – 48, use properties of exponentials to explain why the curve is not exponential.

45.

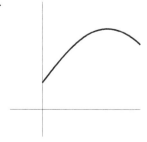

46.

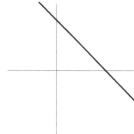

47.

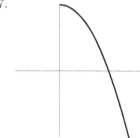

48.

RStudio. In Exercises 49 – 56, use RStudio to graph the exponential on the domain $[-4, 4]$.

49. $y = 3e^{2x}$

50. $y = -3e^{2x}$

51. $y = 3e^{-2x}$

52. $y = -3e^{-2x}$

53. $y = 4 \cdot 10^{x}$

54. $y = -4 \cdot 10^{x}$

55. $y = 4 \cdot 10^{-x}$

56. $y = -4 \cdot 10^{-x}$

RStudio. In Exercises 57 – 62, use RStudio to graph the exponential function and estimate its doubling or halving time, as appropriate.

57. $y = e^{4x}$

58. $y = e^{x/3}$

59. $y = e^{-3x}$

60. $y = e^{-x/4}$

61. $y = 10^{2x}$

62. $y = 10^{-x/2}$

In Exercises 63–66, identify the intervals on which the function is increasing, constant, or decreasing.

63. $f(x) = \begin{cases} -3x+2 & x < 5 \\ 2e^x & 5 \le x \end{cases}$

64. $f(x) = \begin{cases} -4e^{-x} & x < 0 \\ 5x+5 & 0 \le x \end{cases}$

65. $f(x) = \begin{cases} 5e^{-2x} & x \le -2 \\ x-11 & -2 < x \end{cases}$

66. $f(x) = \begin{cases} -2x-3 & x \le 3 \\ 4e^{-3x} & 3 < x \end{cases}$

In Exercises 67–72, consider the exponential function $f(x) = 31.486e^{-0.048x}$ that models the data on United States coal prices (P) in dollars per short ton in each year (Y) since 1989 (Y 0) given in the following table:

Y	0	1	3	7
P	\$31.36	\$30.11	\$27.46	\$22.25

67. Compute $f(5)$ and explain its meaning.

68. Find the vertical intercept of $f(x)$ and explain its meaning.

69. Find the halving time of $f(x)$ and explain its meaning.

70. Plot $f(x)$ and the data on the same axes.

71. Discuss the monotonicity of $f(x)$.

72. Discuss the concavity of $f(x)$.

In Exercises 73–76, consider the exponential function $f(x) = 37.401e^{0.122x}$ that models the data on average birth weight (W) of males in grams each gestational week (T) given in the following table:

T	25	26	30	31	33
W	777	888	1435	1633	2058

73. Compute $f(32)$ and explain its meaning.

74. Find the vertical intercept of $f(x)$ and explain its meaning.

75. Find the doubling time of $f(x)$ and explain its meaning.

76. Plot $f(x)$ and the data on the same axes.

In Exercises 77–80, consider the exponential function $f(x) = 75.567e^{-0.128x}$ that models the data on plasma concentration (C) of Prozac in nanograms per milliliter (ng/mL) as a function of the day (D) for a person who has taken 20 mg of Prozac for 30 days, but stops cold turkey given in the following table. (Note that discontinuing any medication should always be done in consultation with a healthcare provider.)

D	0	5	10	22	27
C	79	40	19.6	4.3	2.5

77. Compute $f(15)$ and explain its meaning.

78. Find the vertical intercept of $f(x)$ and explain its meaning.

79. Find the halving time of $f(x)$ and explain its meaning.

80. Plot $f(x)$ and the data on the same axes.

In Exercises 81–90, express the quantity in the form a^r using the exponent rules.

81. $4^{2x+3} \cdot 4^5$

82. $4^{x+1} \cdot 2^{2x}$

83. $\dfrac{2^x \cdot 3^x}{6^{1-x}}$

84. $\sqrt[7]{x^8}$

85. $\dfrac{1}{\sqrt[5]{x^{13}}}$

86. $\dfrac{\sqrt[5]{x}}{x^6}$

87. $\sqrt{3^{4x} \cdot 9^3}$

88. $2^3 \cdot 3^6 \cdot 5^3$

89. $(16^x)^{0.5} \cdot 4^{x+1}$

90. $\dfrac{(\sqrt{x})^4}{x^{y+2} \cdot 3^y}$

In Your Own Words. In Exercises 91–100, explain the following.

91. Exponential function

92. Vertical intercept

93. Horizontal asymptote

94. Concavity

95. Concave up

96. Concave down

97. Effect of parameter C on $y = Ce^{kx}$

98. Effect of parameter k on $y = Ce^{kx}$

99. Doubling time

100. Halving time

1.5   Inverse Functions

Many aspects of reality can be described in terms of linear and exponential functions. In working with exponential functions, sometimes the action of an exponential needs to be reversed or undone using what is known as a logarithm. A logarithm is able to undo an exponential, and vice versa, because the logarithm and exponential are *inverse functions*. This section studies the general properties of inverse functions, and Section 1.6 studies the specific properties of logarithmic functions.

As an illustration of inverse functions, consider Figure 1, which presents the U.S. Census Bureau's estimates for world population in billions of people per year from two different perspectives. Figure 1(a) has an input of years and an output of billions of people, which addresses the question: "What was the population of the world in a given year?" However, people are also interested in the inverse of this question: "What year did humanity reach a given population?" Based on Figure 1(a), the population of humans on Earth appears to be growing exponentially, and many leaders and researchers would like to know when certain population thresholds will be reached and exceeded. Such inverse questions are explored using Figure 1(b), which interchanges the inputs and outputs in Figure 1(a), giving an input of billions of people and an output of years.

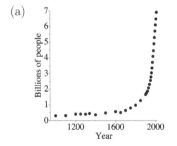
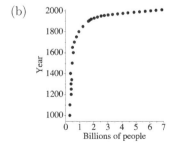

Figure 1: Estimated world population from 1000 to 2015

Many such pairs of inverse questions arise in the study of our economic, physical, and social world. In terms of functions, these questions correspond to thinking about the input-output relationships that result from swapping inputs to outputs and outputs to inputs. Namely, given a function that maps every input to a unique output, the inverse of this function maps each output back to its corresponding input. This idea is formally defined as follows:

Definition. Let $f(x)$ be a function. The **inverse function** of $f(x)$ maps each output from $f(x)$ back to its corresponding input. Symbolically, $f^{-1}(x)$ identifies the inverse of $f(x)$.

The notation $f^{-1}(x)$ for an inverse function can be confusing because the exponent "-1" is often used to indicate the reciprocal of a function or number (that is, its multiplicative inverse). However, $f^{-1}(x)$ is the inverse of $f(x)$ as a function, *not* the multiplicative inverse $1/f(x)$ of $f(x)$. In fact, $f^{-1}(x) \neq 1/f(x)$ save for a few exceptions. Therefore, in order to identify the reciprocal of a function, extra brackets are inserted: $[f(x)]^{-1} = 1/f(x)$.

As a specific example of this behavior, consider the function $f(x) = x + 1$. Working intuitively, the operation of adding one is reversed by subtracting one, which means the inverse of $f(x) = x + 1$ is $f^{-1}(x) = x - 1$. At the same time, the reciprocal of $f(x)$ is $[f(x)]^{-1} = 1/(x+1)$, which happens to be equal to $f^{-1}(x) = x - 1$ for only two inputs. These two different interpretations of the exponent -1 mean that careful attention must be paid to the placement of the -1 in order to distinguish between an inverse $f^{-1}(x)$ or a reciprocal $[f(x)]^{-1}$.

Tabular Inverses

Inverses are first considered from the perspective of tabular data sets, and then studied from a graphical perspective. While not every function has an inverse, for the time being, this section focuses on functions that do have inverses.

◆ **EXAMPLE 1** The following functions have inverses. State the inverse of each function.

(a)

x	1	2	3	4	5	6
$f(x)$	1	4	9	16	25	36

(b)

x	1	2	3	4	2	6
$g(x)$	1	4	9	16	4	36

Solution. The following tables present the inverse of each function. The idea is to map each output back to its corresponding input. Table (a) defining $f(x)$ has $f(1) = 1$, which means that $f^{-1}(1) = 1$. Similarly, $f(2) = 4$, which means that $f^{-1}(4) = 2$. Continuing in this fashion, all of $f^{-1}(x)$ can be obtained as given in table (i).

(i)

x	1	4	9	16	25	36
$f^{-1}(x)$	1	2	3	4	5	6

(ii)

x	1	4	9	16	36
$g^{-1}(x)$	1	2	3	4	6

Table (b) defining $g(x)$ has $g(1) = 1$, which means that $g^{-1}(1) = 1$. Similarly, $g(2) = 4$, which means that $g^{-1}(4) = 2$. For this particular input-output pair of $g(2) = 4$, note that the output of 4 appears twice in the second row of the table, but the corresponding input is 2 in both cases. Such a consistent alignment of inputs and outputs must happen in order for a function to have an inverse. The full inverse of $g(x)$ is given in table (ii).

■

➤ **QUESTION 1** The following functions have inverses. State the inverse of each function.

(a)

x	2	5	6	7	10	20
$a(x)$	42	39	38	37	34	24

(b)

x	-6	-2	-1	4	8
$b(x)$	6	2	1	-4	-8

In addition to these abstract tabular functions, such tabular presentations of functions are encountered when working with various real-life data sets. For such data sets, we want to be able to identify the corresponding inverse functions, while at the same time being aware of the questions that can be addressed by a given tabular function and its inverse.

◆ **EXAMPLE 2** The following table presents the United States annual unemployment rate during each year. This function has an inverse.

Year	2008	2009	2010	2012	2013
Unemployment rate	5	7.8	9.7	8.2	7.9

(a) State a question this function can address.

(b) State a question its inverse can address.

(c) State the inverse function.

Solution.

(a) For a given year, what is the unemployment rate during that year?

(b) For a given unemployment rate, during what year did that unemployment rate occur? Alternatively, during what year did the unemployment rate exceed a given rate?

(c) The inverse is given by

Rate	5	7.8	7.9	8.2	9.7
Year	2008	2009	2013	2012	2010

■

➤ **QUESTION 2** The following table presents how many millions of people used Twitter during each quarter of a year since 2000. For example, 11.25 represents the quarter April to June of the year 2011 during which 85 million people used Twitter. This function has an inverse.

Year	11	11.25	11.5	11.75	12.75
Twitter users	68	85	101	117	185

(a) State a question this function can address.

(b) State a question its inverse can address.

(c) State the inverse function.

Graphical Inverses

Functions and their inverses are now considered from a graphical perspective. When a function $f(x)$ has $f(a) = b$, the point (a, b) is included in the graph of the corresponding curve. If $f(x)$ has an inverse, then a graph of $f^{-1}(x)$ includes the point (b, a), interchanging the input value a of $f(x)$ with its output value b. Graphically, such an interchange of x-coordinates and y-coordinates is obtained by means of a particular *reflection* as follows:

GRAPH OF AN INVERSE. The graph of the inverse $f^{-1}(x)$ is obtained by reflecting the graph of $f(x)$ across the line $y = x$.

◆ **EXAMPLE 3** The functions in Figure 2 have inverses. Graph the inverse of each function.

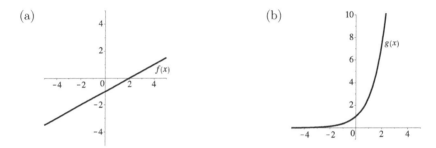

Figure 2: Graphing inverses for Example 3

Solution. The line $y = x$ is sketched, and then the given graph of each function is reflected across this line to obtain the graph of its inverse as shown in Figure 3.

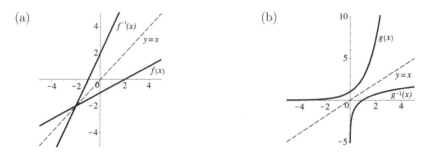

Figure 3: Solution of Example 3

On a practical level, immediately graphing all of an inverse function $f^{-1}(x)$ by reflecting across $y = x$ can be challenging. A helpful intermediate step is to reflect individual points across the line $y = x$ to serve as guides for graphing the entire inverse $f^{-1}(x)$, as Figure 4 illustrates for Example 3(a).

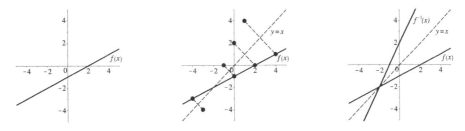

Figure 4: Steps for solving Example 3(a)

Example 3(b) is the graph of the exponential function $g(x) = e^x$. Therefore, Figure 3(b) presents the graph of the inverse of e^x, which is called the **natural logarithm** function and written as $\ln(x)$. The natural logarithm is the focus of study in Section 1.6. The graph of e^x and its inverse $\ln(x)$ illustrate the many parallels that arise between a function and its inverse, to include such properties as that e^x has a vertical intercept of $(0, 1)$ and has the x-axis as a horizontal asymptote, while $\ln(x)$ has a horizontal intercept of $(1, 0)$ and has the y-axis as a vertical asymptote.

➤ **QUESTION 3** The functions in Figure 5 have inverses. Graph the inverse of each function.

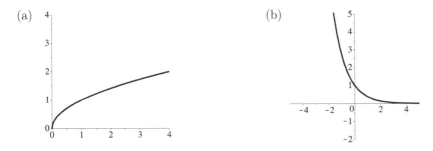

Figure 5: Graphing inverses for Question 3

Existence of Inverses

Thus far, every example and question in this section has explicitly stated that the given function has an inverse. This assertion is included because not every function has an inverse. Recall that, by definition, a function maps every input to exactly one output. When an inverse interchanges inputs and outputs, this inverse is a function only if every output comes from exactly one input. In other words, a function does not have an inverse if some output comes from more than one input, as illustrated in the next example.

◆ **EXAMPLE 4** Explain why each function in Figure 6 does or does not have an inverse.

Solution.

(a) For $a(x)$, each output comes from exactly one input, which means that this function has an inverse. In particular, because $a(1) = 4$ and $a(2) = 9$, the inverse of $a(x)$ is defined by $a^{-1}(4) = 1$ and $a^{-1}(9) = 2$.

(b) For $b(x)$, observe that $b(2) = 9$ and $b(3) = 9$. Therefore, $b(x)$ does not have an inverse, because the output 9 comes from more than one input. Note that even though the output 4 comes from the unique input 1, a single repeated output is enough to make a function not have an inverse.

■

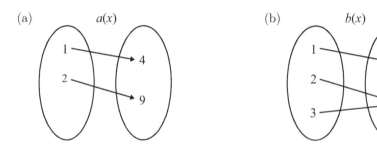

Figure 6: Functions for Example 4

EXISTENCE OF AN INVERSE. A function has an inverse when every output comes from exactly one input. Alternatively, a function does *not* have an inverse if some output comes from more than one input.

Additional functions that do not have inverses are discussed below, first considering tabular presentations of functions and then graphs of functions.

◆ **EXAMPLE 5** Explain why each function does or does not have an inverse and, if so, state its inverse:

(a)
x	-2	-1	1	2
$f(x)$	2	1	1	2

(b)
x	-2	-1	1	2
$g(x)$	2	1	-1	-2

Solution.

(a) The function $f(x)$ does not have an inverse, because the output $y = 2$ comes from multiple distinct inputs. Table (a) shows that $f(-2) = 2$ and $f(2) = 2$. When trying to define an inverse function $f^{-1}(x)$, the supposed inverse would need both $f^{-1}(2) = -2$ and $f^{-1}(2) = 2$, which means that an inverse-input of 2 would map to more than one output and so would not be a function. Alternatively, note that both $f(-1) = 1$ and $f(1) = 1$, and so the output $y = 1$ also ensures that $f(x)$ does not have an inverse for the exact same reason.

(b) The function $g(x)$ does have an inverse. Inspecting the outputs given in the second row of table (b), observe that every output occurs exactly once, which allows the definition of its inverse $g^{-1}(x)$ mapping each of these outputs back to its corresponding unique input as follows.

x	-2	-1	1	2
$g^{-1}(x)$	2	1	-1	-2

∎

As you may recognize, Example 5(a) is a partial tabular presentation of the absolute value function, which shows that $y = |x|$ does not have an inverse. Similarly, Example 5(b) is a partial tabular presentation of the function $y = -x$. While not a conclusive argument, this table provides some partial evidence that $y = -x$ does have an inverse; a complete graphical argument is presented soon. Another feature of Example 5(b) is that the table for $g(x)$ is identical to the table for its inverse $g^{-1}(x)$. In other words, $g(x)$ is its own inverse. While uncommon, such an equality of a function and its inverse function can happen.

➤ **QUESTION 4** Explain why each function does or does not have an inverse and, if so, state its inverse:

(a)
x	-2	-1	1	2
$a(x)$	7	49	81	9

(b)
x	-2	-1	1	2
$b(x)$	4	1	-1	4

In addition to such abstract functions not having inverses, many real-life data sets exhibit these same behaviors. Consider the following example:

◆ **EXAMPLE 6** The following table presents the global gender ratio based on the number of males per 100 females in each year. Explain why this function does not have an inverse.

Year	1970	1980	1990	2000	2010
Males	100.8	101.1	101.5	101.5	101.6

Solution. This function does not have an inverse, because the output 101.5 comes from the inputs of 1990 and 2000. When trying to define an inverse function $f^{-1}(x)$, the supposed inverse would need both $f^{-1}(101.5) = 1990$ and $f^{-1}(101.5) = 2000$, which is not possible for a function.

∎

➤ **QUESTION 5** The following table presents the length of the tornado season in days between the first and last tornado for each year in the warning area of the National Weather Service Office in Goodland, Kansas. Explain why this function does not have an inverse.

Year	1980	1981	1990	1991	2001
Length	89	120	120	35	89

We now consider a graphical approach to determining whether a function does or does not have an inverse. Recall that, by definition, a function must map every input to exactly one output and that the vertical line test is used to decide whether a given curve does or does not define a function. In particular, a curve in the xy-plane is a function when every vertical line intersects the curve at most once, which corresponds to every input mapping to a unique output.

This approach to determining whether a function does or does not have an inverse function is similar, only, instead of the vertical line test, the *horizontal line test* is used. Inverse functions interchange inputs and outputs, and an inverse function exists when each output comes from exactly one input. When a function does not have an inverse, some output b coming from more than one input is recognized by the corresponding horizontal line $y = b$ intersecting the curve more than once.

HORIZONTAL LINE TEST. A function $y = f(x)$ has an inverse function if and only if every horizontal line intersects its graph at most once.

Alternatively, a function does not have an inverse if and only if some horizontal line intersects its graph more than once. Using the horizontal line test to show that a function does not have an inverse requires explicitly stating the horizontal line $y = b$ that serves as a counterexample.

◆ **EXAMPLE 7** Explain why each function in Figure 7 does or does not have an inverse.

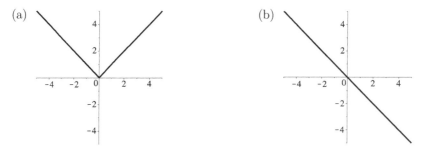

Figure 7: Curves for Example 7

Solution. Each graph is a function, which can be confirmed using the vertical line test. That is, every vertical line intersects the graph in Figure 7(a) exactly once, and similarly for Figure 7(b).

(a) Applying the horizontal line test, Figure 7(a) does not have an inverse, because the horizontal line $y = 2$ intersects the graph twice (so more than once), as shown in Figure 8(a). In fact, any horizontal line $y = b$ with $b > 0$ provides a counterexample.

(b) Every horizontal line intersects Figure 7(b) exactly once, as illustrated by the three example horizontal lines in Figure 8(b). Therefore, this function has an inverse based on the horizontal line test. ■

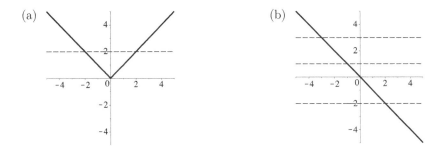

Figure 8: Horizontal line test for Example 7

Figure 7(a) presents the graph of the absolute value function $y = |x|$. This application of the horizontal line test affirms the conclusion based on Example 5(a) that the absolute value function does not have an inverse. In contrast, Figure 7(b) presents the graph of the function $y = -x$, and the horizontal line test provides a complete argument that this function has an inverse. This observation confirms the truth of the conjecture from Example 5(b), which was based on the evidence provided by a partial tabular presentation of $y = -x$. Also, $y = -x$ is its own inverse as indicated by reflecting its graph in Figure 7(b) across the line $y = x$, as shown in Figure 9.

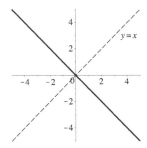

Figure 9: Graph illustrating $y = -x$ is its own inverse

➤ **QUESTION 6** Explain why each function in Figure 10 does or does not have an inverse.

Monotonicity and Inverses

Section 1.3 discussed the monotonic behavior of functions. Recall that the monotonicity of functions falls into one of three different categories: increasing, constant, and decreasing. Reading inputs from left to right, a function is **increasing** when the output values become larger, and a function is **decreasing** when the output values become smaller. A function is **constant** if every input results in the same output, which corresponds to a horizontal line on the plane.

All three types of monotonic behavior are illustrated in Figure 11. This piecewise linear function is: increasing on the interval $(-\infty, -1)$ when its slope is positive, constant

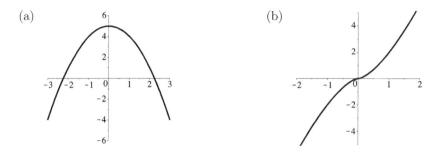

Figure 10: Curves for Question 6

on the interval $(-1, 2)$ when its slope is zero, and decreasing on the interval $(2, \infty)$ when its slope is negative.

Applying the horizontal line test to the function in Figure 11 shows that this function does not have an inverse. For example, the horizontal line $y = 0$ intersects this curve twice (so more than once). Alternatively, $y = 2$ intersects the curve infinitely often on the input interval $[-1, 2]$. On the other hand, focusing only on the increasing part of this function on the input interval $(-\infty, -1)$, observe that this restricted portion of the function passes the horizontal line test and does have an inverse. Similarly, the decreasing part of this function on the interval $(2, \infty)$ passes the horizontal line test and has an inverse.

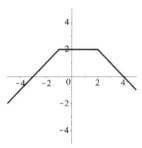

Figure 11: Monotonic behavior of functions

The complete function in Figure 11 does not pass the horizontal line test, both because the function switches from increasing to decreasing and because it is constant on part of its domain. As it turns out, this connection between monotonicity and inverses holds not just for this particular function, but carries over to all functions. In particular, if a function is not exclusively increasing or exclusively decreasing (but switches between them) or if a function is constant on some input interval, then it does not pass the horizontal line test and does not have an inverse. On the other hand, if a function is only increasing or is only decreasing, then it will pass the horizontal line test and does have an inverse. These observations are summarized as follows:

> **MONOTONIC FUNCTIONS HAVE INVERSES**.
> Let $f(x)$ be a function. If $f(x)$ is increasing on its domain or if it is decreasing on its domain, then it has an inverse.

◆ **EXAMPLE 8** Use monotonicity to explain why each function in Figure 12 does or does not have an inverse.

(a)   (b)

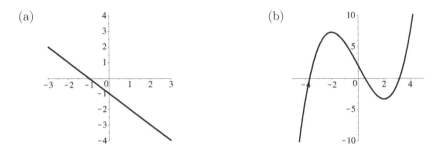

Figure 12: Curves for Example 8

Solution.

(a) The line in Figure 12(a) is decreasing on its domain of $(-\infty, \infty)$, which means that it has an inverse.

(b) The curve in Figure 12(b) is increasing on $(-\infty, 2)$ and $(2, \infty)$, but decreasing on $(-2, 2)$. Therefore, this curve does not have an inverse, because it exhibits more than one type of monotonicity.

                                                                       ■

➤ **QUESTION 7** Use monotonicity to explain why each function in Figure 13 does or does not have an inverse.

(a)   (b)

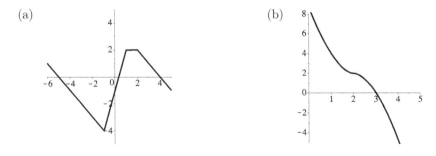

Figure 13: Curves for Question 7

Finding Inverses Algebraically

We know how to find the inverse of a function when the function is presented in a table (swap the input and output rows) and when the function is presented graphically (reflect across $y = x$). In addition to tabular and graphical presentations of functions, algebraically presented functions are also of interest. Determining whether a given algebraic function does or does not have an inverse can be accomplished by graphing such a function with RStudio and then applying the horizontal line test. If such a function does have an inverse, the following two-step process is used to find the algebraic presentation of its inverse function:

FINDING INVERSES OF ALGEBRAIC FUNCTIONS.
If $y = f(x)$ is a function that has an inverse, then $f^{-1}(x)$ is found as follows.

1. Solve $y = f(x)$ for x in terms of y.

2. Interchange the variables x and y.

In the first step of "solve $y = f(x)$ for x in terms of y," an expression is sought that takes an output from the original function, reverses all the function's actions, and returns the output to the corresponding input. In this way, the algebraic expression for the inverse function can be found.

The second step of "interchange the variables x and y" is more of a book-keeping step, because functions are usually presented in terms of input variable x and this step extends this convention to the inverse function. For example, recall the discussion of $f(x) = x+1$ immediately after the definition of an inverse function. Working intuitively, the operation of adding one is reversed by subtracting one, which means that the inverse of $f(x) = x + 1$ is $f^{-1}(x) = x - 1$. This statement of the inverse follows the standard convention of presenting $f^{-1}(x)$ in terms of x, rather than in terms of y as $f^{-1}(y) = y - 1$.

◆ **EXAMPLE 9** Verify each function has an inverse and find its algebraic expression:

(a) $y = 4x - 7$ (b) $y = x^3 + 5$

Solution.

(a) The line $y = 4x - 7$ has a positive slope of $m = 4$ and is increasing on its domain of $(-\infty, \infty)$. Therefore, this function has an inverse because increasing functions have inverses. An algebraic expression for this inverse is found by first solving for x in terms of y.

$$y = 4x - 7 \qquad \text{Given function}$$

$$y + 7 = 4x \qquad \text{Add 7 to both sides}$$

$$\frac{y + 7}{4} = x \qquad \text{Divide both sides by 4}$$

$$\frac{1}{4}y + \frac{7}{4} = x \qquad \text{Simplify}$$

Interchanging the variables, the inverse of $y = 4x - 7$ is the line $y = \dfrac{1}{4}x + \dfrac{7}{4}$ or, equivalently, $y = 0.25x + 1.75$.

(b) The cubic polynomial $y = x^3 + 5$ may be less familiar, so consider the graph of this function given in Figure 14 and apply the horizontal line test. Because every horizontal line intersects the graph exactly once, $y = x^3 + 5$ has an inverse. Alternatively, this function is increasing on its domain of $(-\infty, \infty)$ and has an inverse because of this monotonic behavior.

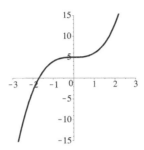

Figure 14: Graph of $y = x^3 + 5$ for Example 9(b)

An algebraic expression is found for the inverse by first solving for x in terms of y:

$$y = x^3 + 5 \qquad \text{Given function}$$
$$y - 5 = x^3 \qquad \text{Subtract 5 from both sides}$$
$$\sqrt[3]{y - 5} = x \qquad \text{Take the cube root of both sides}$$

Interchanging the variables, the inverse of $y = x^3 + 5$ is $y = \sqrt[3]{x - 5}$. ∎

In Example 9(b), the final algebraic step was to take the cube root of both sides of the equation. For square roots and other even-powered roots, recall that such roots of a negative number are undefined as real numbers. For example, if $y = x^2 + 5$ and 5 is subtracted from both sides, then the result is $y - 5 = x^2$. The next step would be to take the square root of both sides, but this step cannot be done because $y - 5$ can be negative. In fact, the parabola $y = x^2 + 5$ fails the horizontal line test (consider $y = 6$) and does not have an inverse, which corresponds to not being able to compute this square root.

As might be expected, some algebraic expressions are more complicated to work with than those in Example 9. The next example considers one such case:

◆ **EXAMPLE 10** Find an algebraic expression for the inverse of $y = \dfrac{3x+4}{2x-1}$.

Solution. First, solve $y = f(x)$ for x in terms of y:

$$y = \frac{3x+4}{2x-1} \qquad\qquad \text{Given function}$$

$$y(2x-1) = 3x+4 \qquad\qquad \text{Multiply both sides by } 2x-1$$

$$y(2x) - y = 3x+4 \qquad\qquad \text{Distribute } y \text{ on the left side}$$

$$(2y)x - 3x = y+4 \qquad\qquad \text{Isolate all } x\text{'s on the left side}$$

$$(2y-3)x = y+4 \qquad\qquad \text{Factor } x \text{ on the left side}$$

$$x = \frac{y+4}{2y-3} \qquad\qquad \text{Divide both sides by } 2y-3$$

Interchanging the variables, the inverse of $y = \dfrac{3x+4}{2x-1}$ is $y = \dfrac{x+4}{2x-3}$. ∎

Working in RStudio

RStudio defines algebraic expressions involving roots by means of fractional powers. For example, $\sqrt[3]{x+1}$ is defined in RStudio with the command (x+1)^(1/3). Similarly, for $\sqrt[5]{(2x-1)^3}$, the exponent rule $\sqrt[s]{a^r} = a^{r/s}$ is used to write $\sqrt[5]{(2x-1)^3} = (2x-1)^{3/5}$, and then this expression is defined in RStudio with the command (2*x-1)^(3/5). The plotFun command is used to graph such functions and decide whether or not they have inverses by means of the horizontal line test.

Examples of Commands:

- (3*x+5)^(5/4)

- plotFun((3*x+5)^(5/4)~x)

▶ **QUESTION 8** Graph each function using RStudio and verify that it has an inverse. Also, find an algebraic expression for its inverse.

(a) $f(x) = \sqrt[5]{3x-7}$

(b) $g(x) = \dfrac{5x-3}{7x+4}$

Summary

- The *inverse function* of a function $f(x)$ maps each output from $f(x)$ back to its corresponding input x. Symbolically, $f^{-1}(x)$ identifies the inverse of $f(x)$.

- The graph of $f^{-1}(x)$ is obtained by reflecting the graph of $f(x)$ across the line $y = x$.

Summary (continued)

- A function has an inverse when every output comes from exactly one input. Alternatively, a function does *not* have an inverse if some output comes from more than one input.

- *Horizontal line test:* A function $y = f(x)$ has an inverse function if and only if every horizontal line intersects its graph at most once. Alternatively, a function does *not* have an inverse if and only if some horizontal line intersects its graph more than once.

- *Monotonic functions have inverses:* Let $f(x)$ be a function. If $f(x)$ is increasing on its domain or if $f(x)$ is decreasing on its domain, then $f(x)$ has an inverse.

- *Finding inverses of algebraic functions:* If $y = f(x)$ has an inverse, $f^{-1}(x)$ can be found by

  1. Solving $y = f(x)$ for x in terms of y.
  2. Interchanging the variables x and y.

Exercises

In Exercises $1-6$, the function has an inverse. State this inverse.

1.

x	1	2	3	4	5
$f(x)$	7	9	6	8	0

2.

x	-3	-1	1	3	5
$f(x)$	4	2	0	6	8

3.

x	2	4	8	16	32
$f(x)$	1	3	9	27	81

4.

x	3	5	7	11	13
$f(x)$	-3	-5	-7	1	3

5.

x	r	s	t	u	v
$f(x)$	d	b	a	e	c

6.

x	a	b	c	d	e
$f(x)$	p	r	s	q	m

In Exercises $7-9$, consider annual e-commerce sales (S) in the United States in billions of dollars for each year (Y).

Y	2009	2010	2011	2012
S	121	143	168	192

7. State a question this function can address.

8. State a question its inverse can address.

9. State the inverse function.

In Exercises $10-12$, consider the S&P 500 stock market closing value (V) on June 1st of each year (Y) in U.S. dollars.

Y	1975	1985	1995	2005	2015
V	95	192	545	1191	2077

10. State a question this function can address.

11. State a question its inverse can address.

12. State the inverse function.

In Exercises 13 – 15, consider the world population growth rate (R) each year (Y).

Y	1995	2000	2005	2010	2015
R	1.41	1.26	1.20	1.13	1.06

13. State a question this function can address.

14. State a question its inverse can address.

15. State the inverse function.

In Exercises 16 – 18, consider the percent of high school graduates (P) enrolled in a two-year or four-year college in each year (Y).

Y	2005	2006	2009	2010	2012
P	68.6	66.0	70.1	68.1	66.2

16. State a question this function can address.

17. State a question its inverse can address.

18. State the inverse function.

In Exercises 19 – 21, consider the average maximum temperature (T) in degrees Fahrenheit each month (M) in New York City.

M	2	4	7	10	12
T	39.4	61.1	88.3	67.8	43.4

19. State a question this function can address.

20. State a question its inverse can address.

21. State the inverse function.

In Exercises 22 – 24, consider the three-year average monthly pollen count (C) in Brooklyn, New York City each month (M).

M	8	9	10	11	12
C	8.4	4.6	1.7	0.3	0.1

22. State a question this function can address.

23. State a question its inverse can address.

24. State the inverse function.

In Exercises 25 – 32, explain why the function does or does not have an inverse and, if it does, state its inverse.

25.
x	1	2	3	4	5
$f(x)$	-1	1	2	-2	3

26.
x	1	2	3	4	5
$f(x)$	0	4	2	3	2

27.
x	1	2	3	4	5
$f(x)$	1	1	1	1	1

28.
x	-2	-1	0	1	2
$f(x)$	4	1	0	-1	-4

29.
x	-2	-1	0	1	2
$f(x)$	-2	-1	0	1	2

30.
x	a	c	e	g	h
$f(x)$	a	a	b	b	a

31.
x	b	k	s	t	u
$f(x)$	d	b	z	e	k

32.
x	a	b	r	p	z
$f(x)$	z	b	a	p	p

In Exercises 33 – 38, explain why the tabular function does or does not have an inverse and, if it does, state its inverse.

33. Annual total retail sales taxes (T) collected in the United States in billions of dollars each year (Y); for example, during 2006, $141,100,000,000 in taxes was collected.

Y	2006	2008	2010	2012
T	$141	$141	$139	$150

34. Average debt (D) in thousands of dollars at the end of the spring term in each year (Y) for bachelor's degree recipients attending public four-year colleges and universities who borrowed money to finance their education.

Y	2001	2003	2005	2006
D	20.4	20.9	21.5	21.8

35. Total United States health expenditures as a percentage of gross domestic product (G) during each year (Y).

Y	2009	2010	2011	2012
G	17.71	17.66	17.68	17.91

36. Floors (F) where a hotel's elevator is located at T minutes past 3 p.m.

T	11	12	13	14	15
F	9	5	1	5	7

37. Average number of goals (G) per game in World Cup Tournaments in each year (Y).

Y	'94	'98	'02	'06	'10
G	2.7	2.7	2.5	2.3	2.3

38. Total annual sales (S) in thousands of hybrid vehicles in the United States in each year (Y).

Y	2010	2011	2012	2013
S	274	266	435	496

Your Turn. In Exercises 39 and 40, give a tabular presentation of a function with the property. Include at least five data points in your table.

39. A function with an inverse.

40. A function without an inverse.

In Exercises 41–50, use the horizontal line test to explain why the function does or does not have an inverse and, if it does, sketch its inverse.

41.
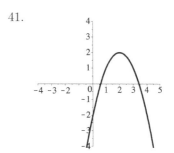

42.
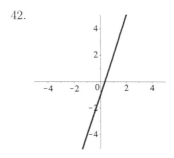

43.
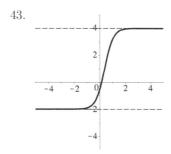

44.
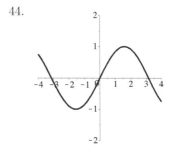

45.

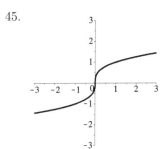

46.

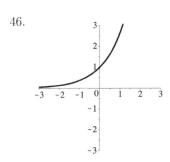

47.

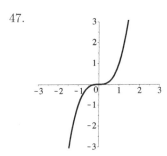

48.

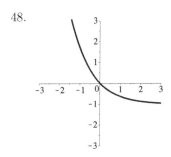

49.

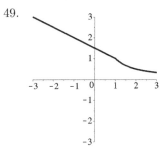

50.

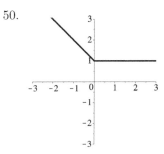

In Exercises 51–60, use monotonicity to explain why the function in Exercises 41–50 does or does not have an inverse and, if so, sketch its inverse.

51. Exercise 41   56. Exercise 46

52. Exercise 42   57. Exercise 47

53. Exercise 43   58. Exercise 48

54. Exercise 44   59. Exercise 49

55. Exercise 45   60. Exercise 50

Your Turn. In Exercises 61–66, graph a function satisfying the set of properties that also does not have an inverse.

61. Decreasing on $(-\infty, \infty)$, concave up on $(-\infty, 0)$, and concave down on $(0, \infty)$.

62. Increasing on $(-\infty, 0)$, decreasing on $(0, \infty)$, concave up on $(-\infty, 0)$, and concave down on $(0, \infty)$.

63. Increasing on $(-\infty, 0)$, decreasing on $(0, \infty)$, and concave up on both $(-\infty, 0)$ and $(0, \infty)$.

64. Decreasing on $(-\infty, 0)$, increasing on $(0, \infty)$, and concave down on both $(-\infty, 0)$ and $(0, \infty)$.

65. Increasing on $(-\infty, 2)$, constant on $(2, \infty)$, and concave up on $(-\infty, 2)$.

66. Decreasing on $(-\infty, 2)$, constant on $(2, \infty)$, and concave down on $(-\infty, 2)$.

In Exercises 67–80, find an algebraic expression for the inverse of the function.

67. $y = 2x - 4$

68. $y = 5x + 15$

69. $y = -3x + 9$

70. $y = -4x - 7$

71. $y = x^7 - 1$

72. $y = x^5 + 3$

73. $y = \sqrt[3]{x + 7}$

74. $y = \sqrt[7]{x^3 - 4}$

75. $y = \dfrac{1}{x - 2}$

76. $y = \dfrac{1}{4x - 5}$

77. $y = \dfrac{x - 3}{x + 2}$

78. $y = \dfrac{x + 5}{9x + 1}$

79. $y = \dfrac{-2x + 1}{3x - 5}$

80. $y = \dfrac{4x - 1}{7x + 2}$

RStudio. In Exercises 81–94, use RStudio to graph the function and use the horizontal line test to explain why the function does or does not have an inverse.

81. $y = x^7 + x$

82. $y = x^6 - x^2$

83. $y = x^4 + x$

84. $y = x^3 - x^2 + 1$

85. $y = |2x - 7|$

86. $y = \sqrt[3]{x + 7}$

87. $y = \sqrt[4]{x^2 + 5}$

88. $y = e^{|x|}$

89. $y = e^{x^2}$

90. $y = -e^{x+1}$

91. $y = 4e^x - 3$

92. $y = xe^{-x}$

93. $y = \dfrac{1}{1 + e^x}$

94. $y = \dfrac{3}{2 - e^x}$

In Your Own Words. In Exercises 95–100, explain the following.

95. Inverse function

96. Graphing an inverse

97. A function without an inverse

98. Horizontal Line Test

99. Monotonicity and inverse functions

100. Finding inverses algebraically

1.6   Logarithmic Functions

Many functions have inverses that map each output of the given function back to its corresponding input. Most of the common modeling functions have inverses, including non-constant linear and exponential functions. Every nonconstant linear function passes the horizontal line test and has an inverse. Such lines are of the form $y = mx + b$ with $m \neq 0$, and the standard algebraic process can be used to find their inverses. Solving $y = f(x)$ for x in terms of y results in the following:

$$y = mx + b \qquad \text{Given function}$$
$$y - b = mx \qquad \text{Subtract } b \text{ from both sides}$$
$$\frac{y}{m} - \frac{b}{m} = x \qquad \text{Divide both sides by } m \neq 0$$

Interchanging the variables, the inverse of $y = mx + b$ is the line $y = \dfrac{x}{m} - \dfrac{b}{m}$.

Exponential functions $y = Ce^{kx}$ with $C \neq 0$ and $k \neq 0$ also have inverses. We focus on the specific case of the natural exponential function $y = e^x$; a similar argument applies for other exponentials. Example 3(b) in Section 1.5 included Figure 1 presenting the graphs of both $f(x) = e^x$ and its inverse $f^{-1}(x) = \ln(x)$, which is obtained by reflecting e^x across the line $y = x$.

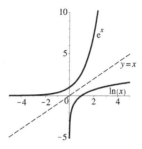

Figure 1: Graph of e^x and its inverse $\ln(x)$

Figure 1 confirms that the natural exponential has an inverse by either the horizontal line test or monotonicity. In particular, every horizontal line intersects $y = e^x$ at most once, so this function passes the horizontal line test. Alternatively, $y = e^x$ is increasing on its domain of all real numbers $(-\infty, \infty)$, and such monotonic functions have inverses.

Once a function is known to have an inverse, finding an expression for the inverse function is the next standard goal. As it turns out, rather than applying the standard algebraic process to find the inverse of the natural exponential function, a new function must be defined.

Definition. The **natural logarithm function**, denoted by $\ln(x)$, is the inverse of the natural exponential function e^x and has domain $(0, \infty)$. In general, natural logarithm functions are of the form $f(x) = C \ln(x - h) + v$ and have a domain of (h, ∞), so they are only defined when $x > h$.

The graphs of e^x and its inverse $\ln(x)$ in Figure 1 illustrate the many parallels that arise between this pair of inverse functions. The natural exponential has a domain of $(-\infty, \infty)$ and a range of $(0, \infty)$, while these intervals are interchanged for its inverse: the natural logarithm has a domain of $(0, \infty)$ and a range of $(-\infty, \infty)$. In addition, e^x has a vertical intercept of $(0, 1)$ and the x-axis as a horizontal asymptote, while $\ln(x)$ has a horizontal intercept of $(1, 0)$ and the y-axis as a vertical asymptote. Both functions are increasing, but e^x is concave up, while $\ln(x)$ is concave down. Furthermore, recall that exponential functions increase more quickly than the other common modeling functions. As the inverse of exponentials, logarithmic functions increase more slowly than the other common modeling functions.

This section continues with a study of the graphical properties of the natural logarithm $f(x) = \ln(x)$, paying particular attention to the effect of its parameters. The algebraic properties of logarithms are then discussed because, in part, they are particularly important to the development of mathematical models.

Logarithms as Parametrized Families of Functions

The simplest parameters to understand in the generalized natural logarithm $y = C \ln(x - h) + v$ are the vertical shift v and the horizontal shift h. The **vertical shift** parameter v moves the graph of $\ln(x)$ vertically up and down. The value of this vertical shift is typically determined by finding the height of the output for input $x = 1 + h$, because $\ln(1) = 0$ and so the height of $f(x) = C \ln(x - h) + v$ is determined at $x = 1 + h$ as follows:

$$f(1 + h) = C \ln[(1 + h) - h] + v = C \ln(1) + v = C \cdot 0 + v = v$$

Figure 2(a) illustrates the effect of the vertical shift parameter on $f(x) = \ln(x) + v$ for various values of v. Curve (a) in Figure 2(a) presents the graph of $y = \ln(x) + 2$ with $v = 2$, curve (b) presents $y = \ln(x)$ with $v = 0$, and curve (c) presents $y = \ln(x) - 2$ with $v = -2$.

The **horizontal shift** parameter h moves the graph of $\ln(x)$ horizontally left and right. The most straightforward way to identify the value of h for $f(x) = C \ln(x - h) + v$ is to determine the location of its vertical asymptote. In its standard position, the vertical asymptote of $\ln(x)$ is $x = 0$, which corresponds to a horizontal shift of $h = 0$. Extending this observation to the general, parametrized form of natural logarithms, the vertical asymptote $x = h$ corresponds to a horizontal shift of h. Figure 2(b) illustrates the effect of the horizontal shift parameter for various values of h. Curve (a) in Figure 2(b) presents the graph of $y = \ln(x + 2) = \ln[x - (-2)]$, with $h = -2$ shifting the graph to the left, curve (b) presents $y = \ln(x)$ with $h = 0$, and curve(c) presents $y = \ln(x - 2)$, with $h = 2$ shifting the graph to the right.

(a) (b)

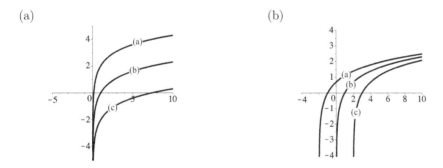

Figure 2: Effect of (a) the vertical shift parameter v and (b) the horizontal shift parameter h on $\ln(x)$

Finally, for a natural logarithm function $y = C \ln(x - h) + v$, the parameter C has two effects on its graph: the magnitude of C determines the relative rate of increase of the graph, and the sign of C determines its monotonicity and concavity. Figure 3(a) presents the graph of $y = 2 \ln(x)$ with $C = 2$, Figure 3(b) presents $y = \ln(x)$ with $C = 1$, and Figure 3(c) presents $y = 0.5 \ln(x)$ with $C = 0.5$. As illustrated in Figure 3, the greater the magnitude of C, the greater the rate of increase of the natural logarithm.

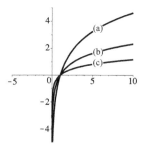

Figure 3: Effect of the magnitude of parameter C on $\ln(x)$

In addition, the sign of the parameter C determines the monotonicity and concavity of the graph of the natural logarithm. When C is positive, $y = C\ln(x - h) + v$ is increasing and concave down; when C is negative, this natural logarithm is decreasing and concave up. Figure 4(a) presents the graph of $y = \ln(x)$ with positive $C = 1$ and so both increasing and concave down; Figure 4(b) presents $y = -\ln(x)$ with negative $C = -1$ and so both decreasing and concave up. This understanding of the effect of these parameters can be used to distinguish among graphs of particular logarithmic functions

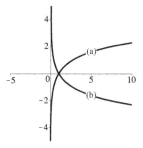

Figure 4: Effect of the sign of parameter C on $\ln(x)$

➤ **QUESTION 1** Identify each curve in Figure 5 as one of $y = 4\ln(x - 2)$ or $y = -2\ln(x + 4)$.

Algebra of Logarithms

The algebraic properties of logarithms motivated the original definition of these functions. In the early 1600s, the study of astronomy was revolutionized by Galileo Galilei's technological improvements to the telescope, which enabled him and others to make previously impossible physical observations and measurements of the solar system. Consequently, astronomy and many other areas of scientific inquiry began to require more computations with substantially larger numbers. In 1614, the Scottish mathematician and physicist John Napier published his *Mirifici Logarithmorum Canonis Descriptio* (*The Description of the Wonderful Canon of Logarithms*), detailing the definition and properties of logarithms, which proved invaluable for enabling these otherwise intractable computations.

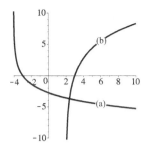

Figure 5: Graphs of logarithms for Question 1

In modern times, calculators and computers can readily handle large-scale computations. In part, this success arises from the underlying programs incorporating logarithms into their code. Beyond simplifying arithmetic computations, the algebraic properties of logarithms are vitally important for mathematical modeling, particularly in working with exponential and power functions. The eight most important properties of the natural logarithm function are given below.

LOGARITHM RULES I. Let a and b be real numbers with a positive:

(1) $e^{\ln(a)} = a$ (2) $\ln(e^b) = b$ (3) $\ln(1) = 0$ (4) $\ln(e) = 1$

These first four rules arise from the fact that the natural logarithm $y = \ln(x)$ is the inverse of the natural exponential $y = e^x$. In addition, rules (3) and (4) can be thought of as special cases of rule (2). In particular, recall that $1 = e^0$ and so $\ln(1) = \ln(e^0) = 0$ using $b = 0$ in rule (2). Similarly, $e = e^1$, which gives $\ln(e) = \ln(e^1) = 1$, using $b = 1$ in rule (2).

◆ **EXAMPLE 1** Simplify each logarithmic expression:

(a) $e^{\ln(4)}$ (b) $\ln\left(\sqrt{e^5}\right)$

Solution.

(a) Applying rule (1) that $e^{\ln(a)} = a$ gives $e^{\ln(4)} = 4$.

(b) First, rewrite the input $\sqrt{e^5} = \sqrt[2]{e^5} = e^{5/2}$ using the exponent rule $\sqrt[s]{a^r} = a^{r/s}$ from Section 1.4. Applying rule (2) that $\ln(e^b) = b$ gives

$$\ln\left(\sqrt{e^5}\right) = \ln\left(e^{5/2}\right) = 5/2 = 2.5$$

➤ **QUESTION 2** Simplify each logarithmic expression:

(a) $e^{\ln(e)}$

(b) $\ln\left(\sqrt[4]{e}\right)$

In addition to these basic inverse properties of $\ln(x)$ with respect to e^x, four other algebraic properties of logarithms are fundamentally important. In essence, these properties enable the exchange of a product (or a quotient) of a pair of numbers for a sum (or a difference) of a pair of related numbers. Addition and subtraction of large numbers are much simpler than multiplication and division, providing the important simplification of computations that facilitated the Scientific Revolution.

LOGARITHM RULES II. Let a and b be positive real numbers:

(5) $\ln(ab) = \ln(a) + \ln(b)$

(7) $\ln(a^r) = r\ln(a)$

(6) $\ln\left(\dfrac{a}{b}\right) = \ln(a) - \ln(b)$

(8) $\ln\left(\dfrac{1}{b}\right) = -\ln(b)$

Observe that rule (6) follows from a combination of rules (5) and (7) because $a/b = ab^{-1}$. Substituting and simplifying gives the following:

$$\ln\left(\frac{a}{b}\right) = \ln(ab^{-1}) = \ln(a) + \ln(b^{-1}) = \ln(a) + (-1)\ln(b) = \ln(a) - \ln(b)$$

Similarly, rule (8) follows from a combination of rules (3) and (6) as follows:

$$\ln\left(\frac{1}{b}\right) = \ln(1) - \ln(b) = 0 - \ln(b) = -\ln(b)$$

As with all skills, practice applying these rules to various expressions is the key for them becoming natural and even second-nature. While working through these abstract examples, the reader is encouraged to keep in mind that these arithmetic techniques are essential to the development and analysis of mathematical models of real-world phenomenon.

◆ **EXAMPLE 2** Expand each expression into a sum and/or difference of logarithms:

(a) $\ln\left(\dfrac{2x}{e}\right)$

(b) $\ln\left[\sqrt{(x^2-1)^3}\right]$

Solution.

(a) $\ln\left(\dfrac{2x}{e}\right) = \ln(2x) - \ln(e)$ $\ln\left(\dfrac{a}{b}\right) = \ln(a) - \ln(b)$

$= \ln(2) + \ln(x) - 1$ $\ln(ab) = \ln(a) + \ln b$ and $\ln(e) = 1$

(b) $\ln\left[\sqrt{(x^2-1)^3}\right] = \ln\left[(x^2-1)^{3/2}\right]$ $\qquad\qquad$ $\sqrt[s]{a^r} = a^{r/s}$

$\qquad\qquad = \dfrac{3}{2}\ln[x^2-1]$ $\qquad\qquad$ $\ln(a^r) = r\ln(a)$

$\qquad\qquad = \dfrac{3}{2}\ln[(x+1)(x-1)]$ $\qquad\qquad$ $x^2-1 = (x+1)(x-1)$

$\qquad\qquad = \dfrac{3}{2}\ln(x+1) + \dfrac{3}{2}\ln(x-1)$ $\qquad\qquad$ $\ln(ab) = \ln(a) + \ln(b)$

$\blacksquare$

➤ **QUESTION 3** Expand each expression into a sum and/or difference of logarithms:

(a) $\ln(3e^2)$ $\qquad\qquad\qquad\qquad\qquad$ (b) $\ln\left(\dfrac{x^2y}{z^3}\right)$

In addition to expanding logarithmic expressions, these rules can be used to combine sums and differences of logarithms into a single quantity.

◆ **EXAMPLE 3** Combine each expression into a logarithm of a single quantity in simplest form:

(a) $2[\ln(x) + \ln(x+1) - \ln(x^2-1)]$ $\qquad$ (b) $\dfrac{3}{2}[\ln(x^2-4) - \ln(x+2)]$

Solution.

(a) $2[\ln(x) + \ln(x+1) - \ln(x^2-1)]$

$\qquad = 2[\ln[x(x+1)] - \ln(x^2-1)]$ $\qquad$ $\ln(ab) = \ln(a) + \ln(b)$

$\qquad = 2\ln\left[\dfrac{x(x+1)}{x^2-1}\right]$ $\qquad$ $\ln\left(\dfrac{a}{b}\right) = \ln(a) - \ln(b)$

$\qquad = 2\ln\left[\dfrac{x(x+1)}{(x+1)(x-1)}\right]$ $\qquad$ $x^2-1 = (x+1)(x-1)$

$\qquad = 2\ln\left[\dfrac{x}{x-1}\right]$ $\qquad$ Cancel $x+1$

$\qquad = \ln\left[\dfrac{x^2}{(x-1)^2}\right]$ $\qquad$ $\ln(a^r) = r\ln(a)$

(b) $\dfrac{3}{2}[\ln(x^2-4) - \ln(x+2)] = \dfrac{3}{2}\ln\left[\dfrac{x^2-4}{x+2}\right]$ $\qquad$ $\ln\left(\dfrac{a}{b}\right) = \ln(a) - \ln(b)$

$\qquad\qquad = \dfrac{3}{2}\ln\left[\dfrac{(x+2)(x-2)}{x+2}\right]$ $\qquad$ $x^2-4 = (x+2)(x-2)$

$\qquad\qquad = \dfrac{3}{2}\ln(x-2)$ $\qquad$ Cancel $x+2$

$\qquad\qquad = \ln\left[(x-2)^{3/2}\right]$ $\qquad$ $\ln(a^r) = r\ln(a)$

$\qquad\qquad = \ln\left[\sqrt{(x-2)^3}\right]$ $\qquad$ $a^{r/s} = \sqrt[s]{a^r}$

$\blacksquare$

▶ **QUESTION 4** Combine each expression into a logarithm of a single quantity in simplest form:

(a) $2\ln(2) + 3\ln(x) - \dfrac{1}{2}\ln(x^2)$

(b) $\ln(x + 8) + 2\ln[(x - 1)^3] - \dfrac{7}{2}\ln[(x - 1)^2]$

Logarithms and Exponential Models

The rules of logarithms are particularly useful when working with exponential models. We often solve for an input variable in an exponential model by applying the natural logarithm to both sides of the equation. Rule (5) $\ln(ab) = \ln(a) + \ln(b)$ and rule (7) $\ln(a^r) = r\ln(a)$ are then used to determine the value of the input variable.

Such processes are illustrated by considering how to find an exponential function when given its vertical intercept and one other point lying on its graph. For an exponential function $f(x) = Ce^{kx}$, the vertical intercept always corresponds to the value of C, because $e^0 = 1$ and $f(0) = Ce^{k \cdot 0} = Ce^0 = C \cdot 1 = C$. The value of the parameter k is determined by substituting the other given point and applying the natural logarithm to solve for k.

◆ **EXAMPLE 4** Find the equation of the exponential function passing through the points $(0, 3)$ and $(2, 12)$. Also, sketch its graph and discuss its monotonicity and concavity.

Solution. For an exponential function $y = Ce^{kx}$, the vertical intercept $(0, 3)$ is $(0, C)$, which gives $C = 3$ and $y = 3e^{kx}$. The coefficient k is obtained by substituting the other point $(2, 12)$ into this intermediate equation and solving for k:

$$y = 3e^{kx} \qquad \text{Intermediate exponential}$$
$$12 = 3e^{k \cdot 2} \qquad \text{Substitute } (x, y) = (2, 12)$$
$$4 = e^{2k} \qquad \text{Divide both sides by 3}$$
$$\ln(4) = \ln(e^{2k}) \qquad \text{Apply } \ln(x) \text{ to both sides}$$
$$\ln(4) = 2k \qquad \ln\left(e^b\right) = b \text{ with } b = 2k$$
$$\frac{\ln(4)}{2} = k \qquad \text{Divide both sides by 2}$$

Rule (7) stating that $\ln(a^r) = r\ln(a)$ gives $\ln(4) = \ln(2^2) = 2\ln(2)$, which means that $k = \ln(4)/2 = 2\ln(2)/2 = \ln(2)$. Therefore, $y = 3e^{\ln(2)x}$ is the desired exponential. Its graph is given in Figure 6 and shows that this exponential $y = 3e^{\ln(2)x}$ is increasing and concave up on its domain of all real numbers $(-\infty, \infty)$. ■

▶ **QUESTION 5** Find the equation of the exponential function passing through the points $(0, -4)$ and $(3, -24)$. Also, sketch its graph and discuss its monotonicity and concavity.

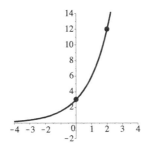

Figure 6: Graph of $y = 3e^{\ln(2)x}$ for Example 4

◆ **EXAMPLE 5** A colony of bacteria begins to grow exponentially on some leftover pizza from a late night study session. One model of this population is $P(t) = 1000e^{\ln(2)t}$, where $P(t)$ denotes the number of bacteria in the colony as a function of the time t in hours after midnight on Thursday, September 15, 2016.

(a) What is the initial population of the colony? What is the population of the colony after 5 hours?

(b) When will the population reach 5000 bacteria?

Solution.

(a) For such an exponential model, the initial population is the parameter $C = P(0) = 1000$, or 1000 bacteria. The population after 5 hours is found by substituting this given time $t = 5$ into $P(t) = 1000e^{\ln(2)t}$ to obtain $P(5) = 1000e^{\ln(2)\cdot5} = 32{,}000$. This result means that 32,000 bacteria are living on the pizza at 5 a.m.

(b) Substituting the given population size $P = 5000$ into $P(t)$ gives

$$5000 = 1000e^{\ln(2)t}$$

Now apply the natural logarithm $\ln(x)$ to both sides of this equation and solve for t:

$$
\begin{aligned}
\ln(5000) &= \ln(1000e^{\ln(2)t}) && \text{Apply } \ln(x) \text{ to both sides} \\
\ln(5000) &= \ln(1000) + \ln(e^{\ln(2)t}) && \ln(ab) = \ln(a) + \ln(b) \\
\ln(5000) - \ln(1000) &= \ln(2) \cdot t && \text{Subtract } \ln(1000) \text{ and } \ln(e^b) = b \\
\frac{\ln(5000) - \ln(1000)}{\ln(2)} &= t && \text{Substitute } \ln(e) = 1 \text{ and divide by } \ln(2) \\
2.322 &= t && \text{Compute the left side}
\end{aligned}
$$

This solution means that 5000 bacteria are living on the pizza 2.322 hours after midnight, or at about 2:19 a.m.

In addition to population growth, exponential models arise in connection with deposits of money earning interest. An initial deposit into an account is called the **principal**, and, for certain types of accounts, this deposit earns **interest**, which the financial institution pays the depositor for the privilege of using their money for other purposes. Periodically, the earned interest is **compounded**, or added into the current account **balance**, at which point the newly added interest begins earning interest as well. Interest is compounded at specified intervals, often monthly, although sometimes annually, quarterly, or even daily. **Continuous compounding** means that the interest is immediately and constantly added to the balance. This type of compounding is modeled using the natural exponential, as illustrated in the following question:

➤ **QUESTION 6** You invest \$2000 earning 1.5% annual interest compounded continuously. The model of the account balance is $B(t) = 2000e^{0.015t}$, where $B(t)$ denotes the amount of money in your account as a function of the number of years t since the initial deposit.

(a) How much money will you have in 2 years?

(b) In how many years will you have \$3000?

Semi-log Plots and Log–Log Plots

When developing models of real-life data sets, graphical presentations of data provide a first rough indicator for a reasonable choice of a model. As will be detailed in Chapter 2, models are typically chosen from among linear, exponential, power, sine, and sigmoidal functions. Looking ahead, observe that the graphs of many exponential and power functions are quite similar at first glance, which can make it difficult to distinguish which model is most appropriate. For example, in Figure 7(a), the data plot corresponds to $y = x^3$, while in Figure 7(b), the data plot corresponds to $y = e^x$. These graphs appear quite similar.

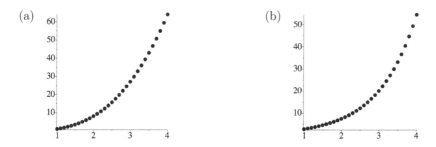

Figure 7: Graphs of (a) $y = x^3$ and (b) $y = e^x$ exhibiting similar shapes

The natural logarithm enables us to distinguish between the graphs of $y = x^3$ and $y = e^x$, as well as between data sets modeled by such functions. Namely, in addition to graphing a given data set, what are known as the *semi-log* plot and the *log–log* plot of the data will also prove useful. A semi-log plot of an input–output relationship is obtained by graphing the natural logarithm of the outputs as a function of the input. In other words, if (a, b) is a point from a given data set or curve, the semi-log plot includes the point $(a, \ln(b))$. A log–log plot extends this modification to the inputs, plotting the natural logarithm of the outputs as a function of the natural logarithms of the inputs. In this case, if (a, b) is a point from a given data set or curve, the log–log plot includes the point $(\ln(a), \ln(b))$.

Definition. If (a, b) is a point from a given data set or curve, its **semi-log plot** includes the point $(a, \ln(b))$, and its **log–log plot** includes the point $(\ln(a), \ln(b))$.

For reasons detailed in Section 2.2, the semi-log plot of an exponential data set is linear, while the semi-log plot of a power data set is not, enabling a familiarity with lines to distinguish between these two types of data sets. For example, in Figure 8(a), the semi-log plot of data corresponding to $y = x^3$ is not linear, while in Figure 8(b), the semi-log plot of data corresponding to $y = e^x$ is linear.

Similarly, log–log plots of power functions and their corresponding data sets are linear, allowing determination of when such a model is appropriate. On a practical level, hand-graphing semi-log and log–log plots point by point is usually impractical, as suggested by Figure 8. Therefore, RStudio is used to graph such plots, using the commands detailed below.

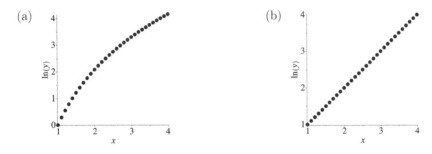

Figure 8: Semi-log plots of (a) $y = x^3$ and (b) $y = e^x$

Working in RStudio

RStudio uses the command `log(x)` for the natural logarithm function $\ln(x)$. This command allows determination of specific values for this function, such as $\ln(1) = 0$ with `log(1)` and $\ln(2) \approx 0.69314782$ with `log(2)`. You may have used the symbolism $\log(x)$ for the common logarithm $\log_{10}(x)$ with a base of 10. However, for RStudio, `log(x)` always refers to the natural logarithm $\ln(x)$. (Note: The RStudio command for the base 10 logarithm is `log10(x)`.) Recall using the RStudio command `exp(x)` for the natural exponential function e^x. This command can be used in combination with `log(x)` to compute such quantities as $\ln(e) = 1$ with `log(exp(1))` and $\ln(\sqrt{e^3}) = 1.5$ with `log(exp(3/2))`.

RStudio graphs functions, including logarithmic functions, using the `plotFun` command. For example, the logarithm $y = \ln(2x)$ is graphed on the interval $[1, 6]$ by entering

```
plotFun(log(2*x)~x,xlim=range(1,6))
```

Recall that the output interval is controlled with the option `ylim=range(,)` and that multiple functions are graphed on the same axes using the option `add=TRUE`.

Examples of Commands:

- `log(5) + log(exp(pi))`

- `plotFun(log(5*x)~x,xlim=range(0.25,10))`

- `plotFun(log(x)~x,add=TRUE)`

◆ **EXAMPLE 6** Simplify each logarithmic expression from Example 1 using RStudio:

(a) $e^{\ln(4)}$ (b) $\ln\left(\sqrt{e^5}\right)$

Solution.

(a) `exp(log(4))`   (b) `log((exp(5))^(1/2))`

    `[1] 4`   `[1] 2.5`

This RStudio output shows that $e^{\ln(4)} = 4$ and $\ln\left(\sqrt{e^5}\right) = 2.5$. ∎

▶ **QUESTION 7** Simplify each logarithmic expression from Question 2 using RStudio:

(a) $e^{\ln(e)}$ (b) $\ln\left(\sqrt[4]{e}\right)$

Using RStudio to graph logarithmic functions can facilitate an exploration of the effects of the parameters on the graph of the natural logarithm.

◆ **EXAMPLE 7** Graph the logarithmic functions $y = 4\ln(x-2)$ and $y = -2\ln(x+4)$ from Question 1 using RStudio and identify the corresponding curve in Figure 5.

Solution. When graphing a logarithm $y = C\ln(x - h) + v$, the vertical asymptote $x = h$ is the leftmost bound that can appear on the range of x-values. Also, when using h as this leftmost bound, bounds for the range of y-values must be provided to obtain a meaningful graph.

- `plotFun(4*log(x-2)~x, xlim=range(2,10), ylim=range(-10,10))`

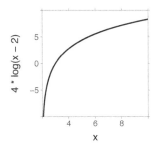

- `plotFun(-2*log(x+4)~x, xlim=range(-4,10), ylim=range(-10,10))`

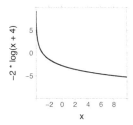

Based on this RStudio output, $y = 4\ln(x - 2)$ corresponds to Figure 5(b) and $y = -2\ln(x + 4)$ to Figure 5(a).

Working in RStudio

RStudio graphs the semi-log plot of a given function by applying the `log(x)` command to the given function inside the `plotFun` command. For example, the semi-log plot of $y = x^5$ is graphed in RStudio by entering

> `plotFun(log(x^5)~x,xlim=range(0,10),ylim=range(-10,10))`

The same practices of specifying ranges on the input values and the output values are needed to obtain a clear, interpretable graph of the given function.

Question 8 considers functional versions of the data sets discussed when introducing the idea of a semi-log plot.

▶ **QUESTION 8** Graph the semi-log plot of each function using RStudio:

(a) $y = e^x$ (b) $y = x^3$

Summary

- The *natural logarithm function* $\ln(x)$ is the inverse of the natural exponential function e^x and has domain $(0, \infty)$. In general, natural logarithm functions are of the form $f(x) = C\ln(x - h) + v$ and have a domain of (h, ∞), so they are only defined when $x > h$.

- For $y = C\ln(x - h) + v$, the parameter v corresponds to a *vertical shift*, h corresponds to a *horizontal shift*, and C affects the relative steepness of its graph, its monotonicity, and its concavity.
  - If C is positive, $y = C\ln(x - h) + v$ is increasing and concave down.
  - If C is negative, $y = C\ln(x - h) + v$ is decreasing and concave up.

- *Logarithm rules* Let a and b be positive real numbers:

  (1) $e^{\ln(a)} = a$ (5) $\ln(ab) = \ln(a) + \ln(b)$

  (2) $\ln(e^b) = b$ (6) $\ln\left(\dfrac{a}{b}\right) = \ln(a) - \ln(b)$

  (3) $\ln(1) = 0$ (7) $\ln(a^r) = r\ln(a)$

  (4) $\ln(e) = 1$ (8) $\ln\left(\dfrac{1}{b}\right) = -\ln(b)$

- If (a, b) is a point from a given data set or curve, its *semi-log plot* includes the point $(a, \ln(b))$, and its *log–log plot* includes the point $(\ln(a), \ln(b))$.

Exercises

In Exercises $1 - 8$, simplify the logarithmic expression by hand.

1. $e^{\ln(5)}$

2. $e^{4\ln(1)}$

3. $e^{3\ln(2)}$

4. $e^{5\ln(e)}$

5. $\ln(e^{-2})$

6. $\ln(e^6)$

7. $\ln\left[\sqrt[4]{e^3}\right]$

8. $\ln\left[\sqrt[8]{e^{11}}\right]$

RStudio. In Exercises $9 - 16$, use RStudio to simplify the logarithmic expression.

9. $e^{\ln(7)}$

10. $e^{6\ln(1)}$

11. $e^{9\ln(\pi)}$

12. $e^{42\ln(e)}$

13. $\ln(e^{-16})$

14. $\ln(e^7)$

15. $\ln\left[\sqrt[17]{e^{22}}\right]$

16. $\ln\left[\sqrt[38]{e^{67}}\right]$

In Exercises $17 - 26$, expand the expression into a sum and/or difference of logarithms.

17. $\ln(3x)$

18. $\ln(4x^2)$

19. $\ln(x^2 - 4)$

20. $\ln(x^2 - 16)$

21. $\ln(x^3 - 9x^2)$

22. $\ln(2x^3 - 50x)$

23. $\ln(9x + 18)$

24. $\ln(2x - 16)$

25. $\ln\left[\dfrac{x(x+1)}{3e}\right]$ 26. $\ln\left[\dfrac{ex}{x^2-1}\right]$

In Exercises 27–34, combine the expression into a logarithm of a single quantity.

27. $\ln(x) + 3\ln(2)$

28. $\ln(5) + 4\ln(x)$

29. $3\ln(x) + 2\ln(x)$

30. $\ln(x) + \ln(x+1)$

31. $2\ln(x+1) - \ln(20) - \ln(x-1)$

32. $3\ln(x) - \ln(x+1) + \ln(e)$

33. $3\ln(e) + \ln(2) + \ln(1) + \ln(0.5)$

34. $2\ln(x^2-9) - 4\ln(x+3) + \ln(x-3)$

Your Turn. In Exercises 35–40, state a specific example of the logarithm rule.

35. $e^{\ln(a)} = a$

36. $\ln(e^b) = b$

37. $\ln(ab) = \ln(a) + \ln(b)$

38. $\ln\left(\dfrac{a}{b}\right) = \ln(a) - \ln(b)$

39. $\ln(a^r) = r\ln(a)$

40. $\ln\left(\dfrac{1}{b}\right) = -\ln(b)$

In Exercises 41–48, solve the equation for the variable x.

41. $\ln(x+1) = 0$

42. $\ln(x-1) = \ln(4)$

43. $\ln(x) + \ln(x-3) = \ln(4)$

44. $\ln(x) + \ln(x-4) = \ln(5)$

45. $\ln(x) - \ln(x-3) = \ln(4)$

46. $\ln(x^2-9) - \ln(x-3) = \ln(8)$

47. $\ln(x^2-25) - \ln(x+5) = \ln(4)$

48. $\ln(x^2) + \ln(x+2) - \ln(x) = \ln(6)$

RStudio. In Exercises 49–56, use RStudio to graph the logarithm on an appropriate domain. Discuss its monotonicity and concavity.

49. $y = \ln(x-2)$

50. $y = -\ln(x-2)$

51. $y = -\ln(x+4)$

52. $y = 7\ln(x+4)$

53. $y = \ln(x) + 2$

54. $y = 4\ln(x) + 2$

55. $y = -3\ln(x) + 1$

56. $y = \ln(x+5) - 6$

In Exercises 57–64, find the exponential function $y = Ce^{kx}$ through the pair of points. Sketch its graph and discuss its monotonicity and concavity.

57. $(0,1), (3,5)$

58. $(0,1), (-3,8)$

59. $(0,8), (-2,1)$

60. $(0,8), (-1,6)$

61. $(0,-3), (5,-8)$

62. $(0,-3), (1,-7)$

63. $(0,-5), (2,-2)$

64. $(0,-5), (-5,1)$

In Exercises 65–68, consider the exponential $B(t) = 1000e^{0.01t}$ that models the balance from an initial deposit of \$1000 earning 1% annual interest compounded continuously for t years.

65. Find the balance in 2 years.

66. Find the balance in 20 years.

67. When will the balance be \$2000?

68. When will the balance be \$50,000?

In Exercises 69–72, consider the exponential $B(t) = 2000e^{0.06t}$ that models the balance from an initial deposit of $2000 earning 6% annual interest compounded continuously for t years.

69. Find the balance in 5 years.

70. Find the balance in 10 years.

71. When will the balance be $10,000?

72. When will the balance be $100,000?

In Exercises 73–76, consider the exponential $A(t) = 10e^{-\ln(2)t/5730}$ that models the amount of radioactive isotope carbon-14 remaining from a 10 gram sample after t years. C-14 is used to estimate the age of many archeological objects because of its half-life of 5730 years.

73. How many grams remain in 2865 years?

74. How many grams remain in 20,000 years?

75. When will the sample be 1 gram?

76. When will the sample be 1 milligram?

In Exercises 77–80, consider the exponential $A(t) = 24e^{-\ln(2)t/30.07}$ that models the amount of the radioactive isotope cesium-137 still in the environment from the nearly 24 kilograms released at Chernobyl t years after 1986.

77. How many kilograms remain in 2015?

78. How many kilograms remain in 2050?

79. When will the sample be 1 kilogram?

80. When will the sample be 1 milligram?

In Exercises 81–84, consider the exponential $P(t) = e^{0.0142(t-1900)+0.3420}$ that models the human population in billions of people in year t.

81. Find $P(2015)$ and explain its meaning.

82. Find $P(2050)$ and explain its meaning.

83. Solve $P(t) = 8$ and explain its meaning.

84. Solve $P(t) = 10$ and explain its meaning.

In Exercises 85–88, consider the exponential $f(x) = 31.486e^{-0.048x}$ that models the data on United States coal prices (P) in dollars per short ton in each year (Y) since 1989 (Year 0) in the following table:

Y	0	1	3	7
P	$31.36	$30.11	$27.46	$22.25

85. Compute $f(5)$ and explain its meaning.

86. Find the vertical intercept of $f(x)$ and explain its meaning.

87. Solve $f(x) = 25$ and explain its meaning.

88. Solve $f(x) = 35$ and explain its meaning.

In Exercises 89–92, consider the exponential $f(x) = 37.401e^{0.122x}$ that models the data on average birth weight (W) of males in grams in each gestational week (T) in the following table:

T	25	26	30	31	33
W	777	888	1435	1633	2058

89. Compute $f(32)$ and explain its meaning.

90. Find the vertical intercept of $f(x)$ and explain its meaning.

91. Solve $f(x) = 1000$ and explain its meaning.

92. Solve $f(x) = 3000$ and explain its meaning.

In Exercises 93–96, consider the exponential $f(x) = 75.567e^{-0.128x}$ that models the data on plasma concentration (C) of Prozac in nanograms per milliliter (ng/mL) as a function of the day (D) for a person who has taken 20 mg of Prozac for 30 days, but stops cold turkey in the following table. (Note that discontinuing any medication should always be done in consultation with a healthcare provider.)

D	0	5	10	22	27
C	79	40	19.6	4.3	2.5

93. Compute $f(15)$ and explain its meaning.

94. Find the vertical intercept of $f(x)$ and explain its meaning.

95. Solve $f(x) = 60$ and explain its meaning.

96. Solve $f(x) = 1$ and explain its meaning.

In Your Own Words. In Exercises 97–102, explain the following.

97. Natural logarithm function

98. Effect of C on $y = C\ln(x - h) + v$

99. Effect of h on $y = C\ln(x - h) + v$

100. Effect of v on $y = C\ln(x - h) + v$

101. Semi-log plot

102. Log–log plot

1.7 Trigonometric Functions

Linear, exponential, and logarithmic functions are commonly used to describe and analyze many different aspects of the real world. This section introduces further important additions to the collection of common models: trigonometric functions. Besides being interesting in and of their own right, trigonometric functions enable the modeling of aspects of reality that cannot be described otherwise.

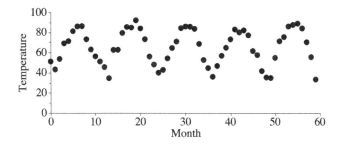

Figure 1: Average maximum temperature each month in Danville, Kentucky

As an example of a phenomenon best modeled by a trigonometric function, consider temperatures in Danville, Kentucky. Figure 1 provides the average maximum temperature in degrees Fahrenheit each month, beginning in January 2006. The inputs along the x-axis are the number of months since January 2006. For example, the input of 12 corresponds to January 2007 and the input of 24 corresponds to January 2008. The outputs along the y-axis provide the average maximum temperatures in degrees Fahrenheit for each month. From studying linear, exponential, and logarithmic functions, we know that none of them can accurately describe this data set. Among other things, all such functions grow toward positive infinity or toward negative infinity as larger x-values are inputted into these functions. In contrast, as Figure 1 illustrates, the average maximum temperature each month oscillates up and down between roughly $30°$ and $90°$ Fahrenheit.

Typical year-to-year experiences of temperature coincide with the oscillating, periodically repeating output values apparent in Figure 1. Many natural phenomena exhibit such **periodic**, or repeating, behavior, including pollen counts, the times of sunrise and sunset, the position of the sun in the sky, and the times of high and low tides each day. Certain social phenomenon also exhibit periodic behavior, such as the prices of textbooks during the academic year and the populations of various vacation towns during a calendar year. Trigonometric functions provide a means for describing this periodic behavior. In fact, the ancient Greeks first defined what are now known as trigonometric functions as part of their study of astronomy.

Recall from Section 1.1 that a **function** provides unambiguous information by mapping each input to exactly one value. Functional input-output relationships are defined in several different ways, including tables, graphs, analytic expressions, and inverses. A distinctive feature of trigonometric functions is the use of geometric objects to define these input-output relationships. Namely, based on right triangles and unit circles, inputs that are angles map to unique numeric outputs and do so in six different ways that give rise to six different trigonometric functions. Working toward these definitions, we introduce two different ways to measure angles and then state both the right triangle and the unit circle definitions of trigonometric functions.

Measuring Angles

Angles are measured in two different ways: degrees and radians. Measuring angles by means of **degrees** dates back to the base 60 number system of the ancient Babylonians in Mesopotamia, which led to 60 minute hours and the $360°$ measure of a circle. Among other things, the number 360 has many nice divisors that enable easy identification of different intermediate angles that result while traveling around a circle.

A complete traversal of any circle is $360°$, but for the sake of this discussion consider a circle centered at the point $(0, 0)$, which is called the **origin**. Moving counterclockwise around such a circle centered at the origin, the positive x-axis corresponds to $0°$, the positive y-axis to $90°$, the negative x-axis to $180°$, the negative y-axis to $270°$, and returning back around to the positive x-axis to $360°$. Figure 2 presents these angles and other intermediate angles measured in degrees on the inside of the circle.

Another approach to measuring angles is in terms of units called *radians*, which correspond to a distance around the unit circle. A circle is uniquely determined by its center and its radius, which is the distance from the center of the circle to each point on the circle. The **unit circle** is centered at the origin and has a radius of one. The radius of one motivates the use of the adjective "unit" for identifying this particular circle.

A **radian** measures the distance around the circumference of the unit circle, beginning from where the unit circle and the positive x-axis intersect at the point $(1, 0)$. As with degrees, the radian measure of an angle is positive when measured in a counterclockwise direction from the positive x-axis; when angles are measured in a clockwise direction from the point $(1, 0)$, the radian measure is negative.

A complete traversal of the unit circle is 2π radians, in parallel with the $360°$ degree measure of a circle, because radians measure the distance traveled around the circumference of the unit circle. Recall that the **circumference** of a circle is $C = 2\pi r$, where

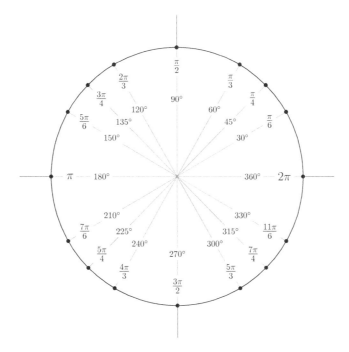

Figure 2: Degree and radian measures of angles

r is the radius of the circle. The unit circle has a radius of $r = 1$, resulting in a circumference of $2\pi \cdot 1 = 2\pi$. The point $(1, 0)$ on the positive x-axis corresponds to 0 radians, the point $(0, 1)$ on the positive y-axis to $\pi/2$ radians, the point $(-1, 0)$ on the negative x-axis to π radians, the point $(0, -1)$ on the negative y-axis to $3\pi/2$ radians, and returning back around to the positive x-axis to 2π radians. The following table presents equivalent angle measures in degrees and radians, which are also illustrated in Figure 2 on the outside of the circle:

degrees	$0°$	$30°$	$45°$	$60°$	$90°$	$120°$	$135°$	$150°$	$180°$	$270°$	$360°$
radians	0	$\pi/6$	$\pi/4$	$\pi/3$	$\pi/2$	$2\pi/3$	$3\pi/4$	$5\pi/6$	π	$3\pi/2$	2π

◆ **EXAMPLE 1** For each radian or degree measure of an angle, state the corresponding degrees or radians, and label each on the unit circle:

(a) $\dfrac{5\pi}{4}$ radians   (b) $\dfrac{3\pi}{2}$ radians   (c) $60°$ (d) $330°$

Solution.

(a) $\dfrac{5\pi}{4}$ radians $= 225°$ (c) $60° = \dfrac{\pi}{3}$ radians

(b) $\dfrac{3\pi}{2}$ radians $= 270°$ (d) $330° = \dfrac{11\pi}{6}$ radians

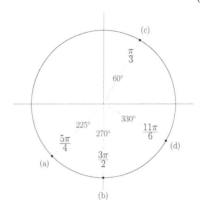

▶ **QUESTION 1** For each radian or degree measure of an angle, state the corresponding degrees or radians, and label each on the unit circle.

(a) $\dfrac{5\pi}{3}$ radians   (b) $\dfrac{7\pi}{6}$ radians   (c) $180°$ (d) $135°$

Right Triangle Definitions of Trigonometric Functions

As the "tri" in trigonometric functions suggests, these functions can be defined using triangles. Recall that a **triangle** is a polygon with three straight sides and that a **right triangle** has an angle with a measure of $90°$ or, equivalently, π radians. The six trigonometric functions are defined in terms of ratios among the various side lengths of a right triangle. The input to a trigonometric function is an angle measure, in degrees or radians, and the sides are referred to based on their relative position to the input angle. For all right triangles, the **hypotenuse** is the side opposite the right angle. In addition, for a given angle x, we refer to the side **opposite** angle x and the side **adjacent** to x, as illustrated in Figure 3.

The most important and fundamental trigonometric functions are sine and cosine, which are defined in terms of the labeled right triangle from Figure 3:

$$\sin(x) = \frac{\text{opposite}}{\text{hypotenuse}} = \frac{\text{opp}}{\text{hyp}} \qquad \cos(x) = \frac{\text{adjacent}}{\text{hypotenuse}} = \frac{\text{adj}}{\text{hyp}}$$

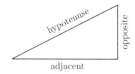

Figure 3: Right triangle for defining trigonometric functions

The other four trigonometric functions include the tangent, which is the ratio of the sine and the cosine, and the reciprocals of these three functions: the cotangent, cosecant, and secant. These functions are defined both in terms of ratios of sine and cosine, and in terms of ratios of side lengths of the labeled right triangle in Figure 3:

$$\tan(x) = \frac{\sin(x)}{\cos(x)} = \frac{\text{opp}}{\text{adj}} \qquad\qquad \cot(x) = \frac{1}{\tan(x)} = \frac{\cos(x)}{\sin(x)} = \frac{\text{adj}}{\text{opp}}$$

$$\csc(x) = \frac{1}{\sin(x)} = \frac{\text{hyp}}{\text{opp}} \qquad\qquad \sec(x) = \frac{1}{\cos(x)} = \frac{\text{hyp}}{\text{adj}}$$

While each of these four functions having two distinct definitions may appear intimidating, keep in mind that they are just functions. These different definitions provide different insights to various questions, but throughout, at their core, these definitions describe particular input-output relationships.

Working with these right triangle definitions of the trigonometric functions often involves using the **Pythagorean theorem**. Recall that for a right triangle $a^2 + b^2 = c^2$, where the hypotenuse has length c and other legs have lengths a and b.

◆ **EXAMPLE 2** Let x be an angle in a right triangle with $\cot(x) = \dfrac{3}{2}$. Find the values of the other five trigonometric functions.

Solution. From its right triangle definition, $\cot(x) = \dfrac{3}{2} = \dfrac{\text{adj}}{\text{opp}}$, which gives an adjacent side length of $a = 3$ and an opposite side length of $b = 2$, as shown in Figure 4. Applying the Pythagorean theorem $a^2 + b^2 = c^2$ gives the length of the hypotenuse $c = \sqrt{3^2 + 2^2} = \sqrt{13}$. The right triangle definitions provide the values of the other five trigonometric functions:

$$\sin(x) = \frac{\text{opp}}{\text{hyp}} = \frac{2}{\sqrt{13}} \qquad \cos(x) = \frac{\text{adj}}{\text{hyp}} = \frac{3}{\sqrt{13}} \qquad \tan(x) = \frac{\text{opp}}{\text{adj}} = \frac{2}{3}$$

$$\csc(x) = \frac{\text{hyp}}{\text{opp}} = \frac{\sqrt{13}}{2} \qquad \sec(x) = \frac{\text{hyp}}{\text{adj}} = \frac{\sqrt{13}}{3}$$

∎

Figure 4: Right triangle for $\cot(x) = \dfrac{3}{2}$ for Example 2

➤ **QUESTION 2** Let x be an angle in a right triangle with $\cos(x) = \dfrac{1}{6}$. Find the values of the other five trigonometric functions.

The Unit Circle Definitions of Trigonometric Functions

The trigonometric functions are defined in terms of two geometric objects: right triangles and the unit circle. These geometric objects are naturally connected to each other, which enables this duality of perspectives on the same functions. As with many aspects of life, developing multiple perspectives enables a richer, more complete understanding.

The unit circle definitions of the trigonometric functions are expressed in terms of radian measures of angles. Recall that radians measure distance around the circumference of the unit circle, where positive distance is measured counterclockwise from the positive x-axis. In the context of unit circle definitions, the Greek letter θ ("theta") will be used to identify angle measurements in order to avoid confusion between references to the x-axis and to the x-coordinates of points. Following the approach with right triangles, sine and cosine are defined first:

Definition.

- $\sin(\theta)$ is the y-coordinate of the point θ radians around the circumference of the unit circle from the point $(1, 0)$ on the positive x-axis

- $\cos(\theta)$ is the x-coordinate of the point θ radians around the circumference of the unit circle from the point $(1, 0)$ on the positive x-axis

---

Alternatively, if (a, b) is the point θ radians around the circumference of the unit circle from the positive x-axis, then $\sin(\theta) = b$ and $\cos(\theta) = a$.

Every point (a, b) on the unit circle corresponds to an ordered pair $(\cos(\theta), \sin(\theta))$. Using known facts about the coordinates of certain points on the unit circle enables the identification of certain values of the sine and cosine. For example, $0°$ and 0 radians both to correspond to the point $(1, 0)$ where the unit circle intersects the positive x-axis. Thus, $\cos(0°) = \cos(0) = 1$ is the x-coordinate of this point of intersection and $\sin(0°) = \sin(0) = 0$ is its y-coordinate. Similarly, $270°$ and $\dfrac{3\pi}{2}$ radians identify the point $(0, -1)$ where the unit circle intersects the negative y-axis. Thus, $\cos(270°) = \cos\left(\dfrac{3\pi}{2}\right) = 0$

is the x-coordinate of this point of intersection and $\sin(270°) = \sin\left(\dfrac{3\pi}{2}\right) = -1$ is its y-coordinate.

In addition to these axis values of the cosine and sine, their values are known for many intermediate "standard reference" angles. The unit circle in Figure 5 presents the standard reference angles and the corresponding values of the cosine and sine

In the first quadrant of this unit circle, observe that the point $\left(\dfrac{1}{2}, \dfrac{\sqrt{3}}{2}\right)$ occurs where the angle is $\theta = \dfrac{\pi}{3} = 60°$. Interpreting this point as $\left(\cos\left(\dfrac{\pi}{3}\right), \sin\left(\dfrac{\pi}{3}\right)\right)$ gives the following:

$$\cos\left(\frac{\pi}{3}\right) = \cos(60°) = \frac{1}{2} \qquad\qquad \sin\left(\frac{\pi}{3}\right) = \sin(60°) = \frac{\sqrt{3}}{2}$$

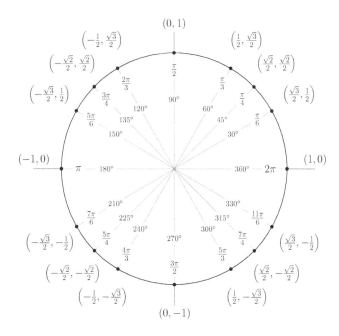

Figure 5: Unit circle values of $\cos(x)$ and $\sin(x)$

A similar process is followed for all the values at the basic reference angles on the unit circle.

The definitions of the other four trigonometric functions are still expressed in terms of the ratios of sine and cosine stated for right triangles:

$$\tan(\theta) = \frac{\sin(\theta)}{\cos(\theta)} \qquad \csc(\theta) = \frac{1}{\sin(\theta)} \qquad \sec(\theta) = \frac{1}{\cos(\theta)} \qquad \cot(\theta) = \frac{\cos(\theta)}{\sin(\theta)}$$

◆ **EXAMPLE 3** Find the values of all six trig functions when $\theta = \dfrac{5\pi}{4} = 135°$.

Solution. From Figure 5, the point $\left(-\dfrac{\sqrt{2}}{2}, \dfrac{\sqrt{2}}{2} \right)$ occurs where the angle is $\theta = \dfrac{5\pi}{4} = 135°$

on the unit circle. Interpreting this point as $\left(\cos\left(\dfrac{5\pi}{4} \right), \sin\left(\dfrac{5\pi}{4} \right) \right)$ gives the following:

$$\cos\left(\frac{5\pi}{4} \right) = \cos(135°) = -\frac{\sqrt{2}}{2} \qquad\qquad \sin\left(\frac{5\pi}{4} \right) = \sin(135°) = -\frac{\sqrt{2}}{2}$$

Substituting these values into the definitions of the other four trigonometric functions gives their values as well:

$$\tan\left(\frac{5\pi}{4} \right) = \frac{\sin\left(\frac{5\pi}{4} \right)}{\cos\left(\frac{5\pi}{4} \right)} = \frac{\frac{\sqrt{2}}{2}}{-\frac{\sqrt{2}}{2}} = -1 \qquad \cot\left(\frac{5\pi}{4} \right) = \frac{\cos\left(\frac{5\pi}{4} \right)}{\sin\left(\frac{5\pi}{4} \right)} = \frac{-\frac{\sqrt{2}}{2}}{\frac{\sqrt{2}}{2}} = -1$$

$$\csc\left(\frac{5\pi}{4} \right) = \frac{1}{\frac{\sqrt{2}}{2}} = \sqrt{2} \qquad\qquad\qquad \sec\left(\frac{5\pi}{4} \right) = \frac{1}{-\frac{\sqrt{2}}{2}} = -\sqrt{2}$$

∎

➤ **QUESTION 3** Find the values of all six trig functions when $\theta = \dfrac{11\pi}{6} = 330°$.

In addition to finding the values of trigonometric functions, solving trigonometric equations will play an important role in the analysis of real-world phenomena, particularly when seeking optimal solutions to certain questions. This work sometimes entails working backward from given output values to the corresponding input angles.

◆ **EXAMPLE 4** Find all solutions of each trigonometric equation on the interval $[0, 2\pi]$:

(a) $\cos(\theta) = 0.5$ (b) $\cos(2\theta) = 0.5$

Solution.

(a) From the unit circle given in Figure 5, $\cos(\theta) = 0.5 = \dfrac{1}{2}$ when $\theta = \dfrac{\pi}{3}$ in the first quadrant and when $\theta = \dfrac{5\pi}{3}$ in the fourth quadrant.

(b) The input to the cosine function is 2θ, which requires consideration of two full traversals of the unit circle. Dividing by 2 shifts these solutions back into $[0, 2\pi]$:

- $2\theta = \dfrac{\pi}{3}$ gives $\theta = \dfrac{\pi}{6}$. · $2\theta = \dfrac{7\pi}{3}$ gives $\theta = \dfrac{7\pi}{6}$.
- $2\theta = \dfrac{5\pi}{3}$ gives $\theta = \dfrac{5\pi}{6}$. · $2\theta = \dfrac{11\pi}{3}$ gives $\theta = \dfrac{11\pi}{6}$.

∎

➤ **QUESTION 4** Find all solutions of each trigonometric equation on the interval $[0, 2\pi]$:

(a) $\tan(\theta) = \sqrt{3}$ (b) $\tan(2\theta) = \sqrt{3}$

Graphs of Trigonometric Functions

Examining the graphs of the six trigonometric functions allows a more complete understanding of these input-output relationships. For these graphs, inputs are angles measured in radians. Recall that radians are distances around the circumference of the circle measured from the positive x-axis, which correspond quite nicely to distances along the x-axis measured from the origin.

To begin, consider the graph of the sine function. Figure 6(a) presents the output values of $\sin(x)$ at the standard reference angles from the unit circle, and Figure 6(b) presents the complete graph of $\sin(x)$.

(a)   (b)

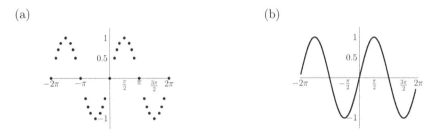

Figure 6: The graph of $\sin(x)$

We highlight two important features of the sine function. First, the **period** of $\sin(x)$ is 2π, which means that $\sin(x)$ completes an entire cycle in the interval $[0, 2\pi]$, returning to its initial output value and then repeating exactly for each successive input interval of length 2π. Second, the graph of $\sin(x)$ oscillates between a maximum output value of 1 and a minimum output value of -1. The **amplitude** of the sine is half the distance between these extreme values, which means that the amplitude of $\sin(x)$ is $\frac{1}{2}[1 - (-1)] = \frac{1}{2} \cdot 2 = 1$.

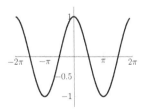

Figure 7: The graph of $\cos(x)$

Figure 7 presents the graph of the cosine function, which is obtained in much the same way and is quite similar to the graph of sine. In fact, shifting the graph of $\sin(x)$

to the left by $\pi/2$ produces the graph of $\cos(x)$. As for $\sin(x)$, the period of $\cos(x)$ is 2π and its amplitude is 1.

The graphs of the other four trigonometric functions are obtained from their defining ratios of sine and cosine. Figure 8(a) presents the graph of the tangent function $\tan(x)$, Figure 8(b) presents $\csc(x)$, Figure 8(c) presents $\sec(x)$, and Figure 8(d) presents $\cot(x)$. The vertical asymptotes of these trigonometric functions occur when the denominator of the corresponding defining ratio is zero. For example, the denominator of both $\tan(x)$ and $\sec(x)$ is $\cos(x)$ and their vertical asymptotes occur when $\cos(x) = 0$ at $x = \dfrac{\pi}{2}$, $x = \dfrac{3\pi}{2}$, $x = -\dfrac{\pi}{2}$, and so on, for every positive and negative odd multiple of $\dfrac{\pi}{2}$.

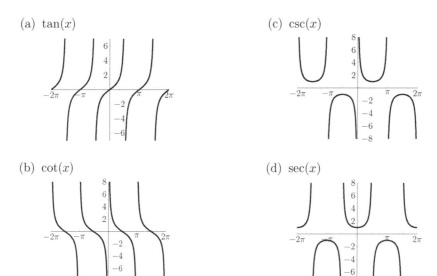

(a) $\tan(x)$

(b) $\cot(x)$

(c) $\csc(x)$

(d) $\sec(x)$

Figure 8: The graphs of the other four trigonometric functions

Trigonometric Identities

Trigonometric functions share various relationships with each other beyond their definitions, many of which are referred to as **trigonometric identities**. The most important of these identities are included here for reference because, among other things, they can prove useful when solving or simplifying trigonometric equations.

PYTHAGOREAN IDENTITIES.

- $\cos^2(x) + \sin^2(x) = 1$

- $1 + \tan^2(x) = \sec^2(x)$

- $\cot^2(x) + 1 = \csc^2(x)$

DOUBLE-ANGLE IDENTITIES.

- $\sin(2x) = 2\sin(x)\cos(x)$

- $\cos(2x) = \cos^2(x) - \sin^2(x)$

- $\cos(2x) = 2\cos^2(x) - 1$

- $\cos(2x) = 1 - 2\sin^2(x)$

PERIODIC IDENTITIES.

- $\sin(x + \pi) = -\sin(x)$

- $\sin(x + 2\pi) = \sin(x)$

- $\sin\left(x + \dfrac{\pi}{2}\right) = \cos(x)$

- $\sin\left(x + \dfrac{3\pi}{2}\right) = -\cos(x)$

Beyond solving equations, the periodic identity $\sin\left(x + \frac{\pi}{2}\right) = \cos(x)$ informs our choice of modeling functions in Chapter 2. Namely, we choose to work exclusively with sine functions when modeling data sets, because every cosine function can be expressed in terms of the sine with an appropriate horizontal shift.

The exercises at the end of this section include graphical verifications of these various identities, by graphing the two functions in each equation on the same pair of axes to show they are equal. While these identities are quite useful and interesting in their own right, a more extensive study is beyond the scope of this book.

Working in RStudio

RStudio uses standard mathematical notation to refer to the three basic trigonometric functions $\cos(x)$, $\sin(x)$, and $\tan(x)$, when defining, evaluating, and graphing these functions. All inputs to trigonometric functions are assumed to be in radians. For example, $\sin(\pi)$ is evaluated by entering the command `sin(pi)`, and $\tan(2\pi)$ by entering `tan(2*pi)`. RStudio works with $\sec(x)$, $\csc(x)$, and $\cot(x)$ by expressing them as multiplicative inverses of the three basic functions. For example, $\sec(\pi)$ is evaluated by entering the command `1/cos(pi)`.

> **Examples of Commands:**
>
> - sin(pi)
>
> - 1/cos(pi)
>
> - plotFun(sin(x)~x,xlim=range(-2*pi,2*pi))

◆ **EXAMPLE 5** Find the values of all six trig functions when $\theta = \dfrac{5\pi}{4} = 135°$ using RStudio.

Solution. For RStudio, all inputs to trigonometric functions are assumed to be in radians, which means that the angle must be entered as $\theta = \dfrac{5\pi}{4}$, not $\theta = 135°$:

sin(5*pi/4)   1/sin(5*pi/4)

[1] -0.707107   [1] -1.41421

cos(5*pi/4)   1/cos(5*pi/4)

[1] -0.707107   [1] -1.41421

tan(5*pi/4)   1/tan(5*pi/4)

[1] 1   [1] 1

■

➤ **QUESTION 5** Find the values of all six trig functions when $\theta = \dfrac{11\pi}{6} = 330°$ using RStudio.

◆ **EXAMPLE 6** Graph $2\sin(x) + 3$ and $-\sin(x) + 3$ on the same pair of axes using RStudio.

Solution.

plotFun(2*sin(x)+3~x,xlim=range(-2*pi,2*pi))
plotFun(-sin(x)+3~x,add=TRUE)

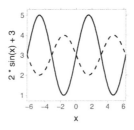

■

➤ **QUESTION 6**   Graph $4\cos(x) - 2$ and $2\cos(2x)$ on the same pair of axes using RStudio.

Summary

- For measuring angles in *degrees*, a complete traversal of a circle is $360°$. For a circle centered at the *origin* $(0,0)$, angles read counterclockwise from the positive x-axis.

- The *unit circle* is centered at the origin $(0,0)$ on the plane and has a radius of one. A *radian* measures distance around the circumference of the unit circle, beginning from where the unit circle and the positive x-axis intersect at the point $(1,0)$. A complete traversal of the unit circle covers 2π radians.

- A *triangle* is a three-sided polygon, a *right triangle* has one angle with a measure of $90°$ (or π radians), and the *hypotenuse* is the side opposite the right angle. The right triangle definitions of the six trigonometric functions refer to the side *opposite* angle x and the side *adjacent* to x as follows:

$$\sin(x) = \frac{\text{opp}}{\text{hyp}} \qquad\qquad \csc(x) = \frac{1}{\sin(x)} = \frac{\text{hyp}}{\text{opp}}$$

$$\cos(x) = \frac{\text{adj}}{\text{hyp}} \qquad\qquad \sec(x) = \frac{1}{\cos(x)} = \frac{\text{hyp}}{\text{adj}}$$

$$\tan(x) = \frac{\sin(x)}{\cos(x)} = \frac{\text{opp}}{\text{adj}} \qquad\qquad \cot(x) = \frac{\cos(x)}{\sin(x)} = \frac{\text{adj}}{\text{opp}}$$

- The unit circle definitions of sine and cosine are as follows:

  ○ $\sin(\theta)$ is equal to the y-coordinate of the point θ radians around the circumference of the unit circle from the positive x-axis;

  ○ $\cos(\theta)$ is equal to the x-coordinate of the point θ radians around the circumference of the unit circle from the positive x-axis.

  If (a, b) is the point θ radians around the circumference of the unit circle from the point $(1,0)$ on the positive x-axis, then $\sin(\theta) = b$ and $\cos(\theta) = a$.

Summary (continued)

- The values of sine and cosine at the standard reference angles are as follows:

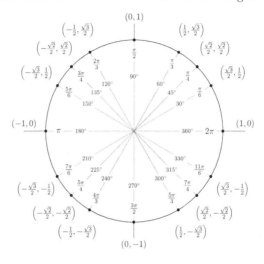

Exercises

In Exercises 1–8, label the degree measure on the unit circle and state the corresponding radians.

1. $x = 45°$

2. $x = 270°$

3. $x = 120°$

4. $x = 300°$

5. $x = 135°$

6. $x = 210°$

7. $x = 60°$

8. $x = 330°$

In Exercises 9–16, label the radian measure on the unit circle and state the corresponding degrees.

9. $x = 0$

10. $x = \dfrac{\pi}{2}$

11. $x = \dfrac{5\pi}{4}$

12. $x = \dfrac{7\pi}{4}$

13. $x = \dfrac{5\pi}{3}$

14. $x = \dfrac{2\pi}{3}$

15. $x = \dfrac{\pi}{6}$

16. $x = \dfrac{5\pi}{6}$

In Exercises 17–24, state the values of all six trigonometric functions for the angle.

17. $x = 45°$

18. $x = 120°$

19. $x = 135°$

20. $x = 60°$

21. $x = 0$

22. $x = \dfrac{5\pi}{4}$

23. $x = \dfrac{5\pi}{3}$

24. $x = \dfrac{5\pi}{6}$

In Exercises 25–32, find the values of the other five trigonometric functions for the angle x between 0 and $\dfrac{\pi}{2}$ radians.

25. $\sin(x) = \dfrac{5}{9}$

26. $\sin(x) = \dfrac{7}{8}$

27. $\cos(x) = \dfrac{3}{7}$

28. $\cos(x) = \dfrac{12}{13}$

29. $\tan(x) = \dfrac{4}{11}$ 31. $\sec(x) = \dfrac{13}{2}$

30. $\csc(x) = \dfrac{12}{5}$ 32. $\cot(x) = \dfrac{5}{3}$

In Exercises 33–44, find all solutions of the equation on the interval $[0, 2\pi]$.

33. $\sin(x) = 0$

34. $\sin(x) = 0.5$

35. $\sin(2x) = 0$

36. $\sin(3x) = -0.5$

37. $2\sin(x) = -1$

38. $2\sin(4x) = \sqrt{3}$

39. $\cos(x) = 0$

40. $\cos(x) = -0.5$

41. $\cos(2x) = -1$

42. $\cos(3x) = 0.5$

43. $2\cos(x) = -1$

44. $2\cos(4x) = \sqrt{3}$

In Exercises 45–56, find all solutions of the equation on the interval $[0, 2\pi]$.

45. $\tan(x) = 0$

46. $\tan(x) = -1$

47. $\tan(2x) = 0$

48. $\tan(3x) = 1$

49. $\tan(x) = -\sqrt{3}$

50. $\tan(4x) = \sqrt{3}$

51. $\cot(x) = 0$

52. $\cot(x) = -0.5$

53. $\cot(2x) = -1$

54. $\cot(3x) = 0$

55. $\cot(x) = -1$

56. $\cot(4x) = \sqrt{3}$

RStudio. In Exercises 57–64, use RStudio to evaluate the expression.

57. $6\sin(\pi)$

58. $-7\sin\left(\dfrac{\pi}{2}\right)$

59. $2\sin\left(\dfrac{3\pi}{4}\right) + 5$

60. $-5\sin\left(\dfrac{\pi}{9}\right) - 11$

61. $4\cos(6)$

62. $-2\cos(7 - \pi)$

63. $2\tan(\pi) + 4$

64. $-8\tan\left(\dfrac{\pi}{4}\right) - 3$

RStudio. In Exercises 65–72, use RStudio to graph the trigonometric function and then discuss its monotonicity and concavity over its first period to the right of the y-axis.

65. $3\sin(x)$

66. $3\sin(x - \pi)$

67. $-2\sin(x + \pi) + 4$

68. $-5\sin(x) - 9$

69. $3\cos(x)$

70. $-3\cos(x - \pi)$

71. $2\tan(x + \pi) + 4$

72. $-5\tan(x) - 9$

RStudio. In Exercises 73–77, use RStudio to study the sine model of the number of minutes (M) after 3 p.m. until sunset at Greenwich in England as a function of the month (T):

$$M = 164.7\sin\left[\dfrac{\pi}{6}(T - 2.6)\right] + 223.6$$

73. Find and explain the meaning of M(5).

74. Find and explain the meaning of M(11).

75. Graph M(T) using RStudio.

76. Use the graph from Exercise 75 to estimate the period of M(T).

77. Use the graph from Exercise 75 to estimate the amplitude of M(T).

RStudio. In Exercises 78–82, use RStudio to study the sine model of the average maximum temperature (T) in degrees Fahrenheit each month (M) in New York City in 2013.

$$T = 24.067 \sin\left[\frac{\pi}{6}(M - 4.3)\right] + 63.317$$

78. Find and explain the meaning of T(3.5).

79. Find and explain the meaning of T(10).

80. Graph T(M) using RStudio.

81. Use the graph from Exercise 80 to estimate the period of T(M).

82. Use the graph from Exercise 80 to estimate the amplitude of T(M).

RStudio. In Exercises 83–95, use RStudio to verify the trigonometric identity by graphing the two functions in the equation on the same pair of axes to show they are equal.

83. $\cos^2(x) + \sin^2(x) = 1$

84. $1 + \tan^2(x) = \sec^2(x)$

85. $\cot^2(x) + 1 = \csc^2(x)$

86. $\sin(2x) = 2\sin(x)\cos(x)$

87. $\cos(2x) = \cos^2(x) - \sin^2(x)$

88. $\cos(2x) = 2\cos^2(x) - 1$

89. $\cos(2x) = 1 - 2\sin^2(x)$

90. $\sin(-x) = -\sin(x)$

91. $\cos(-x) = \cos(x)$

92. $\sin(x + \pi) = -\sin(x)$

93. $\sin(x + 2\pi) = \sin(x)$

94. $\sin\left(x + \frac{\pi}{2}\right) = \cos(x)$

95. $\sin\left(x + \frac{3\pi}{2}\right) = -\cos(x)$

In Your Own Words. In Exercises 96–101, explain the following.

96. Radian

97. Unit circle

98. Right triangle definitions of the six trigonometric functions

99. Unit circle definitions of the six trigonometric functions

100. Period of sine

101. Amplitude of sine

Chapter 2

Mathematical Modeling

Researchers in diverse fields of work and study develop an understanding of reality through a cyclic process known as the **modeling cycle**, as described at the beginning of Chapter 1. In broad strokes, the modeling cycle begins by asking a question about some phenomenon of interest, gathering data relevant to this question, and conjecturing a model describing the data. The accuracy of this model is then tested against our gathered data, possibly leading to modifications that result in a better model that more closely matches the data set. This cycle is repeated as we develop a better understanding of the phenomenon through the model and through additional data collection.

THE MODELING CYCLE. The **modeling cycle** consists of the following five steps:

(1) Ask a question about reality.

(2) Make some observations and collect the corresponding data.

(3) Conjecture a model or modify a known model based on the data.

(4) Test the model against known data (from step (2)) and modify the model as needed.

(5) Repeat steps (2)–(4) to improve the model.

This chapter focuses on the third and fourth steps of the modeling cycle: conjecturing a model and testing the model against known data. We develop a multi-pronged approach to defining a reasonable model for a data set, where models are chosen from among a collection of common functions: linear, exponential, power, sine, and sigmoidal functions. For all data sets, graphs will inform the initial choice of a reasonable model, and then various numerical techniques will be applied both to verify our choice and to conjecture values for the corresponding parameters. In addition, computer software enables the identification of the "best possible" model of a particular type for a given data set.

2.1   Modeling with Linear Functions

This study of modeling begins with linear functions, which are the simplest functions and thus provide the most basic models. Even though they are simple, linear models describe many aspects of reality quite well. A linear function is a function of the form $y = mx + b$, where the slope m of the line through the points (x_1, y_1) and (x_2, y_2) is calculated using the following formulas when $x_1 \neq x_2$:

$$m = \frac{y_2 - y_1}{x_2 - x_1} = \frac{y_1 - y_2}{x_1 - x_2} = \frac{\Delta y}{\Delta x} = \frac{\text{rise}}{\text{run}}$$

A vertical line $x = a$ has an undefined slope. The vertical intercept of a line b is the y-coordinate of the point $(0, b)$ where the line intersects the y-axis, if such a point exists.

For every data set, the first step in the process of identifying a reasonable model is to plot the data points. Visually determining whether or not the data points lie on approximately the same line is relatively straightforward in light of lines being so familiar. If the data points do not appear to lie roughly on the same line, then another type of function might provide a better model.

◆ **EXAMPLE 1** Plot each data set below using appropriate scales on the axes and explain why a linear model is reasonable or not. Table (a) presents how many millions of people used Twitter during each quarter of a year since 2000, where 11 represents the first quarter of 2011 (January–March 2011), 11.25 represents the second quarter of 2011 (April–June 2011), and so on. Table (b) presents the average weight of male fetuses in grams during the given gestational week.

(a)

Year	11	11.25	11.5	11.75	12.75
Twitter users	68	85	101	117	185

(b)

Gestational week	25	26	30	31	33
Average weight	777	888	1435	1633	2058

Solution. Figure 1(a) presents the plot of the data from table (a) and indicates that a linear model is reasonable. Figure 1(b) presents the plot of the data from table (b). The data exhibits a slightly curved behavior as the inputs increase, so a linear function might not provide the best model. ∎

Example 1(b) argues that a linear model is not reasonable for describing the average weight of male fetuses. However, biologists, social scientists, politicians, and others remain interested in obtaining accurate, reasonable models for birth weight and other such data sets. This chapter's continuing study of the most common modeling functions will soon introduce a reasonable model for this data set, expanding our ability to analyze and, hopefully, understand reality.

➤ **QUESTION 1** Plot each data set below using appropriate scales on the axes and explain why a linear model is reasonable or not. Table (a) presents the annual total retail sales taxes collected in the United States in millions of dollars each year; for example, during 2006, $141,179,000,000 was collected in retail sales tax. Table (b) presents the average debt load in thousands of dollars at the end of each year's spring term for

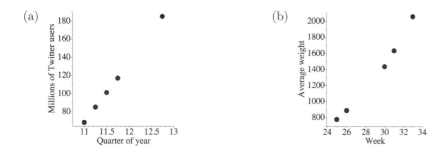

Figure 1: Plots of data sets for Example 1

bachelor's degree recipients who attended public four-year colleges and universities and borrowed money to finance their education.

(a)

Year	2006	2008	2010	2012
Sales tax	\$141,179	\$141,452	\$139,444	\$150,360

(b)

Academic year	2001	2003	2005	2006
Average debt	\$20,400	\$20,900	\$21,500	\$21,800

Numerically Identifying Linear Data Sets

Numerically deciding whether a linear model is reasonable or not for a data set presented in tabular form utilizes the distinguishing feature of a line: the slopes between any two points on the line have the same value. In most real-life settings, the slopes between any two arbitrary data points will usually not be exactly the same, so slopes are inspected for approximate equality. These observations are summarized in the following guide for identifying a reasonable model for a given data set, which will be updated in each section as part of this continuing study of the most common modeling functions.

NUMERICALLY IDENTIFYING REASONABLE MODELS.

(1) Check for a linear model $y = mx + b$ by comparing the slopes of the lines through pairs of data points. Linear data sets have exactly or approximately the same slopes for any two data points. Most often, slopes of lines between successive data points are compared.

Recall that the slope of the line through the points (x_1, x_2) and (y_1, y_2) can be calculated using the following formulas when $x_1 \neq x_2$:

$$m = \frac{y_2 - y_1}{x_2 - x_1} = \frac{y_1 - y_2}{x_1 - x_2} = \frac{\Delta y}{\Delta x} = \frac{\text{rise}}{\text{run}}$$

Applying the above approach to the tables of data given in Example 1 and Question 1 will verify the conclusions of the preceding graphical analysis.

◆ **EXAMPLE 2** For each data set below from Example 1, compute the slopes of the lines through successive data points and explain why a linear model is reasonable or not. Table (a) presents how many millions of people used Twitter during each quarter of the given year. Table (b) presents the average weight of male fetuses in grams during the given gestational week.

(a)

Year	11	11.25	11.5	11.75	12.75
Twitter users	68	85	101	117	185

(b)

Gestational week	25	26	30	31	33
Average weight	777	888	1435	1633	2058

Solution.

(a) First, compute the slopes of the lines through successive data points as shown in the following table. These slopes are interpreted as being approximately equal because the differences between the slopes (of 0 and 4) are relatively small in comparison with the output values (of 68 to 185) and the magnitude of the slopes (of 64 and 68). Therefore, a linear model is reasonable.

Intervals	11 to 11.25	11.25 to 11.5	11.5 to 11.75	11.75 to 12.75
Δx	0.25	0.25	0.25	1
Δy	$85 - 68 = 17$	$101 - 85 = 16$	$117 - 101 = 16$	$185 - 117 = 68$
Slopes	$m_1 = 68$	$m_2 = 64$	$m_3 = 64$	$m_4 = 68$

(b) Again, first compute the slope of the lines through successive data points as shown in the following table. Note that as the inputs increase, the slopes increase as well, and quite significantly relative to the values of the slopes. In contrast, such slopes in a linear setting would have been approximately constant, and so a linear model is not reasonable.

Intervals	25 to 26	26 to 30	30 to 31	31 to 33
Δx	1	4	1	2
Δy	$888 - 777 = 111$	547	198	425
Slopes	$m_1 = 111$	$m_2 = 136.75$	$m_3 = 198$	$m_4 = 212.5$

∎

➤ **QUESTION 2** For each data set below from Question 1, compute the slopes of the lines through successive data points and explain why a linear model is reasonable or not. Table (a) presents the annual total retail sales taxes collected in the United States in millions of dollars each year. Table (b) presents the average debt load in thousands of dollars at the end of each year's spring term for bachelor's degree recipients who attended public four-year colleges and universities and borrowed money to finance their education.

(a)

Year	2006	2008	2010	2012
Sales tax	$141,179	$141,452	$139,444	$150,360

(b)

Academic year	2001	2003	2005	2006
Average debt	$20,400	$20,900	$21,500	$21,800

Conjecturing Linear Models

After determining that a linear model is reasonable for a data set, the next step in the modeling cycle is to conjecture such a model. For linear models $y = mx + b$, this conjecture amounts to identifying values for the slope parameter m and the vertical intercept parameter b. A process for conjecturing parameters for linear models is presented below.

CONJECTURING PARAMETERS FOR LINEAR MODELS.
If a linear model $y = mx + b$ is reasonable for a data set, m and b are conjectured as follows:

- For the slope m, compute the sequence of slopes through successive data points m_1, m_2, ..., m_{n-1} and then average these slopes:

$$m = \frac{m_1 + m_2 + m_3 + \cdots + m_{n-1}}{n - 1}$$

- For the vertical intercept b, substitute each data point into the equation $y = mx + b$ (using the conjectured slope), solve for b_1, b_2, ..., b_n, and then average these intercepts:

$$b = \frac{b_1 + b_2 + b_3 + \cdots + b_n}{n}$$

Example 3 illustrates this process for conjecturing parameters in a specific setting.

◆ **EXAMPLE 3** The table below from Examples 1(a) and 2(a) presents how many millions of people used Twitter during each quarter of the given year. Conjecture a linear model $y = mx + b$ for this data set and graph the model on a plot of the data.

Year	11	11.25	11.5	11.75	12.75
Twitter users	68	85	101	117	185

Solution. Examples 1(a) and 2(a) determined that a linear model $y = mx + b$ is reasonable for this data set. Example 2(a) computed the slopes for the lines through successive data points: $m_1 = 68$, $m_2 = 64$, $m_3 = 64$, and $m_4 = 68$. Following the above process, the slope m of the linear model is conjectured by averaging these computed slopes together:

$$m = \frac{m_1 + m_2 + m_3 + m_4}{4} = \frac{68 + 64 + 64 + 68}{4} = 66$$

Substituting into the general formula gives a linear model of the form $y = 66x + b$. In a similar fashion, the vertical intercept b is obtained by substituting each of the five given data points into this equation, solving for the vertical intercept in each case, and then averaging the results b_1, b_2, b_3, b_4, and b_5. For example, substituting the first data point $(11, 68)$ into the linear model gives $b_1 = -658$ as follows:

$$y = 66 \cdot x + b \qquad \text{Intermediate model}$$
$$68 = 66 \cdot 11 + b_1 \qquad \text{Substitute } (x, y) = (11, 68)$$
$$68 - 726 = b_1 \qquad \text{Subtract } 66 \cdot 11 = 726 \text{ from both sides}$$
$$-658 = b_1 \qquad \text{Simplify}$$

Similar computations with the other data points yield: $b_2 = -657.5$, $b_3 = -658$, $b_4 = -658.5$, and $b_5 = -656.5$, which average to $b = -657.7$. Substituting into the intermediate equation gives the linear model $y = 66x - 657.7$ for this data set. In such real-life settings, descriptive names are often chosen for the names of the variables appearing in models, such as Users $= 66 \cdot$ Year $- 657.7$.

As part of the modeling cycle, the conjectured model is compared against the known data to ascertain its accuracy and reliability, and to inform any adjustments to the parameters. Figure 2 presents the graph of this linear model on a plot of the data. While not every data point lies exactly on the line determined by the model Users $= 66 \cdot$ Year $- 657.7$, this linear model does appear to provide a pretty good match.

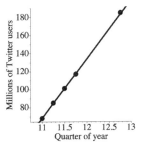

Figure 2: Graph of the data and a linear model for Example 3

In Example 3, the conjectured vertical intercept of $b = -657.7$ is negative, which might seem a bit strange in the given context. What would it mean to have -657.5 million Twitter users in the first quarter of 2000? As you may know, Twitter was first launched in March 2006, which means that because quarters are being counted since the beginning of 2000, the linear model must have a negative vertical intercept. In addition, this model should only be considered for inputs greater than 6.25, which corresponds to the April, May, and June quarter of 2006.

The verification of models is a key step in the modeling cycle, and modifying parameters to obtain better models is an iterative process. A graphical comparison, as demonstrated in Example 3 and Figure 2, provides a rough indication of validity. Section 2.6 discusses why and how to adjust conjectured parameters to obtain better models.

➤ **QUESTION 3** The table below from Questions 1(b) and 2(b) presents the average debt load in thousands of dollars at the end of each year's spring term for bachelor's degree recipients who attended public four-year colleges and universities and borrowed money to finance their education. Conjecture a linear model $y = mx + b$ for this data set and graph the model on a plot of the data.

Academic year	2001	2003	2005	2006
Average debt	$20,400	$20,900	$21,500	$21,800

Best Possible Linear Models

The techniques developed in this section enable the conjecture of a reasonable linear model for a given data set. However, these methods are not guaranteed to produce the best possible linear model. Rather, the only claim that can be made is that a reasonable model has been identified and that a visual check does or does not affirm its apparent validity. Chapter 3 studies the method of least squares, which determines the best linear model for a given set of data. The RStudio command `fitModel` implements a version of the method of least squares and provides a way to find the best possible linear model for a given data set.

Working in RStudio

Before plotting a set of data points or using `fitModel` to determine its best possible linear model, the data must be stored in RStudio. Data is stored and named with the command `var1=c(,)`, where the data points are listed inside the parentheses in order and separated by commas. Giving data contextually meaningful names becomes particularly helpful when using these names in later commands and when interpreting the results of an analysis.

In order to determine if a linear model is reasonable for a given data set, plot the data points using the RStudio command `plotPoints`. The `plotPoints` command needs to know which variable identifies the output and which variable identifies the input. The format of the command is similar to that of other commands already introduced. Namely, if the output variable is `var2` and the input variable is `var1`, then the data set is plotted by entering `plotPoints(var2~var1)`.

The `mosaic` package in RStudio contains a command called `fitModel` that can be used to find the best possible model of a specified type. The `fitModel` command expects three things: the output variable, followed by the tilde symbol ~, followed by the form of the model (e.g., $mx+b$). In addition, `fitModel` can be provided with an initial guess for the values of the parameters, which will be relevant for models that are highly sensitive to changes in their parameters; this option is generally not needed for linear models.

While the parameters can have any name, the names of the input and output variables must exactly match the names for the data. For example, suppose the relationship between the variable quantities `time` and `location` is modeled by a linear function of the form location $= m \cdot$ time $+ b$. The appropriate syntax for the `fitModel` command is:

```
bestLinModel=fitModel(location~m*time+b).
```

In this case, the output of the `fitModel` command is stored in `bestLinModel`, which allows the resulting model to be used easily in subsequent commands. The coefficients determined by `fitModel` can be viewed by entering the command `coef(bestLinModel)`.

Examples of Commands

- `input=c(1,2,3,4)`

- `output=c(8,12,15,20)`

- `plotPoints(Users~Year)`

- `bestLinModel=fitModel(output~m*input+b)`

- `coef(bestLinModel)`

◆ **EXAMPLE 4** The linear data set below from Examples 1(a), 2(a), and 3 presents how many millions of people used Twitter during each quarter of the given year.

Year	11	11.25	11.5	11.75	12.75
Twitter users	68	85	101	117	185

(a) Verify that a linear model is reasonable using a plot of the data created by `plotPoints`.

(b) Use `fitModel` to find the best linear model for this data set and graph the model on a plot of the data.

Solution.

(a) First store the data in RStudio and then plot the data set. The data points appear to fall approximately the same line on the plane, indicating that a linear model is reasonable.

```
Year=c(11, 11.25, 11.5, 11.75, 12.75)
Users=c(68, 85, 101, 117, 185)
plotPoints(Users~Year)
```

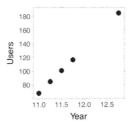

(b) The best linear model for this data set is found using `fitModel`.

```
bestModel=fitModel(Users~m*Year+b)
coef(bestModel)

      m          b
 66.7671 -666.6370
```

Note that RStudio presents the values of the two parameters in two columns; despite the close spacing, the negative sign with parameter b does not express a difference. From this output, the parameters of the line are $m = 66.7671$ and $b = -666.6370$, which means that the best linear model of this data is Users $= 66.7671 \cdot$ Year $- 666.637$. This model is relatively similar to the line Users $= 66 \cdot$ Year $- 657.7$ that was conjectured in Example 3. The following graph of the best linear model on a plot of the data indicates its relatively close fit with the data:

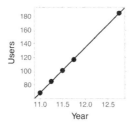

➤ **QUESTION 4** The linear data set below from Questions 1(b), 2(b), and 3 presents the average debt load in thousands of dollars at the end of each year's spring term for bachelor's degree recipients who attended public four-year colleges and universities and borrowed money to finance their education.

(a) Verify that a linear model is reasonable using a plot of the data created by `plotPoints`.

(b) Use `fitModel` to find the best linear model for this data set and graph the model on a plot of the data.

Academic Year	2001	2003	2005	2006
Average Debt	$20,400	$20,900	$21,500	$21,800

One major advantage of using RStudio for this type of analysis is the relative ease of working with large data sets. The approach of using `fitModel` to determine the best linear model for a given data set is identical for both small and large sets of data, as illustrated in Example 5. The first step in this direction is to learn how to load data sets into RStudio from existing electronic files.

Working in RStudio

Entering large data sets is not feasible using the `var1=c(,)` command described above. The large data sets used throughout this book are included in the MMAC package. The `names` command can be used to output the variable names in a data set. For example, for the data set `HealthExpenditure`, the command `names(Health Expenditure)` outputs the names of the two variables in this data set. Both `plotPoints` and `fitModel` work exactly as before with such data sets, with one required addition: the inclusion of the argument `data=myData`, where `myData` is the name under which the data was stored. The following examples illustrate this addition to these commands:

Examples of Commands

- `plotPoints(PercentGDP~Year,data=HealthExpenditure)`

- `fitModel(PercentGDP~m*Year+b,data=HealthExpenditure)`

◆ **EXAMPLE 5** The data set stored in `HealthExpenditure` contains the World Bank's data for total U.S. health expenditures as a percentage of U.S. gross domestic product (or GDP) from 1995 to 2012.

(a) Verify that a linear model is reasonable using a plot of the data.

(b) Use `fitModel` to find the best linear model for this data set and graph the model on a plot of the data.

Solution.

(a) First, determine the variables of the data stored in `HealthExpenditure`:

```
names(HealthExpenditure)
```

```
[1] "Year"        "PercentGDP"
```

The input variables name is `Year` and the output variable name is `PercentGDP`, which are used to plot the data set and determine if a linear model is reasonable:

```
plotPoints(PercentGDP~Year,data=HealthExpenditure,ylim=range(10,20))
```

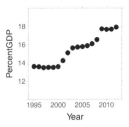

While the data is not perfectly linear, a linear model does appear to be somewhat reasonable. While less obviously linear than the other data sets considered in this section, linear models are widespread in economic analysis because of their relative simplicity and this data set is in the right ballpark for such a model.

(b) The best linear model for this data set is found using `fitModel`:

```
healthExpendModel=fitModel(PercentGDP~m*Year+b,
                            data=HealthExpenditure)
coef(healthExpendModel)

       m         b
 0.29978 -585.21396
```

As in Example 4, RStudio presents the values of the two parameters in two columns; despite the close spacing, the negative sign with parameter b does not express a difference. From this output, the parameters of the line are $m = 0.29978$ and $b = -585.21396$, which gives the following best linear model of this data:

$$\text{PercentGDP} = 0.29978 \cdot \text{Year} - 585.21396$$

The graph of this model on a plot of the data indicates its relative goodness of fit in this setting where the data is somewhat linear. Namely, as noted above, the data is not perfectly linear, but the model does provide a pretty reasonable approximation of the overall behavior of the data set.

```
plotPoints(PercentGDP~Year,data=HealthExpenditure,ylim=range(10,20))
plotFun(healthExpendModel(Year)~Year,add=TRUE)
```

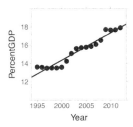

▶ **QUESTION 5** The data stored in `WeightChange` contains the change in one woman's weight during a pregnancy from the beginning of her second trimester until birth.

(a) Verify that a linear model is reasonable for this data set.

(b) Use `fitModel` to find the best linear model for this data set and graph the model on a plot of the data.

Summary

- A *linear function* is of the form $y = mx + b$. The *slope* m of a line through the points (x_1, x_2) and (y_1, y_2) is calculated using the following formulas when $x_1 \neq x_2$:
$$m = \frac{y_2 - y_1}{x_2 - x_1} = \frac{y_1 - y_2}{x_1 - x_2} = \frac{\Delta y}{\Delta x} = \frac{\text{rise}}{\text{run}}.$$
A vertical line $x = a$ has an *undefined* slope. The *vertical intercept* of a line b is the y-coordinate of the point $(0, b)$ where the line intersects the y-axis, if such a point exists.

- *Numerically Identifying Linear Models*: Check for a linear model $y = mx + b$ by comparing the slopes of the lines through pairs of data points. Linear data sets have exactly or approximately the same slopes for any two data points. Most often, slopes of lines between successive data points are compared.

- *Conjecturing Linear Parameters*: If a linear model $y = mx + b$ is reasonable for a set of data, then m and b are conjectured as follows.
  - For slope m, compute the sequence of slopes through successive data points $m_1, m_2, \ldots, m_{n-1}$ and then average these slopes:
$$m = \frac{m_1 + m_2 + m_3 + \cdots + m_{n-1}}{n - 1}$$
  - For vertical intercept b, substitute each data point into the equation $y = mx + b$ (using the conjectured slope), solve for $b_1, b_2, \ldots, b_n$, and then average these intercepts:
$$b = \frac{b_1 + b_2 + b_3 + \cdots + b_n}{n}$$

Exercises

In Exercises 1–8, explain why a linear model $y = mx + b$ of the data set is reasonable or not. If so, estimate its slope m and vertical intercept b.

1.

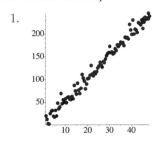

2.

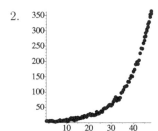

3.

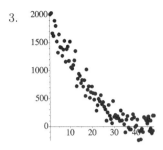

7.

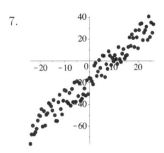

8.

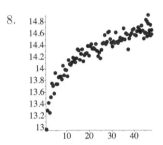

4.

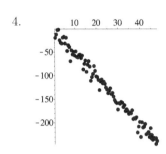

RStudio. In Exercises 9–16, use RStudio to plot the data set and explain why a linear model of the data is reasonable or not.

9.

x	1	2	3	4	5
y	111	69	26	−15	−57

10.

x	8	10	12	14	16
y	−2.5	4	11.5	18.5	25.5

11.

x	1	2	4	5	6
y	78.4	91.8	99.5	93.7	81.6

12.

x	1	2	4	5	6
y	40.5	32.8	21.5	17.4	14.1

5.

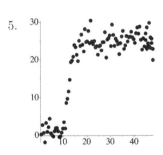

13.

x	2	8	12	16	18
y	33.7	1.9	−19.3	−40.5	−51.1

14.

x	−4	−2	1	3	7
y	−86.6	−42.9	−7.4	−3.7	−44.4

15.

x	0	5	10	15	20
y	5.1	4.42	3.87	3.42	2.97
x	30	35	40	45	50
y	1.84	1.38	1.06	0.61	−0.04

6.

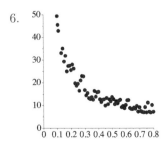

16.

x	0	1	2	3	4
y	7.88	10.03	13.1	12.2	9.3
x	5	6	7	8	9
y	13.7	19.2	21.9	23.8	26.1

RStudio. In Exercises 17–22, use RStudio to plot the data set and explain why a linear model of the data is reasonable or not.

17. Annual e-commerce sales (S) in the United States in billions of dollars for each year (Y)

Y	2009	2010	2011	2012
S	121	143	168	192

18. High value (H) of Facebook stock each month (M) in 2014

M	1	2	3	4	6
H	63.4	71.4	72.6	63.9	68

19. World population growth rate (R) by year (Y) since 1995

Y	0	5	10	15	20
R	1.41	1.26	1.20	1.13	1.06

20. Average number of goals (G) per game in World Cup tournaments each year (Y) since 1994

Y	0	4	8	12	16
G	2.7	2.7	2.5	2.3	2.3

21. Percent (P) of each year's high school graduates enrolling each year (Y) since 2005 in a two-year or four-year college

Y	0	1	4	5	6
P	68.6	66.0	70.1	68.1	66.2

22. Average monthly maximum temperature (T) in degrees Fahrenheit each month (M) in New York City

M	2	4	7	10	12
T	39.4	61.1	88.3	67.8	43.4

In Exercises 23–28, verify that a linear model $y = mx + b$ is reasonable for the data set by computing the slopes of lines through successive data points. Also, conjecture its slope m.

23.

x	1	2	3	4	5
y	5.4	7.5	11.5	14.4	16.5

24.

x	−2	−1	1	2	4
y	10.9	6.8	−1.3	−4.5	−12.6

25.

x	−5	−4	−3	−2	−1
y	−14.2	−10.9	−8.4	−5.9	−2.9

26.

x	−2	2	6	10	14
y	3.6	1.3	−1.8	−4.4	−7.4

27.

x	−5	−4	−3	−2	−1
y	−28.0	−20.7	−16.2	−8.4	−3.4
x	0	1	2	3	4
y	3.5	10.3	16.5	21.7	27.3

28.

x	−10	−5	0	5
y	65.9	23.4	−12.4	−49.2
x	15	20	25	35
y	−113.9	−154.8	−185.8	−270.5

In Exercises 29–36, for the data sets in Exercises 9–16, compute the slopes between successive data points and explain why a linear model $y = mx + b$ of the data set is reasonable or not. If so, conjecture its slope m.

29. Exercise 9   33. Exercise 13

30. Exercise 10   34. Exercise 14

31. Exercise 11   35. Exercise 15

32. Exercise 12   36. Exercise 16

In Exercises 37–40, conjecture the vertical intercept b for the data set and given intermediate linear model.

37. $y = -4x + b$

x	−2	1	2	4
y	17.5	5.8	1.0	−6.4

38. $y = -0.5x + b$

x	1	3	7	9
y	-27.9	-28.9	-30.7	-31.8

39. $y = 19x + b$

x	4	5	6	7
y	-26	-10	8	29

40. $y = 3.1x + b$

x	-3	-1	0	3
y	5.8	12.1	14.9	23.9

RStudio. In Exercises 41–46, use RStudio to find the best possible linear model for the data sets in Exercises 23–28 using the `fitModel` command.

41. Exercise 23

42. Exercise 24

43. Exercise 25

44. Exercise 26

45. Exercise 27

46. Exercise 28

In Exercises 47–52, consider the length (L) of the tornado season in days between the first and last tornado for each year (Y) in the warning area of the National Weather Service Office in Goodland, Kansas.

Y	1990	2000	2005	2010
L	120	142	149	155

47. Graphically verify that a linear model is reasonable by plotting the data.

48. Numerically verify that a linear model is reasonable by computing the slopes of lines through successive data points.

49. Conjecture a linear model $y = mx + b$ by averaging the slopes from Exercise 48 and the vertical intercepts from each data point.

50. Use `fitModel` to obtain a linear model $y = mx + b$ for the data set.

51. Evaluate the linear models from Exercises 49 and 50 at $x = 2008$ and explain the meaning of these answers.

52. Evaluate the linear models from Exercises 49 and 50 at $x = 2015$ and explain the meaning of these answers.

In Exercises 53–58, consider U.S. retail prescription drug sales (S) in billions of dollars for each year (Y); that is, in 2006, $262.7 billion was spent on prescriptions in the United States.

Y	2006	2007	2009	2010
S	$262.7	$261.6	$266.8	$266.4

53. Graphically verify that a linear model is reasonable by plotting the data.

54. Numerically verify that a linear model is reasonable by computing the slopes of lines through successive data points.

55. Conjecture a linear model $y = mx + b$ by averaging the slopes from Exercise 54 and the vertical intercepts from each data point.

56. Use `fitModel` to obtain a linear model $y = mx + b$ for the data set.

57. Evaluate the linear models from Exercises 55 and 56 at $x = 2008$ and explain the meaning of these answers.

58. Evaluate the linear models from Exercises 55 and 56 at $x = 2015$ and explain the meaning of these answers.

In Exercises 59–64, consider the total number of prescription (P) drugs sold in the United States in millions per year (Y); that is, in 2006, $3,419,000,000$ prescriptions were sold.

Y	2006	2007	2009	2010
P	3419	3530	3633	3676

59. Graphically verify that a linear model is reasonable by plotting the data.

60. Numerically verify that a linear model is reasonable by computing the slopes of lines through successive data points.

61. Conjecture a linear model $y = mx + b$ by averaging the slopes from Exercise 60 and the vertical intercepts from each data point.

62. Use `fitModel` to obtain a linear model $y = mx + b$ for the data set.

63. Evaluate the linear models from Exercises 61 and 62 at $x = 2008$ and explain the meaning of these answers.

64. Evaluate the linear models from Exercises 61 and 62 at $x = 2015$ and explain the meaning of these answers.

In Exercises 65 – 70, consider the total number of burgers (B) sold by McDonald's in billions as of each year (Y).

Y	1976	1987	1990	1994
B	20	65	80	100

65. Graphically verify that a linear model is reasonable by plotting the data.

66. Numerically verify that a linear model is reasonable by computing the slopes of lines through successive data points.

67. Conjecture a linear model $y = mx + b$ by averaging the slopes from Exercise 66 and the vertical intercepts from each data point.

68. Use `fitModel` to obtain a linear model $y = mx + b$ for the data set.

69. Evaluate the linear models from Exercises 67 and 68 at $x = 1992$ and explain the meaning of these answers.

70. Evaluate the linear models from Exercises 67 and 68 at $x = 2010$ and explain the meaning of these answers.

In Exercises 71 – 76, consider the global gender ratio based on the number of males per 100 females (M) in each year (Y).

Y	1990	1995	2005	2010
M	101.5	101.5	101.6	101.6

71. Graphically verify that a linear model is reasonable by plotting the data.

72. Numerically verify that a linear model is reasonable by computing the slopes of lines through successive data points.

73. Conjecture a linear model $y = mx + b$ by averaging the slopes from Exercise 72 and the vertical intercepts from each data point.

74. Use `fitModel` to obtain a linear model $y = mx + b$ for the data set.

75. Evaluate the linear models from Exercises 73 and 74 at $x = 2000$ and explain the meaning of these answers.

76. Evaluate the linear models from Exercises 73 and 74 at $x = 2015$ and explain the meaning of these answers.

RStudio. In Exercises 77 – 80, use RStudio to find a linear model of the United States monthly unemployment rate from January 2010 to December 2014 stored in `MonthlyUnemployment`.

77. Graphically verify that a linear model $y = mx + b$ is reasonable by plotting the data. Also, conjecture its slope m and vertical intercept b.

78. Use `fitModel` to obtain a linear model $y = mx + b$ for the data set.

79. Evaluate the model from Exercise 78 at $x = 36$. Explain the answer's meaning.

80. Evaluate the model from Exercise 78 at $x = 65$. Explain the answer's meaning.

RStudio. In Exercises 81 – 84, use RStudio to find a linear model of the total midyear population for the world from 1950 to 2015 stored in `WorldPopulation`.

81. Graphically verify that a linear model $y = mx + b$ is reasonable by plotting the data. Also, conjecture its slope m and vertical intercept b.

82. Use `fitModel` to obtain a linear model $y = mx + b$ for the data set.

83. Evaluate the model from Exercise 82 at $x = 1988$. Explain the answer's meaning.

84. Evaluate the model from Exercise 82 at $x = 2020$. Explain the answer's meaning.

RStudio. In Exercises 85 – 88, use RStudio to find a linear model of interest rates on 15-year, fixed-rate conventional home mortgages annually from 1992 to 2014 stored in `Mortgage15YrAnnual`.

85. Graphically verify that a linear model $y = mx + b$ is reasonable by plotting the data. Also, conjecture its slope m and vertical intercept b.

86. Use `fitModel` to obtain a linear model $y = mx + b$ for the data set.

87. Evaluate the model from Exercise 86 at $x = 2007$. Explain the answer's meaning.

88. Evaluate the model from Exercise 86 at $x = 2015$. Explain the answer's meaning.

RStudio. In Exercises 89 – 92, use RStudio to find a linear model of the number of Facebook users in millions of people from 2009 through 2012 stored in `FacebookUsers`.

89. Graphically verify that a linear model $y = mx + b$ is reasonable by plotting the data. Also, conjecture its slope m and vertical intercept b.

90. Use `fitModel` to obtain a linear model $y = mx + b$ for the data set.

91. Evaluate the model from Exercise 90 at $x = 13$. Explain the answer's meaning.

92. Evaluate the model from Exercise 90 at $x = 60$. Explain the answer's meaning.

RStudio. In Exercises 93 – 96, use RStudio to find a linear model of the high school dropout rate in the United States from 1970 through 2012 stored in `HSDropoutRate`.

93. Graphically verify that a linear model $y = mx + b$ is reasonable by plotting the data. Also, conjecture its slope m and vertical intercept b.

94. Use `fitModel` to obtain a linear model $y = mx + b$ for the data set.

95. Evaluate the model from Exercise 94 at $x = 2005$. Explain the answer's meaning.

96. Evaluate the model from Exercise 94 at $x = 2015$. Explain the answer's meaning.

RStudio. In Exercises 97 – 100, use RStudio to find a linear model of U.S. carbon dioxide emissions in kT (energy) annually from 1980 to 2008 according to the World Bank stored in `USCO2Emissions`.

97. Graphically verify that a linear model $y = mx + b$ is reasonable by plotting the data. Also, conjecture its slope m and vertical intercept b.

98. Use `fitModel` to obtain a linear model $y = mx + b$ for the data set.

99. Evaluate the model from Exercise 98 at $x = 2000$. Explain the answer's meaning.

100. Evaluate the model from Exercise 98 at $x = 2020$. Explain the answer's meaning.

In Your Own Words. In Exercises 101 – 104, explain the following.

101. Graphically determining if a linear model is reasonable or not

102. Numerically determining if a linear model is reasonable or not

103. Conjecturing the slope of a linear model

104. Conjecturing the vertical intercept of a linear model

2.2   Modeling with Exponential Functions

Exponential models provide another important option in the third and fourth steps of the modeling cycle: conjecturing a model and testing the model against known data. Given a data set, always plot the data first and then use this graph to inform an initial choice of a reasonable model. Various numerical techniques allow both verifying this choice and conjecturing values for the corresponding parameters. In addition, RStudio can be used to identify the "best possible" exponential model for a given data set. In practice, a natural exponential function $y = Ce^{k(x-h)}$ with base $e = 2.71828182\ldots$ is primarily used for exponential models.

Section 1.4 discussed exponential functions of the form $y = Ce^{kx}$ in some detail, with particular attention to the effect of the parameters C and k. When modeling real-life data sets, modifying the presentation of exponentials to $y = Ce^{k(x-h)}$ to include a horizontal shift parameter h in the exponent becomes necessary for practical, computational reasons. The key idea is to set h equal to the least input value. Further details about h are discussed later in this section as needed.

The graphical approach to linear data sets relies upon a familiarity with graphs of lines, which allows a visual determination of whether a data set is linear or not. While graphs of exponential functions are also familiar, many other functions are similarly shaped for portions of their graphs. These similarities can make it difficult to visually distinguish between plots of exponential and non-exponential data. For example, the data plot in Figure 1(a) corresponds to $y = e^x$, while the data plot in Figure 1(b) corresponds to $y = x^3$.

Section 1.4 introduced the notions of doubling time and halving time as a means to identifying exponential functions. The graphs in Figure 1 contain enough points to use doubling time to distinguish between $y = e^x$ and $y = x^3$. However, many data sets do not contain enough information to make such a determination. In such cases, a semi-log plot provides the key. Therefore, given a data set, always examine both its standard plot and its semi-log plot.

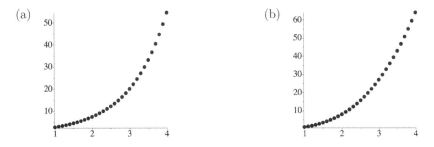

Figure 1: Graphs of (a) $y = e^x$ and (b) $y = x^3$ with similar shapes

When deciding whether an exponential model is reasonable or not for a data set, first plot the given data points (x, y) to determine if its general shape is exponential. Then, create a **semi-log plot** consisting of the points $(x, \ln(y))$, in which the logarithms of the output values $\ln(y)$ are plotted as a function of the given input values x. For reasons discussed shortly, the semi-log plot of an exponential data set is linear, which allows the use of a familiarity with lines to identify an exponential function. For example, the semi-log plot in Figure 2(a) of data corresponding to $y = e^x$ is linear, while the semi-log plot in Figure 2(b) of data corresponding to $y = x^3$ is curved and not linear.

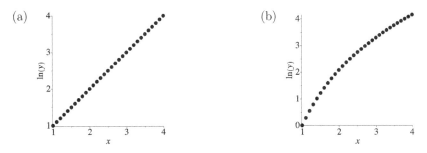

Figure 2: Semi-log plots of (a) $y = e^x$ and (b) $y = x^3$

GRAPHICALLY IDENTIFYING REASONABLE MODELS.

(1) If the standard plot (x, y) of a given data set is linear, then a linear model $y = mx + b$ is reasonable.

(2) If the semi-log plot $(x, \ln(y))$ of a given data set is nonconstant linear, then an exponential model $y = Ce^{k(x-h)}$ is reasonable.

Stipulating that the semi-log plot must be *nonconstant* linear in order for an exponential model to be reasonable might initially seem odd. However, if the semi-log plot is constant, then its slope is $m = 0$. As discussed soon, this slope m is equal to the parameter k in the corresponding exponential model $y = Ce^{k(x-h)}$. When $k = m = 0$, the corresponding "exponential" model is $y = Ce^{k(x-h)} = Ce^{0 \cdot (x-h)} = Ce^0 = C \cdot 1 = C$, which is a constant linear function and not exponential.

◆ **EXAMPLE 1** The table below presents the average weight of male fetuses in grams during the given gestational week. Graph the standard plot and the semi-log plot of this data set, and explain why an exponential model of the given data is reasonable or not.

Gestational week	25	26	30	31	33
Average weight	777	888	1435	1633	2058

Solution. First, transform the data to obtain the following table, which presents the logarithm of the output values as a function of the given input values:

Gestational week	25	26	30	31	33
ln(Average weight)	6.655	6.789	7.269	7.398	7.629

Figure 3(a) is the standard plot of this data, and its curved shape suggests that an exponential model is possible. Figure 3(b) is the semi-log plot of the given data, and its linear shape indicates that an exponential model is reasonable.

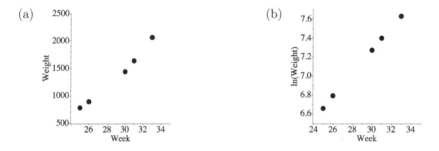

Figure 3: Standard and semi-log plots of the data for Example 1

◆ **EXAMPLE 2** The table below presents the annual U.S. Federal Funds Interest Rate in each year, which is the annual average of the interest rates at which banks and credit unions lend money to each other overnight. Graph the standard plot and the semi-log plot of this data set, and explain why an exponential model of the given data is reasonable or not.

Year	2000	2002	2004	2006	2008
Rate	6.24	1.67	1.35	4.97	1.92

Solution. First, transform the data to obtain the following table, which presents the logarithm of the output values as a function of the given input values:

Year	2000	2002	2004	2006	2008
ln(Rate)	1.831	0.513	0.300	1.603	0.652

Figure 4(a) is the standard plot of this data, and, because the data is oscillating up and down, an exponential model is not reasonable. Even more, Figure 3(b) is the semi-log plot of the given data, and its nonlinear shape also indicates that an exponential model is not reasonable.

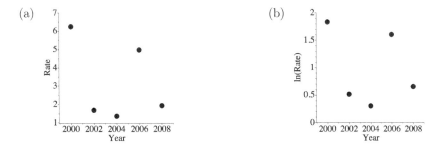

Figure 4: Standard and semi-log plots of the data for Example 2

➤ **QUESTION 1** The table below presents the plasma concentration of Prozac in nanograms per milliliter (ng/mL) for a person who has taken 20 mg of Prozac for 30 days, but stops cold turkey. (Note that discontinuing any medication should always be done in consultation with a healthcare provider.) Graph the standard plot and the semi-log plot of this data set, and explain why an exponential model of the given data is reasonable or not.

Day	0	5	10	22	27
Concentration	79	40	19.6	4.3	2.5

➤ **QUESTION 2** The table below presents the United Nations global estimates of the number of males per 100 females at the beginning of the last five decades. Graph the standard plot and the semi-log plot of this data set, and explain why an exponential model of the given data is reasonable or not.

Year	1970	1980	1990	2000	2010
Males per 100 females	100.8	101.1	101.5	101.5	101.6

This graphical approach to deciding whether an exponential model is reasonable or not relies on the linearity of the semi-log plot of the given data set for two reasons. First, linear plots are usually, quickly and easily recognized, and we want to capitalize on this expertise. Second, the natural logarithm of the exponential model $y = Ce^{k(x-h)}$ results in a linear model, which provides the necessary connection to these simplest models.

The following calculations demonstrate this exponential-linear connection by means of the natural logarithm:

$$y = Ce^{k(x-h)}$$ Exponential model

$$\ln(y) = \ln(Ce^{k(x-h)})$$ Take $\ln(x)$ of both sides

$$\ln(y) = \ln(C) + \ln(e^{k(x-h)})$$ $\ln(ab) = \ln(a) + \ln(b)$

$$\ln(y) = \ln(C) + k(x - h)$$ $\ln(e^b) = b$

$$\ln(y) = k \cdot (x - h) + \ln(C)$$ Rearrange terms

$$\ln(y) = k \cdot x + (-kh + \ln(C))$$ Rearrange terms some more

Based on this analysis of an exponential function, the semi-log plot of points $(x, \ln(y))$ gives $\ln(y)$ as a linear function of x. Furthermore, the semi-log plot has slope $m = k$ and passes through the point $(h, \ln(C))$, which provides the parameters of an exponential model $y = Ce^{k(x-h)}$ of the given data set.

While not common, some exponential data sets have outputs that are all negative. In this case, the corresponding model $y = Ce^{k(x-h)}$ has a negative parameter C. These negative outputs are problematic for graphing the semi-log plot because the domain of the natural logarithm $\ln(x)$ consists of all positive real numbers, which means the logarithm of such a negative output is undefined. In such cases, first take the absolute values of the given negative outputs (which will be positive) and then examine the resulting semi-log plot. An exponential model can be conjectured based on these absolute values of the outputs and C made negative as the last step.

Numerically Identifying Exponential Data Sets

A graphical analysis of reasonable models for a given data set is complemented with a parallel numerical assessment. Section 2.1 presented a numerical approach to identifying linear data sets based on a table of data. The following process extends this approach to include a check for exponential data sets:

NUMERICALLY IDENTIFYING REASONABLE MODELS.

(1) Check for a linear model $y = mx + b$ by comparing the slopes of the lines through pairs of data points. Linear data sets have exactly or approximately the same slopes for any two data points. Most often, slopes of lines between successive data points are compared.

(2) Check for an exponential model $y = Ce^{k(x-h)}$ by determining if the corresponding semi-log data is nonconstant linear by applying (1) to the transformed data $(x, \ln(y))$. Exponential data sets have nonconstant linear semi-log data. If the semi-log data has a linear model $y = m(x - h) + b$ with $m \neq 0$, then the exponential model has parameters $k = m$ and $C = e^b$.

◆ **EXAMPLE 3** The table below presents United States coal prices in dollars per short ton in the given year. Determine if the corresponding semi-log data is nonconstant linear and explain why an exponential model of the given data is reasonable or not.

Year	1989	1990	1992	1996	1998
Price	$31.36	$30.11	$27.46	$22.25	$20.65

Solution. Because this example focuses on the reasonableness of an exponential model, first create the following extended table by taking the natural logarithm of each output:

Year	1989	1990	1992	1996	1998
Price	$31.36	$30.11	$27.46	$22.25	$20.65
ln(Price)	3.446	3.405	3.313	3.102	3.028

Then compute the slopes of lines between successive data points:

Intervals	1989 to 1990	1990 to 1992	1992 to 1996	1996 to 1998
ΔYear	$1990 - 1989 = 1$	2	4	2
Δ ln(Price)	-0.041	-0.092	-0.211	-0.074
Slopes	$m_1 = -0.04$	$m_2 = -0.05$	$m_3 = -0.05$	$m_4 = -0.04$

These four slopes are within 0.01 of each other and so are considered approximately equal in the context of the modified output values of ln(Price), which vary between 3.028 and 3.446. Therefore, the semi-log data is approximately nonconstant linear, which means that an exponential model is reasonable for the given data set. ◼

◆ **EXAMPLE 4** The table below from Example 2 presents the annual U.S. Federal Funds Interest Rate. Determine if the corresponding semi-log data is nonconstant linear and explain why an exponential model of the given data is reasonable or not.

Year	2000	2002	2004	2006	2008
Rate	6.24	1.67	1.35	4.97	1.92

Solution. Because this example focuses on the reasonableness of an exponential model, first create the following extended table by taking the natural logarithm of each output:

Year	2000	2002	2004	2006	2008
Rate	6.24	1.67	1.35	4.97	1.92
ln(Rate)	1.831	0.513	0.300	1.603	0.652

Then compute the slopes of lines between successive data points:

Intervals	2000 to 2002	2002 to 2004	2004 to 2006	2006 to 2008
ΔYear	$2002 - 2000 = 2$	2	2	2
Δ ln(Rate)	-1.318	-0.213	1.303	-0.951
Slopes	$m_1 = -0.659$	$m_2 = -0.107$	$m_3 = 0.652$	$m_4 = -0.476$

These four slopes are not approximately equal. They differ by as much as 1.311, while the modified output values of ln(Rate) are between 0.300 and 1.831. Therefore, a linear model is not reasonable for the semi-log data, which means that an exponential model is not reasonable for the given data set. ◼

▶ **QUESTION 3** The table below from Example 1 presents the average weight of male fetuses in grams during the given gestational week. Determine if the corresponding semi-log data is nonconstant linear and explain why an exponential model of the given data is reasonable or not.

Gestational week	25	26	30	31	33
Average weight	777	888	1435	1633	2058

▶ **QUESTION 4** The table below from Question 2 presents the United Nations global estimates of the number of males per 100 females at the beginning of the last five decades. Determine if the corresponding semi-log data is nonconstant linear and explain why an exponential model of the given data is reasonable or not.

Year	1970	1980	1990	2000	2010
Males per 100 females	100.8	101.1	101.5	101.5	101.6

Conjecturing Exponential Models

After determining that an exponential model is reasonable for a given data set, the next step is to conjecture the parameters of such a model $y = Ce^{k(x-h)}$. The first step is to identify the value of the horizontal shift parameter h, which essentially "shifts" the data so that the first data point corresponds to the vertical axis. This shift is accomplished by replacing the "x" in $y = Ce^{kx}$ with "$x - h$" in $y = Ce^{k(x-h)}$ and setting h equal to the least (or leftmost) input value from the data set. For example, suppose $(100, 85)$ is the first data point in the sense that all other data points have an input value greater than 100. In this setting, let $h = 100$ and work with the intermediate exponential model $y = Ce^{k(x-100)}$ to identify values for the parameters C and k.

CONJECTURING PARAMETERS FOR EXPONENTIAL MODELS.
If an exponential model $y = Ce^{k(x-h)}$ is reasonable for a data set, conjecture h, k, and C as follows.

- For horizontal shift parameter h, let h equal the least (or leftmost) input value from the data set.

- For exponent parameter k, compute the sequence of slopes through successive data points $m_1, m_2, \ldots, m_{n-1}$ for the semi-log data and then average these slopes:
$$k = m = \frac{m_1 + m_2 + m_3 + \cdots + m_{n-1}}{n - 1}$$

- For the coefficient parameter C, substitute each point from the original data set into the equation $y = Ce^{k(x-h)}$ (using the conjectured values of h and k), solve for $C_1, C_2, \ldots, C_n$, and then average these values:
$$C = \frac{C_1 + C_2 + C_3 + \cdots + C_n}{n}$$

◆ **EXAMPLE 5** The table below from Example 3 presents U.S. coal prices in dollars per short ton in the given year. Conjecture an exponential model $y = Ce^{k(x-h)}$ for this data set and graph the model on a plot of the data.

Year	1989	1990	1992	1996	1998
Price	$31.36	$30.11	$27.46	$22.25	$20.65

Solution. Example 3 demonstrated that an exponential model is reasonable for this data set. In this setting, let "Year" denote the input, let "Price" denote the output, and let the horizontal shift parameter be $h = 1989$ based on the leftmost data point $(1989, 31.36)$ with the least input of 1989. Therefore, the sought-for model is of the form

$$\text{Price} = Ce^{k(\text{Year}-1989)}$$

Example 3 computed slopes $m_1 = -0.04$, $m_2 = -0.05$, $m_3 = -0.05$, and $m_4 = -0.04$ for the semi-log data. These slopes average to $m = -0.045$, which is the conjectured value of the exponent parameter k and gives an intermediate exponential model of $\text{Price} = Ce^{-0.045(\text{Year}-1989)}$. The coefficient parameter C is conjectured by substituting each data point into this intermediate model, solving for the corresponding C, and then averaging the results. For example, substituting the second data point $(1990, 30.11)$ produces $C_2 = 31.496$ as follows:

$$\text{Price} = Ce^{-0.045(\text{Year}-1989)} \qquad \text{Intermediate model}$$
$$30.11 = C_1 e^{-0.045(1990-1989)} \qquad \text{Substitute (Year, Price)} = (1990, 30.11)$$
$$\frac{30.11}{e^{-0.045}} = C_1 \qquad \text{Divide both sides by } e^{-0.045(1990-1989)} = e^{-0.045}$$
$$31.496 = C_1 \qquad \text{Simplify}$$

Similar computations with the remaining four data points yield $C_2 = 31.36$, $C_3 = 31.429$, $C_4 = 30.488$, and $C_5 = 30.961$, which average to $C = 31.147$. Substituting into the intermediate model $\text{Price} = Ce^{-0.045(\text{Year}-1989)}$ gives the conjectured exponential model:

$$\text{Price} = 31.147e^{-0.045(\text{Year}-1989)}$$

As part of the modeling cycle, this conjectured model is compared against the known data to assess its accuracy and reliability, and to inform any adjustments to the parameters. Figure 5 presents the graph of this exponential model on a plot of the data. While not every data point lies exactly on the corresponding curve, the model appears quite good.

∎

➤ **QUESTION 5** The table below from Example 1 and Question 3 presents the average weight of male fetuses in grams during the given gestational week. Conjecture an exponential model $y = Ce^{k(x-h)}$ for this data set and graph the model on a plot of the data.

Gestational week	25	26	30	31	33
Average weight	777	888	1435	1633	2058

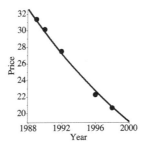

Figure 5: Graph of the data and an exponential model for Example 5

Best Possible Exponential Models

The techniques developed in this section enable us to conjecture a reasonable exponential model for a given data set. However, these methods are not guaranteed to produce the best possible exponential model. Rather, the only claim that can be made is that a reasonable model has been identified and that a visual check does or does not affirm its apparent validity. As with linear models, the RStudio command `fitModel` implements a version of the method of least squares and provides a way to find the "best possible" exponential model for a given data set.

Working in RStudio

As with linear models, the `fitModel` command in the `mosaic` package of RStudio finds the best possible exponential model for a given data set. Recall that before using `fitModel`, the data must first be stored in RStudio. Small data sets can be entered by hand, storing the data with the command `var1=c(,)`, where the data points are listed inside the parentheses and separated by commas; large data sets used in this book are included in the `MMAC` package

For an exponential model $y = Ce^{k(x-h)}$, the horizontal shift parameter h is identified by inspecting the data. Because exponentials grow so rapidly, `fitModel` sometimes has difficulty finding the best values for the other parameters C and k directly. Therefore, `fitModel` is applied to the corresponding semi-log data $(x, \ln(y))$ to find the corresponding best linear model $\ln(y) = m(x - h) + b$. From such a model, the slope m is equal to the parameter k, and the vertical intercept $b = \ln C$, which is used to compute $C = e^b$. Recall that RStudio uses the command `log(y)` to compute $\ln(y)$.

For example, suppose an exponential model is reasonable for data with variable names `input` and `output`. The parameters for the exponential model are determined with the command `fitModel(log(output)~m*(input-h)+b)`.

Examples of Commands

- `input=c(1,2,3,4)`
- `output=c(2.7,7.4,20.1,54.6)`
- `bestExpModel=fitModel(log(output)~m*(input-1)+b)`
- `coef(bestExpModel)`
- To determine C: `exp(b)`

◆ **EXAMPLE 6** The table below from Example 1 and from Questions 3 and 5 presents the average weight of male fetuses in grams during the given gestational week. Use `fitModel` to find the best exponential model $y = Ce^{k(x-h)}$ for this data set and graph the model on a plot of the data.

Gestational week	25	26	30	31	33
Average weight	777	888	1435	1633	2058

Solution. First, store the data set in RStudio. Example 1 and Question 3 demonstrated that an exponential model is reasonable for this data set.

```
Week = c(25,26,30,31,33)
Weight = c(777,888,1435,1633,2058)
```

In this setting, "Week" denotes the input, "Weight" denotes the output, and let the horizontal shift parameter be $h = 25$ based on the leftmost data point $(25, 777)$ with the least input of 25. Therefore, the sought-for model is of the form Weight $= Ce^{k(\text{Week}-25)}$. The parameters C and k are found by working with the corresponding semi-log data and using `fitModel` to determine the best possible linear model $\ln(\text{Weight}) = m(\text{Week} - 25) + b$ for the semi-log data:

```
bestWeightModel = fitModel(log(Weight)~m*(Week-25)+b)
coef(bestWeightModel)

      m        b
0.121621 6.661717
```

From `fitModel`, the linear model for the corresponding semi-log data has slope $m = 0.1216$ and vertical intercept $b = 6.6617$. Thus, the parameters of exponential model are $k = m = 0.1216$ and $C = e^b = e^{6.6617}$, which can be approximated in RStudio:

```
exp(6.6617)
```

`[1] 781.879`

These results provide the best possible exponential model for the average weight of male fetuses in grams during the given gestational week based on the given data:

$$\text{Weight} = e^{6.6617} \cdot e^{0.1216(\text{Week}-25)} \approx 781.9e^{0.1216(\text{Week}-25)}$$

The graph of the model on a plot of the data, while not exact, indicates its relative goodness of fit in the sense that the graph of the model matches the data relatively closely.

```
plotPoints(Weight~Week)
plotFun(exp(6.6617)*exp(0.1216*(Week-25))~Week,add=TRUE)
```

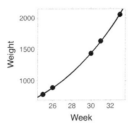

The `fitModel` command produces the best model of a particular type for a given set of data. This means that the following model from Example 6 is the best possible exponential model for the data set about the average weight of male fetuses in grams during the given gestational week:

$$\text{Weight} = e^{6.6617} \cdot e^{0.1216(\text{Week}-25)} \approx 781.9 e^{0.1216(\text{Week}-25)}$$

Question 5 introduced this same data set and the numerical process for conjecturing exponential model results $\text{Weight} = 770 e^{0.125(\text{Week}-25)}$. These conjectured parameters of $k = 0.125$ and $C = 770$ are quite close to the best possible parameters of $k = 0.1216$ and $C = 781.9$. For the sake of comparison, these two models are graphed side by side on a plot of the data: the `fitModel` exponential from Example 6 on the left and the conjectured model from Question 5 on the right. As the graphs suggest, `fitModel` produces a better model that more closely approximates each data point, particularly including the data points at the extreme ends of the range of data.

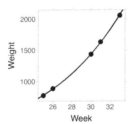
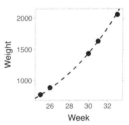

Finally, while this model works quite well for the given data set, additional data might lead to new, different parameters for an exponential model or even to a non-exponential model. Section 2.6 studies this iterative aspect of the fifth step of the modeling cycle.

▶ **QUESTION 6** The table below from Question 1 presents the plasma concentration of Prozac in nanograms per milliliter (ng/mL) for a person who has taken 20 mg of Prozac for 30 days, but stops cold turkey. (Note that discontinuing any medication should always be done in consultation with a healthcare provider.) Use `fitModel` to find the best exponential model $y = Ce^{k(x-h)}$ of this data set and graph the model on a plot of the data.

Day	0	5	10	22	27
Concentration	79	40	19.6	4.3	2.5

One major advantage of using RStudio for this type of analysis is the relative ease of working with large data sets. The `fitModel` approach to determining the best exponential model for a given data set is essentially identical for both small and large sets of data, as illustrated in Example 7.

Working in RStudio

Similar to the **names** command, the command **head** provides both the variable names and the first few data points. For example `head(USPopulation)` would provide the variables names and the first few data points for the data set `USPopulation`.

Example of Command

- `head(USPopulation)`

◆ **EXAMPLE 7** The data set stored in `WorldPopulation` contains the U.S. Census Bureau's estimate of world population in billions of people every five years since 1950.

(a) Verify that an exponential model is reasonable using a semi-log plot of the data.

(b) Use `fitModel` to find the best exponential model for this data set and then graph the model on a plot of the data.

Solution.

(a) First, determine the variables and first few data points using **head**:

```
head(WorldPopulation)
```

```
   Year People
1 1950   2.55
2 1955   2.78
3 1960   3.04
4 1965   3.35
5 1970   3.71
6 1975   4.09
```

Using the variables named **Year** and **People**, create (a) the standard plot of the data and (b) the semi-log plot of the data set to verify that an exponential model is reasonable.

(a) `plotPoints(People~Year,`
     `data=WorldPopulation)`

(b) `plotPoints(log(People)~Year,`
     `data=WorldPopulation)`

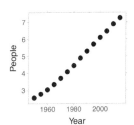

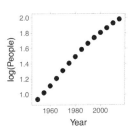

The curved shape of the standard plot (a) suggests that an exponential model is possible and, while not perfectly linear, the semi-log plot of the data appears roughly linear for most of the data points and suggests that an exponential model is reasonable. At the same time, the exponential model will have limited accuracy because the semi-log plot is only roughly linear.

(b) Let the horizontal shift parameter be $h = 1950$ based on the data point $(1950, 2.55)$ with the least input of 1950. Therefore, the sought-for model is of the form People $= Ce^{k(\text{Year}-1950)}$. The parameters C and k are found by working with the corresponding semi-log data and using `fitModel` to determine the best possible linear model $\ln(\text{People}) = m \cdot (\text{Year} - 1950) + b$ for the semi-log data:

```
worldPopModel=fitModel(log(People)~m*(Year-1950)+b,
                        data=WorldPopulation)
coef(worldPopModel)


        m        b
0.016511 0.968407
```

Based on the RStudio output from `fitModel`, the linear model for the corresponding semi-log data has slope $m = 0.016511$ and vertical intercept $b = 0.968407$, which means that the parameters of the exponential model are $k = m = 0.1216$ and $C = e^b = e^{0.968407} \approx 2.633746$. Thus, the given data provides the following best possible exponential model of world population:

$$\text{People} = e^{0.968407} \cdot e^{0.016511 \cdot (\text{Year}-1950)} \approx 2.633746 \cdot e^{0.016511 \cdot (\text{Year}-1950)}$$

The following graph of this model on a plot of the data indicates its relative goodness of fit. As noted above in part (a), the model does not provide an exact match to the data, because the semi-log plot of the data is only roughly linear. However, the exponential model does appear reasonable for the given data set.

```
plotPoints(People~Year,data=WorldPopulation)
plotFun(exp(0.968407)*exp(0.016511*(Year-1950))~Year,add=TRUE)
```

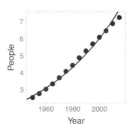

This book's continuing study of models will introduce other functions that provide a better match for this world population data set. In addition, Section 2.6 studies the iterative aspect of the modeling cycle, which will enable further refinements of this model through the inclusion of additional data.

➤ **QUESTION 7** The data set stored in `DJIACloseQuarterly` contains the closing stock market value of the Dow Jones Industrial Average on the last day of each quarter from March 31, 1935 (quarter 1) through December 31, 2014 (quarter 320).

(a) Verify that an exponential model is reasonable using a semi-log plot of the data.

(b) Use `fitModel` to find the best exponential model for this data set and then graph the model on a plot of the data.

Summary

- The *natural exponential function* is of the form $y = Ce^{k(x-h)}$ with base $e = 2.71828182\ldots$.

- If (a, b) is a point from a data set or curve, its *semi-log plot* includes the point $(a, \ln(b))$.

- *Graphically Identifying Exponential Models*: If the semi-log plot $(x, \ln(y))$ of a data set is nonconstant linear, then an exponential model is reasonable.

- *Numerically Identifying Exponential Models*: Check for an exponential model $y = Ce^{k(x-h)}$ by determining if the corresponding semi-log data is nonconstant linear by determining if the slopes of the lines through pairs of transformed data points $(x, \ln(y))$ are exactly or approximately equal. Exponential data sets have nonconstant linear semi-log data. If the semi-log data has a linear model $y = mx + b$ with $m \neq 0$, then the exponential model has parameters $k = m$ and $C = e^b$.

Summary (continued)

- *Conjecturing Parameters for Exponential Models*: If an exponential model $y = Ce^{k(x-h)}$ is reasonable for a set of data, then conjecture h, k, and C as follows.

  ○ For horizontal shift parameter h, let h equal the least (or leftmost) input value from the data set.

  ○ For exponent parameter k, compute the sequence of slopes through successive data points m_1, m_2, ..., m_{n-1} for the semi-log data and then average these slopes:
  $$k = m = \frac{m_1 + m_2 + m_3 + \cdots + m_{n-1}}{n-1}$$

  ○ For coefficient parameter C, substitute each data point from the original data set into the equation $y = Ce^{k(x-h)}$ (using the conjectured values of h and k), solve for C_1, C_2, ..., C_n, and then average these values:
  $$C = \frac{C_1 + C_2 + C_3 + \cdots + C_n}{n}$$

Exercises

In Exercises 1–8, explain why an exponential model $y = Ce^{k(x-h)}$ of the data set is reasonable or not based on a plot of the data points (left-hand plot) and a semi-log plot of the data points (right-hand plot). If so, estimate its parameters C and k.

1.
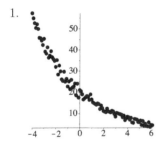 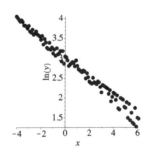

2.
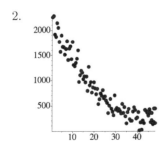 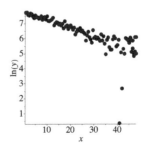

3.

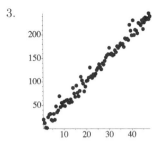
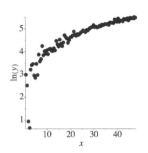

4.

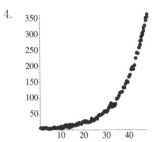
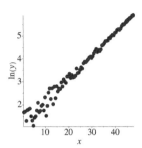

5.

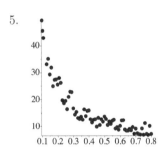
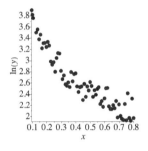

6.

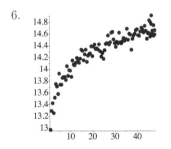
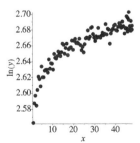

7.

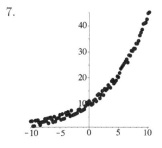
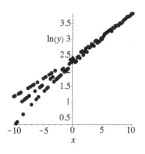

8.

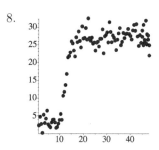
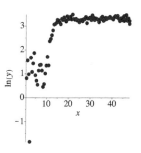

RStudio. In Exercises 9–16, use RStudio to graph the standard plot and the semi-log plot of the data set, and explain why an exponential model of the given data is reasonable or not. If so, estimate its parameters C and k.

9.

x	15	18	24	30	31
y	7.3	5.4	3	1.6	1.5

10.

x	15	18	24	30	31
y	2	3	6	10	11

11.

x	5	20	25	55	60
y	900	625	400	400	225

12.

x	5	20	25	55	60
y	52	59	43	35	40

13.

x	−0.5	0.4	2.9	5.2	7.7
y	0.2	1.5	18	181	2208

14.

x	−0.5	0.4	2.9	5.2	7.7
y	1.3	0.9	0.6	0.4	0.2

15.

x	−2	0	3	5	9
y	1	27	216	512	1728

16.

x	1	2	3	4	5
y	34	37	50	85	182

RStudio. In Exercises 17–22, use RStudio to graph the standard plot and the semi-log plot of the data set. Explain why an exponential model of the given data is reasonable or not and, if so, estimate its parameters C and k.

17. S&P 500 stock market closing value (V) on June 1st of each year (Y) since 1975 in U.S. dollars

Y	0	10	20	30	40
V	95	192	545	1191	2077

18. Annual U.S. unemployment rate (R) each year (Y)

Y	2007	2008	2009	2010
R	4.6	5	7.8	9.7

19. Total annual sales (S) in thousands of hybrid vehicles in the United States each year (Y)

Y	2010	2011	2012	2013
S	274	266	435	496

20. U.S. field production of crude oil (O) in billions of barrels in each year (Y) since 1970

Y	0	10	20	30	35
O	3.5	3.1	2.7	2.1	1.9

21. Three-year average monthly pollen count (C) each month (M) in Brooklyn, New York City

M	8	9	10	11	12
C	8.4	4.6	1.7	0.3	0.1

22. Average monthly maximum temperature (T) in degrees Fahrenheit each month (M) in New York City

M	2	4	7	10	12
T	39.4	61.1	88.3	67.8	43.4

In Exercises 23–28, numerically verify that an exponential model $y = Ce^{k(x-h)}$ is reasonable for the data set by showing the corresponding semi-log data is nonconstant linear. Also, conjecture its parameter k.

23.
x	1	2	3	4	5
y	3.6	4.5	5.5	6.7	8.2

24.
x	−2	−1	1	2	4
y	7.3	5.4	2.9	2.2	1.2

25.
x	−5	−4	−3	−2	−1
y	8113	1330	220	35	6

26.
x	−2	2	6	10	14
y	1.4	6.6	33.1	163.7	811.6

27.
x	−4	−3	−2	−1
y	0.058	0.145	0.351	0.891
x	0	1	2	3
y	2.146	5.281	12.999	31.927

28.
x	−15	−5	0	5
y	0.001	0.003	0.007	0.011
x	10	20	25	35
y	0.018	0.050	0.082	0.223

In Exercises 29–36, for the data sets in Exercises 9–16, numerically determine if the semi-log data for each data set is nonconstant linear and explain why an exponential model $y = Ce^{k(x-h)}$ is reasonable or not. If so, conjecture k and C.

29. Exercise 9

30. Exercise 10

31. Exercise 11

32. Exercise 12

33. Exercise 13

34. Exercise 14

35. Exercise 15

36. Exercise 16

In Exercises 37–40, conjecture the coefficient parameter C for the data set and given intermediate exponential model.

37. $y = Ce^{-(x+2)}$

x	−2	1	2	4
y	14.78	0.74	0.27	0.04

38. $y = Ce^{0.333(x-1)}$

x	1	3	7	9
y	7.53	14.66	55.55	108.13

39. $y = Ce^{2.1(x-4)}$

x	4	5	6	7
y	4.4	36.3	296.6	2421.7

40. $y = Ce^{-1.25(x+6)}$

x	−6	−3	−1	3
y	−723.22	−17.01	−1.40	−0.01

RStudio. In Exercises 41–46, use RStudio to find the best possible exponential model for each of the data sets in Exercises 23–28 using the `fitModel` command.

41. Exercise 23

42. Exercise 24

43. Exercise 25

44. Exercise 26

45. Exercise 27

46. Exercise 28

In Exercises 47–52, consider the number of red cards (C) given per men's World Cup tournament in each year (Y).

Y	1978	1982	1986	1994	1998
C	3	5	8	15	22

47. Graphically verify that an exponential model is reasonable using a semi-log plot.

48. Numerically verify that an exponential model is reasonable using slopes of lines through successive semi-log data points.

49. Conjecture an exponential model using the first data point, and by averaging the slopes from Exercise 48 and the results of substituting each data point.

50. Use `fitModel` to obtain an exponential model $y = Ce^{k(x-h)}$ for the data set.

51. Evaluate the exponential models from Exercises 49 and 50 at $x = 2006$ and explain the meaning of these answers.

52. Evaluate the exponential models from Exercises 49 and 50 at $x = 2014$ and explain the meaning of these answers.

In Exercises 53–58, consider the average total debt load (D) in thousands of dollars of bachelor's degree recipients who attended U.S. public colleges and universities and borrowed money to finance their education each year (Y).

Y	2008	2010	2011	2012
D	$22	$23.6	$24.6	$25.3

53. Graphically verify that an exponential model is reasonable using a semi-log plot.

54. Numerically verify that an exponential model is reasonable using slopes of lines through successive semi-log data points.

55. Conjecture an exponential model using the first data point, and by averaging the slopes from Exercise 54 and the results of substituting each data point.

56. Use `fitModel` to obtain an exponential model $y = Ce^{k(x-h)}$ for the data set.

57. Evaluate the exponential models from Exercises 55 and 56 at $x = 2009$ and explain the meaning of these answers.

58. Evaluate the exponential models from Exercises 55 and 56 at $x = 2013$ and explain the meaning of these answers.

In Exercises 59–64, consider the highest value (V) of Facebook stock each month (M) in 2013, where the number identifies the month; for example, 8 corresponds to August, the 8th month of the year.

M	5	7	8	9	10
V	29.1	38.3	42.5	51.6	54.8

59. Graphically verify that an exponential model is reasonable using a semi-log plot.

60. Numerically verify that an exponential model is reasonable using slopes of lines through successive semi-log data points.

61. Conjecture an exponential model using the first data point, and by averaging the slopes from Exercise 60 and the results of substituting each data point.

62. Use `fitModel` to obtain an exponential model $y = Ce^{k(x-h)}$ for the data set.

63. Evaluate the exponential models from Exercises 61 and 62 at $x = 6$ and explain the meaning of these answers.

64. Evaluate the exponential models from Exercises 61 and 62 at $x = 12$ and explain the meaning of these answers.

In Exercises 65–70, consider the total number of tornadoes (T) each year (Y) in the warning area of the National Weather Service Office in Goodland, Kansas.

Y	1975	1985	1995	2000	2005
T	5	8	18	28	33

65. Graphically verify that an exponential model is reasonable using a semi-log plot.

66. Numerically verify that an exponential model is reasonable using slopes of lines through successive semi-log data points.

67. Conjecture an exponential model using the first data point, and by averaging the slopes from Exercise 66 and the results of substituting each data point.

68. Use `fitModel` to obtain an exponential model $y = Ce^{k(x-h)}$ for the data set.

69. Evaluate the exponential models from Exercises 67 and 68 at $x = 2006$ and explain the meaning of these answers.

70. Evaluate the exponential models from Exercises 67 and 68 at $x = 2011$ and explain the meaning of these answers.

In Exercises 71–76, consider the population (P) of Ireland in millions of people each year (Y).

Y	1834	1847	1854	1867	1881
P	7.89	8.02	6.08	5.64	3.87

71. Graphically verify that an exponential model is reasonable using a semi-log plot.

72. Numerically verify that an exponential model is reasonable using slopes of lines through successive semi-log data points.

73. Conjecture an exponential model using the first data point, and by averaging the slopes from Exercise 72 and the results of substituting each data point.

74. Use `fitModel` to obtain an exponential model $y = Ce^{k(x-h)}$ for the data set.

75. Evaluate the exponential models from Exercises 73 and 74 at $x = 1841$ and explain the meaning of these answers.

76. Evaluate the exponential models from Exercises 73 and 74 at $x = 1891$ and explain the meaning of these answers.

RStudio. In Exercises 77–80, use RStudio to find an exponential model of the closing price of Apple Inc. stock in U.S. dollars adjusted for dividends and splits at the beginning of each month from January 1981 (month 1) through December 2014 (month 408) stored in `AAPLStockMonthly`.

77. Graphically verify that an exponential model $y = Ce^{k(x-h)}$ is reasonable using a semi-log plot. Also, conjecture its parameters h, k, and C.

78. Use `fitModel` to obtain an exponential model $y = Ce^{k(x-h)}$ for the data set.

79. Evaluate the model from Exercise 78 at $x = 380$. Explain the answer's meaning.

80. Evaluate the model from Exercise 78 at $x = 415$. Explain the answer's meaning.

RStudio. In Exercises 81–84, use RStudio to find an exponential model of the closing NASDAQ stock market value in U.S. dollars at the end of each quarter from March 1938 (quarter 1) through December 2014 (quarter 308) stored in `NASDAQQuarterly`.

81. Graphically verify that an exponential model $y = Ce^{k(x-h)}$ is reasonable using a semi-log plot. Also, conjecture its parameters h, k, and C.

82. Use `fitModel` to obtain an exponential model $y = Ce^{k(x-h)}$ for the data set.

83. Evaluate the model from Exercise 82 at $x = 12$. and explain the meaning of the answer.

84. Evaluate the model from Exercise 82 at $x = 310$ and explain the meaning of the answer.

RStudio. In Exercises 85–88, use RStudio to find an exponential model of average interest rate for conventional 30-year mortgages each year from 1981 to 2012 stored in `Mortgage30YrAnnual`.

85. Graphically verify that an exponential model $y = Ce^{k(x-h)}$ is reasonable using a semi-log plot. Also, conjecture its parameters h, k, and C.

86. Use `fitModel` to obtain an exponential model $y = Ce^{k(x-h)}$ for the data set.

87. Evaluate the model from Exercise 86 at $x = 1993$. Explain the answer's meaning.

88. Evaluate the model from Exercise 86 at $x = 2014$. Explain the answer's meaning.

RStudio. In Exercises 89–92, use RStudio to find an exponential model of the population of the Netherlands each decade from 1700 through 2010 stored in `NetherlandsPopulation`.

89. Graphically verify that an exponential model $y = Ce^{k(x-h)}$ is reasonable using a semi-log plot. Also, conjecture its parameters h, k, and C.

90. Use `fitModel` to obtain an exponential model $y = Ce^{k(x-h)}$ for the data set.

91. Evaluate the model from Exercise 90 at $x = 2002$. Explain the answer's meaning.

92. Evaluate the model from Exercise 90 at $x = 2020$. Explain the answer's meaning.

RStudio. In Exercises 93–96, use RStudio to find an exponential model of the number of physicians per 1000 people as a function of average life expectancy in different countries in 2010 stored in `LifeExpectancyPhysicians`.

93. Graphically verify that an exponential model $y = Ce^{k(x-h)}$ is reasonable using a semi-log plot. Also, conjecture its parameters h, k, and C.

94. Use `fitModel` to obtain an exponential model $y = Ce^{k(x-h)}$ for the data set.

95. Evaluate the model from Exercise 94 at $x = 75$. Explain the answer's meaning.

96. Evaluate the model from Exercise 94 at $x = 90$. Explain the answer's meaning.

RStudio. In Exercises 97–100, use RStudio to find an exponential model of the total number of AP Calculus exams taken each year from 1955 to 2015 stored in `APCalculus`.

97. Graphically verify that an exponential model $y = Ce^{k(x-h)}$ is reasonable using a semi-log plot. Also, conjecture its parameters h, k, and C.

98. Use `fitModel` to obtain an exponential model $y = Ce^{k(x-h)}$ for the data set.

99. Evaluate the model from Exercise 98 at $x = 2000$. Explain the answer's meaning.

100. Evaluate the model from Exercise 98 at $x = 2017$. Explain the answer's meaning.

In Your Own Words. In Exercises 101–105, explain the following.

101. Graphically determining if an exponential model is reasonable or not

102. Numerically determining if an exponential model is reasonable or not

103. Conjecturing the horizontal shift parameter h of an exponential model $y = Ce^{k(x-h)}$

104. Conjecturing the exponent parameter k of an exponential model $y = Ce^{k(x-h)}$

105. Conjecturing the coefficient parameter C of an exponential model $y = Ce^{k(x-h)}$

2.3 Modeling with Power Functions

We have learned how to model various aspects of reality using linear and exponential functions. This section studies power functions, a third important category of mathematical models that correspond to diverse natural and social phenomena. While the name "power" function may be new, this class is already quite familiar and includes such functions as $y = x^2$, $y = x^3$, and $y = \sqrt{x} = x^{1/2}$. Roughly speaking, power functions are defined by exponential expressions a^r, where the base a is a variable x and the exponent r (or power) is a nonzero real number. Multiplication by a nonzero coefficient is also allowed, as indicated in the following definition:

> **Definition.** A **power function** is of the form $y = Cx^k$, where C and k are nonzero constants.

Power functions encompass a wide variety of functions whose graphs exhibit a myriad of different shapes, features, and overall behaviors, many of which will already be familiar. For example, when n is a natural number, $y = x^n$ gives us the power functions $y = x$, $y = x^2$, $y = x^3$, $y = x^4$, and so on. Figure 1 presents the graphs of three such power functions: (a) is $y = x$, (b) is $y = x^2$, and (c) is $y = x^3$.

In addition to these familiar functions, sometimes exponents of power functions are not natural numbers. For example, Figure 2(a) presents the graph of the power function $y = x^{0.135}$ and (b) is $y = x^{-1.742}$.

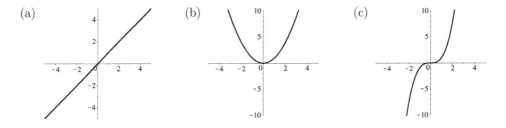

Figure 1: Graphs of power functions (a) $y = x$, (b) $y = x^2$, and (c) $y = x^3$

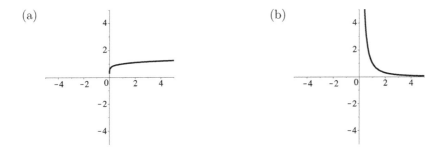

Figure 2: Graphs of power functions (a) $y = x^{0.135}$ and (b) $y = x^{-1.742}$

While these functions may appear to be quite different, they are all of the same basic form $y = Cx^k$. This commonality enables us to follow the same standard process when modeling a data set with a power function, regardless of the value of the exponent k.

Power functions arise in many contexts, some of which you already know well. For example, the area of a circle with radius r is often found using the function Area$(r) = \pi r^2$, which is a power function with coefficient $C = \pi$ and exponent $k = 2$. Similarly, the area of a square with side length s is found using the power function Area$(s) = s^2$ with $C = 1$ and $k = 2$.

Beyond such geometric settings, power functions are implicitly present in many aspects of our daily lives. As an example, suppose you are filling up the gas tank in a car and a gallon of 87 octane unleaded gasoline costs \$2.96. The price of a fill-up is a function of the price per gallon times the number of gallons pumped into the gas tank and is modeled by the power function Price(gallon) $= 2.96 \cdot$ gallon. In this power function $y = Cx^k$, the coefficient is $C = 2.96$, the variable is $x =$ gallons, and the exponent is $k = 1$.

Another, perhaps less familiar example of a power function is Zipf's Law, which states that the probability of encountering the xth most common word in the English language is approximately $y = 0.1x^{-1}$. This power function has coefficient $C = 0.1$ and exponent $k = -1$. As it turns out, Zipf's Law is valid for almost all of the top one-thousand most commonly used English words.

An alternative to this global model is Goetz's Law, which provides the probability of encountering the xth most common word in the English language based on how many words R are read or spoken in some given interval of time. Goetz's Law asserts that this probability is given by the power function $y = 1/(x \ln(1.78R)) = 1/\ln(1.78R)x^{-1}$ with coefficient $C = 1/\ln(1.78R)$ and exponent $k = -1$.

As suggested by Figures 1 and 2, power functions behave in quite different ways. The exponent k plays the dominant role in producing these diverse types of behavior. The following example illustrates such differences among power functions based on the exponents k:

◆ **EXAMPLE 1** In 1957, Stanley Smith Stevens published a body of psychophysical data showing that the power function $\psi(I) = CI^k$ models the magnitude ψ of a subjective sensation as a function of a physical stimulus of magnitude I. This power function is now known as Steven's Power Law, and its exponent k varies depending on the physical stimulus and subjective sensation.

(a) When the physical stimulus is the sound pressure of a 3000 Hz tone and the sensation is loudness, the exponent in Steven's Power Law is $k = 0.67$. If the magnitude of the sound pressure is doubled, how much greater is the sensation of loudness?

(b) When the physical stimulus is an electric current through a person's finger and the sensation is electric shock, the exponent in Steven's Power Law is $k = 3.5$. If the magnitude of the current is doubled, how much greater is the sensation of electric shock?

Solution.

(a) When a sound pressure has magnitude P, the magnitude of the sensation of loudness is given by $\psi(P) = CP^{0.67}$. If the magnitude of the sound pressure is doubled to $2P$, then the magnitude of the sensation of loudness is found as follows:

$$\psi(2P) = C(2P)^{0.67} = C2^{0.67}P^{0.67} = 2^{0.67}CP^{0.67} \approx 1.59CP^{0.67}$$

Thus, when the magnitude of the sound pressure is doubled, the magnitude of the sensation of loudness increases by a factor of 1.59.

(b) When an electric current through a person's finger has magnitude E, the magnitude of the sensation of electric shock is given by $\psi(E) = CE^{3.5}$. If the magnitude of the current is doubled to $2E$, then the magnitude of the shock sensation is found as follows:

$$\psi(2E) = C(2E)^{3.5} = C2^{3.5}E^{3.5} = 2^{3.5}CE^{3.5} \approx 11.31CE^{3.5}$$

This result means that when the magnitude of the current is doubled, the magnitude of the sensation of electric shock increases by a factor of 11.31. ◼

Both parts of Example 1 addressed the exact same question: If the input value is doubled, what is the change in the magnitude of the output? In part (a) with exponent $k = 0.67$, the output increases by a factor of 1.59. However, in part (b) with exponent $k = 3.5$, the output increases by a factor of 11.31. As illustrated by this example, even when two power functions have the same form, their different exponents can result in quite different behaviors.

Graphically Identifying Power Functional Data Sets

The study of exponential functions in Section 2.2 discussed the difficulty of distinguishing among different curved graphs. Recall that the data plot in Figure 3(a) corresponds to $y = e^x$, while the data plot in Figure 3(b) corresponds to $y = x^3$.   This difficulty was

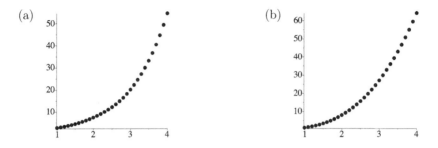

Figure 3: Graphs of (a) $y = e^x$ and (b) $y = x^3$ with similar shapes

overcome by considering the logarithms of the outputs as a function of the given inputs. For an exponential data set, the semi-log data is exactly or approximately linear, which can be identified graphically or numerically by computing slopes between successive semi-log data points. This section extends this approach to the case of power functions and, once again, logarithms play a central role. Working in this direction, consider the following definitions:

Definition. If (a, b) is a point from a given data set or on a curve, its **semi-log plot** includes the point $(a, \ln(b))$, and its **log–log plot** includes the point $(\ln(a), \ln(b))$.

When deciding whether a power function model is reasonable for a given data set, the key step is to plot the logarithm of the outputs as a function of the logarithm of the inputs. For reasons discussed shortly, such plots are always linear for power functions. Such a plot can be made in two different ways. The first option is to actually compute the logarithms $\ln(x)$ of the inputs x and the logarithms $\ln(y)$ of the outputs y, and then graph the corresponding log–log plot of the ordered pairs $(\ln(x), \ln(y))$ on a standard xy-axis. The second option is to create a **logarithmic scale plot**, which graphs the function on a plane using logarithmic scale on both the horizontal and vertical axes; this second approach is discussed in more detail later in this section.

Regardless of whether a log–log plot or a logarithmic scale plot is being created, such a graph for a power function or a data set modeled by a power function is exactly or approximately linear. Furthermore, the slope of this line is equal to the exponent k of the corresponding power function. This approach is applied in the next example.

◆ **EXAMPLE 2** The revolutions per minute (or RPM) of an engine varies according to the size of the engine. The table below presents the RPMs of five different engines as a function of the mass of each engine in kilograms. Graph the log–log plot of the data set and explain why a power function model of the given data is reasonable or not.

Engine mass	0.135	0.67	2.45	75	1775
RPM	22,000	13,000	8000	2550	900

Solution. First, transform the data to obtain a new table presenting the logarithms of both the input values and the output values.

ln(Engine mass)	−2.00	−0.40	0.90	4.32	7.48
ln(RPM)	10.00	9.47	8.99	7.84	6.80

Now, graph the log–log plot of the given data as shown in Figure 4, which presents the logarithms of the outputs as a function of the logarithms of the inputs. The linear shape of this log–log plot indicates that a power function is reasonable for the given data set.

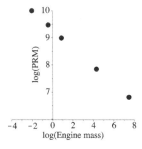

Figure 4: log–log plot of data for Example 2

This approach to identifying when a power function provides a reasonable model can be incorporated into the overall process for graphically identifying reasonable models:

GRAPHICALLY IDENTIFYING REASONABLE MODELS.

(1) If the standard plot (x, y) of a given data set is linear, then a linear model $y = mx + b$ is reasonable.

(2) If the semi-log plot $(x, \ln(y))$ of a given data set is nonconstant linear, then an exponential model $y = Ce^{k(x-h)}$ is reasonable.

(3) If the log–log plot $(\ln(x), \ln(y))$ of a given data set is nonconstant linear, then a power function model $y = Cx^k$ is reasonable.

➤ **QUESTION 1** The table below presents the maximum observed flying speed of five different species of animals in centimeters per second as a function of their length in centimeters. Graph the log–log plot of the data set and explain why a power function model of the given data is reasonable or not.

Length	1.3	8.1	11	34	160
Speed	660	1120	1200	1560	2280

Two primary reasons motivate this graphical approach to deciding whether a power function model is reasonable based on the linearity of the log–log plot (and logarithmic scale plot) of the given data set. First, lines can be quickly and easily recognized. Second, applying the natural logarithm to a power function $y = Cx^k$ results in a linear expression. The following calculations demonstrate this connection between power functions and linear functions by means of the natural logarithm:

$$y = Cx^k \qquad\qquad\qquad \text{Power function model}$$
$$\ln(y) = \ln(Cx^k) \qquad\qquad \text{Take } \ln(x) \text{ of both sides}$$
$$\ln(y) = \ln(C) + \ln(x^k) \qquad \ln(ab) = \ln(a) + \ln(b)$$
$$\ln(y) = \ln(C) + k\ln(x) \qquad \ln(a^r) = r\ln(a)$$
$$\ln(y) = k \cdot \ln(x) + \ln(C) \qquad \text{Rearrange terms}$$

Based on this analysis, a power function relationship $y = Cx^k$ ensures that $\ln(y)$ is a linear function of $\ln(x)$. Even more, the log–log plot has slope $m = k$ and vertical intercept $b = \ln(C)$ and so provides the parameters of the sought for power function model $y = Cx^k$ of the given data set.

Finally, note that this approach cannot be applied when either the input or output values include negative numbers, because the domain of the natural logarithm $\ln(x)$ is the set of positive real numbers. For exponential functions, one version of this issue could be dealt with by taking the absolute value of the outputs because of the particular properties of exponentials. Power functions are not so uniformly behaved and such a simple approach will not apply. As it turns out, the methods of this section can be adapted for data with negative values, but such adaptations are beyond the scope of this book.

Numerically Identifying Power Functional Data Sets

As with other common modeling functions, a graphical assessment of the reasonableness of a power function model for a given data set is complemented with a parallel numerical assessment.

NUMERICALLY IDENTIFYING REASONABLE MODELS.

(1) Check for a linear model $y = mx + b$ by comparing the slopes of the lines through pairs of data points. Linear data sets have exactly or approximately the same slopes for any two data points. Most often slopes of lines between successive data points are compared.

(2) Check for an exponential model $y = Ce^{k(x-h)}$ by determining if the corresponding semi-log data is nonconstant linear by applying (1) to the transformed data $(x, \ln(y))$. Exponential data sets have nonconstant linear semi-log data. If the semi-log data has a linear model $y = m(x - h) + b$ with $m \neq 0$, then the exponential model has parameters $k = m$ and $C = e^b$.

**NUMERICALLY IDENTIFYING REASONABLE MODELS.
(CONTINUED)**

(3) Check for a power function model $y = Cx^k$ by determining if the corresponding log–log data is nonconstant linear by applying (1) to the transformed data $(\ln(x), \ln(y))$. Power function data sets have nonconstant linear log–log data. If the log–log data has a linear model $y = mx + b$ with $m \neq 0$, then the power function has parameters $k = m$ and $C = e^b$.

◆ **EXAMPLE 3** The table below from Example 2 presents the RPMs of five different engines as a function of each engine's mass in kilograms. Determine if the corresponding log–log data is nonconstant linear and explain why a power function model of the given data is reasonable or not.

Engine mass	0.135	0.67	2.45	75	1775
RPM	22,000	13,000	8000	2550	900

Solution. Example 2 computed the following table with the corresponding log–log data:

ln(Engine mass)	−2.00	−0.40	0.90	4.32	7.48
ln(RPM)	10.00	9.47	8.99	7.84	6.80

Now, calculate the slopes between successive points in this transformed data set:

Interval	−2.00 to −0.40	−0.40 to 0.90	0.90 to 4.32	4.32 to 7.48
$\Delta \ln(\text{Mass})$	1.60	1.30	3.42	3.16
$\Delta \ln(\text{RPM})$	−0.53	−0.48	−1.15	−1.04
Slope	$m_1 = -0.33$	$m_2 = -0.37$	$m_3 = -0.34$	$m_4 = -0.33$

These four slopes are within 0.04 of each other, which can be considered approximately equal in the context of the values of ln(RPM), which vary between 6.80 and 10.00. Therefore, the log–log data is nonconstant linear and a power function model is reasonable for the given data set. ■

▶ **QUESTION 2** The table below from Question 1 presents the maximum observed flying speed of five different species of animals in centimeters per second as a function of their length in centimeters. Determine if the corresponding log–log data is nonconstant linear and explain why a power function model of the given data is reasonable or not.

Length	1.3	8.1	11	34	160
Speed	660	1120	1200	1560	2280

Conjecturing Power Function Models

After determining that a power function is reasonable for a given data set, the next step is to conjecture the parameters of such a model. The conjecturing process for power functions is essentially the same as that for exponential functions.

CONJECTURING POWER FUNCTION MODEL PARAMETERS.
If a power function $y = Cx^k$ is reasonable for a data set, conjecture k and C as
follows:

- For exponent k, compute the sequence of slopes through successive data points
  $m_1, m_2, \ldots, m_{n-1}$ for the log–log data and then average these slopes:

$$k \; = \; m \; = \; \frac{m_1 + m_2 + m_3 + \cdots + m_{n-1}}{n-1}$$

- For coefficient C, substitute each data point from the original data set into
  the equation $y = Cx^k$ (using the conjectured value of k), solve for C_1, C_2,
  $\ldots$, C_n, and then average these values:

$$C \; = \; \frac{C_1 + C_2 + C_3 + \cdots + C_n}{n}$$

◆ **EXAMPLE 4** The table below from Examples 2 and 3 presents the RPMs of five
different engines as a function of each engine's mass given in kilograms. Conjecture a
power function model $y = Cx^k$ for this data set and graph the model on a plot of the
data.

Engine mass	0.135	0.67	2.45	75	1775
RPM	22,000	13,000	8000	2550	900

Solution. Examples 2 and 3 determined that the model RPM $= C \cdot (\text{Mass})^k$ is reasonable for this data set. The value of the exponent parameter k is conjectured by averaging
the four slopes from the log–log data set in Example 3:

$$k \; = \; \frac{(-0.33) + (-0.37) + (-0.34) + (-0.33)}{4} \; = \; -0.34$$

Working with the intermediate model of RPM $= C \cdot (\text{Mass})^{-0.34}$, the coefficient parameter C is conjectured by substituting each data point into this intermediate model and
solving for the corresponding C, and then averaging the results. For example, substituting the data point $(\text{RPM}, \text{Mass}) = (0.135, 22{,}000)$ into the intermediate model gives
$C_1 = 11{,}136.2$ as follows:

$$\text{RPM} = C_1 \cdot (\text{Mass})^{-0.34} \qquad \text{\small The intermediate model}$$
$$22{,}000 = C_1 \cdot (0.135)^{-0.34} \qquad \text{\small Substitute (RPM, Mass)} = (0.135, 22{,}000)$$
$$\frac{22{,}000}{0.135^{-0.34}} = C_1 \qquad \text{\small Divide by } 0.135^{-0.34}$$
$$11{,}136.2 = C_1 \qquad \text{\small Simplify}$$

Similar calculations with the remaining four data points yields $C_2 = 11{,}345.1$, $C_3 =
10{,}849.4$, $C_4 = 11{,}067.8$, and $C_5 = 11{,}454.3$, which average to $C = 11{,}170.6$. Substi-
tuting into the intermediate model RPM $= C \cdot (\text{Mass})^{-0.34}$ provides the following final
conjectured model:

$$\text{RPM} \; = \; 11{,}170.6 \cdot (\text{Mass})^{-0.34}$$

As part of the modeling cycle, the conjectured model is compared against the given data. Figure 5 presents the graph of this power function model on a plot of the data and indicates how well this model matches the data.

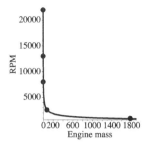

Figure 5: Plot of data set and model for Example 4

➤ **QUESTION 3** The table below from Questions 1 and 2 presents the maximum observed flying speed of five different species of animals in centimeters per second as a function of their length in centimeters. Conjecture a power function model $y = Cx^k$ for this data set and graph the model on a plot of the data.

Length	1.3	8.1	11	34	160
Speed	660	1120	1200	1560	2280

Best Possible Power Function Models

The techniques developed in this section enable us to conjecture a reasonable power function model for a given data set. However, these methods are not guaranteed to produce the best possible power function model. Rather, the only claim that can be made is that a reasonable model has been identified and that a visual check does or does not affirm its apparent validity. As with linear and exponential models, the RStudio command `fitModel` implements a version of the method of least squares and provides a way to find the "best possible" power function model for a given data set. We discuss the modifications needed to apply the `fitModel` command in the context of power functions and data sets, and then consider some specific examples.

Working in RStudio

RStudio can be used to create log–log plots to determine whether a power function is a reasonable model for a data set. The reasonableness of a power function is determined by plotting the logarithms of the input and output values, and then deciding if the resulting graph appears exactly or approximately nonconstant linear. In RStudio, `plotPoints` is used to graph the ordered pairs of the logarithm-modified input and output values. If the input values are stored in `var1` and the output values are stored in `var2`, then the corresponding log–log plot of the inputs and outputs is plotted by entering the command `plotPoints(log(var2)~log(var1))`.

When it is known that a power function model $y = Cx^k$ is reasonable for the given data set, the parameters C and k for the best possible power function model are determined by the `fitModel` command. The input and output values to `fitModel` are modified in a fashion paralleling that for `plotPoints`. Namely, the best linear model for the logarithm of the input and output values is obtained with the command `bestModel=fitModel(log(var2)~m*log(var1)+b)`. The slope m of this linear model is the exponent k for the corresponding power function and C is found by taking the exponential of the vertical intercept b of this linear model.

Examples of Commands

- `plotPoints(log(var2)~log(var1))`

- `bestModel=fitModel(log(var2)~m*log(var1)+b)`

- `coef(bestModel)`

◆ **EXAMPLE 5** The following table from Examples 2–4 presents the RPMs of five different engines as a function of each engine's mass in kilograms:

Engine mass	0.135	0.67	2.45	75	1775
RPM	22,000	13,000	8000	2550	900

(a) Verify that a power function model is reasonable using a log–log plot of the data.

(b) Use `fitModel` to find the best power function model for this data set and graph the model on a plot of the data.

Solution.

(a) First, store the data set in RStudio and then graph its log–log plot. The approximately nonconstant linear shape of the resulting log–log plot indicates that a power function model is reasonable.

```
Mass = c(0.135, 0.67, 2.45, 75, 1775)
RPM = c(22000,13000,8000,2550,900)
plotPoints(log(RPM)~log(Mass))
```

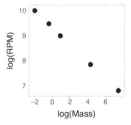

(b) Working toward finding the best power function model for the given data set, use `fitModel` to identify a linear model $\ln(\text{RPM}) = m \cdot \ln(\text{Mass}) + b$ for the corresponding log–log data:

```
bestModel = fitModel(log(RPM)~m*log(Mass)+b)
coef(bestModel)
```

```
        m           b
-0.337886   9.316505
```

From this RStudio output, the linear model for the log–log data has slope $m = -0.337886$ and vertical intercept $b = 9.316505$. As a result, the parameters of the power function model for the given data are $k = m = -0.337886$ and $C = e^b = e^{9.316505}$, which gives the following best possible power function model for the RPM of an engine as a function of the engine's mass:

$$\text{RPM} = e^{9.316505} \cdot \text{Mass}^{-0.337886} \approx 11{,}120 \cdot \text{Mass}^{-0.337886}$$

The following graph of this model on a plot of the data indicates that the model appears to matches up quite well with the data:

```
plotPoints(RPM~Mass)
plotFun(exp(9.316505)*Mass^(-0.337886)~Mass,add=TRUE)
```

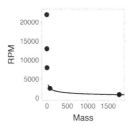

► **QUESTION 4** The following table from Questions 1–3 presents the maximum observed flying speed of five different species of animals in centimeters per second as a function of their length in centimeters:

Length	1.3	8.1	8.5	11	41
Speed	660	1120	1000	1200	2320

(a) Verify that a power function model is reasonable using a log–log plot of the data.

(b) Use `fitModel` to find the best power function model for this data set and graph the model on a plot of the data.

One major advantage of using RStudio for this type of analysis is the relative ease of working with large data sets. The `fitModel` approach to determining the best power function model for a given data set is identical for both small and large sets of data. Example 6 considers a larger version of the data set on RPM and engine size and illustrates the equivalent work needed to find the best power function model using the `fitModel` command, independent of the size of the data set.

◆ **EXAMPLE 6** The data set stored in `EngineRPM` contains data on the revolutions per minute and size (in kg) of different engines.

(a) Verify that a power function model is reasonable using a log–log plot of the data.

(b) Use `fitModel` to find the best power function model for this data set and graph the model on a plot of the data.

Solution.

(a) First, use `names` to determine the variables in `EngineRPM`:

```
names(EngineRPM)
```

```
[1] "Mass" "RPM"
```

Now, using the variables names `Mass` and `RPM`, create a log–log plot of the data set to verify a power function model is reasonable. While not perfectly linear, this log–log plot is generally linear in shape and indicates that a power function is reasonable. At the same time, the power function model will have limited accuracy because the log–log plot is only roughly linear.

```
plotPoints(log(RPM)~log(Mass),
           data=EngineRPM)
```

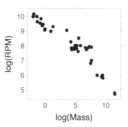

(b) Working toward finding the best power function model for the given data set, use `fitModel` to identify a linear model $\ln(\text{RPM}) = m \cdot \ln(\text{Mass}) + b$ for the corresponding log–log data:

```
bestModel=fitModel(log(RPM)~m*log(Mass)+b,data=EngineRPM)
coef(bestModel)
```

```
        m         b
-0.339622  9.513810
```

From this RStudio output, the linear model for the log–log data has slope $m = -0.339622$ and vertical intercept $b = 9.513810$. As a result, the parameters of the power function model for the given data are $k = m = -0.339622$ and $C = e^b = e^{9.513810}$, which gives the following best possible power function model for the RPM of an engine as a function of the engine's mass:

$$\text{RPM} = e^{9.513810} \cdot \text{Mass}^{-0.339622} \approx 13{,}545.5 \cdot \text{Mass}^{-0.339622}$$

The graph of this model on a plot of the data indicates its goodness of fit. As noted in part (a), the model does not provide an exact match to the data because the log–log plot of the data is only roughly linear. However, the power function model does appear pretty reasonable for the given data set.

```
plotPoints(RPM~Mass,data=EngineRPM)
plotFun(exp(9.513810)*Mass^(-0.339622)~Mass,add=TRUE)
```

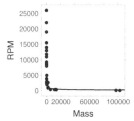

Logarithmic Scale Plots

As promised, this section concludes by considering another way to graphically display the logarithms of outputs as a function of the logarithms of inputs, and so determine if a power function model is reasonable for a data set with a curved plot. Instead of creating a log–log plot, the given data or function can be graphed using a *logarithmic scale* on both axes. This book almost always uses a log–log plot to determine whether a power function model is reasonable for a given data set. However, logarithmic scale plots are also commonly used and appear widely in the literature of many disciplines, so learning how to read and work with them can prove helpful.

> **Definition.** A **logarithmic scale plot** has a scale on both its axes where the spacing between numbers changes logarithmically, which means that integer powers of the base of the logarithm are spaced evenly on both axes.

For example, consider a logarithmic scale plot using a base 10 logarithm. In this setting, the spacing on the axes between 1 and 10 is the same as the spacing between 10 and 100, because both $\log_{10}(10) - \log_{10}(1) = 1 - 0 = 1$ and $\log_{10}(100) - \log_{10}(10) = \log_{10}(10^2) - \log_{10}(10) = 2 - 1 = 1$. Figure 6(a) presents the graph of the power function $y = x^{1/2}$ on standard scale axes and Figure 6(b) presents the corresponding logarithmic scale plot of $y = x^{1/2}$ using a base 10 logarithm.

On the logarithmic scale plot in Figure 6(b), notice that the distance between 1 and 10 on the x-axis is the same as the distance between 10 and 100. This same spacing would appear on the y-axis if the graph were extended further. Notice the logarithmic scale spacing among the even integers between 1 and 10 appearing on the y-axis in Figure 6(b).

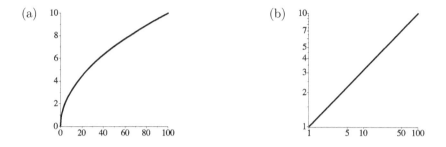

Figure 6: Standard plot and logarithmic scale plot of $y = x^{1/2}$

In fact, a distinguishing identifier of a logarithmic scale plot is the spacing between the numeric labels on the axes. If the spacing between numbers on the axes is not the uniform spacing usually presented on plots, then you are most likely working with a logarithmic scale plot. Often the axes of a logarithmic scale plot are labeled with powers of the base of the logarithm. For example, if a logarithmic scale plot uses a base 10 logarithm, then the marks on the axes would indicate powers of 10, such as 0.001, 0.01, 0.1, 1, 10, 100, 1000, and so on.

Beyond providing an example of a logarithmic scale plot, Figure 6(b) illustrates a key property: the logarithmic scale plot of $y = x^{1/2}$ is linear. In fact, every graph of a power function graph on any logarithmic scale plot is linear, independent of the base of the logarithm. In more detail, the base of the logarithm (e.g., 2, e, 10, etc.) does not impact the linearity of a logarithmic scale plot of a power function, nor do the values of C and k. Different choices for the base might be made depending on the context, but the result will always be a linear logarithmic scale plot for a power function. The most commonly used bases for logarithmic scale plots are base 10 and base e.

As with log–log plots, the parameters of the corresponding power function $y = Ce^{kx}$ can be determined using a logarithmic scale plot. The exponent parameter is equal to the slope of the line on the logarithmic scale plot and the coefficient parameter C is found by substituting a point (or points and averaging). More details are given below.

SLOPE OF A LINE ON A LOGARITHMIC SCALE PLOT.
If (x_1, y_1) and (x_2, y_2) are points on a linear logarithmic scale plot using a base b logarithm, then the slope k of the line through these points is computed as follows:

$$k = \frac{\log_b(y_2) - \log_b(y_1)}{\log_b(x_2) - \log_b(x_1)} = \frac{\log_b(y_2/y_1)}{\log_b(x_2/x_1)}$$

◆ **EXAMPLE 7**   For the logarithmic scale plots in Figure 7, explain why the curve corresponds to a power function $y = Cx^k$ or not. If so, conjecture its parameters.

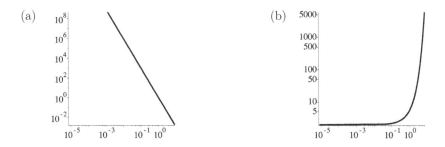

Figure 7: Logarithmic scale plots for Example 7

Solution.

(a) The logarithmic scale plot is linear, which means that the curve corresponds to a power function $y = Cx^k$. For this example, the slope of the line is calculated using the points $(10^{-4}, 10^{12})$ and $(10^0, 10^0)$; other choices would produce the same slope. The axes of this logarithmic scale plot are labeled with powers of 10, indicating that a base $b = 10$ logarithm should be used in the slope formula as follows:

$$ k = \frac{\log_{10}(10^0/10^{12})}{\log_{10}(10^0/10^{-4})} = \frac{\log_{10}(10^{-12})}{\log_{10}(10^4)} = \frac{-12\log_{10}(10)}{4\log_{10}(10)} = \frac{-12}{4} = -3 $$

Therefore, the corresponding power function is of the form $y = Cx^{-3}$. The coefficient C is determined by substituting a point on the line into this intermediate form of the power function. In particular, substituting the data point $(10^0, 10^0) = (1, 1)$ gives $1 = C1^{-3} = C$. Therefore, the sought-for corresponding power function is $y = x^{-3}$.

(b) The logarithmic scale plot is not linear, which means that the curve does not correspond to a power function. ■

➤ **QUESTION 5**   For each logarithmic scale plot in Figure 8, determine whether the function corresponds to a power function and, if so, state the corresponding power function.

(a) 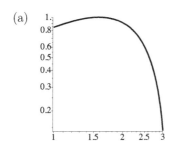   (b)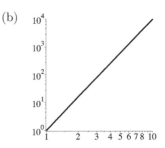

Figure 8: Logarithmic scale plots for Question 5

Summary

- A *power function* is of the form $y = Cx^k$, where C and k are nonzero constants.

- If (a, b) is a point from a data set or curve, its *log–log plot* includes the point $(\ln(a), \ln(b))$.

- *Graphically Identifying Power Function Models*: If the log–log plot $(\ln(x), \ln(y))$ of a given data set is nonconstant linear, then a power function model $y = Cx^k$ is reasonable.

- *Numerically Identifying Power Function Models*: Check for a power function model $y = Cx^k$ by determining if the corresponding log–log data is nonconstant linear by applying (1) to the transformed data $(\ln(x), \ln(y))$. Power function data sets have linear log–log data. If the log–log data has a linear model $y = mx + b$, then the power function has parameters $k = m$ and $C = e^b$.

- *Conjecturing Parameters for Power Function Models*: If a power function model $y = Cx^k$ is reasonable for a set of data, conjecture k and C as follows:

  - For exponent k, compute the sequence of slopes through successive data points m_1, m_2, ..., m_{n-1} for the log–log data and then average these slopes:
  $$k = m = \frac{m_1 + m_2 + m_3 + \cdots + m_{n-1}}{n-1}$$

  - For coefficient C, substitute each data point from the original data set into the equation $y = Cx^k$ (using the conjectured value of k), solve for $C_1, C_2, \ldots, C_n$, and then average these values:
  $$C = \frac{C_1 + C_2 + C_3 + \cdots + C_n}{n}$$

Summary (continued)

- A *logarithmic scale plot* has a scale on its axes where the spacing between numbers changes logarithmically, which means that integer powers of the base of the logarithm are spaced evenly on the axes.

- *Slope of a Line on a Logarithmic Scale Plot*: If (x_1, y_1) and (x_2, y_2) are points on a linear logarithmic scale plot using a base b logarithm, then the slope of the line is computed as follows:

$$k = \frac{\log_b(y_2) - \log_b(y_1)}{\log_b(x_2) - \log_b(x_1)} = \frac{\log_b(y_2/y_1)}{\log_b(x_2/x_1)}$$

Exercises

In Exercises 1 – 10, explain why the given function is a power function or not.

1. $y = x$

2. $y = x + 4$

3. $y = x^2 + 2x$

4. $y = x^{-1.327}$

5. $y = \dfrac{1}{x^2}$

6. $y = [\sin(x)]^2$

7. $y = 3\sqrt{x}$

8. $y = 7x^{\sqrt{2}}$

9. $y = 2^x$

10. $y = x^x$

In Exercises 11 – 18, identify which type of model is reasonable for the data set from among the known models: linear, exponential, power function, or none of these. More than one type of model may be reasonable.

12.

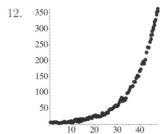

13.

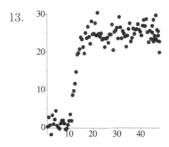

11.

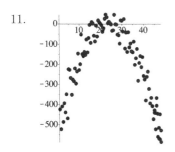

14.

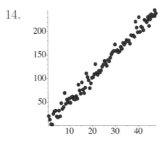

15.

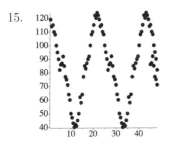

16.

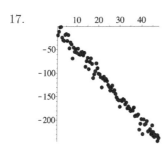

17.

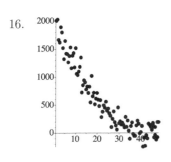

18.

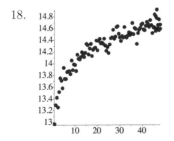

In Exercises 19 – 26, explain why the logarithmic scale plot corresponds to a power function $y = Cx^k$ or not. If so, estimate k and C.

19.

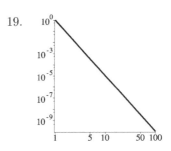

20.

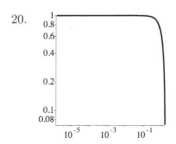

21.

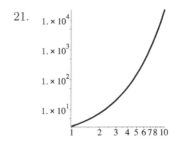

22.

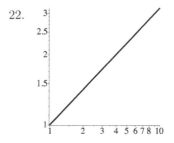

23.

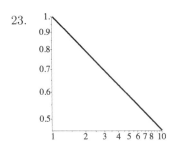

In Exercises 27–32, find the power function $y = Cx^k$ corresponding to the logarithmic scale plot.

27.

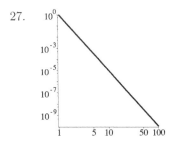

24.

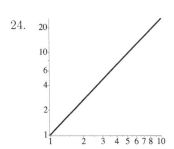

28.

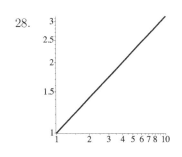

25.

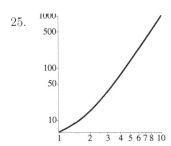

29.

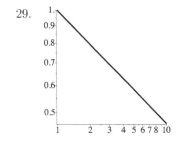

26.

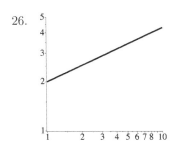

30.

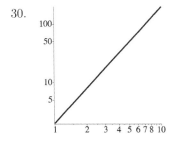

31.

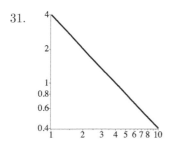

32.

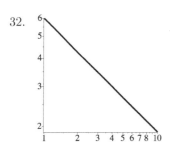

RStudio. In Exercises 33 – 38, use RStudio to graph the log–log plot of the data set and explain why a power function model $y = Cx^k$ is reasonable or not. If so, conjecture k and C.

33.

x	1.1	2.2	2.9	3.6	4.5
y	8.7	5.3	3.4	2.6	1.5

34.

x	1	2	3
y	1.000	0.126	0.037
x	4	5	
y	0.015	0.009	

35.

x	2	2.5	3.1	4.5	6.2
y	4.02	6.00	9.73	19.93	38.12

36.

x	1.1	2.2	2.9	3.6	4.5
y	0.5	5.2	8.4	10.8	14.3

37.

x	0.5	1	3	4	8
y	1.43	1.01	0.59	0.49	0.35

38.

x	1	3	5	9	12
y	3.6	11.8	20.3	36.5	47

RStudio. In Exercises 39 – 44, use RStudio to graph the log–log plot of the data set. Explain why a power function model is reasonable or not. If so, estimate both k and C.

39. S&P 500 stock market closing value (V) on June 1st of each year (Y) since 1975 in U.S. dollars

Y	15	25	35	45
V	192	545	1191	2077

40. Percent of children in 1990 with telephone service (P) by parental income percentile (I)

I	5	15	30	50
P	68.7	79.7	90.8	96.5

41. Total annual sales (S) in thousands of hybrid vehicles each year (Y) in the United States

Y	2010	2011	2012	2013
S	274	266	435	496

42. U.S. field production of crude oil (O) in billions of barrels in each year (Y)

Y	1970	1980	2000	2005
O	3.5	3.1	2.1	1.9

43. Average number of goals (G) per game in World Cup tournaments each year (Y) since 1994

Y	4	8	12	16
G	2.7	2.5	2.3	2.3

44. Average monthly maximum temperature (T) in degrees Fahrenheit each month (M) in New York City

M	2	4	7	10	12
T	39.4	61.1	88.3	67.8	43.4

In Exercises 45 – 50, numerically verify that a power function model $y = Cx^k$ is reasonable for the data set by showing the corresponding log–log data is nonconstant linear. Also, conjecture k.

45.

x	2	3	4	5	6
y	1.41	2.6	4	5.59	7.35

46.

x	2	3	5	7	8
y	1.2	2.7	7.5	14.7	19.2

47.

x	2	3	4	5	6
y	163	177	187	196	203

48.

x	1	3	6	10	14
y	8.3	29.1	64.3	115.4	169.4

49.

x	1	2	3	4
y	0.324	0.339	0.35	0.359

50.

x	10	20	25	35
y	0.215	0.271	0.292	0.327

In Exercises 51–56, for the data sets in Exercises 33–38, numerically determine if the corresponding log–log data is nonconstant linear and explain why a power function model $y = Cx^k$ is reasonable or not. If so, conjecture k and C.

51. Exercise 33   54. Exercise 36

52. Exercise 34   55. Exercise 37

53. Exercise 35   56. Exercise 38

In Exercises 57–60, conjecture the coefficient C for the data set and given intermediate power function model.

57. $y = Cx^{-1}$

x	-2	1	2	4
y	-1.21	2.6	1.3	0.63

58. $y = Cx^{0.35}$

x	1	3	7	9
y	1.4	2.1	2.8	3.1

59. $y = Cx^{2.1}$

x	4	5	6	7
y	-16.5	-26.4	-38.7	-53.5

60. $y = Cx^{-0.6}$

x	-6	-3	-1	3
y	-2.8	-4.2	-5.4	4.3

RStudio. In Exercises 61–66, for the data sets in Exercises 45–50, use RStudio to find the best possible power function model for each data set using the `fitModel` command.

61. Exercise 45   64. Exercise 48

62. Exercise 46   65. Exercise 49

63. Exercise 47   66. Exercise 50

RStudio. In Exercises 67–70, use RStudio to find a power function model of the swimming speed in centimeters per second as a function of the length in centimeters of various animals stored in `SwimmingSpeed`.

67. Verify that a power function model $y = Cx^k$ is reasonable using a log–log plot of the data. Also, conjecture its parameters k and C.

68. Use `fitModel` to find a power function model $y = Cx^k$ for the data set.

69. Evaluate the model from Exercise 68 at $x = 25$. Explain the answer's meaning.

70. Evaluate the model from Exercise 68 at $x = 150$. Explain the answer's meaning.

RStudio. In Exercises 71–74, use RStudio to find a power function model of the running speed in centimeters per second as a function of the length in centimeters of various animals stored in `RunningSpeed`.

71. Verify that a power function model $y = Cx^k$ is reasonable using a log–log plot of the data. Also, conjecture its parameters k and C.

72. Use `fitModel` to find a power function model $y = Cx^k$ for the data set.

73. Evaluate the model from Exercise 72 at $x = 10$. Explain the answer's meaning.

74. Evaluate the model from Exercise 72 at $x = 125$. Explain the answer's meaning.

RStudio. In Exercises 75 – 79, use RStudio to find a power function model of atmospheric carbon dioxide from Mauna Loa in ppmv (or parts per million by volume) as a function of years from 1958 to 2008 stored in `MaunaLoaCO2`.

75. Verify that a power function model $y = Cx^k$ is reasonable using a log–log plot of the data. Also, conjecture its parameters k and C.

76. Use `fitModel` to find a power function model $y = Cx^k$ for the data set.

77. Evaluate the model from Exercise 76 at $x = 1960$. Explain the answer's meaning.

78. Evaluate the model from Exercise 76 at $x = 2000$. Explain the answer's meaning.

79. Evaluate the model from Exercise 76 at $x = 2040$. Explain the answer's meaning.

RStudio. In Exercises 80 – 83, use RStudio to find power function model of the field metabolic rate of individual birds and mammals in kilojoules per day as a function of body mass in kilograms stored in `BodyMassMetabolicRate`.

80. Verify that a power function model $y = Cx^k$ is reasonable using a log–log plot of the data. Also, conjecture its parameters k and C.

81. Use `fitModel` to find a power function model $y = Cx^k$ for the data set.

82. Evaluate the model from Exercise 81 at $x = 10$. Explain the answer's meaning.

83. Evaluate the model from Exercise 81 at $x = 75$. Explain the answer's meaning.

In Your Own Words. In Exercises 84 – 90, explain the following.

84. Power function

85. Log–log plot

86. Graphically determining if a power function model is reasonable or not

87. Numerically determining if a power function model is reasonable or not

88. Conjecturing the exponent k of a power function model $y = Cx^k$

89. Conjecturing the coefficient C of a power function model $y = Cx^k$

90. Logarithmic scale plot

2.4   Modeling with Sine Functions

The previous sections introduced the three most common models for data sets: linear, exponential, and power function models. This section adds sine functions to the collection of possible models. These important functions help model aspects of reality that are periodic (or repeating) in nature and that cannot be described using any of the other models.

As an example of one phenomenon best described by a sine function, consider average temperatures in Danville, Kentucky. Figure 1 provides the average maximum temperature in degrees Fahrenheit each month beginning in January 2006. The inputs along the x-axis are the number of months since January 2006, which means, for example, that 24 identifies January 2008. The outputs along the y-axis are the average maximum temperatures in degrees Fahrenheit in each month.

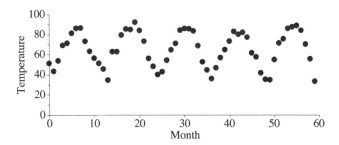

Figure 1: Average maximum temperature each month in Danville, Kentucky

Our year-to-year experience of temperature coincides with the oscillating, periodically repeating output values illustrated in Figure 1. None of linear, exponential, or power functions can accurately describe these temperatures alternating up and down over time. Fortunately, sine functions allow the description of such periodic behavior.

The Sine Function

The sine function provides the most useful trigonometric model. Recall from Section 1.7 that trigonometric functions are defined in terms of right triangles and unit circles, mapping inputs that are angles to unique numeric outputs. The two common units for measuring angles are degrees and radians. A complete traversal of any circle has a degree measure of $360°$, which in the case of a circle centered at the origin $(0,0)$ is read counterclockwise from the positive x-axis. A radian measures distance around the circumference of the unit circle, beginning from the positive x-axis. A complete traversal of the unit circle covers 2π radians. Beginning from the positive x-axis, the radian measure of an angle is positive when measured in a counterclockwise direction and is negative when measured in a clockwise direction.

The right triangle definitions of the trigonometric functions consist of ratios between the lengths of its three sides. The input to a trigonometric function is an angle measure, in degrees or radians, and the sides are referred to as hypotenuse, opposite, or adjacent based on their relative position to the input angle, as illustrated in Figure 2.

The two fundamental trigonometric functions are sine and cosine, which are defined in terms of the labeled right triangle in Figure 2 as follows:

$$\sin(x) = \frac{\text{opposite}}{\text{hypotenuse}} \qquad\qquad \cos(x) = \frac{\text{adjacent}}{\text{hypotenuse}}$$

The unit circle definitions of sine and cosine are expressed in terms of radian measures of angles, where the Greek letter θ ("theta") is used to identify angle measurements in order to avoid confusion between references to the x-axis and to the x-coordinates of points. The sine and cosine functions are defined as follows:

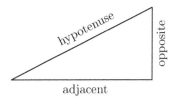

Figure 2: Right triangle for defining trigonometric functions

- $\sin(\theta)$ is the y-coordinate of the point θ radians around the circumference of the unit circle from the point $(1, 0)$ on the positive x-axis;

- $\cos(\theta)$ is the x-coordinate of the point θ radians around the circumference of the unit circle from the point $(1, 0)$ on the positive x-axis.

Alternatively, if (a, b) is the point θ radians around the circumference of the unit circle from the point $(1, 0)$ on the positive x-axis, then $\sin(\theta) = b$ and $\cos(\theta) = a$.

The graph of $\sin(x)$ is shown in Figure 3 and highlights two important features of the sine function. First, the period of $\sin(x)$ is 2π, which means that $\sin(x)$ completes an entire cycle in the interval $[0, 2\pi]$, returning to its initial output value and then repeating itself exactly. Second, the graph of $\sin(x)$ oscillates between a maximum output value of 1 and a minimum output value of -1. The amplitude of $\sin(x)$ is half the distance between these extreme values and has numerical value 1 because $A = \frac{1}{2}[1 - (-1)] = \frac{1}{2} \cdot 2 = 1$.

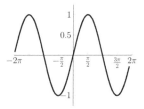

Figure 3: The graph of $\sin(x)$

Parameters of Sine Functions

The sine function is the most useful trigonometric function for modeling real-life phenomena. The main indicator of sinusoidal data is the distinctive shape of the graph of sine functions with their regular oscillations from one output to another and back. The basic shape of the $\sin(x)$ graph can be altered in four different ways: adjusting its amplitude and period, as well as making horizontal and vertical shifts. These four transformations of the $\sin(x)$ graph are expressed in the following parametrized version of a sine function, where A is the amplitude, P is the period, h is the horizontal shift, and v is the vertical shift:

$$y = A \sin\left[\frac{2\pi}{P}(x - h)\right] + v$$

While this form of the sine function may appear quite daunting, don't panic! We immediately explore the effect of each parameter on the graph of sine and then develop an expertise in conjecturing their numerical values.

The parameter A is called the **amplitude** of a sine function and is equal to half the distance between the maximum and minimum output values of the function; in other words, the amplitude is half the total height of the sine curve. Figure 4 presents $2\sin(x)$ with amplitude $A = 2$ in graph (a), $2\sin(x) + 2$ with $A = 2$ in graph (b), and $0.5\sin(x)$ with $A = 0.5$ in graph (c).

In Figure 4(a), the amplitude of the sine function $2\sin(x)$ is $A = 2$ because the curve oscillates between a maximum value of 2 and a minimum value of -2, and half the distance between these extreme values is $A = \frac{1}{2}[2 - (-2)] = \frac{1}{2} \cdot 4 = 2$. Similarly, in Figure 4(b), the amplitude is also $A = 2$ because $2\sin(x) + 2$ oscillates between 4 and 0. As illustrated by these two graphs, the amplitude of a sine function is independent of its vertical location on the plane. Finally, in Figure 4(c), the amplitude of $0.5\sin(x)$ is $A = 0.5$ because the curve oscillates between 0.5 and -0.5, which gives $A = \frac{1}{2}[0.5 - (-0.5)] = \frac{1}{2} \cdot 1 = 0.5$.

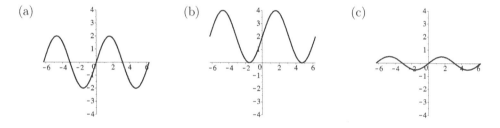

Figure 4: Effect of the amplitude parameter A on a sine function

The parameter P is called the **period** of a sine function and is equal to the length of the input interval needed for a sine function to complete an entire oscillation; in other words, the period is the length of the smallest input interval after which a sine curve repeats itself. The standard sine function $\sin(x)$ has period $P = 2\pi$, as illustrated in Figure 3 and as can be verified by substituting $P = 2\pi$ into the parametrized version of sine:

$$\sin\left[\frac{2\pi}{P} \cdot x\right] = \sin\left[\frac{2\pi}{2\pi} \cdot x\right] = \sin(1 \cdot x) = \sin(x)$$

Figure 5 presents $\sin(2\pi x)$ with period $P = 1$ in graph (a), $\sin(\pi x)$ with $P = 2$ in graph (b), and $\sin(0.5\pi x)$ with $P = 4$ in graph (c). You are encouraged to visually confirm the correspondence between the period and the input interval needed for one complete oscillation of sine for all three of the example graphs in Figure 5.

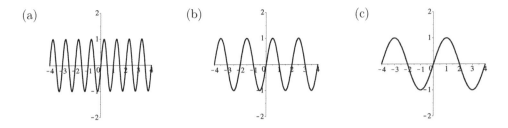

Figure 5: Effect of the period parameter P on a sine function

The **vertical shift** parameter v adjusts the graph of a sine function vertically up and down on the plane. The vertical shift is typically determined by finding the height of the output that is halfway between its maximum and minimum output values; in other words, the vertical shift is the height halfway between its extreme values. Figure 6 presents $\sin(x) + 2$ with vertical shift $v = 2$ in graph (a), $2\sin(x) + 2$ with $v = 2$ in graph (b), and $\sin(x) - 2$ with $v = -2$ in graph (c). Figures 6(a) and 6(b) show that the vertical shift v of a sine function is independent of its amplitude. Examining Figure 6(a) in more detail, the maximum output value of $\sin(x) + 2$ is $y = 3$ and its minimum output value is $y = 1$. The halfway point of $y = 2$ corresponds to the vertical shift $v = 2$ for $\sin(x) + 2$ in graph (a). You are encouraged to confirm that similar results hold for the other two graphs of $2\sin(x) + 2$ and $\sin(x) - 2$ in Figure 6.

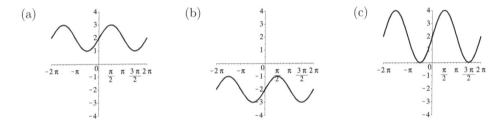

Figure 6: Effect of vertical shift parameter v on a sine function

The **horizontal shift** parameter h adjusts the graph of a sine function horizontally left or right on the plane. First, observe that when $x = 0$, the output of $\sin(x)$ is $y = 0$. Also, this output of $y = 0$ both lies halfway between its maximum output of $y = 1$ and its minimum output of $y = -1$ and occurs along an increasing part of the sine curve. The value of h identifies how far this vertical midpoint between the extreme values along an increasing part of the curve has shifted from the input $x = 0$. In order to keep things as simple as possible, always think of the horizontal shift as a shift to the right. Figure 7 presents sine functions $\sin[0.5\pi(x - h)]$ with period $P = 4$ and three different horizontal shifts: $\sin[0.5\pi(x - 1)]$ with horizontal shift $h = 1$ in graph (a), $\sin[0.5\pi(x - 2)]$ with $h = 2$ in graph (b), and $\sin[0.5\pi(x - 3)]$ with $h = 3$ in graph (c). In each case, h is measured by moving right from an input of $x = 0$ to the first occurrence of the vertical midpoint along an increasing part of the sine curve. You are encouraged to visually confirm all three of these horizontal shifts in Figure 7.

This analysis of the effects of the parameters of a sine function on the graphs of such functions can be summarized as follows. Figure 8 provides a graphical summary of the effects of the four parameters on the graph of a sine function.

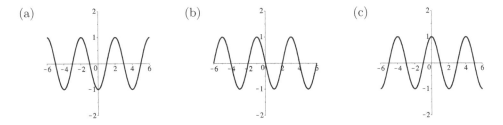

Figure 7: Effect of horizontal shift parameter h on a sine function

---

EFFECTS OF THE PARAMETERS OF A SINE FUNCTION.

A sine function is of the form $y = A \sin \left[\dfrac{2\pi}{P}(x - h) \right] + v$, where

- *amplitude* A is equal to half the distance between the maximum and the minimum output values of the sine function;

- *period* P is equal to the length of the minimum input interval needed for the sine function to complete an entire oscillation;

- *horizontal shift* h is equal to the distance from $x = 0$ to the first input $x = h$ to the right providing the vertical midpoint along an increasing part of the sine function; and

- *vertical shift* v is the height of the output that is halfway between the maximum and the minimum output values of the sine function.

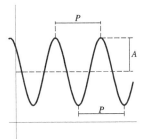
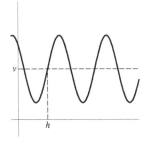

Figure 8: Parameters for sine functions

Conjecturing Sine Models

Working toward solidifying an understanding of the effect of the parameters of a sine function, the next example and question conjecture their values for specific graphs. After considering complete graphs of sine functions, we conjecture parameters for sine models of data sets of periodic phenomenon. As illustrated in Example 1, the values of parameters are typically determined in the following order: amplitude A, vertical shift v, horizontal shift h, and period P. This ordering reflects the relative ease of finding the values of these parameters and their interdependence.

◆ **EXAMPLE 1**   Conjecture the parameters A, P, h, and v of each sine function $y = A \sin\left[\dfrac{2\pi}{P}(x - h)\right] + v$ in Figure 9.

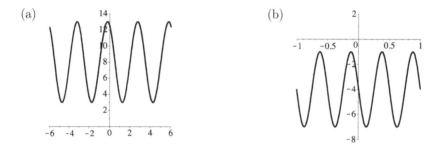

Figure 9: Sine functions for Example 1

Solution.

(a) The maximum output value is $y = 13$ and the minimum output is $y = 3$. The distance between them is $13 - 3 = 10$, and dividing this distance in half provides its amplitude of $A = 5$. The midpoint between these extreme values is 8, which corresponds to its vertical shift of $v = 8$. The horizontal shift is determined by the first occurrence when $x = 2$ of the midpoint $y = 8$ on an increasing part of the sine curve to the right of the y-axis, which gives $h = 2$. Finally, from a "beginning" when $x = 2$, the given sine curve completes an entire oscillation when $x = 5$. Therefore, its period is $P = 5 - 2 = 3$. In summary, graph (a) is the plot of the following sine function:

$$y \;=\; A \sin\left[\frac{2\pi}{P}(x - h)\right] + v \;=\; 5 \sin\left[\frac{2\pi}{3}(x - 2)\right] + 8$$

(b) The maximum output value is $y = -1$ and the minimum output is $y = -7$, which gives an amplitude of $A = \frac{1}{2}[-1 - (-7)] = \frac{1}{2} \cdot 6 = 3$. The midpoint between these extreme values is -4, which corresponds to a vertical shift of $v = -4$. The horizontal shift is determined by the first occurrence when $x = 0.25$ of the midpoint $y = -4$ on an increasing part of the sine curve to the right of the y-axis, which gives $h = 0.25$. Finally, from a "beginning" when $x = 0.25$, the given sine curve completes an entire oscillation when $x = 0.75$, which means that its period is $P = 0.75 - 0.25 = 0.5$. In summary, graph (b) is the plot of the following sine function:

$$y = A \sin\left[\frac{2\pi}{P}(x - h)\right] + v = 3 \sin\left[\frac{2\pi}{0.5}(x - 0.25)\right] - 4$$

∎

➤ **QUESTION 1** Conjecture the parameters A, P, h, and v of each sine function $y = A \sin\left[\frac{2\pi}{P}(x - h)\right] + v$ Figure 10.

(a) (b)

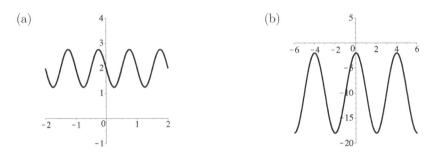

Figure 10: Sine functions for Question 1

This facility with identifying the parameters of graphs of continuous sine functions extends to sinusoidal data sets. Values for the parameters A, P, h, and v are conjectured in the same fashion, although these informed guesses will usually be less accurate. In particular, less information is available in a data set as opposed to an entire graph, which influences our ability to accurately determine the parameters. Also, data exhibiting sinusoidal behavior is often not exactly sinusoidal, because actual phenomenon and data measurements tend to be somewhat noisy. Even so, reasonable conjectures can be made to obtain good models of data sets.

◆ **EXAMPLE 2** The data set stored in `TemperaturesDanville` contains the average maximum temperature in Danville, Kentucky at the beginning of each month since January 2006, which was presented in Figure 1.

(a) Verify that a sine model is reasonable using a plot of the data.

(b) Conjecture a sine model $y = A \sin\left[\frac{2\pi}{P}(x - h)\right] + v$ and graph the model on a plot of the data.

Solution.

(a) First, use **names** to determine the variables in `TemperaturesDanville`:

```
names(TemperaturesDanville)
```

```
[1] "Month"        "Temperature"
```

Then, using the variable names **Month** for the input and **Temperature** for the output, plot the data set to observe the regular oscillations between two extreme outputs that are characteristic of a sinusoidal data set.

```
plotPoints(Temperature~Month,data=TemperaturesDanville)
```

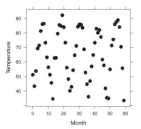

(b) Based on the plot of the data set, a reasonable maximum value is 88 and a reasonable minimum value is 32. These extreme values result in an amplitude of $A = \frac{1}{2}(88 - 32) = 28$ and a vertical shift of $v = 60$. Based on both the graph and our real-life experiences with temperatures, the period is $P = 12$ months. The first time the vertical midpoint of the sine curve occurs on an increasing part of the graph to the right of input Month= 0 appears to be about Month= 3.5, which gives horizontal shift $h = 3.5$. Thus, one conjecture of the parameters for a sine model of this data set is: $A = 28$, $P = 12$, $h = 3.5$, and $v = 60$. In light of the partial information provided by the data set, other conjectures are possible and reasonable, although they will not differ radically from these values. Substituting into the parametrized sine function, provides the following conjectured sine model:

$$\text{Temperature} = 28 \sin\left[\frac{2\pi}{12}(\text{Month} - 3.5)\right] + 60$$

The graph of this model on a plot of the data indicates its relative goodness of fit. The data is not exactly sinusoidal, and so the model does not match exactly with every data point. However, the model does capture the overall trends of the data set and reasonably approximates many of the data points quite well, particularly in light of just following the rough approximation process for the parameters.

```
plotPoints(Temperature~Month,data=TemperaturesDanville)
plotFun(28*sin(2*pi/12*(Month-3.5))+60~Month,add=TRUE)
```

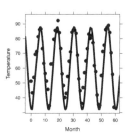

▶ **QUESTION 2** The data set stored in SunsetGreenwich contains the number of minutes after 3 p.m. until sunset at Greenwich, England since January 2010.

(a) Verify that a sine model is reasonable using a plot of the data.

(b) Conjecture a sine model and graph the model on a plot of the data.

Best Possible Sine Models

We have learned how to conjecture parameters for a sine model based on a plot of a given data set. However, this approach is not guaranteed to produce the best possible sine model. Rather, we can only say that a reasonable model has been identified based on a visual check of its validity. The RStudio command `fitModel` provides a way to find the best possible sine model for a given data set.

Working in RStudio

The syntax for the `fitModel` command remains the same for sine models as for the other models. However, additional arguments must be included with `fitModel` to provide an initial guess for the parameters of the model. The argument `start=` `list(,)` is added to `fitModel`, where the conjectured values of the parameters from the plot of the data set are included inside the parentheses separated by commas. For example, the argument `start=list(A=4,P=3,h=2,v=4)` might be added to `fitModel`, as illustrated below.

Examples of Commands

- `bestSineModel=fitModel(y~A*sin(2*pi/P*(x-h))+v,data=sineData,`
        `start=list(A=4,P=3,h=2,v=4))`

- `coef(bestSineModel)`

◆ **EXAMPLE 3** The data set stored in `TemperaturesDanville` contains the average maximum temperature in Danville, Kentucky at the beginning of each month since January 2006. Use `fitModel` with the conjectured parameters from Example 2 to find the best sine model for this data set and graph the model on a plot of the data.

Solution. Example 2 determined that the variable names of `TemperaturesDanville` are `Month` for the input and `Temperature` for the output, and from a plot of the data identified parameters $A = 28$, $P = 12$, $h = 3.5$, and $v = 60$ to conjecture the following sine model:

$$\text{Temperature} = 28 \sin\left[\frac{2\pi}{12}(\text{Month} - 3.5)\right] + 60$$

These conjectured parameters are used for the **start** values in the `fitModel` command to find the best sine model for this data set:

```
bestTempsModel=fitModel(Temperature~A*sin((2*pi/P)*(Month-h))+v,
        data=TemperaturesDanville,start=list(A=28,P=12,h=3.5,v=60))
coef(bestTempsModel)

        A        P        h        v
23.28683 11.98210  3.16363 65.35828
```

From this RStudio output, the best sine model has amplitude $A = 23.28683$, period $P = 11.98210$, horizontal shift $h = 3.16363$, and vertical shift $v = 65.35828$ as follows:

$$\text{Temperature} = 23.28683 \sin\left[\frac{2\pi}{11.98210}(\text{Month} - 3.16363)\right] + 65.35828$$

As an indication of its relative goodness of fit, the graph of this best possible model is added to a plot of the data. Because the data set is only roughly sinusoidal, the model does not provide an exact match for every data point, but it does provide a relatively close match for many data points and certainly captures the overall trends in the data quite well.

```
plotPoints(Temperature~Month, data=TemperaturesDanville)
plotFun(bestTempsModel(Month)~Month,add=TRUE)
```

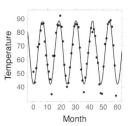

For further comparison, the graph of this best possible model is added to the plot of the data on the left and a graph of the conjectured model from Example 2 on the right.

```
plotPoints(Temperature~Month, data=TemperaturesDanville)
plotFun(28*sin((2*pi/12)*(Month-3.5))+60~Month,add=TRUE)
plotFun(bestTempsModel(Month)~Month,add=TRUE)
```

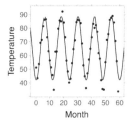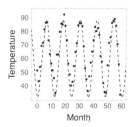

■

➤ **QUESTION 3** The data set stored in `SunsetGreenwich` contains the number of minutes after 3 p.m. until sunset at Greenwich, England since January 2010. Use `fitModel` with the conjectured parameters from Question 2 to find the best sine model for this data set and graph the model on a plot of the data.

Summary

- A sine function is of the form $y = A \sin\left[\dfrac{2\pi}{P}(x - h)\right] + v$, where

  - *amplitude* A is equal to half the distance between the maximum and the minimum output values of the sine function;

  - *period* P is equal to the length of the minimum input interval needed for the sine function to complete an entire oscillation;

  - *horizontal shift* h is equal to the distance from $x = 0$ to the first input $x = h$ to the right providing the vertical midpoint along an increasing part of the sine function; and

  - *vertical shift* v the height of the output that is halfway between the maximum and the minimum output values of the sine function.

- The following plots provide a graphical summary of the effects of the four parameters on the graph of a sine function:

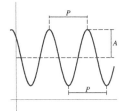 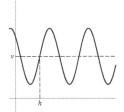

Exercises

In Exercises 1–8, identify which type of model is reasonable for the data set from among the known models: linear, exponential, power function, sine, or none of these. More than one type of model may be reasonable.

1.
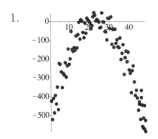

2.
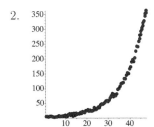

3.

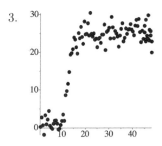

7.

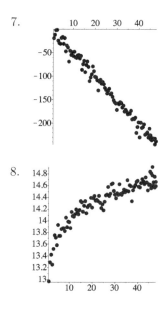

4.

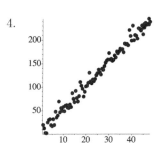

8.

In Exercises 9 – 16, conjecture the amplitude A and the vertical shift v for the sine function.

5.

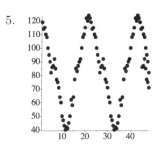

9.

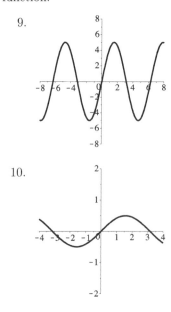

10.

6.

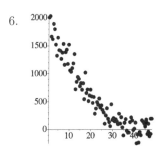

11.

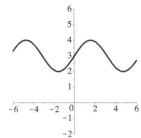

15.

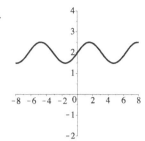

12.

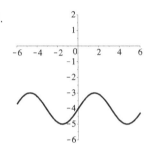

16.

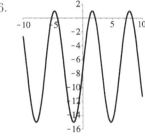

In Exercises 17 – 24, conjecture the period P and the horizontal shift h for the sine function.

13.

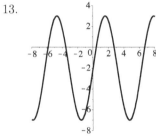

17.

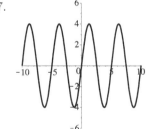

14.

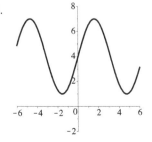

18.

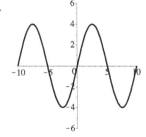

19.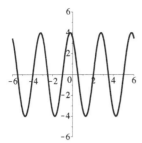

20.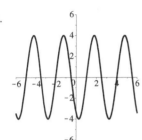

21.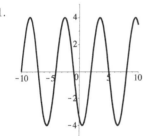

22.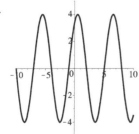

23.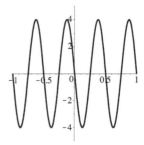

24.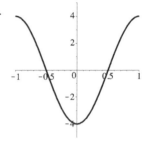

In Exercises 25–32, conjecture the parameters for the sine function $y = A \sin\left[\dfrac{2\pi}{P}(x - h)\right] + v.$

25.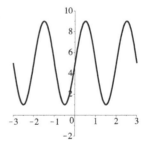

26.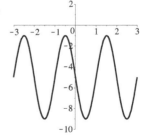

27.

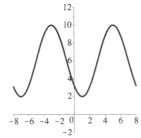

31.

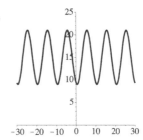

28.

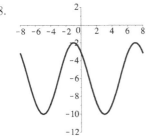

32.

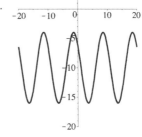

In Exercises 33–40, estimate the parameters for a sine model of the data set.

29.

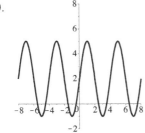

33.

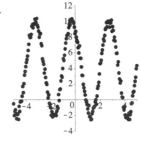

30.

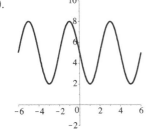

34.

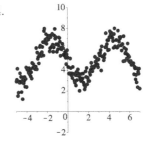

35.

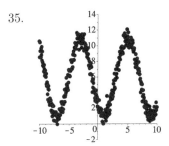

39.

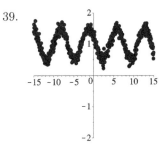

36.

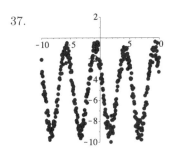

40.

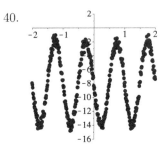

37.

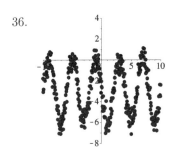

RStudio. In Exercises 41–45, use RStudio to find a sine model of the number of minutes after 4 a.m. until sunrise in Los Angeles, California, adjusted for Daylight Savings Time, from January 2010 (month 1) through December 2011 (month 24) stored in `SunriseLA`.

41. Verify that a sine model is reasonable by plotting the data. Also, conjecture its parameters A, P, h, and v.

42. Graph a sine model with the parameters from Exercise 41 on the data plot.

43. Use `fitModel` with `start` values from Exercise 41 to obtain a sine model for the data set.

44. Evaluate the models from Exercises 41 and 43 at $x = 15$. Explain the answers' meaning and discuss their accuracy.

38.

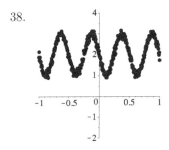

45. Evaluate the models from Exercises 41 and 43 at $x = 30$. Explain the answers' meaning and discuss their accuracy.

RStudio. In Exercises 46–50, use RStudio to find a sine model of the number of minutes after 4 p.m. until sunset in Los Angeles, California, adjusted for Daylight Savings time, from January 2010 (month 1) through December 2013 (month 48) stored in `SunsetLA`.

46. Verify that a sine model is reasonable by plotting the data. Also, conjecture its parameters A, P, h, and v.

47. Graph a sine model with the parameters from Exercise 46 on the data plot.

48. Use `fitModel` with `start` values from Exercise 46 to obtain a sine model for the data set.

49. Evaluate the models from Exercises 46 and 48 at $x = 30$. Explain the answers' meaning and discuss their accuracy.

50. Evaluate the models from Exercises 46 and 48 at $x = 60$. Explain the answers' meaning and discuss their accuracy.

RStudio. In Exercises 51–56, use RStudio to find a sine model of the altitude angle of the sun in Anchorage, Alaska, each hour from midnight on June 29, 2014 (hour 0) until midnight on June 30, 2014 (hour 24) stored in `SunPositionAlaska`.

51. Verify that a sine model is reasonable by plotting the data. Also, conjecture its parameters A, P, h, and v.

52. Graph a sine model with the parameters from Exercise 51 on the data plot.

53. Use `fitModel` with `start` values from Exercise 51 to obtain a sine model for the data set.

54. Evaluate the models from Exercises 51 and 53 at $x = 20.5$. Explain the answers' meaning and discuss their accuracy.

55. Evaluate the models from Exercises 51 and 53 at $x = 36$. Explain the answers' meaning and discuss their accuracy.

56. Evaluate the models from Exercises 51 and 53 at $x = 2160$. Explain the answers' meaning and discuss their accuracy.

RStudio. In Exercises 57–61, use RStudio to find a sine model of U.S. retail tax in millions of dollars each year from 2005 through 2011 stored in `USRetailTax`.

57. Verify that a sine model is reasonable by plotting the data. Also, conjecture its parameters A, P, h, and v.

58. Graph a sine model with the parameters from Exercise 57 on the data plot.

59. Use `fitModel` with `start` values from Exercise 57 to obtain a sine model for the data set.

60. Evaluate the models from Exercises 57 and 59 at $x = 2010$. Explain the answers' meaning and discuss their accuracy.

61. Evaluate the models from Exercises 57 and 59 at $x = 2020$. Explain the answers' meaning and discuss their accuracy.

RStudio. In Exercises 62–66, use RStudio to find a sine model of the electric bill of a single-family home in Minnesota from 2000 through 2003 stored in `ElectricBill`.

62. Verify that a sine model is reasonable by plotting the data. Also, conjecture the value of its parameters A, P, h, and v.

63. Graph a sine model with the parameters from Exercise 62 on the data plot.

64. Use `fitModel` with `start` values from Exercise 62 to obtain a sine model for the data set.

65. Evaluate the models from Exercises 62 and 64 at $x = 24$. Explain the answers' meaning and discuss their accuracy.

66. Evaluate the models from Exercises 62 and 64 at $x = 43$. Explain the answers' meaning and discuss their accuracy.

RStudio. In Exercises 67–71, use RStudio to find a sine model for tidal measurements at Pearl Harbor, Hawaii stored in `Hawaii`.

67. Verify that a sine model is reasonable by plotting the data. Also, conjecture its parameters A, P, h, and v.

68. Graph a sine model with the parameters from Exercise 67 on the data plot.

69. Use `fitModel` with `start` values from Exercise 67 to obtain a sine model for the data set.

70. Evaluate the models from Exercises 67 and 69 at $x = 50$. Explain the answers' meaning and discuss their accuracy.

71. Evaluate the models from Exercises 67 and 69 at $x = 115$. Explain the answers' meaning and discuss their accuracy.

In Your Own Words. In Exercises 72–79, explain the following.

72. Sine function

73. Parametrized sine function

74. Graphically determining a sine model is reasonable or not

75. Effect of the parameters A, P, h, and h

76. Conjecturing the amplitude A of a sine model

77. Conjecturing the vertical shift v of a sine model

78. Conjecturing the horizontal shift h of a sine model

79. Conjecturing the period P of a sine model

2.5   Modeling with Sigmoidal Functions

Linear, exponential, power function, and sine models are commonly used to describe and analyze many different aspects of our world. This section makes a final addition to this collection of common models: sigmoidal functions. While interesting in and of their own right, these functions are included to help us understand aspects of reality that cannot be described using the other models.

As an example, such models arise when considering the changing sizes of populations. During their initial stages of growth, populations often increase exponentially. However, this exponential growth is eventually constrained by the limited resources available in any environment, such as space, food, water, and shelter. These constraints means that a population can only become so large in any given environment. This limit on how large a population can grow is called the **carrying capacity** of the environment. As a specific example of this phenomenon, Figure 1 presents the population of the Netherlands measured in millions of people at the beginning of each decade since 1700.

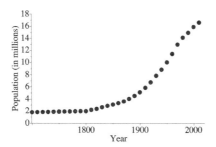

Figure 1: Population of the Netherlands in millions of people each decade

Figure 1 illustrates three distinct types of functional behavior over different input intervals in that this data set appears to be roughly:

- linear from 1700 to 1800, and from 1930 to 1970,

- exponential or power functional from 1860 to 1930, and

- power functional, more recently, from 1960 to 2010.

As such, none of these types of functions can, by themselves, provide a satisfactory model of the entire data set. Such realities led to the study of sigmoidal models, which blend exponential functions with **asymptotic functions**, which level off and approach some fixed, horizontal limiting line for the height of their outputs. These sigmoidal models enable us to describe and analyze data sets exhibiting behaviors like that of the population of the Netherlands as well as many other phenomena, including some that decay (or decrease) toward a limiting value.

You may have never heard of a sigmoidal function before. These functions are called "sigmoidal" because they usually resemble an elongated "S" and the Greek name for the letter "S" is "sigma." The primary method for identifying a sigmoidal function as a reasonable model for a given data set is recognizing their characteristic elongated "S" shape as illustrated in Figure 2.

In Figure 2, the value L corresponds to the limiting value the function approaches asymptotically as the inputs increase toward infinity. In the context of population growth, L is often referred to as the carrying capacity parameter because it identifies the limit on population growth as a result of environmental constraints. In more general settings, this horizontal asymptote is equal to a sum of the value of L and a vertical shift parameter v, as will be discussed in more detail shortly.

The value $y(h)$ of the sigmoidal function at input $x = h$ corresponds to the **point of inflection** where the function transitions from being concave up to concave down. Recall from Section 1.4, that a concave up function looks like a "cup" and that a concave down function looks like a "frown." As illustrated in Figure 2, sigmoidal functions exhibit both of these behaviors, and the point of inflection at the transition between these two behaviors is important in the modeling process.

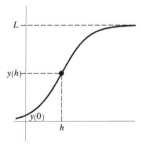

Figure 2: A typical sigmoidal function

The initial exponential growth (or decay) of a sigmoidal function depends on an exponent parameter k and a vertical shift parameter v adjusts the graphs of some sigmoidal functions up and down on the plane, as incorporated into the following parametrized definition of these functions:

Definition. A sigmoidal function is of the form

$$y = \frac{L}{1 + Ce^{-k(x-h)}} + v$$

where v, L, h, C, and k are constants with L, C, and k all nonzero and with

$$C = \frac{L}{y(h) - v} - 1$$

While this general form for a sigmoidal function may appear quite daunting, we examine the effect of these parameters on the graph of such functions and learn how to conjecture their values for specific examples of sigmoidals.

Parameters of Sigmoidal Functions

Following our usual approach, we explore the effect of each parameter on the graph of a sigmoidal function, beginning with v, and then considering L, h, C, and k in this order. Each parameter has a particular effect on the graph of a sigmoidal, although, as suggested by the above parametrized definition, the parameters of a sigmoidal function are more interconnected than those of the other common modeling functions.

The **vertical shift** parameter v corresponds to the output value that the graph of a sigmoidal function approaches asymptotically as the input values decrease toward negative infinity. Figure 3(a) presents the graph of the sigmoidal function $y = 5/(1 + e^{-x}) - 5$, Figure 3(b) the graph of $y = 5/(1 + e^{-x})$, and Figure 3(c) the graph of $y = 5/(1 + e^{-x}) + 5$.

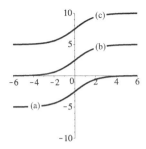

Figure 3: Effect of the vertical shift parameter v on a sigmoidal function

For all three of these example sigmoidal functions, the other parameters $L = 5$, $h = 0$, $C = 1$, and $k = 1$ are held constant so that the behavior of each function depends only on v. And, as illustrated in Figure 3, the value of the vertical shift parameter v determines the horizontal asymptote $y = v$ of each sigmoidal function as the inputs decrease toward negative infinity.

When $v = 0$, the **numerator** parameter L corresponds to the output value that its graph approaches asymptotically as the input values increase toward positive infinity. Figure 4(a) presents the graph of the sigmoidal function $y = 1/(1 + e^{-x})$, Figure 4(b) the graph of $y = 5/(1 + e^{-x})$, and Figure 4(c) the graph of $y = 10/(1 + e^{-x})$.

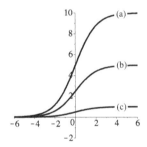

Figure 4: Effect of the carrying capacity parameter L on a sigmoidal function

For all three of these example sigmoidal functions, the other parameters $v = 0$, $h = 0$, $C = 1$, and $k = 1$ are held constant so that the behavior of each function depends only on L. Because $v = 0$, the value of L determines the horizontal asymptote $y = L$ of each sigmoidal function as the inputs increase toward infinity. When $v \neq 0$, the horizontal asymptote of the sigmoidal is $y = v + L$ and the value of L can be solved for after determining v; such cases are studied in more detail soon. In addition, the value of L is connected both to the vertical height of its point of inflection $(h, y(h))$ and to the intercept with the vertical axis, although these relationships are more subtle and interconnected with the value of C.

The **horizontal shift** parameter h corresponds to the input value $x = h$ for which $(h, y(h))$ is the point of inflection of the sigmoidal function. It turns out that this point of inflection is also connected to the coefficient parameter C in the denominator. Namely,

C is determined by the carrying capacity L and the vertical height $y(h)$ of the point of inflection via the formula

$$C = \frac{L}{y(h) - v} - 1$$

We do not explore the graphical effect of the horizontal shift parameter beyond its relationship to C and L. The key idea here is to be able to recognize the point of inflection on the graph of a sigmoidal function.

The **denominator coefficient** parameter C can have multiple effects on the shape of the graph of a sigmoidal function. One effect amounts to a horizontal shift left or right on the plane. Figure 5 presents the graph of the following three sigmoidal functions with the other parameters $v = 0$, $L = 10$, $h = 0$, and $k = 1$ held constant so that the behavior of each function depends only on C:

(a) $C = 1$ (b) $C = 10$ (c) $C = 0.1$

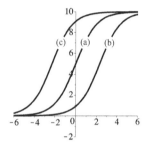

Figure 5: Effect of the coefficient parameter C on a sigmoidal function

As can be seen, when $0 < C < 1$, the graph retains the same shape but shifts to the left, while for $C > 1$, the graph shifts to the right. Consequently, the value of the vertical intercept changes the point of inflection as the graph shifts left or right accordingly. While rarely encountered, sometimes the coefficient parameter C is negative. The difference between C being positive and negative is illustrated in Figure 6, where $2/(1 + e^{-x})$ with $C = 1 > 0$ is shown by a solid line and $2/(1 - e^{-x})$ with $C = -1 < 0$ by a dashed line.

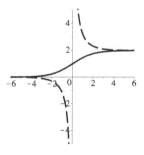

Figure 6: Effect of the sign of the parameter C on a sigmoidal function

The **exponent** parameter k determines the initial exponential growth or decay of a sigmoidal function. Figure 7 presents the graphs of the following three sigmoidal functions with the other parameters $v = 0$, $L = 2$, $h = 0$, and $C = 1$ held constant so that the behavior of each function depends only on k:

(a) $k = 1$ (b) $k = 4$ (c) $k = 0.5$

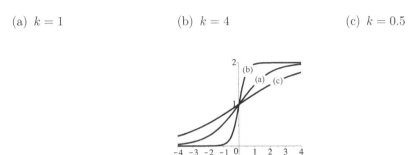

Figure 7: Effect of the exponent parameter k on a sigmoidal function

These three graphs indicate that the value of k affects the growth rate of the function from its vertical shift asymptote $y = v$ and its carrying capacity $y = v + L$. For positive k, the larger values of k result in steeper growth, while smaller values of k result in slower growth. In addition, k can be negative. When $k > 0$, the function grows towards its carrying capacity $y = v + L$ as shown in graph (a) of Figure 8, but when $k < 0$, the function decays towards its vertical shift $y = v$ as shown in graph (b).

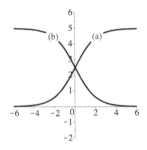

Figure 8: Effect of the sign of k on the graph of a sigmoidal function

Conjecturing Sigmoidal Models

Working toward solidifying an understanding of the effect of the parameters of a sigmoidal function, the next example and question conjecture their values for specific graphs. After considering complete graphs of such functions, we employ these invaluable skills in identifying models for sigmoidal data sets. As illustrated in Example 1, the asymptotes $y = v$ and $y = v + L$, and the point of inflection $(h, y(h))$, are conjectured based on a sigmoidal graph or data plot. These values are then used to compute C. Finally, a point from the curve or data (other than the point of inflection $(h, y(h))$) is substituted into this intermediate model to solve for k and obtain a final conjectured sigmoidal model.

◆ **EXAMPLE 1** The graph of a sigmoidal function $y = \dfrac{L}{1 + Ce^{-k(x-h)}} + v$ is given in Figure 9.

(a) Conjecture values for the parameters v, L, and $(h, y(h))$, and then compute C.

(b) Use a point on the curve to determine a value for exponent parameter k.

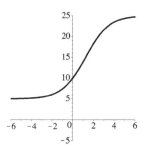

Figure 9: Sigmoidal function for Example 1

Solution.

(a) Based on the horizontal asymptote as inputs decrease toward negative infinity, a vertical shift of $v = 5$ is conjectured. The horizontal asymptote as inputs increase to positive infinity is $y = 25$, which means that $v + L = 25$. Substituting $v = 5$ and solving $5 + L = 25$ gives $L = 20$. The point of inflection appears to be about $(h, y(h)) = (1, 14)$. Using these parameters, calculate C as follows:

$$C = \frac{L}{y(h) - v} - 1 = \frac{20}{14 - 5} - 1 = \frac{20}{9} - 1 = \frac{11}{9}$$

(b) Any point on the curve (other than the point of inflection $(h, y(h)) = (1, 14)$) can be used to determine the exponent parameter k. This example works with $(0, 10)$, substituting this point into the intermediate model from part (a) and solving for k:

$$y = \frac{20}{1 + \frac{11}{9}e^{-k(x-1)}} + 5 \qquad \text{Intermediate model}$$

$$10 = \frac{20}{1 + \frac{11}{9}e^{k}} + 5 \qquad \text{Substitute } (x, y) = (0, 10) \text{ and } -k \cdot (0 - 1) = k$$

$$5 = \frac{20}{1 + \frac{11}{9}e^{k}} \qquad \text{Subtract 5 from both sides}$$

$$1 + \frac{11}{9}e^{k} = \frac{20}{5} \qquad \text{Multiply both sides by } 1 + \frac{11}{9}e^{k} \text{ and } \frac{1}{5}$$

$$\frac{11}{9}e^k = 4 - 1 \qquad\qquad \text{Simplify } \frac{20}{5} = 4 \text{ and subtract 1 from both sides}$$

$$e^k = 3 \cdot \frac{9}{11} \qquad\qquad \text{Simplify } 4 - 1 = 3 \text{ and multiply both sides by } \frac{9}{11}$$

$$k = \ln\left(\frac{27}{11}\right) \qquad\qquad \text{Simplify, take } \ln(x) \text{ of both sides, and } \ln(e^b) = b$$

$$k \approx -0.898 \qquad\qquad \text{Simplify}$$

Thus, the conjectured model is

$$y \;=\; \frac{20}{1 + \frac{11}{9}e^{0.898(x-1)}} + 5 \;\approx\; \frac{20}{1 + 1.222 \cdot e^{0.898(x-1)}} + 5$$

▶ **QUESTION 1** The graph of a sigmoidal function $y = \dfrac{L}{1 + Ce^{-k(x-h)}} + v$ is given in Figure 10.

(a) Conjecture values for the parameters v, L, and $(h, y(h))$, and then compute C

(b) Use a point on the curve to determine a value for exponent parameter k.

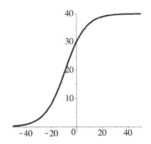

Figure 10: Sigmoidal function for Question 1

◆ **EXAMPLE 2** The data set in Figure 11 presents the number of monthly active Twitter users worldwide by quarter (e.g., 10.25 represents April to June 2010) in millions. For example, from January to March 2010 (or input 10), 30 million people used Twitter and, from January to March 2014 (or input 14), 255 million people used Twitter.

(a) Verify that a sigmoidal model is reasonable using the data plotted in Figure 11.

(b) Conjecture a sigmoidal model Users $= \dfrac{L}{1 + Ce^{-k(\text{Year}-h)}} + v$ for this data set.

Solution.

(a) The data set in Figure 11 can be interpreted as exhibiting the elongated "S" shape characteristic of sigmoidal data. As a result of the limited number of data points, the long-term asymptotic behavior of the data points approaching a horizontal limiting value is less apparent, although it is still suggested by the relative output values at the two ends of the data set beginning to level off.

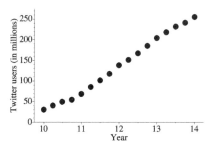

Figure 11: Millions of Twitter Users for Example 2

(b) Based on the data and the knowledge that Twitter began with zero users, a vertical shift of $v = 0$ is reasonable. The horizontal asymptote as inputs increase toward infinity is more difficult to conjecture, because the outputs are still increasing, but a reasonable assumption is that $L = 275$. The point of inflection appears to be approximately $(h, y(h)) = (12.25, 150)$, which gives $h = 12.25$. Now use this point to compute C as follows:

$$C = \frac{L}{y(12) - v} - 1 = \frac{275}{150 - 0} - 1 \approx 0.83.$$

Next, solve for k using the data point $(\text{Year}, \text{Users}) = (11.5, 100)$:

$$\text{Users} = \frac{275}{1 + 0.83e^{-k(\text{Year} - 12.25)}} \qquad \text{Intermediate model}$$

$$100 = \frac{275}{1 + 0.83e^{-k(11.5 - 12.25)}} \qquad \text{Substitute } (\text{Year}, \text{Users}) = (11.5, 100)$$

$$100(1 + 0.83e^{0.75k}) = 275 \qquad \text{Multiply both sides by } 1 + 0.83e^{0.75k}$$

$$1 + 0.83e^{0.75k} = \frac{275}{100} \qquad \text{Divide both sides by } 100$$

$$0.83e^{0.75k} = 2.75 - 1 \qquad \text{Subtract 1 from both sides}$$

$$e^{0.75k} = 1.823 \qquad \text{Simplify and divide both sides by } 0.83$$

$$0.75k = \ln(2.108) \qquad \text{Take } \ln(x) \text{ of both sides and } \ln(e^b) = b$$

$$k = \frac{1}{0.75} \cdot \ln(2.108) \qquad \text{Divide both sides by } 0.75$$

$$k \approx 0.99 \qquad \text{Simplify}$$

Thus, the conjectured model is

$$\text{Users} = \frac{275}{1 + 0.83e^{-0.99(\text{Year} - 12.25)}}$$

A graph of this model on a plot of the data in Figure 12 indicates its reasonableness, at least up through 2013. The misalignment of this model and the latter data points might suggest revising the values of the conjectured parameters or using `fitModel` in RStudio to find the best possible sigmoidal model.

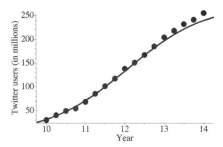

Figure 12: Model of Twitter users for Example 2

➤ **QUESTION 2** The data set in Figure 13 presents the population of the Netherlands measured in millions of people at the beginning of each decade since 1700. For example, in 1990, the population of the Netherlands was approximately 14.892 million people.

(a) Verify that a sigmoidal model is reasonable using the data plotted in Figure 13.

(b) Conjecture a sigmoidal model Population $= \dfrac{L}{1 + Ce^{-k(\text{Year}-h)}} + v$ for this data set.

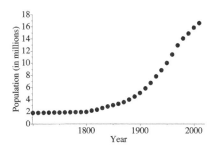

Figure 13: Population of the Netherlands for Question 2

Best Possible Sigmoidal Models

We have learned how to conjecture a sigmoidal model based on a plot of a given data set. However, this approach is not guaranteed to produce the best possible sigmoidal model. Rather, we can only say that a reasonable model has been identified based on a visual check of its validity. The RStudio command `fitModel` provides a way to find the best possible sigmoidal model for a given data set.

Working in RStudio

The syntax for the `fitModel` command for sigmoidal models is identical to that for sine models. In particular, the additional argument `start=list(,)` must be included in the `fitModel` command to provide initial guesses for the parameters of the model, where these are listed inside the parentheses separated by commas. For example, the argument `start=list(v=1,L=2,C=3,k=4)` might be added to `fitModel`, as illustrated below.

Examples of Commands

- `bestSigModel=fitModel(y~L/(1+C*exp(-k*(x-10))+v),data=sigData,`
  `start=list(v=1,L=2,C=3,k=4))`

- `coef(bestSigModel)`

◆ **EXAMPLE 3** The data set stored in `TwitterUsers` contains the number of monthly active Twitter users worldwide by quarter (e.g., 10.25 represents April to June 2010) in millions of people. Use `fitModel` with the conjectured parameters from Example 2 to find the best possible sigmoidal model for this data set and graph the model on a plot of the data.

Solution. First, use `names` to determine the variables in `TwitterUsers`:

```
names(TwitterUsers)
```

```
[1] "Year"  "Users"
```

Figure 11 in Example 2 presented the graph of the data and led to the parameters $v = 0$, $L = 275$, $(h, y(h)) = (12.25, 150)$, $C = 0.83$, and $k = 0.99$, and a conjectured sigmoidal model of

$$\text{Users} = \frac{275}{1 + 0.83e^{-0.99(\text{Year}-12.25)}}$$

Now, use these conjectured parameters for the **start** values of v, L, C, and k in the `fitModel` command to obtain the best sigmoidal model for this data set:

```
bestModel=fitModel(Users~L/(1+C*exp(-k*(Year-12.25)))+v,
  data=TwitterUsers,start=list(v=0,L=275,C=0.83,k=0.99))
coef(bestModel)
```

```
        v           L           C           k
 -4.698342 306.975542    0.963274    0.930132
```

From this RStudio output, the best sigmoidal model has parameters $v = -4.698342$, $L = 306.975542$, $C = 0.963274$, and $k = 0.930132$, which results in the following when substituted into the standard form of the model:

$$\text{Users} = \frac{306.975542}{1 + 0.963274 \cdot e^{-0.930132(\text{Year}-12.25)}} - 4.698342$$

As an indication of goodness of fit, the graph of this best possible model is graphed as a solid line on a plot of the data in the rightmost graph and the graph of the conjectured model from Example 2 as a dashed line on the leftmost graph. As the graphs indicate, the best possible model provided by `fitModel` provides a much closer match for more of the data set.

```
plotPoints(Users~Year,data=TwitterUsers)
plotFun(275/(1+0.83*exp(-0.99*(Year-12.25)))~Year,add=TRUE)
plotFun(bestModel(Year)~Year,add=TRUE)
```

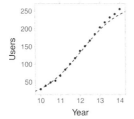 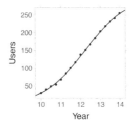

▶ **QUESTION 3** The data set stored in `NetherlandsPopulation` contains the population of the Netherlands measured in millions of people at the beginning of each decade since 1700. Use `fitModel` with the conjectured parameters from Question 2 to find the best possible sigmoidal model for this data set and graph the model on a plot of the data.

Summary

- A *sigmoidal* function is of the form $y = \dfrac{L}{1 + Ce^{-k(x-h)}} + v$, where v, L, h, C, and k are constants with L, C, and k all nonzero and with $C = \dfrac{L}{y(h) - v} - 1$.

- In the parametrized form of a sigmoidal function v gives the vertical shift, $v + L$ gives the carrying capacity, $(h, y(h))$ the coordinates of the point of inflection, C shifts the sigmoidal left and right, and k effects the steepness of the growth.

- To conjecture the parameters, first estimate v, $v + L$, and $(h, y(h))$ from the graph of the sigmoidal function or data set. Then calculate C using the above formula. Finally, compute k by substituting any point (other than the point of inflection $(h, y(h))$) into the intermediate sigmoidal and solving.

Exercises

In Exercises 1–10, identify which type of model is reasonable for the data set from among the known models: linear, exponential, power function, sine, sigmoidal, or none of these. More than one type of model may be reasonable.

1.

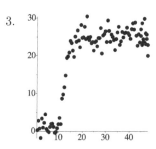

2.

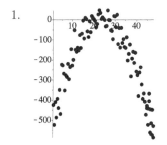

3.

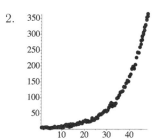

4.

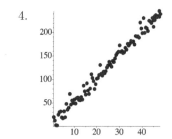

5.

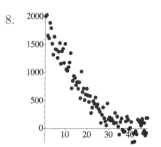

6.

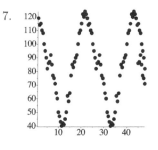

7.

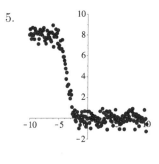

8.

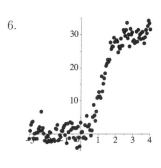

9.

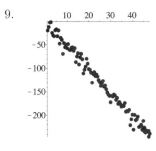

10.

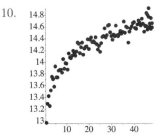

In Exercises $11-14$, find L for the sigmoidal function through the given point.

11. $y(x) = \dfrac{L}{1 + 8e^{-2(x-4)}}$; $(4, 5)$

12. $y(x) = \dfrac{L}{1 + 4e^{-0.5(x-20)}}$; $(10, 15)$

13. $y(x) = \dfrac{L}{1 + 2e^{-(x-100)}}$; $(108, 25)$

14. $y(x) = \dfrac{L}{1 + 0.5e^{-4(x+10)}}$; $(1, 8)$

In Exercises $15-18$, find C for the sigmoidal function through the given point.

15. $y(x) = \dfrac{10}{1 + Ce^{-2(x-5)}}$; $(5, 4)$

16. $y(x) = \dfrac{5}{1 + Ce^{-0.5(x-20)}}$; $(10, 5)$

17. $y(x) = \dfrac{30}{1 + Ce^{-(x-100)}}$; $(104, 60)$

18. $y(x) = \dfrac{100}{1 + Ce^{-4(x+10)}}$; $(1, 80)$

In Exercises $19-22$, find k for the sigmoidal function through the given point.

19. $y(x) = \dfrac{10}{1 + 3e^{-k(x-4)}}$; $(5, 4)$

20. $y(x) = \dfrac{5}{1 + 4e^{-k(x-20)}}$; $(10, 2)$

21. $y(x) = \dfrac{30}{1 + 2e^{-k(x-100)}}$; $(102, 15)$

22. $y(x) = \dfrac{100}{1 + 0.5e^{-k(x+10)}}$; $(1, 80)$

In Exercises $23-30$, conjecture values for the parameters v, L, and $(h, y(h))$, and use them to compute C for the sigmoidal function.

23.

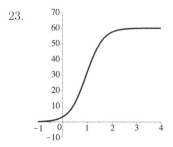

24.

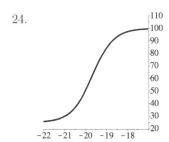

25.

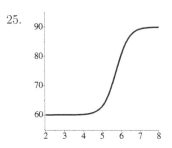

26.

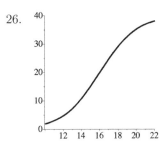

30.

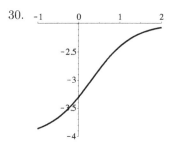

In Exercises $31-34$, compute k using a point on the sigmoidal curve and given intermediate model.

31. $y(x) = \dfrac{20}{1 + 8e^{-k(x-4)}} + 10$

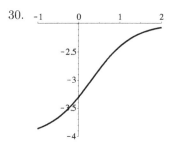

27.

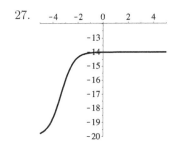

28.

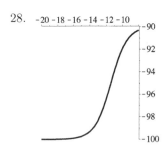

32. $y(x) = \dfrac{2}{1 + e^{-k(x+3)}} - 15$

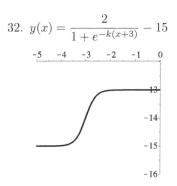

33. $y(x) = \dfrac{200}{1 + 0.5e^{-k(x+30)}} - 150$

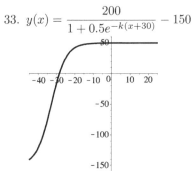

29.

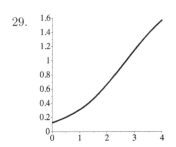

34. $y(x) = \dfrac{-12}{1 + 0.1e^{-k(x-1)}} + 8$

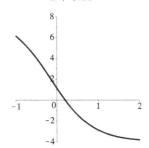

In Exercises 35–40, conjecture a sigmoidal model for the curve.

35.

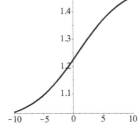

36.

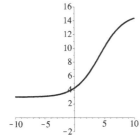

37.

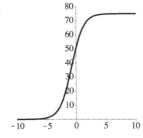

38.

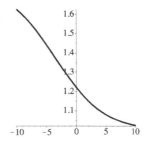

39.

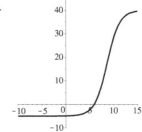

40.

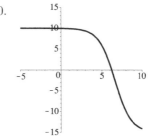

In Exercises 41–46, conjecture a sigmoidal model for the data set.

41.

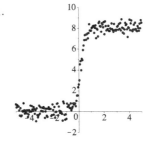

42.

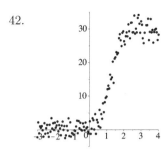

46.

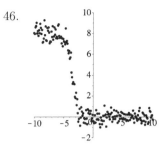

RStudio. In Exercises 47–51, use RStudio to find a sigmoidal model of the population of Belgium in millions of people stored in `PopulationBelgium`.

47. Verify that a sigmoidal model is reasonable using a plot of the data. Also, conjecture its parameters v, L, h, C, and k.

48. Graph a sigmoidal model with the parameters from Exercise 47 on the data plot.

43.

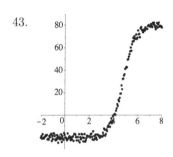

49. Use `fitModel` with `start` values from Exercise 47 to obtain a sigmoidal model for the data set.

50. Evaluate the models from Exercises 47 and 49 at $x = 1950$. Explain the answers' meaning and discuss their accuracy.

44.

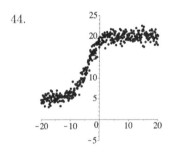

51. Evaluate the models from Exercises 47 and 49 at $x = 2020$. Explain the answers' meaning and discuss their accuracy.

RStudio. In Exercises 52–56, use RStudio to find a sigmoidal model of the number of Facebook users in millions from 2009 through 2012 stored in `FacebookUsers`.

52. Verify that a sigmoidal model is reasonable using a plot of the data. Also, conjecture its parameters v, L, h, C, and k.

45.

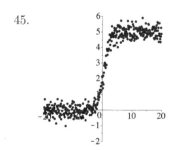

53. Graph a sigmoidal model with the parameters from Exercise 52 on the data plot.

54. Use `fitModel` with `start` values from Exercise 52 to obtain a sigmoidal model for the data set.

55. Evaluate the models from Exercises 52 and 54 at $x = 13$. Explain the answers' meaning and discuss their accuracy.

56. Evaluate the models from Exercises 52 and 54 at $x = 60$. Explain the answers' meaning and discuss their accuracy.

RStudio. In Exercises 57 – 61, use RStudio to find a sigmoidal model of the average SAT math score in Kentucky each year from 1980 to 2013 stored in `SATMathKentucky`.

57. Verify that a sigmoidal model is reasonable using a plot of the data. Also, conjecture its parameters v, L, h, C, and k.

58. Graph a sigmoidal model with the parameters from Exercise 57 on the data plot.

59. Use `fitModel` with `start` values from Exercise 57 to obtain a sigmoidal model for the data set.

60. Evaluate the models from Exercises 57 and 59 at $x = 1998$. Explain the answers' meaning and discuss their accuracy.

61. Evaluate the models from Exercises 57 and 59 at $x = 2020$. Explain the answers' meaning and discuss their accuracy.

RStudio. In Exercises 62 – 66, use RStudio to find a sigmoidal model of the percent of each year's high school graduates to enroll in either a two-year or four-year college from 1972 to 2012 stored in `HSGradsinCollege`.

62. Verify that a sigmoidal model is reasonable using a plot of the data. Also, conjecture its parameters v, L, h, C, and k.

63. Graph a sigmoidal model with the parameters from Exercise 62 on the data plot.

64. Use `fitModel` with `start` values from Exercise 62 to obtain a sigmoidal model for the data set.

65. Evaluate the models from Exercises 62 and 64 at $x = 1998$. Explain the answers' meaning and discuss their accuracy.

66. Evaluate the models from Exercises 62 and 64 at $x = 2020$. Explain the answers' meaning and discuss their accuracy.

RStudio. In Exercises 67 – 71, use RStudio to find a sigmoidal model of the cumulative number of Ebola cases in Sierra Leone from May 1, 2014 to December 16, 2015 stored in `EbolaSierraLeone`.

67. Verify that a sigmoidal model is reasonable using a plot of the data. Also, conjecture its parameters v, L, h, C, and k.

68. Graph a sigmoidal model with the parameters from Exercise 67 on the data plot.

69. Use `fitModel` with `start` values from Exercise 67 to obtain a sigmoidal model for the data set.

70. Evaluate the models from Exercises 67 and 69 at $x = 300$. Explain the answers' meaning and discuss their accuracy.

71. Evaluate the models from Exercises 67 and 69 at $x = 700$. Explain the answers' meaning and discuss their accuracy.

RStudio. In Exercises 72–76, use RStudio to find a sigmoidal model of the number of yellow cards given per men's World Cup tournament from 1970 to 2010 stored in `YellowCards`.

72. Verify that a sigmoidal model is reasonable using a plot of the data. Also, conjecture its parameters v, L, h, C, and k.

73. Graph a sigmoidal model with the parameters from Exercise 72 on the data plot.

74. Use `fitModel` with `start` values from Exercise 72 to obtain a sigmoidal model for the data set.

75. Evaluate the models from Exercises 72 and 74 at $x = 1998$. Explain the answers' meaning and discuss their accuracy.

76. Evaluate the models from Exercises 72 and 74 at $x = 2018$. Explain the

answers' meaning and discuss their accuracy.

In Your Own Words. In Exercises 77–84, explain the following.

77. Sigmoidal function

78. Graphically determining a sigmoidal model is reasonable or not

79. Effect of the parameters L, v, h, C, and k

80. Conjecturing the vertical shift v of a sigmoidal model

81. Conjecturing the numerator L of a sigmoidal model

82. Conjecturing the horizontal shift h of a sigmoidal model

83. Finding the denominator coefficient C of a sigmoidal model

84. Conjecturing the exponent k of a sigmoidal model

2.6   Single-Variable Modeling

We have learned about various functions that are commonly used to model data sets, focusing primarily on each modeling function one at a time. This section examines these diverse functions simultaneously, uniting this study into a single, coherent approach to modeling data sets. We also expand our abilities to work with the last three steps of the modeling cycle as well as explore the limitations of these models. Recall the five steps of this process:

THE MODELING CYCLE.

The five steps of the **modeling cycle** are:

(1) Ask a question about reality.

(2) Make some observations and collect the corresponding data.

(3) Conjecture a model or modify a known model based on the data.

THE MODELING CYCLE. (CONTINUED)

(4) Test the model against known data (from step (2)) and modify the model as needed.

(5) Repeat steps (2), (3), and (4) to improve the model.

Part of conjecturing a model in step (3) of the modeling cycle is determining which type of model appears to provide the best match for a given data set. Making this determination always begins with visually examining standard, semi-log, and log-log plots of the data set to identify which model appears most reasonable. These plots are also used to conjecture the parameters of this most reasonable model. Sections 2.1, 2.2, and 2.3 introduced numerical approaches to assessing the reasonableness of linear, exponential, and power function models, respectively. In contrast, this section relies on RStudio when determining the most reasonable model, rather than plotting data sets by hand or performing hand computations. This choice enables a nimble, robust approach to identifying, developing, and analyzing models of real-life phenomena.

Graphically Identifying Reasonable Models

We begin by summarizing into a single step-by-step process our approach to deciding which model appears most reasonable for a given data set based on a graphical examination of the data.

GRAPHICALLY IDENTIFYING REASONABLE MODELS.

Graph the standard plot (x, y) of the given data set.

(1) If the standard plot is approximately linear, then a linear model $y = mx + b$ is reasonable.

(2) If the standard plot exhibits periodic oscillations between two extreme values, then a sine model $y = A \sin \left[\dfrac{2\pi}{P}(x - h) \right] + v$ is reasonable.

(3) If the standard plot has two horizontal asymptotes $y = v$ and $y = v + L$ with initial exponential growth from $y = v$ followed by exponential decay toward $y = v + L$, then a sigmoidal model $y = \dfrac{L}{1 + Ce^{-k(x-h)}} + v$ where $C = \dfrac{L}{y(h) - v} - 1$ is reasonable.

GRAPHICALLY IDENTIFYING REASONABLE MODELS.
(CONTINUED)
If the standard plot does not identify a reasonable model, then graph both the
semi-log plot $(x, \ln(y))$ and the log-log plot $(\ln(x), \ln(y))$ of the given data set.

(4) If the semi-log plot is approximately linear and the log-log plot is not approximately linear, then an exponential model $y = Ce^{k(x-h)}$ is reasonable.

(5) If the log-log plot is approximately linear and the semi-log plot is not approximately linear, then a power function model $y = Cx^k$ is reasonable.

(6) If both the semi-log plot and the log-log plot are linear, then determine which
plot has the least variance in slopes and which of the corresponding models
from (4) or (5) is most reasonable.

◆ **EXAMPLE 1** The data set stored in `APCalculus2` contains the number of students
who took the Advanced Placement AB Calculus exam each year from 1973 to 2010.

(a) Graphically identify the most reasonable model for this data set.

(b) Use `fitModel` to determine the best model of this type for the data set.

(c) Discuss the model's quality of fit based on a graph of the model on a plot of the
data.

Solution.

(a) Recall that the `head` command can be used to view the first few data points along
with the variable names:

```
head(APCalculus2)
```

```
   Year Exams
1  1973 14310
2  1974 16038
3  1975 17090
4  1976 19065
5  1977 20317
6  1978 22510
```

Therefore, the input variable is `Year` and the output variable is `Exams`. Now graph
a standard plot (x, y) of the data set using `plotPoints` in order to check whether
a linear, sine, or sigmoidal model might be reasonable.

```
plotPoints(Exams~Year,data=APCalculus2)
```

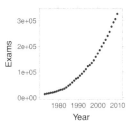

The data is not approximately linear (ruling out a linear model), does not exhibit periodic oscillations (ruling out a sine model), and does not appear to have two horizontal asymptotes (ruling out a sigmoidal model). Therefore, graph a semi-log plot $(x, \ln(y))$ and a log-log plot $(\ln(x), \ln(y))$ of the given data set to check whether an exponential or power function model might be reasonable:

```
plotPoints(log(Exams)~Year,data=APCalculus2)
plotPoints(log(Exams)~log(Year),data=APCalculus2)
```

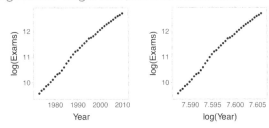

An initial examination of the semi-log and log-log plots indicates that either an exponential or a power function model would appear to be reasonable, because both are roughly linear. In order to choose between them, identify which plot has the least variation in the slopes of consecutive data points. The scale on the vertical axes is the same for both plots and does not distinguish between these two models. However, the scale on the horizontal axis is much greater for the semi-log plot (resulting in less variation in the slopes) than for the log-log plot (resulting in more variation in the slopes). Therefore, the semi-log plot is more linear, which indicates that an exponential function provides the most reasonable model for this data set.

(b) Using the first data point obtained using **head** in part (a), let the horizontal shift parameter be $h = 1973$ based on the data point $(1973, 14, 310)$ with the least input of 1973, which gives a sought-for model of the form $\text{Exams} = Ce^{k(\text{Year}-1973)}$. Now, the best exponential model for the given data set is found by determining the best linear model for the corresponding semi-log data with the **fitModel** command:

```
expAPModel = fitModel(log(Exams)~m*(Year-1973)+b,data=APCalculus2)
coef(expAPModel)
```

```
        m          b
0.0869263 9.6655055
```

From this RStudio output, the linear model for the semi-log data has slope $m = 0.08693$ and vertical intercept $b = 9.66551$, which means that the parameters of the exponential model for the given data set are $k = m = 0.08693$ and $C = e^b = e^{9.66551} \approx 15,764.41$. Therefore, the following function provides the best exponential model for the number of students who took the Advanced Placement AB Calculus exam between 1973 to 2010:

$$\text{Exams} = e^{9.66551} \cdot e^{0.08693(\text{Year}-1973)} \approx 15,764.41 \cdot e^{0.08693(\text{Year}-1973)}$$

(c) The quality of fit of this model can be discussed in light of the following graph of the model on a plot of the data:

```
plotPoints(Exams~Year,data=APCalculus2)
plotFun(exp(9.66551)*exp(0.08693*(Year-1973))~Year,add=TRUE)
```

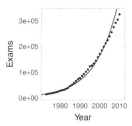

While the model does not fit the data perfectly, it certainly captures the overall, general trend of the data. Because the semi-log plot is only roughly linear, some differences between the model and the actual data should be expected. Even so, the model closely approximates most of the data, particularly early in the data set. The model is less accurate for the latter portion of the data set, and this model should not be expected to predict the number of students who will take the AP Calculus exam some indefinite number of years in the future. In order to conjecture future results, the model needs to be further refined, which might be accomplished by extending the data set to include the number of students who took the Advanced Placement AB Calculus exam in 2011 and beyond.

◼

◆ **EXAMPLE 2** The data set stored in `EbolaSierraLeone` contains the cumulative number of Ebola cases in Sierra Leone from May 1, 2014 to December 16, 2015.

(a) Graphically identify the most reasonable model for this data set.

(b) Use `fitModel` to determine the best model of this type for the data set.

(c) Discuss the model's quality of fit based on a graph of the model on a plot of the data.

Solution.

(a) The variables and first few data points in `EbolaSierraLeone` are found using `head`:

```
head(EbolaSierraLeone)
```

```
  Day Cases
1   0    0
2   9    0
3  13    1
4  14   16
5  19   50
6  22   81
```

The input variable is `Day` and the output variable is `Cases`. Now graph a standard plot (x, y) of the data set using `plotPoints` in order to check whether a linear, sine, or sigmoidal model might be reasonable:

```
plotPoints(Cases~Day,data=EbolaSierraLeone)
```

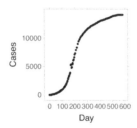

The standard plot indicates that a sigmoidal model is the most reasonable, because the data appears to have two horizontal asymptotes $y = 0$ and $y \approx 14,000$, with initial exponential growth from $y = 0$ followed by exponential decay toward $y = 14,000$.

(b) A sigmoidal model is of the form

$$y(x) = \frac{L}{1 + Ce^{-k(x-h)}} + v \quad \text{with} \quad C = \frac{L}{y(h) - v} - 1$$

When using RStudio to find the best possible sigmoidal model, `fitModel` must be provided with conjectured values for its five parameters. From the standard plot of the data in part (a), observe that $v = 0$, $L \approx 14,000$, and $(h, y(h)) \approx (200, 7200)$. Substituting these values into the standard formula, compute C as follows:

$$C = \frac{L}{y(h) - v} - 1 = \frac{14000}{7200 - 0} - 1 \approx 0.94444$$

Next, solve for k by substituting a point (other than the already used point of inflection) into the current intermediate model. This example conjectures k by substituting the fourth point $(\text{Day}, \text{Cases}) = (14, 16)$ from the data set into the intermediate model and solving for k:

$$\text{Cases} = \frac{14000}{1 + 0.94444e^{-k(\text{Day}-200)}} \qquad \text{Intermediate model}$$

$$16 = \frac{14000}{1 + 0.94444e^{-k(14-200)}} \qquad \text{Substitute } (\text{Day}, \text{Cases}) = (14, 16)$$

$$1 + 0.94444e^{186k} = \frac{14000}{16} \qquad \text{Simplify and cross multiply}$$

$$0.94444e^{186k} = \frac{14000}{16} - 1 = 874 \qquad \text{Subtract 1 and simplify}$$

$$e^{186k} = \frac{874}{0.94444} = 925.4161 \qquad \text{Divide by 0.94444 and simplify}$$

$$k = \frac{\ln(925.4161)}{186} \approx 0.03672174 \qquad \text{Solve for } k \text{ using the natural logarithm}$$

Now find the best sigmoidal model for this data set using `fitModel` and these initial conjectures for the values of its parameters:

```
sigModel = fitModel(Cases~L/(1+C*exp(-k*(Day-200)))+v,
                    data=EbolaSierraLeone,
                    start=list(v=0,L=14000,C=0.944, k=0.0367))
coef(sigModel)
```

```
        v            L            C            k
-5.00361e+02  1.40155e+04  9.52042e-01  2.10288e-02
```

RStudio outputs the values of the parameters using scientific notation, with $v = $ `-5.0036e+02` $= -500.36$, $L = $ `1.4016e+04` $= 14016$, $C = $ `9.52e-01` $= 9.52 \times 10^{-1} = 0.952$, and $k = $ `2.103e-02` $= 2.103 \times 10^{-2} = 0.02103$. Substituting these values into the intermediate model provides the following best possible sigmoidal model:

$$\text{Cases} = \frac{14016}{1 + 0.952e^{-0.02103(\text{Day}-200)}} - 500.36$$

(c) The quality of fit of this model can be discussed in light of the following graph of the model on a plot of the data:

```
plotPoints(Cases~Day,data=EbolaSierraLeone)
plotFun(sigModel(Day)~Day,add=TRUE)
```

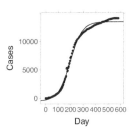

While the model does not perfectly capture all features of the data, it does express the initial exponential growth quite well. The model approaches the horizontal asymptote $y = 13{,}420$ more quickly than the data. This behavior occurs because the initial exponential growth in the data is more rapid than the exponential decay toward the top horizontal asymptote.

■

▶ **QUESTION 1** In 1945, the United States detonated the first atomic bomb in the Trinity Test at White Sands, New Mexico. The data set stored in `BlastData` contains the blast radius in meters as a function of the time in seconds after the initial blast.

  (a) Graphically identify the most reasonable model for this data set.

  (b) Use `fitModel` to determine the best model of this type for the data set.

  (c) Discuss the model's quality of fit based on a graph of the model on a plot of the data.

▶ **QUESTION 2** The data set stored in `WaterLevelsEastportMaine` contains the water level in Eastport, Maine from May 26, 2016 to May 27, 2016 measured in feet above the mean lower water level as a function of the time after 12:00 a.m. on May 26, 2016.

  (a) Graphically identify the most reasonable model for this data set.

  (b) Use `fitModel` to determine the best model of this type for the data set.

  (c) Discuss the model's quality of fit based on a graph of the model on a plot of the data.

▶ **QUESTION 3** The data set stored in `NaturalGasConsumption` contains the total number of millions of cubic feet of natural gas consumed in the United States from 1950 to 1970.

  (a) Graphically identify the most reasonable model for this data set.

  (b) Use `fitModel` to determine the best model of this type for the data set.

  (c) Discuss the model's quality of fit based on a graph of the model on a plot of the data.

Context and Choosing Models

Another consideration when identifying reasonable models is that the selected modeling function needs to make sense in context of the physical or social situation being analyzed. In some settings, several different models might fit the data quite well, but they may produce diverse predictions that may correspond more or less to what might be expected. Therefore, careful reflection must be invested in whether a particular type of model would provide a reasonable picture of the situation at hand. The following example illustrates how the phenomenon being modeled plays a role in determining the choice of a model:

◆ **EXAMPLE 3** Article I of the U.S. Constitution mandates that a census be conducted every ten years. The data set stored in **USTotalPopulation** contains the U.S. population in millions of people each decade from 1900 to 2010 based on the census.

(a) Graphically identify the most reasonable model for this data set.

(b) Use **fitModel** to determine the best model of this type for the data set.

(c) Discuss the model's quality of fit based on a graph of the model on a plot of the data.

Solution.

(a) First, determine the variables of **USTotalPopulation** using **names**:

```
names(USTotalPopulation)
```

```
[1] "Year"        "Population" "RelGrowth"
```

This data set contains three columns of data. Guided by the motivating question in the example prompt, the input variable is **Year** and the output variable is **Population**. Now, graph its standard plot, semi-log plot, and log-log plot as follows:

```
plotPoints(Population~Year,data=USTotalPopulation)
plotPoints(log(Population)~Year,data=USTotalPopulation)
plotPoints(log(Population)~log(Year),data=USTotalPopulation)
```

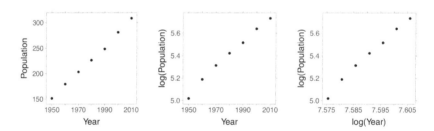

Based on these graphs, any of a linear, exponential, or power function model would seem to be reasonable. However, thinking about the phenomena being modeled, the population of the United States would not be expected to continue to increase at a linear, exponential, or power function rate far into the future. Rather, populations exhibit sigmoidal behavior over the long run because the total number of people in a region levels off over time in response to its carrying capacity.

(b) The best possible sigmoidal model can be found using **fitModel**, where the conjectured parameters of $v = 0$, $L = 525$, $h = 1950$, $C = 2.4$, and $k = 0.021$ are determined as in Example 2. The following version of this command forces the model to have a horizontal asymptote of $v = 0$, which corresponds to an initial U.S. population of zero people:

```
popModel=fitModel(Population~L/(1+C*exp(-k*(Year-1950)))),
                  data=USTotalPopulation,
                  start=list(L=525,C=2.4,k=0.021))
coef(popModel)
```

```
           L            C             k
591.1733348     2.8305085     0.0187521
```

This RStudio output yields the following best possible sigmoidal model:

$$\text{Population} \;=\; \frac{591.17332}{1 + 2.83051e^{-0.01875(\text{Year}-1950)}}$$

(c) The graph of this model on a plot of the data given below affirms its quality of fit, since the model both expresses the general trend of the data and closely approximates the actual value of all the data points. As such, this good model for the data also makes sense in the context of the phenomenon being analyzed.

```
plotPoints(Population~Year,data=USTotalPopulation)
plotFun(591.17332/(1+2.83051*exp(-0.01875*(Year-1950)))~Year,
        add=TRUE)
```

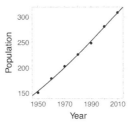

Refining Models Using More Data

Often a model can provide good information in response to input values that lie within the input interval used to create the model. However, considering values outside this input interval can produce quite diverse results, sometimes yielding outputs that are reasonable and sometimes outputs that are quite the opposite and not reasonable in the least. By collecting more data to enlarge the available interval of inputs with known outputs, the model can be refined and, hopefully, produce accurate predictions for a wider input interval. The following example illustrates the need to refine a model of a given phenomenon based on additional data. This approach implements the fourth step of the modeling cycle: testing the model against known data and modifying it as needed.

◆ **EXAMPLE 4** The following three data sets contain how many millions of people used Twitter during each quarter of a year for various years. For all three data sets, the input variable is Year and the output variable is Users.

(a) The data set stored in **TwitterUsers1** contains how many millions of people used Twitter from the first quarter of 2012 to the end of the third quarter of 2013. Graphically identify a reasonable model for this data set.

(b) The data set stored in **TwitterUsers2** contains how many millions of people used Twitter from the first quarter of 2010 to the end of the third quarter of 2013. Graphically determine whether the type of model identified in part (a) is still reasonable. If not, identify a more reasonable model.

(c) The data set stored in **TwitterUsers3** contains how many millions of people used Twitter from the first quarter of 2010 to the beginning of 2016. Graphically determine whether the type of model identified in part (b) is still reasonable. If not, identify a more reasonable model.

Solution. After loading each data set, the three plots used to identify models are graphed: a standard plot (Year, Users) to determine whether a linear, sine, or sigmoidal model might be reasonable; and a semi-log plot (Year, ln(Users)) and a log-log plot (ln(Year), ln(Users)) to determine whether an exponential or a power function model might be reasonable.

(a) Use **plotPoints** to generatre the three plots using **TwitterUsers1** as shown below. Based on the standard plot, a linear function Users $= m \cdot$ Year $+ b$ appears to provide the most reasonable model for this first data set.

```
plotPoints(Users~Year,data=TwitterUsers1)
plotPoints(log(Users)~Year,data=TwitterUsers1)
plotPoints(log(Users)~log(Year),data=TwitterUsers1)
```

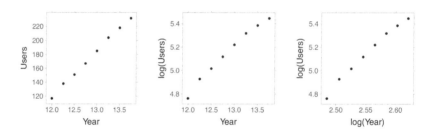

(b) Use **plotPoints** to generatre the three plots using **TwitterUsers2** as shown below. Based on the standard plot, a linear model might be reasonable. Although, in light of the additional data provided with this second data set, the best such model would need to be piecewise linear because of the two distinct slopes apparent before and after the first quarter of 2011. Examining the other two plots shows that the log-log plot appears to be approximately linear (or at least more so than the semi-log plot), which suggests that a power function model might be the most reasonable. Furthermore, a power function would provide a good model of the slow initial growth in the number of Twitter users up until 2011. Therefore, a power function Users $= C \cdot$ Year^k provides the most reasonable model for this second data set.

```
plotPoints(Users~Year,data=TwitterUsers2)
plotPoints(log(Users)~Year,data=TwitterUsers2)
plotPoints(log(Users)~log(Year),data=TwitterUsers2)
```

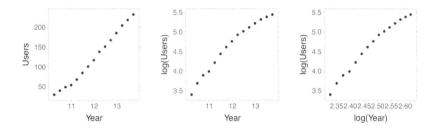

(c) Use `plotPoints` to generatre the three plots using `TwitterUsers2` as shown below. Based on the standard plot with this additional data, a sigmoidal model appears to provide the best model. In fact, a sigmoidal function might be expected to provide a good model for this phenomenon, because the total number of users of such social media typically levels off or even decreases over time. In addition, the continued growth toward infinity characteristic of linear, exponential, and power functions cannot happen in reality because of the limited population size of humanity. Therefore, a sigmoidal function $\text{Users} = \dfrac{L}{1 + Ce^{-k(\text{Year} - h)}} + v$ appears to provide the most reasonable model for this third data set.

```
plotPoints(Users~Year,data=TwitterUsers3)
plotPoints(log(Users)~Year,data=TwitterUsers3)
plotPoints(log(Users)~log(Year),data=TwitterUsers3)
```

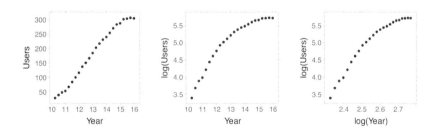

◆ **EXAMPLE 5** Question 3 examined the data set stored in `NaturalGasConsumption`, which contains the total number of millions of cubic feet of natural gas consumed in the United States from 1950 to 1970, and identified a linear model as the most reasonable. The extended data set stored in `NaturalGasConsumption2` contains the total number of millions of cubic feet of natural gas consumed in the United States from 1949 to 2015. Graphically determine whether a linear model is still reasonable and, if not, identify a more reasonable model.

Solution. First, use **names** to determine the data set's variables and then graph its standard plot as shown below. As can be seen, the extended data set is not linear. While a linear model would still capture the general trend of the data, such a function would not model the interesting fluctuations in the data. In fact, none of the common modeling functions can express the mixed increasing and decreasing behavior of the data in the midst of its overall linear trend.

```
names(NaturalGasConsumption2)
```

```
[1] "Year"        "CubicFeet"
```

```
plotPoints(CubicFeet~Year,data=NaturalGasConsumption2)
```

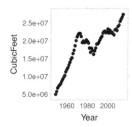

Sometimes the kind of "nonsolution" illustrated in Example 5 arises as different aspects of reality are studied and analyzed. Further studies in modeling include identifying additional functions for describing such data sets with more complicated behaviors. Also, arithmetic and other combinations of modeling functions can sometimes be created to describe such phenomenon more accurately. This more sophisticated work with models lies beyond the scope of this book and is left for the reader's future studies.

A Limitation of Mathematical Models

One big advantage of mathematical models is that they provide a formula that represents a given data set reasonably well. Such a formula enables two important tasks:

(1) conjecturing output values for various inputs, and

(2) analyzing the model with the tools of calculus (studied in Chapters 4–6).

A modeling function often provides a reasonable prediction of the corresponding output when the selected input lies within the input interval used to create the model. However, when trying to predict outputs for inputs that lies outside the input interval provided by the data, the resulting prediction may or may not be accurate. This process of seeking such outputs is called *extrapolation*, which is defined as follows:

> **Definition. Extrapolation** uses a mathematical model to attempt to predict an output value for an input that lies outside the input interval used to create the model.

In many settings, extrapolating with inputs immediately outside of the given data set might be expected to provide reasonable outputs. However, we cannot be completely certain that the trends observed in the data will continue or that a model will accurately describe the phenomenon being studied for any inputs outside the given input interval of the data. Indeed, our real-life experiences include settings with precipitous changes in output values that may or may not be expressed by a model. Furthermore, people are sometimes tempted to extrapolate to outputs that lie far outside the input interval of the data used to create the model. For such inputs, we should be far less confident that the observed trends in the data will continue and so less confident in the predictive accuracy of the model.

> **Note.** In general, a mathematical model can only be expected to provide reasonable outputs for inputs that lie inside the input interval used to create the model.

The following example and question explore these ideas in further detail:

◆ **EXAMPLE 6** The data set stored in `WorldPopulationChange` contains the percent growth of the world's population as a function of the year from 1970 to 2015. Find the best linear model for this data set using `fitModel` and then graph the model on a plot of the data to check its quality of fit. Using this model, conjecture the percent growth of the world's population in each of the following years and discuss the reasonableness of each prediction:

   (a) 1998   (b) 2025   (c) 1950

Solution. First, determine the variables using **names**, and then find the best possible linear model using `fitModel`:

```
names(WorldPopulationChange)
```

```
[1] "Year"    "Growth"
```

```
worldChangeModel=fitModel(Growth~m*Year+b,data=WorldPopulationChange)
coef(worldChangeModel)
```

```
        m           b
-0.0223006 45.9359576
```

Based on this RStudio output, the best possible linear model for the percent growth of the world's population is Growth $= -0.0223 \cdot$ Year $+ 45.936$. Now, graph this model on a plot of the data to check its quality of fit:

```
plotPoints(Growth~Year,data=WorldPopulationChange)
plotFun(-0.0223*Year+45.936~Year,add=TRUE)
```

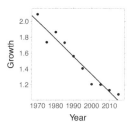

Because the data set is not perfectly linear, no linear model can perfectly match all of the data points. However, as can be seen, the line captures the overall trend of the data quite well and provides reasonable approximations for most of the data points. This model is now used to conjecture the percent growth of the world's population in the given years, and the reasonableness of each prediction is discussed.

(a) For 1998, the growth rate of the world's population is found by substituting Year = 1998 into worldChangeModel. The resulting growth rate of approximately 1.38% provided by the model is reasonably close to the actual growth rate of 1.295% reported by the U.S. Census Bureau. This relatively close match between predicted and observed growth rates is not unexpected, because 1998 lies within the input interval of the given data set.

```
worldChangeModel(1998)
```

```
[1] 1.37935
```

(b) For 2025, the growth rate of the world's population is found by substituting Year = 2025 into worldChangeModel. The resulting growth rate of 0.78% based on the model may or may not provide an accurate prediction of what will happen in 2025. The reliability of this extrapolation hinges on the trends in observed population growth rate from 1970 to 2015 continuing into the future, which may or may not happen.

```
worldChangeModel(2025)
```

```
[1] 0.77723
```

(c) For 1950, the growth rate of the world's population is found by substituting Year = 1950 into worldChangeModel. The resulting growth rate of 2.45% based on the model is much greater than the actual growth rate of 1.459% reported by the U.S. Census Bureau. As in part (b), this result is an extrapolation. The year 1950 lies far outside of the input interval of 1970 to 2015 used to create this linear model, and this model cannot be expected to necessarily provide accurate predictions for inputs such as 1950.

```
worldChangeModel(1950)
```

```
[1] 2.44978
```

■

➤ **QUESTION 4** The data set stored in `NaturalGasConsumption` contains the total number of millions of cubic feet of natural gas consumed in the United States from 1950 to 1970. Question 3 found that CF = $723,130 \cdot$ Year $- 1,404,932,269$ provides the best possible linear model for this data set. Using this linear model, conjecture U.S. natural gas consumption in each year and discuss the reasonableness of each prediction.

(a) 1949   (b) 1971   (c) 1986

Summary

- In order to identify a reasonable model for a given data set, first graph its standard plot (x, y) and check whether a linear, sine, or sigmoidal model is reasonable. If the standard plot does not identify a model, then graph both its semi-log plot $(x, \ln(y))$ and its log-log plot $(\ln(x), \ln(y))$ and check whether an exponential or a power function model is most reasonable.

- When identifying reasonable models, one must consider whether the selected function makes sense in the context of the situation being studied.

- Expanding the input interval with known outputs used to determine a model not only increases its range of accuracy, but also requires one to reaffirm whether a model is still reasonable or, if not, to identity a new type of model.

- *Extrapolation* is when a mathematical model is used to predict an output value for an input value that lies outside the input interval used to create the model. In general, a mathematical model is only expected to provide reasonable outputs for inputs inside of the input interval used to create the model.

Exercises

In Exercises 1–10, identify which type of model is most reasonable for the data set from among: linear, exponential, power function, sine, sigmoidal, or none of these. More than one type of model may be reasonable.

1.

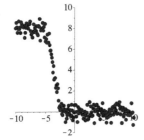

2.

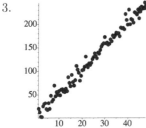

3.

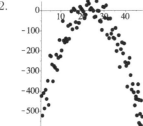

4.

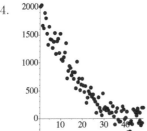

5.

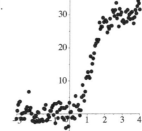

6.

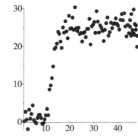

7.

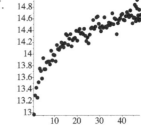

8.

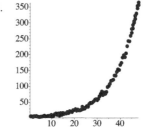

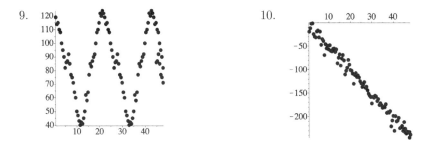

In Exercises 11–18, identify which type of model is most reasonable for the data set from among: linear, exponential, power function, sine, sigmoidal, or none of these. More than one type of model may be reasonable.

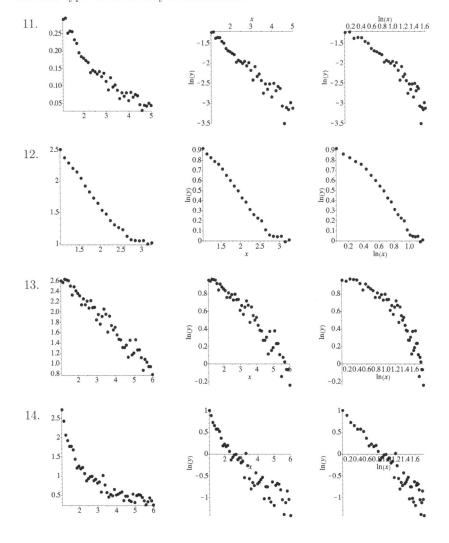

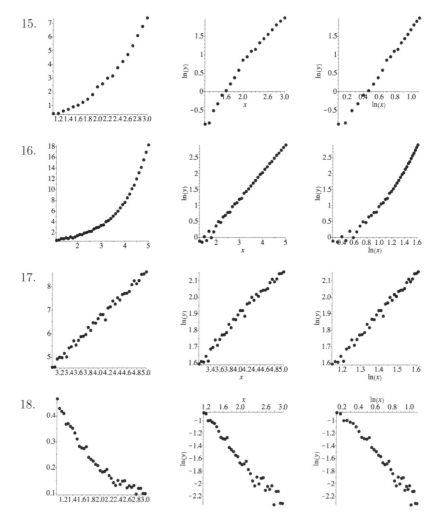

RStudio. In Exercises 19–30, use RStudio to identify which type of model is most reasonable for the data set from among: linear, exponential, power function, sine, sigmoidal, or none of these. More than one type of model may be reasonable.

19. Annual U.S. unemployment rate (R) each year (Y)

Y	2007	2008	2009	2010
R	4.6	5	7.8	9.7

20. Annual e-commerce sales (S) in the United States in billions of dollars for each year (Y)

Y	2009	2010	2011	2012
S	121	143	168	192

21. S&P 500 stock market closing value (V) on June 1 of each year (Y) since 1975 in U.S. dollars

Y	0	10	20	30	40
V	95	192	545	1191	2077

22. Percent of children in 1990 with telephone service (P) by parental income percentile (I)

I	5	15	30	50
P	68.7	79.7	90.8	96.5

23. High value of Facebook stock (H) each month (M) in 2014

M	1	2	3	4	6
H	63.4	71.4	72.6	63.9	68

24. Total annual sales (S) in thousands of hybrid vehicles each year (Y) in the United States

Y	2010	2011	2012	2013
S	274	266	435	496

25. Three-year average monthly pollen count (C) each month (M) in Brooklyn, New York City

M	8	9	10	11	12
C	8.4	4.6	1.7	0.3	0.1

26. U.S. field production of crude oil (O) in billions of barrels in each year (Y)

Y	1970	1980	2000	2005
O	3.5	3.1	2.1	1.9

27. Average number of goals (G) per game in World Cup tournaments each year (Y) since 1994

Y	0	4	8	12	16
G	2.7	2.7	2.5	2.3	2.3

28. Average monthly maximum temperature (T) in degrees Fahrenheit each month (M) in New York City

M	2	4	7	10	12
T	39.4	61.1	88.3	67.8	43.4

29. Percent (P) of each year's high school graduates enrolled in a two-year or four-year college each year (Y) since 2005

Y	0	1	4	5	7
P	68.6	66.0	70.1	68.1	66.2

30. world population growth rate (R) by year (Y) since 1995

Y	0	5	10	15	20
R	1.41	1.26	1.20	1.13	1.06

RStudio. In Exercises 31–38, use RStudio to examine the total U.S. electronic and mail-order shopping sales in billions of dollars each year from 1999 to 2012 stored in `ElectronicMailOrderSales`.

31. Graphically identify the most reasonable type of model for the data set.

32. Find the best model of the type from Exercise 31 for the data set using `fitModel`.

33. Graph the model from Exercise 32 on a plot of the data, and discuss its quality of fit.

34. Find sales for 2011 using the model from Exercise 32 and discuss the reasonableness of the result.

35. Find sales for 2001 using the model from Exercise 32 and discuss the reasonableness of the result.

36. Find sales for 1982 using the model from Exercise 32 and discuss the reasonableness of the result.

37. Find sales for 2016 using the model from Exercise 32 and discuss the reasonableness of the result.

38. Does the model from Exercises 31 and 32 make sense in the context of the data? Why or why not?

RStudio. In Exercises 39 – 46, use RStudio to examine the closing NASDAQ stock market value in U.S. dollars at the end of each quarter from March 1938 (quarter 1) through December 2014 (quarter 308) stored in `NASDAQQuarterly`.

39. Graphically identify the most reasonable type of model for the data set.

40. Find the best model of the type from Exercise 39 for the data set using `fitModel`.

41. Graph the model from Exercise 40 on a plot of the data, and discuss its quality of fit.

42. Find the NASDAQ value for quarter 225 (January to March 1994) using the model from Exercise 40 and discuss the reasonableness of the result.

43. Find the NASDAQ value for quarter 24 (October to December 1943) using the model from Exercise 40 and discuss the reasonableness of the result.

44. Find the NASDAQ value for quarter 312 (October to December 2016) using the model from Exercise 40 and discuss the reasonableness of the result.

45. Find the NASDAQ value for quarter −16 (January to March 1934) using the model from Exercise 40 and discuss the reasonableness of the result.

46. Does the model from Exercises 39 and 40 make sense in the context of the data? Why or why not?

RStudio. In Exercises 47 – 54, use RStudio to examine the interest rates on 15-year, fixed-rate conventional home mortgages annually from 1992 to 2014 stored in `Mortgage15YrAnnual`.

47. Graphically identify the most reasonable type of model for the data set.

48. Find the best model of the type from Exercise 47 for the data set using `fitModel`.

49. Graph the model from Exercise 48 on a plot of the data, and discuss its quality of fit.

50. Find the interest rate in 2007 using the model from Exercise 48 and discuss the reasonableness of the result.

51. Find the interest rate in 2009 using the model from Exercise 48 and discuss the reasonableness of the result.

52. Find the interest rate in 1963 using the model from Exercise 48 and discuss the reasonableness of the result.

53. Find the interest rate in 1981 using the model from Exercise 48 and discuss the reasonableness of the result.

54. Does the model from Exercises 47 and 48 make sense in the context of the data? Why or why not?

RStudio. In Exercises 55 – 62, use RStudio to examine the tidal measurements in Pearl Harbor, Hawaii stored in `Hawaii`.

55. Graphically identify the most reasonable type of model for the data set.

56. Find the best model of the type from Exercise 55 for the data set using `fitModel`.

57. Graph the model from Exercise 56 on a plot of the data, and discuss its quality of fit.

58. Find the depth at time 65 using the model from Exercise 56 and discuss the reasonableness of the result.

59. Find the depth at time 15 using the model from Exercise 56 and discuss the reasonableness of the result.

60. Find the depth at time 100 using the model from Exercise 56 and discuss the reasonableness of the result.

61. Find the depth at time 115 using the model from Exercise 56 and discuss the reasonableness of the result.

62. Does the model from Exercises 55 and 56 make sense in the context of the data? Why or why not?

RStudio. In Exercises 63 – 70, use RStudio to examine the field metabolic rate of individual birds and mammals in kilojoules per day as a function of body mass in kilograms stored in `BodyMassMetabolicRate`.

63. Graphically identify the most reasonable type of model for the data set.

64. Find the best model of the type from Exercise 63 for the data set using `fitModel`.

65. Graph the model from Exercise 64 on a plot of the data, and discuss its quality of fit.

66. Find the rate for a mass of 8 kg using the model from Exercise 64 and discuss the reasonableness of the result.

67. Find the rate for a mass of 12 kg using the model from Exercise 64 and discuss the reasonableness of the result.

68. Find the rate for a mass of 300 kg using the model from Exercise 64

and discuss the reasonableness of the result.

69. Find the rate for a mass of 200 kg using the model from Exercise 64 and discuss the reasonableness of the result.

70. Does the model from Exercises 63 and 64 make sense in the context of the data? Why or why not?

RStudio. In Exercises 71 – 78, use RStudio to examine U.S. carbon dioxide emissions in kT annually from 1980 to 2008 according to the World Bank stored in `USCO2Emissions`.

71. Graphically identify the most reasonable type of model for the data set.

72. Find the best model of the type from Exercise 71 for the data set using `fitModel`.

73. Graph the model from Exercise 72 on a plot of the data, and discuss its quality of fit.

74. Find the emissions in 1999 using the model from Exercise 72 and discuss the reasonableness of the result.

75. Find the emissions in 2005 using the model from Exercise 72 and discuss the reasonableness of the result.

76. Find the emissions in 2016 using the model from Exercise 72 and discuss the reasonableness of the result.

77. Find the emissions in 1972 using the model from Exercise 72 and discuss the reasonableness of the result.

78. Does the model from Exercises 71 and 72 make sense in the context of the data? Why or why not?

RStudio. In Exercises 79 – 86, use RStudio to examine the percent of high school completers to enroll in either a two-year or four-year college each year from 1972 to 2012 stored in `HSGradsinCollege`.

79. Graphically identify the most reasonable type of model for the data set.

80. Find the best model of the type from Exercise 79 for the data set using `fitModel`.

81. Graph the model from Exercise 80 on a plot of the data, and discuss its quality of fit.

82. Find the percentage in 2000 using the model from Exercise 80 and discuss the reasonableness of the result.

83. Find the percentage in 1982 using the model from Exercise 80 and discuss the reasonableness of the result.

84. Find the percentage in 1960 using the model from Exercise 80 and discuss the reasonableness of the result.

85. Find the percentage in 2020 using the model from Exercise 80 and discuss the reasonableness of the result.

86. Does the model from Exercises 79 and 80 make sense in the context of the data? Why or why not?

RStudio. In Exercises 87 – 94, use RStudio to examine the atmospheric carbon dioxide from Mauna Loa in ppmv (or parts per million by volume) as a function of years from 1958 to 2008 stored in `MaunaLoaCO2`.

87. Graphically identify the most reasonable type of model for the data set.

88. Find the best model of the type from Exercise 87 for the data set using `fitModel`.

89. Graph the model from Exercise 88 on a plot of the data, and discuss its quality of fit.

90. Find the atmospheric carbon dioxide in 2000 using the model from Exercise 88 and discuss the reasonableness of the result.

91. Find the atmospheric carbon dioxide in 1960 using the model from Exercise 88 and discuss the reasonableness of the result.

92. Find the atmospheric carbon dioxide in 2020 using the model from Exercise 88 and discuss the reasonableness of the result.

93. Find the atmospheric carbon dioxide in 1900 using the model from Exercise 88 and discuss the reasonableness of the result.

94. Does the model from Exercises 87 and 88 make sense in the context of the data? Why or why not?

RStudio. In Exercises 95 – 102, use RStudio to examine the altitude angle of the sun in Anchorage, Alaska, each hour from midnight on June 29, 2014 (hour 0) until midnight on June 30, 2014 (hour 24) stored in `SunPositionAlaska`.

95. Graphically identify the most reasonable type of model for the data set.

96. Find the best model of the type from Exercise 95 for the data set using `fitModel`.

97. Graph the model from Exercise 96 on a plot of the data, and discuss its quality of fit.

98. Find the altitude angle in hour 12 using the model from Exercise 96 and discuss the reasonableness of the result.

99. Find the altitude angle in hour 3 using the model from Exercise 96 and discuss the reasonableness of the result.

100. Find the altitude angle in hour 26 using the model from Exercise 96 and discuss the reasonableness of the result.

101. Find the altitude angle in hour 120 using the model from Exercise 96 and discuss the reasonableness of the result.

102. Does the model from Exercises 95 and 96 make sense in the context of the data? Why or why not?

RStudio. In Exercises 103 – 110, use RStudio to examine the number of physicians per 1000 people as a function of average life expectancy in different countries in 2010 stored in `LifeExpectancyPhysicians`.

103. Graphically identify the most reasonable type of model for the data set.

104. Find the best model of the type from Exercise 103 for the data set using `fitModel`.

105. Graph the model from Exercise 104 on a plot of the data, and discuss its quality of fit.

106. Find the number of physicians for a life expectancy of 50 years using the model from Exercise 104 and discuss the reasonableness of the result.

107. Find the number of physicians for a life expectancy of 75 years using the model from Exercise 104 and discuss the reasonableness of the result.

108. Find the number of physicians for a life expectancy of 95 years using the model from Exercise 104 and discuss the reasonableness of the result.

109. Find the number of physicians for a life expectancy of 35 years using the model from Exercise 104 and discuss the reasonableness of the result.

110. Does the model from Exercises 103 and 104 make sense in the context of the data? Why or why not?

RStudio. In Exercises 111 – 113, use RStudio to examine the average cumulative debt of bachelor's degree students enrolled in public colleges and universities.

111. Graphically identify the most reasonable type of model for the data stored in `StudentDebt1`.

112. Using the additional data stored in `StudentDebt2`, determine if the model in Exercise 111 is still reasonable or identify a more reasonable model.

113. Do the models in Exercises 111 and 112 make sense in the context of the data? Why or why not?

RStudio. In Exercises 114 – 116, use RStudio to examine the number of males per 100 females in the word population.

114. Graphically identify the most reasonable type of model for the data stored in `GenderRatio1`.

115. Using the additional data stored in `GenderRatio2`, determine if the model in Exercise 114 is still reasonable or identify a more reasonable model.

116. Do the models in Exercises 114 and 115 make sense in the context of the data? Why or why not?

RStudio. In Exercises 117–119, use RStudio to examine the number of burgers (in billions) sold by McDonald's since 1955.

117. Graphically identify the most reasonable type of model for the data stored in McDBurgers1.

118. Using the additional data stored in McDBurgers2, determine if the model in Exercises 117 is still reasonable or identify a more reasonable model.

119. Do the models in Exercises 117 and 118 make sense in the context of the data? Why or why not?

RStudio. In Exercises 120–122, use RStudio to examine the interest rate on a 30-year fixed-rate conventional home mortgage.

120. Graphically identify the most reasonable type of model for the data stored in Mortgage30YrMonthly1.

121. Using the additional data stored in Mortgage30YrMonthly2, determine if the model in Exercises 120 is still reasonable or identify a more reasonable model.

122. Do the models in Exercises 120 and 121 make sense in the context of the data? Why or why not?

RStudio. In Exercises 123–125, use RStudio to examine the volume of Ford Motor Company stock shares traded per quarter since January 1, 2007.

123. Graphically identify the most reasonable type of model for the data stored in FordMarketVolume1.

124. Using the additional data stored in FordMarketVolume2, determine if the model in Exercises 123 is still reasonable or identify a more reasonable model.

125. Do the models in Exercises 123 and 124 make sense in the context of the data? Why or why not?

RStudio. In Exercises 126–128, use RStudio to examine the number of barrels (in thousands) of crude oil produced per year in the United States.

126. Graphically identify the most reasonable type of model for the data stored in OilProductionAnnual1.

127. Using the additional data stored in OilProductionAnnual2, determine if the model in Exercises 126 is still reasonable or identify a more reasonable model.

128. Do the models in Exercises 126 and 127 make sense in the context of the data? Why or why not?

RStudio. In Exercises 129–130, use RStudio to determine a model that fits the data reasonably well and makes sense in the context of the data.

129. Consider the millions of Latinos living in the United States stored in HispanicPopulation.

130. Consider the pollen count each day in Los Angeles from April to July of 2014 stored in PollenCountLA.

In Your Own Words. In Exercises 131–136, explain the following.

131. The modeling cycle

132. Graphically identifying reasonable models

133. Influence of context on choosing models

134. Impact of new data on a model

135. Extrapolation

136. Inputs for which a model is reliable

2.7   Dimensional Analysis

Humans care about many aspects of reality, from the price of textbooks to interpersonal relationships, from weather events to changes in population. Many aspects of reality can be measured and described in terms of numerical quantities, most of which are expressed using some kind of units. In such diverse settings as biology, economics, medicine, and psychology, dimensions and their associated units play an essential role in understanding our world. Dimensions and units are pivotal in the study and development of mathematical models as well. This section discusses dimensions and units, and uses them to answer certain modeling questions.

Length is a familiar example of a *dimension*. Length is a dimension because it provides a means for measuring a quantity. Many different *units* are used to measure length, including: feet, inches, meters, centimeters, kilometers, versts, leagues, and sazhen. These diverse choices for units mean that many different numerical values can be assigned to the same measurement. For example, a distance of one mile can be expressed as follows, among many possible options:

1 mile = 5280 feet = 63,360 inches = 1609.34 meters = 1.60934 kilometers

This example illustrates the distinction between the two concepts of dimension and unit, which are defined as follows:

Definition.

- A **dimension** is an inherent way of measuring a quantity.

- A **unit** is a way to assign a numerical value to the measurement of a quantity.

➤ **QUESTION 1**

(a) Name three additional units associated with the dimension of length.

(b) State another dimension and name three units associated with that dimension.

In many ways, the concept of dimension is familiar. At the same time, many find this notion unusual because, more often than not, we refer to the units associated with a particular quantity. A numerical value of a measurement and its corresponding units are important, but are also very specific. On a more general level, dimension can be thought of as an abstraction of specific measurements and their units. As will be shown, working with this broader notion of dimension enables an effective analysis of otherwise intractable questions.

Fundamental and Derived Dimensions

People have identified many distinct dimensions, and these are placed into one of two categories. The most basic measurable quantities, such as length and time, are known as *fundamental dimensions*. In addition, we often work with *derived dimensions*, such as velocity and acceleration, which are dimensions that can be expressed in terms of fundamental dimensions.

Definition.

- A **fundamental dimension** is a dimension that cannot be expressed in terms of other dimensions.

- A **derived dimension** is a dimension that can be expressed as a product of integer powers of fundamental dimensions.

Scientists recognize seven physical fundamental dimensions, with the most familiar being length, mass, and time. The other four are given in Table 1, along with both the symbol used to represent the dimension and some examples of their associated units. These seven physical fundamental dimensions are the only such dimensions. Non-physical fundamental dimensions are discussed shortly.

Dimension	*Symbol*	*Examples of units*
Length	L	foot, mile, meter, kilometer
Mass	M	gram, pound, ton, stone
Time	T	second, hour, day, century
Amount of substance	N	mole
Temperature	Θ	kelvin, degree F, degree C
Electric current	I	ampere, coulombs per second
Luminous intensity	J	candela, candlepower, Hefnerkerze

Table 1: The seven physical fundamental dimensions

As might be expected when studying mathematics, dimensions are regularly identified by means of their symbols. The dimension of a quantity is expressed using the notation "[quantity] = dimension symbol," such as [length] $= L$ and [time] $= T$. This same notation is used for specific numeric examples, such as [3 feet] $= L$ and [2 days] $= T$. With practice, this symbolism will become familiar and natural.

Working with the fundamental dimensions as the basic building blocks, a wide variety of derived dimensions can be specified. By definition, a derived dimension can be expressed as a product of integer powers of fundamental dimensions, where the integers are $\ldots, -3, -2, -1, 0, 1, 2, 3, \ldots$ and consist of all the whole numbers together with their negatives. Table 2 presents some examples of physical derived dimensions. The symbol associated with each example illustrates the definition of derived dimension in action. The same bracket notation "[quantity] = dimension symbol" is used to express both fundamental and derived dimensions. For example, [area] $= L^2$ and [4 square meters] $= L^2$ as well as [velocity] $= LT^{-1}$ and [35 miles per hour] $= LT^{-1}$.

Dimension	Intuition	Symbol	Examples of units
Area	length × width	L^2	square foot
Volume	length × width × height	L^3	meter cubed
Density	mass per volume	ML^{-3}	ounces per cubic inch
Frequency	per time (how often)	T^{-1}	per minute
Velocity	distance per time	LT^{-1}	kilometers per hour
Acceleration	velocity per time	LT^{-2}	feet per square second
Force	$F = ma$	MLT^{-2}	gram meters per square hour
Pressure	force per area	$ML^{-1}T^{-2}$	grams per meter square second
Energy	force × displacement	ML^2T^{-2}	joule, newton-meter

Table 2: Some physical derived dimensions

Note. A parallel can be drawn between the concepts of fundamental and derived dimensions on the one hand, and the concepts of prime numbers and positive rational numbers on the other. Recall that a prime number's only positive divisors are one and itself. For example, 2 can only be divided by 1 and 2, and 5 can only be divided by 1 and 5. Positive rational numbers can be written as fractions $\frac{p}{q} = p \cdot q^{-1}$ where p and q are products of integer powers of primes. Examples include the following:

$$\frac{12}{5} = \frac{2^2 \cdot 3}{5} = 2^2 \cdot 3^1 \cdot 5^{-1} \qquad\qquad \frac{49}{18} = \frac{7^2}{2 \cdot 3^2} = 7^2 \cdot 2^{-1} \cdot 3^{-2}$$

In essentially the same way, derived dimensions are expressed as products of integer powers of fundamental dimensions.

In addition to the physical fundamental dimensions, widespread use is made of other nonphysical fundamental dimensions, such as those in Table 3.

Dimension	Symbol	Examples of units
Angle	A	degree, radian
Money	B	dollar, euro, yen
Population	P	humans, dogs, mice

Table 3: Some nonphysical fundamental dimensions

Similarly, various nonphysical derived dimensions are used, such as those in Table 4. Finally, some quantities are pure numbers and are said to be dimensionless.

Definition. A pure number that does not have a dimension is said to be **dimensionless**. The symbol "1" denotes a dimensionless quantity and is written as [number] = 1.

Dimension	Intuition	Symbol	Examples of units
Wages	money per time	BT^{-1}	dollars per hour
Population growth	population per time	PT^{-1}	people per year
Population density	population per area	PL^{-2}	people per square mile
Rotational speed	angle per time	AT^{-1}	degrees per second

Table 4: Some nonphysical derived dimensions

For example, the number "6" is dimensionless, but "6 inches" has the dimension of length. The bracket notation for dimensions extends to pure numbers by writing [number] = 1. Continuing with the example with the number "6", one writes [6] = 1.

In many mathematics courses, relatively little attention is paid to units in the midst of focusing on learning new ideas and methods. However, most quantities have dimensions and units associated with them, and these play an important role in exploring and understanding this world and other worlds as well. One infamous example of a units mismatch resulted in a NASA satellite sent to orbit Mars dipping too far down into the Martian atmosphere, which overheated its propulsion system and sent it careering off into space. Following the loss of the $125 million Mars Climate Orbiter on September 23, 1999, investigators learned that NASA was using the metric system, which is standard in the scientific community, while one of its contractors used English units of measurement. Among the many possible lessons here, careful attention must be paid to dimensions and units.

Arithmetic with Dimensions and Units

Using dimensions to analyze mathematical models hinges on understanding how to compute quantities with dimension and units. A number of rules have been established for combining dimensional quantities, which means that some computations are possible (and are referred to as being **valid**), while others are not possible (and are referred to as being **invalid**). Four of these rules are introduced along with examples of their application.

RULE 1 OF DIMENSIONAL ARITHMETIC.
When adding, subtracting, or comparing (e.g., with =, ≥, or <) two quantities, they must have the same dimension, although their units may be different. The resulting dimension is equal to that of the quantities being added, subtracted, or compared.

◆ **EXAMPLE 1** Identify each computation or comparison as valid or invalid using rule 1:

(a) 3.2 meters + 4.5 inches

(b) 2.1 hours − 3.675 meters

(c) 18 feet/hour < 22 miles/second

(d) 5° Fahrenheit = $2.18

Solution.

(a) The quantities 3.2 meters and 4.5 inches both have the fundamental dimension of length, which is written as [3.2 meters] = L and [4.5 inches] = L. Therefore, this computation is valid. The difference of units between meters and inches does not affect the validity of the computation.

(b) These quantities being subtracted have the fundamental dimensions of [2.1 hours] = T for time and [3.675 meters] = L for length. The dimensions of these quantities are different, which means the computation is invalid and this difference cannot be computed.

(c) The comparison is valid because the quantities have the same derived dimension of LT^{-1} for velocity. As in (a), the different units for velocity do not affect the validity of this comparison.

(d) The equality is stated between [5° Fahrenheit] = Θ with a dimension of temperature and [$2.18] = B with a dimension of money, which are different fundamental dimensions. Therefore, this comparison is invalid.

▮

➤ **QUESTION 2** Identify each computation or comparison as valid or invalid using rule 1. Justify your answers as shown in Example 1:

(a) 4.9 square feet + 8.5 cubic centimeters

(b) 2.25 hours − 5.3 seconds

(c) 0 kelvin < 0° centigrade

(d) 8.3 grams/minute = 1.6 lbs/m^3

In addition to determining the validity of adding, subtracting, and comparing quantities, rule 1 of dimensional arithmetic sometimes enables the determination of unknown dimensions.

◆ **EXAMPLE 2** Determine the dimensions of the quantities a and b in each setting. Recall that M is the symbol for the fundamental dimension of mass, B is the symbol for money, and T is the symbol for time.

(a) $a + b − c$, when $[c] = M$

(b) $a > b + c$, when $[c] = BT^{-1}$

Solution.

(a) According to rule 1, quantities can only be added and subtracted when they have the same dimension. Therefore, both $[a] = [c] = M$ and $[b] = [c] = M$.

(b) Applying rule 1 to the sum $b + c$ gives $[b] = [c] = BT^{-1}$ and so $[b + c] = BT^{-1}$ because the resulting dimension is equal to the dimension of the quantities being added. Also, quantities must have the same dimension to be compared, which means that $[a] = [b + c] = BT^{-1}$.

■

➤ **QUESTION 3** Determine the dimensions of the quantities a and b in each setting. Recall that P is the symbol for the fundamental dimension of population.

(a) $a + b \leq c$, when $[c] = P$ (b) $a - b + 3$

Rule 1 of dimensional arithmetic asserts strict dimension requirements for sums, differences, and comparisons of quantities. In contrast, rule 2 observes that products and quotients of quantities can always be computed independent of their dimensions.

RULE 2 OF DIMENSIONAL ARITHMETIC.
When multiplying and dividing quantities, they can have any dimensions and any units. The resulting dimension is equal to the product or quotient of the dimensions of the quantities being multiplied or divided.

◆ **EXAMPLE 3** Identify the dimension and units associated with each quantity using rule 2:

(a) The distance traveled by a car going 55 miles per hour for half an hour.

(b) The population density of Chicago, Illinois in 2010 when 2.696 million people lived in 227.63 square miles.

Solution.

(a) The velocity of 55 miles per hour has dimension LT^{-1} and the time of 0.5 hours has dimension T. The distance the car travels is found by taking the product of these two quantities to obtain the dimension of $LT^{-1} \cdot T = L$, or length. The units are miles per hour multiplied by hour, which simplifies to miles.

(b) The population of 2.696 million people has dimension P and the area of 227.63 square miles has dimension L^2. The population density is equal to the size of the population divided by the area of the city, producing a dimension of $P/L^2 = PL^{-2}$ with units of millions of people per square mile.

■

➤ **QUESTION 4** Identify the dimension and units associated with each quantity using rule 2. Justify your answers as shown in Example 3.

(a) The average force acting on a 3200 pound car that accelerates from 0 to 60 miles per hour in 9 seconds.

(b) The concentration of the chlorine in a 50 meter Olympic size swimming pool that contains 2500 cubic meters of water and 5000 grams of chlorine.

A combined application of rules 1 and 2 of dimensional arithmetic enables the determination of unknown dimensions in even more settings.

◆ **EXAMPLE 4** Determine the dimension of the quantity a in each setting. Recall that T is the symbol for time, L is the symbol for length, and B is the symbol for money.

(a) $a = 4bc$, when $[b] = T$ and $[c] = LT^{-1}$

(b) $ab \leq c^2$, when $[b] = B$ and $[c] = BT^{-1}$

Solution.

(a) According to rule 2, the dimension of $4bc$ is the product of the dimensions of the quantities being multiplied. Because 4 is a dimensionless pure number, rule 2 yields the following:

$$[4bc] = [4] \cdot [b] \cdot [c] = 1 \cdot T \cdot LT^{-1} = L \cdot T \cdot T^{-1} = L \cdot 1 = L$$

Applying the rule 1 requirement that quantities must have the same dimension in order to be compared yields $[a] = [4bc] = L$.

(b) The dimension of a is $[a] = BT^{-2}$ as found by the following procedure:

$$[ab] = [c^2] \qquad \text{Rule 1: compared quantities have the same dimensions}$$
$$[a] \cdot [b] = [c] \cdot [c] \qquad \text{Rule 2: the dimension of a product is the product of the dimensions}$$
$$[a] \cdot B = BT^{-1} \cdot BT^{-1} \qquad \text{Substitute known dimensions } [b] = B \text{ and } [c] = BT^{-1}$$
$$[a] = T^{-1} \cdot BT^{-1} \qquad \text{Divide both sides by } B$$
$$[a] = BT^{-2} \qquad \text{Apply exponent rule } a^r a^s = a^{r+s}$$

■

➤ **QUESTION 5** Determine the dimension of the quantity a in each setting. Recall that P is the symbol for the fundamental dimension of population.

(a) $ab^2 = 3c$, when $[b] = P$ and $[c] = P^2 L^{-2}$

(b) $ab \geq 5 + c$, when $[b] = ML^{-3}$

RULE 3 OF DIMENSIONAL ARITHMETIC.
In an exponential expression a^r, the exponent r must be dimensionless.

◆ **EXAMPLE 5** Identify each computation as valid or invalid using rule 3:

(a) $5^{8\,\mathrm{ft}}$ (b) $e^{7\,\mathrm{ft}/6\,\mathrm{in}}$

Solution.

(a) The quantity "8 ft" has dimension of length or, written symbolically, $[8\,\mathrm{ft}] = L$. Because rule 3 asserts that the exponent must be dimensionless, the computation is invalid.

(b) Both "7 ft" and "6 in" have dimension of length, or, written symbolically, $[7\,\mathrm{ft}] = [6\,\mathrm{in}] = L$. Therefore, using rule 2, the dimension of 7 ft/6 in is $L/L = 1$, which means that the exponent is dimensionless. Thus, according to rule 3, the computation is valid.

                                                                                              ■

➤ **QUESTION 6** Identify each computation as valid or invalid using rule 3:

(a) $5^{3\,\mathrm{hr}/6\,\mathrm{min}}$ (b) $e^{3\,\mathrm{kg}}$

Rule 3 of dimensional arithmetic can also be used to determine unknown dimensions.

◆ **EXAMPLE 6** Suppose $y = a \cdot 2^{bt} + cd$ where $[b] = T^{-1}$, $[y] = B$, and $[c] = BT^{-1}$. Determine the dimensions $[a]$, $[t]$, and $[d]$.

Solution. First, find the dimension $[t]$:

$[bt] = 1$	Rule 3: exponent must be dimensionless
$[b] \cdot [t] = 1$	Rule 2: the dimension of a product is the product of the dimensions
$T^{-1} \cdot [t] = 1$	Substitute known dimension $[b] = T^{-1}$
$[t] = 1/T^{-1} = T$	Solve for $[t]$

From rule 1, $[y] = [a \cdot 2^{bt}] = [cd] = B$, and then applying rule 2 gives $[y] = [a] \cdot [2^{bt}]$. Because 2 is dimensionless, $[2] = 1$ and $[2^{bt}] = 1$. Thus, $[y] = [a] \cdot [2^{bt}] = [a] \cdot 1 = [a] = B$. Finally, another application of rule 2 yields $[y] = [cd] = [c] \cdot [d] = BT^{-1} \cdot [d]$. Substituting $[y] = B$ provides $B = BT^{-1} \cdot [d]$. Dividing both sides of this equation by BT^{-1} and simplifying gives $[d] = B/(BT^{-1}) = BB^{-1}T = T$.

                                                                                              ■

➤ **QUESTION 7** Suppose $y = ke^m + c$, where the numeric base e is dimensionless and $[c] = MLT$. Determine the dimensions $[y]$, $[k]$, and $[m]$.

Solving for Unknown Dimensions

An important element of dimensional analysis is manipulating dimensional quantities based on the rules of dimensional arithmetic. In addition to specific numeric computations, these rules can be applied in a more general fashion. While working with models in some settings, the dimensions of some, but not all, of the relevant quantities might be known. These rules enable the determination of unknown dimensions based on the known dimensions of the other quantities in the computation, as illustrated in the following examples.

◆ **EXAMPLE 7** Suppose $y = mx + b$, where the dimension of y is length, or $[y] = L$, and the dimension of m is force, or $[m] = MLT^{-2}$. Determine the dimensions $[x]$ and $[b]$.

Solution. First, observe that $[y] = [mx] = [b]$ because rule 1 requires quantities being added together to have the same dimensions. Therefore, the given information that $[y] = L$ yields both $[mx] = L$ and $[b] = L$. Applying rule 2 that the dimension of a product is the product of the corresponding dimensions gives $[mx] = [m] \cdot [x]$. Substituting $[mx] = L$ and $[m] = MLT^{-2}$ into this equality provides the dimension $[x]$ as follows:

$$[mx] = [m] \cdot [x] \qquad \text{Rule 2: the dimension of a product is the product of the dimensions}$$
$$L = MLT^{-2} \cdot [x] \qquad \text{Substitute known dimensions } [mx] = L \text{ and } [m] = MLT^{-2}$$
$$\frac{L}{MLT^{-2}} = [x] \qquad \text{Divide by } MLT^{-2} \text{ to solve for } [x]$$
$$M^{-1}T^2 = [x] \qquad \text{Cancel } L\text{'s and apply exponent rule } \frac{1}{a^r} = a^{-r}$$

◆ **EXAMPLE 8** Suppose that $p = c^3 t + r$ and that the dimensions of both t and r are money per time, or $[t] = [r] = BT^{-1}$. Determine the dimensions $[p]$ and $[c]$.

Solution. Rule 1 for sums yields $[p] = [c^3 t] = [r]$. Because $[r] = BT^{-1}$ both $[p] = BT^{-1}$ and $[c^3 t] = BT^{-1}$. Finally, $[c] = 1$ based on the following:

$$[c^3 t] = [c^3] \cdot [t] \qquad \text{Rule 2: the dimension of a product is the product of the dimensions}$$
$$[c^3 t] = [c]^3 \cdot [t] \qquad \text{Rule 2: the dimension of a product is the product of the dimensions}$$
$$BT^{-1} = [c]^3 \cdot BT^{-1} \qquad \text{Substitute known dimensions } [c^3 t] = BT^{-1} \text{ and } [t] = BT^{-1}$$
$$1 = [c]^3 \qquad \text{Divide by } BT^{-1} \text{ to solve for } [c]^3$$
$$1 = [c] \qquad \text{Rule 2: the dimension of a product is the product of the dimensions}$$

➤ **QUESTION 8** Let $r = a \cdot q^2 + k$. Use the rules of dimensional arithmetic and the given dimensions to find the unknown dimensions.

(a) Let $[r] = P$ and $[q] = N$, and determine the dimensions $[a]$ and $[k]$.

(b) Let $[a] = T^2$ and $[k] = 1$, and determine the dimensions $[r]$ and $[q]$.

◆ **EXAMPLE 9** The Ideal Gas Law states that $pv = nrt$, where p is the absolute pressure of the gas, v is the volume of the gas, n is the amount of substance of gas (measured in moles), r is the ideal gas constant, and t is the absolute temperature of the gas. Determine the dimensions $[p]$, $[v]$, $[n]$, $[r]$, and $[t]$.

Solution. The dimensions of $[p]$, $[v]$, $[n]$, and $[t]$ are determined by the physical quantities they represent. Pressure is force per area and has dimension $[p] = MLT^{-2}/L^2 = ML^{-1}T^{-2}$. Similarly, volume has dimension $[v] = L^3$, $[n] = N$, and $[t] = \theta$. The dimension of the ideal gas constant r is found using the rules of dimensional arithmetic as follows:

$$[pv] = [nrt] \qquad \text{Rule 1}$$
$$[p] \cdot [v] = [n] \cdot [r] \cdot [t] \qquad \text{Rule 2}$$
$$ML^{-1}T^{-2} \cdot L^3 = N \cdot [r] \cdot \theta \qquad \text{Substitute known dimensions}$$
$$ML^2T^{-2} = N\theta \cdot [r] \qquad \text{Simplify the exponent of } L \text{ with exponent rule } a^r a^s = a^{r+s}$$
$$\frac{ML^2T^{-2}}{N\theta} = [r] \qquad \text{Divide both sides by } N\theta$$
$$ML^2T^{-2}N^{-1}\theta^{-1} = [r] \qquad \text{Apply exponent rule } \frac{1}{a^r} = a^{-r}$$

■

Generalized Products

The goal of dimensional analysis is to develop a model of some phenomenon of interest using the dimensions of the quantities relevant to the phenomenon. This process assumes a particular form for the usually multivariable model that consists of setting the output variable equal to a *generalized product* of the input variables.

Working with derived dimensions included specific instances of generalized products, because derived dimensions are equal to a product of fundamental dimensions raised to integer exponents. Perhaps the most familiar examples are velocity, which is expressed as $[\text{velocity}] = LT^{-1}$, and acceleration, which is expressed as $[\text{acceleration}] = LT^{-2}$. These specific examples extend to the notion of generalized products defined as follows:

> **Definition.** The **generalized product** of dimensional quantities a and b with output c is written $c = k\,a^x b^y$, where k is a dimensionless constant and x and y are chosen so that the equality is valid. The generalized product of dimensional quantities $a_1, a_2, \ldots, a_n$ with output c is written $c = ka_1^{x_1} a_2^{x_2} \cdots a_n^{x_n}$, where k is a dimensionless constant and $x_1, x_2, \ldots, x_n$ are chosen so that the equality is valid.

Recall from rule 1 of dimensional arithmetic that an equality is valid exactly when the dimensions of the quantities on both sides of the equation are the same.

◆ **EXAMPLE 10** State the generalized product of the dimensional quantities:

(a) Output of distance (d) from inputs of mass (m), time (t), and velocity (v).

(b) Output of time (t) from inputs of weight (w), length (ℓ), and cost (c).

Solution.

(a) The generalized product is $d = k\,m^x\,t^y\,v^z$.

(b) The generalized product is $t = k\,w^x\,\ell^y\,c^z$.

■

As suggested by Example 10, the variable associated with each quantity is used when stating these generalized products, rather than the variable's corresponding dimensional symbol. Dimensional analysis involves working with such substitutions. However, the starting point is always stating the generalized product using a given or appropriate variable for each input and output appearing in the situation being modeled.

➤ **QUESTION 9** State the generalized product of the dimensional quantities:

(a) Output of cost (c) from inputs of amount (a) and temperature (t).

(b) Output of time (t) from inputs of population (p), frequency (f), and cost (c).

The process of dimensional analysis continues by substituting the dimensional symbols for the associated quantities and then trying to find values for the variable exponents that make the equation dimensionally compatible. Solving for the exponents involves isolating the dimensions of the known quantities from the unknown exponent, which is accomplished via rule 4 of dimensional arithmetic.

RULE 4 OF DIMENSIONAL ARITHMETIC.
The equation $c = k\,a_1^{x_1}a_2^{x_2}\dots a_n^{x_n}$ is dimensionally valid when

$$
\begin{aligned}
[c] &= [k\,a_1^{x_1}a_2^{x_2}\dots a_n^{x_n}] \\
&= [k]\cdot[a_1]^{x_1}\cdot[a_2]^{x_2}\cdots[a_n]^{x_n}.
\end{aligned}
$$

As an illustration of rule 4, consider the generalized products in Example 10. Using the fact that k is dimensionless (and so $[k] = 1$) and applying rule 4 to these generalized products provides the following results:

(a) $[d] = [k\,m^x\,t^y\,v^z] = [k]\cdot[m]^x\cdot[t]^y\cdot[v]^z = 1\cdot[m]^x\cdot[t]^y\cdot[v]^z = [m]^x\cdot[t]^y\cdot[v]^z$

(b) $[t] = [k\,w^x\,\ell^y\,c^z] = [k]\cdot[w]^x\cdot[\ell]^y\cdot[c]^z = 1\cdot[w]^x\cdot[\ell]^y\cdot[c]^z = [w]^x\cdot[\ell]^y\cdot[c]^z$

For additional immediate practice, you might apply rule 4 to the generalized products that are the answers to Question 9.

Dimensional Analysis

Mathematicians and scientists in diverse disciplines study physical and social phenomena and then try to describe them using the language of mathematics. Another powerful tool when modeling is called *dimensional analysis*. The underlying goal in dimensional analysis is the same as that for single-variable modeling studied in Sections 2.1–2.6: find a function that accurately expresses a phenomenon of interest. The difference between these two methods is the information available to create the model. For single-variable modeling, a data set serves as the basis for the model. For dimensional analysis, assumptions are made about which dimensional quantities are involved in the model, and then a dimensionally compatible way to combine these quantities is sought.

The first step in dimensional analysis is to identify the relevant quantities, which requires us to make assumptions about which quantities are relevant and which are irrelevant to the situation being modeled. Ideally, all of the relevant quantities will be included in the initial assumptions, leaving none out and including no extras. However, sometimes, these assumptions may result in a dimensionally incompatible model, which will require revisiting and modifying the assumptions. While frustrating, making such revisions is common during the modeling process and provides important insights that enable refinements and improvements of a model. The process of dimensional analysis involves the following six steps:

DIMENSIONAL ANALYSIS.

(1) Identify the relevant input and output quantities.

(2) Write the output as a generalized product of the inputs.

(3) Find a dimensional version of the equation from step (2).

(4) Use rule 1 of dimensional arithmetic to solve for the unknown exponents.

(5) State the model based on steps (2) and (4).

(6) Interpret the resulting model.

A couple of features of this process of dimensional analysis are highlighted before diving into some examples. When writing the output as a generalized product of the inputs in step (2), the actual form of the desired model is explicitly being stated. Almost always, this model is a multivariable function. The generalized product form used by dimensional analysis is a choice, and there might be other relevant choices. However, this type of choice covers many relevant situations and is relatively simple in nature.

Steps (3) and (4) determine the specific, numeric exponents in the generalized product that lead to the actual model in step (5). Finally, step (6) involves interpreting the resulting model. This interpretation and analysis of the model can take on many different forms, depending on the resulting model and how one plans to use the model.

The process of dimensional analysis is illustrated by working through two specific examples in detail. Each step of the dimensional analysis process of modeling is clearly identified in these examples.

◆ **EXAMPLE 11** Suppose that a ball of mass m is thrown directly upwards with an initial velocity v. Find an equation for the height of the ball using dimensional analysis

Solution.
Step (1): Identify the relevant input and output quantities.
The relevant quantities involved in finding the height of the ball are assumed to be the mass of the ball, its initial velocity, gravity, and its height. Furthermore, assuming the

height of the ball depends on the other three quantities, the output is height (h) from inputs of mass (m), initial velocity (v), and gravity (g).

Step (2): Write the output as a generalized product of the inputs.
The generalized product is $h = k\,m^x\,v^y\,g^z$, where k is a dimensionless constant and the exponents x, y, and z are numbers that will be identified in step (4).

Step (3): Find a dimensional version of the equation from step (2).
First, use rule 4 of dimensional arithmetic and the fact that k is dimensionless (and so $[k] = 1$) to obtain the following equation involving the dimensions $[h]$, $[v]$, $[g]$, and $[m]$:

$$[h] = [k\,m^x\,v^y\,g^z] = [k]\cdot[m]^x\cdot[v]^y\cdot[g]^z = 1\cdot[m]^x\cdot[v]^y\cdot[g]^z = [m]^x\cdot[v]^y\cdot[g]^z$$

Now, substitute the dimensions of the variables into this equation and simplify the resulting expression using exponent rules. The dimensions of the variables are $[h] = L$, $[m] = M$, $[v] = LT^{-1}$, and $[g] = LT^{-2}$ (because gravity is an acceleration). Substitute and simplify as follows:

$$
\begin{aligned}
[h] &= [m]^x \cdot [v]^y \cdot [g]^z && \text{Dimensional equation} \\
L &= M^x\,(LT^{-1})^y\,(LT^{-2})^z && \text{Substitute the dimensions of } m,\ v,\text{ and } g \\
&= M^x\,L^y T^{-y}\,L^z T^{-2z} && \text{Simplify using exponent rule } (ab)^r = a^r b^r \\
&= M^x\,(L^y L^z)\,(T^{-y} T^{-2z}) && \text{Group like terms together} \\
&= M^x\,L^{y+z}\,T^{-y-2z} && \text{Combine like terms using exponent rule } a^r a^s = a^{r+s}
\end{aligned}
$$

Step (4): Use rule 1 of dimensional arithmetic to solve for the unknown exponents.
In order for the equation from step (3) to be dimensionally compatible, rule 1 requires that the dimensions on the left and right sides of the equality must be the same. Because only L appears on the left and M, L, and T appear on the right, use the fact that $a^0 = 1$ for any $a > 0$ to rewrite the dimensional equation as follows:

$$L = M^x \cdot L^{y+z} \cdot T^{-y-2z}$$
$$M^0 \cdot L^1 \cdot T^0 = M^x \cdot L^{y+z} \cdot T^{-y-2z}$$

Now, set the corresponding exponents on both sides of the equation equal to each other:

$$
\begin{aligned}
M &: & 0 &= x \\
L &: & 1 &= y + z \\
T &: & 0 &= -y - 2z
\end{aligned}
$$

These three linear equations can be used to solve for the three unknown exponents x, y, and z. The M equation immediately gives $x = 0$. Rearranging the T equation provides $y = -2z$. Substituting $-2z$ for y into the L equation gives $1 = -2z + z$, and so $z = -1$. Finally, substituting $z = -1$ into the L equation yields $1 = y - 1$, or $y = 2$. Thus, the three exponents are $x = 0$, $y = 2$, and $z = -1$.

Step (5): State the model based on steps (2) and (4).
Step (2) stated that the generalized product is $h = k\,m^x\,v^y\,g^z$ and step (4) determined

that $x = 0$, $y = 2$, and $z = -1$. Thus, the resulting model for the height of the ball is $h = k\, m^0\, v^2\, g^{-1}$, which simplifies to $h = k\, v^2\, g^{-1}$.

Step (6): Interpret the resulting model.
Several observations can be made based on this model, although this handful of comments provides just the tip of the proverbial iceberg:

- Launch velocity and gravity affect the height of the ball.

- The mass of the ball does *not* affect the height of the ball, as indicated by mass not appearing in the final model.

- As the launch velocity increases, the height of the ball also increases.

- The height of the ball increases by a factor of four if the launch velocity is doubled.

■

Step (1) makes certain assumptions about the relevant quantities. When making such assumptions, we must keep in mind that quantities may have been identified that are not relevant or quantities that are relevant may have been ignored. Example 11 conjectured that mass is relevant to the height of a ball thrown upward, when it is not. Such mindful reflection is extremely important whenever modeling a phenomenon of interest, because these modeling functions are being developed under a specific set of assumptions that may need to be modified as more is learned about the phenomenon.

◆ **EXAMPLE 12** In 1945, the United States detonated the first atomic bomb in the Trinity Test at White Sands, New Mexico. The details of this test were highly classified, but the pictures of the blast in Figure 1 were shared with the public. Find an equation for the radius of the blast using dimensional analysis.

Figure 1: Photos of the Trinity Test at White Sands, New Mexico for Example 12

Solution.
Step (1): Identify the relevant input and output quantities.
The relevant quantities involved in finding the blast radius are assumed to be time, the density of air, and the explosive energy from the bomb. Furthermore, assuming the blast radius depends on the other three quantities, the output is blast radius (r) from inputs of air density (d), explosive energy (n), and time (t).

Step (2): Write the output as a generalized product of the inputs.
The generalized product is $r = k\, d^x\, n^y\, t^z$, where k is a dimensionless constant and the exponents x, y, and z are numbers that will be identified in step (4).

Step (3): Find a dimensional version of the equation from step (2).
First, use rule 4 of dimensional arithmetic and the fact that k is dimensionless (and so $[k] = 1$) to obtain the following equation involving the dimensions $[b]$, $[d]$, $[n]$, and $[t]$:

$$[r] = [k\, d^x\, n^y\, t^z] = [k] \cdot [d]^x \cdot [n]^y \cdot [t]^z = 1 \cdot [d]^x \cdot [n]^y \cdot [t]^z = [d]^x \cdot [n]^y \cdot [t]^z$$

Now, substitute the dimensions of the variables into this equation and simplify the resulting expression using exponent rules. The dimensions of the variables are $[r] = L$, $[d] = ML^{-3}$, $[n] = ML^2T^{-2}$, and $[t] = T$ (see Tables 1 and 2 of physical dimensions). Substitute and simplify as follows:

$$[r] = [d]^x \cdot [n]^y \cdot [t]^z \qquad \text{Dimensional equation}$$
$$L = (ML^{-3})^x \cdot (ML^2T^{-2})^y \cdot T^z \qquad \text{Substitute the dimensions of } r,\, d,\, n,\, \text{and } t$$
$$= M^x L^{-3x} \cdot M^y L^{2y} T^{-2y} \cdot T^z \qquad \text{Apply exponent rule } (ab)^r = a^r b^r$$
$$= L^{-3x} L^{2y} \cdot M^x M^y \cdot T^{-2y} T^z \qquad \text{Group like terms together}$$
$$= L^{-3x+2y} \cdot M^{x+y} \cdot T^{-2y+z} \qquad \text{Combine like terms using exponent rule } a^r a^s = a^{r+s}$$

Step (4): Use rule 1 of dimensional arithmetic to solve for the unknown exponents.
In order for the equation from step (3) to be dimensionally compatible, rule 1 requires that the dimensions on the left and right sides of the equality must be the same. Because T and M do not appear on the left-hand side of the equation, use the fact that $a^0 = 1$ for any $a > 0$ to rewrite the dimensional equation as follows:

$$L^1 = L^{-3x+2y} \cdot M^{x+y} \cdot T^{-2y+z}$$
$$L^1 \cdot M^0 \cdot T^0 = L^{-3x+2y} \cdot M^{x+y} \cdot T^{-2y+z}$$

Now, set the corresponding exponents on both sides of the equation equal to each other:

$$L: \quad 1 \ = \ -3x + 2y$$
$$M: \quad 0 \ = \ x + y$$
$$T: \quad 0 \ = \ -2y + z$$

These three linear equations can be used to solve for the three unknown exponents x, y, and z. The M equation provides $x = -y$, and substituting into the L equation yields the following:

$$L: \quad 1 = -3x + 2y = -3(-y) + 2y = 3y + 2y = 5y \ \Rightarrow \ y = \frac{1}{5}.$$

The identity $x = -y$ gives $x = -1/5$. Finally, substituting $z = 1/5$ into the T equation provides $0 = -2/5 + z$ or $z = 2/5$. Therefore, the three exponents are $x = -1/5$, $y = 1/5$, and $z = 2/5$.

Step (5): State the model based on steps (2) and (4).
Step (2) stated that the generalized product is $r = k\, d^x\, n^y\, t^z$ and step (4) determined

that $x = -1/5$, $y = 1/5$, and $z = 2/5$. Substituting these numeric exponents into the generalized product provides a model for the radius of the blast:

$$r = k\, d^{-1/5}\, n^{1/5}\, t^{2/5} = \frac{k\, \sqrt[5]{n} \cdot \sqrt[5]{t^2}}{\sqrt[5]{d}} = k \sqrt[5]{\frac{n \cdot t^2}{d}}.$$

Step (6): Interpret the resulting model.
An interesting application of this model is provided in Example 13.   ■

Despite the highly classified nature of almost all details around the Trinity Test, a few publicly released photos enabled the British physicist G. I. Taylor to discover certain key facts. Working with these photos, including those given in Figure 1, Taylor measured the blast radius r as a function of time t, producing the data contained in BlastData. As illustrated in Example 13, Taylor famously used dimensional analysis to determine the explosive energy of this first atomic bomb.

◆ **EXAMPLE 13** Use the model developed in Example 12 and the data in BlastData gathered by G. I. Taylor to estimate the explosive energy n of the atomic bomb detonated in the Trinity Test.

Solution. After careful study, physicists determined that the constant $k \approx 1$ and the air density $d \approx 1$ are reasonable assumptions, which allows the model to be simplified as follows:
$$r = k\, d^{-1/5}\, n^{1/5}\, t^{2/5} = 1 \cdot 1^{-1/5} \cdot n^{1/5} \cdot t^{2/5} = n^{1/5} \cdot t^{2/5}.$$

Because the explosive energy n is a constant, proceed in parallel with the process of identifying power function models. Namely, rewrite the modeling equation by taking the natural logarithm of both sides of the model.:

$$r = n^{1/5} \cdot t^{2/5} \qquad \text{Simplified modeling equation}$$
$$\ln(r) = \ln(n^{1/5} \cdot t^{2/5}) \qquad \text{Apply natural logarithm function to both sides}$$
$$\ln(r) = \ln(n^{1/5}) + \ln(t^{2/5}) \qquad \text{Use logarithm property } \ln(ab) = \ln(a) + \ln(b)$$
$$\ln(r) = \frac{1}{5}\ln(n) + \frac{2}{5}\ln(t) \qquad \text{Use logarithm property } \ln(a^r) = r\ln(a)$$

Now, approximate n by using fitModel to find the best linear model with slope $\frac{2}{5}$ for the data points $(\ln(r), \ln(t))$:

```
names(BlastData)

[1] "X"        "time"     "radius"

energyModel = fitModel(log(radius)~2/5*log(time)+1/5*log(n),
                        data=BlastData)
coef(energyModel)

          n
6.8572e+13
```

Thus, based on the model, the explosive energy of the bomb was $n \approx 6.86 \times 10^{13}$ joules. The actual explosive energy of the atomic bomb denoted in the Trinity Test was 9×10^{13} joules.

∎

In this way, dimensional analysis and a handful of data points allowed G. I. Taylor (and us!) to approximate the explosive energy of an atomic bomb to the correct order of magnitude. While the multi-step process of dimensional analysis can require lots of algebraic effort, the facts that can be determined from relatively little information are really quite striking.

➤ **QUESTION 10** Find an equation for the frequency f of the sound produced by a human's vocal chords as a function of their length (ℓ), tension (s), and mass density (d) using dimensional analysis. Tension is a force and mass density is mass per unit length.

Summary

- A *dimension* is an inherent way of measuring a quantity. A *unit* is a way to assign a numerical value to the measurement of a quantity

- A *fundamental dimension* is a basic, inherent way of measuring a quantity that cannot be expressed in terms of other dimensions; for example, length with symbol [length] $= L$. A *derived dimension* is a dimension that can be expressed as a product of integer powers of fundamental dimensions; for example, area with symbol [area] $= L^2$. A pure number is *dimensionless*; for example, the number 2 is dimensionless, which is written as $[2] = 1$.

- *Rules of dimensional arithmetic*

  - *Rule 1*: When adding, subtracting, or comparing (e.g., $=$, $\geq$, or $<$) two quantities, they must have the same dimension, although their units may be different.

  - *Rule 2*: When multiplying and dividing quantities, they can have any dimensions and any units.

  - *Rule 3*: In an exponential expression a^r, the exponent must be dimensionless.

  - *Rule 4*: The equation $c = k\, a_1^{x_1} a_2^{x_2} \dots a_n^{x_n}$ is dimensionally valid when
    $$\begin{aligned}[c] &= [k\, a_1^{x_1} a_2^{x_2} \dots a_n^{x_n}] \\ &= [k] \cdot [a_1]^{x_1} \cdot [a_2]^{x_2} \dots [a_n]^{x_n}.\end{aligned}$$

Summary (continued)

- The *generalized product* of dimensional quantities a and b with output c is written $c = k\,a^x b^y$ where k is a dimensionless constant and x and y are chosen so that the equality is valid. The generalized product of dimensional quantities $a_1, a_2, \ldots, a_n$ with output c is written $c = k a_1^{x_1} a_2^{x_2} \cdots a_n^{x_n}$, where k is a dimensionless constant and $x_1, x_2, \ldots, x_n$ are chosen so that the equality is valid. From rule 1 of dimensional arithmetic, an equality is valid exactly when the dimensions of the quantities on both sides of the equation are the same.

- *Dimensional Analysis*:

  (1) Identify the relevant input and output quantities.

  (2) Write the output as a generalized product of the inputs.

  (3) Find a dimensional version of the equation from step (2).

  (4) Use rule 1 of dimensional arithmetic to solve for the unknown exponents.

  (5) State the model based on steps (2) and (4).

  (6) Interpret the resulting model.

Exercises

In Exercises $1-10$, state one unit in the English system and one unit in the metric system for the dimension, and the relationship between them. For example, length has units of 1 mile $= 1.60934$ kilometers. The Internet is a useful resource.

1. $[\text{length}] = L$

2. $[\text{mass}] = M$

3. $[\text{area}] = L^2$

4. $[\text{volume}] = L^3$

5. $[\text{velocity}] = LT^{-1}$

6. $[\text{acceleration}] = LT^{-2}$

7. $[\text{force}] = MLT^{-2}$

8. $[\text{pressure}] = ML^{-1}T^{-2}$

9. $[\text{density}] = ML^{-3}$

10. $[\text{energy}] = ML^2T^{-2}$

In Exercises $11-16$, state two different units for the dimension and the relationship between them. The Internet is a useful resource.

11. $[\text{time}] = T$

12. $[\text{frequency}] = T^{-1}$

13. $[\text{angle}] = A$

14. $[\text{money}] = B$

15. $[\text{population density}] = PL^{-2}$

16. $[\text{wages}] = BT^{-1}$

In Exercises $17-30$, identify the computation or comparison as valid or invalid based on the rules for dimensional arithmetic. Justify your answer.

17. $4 \text{ m} - 3 \text{ ft}$

18. $12.1 \text{ g/L} + 7 \text{ lbs/ft}^2$

19. $\$22.14 = £4.1$

20. $5 \text{ g} > 3 \text{ hr}$

21. $12°\text{C/hr} \times 15 \text{ min}$

22. $6 \text{ kg} \times 2 \text{ m}$

23. $(8 \text{ mph})^3$

24. $(3 \text{ hr})^2 \times 5 \text{ km}$

25. $\dfrac{7 \text{ m} - 0.25 \text{ km}}{4 \text{ min} - 25 \text{ sec}}$

26. $\dfrac{3 \text{ g} \times 2 \text{ m}}{5 \text{ km}^2}$

27. $2^{4\,\text{ft}/\,3\,\text{g}}$

28. $7^{4\,\text{hr}/\,3\,\text{min}}$

29. $5 \text{ g} \times 3 \text{ kg} - 7 \text{ lbs}$

30. $\sqrt[3]{8 \text{ m}^3 + 27 \text{ ft}^2}$

Your Turn. In Exercises 31–36, state a specific example of a valid and an invalid computation or comparison, if possible, based on the rules of dimensional arithmetic.

31. Rule 1 for $a + b$

32. Rule 1 for $a - b$

33. Rule 1 for $a = b$

34. Rule 1 for $a < b$

35. Rule 2 for $a \times b$

36. Rule 2 for a/b

In Exercises 37–50, determine the dimensions of b and c using the given information.

37. $a + b + c$, when $[a] = L$

38. $a - b + 3c$, when $[a] = M$

39. $a + b = c$, when $[a] = T$

40. $a > b - c$, when $[a] = P$

41. $5 = ab + c$, when $[a] = M$

42. $(a + b)^2 + c$, when $[a] = 1$

43. $2^{ab} - 4c$, when $[a] = T^{-1}$

44. 1.1^{ab+c}, when $[a] = B$

45. $x + (ab + c)$, when $[a] = T$, $[x] = L$

46. $a + (bx + c)^3$, when $[a] = L^6$, $[x] = M$

47. $y \leq ab - c$, when $[a] = T$, $[y] = L$

48. $y^2 \geq a^2 b + c$, when $[a] = N$, $[y] = B$

49. $z = b\,4^{ac}$, when $[a] = LT^{-1}$, $[z] = M$

50. $z = ab\,2^{bc}$, when $[a] = T$, $[z] = B$

In Exercises 51–61, determine the dimensions of all quantities identified in the given setting.

51. The total amount of money earned in interest i for a particular time period is given by $i = prt$, where p is the principal (or initial) amount of money invested, r is the interest rate, and t is the length of the time period. Determine $[i]$, $[p]$, $[t]$, and $[r]$.

52. Populations of organisms increase very rapidly in the presence of abundant environmental resources. Such population growth is described with the exponential model $n = c \cdot 2^{kt}$, where n is the size of the population, c and k are organism-specific constants, and t is time. Determine $[n]$, $[c]$, $[t]$, and $[k]$.

53. When populations of organisms live in an environment with limited resources, the population grows exponentially at first, but eventually grows more slowly as the population reaches the carrying capacity of the environment. Such growth is described using a sigmoidal model

$$n = \frac{\ell}{1 + c \cdot 2^{kt}},$$

where n is the population size, ℓ is the population size at the carrying capacity of the environment, c and k are organism-specific constants, and t is time. Determine $[n]$, $[\ell]$, $[c]$, $[t]$, and $[k]$.

54. Organisms need resources to survive, such as water, oxygen, and food. The available resources inside an organism are given by the difference between the resources absorbed from the environment and the resources used inside the organism. For example, bloodstream oxygen level is the difference between how much is absorbed through respiration and how much the body uses. One model for available resources in a simple organism is $a = bw^2 - cw^3$, where a is the mass of the available resources and w is the length of the organism. In this case, bw^2 provides the resources taken in and cw^3 provides the resources used. Determine $[a]$, $[w]$, $[b]$, and $[c]$.

55. The available resource model in Exercise 54 can be modified for "wormlike" organisms such as *Paramecium*. One such model is $a = b \cdot r^2 + c \cdot r \cdot h - d \cdot r^3$, where a is the mass of the surplus resources, r is the radius of the organism, and h is the length of the organism. Determine $[a]$, $[r]$, $[b]$, $[h]$, $[c]$, and $[d]$.

56. The SIR disease model describes the spread of infectious disease in a constant population of organisms. In this model, s is the size of the susceptible population, i is the size of the infected population, and the two-letter symbolism ds is the rate of change in the size of the susceptible population. Example units for ds are people per day or fish per hour. The SIR model states

$ds = -a \cdot s \cdot i$. Determine $[ds]$, $[s]$, $[i]$, and $[a]$.

57. A second component of the SIR disease model introduced in Exercise 56 defines di, the rate of change in the size of the infected population, which has units such as people per day or fish per hour. The SIR model states that $di = a \cdot s \cdot i - b \cdot i$. Determine $[di]$, $[s]$, $[i]$, $[a]$, and $[b]$.

58. A third component of the SIR disease model introduced in Exercise 56 defines dr, the rate of change in the size of the recovered population, which has units such as people per day or fish per hour. The SIR model states that $dr = b \cdot i$. Determine $[dr]$, $[i]$, and $[b]$.

59. After taking an oral medication, the concentration of the medication in the blood increases and then tapers off. This behavior is modeled by $c = ate^{-bt}$, where c is the concentration of the medication (with units such as milligrams per liter), a and b are constants particular to the medication, and t is time. Determine $[c]$, $[t]$, $[a]$, and $[b]$.

60. Radiocarbon dating is often used to approximate the age of objects less than 50,000 years old. The radioactive isotope carbon-14 begins to decay when an organism dies, as modeled by the equation $a = c \cdot 2^{-kt}$, where a is the mass of the carbon-14 present, c and k are carbon-14 specific constants, and t is time. Determine $[a]$, $[c]$, $[t]$, and $[k]$.

61. Newton's law of gravitation states that
$$F = \frac{Gm_1 m_2}{r},$$
where F is a force, G is the universal gravitational constant, m_1 and

m_2 are masses, and r is a distance. Determine $[F]$, $[m_1]$, $[m_2]$, $[r]$, and $[G]$. Note that G is different from Earth's gravitational constant (the acceleration due to gravity) $g = 9.8$ meters per second squared.

In Exercises 62–71, use exponent rules to combine like terms as in step (3) of dimensional analysis.

62. $LL^{-1}M$

63. L^2PL^{-2}

64. $L^2T^{-3}LT$

65. $LT^{-1}MT^2$

66. $BT^{-1}TB$

67. $PLT^{-2}P^2L^{-2}$

68. $M^2L^{-3}M^{-2}L^3$

69. $MB^2P^{-2}M^2B$

70. $LT^{-3}MLT^{-2}$

71. $NT^{-2}MT^{-1}N$

In Exercises 72–81, complete steps (2) - (5) of dimensional analysis to express the output as a generalized product of the inputs with specific numeric exponents.

72. Inputs: $[\ell]=L$, $[t]=T$, $[d]=ML^{-1}T^{-1}$
Output: $[w] = M^2L$

73. Inputs: $[u] = LT^{-2}$, $[v] = TL^{-1}$
Output: $[w] = L^{-2}$

74. Inputs: $[c] = PL^{-1}$, $[d] = L^2T^{-1}$
Output: $[f] = P^2T^{-1}$

75. Inputs: $[a] = M$, $[b] = LM^{-1}$
Output: $[c] = M^3L^2$

76. Inputs: $[\ell]=P$, $[f]=BP^{-1}$, $[n]=NB^{-1}$
Output: $[p] = P^2N^2$

77. Inputs: $[p]=L$, $[q]=L^{-1}M$, $[r]=LT$
Output: $[s] = M^2T^3$

78. Inputs: $[d]=ML^{-1}$, $[t]=T$, $[v]=LT^{-1}$
Output: $[a] = L^{-2}M^3T^{-2}$

79. Inputs: $[p]=ML$, $[q]=LP^{-1}$, $[r]=P$
Output: $[s] = LM^{-2}P^{-1}$

80. Inputs: $[f]=BL$, $[\ell]=BM$, $[h]=ML^{-1}$
Output: $[p] = 1$

81. Inputs: $[a]=M$, $[b]=LT$, $[c]=M^{-1}T^{-1}$
Output: $[d] = L^4MT$

In Exercises 82–90, find a generalized product of the given quantities to model the given setting using dimensional analysis.

82. The potential energy stored in a compressed spring is a function of the dimensional spring constant (with units such as newtons per meter) and the distance the spring has been compressed.

83. The period of a pendulum is a function of the length of the pendulum and the acceleration due to gravity g.

84. The magnitude of the centripetal force of an object traveling along a curved path is a function of the mass of the object, its velocity, and the radius of curvature of the path.

85. Consider a car that slams on its brakes and skids to a stop. The speed of the car when it slams on the brakes is a function of the length of the skid and the acceleration due to gravity g.

86. The drag force on a disk of diameter ℓ created by air blown on the disk is a function of the density of the air, the speed of the air, and the diameter of the disk ℓ.

87. The electrostatic force between two charged particles is a function of the magnitude of the two charges

q_1 and q_2, the distance between the particles, and Coulomb's constant k_e. Charge has dimension IT and Coulomb's constant has dimension $ML^3T^{-4}I^{-2}$.

88. In humans, the hydrostatic pressure of blood contributes to total blood pressure. Express the hydrostatic pressure of blood as a function of blood density, the height of the blood column between the heart and some lower point of the body, and the acceleration due to gravity g.

89. The speed of a vehicle is affected by the wind force on a vehicle driving down the road. Determine whether or not there is a dimensionally compatible relationship for the speed of the vehicle as a function of the surface area of the vehicle and the force of the wind.

90. The speed of a vehicle is affected by the wind force on a vehicle driving down the road. Determine whether or not there is a dimensionally compatible relationship for the speed of the vehicle as a function of the surface area of the vehicle, the force of the wind, and the density of air.

In Your Own Words. In Exercises 91–100, explain the following.

91. Relationship between dimension and units

92. Fundamental dimension

93. Derived dimension

94. Dimensionless quantity

95. Rule 1 of dimensional arithmetic

96. Rule 2 of dimensional arithmetic

97. Rule 3 of dimensional arithmetic

98. Rule 4 of dimensional arithmetic

99. Generalized product

100. Dimensional analysis

Chapter 3

The Method of Least Squares

Chapters 1 and 2 introduced the modeling cycle and how to find a mathematical model for a given data set of data, chosen from among linear, exponential, power, sine, and sigmoidal functions. This study included using the `fitModel` command in RStudio to find the best possible model of a particular type for a given data set. This chapter explores the underlying mathematical framework that produces this best model. In particular, we study the *method of least squares*, which is the mathematical method encoded in the RStudio command `fitModel`.

More than just being programmed into `fitModel`, the method of least squares is the go-to approach of biologists, economists, psychologists, and other physical and social scientists in developing mathematical models of data sets. Learning these ideas will prove incredibly useful as you develop and interpret such models in whatever field you choose to pursue.

3.1   Vectors and Vector Operations

This study of the method of least squares begins with the idea of a *vector*. Vectors are used to describe a wide variety of phenomena because they capture both the strength of a quantity and the direction in which it acts. In many cases, the strength and direction of a quantity are different at various points, and so a vector is associated with each point. This identification of vectors with points is called a *vector field*. Graphically, an arrow represents a vector, with the length of the arrow indicating the strength (or magnitude) of the vector and the direction of the arrow indicating the direction of the vector.

You have probably encountered vector fields before, even if you did not use such terminology to describe them. Figure 1 provides an example of a vector field that represents the flow of blood through a particular cross-section of an artery in the human circulatory system. In this setting, observe the clockwise rotation in the blood flowing through the artery and the varying lengths of the arrows representing the relative strength of the flow. This vector field enables scientists to better understand the corresponding system and to make advances in their work.

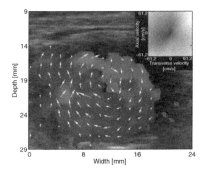

Figure 1: Example of a vector field representing blood flow in the human body

Vector fields are used to represent and understand many different phenomena. Weather maps often represent wind speed and direction by means of arrows (or vectors) overlaid on a map. Similarly, ocean and river currents, blood flow in organs and veins, dispersion of medication, the magnetic field around the earth, the gravitational field of the sun, and air flow around planes, trains, and automobiles can all be represented using vector fields, enabling us to model and better understand these systems.

The simplest way to think about a vector is to picture the directed line segment between two points. A vector efficiently captures two pieces of information: the distance between the two points and the direction of the line segment. Recall that the slope of a nonhorizontal line is the ratio of its change in the vertical direction (often called "rise" and denoted by Δy) and its change in the horizontal direction (often called "run" and denoted by Δx). In a two-dimensional setting, a vector keeps track of vertical change and horizontal change separately by means of two components. The following definition extends this notion to multiple dimensions:

Definition. The **vector** $\overline{PQ}$ starting at point $P = (P_1, P_2)$ and ending at point $Q = (Q_1, Q_2)$ is

$$\overline{PQ} = \begin{pmatrix} Q_1 - P_1 \\ Q_2 - P_2 \end{pmatrix} = \begin{pmatrix} \Delta x \\ \Delta y \end{pmatrix}$$

In general, the vector $\overline{PQ}$ starting at point $P = (P_1, P_2, \ldots, P_n)$ and ending at point $Q = (Q_1, Q_2, \ldots, Q_n)$ is

$$\overline{PQ} = \begin{pmatrix} Q_1 - P_1 \\ Q_2 - P_2 \\ \vdots \\ Q_n - P_n \end{pmatrix}$$

Each entry in the column is called a **component**. The first number is called the first component of the vector, the second number is called its second component, and so on.

For a vector with only two components, the first component is often called the **x-component** because it represents the change in the x-direction. Similarly, the second component of a two-component vector is often called the **y-component** because it measures the change in the y-direction.

◆ **EXAMPLE 1** Compute the vector starting at point P and ending at point Q and, if possible, graph the resulting vector:

(a) $P = (4, 3)$ and $Q = (6, 7)$

(b) $P = (0, 0)$ and $Q = (2, 4)$

(c) $P = (3, 0, 0)$ and $Q = (3, 4, 4)$

(d) $P = (3, 1, 5, -4, 2)$ and $Q = (7, 8, -2, 1, 0)$

Solution.

(a) $\overline{PQ} = \begin{pmatrix} 6 - 4 \\ 7 - 3 \end{pmatrix} = \begin{pmatrix} 2 \\ 4 \end{pmatrix}$

(b) $\overline{PQ} = \begin{pmatrix} 2 - 0 \\ 4 - 0 \end{pmatrix} = \begin{pmatrix} 2 \\ 4 \end{pmatrix}$

(c) $\overline{PQ} = \begin{pmatrix} 3 - 3 \\ 4 - 0 \\ 4 - 0 \end{pmatrix} = \begin{pmatrix} 0 \\ 4 \\ 4 \end{pmatrix}$

(d) $\overline{PQ} = \begin{pmatrix} 7 - 3 \\ 8 - 1 \\ -2 - 5 \\ 1 - (-4) \\ 0 - 2 \end{pmatrix} = \begin{pmatrix} 4 \\ 7 \\ -7 \\ 5 \\ -2 \end{pmatrix}$

The two-component vectors from parts (a) and (b) are graphed in the left-hand plot of Figure 2 using the standard pair of axes on the plane with the identifying labels of $\overline{PQ}_a$ and $\overline{PQ}_b$. The three-component vector from part (c) is graphed in the right-hand plot of Figure 2 using the standard three axes in space and is labeled $\overline{PQ}_c$. The five-component vector from part (d) cannot be graphed, because humans cannot visually present five dimensions simultaneously.

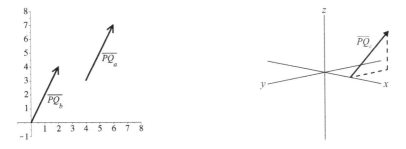

Figure 2: Graphs of vectors for Example 1

➤ **QUESTION 1** Compute the vector starting at point P and ending at point Q and, if possible, graph the resulting vector:

(a) $P = (3, 5)$ and $Q = (4, 7)$

(b) $P = (0, 0)$ and $Q = (-2, 3)$

(c) $P = (1, 2, 4, 1)$ and $Q = (-3, 5, 5, 6)$

(d) $P = (-3, 5, 1, 1, 1)$ and $Q = (9, 1, 2, 0, 3)$

Often a vector is computed between a given pair of points. However, sometimes one point and the resulting vector are given, and we will be interested in determining the other point.

◆ **EXAMPLE 2** Find either the starting point P or the ending point Q of each vector:

(a) Find the starting point P when $Q = (3, -5)$ and $\overline{PQ} = \begin{pmatrix} -1 \\ -2 \end{pmatrix}$.

(b) Find the ending point Q when $P = (2, -4, 6)$ and $\overline{PQ} = \begin{pmatrix} 3 \\ 1 \\ 2 \end{pmatrix}$.

Solution.

(a) Point $P = (x, y)$ is found by setting up a difference equal to each component of $\overline{PQ}$:

$$\overline{PQ} = \begin{pmatrix} Q_1 - P_1 \\ Q_2 - P_2 \end{pmatrix} \quad \Rightarrow \quad \begin{pmatrix} -1 \\ -2 \end{pmatrix} = \begin{pmatrix} 3 - x \\ -5 - y \end{pmatrix} \quad \begin{array}{l} \Rightarrow \quad -1 = 3 - x \\ \Rightarrow \quad -2 = -5 - y \end{array}$$

For the x-coordinate of P, $3 - x = -1$ and solving gives $-x = -1 - 3 = -4$, or $x = 4$. Similarly, for the y-coordinate of P, $-5 - y = -2$ and solving gives $y = -3$. Thus, the starting point is $P = (4, -3)$.

(b) Point $Q = (x, y, z)$ is found by setting up a difference equal to each component of $\overline{PQ}$:

$$\overline{PQ} = \begin{pmatrix} Q_1 - P_1 \\ Q_2 - P_2 \\ Q_3 - P_3 \end{pmatrix} \quad \Rightarrow \quad \begin{pmatrix} 3 \\ 1 \\ 2 \end{pmatrix} = \begin{pmatrix} x - 2 \\ y - (-4) \\ z - 6 \end{pmatrix} \quad \begin{array}{l} \Rightarrow \quad 3 = x - 2 \\ \Rightarrow \quad 1 = y - (-4) \\ \Rightarrow \quad 2 = z - 6 \end{array}$$

Now solve for the unknown coordinates of Q as follows:

$$\begin{array}{lll}
x: \quad \begin{aligned} x - 2 &= 3 \\ x &= 3 + 2 \\ x &= 5 \end{aligned} &
y: \quad \begin{aligned} y - (-4) &= 1 \\ y &= 1 + (-4) \\ y &= -3 \end{aligned} &
z: \quad \begin{aligned} z - 6 &= 2 \\ z &= 2 + 6 \\ z &= 8 \end{aligned}
\end{array}$$

Thus, the ending point is $Q = (5, -3, 8)$.

■

▶ **QUESTION 2** Find either the starting point P or the ending point Q of each vector:

(a) Find the starting point P when $Q = (4, 8)$ and $\overline{PQ} = \begin{pmatrix} 1 \\ -5 \end{pmatrix}$.

(b) Find the ending point Q when $P = (7, 5, 3)$ and $\overline{PQ} = \begin{pmatrix} 9 \\ 0 \\ 1 \end{pmatrix}$.

Vectors do not uniquely identify the directed line segment between two points. In other words, different directed line segments between different pairs of points can have the same vector representation. Example 1 found that the vector with x-component 2 and y-component 4 represents both the directed line segment between (a)'s points $P = (4, 3)$ and $Q = (6, 7)$ and the directed line segment between (b)'s points $P = (0, 0)$ and $Q = (2, 4)$, even though they are distinct. The x-component of 2 indicates moving two units in the horizontal direction from the starting point, and the y-component of 4 indicates moving four units in the vertical direction from the starting point. Because the starting points are different, the ending points are also different, even though the same distance is moved in the same direction in both cases.

Because many pairs of points result in the same vector, this discussion will be more focused and clear by letting $P = (0, 0)$ unless otherwise specified. In other words, the implicitly assumed starting point of every vector is located at the origin. This choice results in an immediate connection between the components of a vector and its ending point Q. In particular, if $P = (0, 0)$ and $Q = (a, b)$, then

$$\overline{PQ} = \begin{pmatrix} a - 0 \\ b - 0 \end{pmatrix} = \begin{pmatrix} a \\ b \end{pmatrix}$$

which corresponds nicely with the ending point $Q = (a, b)$.

KEY FACT.
From this point forward, assume that all vectors begin at the origin unless otherwise specified. Typically, the symbolism $\overline{u}$, $\overline{v}$, and $\overline{w}$ is used to denote vectors.

Vectors can be used to find the direction and the length of the directed line segment between two points. In order that we can use our visual intuitions to facilitate this study, two-component vectors will be the focus of this discussion. However, these ideas extend to vectors with any number of components, and the definitions are stated in such a way as to reflect this generality.

The **direction** of a vector is determined by the x- and y-components of the vector. In particular, each component dictates how far to move in the x-direction and how far to move in the y-direction, respectively. Furthermore, observe that any direction can be expressed by an appropriate choice of x- and y-components, because these numbers can be any of positive, zero, or negative.

The *length* of a vector is determined using the Pythagorean theorem. Recall that $a^2 + b^2 = c^2$ when a right triangle has side lengths a and b, and hypotenuse length c. For a vector, the change in the x-direction and the change in the y-direction form the two sides of a right triangle, and the length of the hypotenuse of this triangle is equal to the length of the vector. Applying the Pythagorean theorem provides the length of the hypotenuse and so the length of the vector.

◆ **EXAMPLE 3** Determine the length of the vector from $P = (1, 1)$ to $Q = (4, 4)$.

Solution. As illustrated in Figure 3, the length of the x-component of this vector is $4 - 1 = 3$ and the length of the y-component is $4 - 1 = 3$. Applying the Pythagorean theorem yields $c^2 = a^2 + b^2 = 3^2 + 3^2 = 9 + 9 = 18$. Taking the square root determines that $c = \sqrt{18} = 3\sqrt{2}$, and so the length of the vector $\begin{pmatrix} 3 \\ 3 \end{pmatrix}$ from $(1, 1)$ to $(4, 4)$ is $3\sqrt{2} \approx 4.243$.

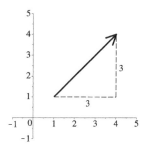

Figure 3: Graph of the vector from $P = (1, 1)$ to $Q = (4, 4)$ for Example 3 ■

▶ **QUESTION 3** Determine the length of the vector from $P = (2, 10)$ to $Q = (5, 8)$.

Operating under the standard assumption that a vector has a starting point at the origin $P = (0, 0)$ and an ending point $Q = (a, b)$ yields

$$\overline{PQ} = \begin{pmatrix} a - 0 \\ b - 0 \end{pmatrix} = \begin{pmatrix} a \\ b \end{pmatrix}$$

This vector $\overline{PQ}$ changes a in the x-direction and b in the y-direction, which means that applying the Pythagorean theorem in this case gives a vector length of $\sqrt{a^2 + b^2}$. The following definition summarizes this discussion of the length of a vector:

Definition. The **length** of a vector $\overline{v} = \begin{pmatrix} v_1 \\ v_2 \end{pmatrix}$ is $\|\overline{v}\| = \sqrt{v_1^2 + v_2^2}$.

In general, the length of a vector $\overline{v} = \begin{pmatrix} v_1 \\ v_2 \\ \vdots \\ v_n \end{pmatrix}$ is $\|\overline{v}\| = \sqrt{v_1^2 + v_2^2 + \cdots + v_n^2}$.

As indicated in this definition, the symbolism $\|\overline{v}\|$ denotes the length of a vector. Also, the length of $\overline{v}$ is sometimes called the **magnitude** of $\overline{v}$.

◆ **EXAMPLE 4** Find the length of each vector:

(a) $\bar{v} = \begin{pmatrix} -2 \\ 5 \end{pmatrix}$
(b) $\bar{v} = \begin{pmatrix} 4 \\ 1 \\ 2 \end{pmatrix}$

Solution.

(a) $\|\bar{v}\| = \sqrt{(-2)^2 + 5^2} = \sqrt{4 + 25} = \sqrt{29} \approx 5.385$

(b) $\|\bar{v}\| = \sqrt{4^2 + 1^2 + 2^2} = \sqrt{16 + 1 + 4} = \sqrt{21} \approx 4.583$ ∎

▶ **QUESTION 4** Find the length of each vector:

(a) $\bar{v} = \begin{pmatrix} 4 \\ -3 \end{pmatrix}$
(b) $\bar{w} = \begin{pmatrix} 1 \\ -5 \\ 7 \end{pmatrix}$

Three Vector Operations

The most basic arithmetic operations include addition, subtraction, and multiplication. As defined below, vector addition and vector subtraction are carried out by adding or subtracting the corresponding components and are pretty much all of what might be hoped and expected. Vector multiplication is more complicated. In fact, there is no way to define a multiplication of vectors that has all the desirable properties of multiplication, such as commutativity, which is illustrated by $2 \times 3 = 3 \times 2$, and associativity, which is illustrated by $2 \times (3 \times 4) = (2 \times 3) \times 4$.

Mathematicians have developed three different types of "multiplication," each of which preserves some properties of standard multiplication as well as having important physical interpretations. These three different types of multiplication are called scalar multiplication, dot product, and cross product. This section focuses on scalar multiplication and Section 3.4 introduces the dot product, but the study of the cross product is beyond the scope of this book.

In the context of working with vectors, real numbers are called **scalars** for reasons that are discussed soon. For example, 2, 5, -8, π, and $-e$ are all scalars because they are all real numbers. Scalar multiplication is carried out by multiplying each component of the vector by the same scalar.

Definition.

- The **addition** and **subtraction** of vectors $\bar{u}$ and $\bar{v}$ are defined by

$$\bar{u} \pm \bar{v} = \begin{pmatrix} u_1 \\ u_2 \\ \vdots \\ u_n \end{pmatrix} \pm \begin{pmatrix} v_1 \\ v_2 \\ \vdots \\ v_n \end{pmatrix} = \begin{pmatrix} u_1 \pm v_1 \\ u_2 \pm v_2 \\ \vdots \\ u_n + v_n \end{pmatrix}$$

Definition. (continued)

- The **scalar multiplication** of a vector $\bar{u}$ by a scalar m is defined by

$$m\bar{u} = m \begin{pmatrix} u_1 \\ u_2 \\ \vdots \\ u_n \end{pmatrix} = \begin{pmatrix} m\,u_1 \\ m\,u_2 \\ \vdots \\ m\,u_n \end{pmatrix}$$

Often $(-1)\bar{v}$ is denoted by $-\bar{v}$.

Note that in order to add or subtract two vectors, they must have the same number of components, because these operations on vectors are based on combining corresponding components. If a component does not have a "partner" so to speak, then the addition or subtraction is undefined. In addition, observe that all three of these vector operations result in a vector. As will be seen in Section 3.4, the dot product of two vectors results in a scalar, and so not all vector operations produce vectors. Some examples of vector addition, vector subtraction, and scalar multiplication are worked through, followed by a discussion of the geometric interpretations associated with these operations.

◆ **EXAMPLE 5** Let $\bar{u} = \begin{pmatrix} -5 \\ 3 \\ -2 \\ 4 \end{pmatrix}$ and $\bar{v} = \begin{pmatrix} 1 \\ 2 \\ 7 \\ -8 \end{pmatrix}$, and compute each vector:

(a) $\bar{u} + \bar{v}$ (b) $\bar{u} - \bar{v}$ (c) $5\bar{u}$ (d) $3\bar{u} - 2\bar{v}$

Solution.

(a) $\bar{u} + \bar{v} = \begin{pmatrix} -5 \\ 3 \\ -2 \\ 4 \end{pmatrix} + \begin{pmatrix} 1 \\ 2 \\ 7 \\ -8 \end{pmatrix} = \begin{pmatrix} -5 + 1 \\ 3 + 2 \\ -2 + 7 \\ 4 + (-8) \end{pmatrix} = \begin{pmatrix} -4 \\ 5 \\ 5 \\ -4 \end{pmatrix}$

(b) $\bar{u} - \bar{v} = \begin{pmatrix} -5 \\ 3 \\ -2 \\ 4 \end{pmatrix} - \begin{pmatrix} 1 \\ 2 \\ 7 \\ -8 \end{pmatrix} = \begin{pmatrix} -5 - 1 \\ 3 - 2 \\ -2 - 7 \\ 4 - (-8) \end{pmatrix} = \begin{pmatrix} -6 \\ 1 \\ -9 \\ 12 \end{pmatrix}$

(c) $5\bar{u} = 5 \begin{pmatrix} -5 \\ 3 \\ -2 \\ 4 \end{pmatrix} = \begin{pmatrix} 5 \cdot (-5) \\ 5 \cdot 3 \\ 5 \cdot (-2) \\ 5 \cdot 4 \end{pmatrix} = \begin{pmatrix} -25 \\ 15 \\ -10 \\ 20 \end{pmatrix}$

(d) $3\bar{u} - 2\bar{v} = \begin{pmatrix} 3 \cdot (-5) \\ 3 \cdot 3 \\ 3 \cdot (-2) \\ 3 \cdot 4 \end{pmatrix} - \begin{pmatrix} 2 \cdot 1 \\ 2 \cdot 2 \\ 2 \cdot 7 \\ 2 \cdot (-8) \end{pmatrix} = \begin{pmatrix} -15 - 2 \\ 9 - 4 \\ -6 - 14 \\ 12 - (-16) \end{pmatrix} = \begin{pmatrix} -17 \\ 5 \\ -20 \\ 28 \end{pmatrix}$

▶ **QUESTION 5** Let $\bar{u} = \begin{pmatrix} 4 \\ 8 \\ -1 \\ 0 \end{pmatrix}$ and $\bar{v} = \begin{pmatrix} -1 \\ 6 \\ 2 \\ 1 \end{pmatrix}$, and compute each vector:

(a) $\bar{u} + \bar{v}$ (b) $\bar{u} - \bar{v}$ (c) $2\bar{u} + \bar{v}$ (d) $\dfrac{1}{2}\bar{u} - 3\bar{v}$

A Geometric Interpretation of Scalar Multiplication

The operation of scalar multiplication has a nice geometric interpretation. Multiplying a vector by a scalar changes the length of the vector by the size of the scalar. For example, if a vector $\bar{v}$ is scalar multiplied by $m = 2$, then $2\bar{v}$ is twice as long as $\bar{v}$. Similarly, if $\bar{v}$ is scalar multiplied by $m = 0.5$, then $0.5\bar{v}$ is half as long as $\bar{v}$. This scaling of a vector's length is the reason why real numbers are called "scalars" in the context of working with vectors.

In addition to changing a vector's length, scalar multiplication can also reverse a vector's direction. For positive scalars, the direction remains unchanged. For example, $2\bar{v}$ points in the same direction from the origin as $\bar{v}$. However, multiplying by a negative scalar reverses the direction. For example, $(-2)\bar{v} = -2\bar{v}$ points in the opposite direction from the origin to $\bar{v}$. These properties of scalar multiplication are illustrated in the following example:

◆ **EXAMPLE 6** Let $\bar{v} = \begin{pmatrix} 3 \\ 4 \end{pmatrix}$. Compute each vector and its length, and then graphically illustrate their geometric relationships:

(a) $\bar{v}$ (b) $2\bar{v}$ (c) $-\bar{v}$

Solution.

(a) The vector $\bar{v} = \begin{pmatrix} 3 \\ 4 \end{pmatrix}$ has length $\bar{v}$ is $\|\bar{v}\| = \sqrt{3^2 + 4^2} = \sqrt{9 + 16} = \sqrt{25} = 5$.

(b) The vector $2\bar{v} = \begin{pmatrix} 2 \cdot 3 \\ 2 \cdot 4 \end{pmatrix} = \begin{pmatrix} 6 \\ 8 \end{pmatrix}$ has length $2\bar{v}$ is $\|2\bar{v}\| = \sqrt{6^2 + 8^2} = \sqrt{100} = 10$.

(c) Recall that $-\bar{v}$ denotes $(-1)\bar{v}$, so

$$-\bar{v} = (-1)\bar{v} = \begin{pmatrix} (-1) \cdot 3 \\ (-1) \cdot 4 \end{pmatrix} = \begin{pmatrix} -3 \\ -4 \end{pmatrix}$$

which has length $-\bar{v}$ is $\|-\bar{v}\| = \sqrt{(-3)^2 + (-4)^2} = \sqrt{9 + 16} = \sqrt{25} = 5$.

The graphs in Figure 4 illustrate the geometric relationships among these vectors. As illustrated in Figure 4(a), $2\bar{v}$ is twice as long as $\bar{v}$ and points in the same direction as $\bar{v}$. Similarly, as illustrated in Figure 4(b), $-\bar{v}$ has the same length as $\bar{v}$, but points in the opposite direction to $\bar{v}$.   ■

(a)   (b)

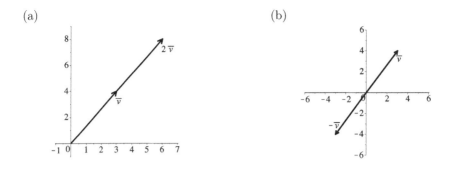

Figure 4: The effect of scalar multiplication for Example 6

➤ **QUESTION 6** Let $\bar{v} = \begin{pmatrix} -2 \\ 4 \end{pmatrix}$. Compute each vector and its length, and then graphically illustrate their geometric relationships:

(a) $2\bar{v}$ (b) $\dfrac{1}{2}\bar{v}$ (c) $-\dfrac{3}{2}\bar{v}$

In summary, scalar multiplication changes the length of the vector by the magnitude of the scalar. Also, if the scalar is positive, the resulting vector points in the same direction as the original; if the scalar is negative, the resulting vector points in the opposite direction to the original vector.

A Geometric Interpretation of Vector Addition

Example 7 below shows that for a given pair of vectors $\bar{u}$ and $\bar{v}$, the sums $\bar{u} + \bar{v}$ and $\bar{v} + \bar{u}$ are identical. This equality raises the question of whether or not vector addition is **commutative**, which means that any two vectors can be added in any order. As it turns out, this property of commutativity holds not just for the two vectors in Example 7, but for all vectors. Symbolically, the commutativity of vector addition is expressed by stating that for all possible vectors $\bar{u}$ and $\bar{v}$ with the same number of components, $\bar{u} + \bar{v} = \bar{v} + \bar{u}$.

The addition of two vectors also has a very nice geometric interpretation. Geometrically, $\bar{u} + \bar{v}$ is computed by placing the starting point of $\bar{v}$ at the tip of $\bar{u}$ and then drawing the vector from the starting point of $\bar{u}$ to the tip of $\bar{v}$ to obtain the vector $\bar{u} + \bar{v}$. Since vector addition is commutative, placing the starting point of $\bar{u}$ at the tip of $\bar{v}$ and then drawing the vector from the starting point of $\bar{v}$ to the tip of $\bar{u}$ must produce the vector $\bar{u} + \bar{v}$ as well. In this way, $\bar{u} + \bar{v}$ corresponds to one of the diagonals of the parallelogram formed by $\bar{u}$ and $\bar{v}$. Example 7 illustrates this geometric relationship.

◆ **EXAMPLE 7** Let $\overline{u} = \begin{pmatrix} 3 \\ -1 \end{pmatrix}$ and $\overline{v} = \begin{pmatrix} 2 \\ 3 \end{pmatrix}$. Compute each vector addition and then graphically illustrate their geometric relationships:

(a) $\overline{u} + \overline{v}$ (b) $\overline{v} + \overline{u}$

Solution.

(a) $\overline{u} + \overline{v} = \begin{pmatrix} 3+2 \\ (-1)+3 \end{pmatrix} = \begin{pmatrix} 5 \\ 2 \end{pmatrix}$ (b) $\overline{v} + \overline{u} = \begin{pmatrix} 2+3 \\ 3+(-1) \end{pmatrix} = \begin{pmatrix} 5 \\ 2 \end{pmatrix}$

Figure 5(a) provides the graph of $\overline{u} + \overline{v}$ and Figure 5(b) provides the graph of $\overline{v} + \overline{u}$. Both graphs illustrate the characteristic diagonal of the parallelogram relationship that results from vector addition.   ◼

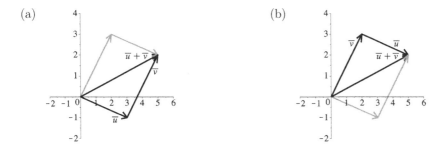

Figure 5: Vector sums for Example 7

➤ **QUESTION 7** Consider the vectors $\overline{u} = \begin{pmatrix} -2 \\ 3 \end{pmatrix}$, $\overline{v} = \begin{pmatrix} 0 \\ 3 \end{pmatrix}$, and $\overline{w} = \begin{pmatrix} 1 \\ 2 \end{pmatrix}$. Compute each vector addition and then graphically illustrate their geometric relationships:

(a) $\overline{u} + \overline{v}$ (b) $\overline{u} + \overline{w}$ (c) $\overline{w} + \overline{v}$

An Introduction to Vector Fields

This section began by mentioning the idea of a *vector field*, which is a mathematical way to describe a phenomenon in which the magnitude and direction of a quantity differ at each input. Recall that Figure 1(a) represents the flow of water over a short falls in a river, in which the speed and the direction of the water varies from point to point in the river. Similarly, Figure 1(b) represents the rotational flow of blood in a cross-section of an artery, which again varies from point to point. After discussing vectors, this idea of a vector field can be defined more carefully and examined from both an analytical and a graphical point of view.

> **Definition.** A **vector field** is a vector-valued function that has two or more inputs uniquely mapped to two or more outputs. A two-dimensional vector field $\overline{F}(x, y)$ assigns each input (x, y) to exactly one output $\overline{F}(x, y) = \begin{pmatrix} f(x, y) \\ g(x, y) \end{pmatrix}$.

Recall from Section 1.1 that a vector-valued function is a function that has one or more inputs and two or more outputs. A key point of the preceding definition is that a vector field is a type of vector-valued function. The input to the function is a point (x, y) in the plane and the output to the function is a vector unique to that point. The first component of the vector is a function $f(x, y)$ that depends on both of the inputs x and y, and the second component is a (possibly) different function $g(x, y)$ that also depends on both of the input values.

◆ **EXAMPLE 8** Consider the vector field $\overline{F}(x, y) = \begin{pmatrix} \cos(x) \\ \cos(y) \end{pmatrix}$.

(a) Evaluate $\overline{F}(x, y)$ at the inputs $(0, 0)$, $(\pi/2, 0)$, $(\pi, 0)$, $(0, \pi/2)$, $(\pi, \pi/2)$, $(0, \pi)$, $(\pi/2, \pi)$, (π, π).

(b) Graph part of vector field $\overline{F}(x, y)$ by plotting each output vector in (a) at the associated input point.

Solution.

(a) Substitute each input point into the two components of the vector field $\overline{F}(x, y)$ to determine the corresponding output vector:

$$\overline{F}(0, 0) = \begin{pmatrix} \cos(0) \\ \cos(0) \end{pmatrix} = \begin{pmatrix} 1 \\ 1 \end{pmatrix} \qquad \overline{F}(\pi/2, 0) = \begin{pmatrix} \cos(\pi/2) \\ \cos(0) \end{pmatrix} = \begin{pmatrix} 0 \\ 1 \end{pmatrix}$$

$$\overline{F}(\pi, 0) = \begin{pmatrix} \cos(\pi) \\ \cos(0) \end{pmatrix} = \begin{pmatrix} -1 \\ 1 \end{pmatrix} \qquad \overline{F}(0, \pi/2) = \begin{pmatrix} \cos(0) \\ \cos(\pi/2) \end{pmatrix} = \begin{pmatrix} 1 \\ 0 \end{pmatrix}$$

$$\overline{F}(0, \pi) = \begin{pmatrix} \cos(0) \\ \cos(\pi) \end{pmatrix} = \begin{pmatrix} 1 \\ -1 \end{pmatrix} \qquad \overline{F}(\pi, \pi/2) = \begin{pmatrix} \cos(\pi) \\ \cos(\pi/2) \end{pmatrix} = \begin{pmatrix} -1 \\ 0 \end{pmatrix}$$

$$\overline{F}(\pi, \pi) = \begin{pmatrix} \cos(\pi) \\ \cos(\pi) \end{pmatrix} = \begin{pmatrix} -1 \\ -1 \end{pmatrix} \qquad \overline{F}(\pi/2, \pi) = \begin{pmatrix} \cos(\pi/2) \\ \cos(\pi) \end{pmatrix} = \begin{pmatrix} 0 \\ -1 \end{pmatrix}$$

(b) Plot each output vector, placing the starting point of the vector at its associated input. ∎

➤ **QUESTION 8** Consider the vector field $\overline{G}(x, y) = \begin{pmatrix} \dfrac{x}{\sqrt{x^2 + y^2 + 4}} \\ \dfrac{-y}{\sqrt{x^2 + y^2 + 4}} \end{pmatrix}$.

(a) Evaluate $\overline{G}(x, y)$ at the inputs $(-1, -1)$, $(0, -1)$, $(1, -1)$, $(-1, 0)$ $(0, 0)$, $(1, 0)$, $(-1, 1)$, $(0, 1)$, $(1, 1)$.

(b) Graph part of vector field $\overline{G}(x, y)$ by plotting each output vector in (a) at the associated input point.

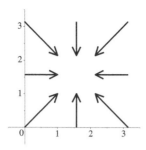

Figure 6: Vector field $\overline{F}(x, y)$ for Example 8

Working in RStudio

A vector is defined in RStudio by listing its components separated by commas inside
c(,) in the exact same fashion as when storing small data sets in RStudio. For
example, define the vector

$$\overline{u} = \begin{pmatrix} 3 \\ -6 \\ 5 \end{pmatrix}$$

in RStudio by entering the command u=c(3,-6,5). The components of a vector
u are obtained with the commands u[1], u[2], etc., where the numbered position
of the desired component is listed between the square brackets. For example, the
value of the third component of u defined by u=c(3,-6,5) is viewed by entering
the command u[3], which produces an output of 5. The operations of scalar multiplication and vector addition are performed using the standard symbols $*$ and $+$.
For example, compute

$$2\overline{u} = 2 \begin{pmatrix} 3 \\ -6 \\ 5 \end{pmatrix} \quad \text{and} \quad \overline{u} + 3\overline{v} = \begin{pmatrix} 3 \\ -6 \\ 5 \end{pmatrix} + 3 \begin{pmatrix} 4 \\ 1 \\ -8 \end{pmatrix}$$

by first defining the vectors with the commands u=c(3,-6,5) and v=c(4,1,-8),
and then entering the commands 2*u and u+3*v.

Examples of Commands

- u=c(4,1,2,1)

- u[2]

- v=c(1,5,7,-2)

- 3*u-8*v

◆ **EXAMPLE 9** Let $\bar{u} = \begin{pmatrix} 1 \\ 7 \\ 3 \\ 0 \end{pmatrix}$ and $\bar{v} = \begin{pmatrix} 2 \\ -1 \\ -4 \\ -1 \end{pmatrix}$. Compute each vector using RStudio:

(a) $3\bar{v}$ (b) $\bar{u} - 2\bar{v}$ (c) $2\bar{u} - \bar{v}$

Solution.

```
u = c(1,7,3,0)                    u-2*v
v = c(2,-1,-4,-1)
3*v                              [1] -3  9 11   2

[1]    6  -3 -12  -3             2*u-v

                                 [1]   0 15 10   1
```

This output from RStudio provides the following answers:

(a) $3\bar{v} = \begin{pmatrix} 6 \\ -3 \\ -12 \\ -3 \end{pmatrix}$ (b) $\bar{u} - 2\bar{v} = \begin{pmatrix} -3 \\ 9 \\ 11 \\ 2 \end{pmatrix}$ (c) $2\bar{u} - \bar{v} = \begin{pmatrix} 0 \\ 15 \\ 10 \\ 1 \end{pmatrix}$

■

➤ **QUESTION 9** Let $\bar{u} = \begin{pmatrix} 4 \\ 2 \\ 2 \end{pmatrix}$ and $\bar{v} = \begin{pmatrix} 1 \\ 1 \\ -1 \end{pmatrix}$. Compute each vector using RStudio:

(a) $2\bar{v}$ (b) $3\bar{u} - 4\bar{v}$ (c) $6\bar{u} - \dfrac{1}{2}\bar{v}$

Summary

- The *vector* $\overline{PQ}$ starting at point $P = (P_1, P_2)$ and ending at point $Q = (Q_1, Q_2)$ is

$$\overline{PQ} = \begin{pmatrix} Q_1 - P1 \\ Q_2 - P_2 \end{pmatrix} = \begin{pmatrix} \Delta x \\ \Delta y \end{pmatrix}$$

  Each entry in the column of numbers is called a *component*. Vectors with more components are defined similarly.

- *Key Fact:* Assume that all vectors start at the origin unless otherwise specified.

Summary (continued)

- The *length* (or *magnitude*) of a vector $\overline{v} = \begin{pmatrix} v_1 \\ v_2 \\ \vdots \\ v_n \end{pmatrix}$ is $\|\overline{v}\| = \sqrt{v_1^2 + v_2^2 + \cdots + v_n^2}$

- *Scalar* is another name for a real number.

- The *addition*, *subtraction*, and *scalar multiplication* of vectors are defined as follows:

$$\overline{u} \pm \overline{v} = \begin{pmatrix} u_1 \\ u_2 \\ \vdots \\ u_n \end{pmatrix} \pm \begin{pmatrix} v_1 \\ v_2 \\ \vdots \\ v_n \end{pmatrix} = \begin{pmatrix} u_1 \pm v_1 \\ u_2 \pm v_2 \\ \vdots \\ u_n \pm v_n \end{pmatrix} \qquad m\,\overline{u} = m \begin{pmatrix} u_1 \\ u_2 \\ \vdots \\ u_n \end{pmatrix} = \begin{pmatrix} m\,u_1 \\ m\,u_2 \\ \vdots \\ m\,u_n \end{pmatrix}$$

- *Scalar multiplication* changes the length of the vector by the magnitude of the scalar. If the scalar is positive, the resulting vector points in the same direction as the original; if the scalar is negative, the resulting vector points in the opposite direction.

- A *vector addition* $\overline{u} + \overline{v}$ produces a vector corresponding to one of the diagonals of the parallelogram formed by $\overline{u}$ and $\overline{v}$.

- A *vector field* is a vector-valued function that has two or more inputs uniquely mapped to two or more outputs. A vector field $\overline{F}(x, y)$ assigns each input (x, y) to a unique vector output $\overline{F}(x, y) = \begin{pmatrix} f(x, y) \\ g(x, y) \end{pmatrix}$.

Exercises

In Exercises $1-6$, compute both the vector $\overline{PQ}$ from P to Q and the vector $\overline{QP}$ from Q to P.

1. $P = (1, 1)$ and $Q = (7, 3)$

2. $P = (-2, 3)$ and $Q = (5, 0)$

3. $P = (3, 4)$ and $Q = (-3, 2)$

4. $P = (6, -2)$ and $Q = (0, -3)$

5. $P = (0, 0)$ and $Q = (4, -9)$

6. $P = (-3, -4)$ and $Q = (-1, 8)$

In Exercises $7-12$, compute both the vector $\overline{PQ}$ from P to Q and the vector $\overline{QP}$ from Q to P.

7. $P = (0, 7, 2)$ and $Q = (7, 1, 3)$

8. $P = (-2, 3, 1)$ and $Q = (4, -8, 2)$

9. $P = (2, 4, 8)$ and $Q = (-16, -8, -4)$

10. $P = (2, -3, -5)$ and $Q = (8, 13, 21)$

11. $P = (1, 7, 2, 8)$ and $Q = (9, 3, 5, 4)$

12. $P = (9, -2, 2, -3)$ and
$Q = (-7, 1, 3, 4)$

Your Turn. In Exercises 13–16, choose your own points with the specified number of coordinates and compute both the vector $\overline{PQ}$ from P to Q and the vector $\overline{QP}$ from Q to P.

13. P and Q with two coordinates

14. P and Q with three coordinates

15. P and Q with four coordinates

16. P and Q with five coordinates

In Exercises 17 and 18, show that the directed line segments defined by each pair of points in the table are all represented by the same vector $\overline{PQ}$. Sketch the three directed line segments on the same pair of axes.

17.

P	Q	$\overline{PQ}$
$(-1, 2)$	$(5, 3)$	
$(-2, 1)$	$(4, 2)$	
$(0, 5)$	$(6, 6)$	

18.

P	Q	$\overline{PQ}$
$(3, 7)$	$(6, 2)$	
$(-2, 4)$	$(1, -1)$	
$(8, 0)$	$(11, -5)$	

Your Turn. In Exercises 19 and 20, find another pair of points P and Q such that the directed line segment from P to Q is represented by the same vector $\overline{PQ}$ from the given exercise.

19. Exercise 17   20. Exercise 18

In Exercises 21–24, find the starting point P for the vector $\overline{PQ}$ with ending point Q.

21. $\overline{PQ} = \begin{pmatrix} 2 \\ 1 \end{pmatrix}$; $Q = (-1, 1)$

22. $\overline{PQ} = \begin{pmatrix} 3 \\ -2 \end{pmatrix}$; $Q = (-3, -4)$

23. $\overline{PQ} = \begin{pmatrix} 4 \\ 6 \end{pmatrix}$; $Q = (8, -1)$

24. $\overline{PQ} = \begin{pmatrix} -9 \\ 7 \end{pmatrix}$; $Q = (8, -2)$

In Exercises 25–28, find the ending point Q for the vector $\overline{PQ}$ with starting point P.

25. $\overline{PQ} = \begin{pmatrix} 2 \\ 1 \end{pmatrix}$; $P = (4, 3)$

26. $\overline{PQ} = \begin{pmatrix} 3 \\ -2 \end{pmatrix}$; $P = (-6, -7)$

27. $\overline{PQ} = \begin{pmatrix} -4 \\ 3 \end{pmatrix}$; $P = (1, 2)$

28. $\overline{PQ} = \begin{pmatrix} -9 \\ 7 \end{pmatrix}$; $P = (-2, 16)$

In Exercises 29–36, compute the length of the vector.

29. $\overline{v} = \begin{pmatrix} 4 \\ -3 \end{pmatrix}$

30. $\overline{v} = \begin{pmatrix} 2 \\ 6 \end{pmatrix}$

31. $\overline{v} = \begin{pmatrix} -1 \\ -3 \end{pmatrix}$

32. $\overline{v} = \begin{pmatrix} 1 \\ 1 \\ 1 \end{pmatrix}$

33. $\overline{v} = \begin{pmatrix} 4 \\ 1 \\ -3 \end{pmatrix}$

34. $\overline{v} = \begin{pmatrix} -1 \\ 5 \\ 7 \\ -2 \end{pmatrix}$

35. $\overline{v} = \begin{pmatrix} 1 \\ 1 \\ 1 \\ 0 \end{pmatrix}$

36. $\overline{v} = \begin{pmatrix} -2 \\ 3 \\ 5 \\ 1 \end{pmatrix}$

Your Turn. In Exercises 37–40, choose your own vector $\overline{v}$ with the specified number of components and compute the length of $\overline{v}$.

37. Two components

38. Three components

39. Four components

40. Five components

In Exercises 41 – 48, let

$$\overline{u} = \begin{pmatrix} 4 \\ -3 \end{pmatrix} \text{ and } \overline{v} = \begin{pmatrix} 2 \\ 2 \end{pmatrix}$$

and compute the vector.

41. $3\overline{u}$

42. $-2\overline{u}$

43. $-\overline{v}$

44. $14\overline{v}$

45. $\overline{u} + \overline{v}$

46. $\overline{v} + \overline{u}$

47. $8\overline{u} + 3\overline{v}$

48. $2\overline{u} - 5\overline{v}$

RStudio. In Exercises 49 – 56, use RStudio to compute the vectors from Exercises 41 – 48.

49. Exercise 41

50. Exercise 42

51. Exercise 43

52. Exercise 44

53. Exercise 45

54. Exercise 46

55. Exercise 47

56. Exercise 48

In Exercises 57 – 64, let

$$\overline{u} = \begin{pmatrix} 1 \\ -7 \\ 2 \end{pmatrix} \text{ and } \overline{v} = \begin{pmatrix} 0 \\ -3 \\ 9 \end{pmatrix},$$

and compute the vector.

57. $7\overline{u}$

58. $-2\overline{u}$

59. $-3\overline{v}$

60. $5\overline{v}$

61. $\overline{u} + \overline{v}$

62. $-\overline{u} - \overline{v}$

63. $7\overline{u} + 5\overline{v}$

64. $-2\overline{v} - 3\overline{u}$

RStudio. In Exercises 65 – 72, use RStudio to compute the vectors from Exercises 57 – 64.

65. Exercise 57

66. Exercise 58

67. Exercise 59

68. Exercise 60

69. Exercise 61

70. Exercise 62

71. Exercise 63

72. Exercise 64

In Exercises 73 – 75, let $\overline{v} = \begin{pmatrix} 1 \\ 3 \end{pmatrix}$.

73. Sketch the vectors $\overline{v}$ and $-\overline{v}$ on the same pair of axes.

74. How does the length of $\overline{v}$ relate to the length of $-\overline{v}$?

75. How does the direction of $\overline{v}$ relate to the direction of $-\overline{v}$?

In Exercises 76 – 78, let $\overline{v} = \begin{pmatrix} -1 \\ 0 \end{pmatrix}$.

76. Sketch the vectors $\overline{v}$ and $3\overline{v}$ on the same pair of axes.

77. How does the length of $\overline{v}$ relate to the length of $3\overline{v}$?

78. How does the direction of $\overline{v}$ relate to the direction of $3\overline{v}$?

Your Turn. For Exercises 79 – 81, choose your own vector $\overline{v}$ and scalar m.

79. Sketch the vectors $\overline{v}$ and $m\overline{v}$ on the same pair of axes.

80. How does the length of $\overline{v}$ relate to the length of $m\overline{v}$?

81. How does the direction of $\overline{v}$ relate to the direction of $m\overline{v}$?

In Exercises 82 – 85, let

$$\overline{u} = \begin{pmatrix} 2 \\ 3 \end{pmatrix} \text{ and } \overline{v} = \begin{pmatrix} -1 \\ 1 \end{pmatrix}.$$

82. Compute $\overline{u} + \overline{v}$ algebraically.

83. Sketch the vectors $\overline{u}$, $\overline{v}$, and $\overline{u} + \overline{v}$ as directed line segments starting at $(0, 0)$.

84. On the plot from Exercise 83, draw the other two sides of the parallelogram with edges $\overline{u}$ and $\overline{v}$, with the directed line segments pointing in the same direction. Also, label the sides represented by $\overline{u}$ and the sides represented by $\overline{v}$.

85. The vector $\overline{u} + \overline{v}$ represents one diagonal of the parallelogram determined by the vectors $\overline{u}$ and $\overline{v}$. Compute a vector representing the other diagonal and write this vector in the form $m\overline{u} + n\overline{v}$, for appropriate scalars m and n. Note that there are two possible pairs of answers for m and n owing to the two possible directions of the vector representing the other diagonal.

Your Turn. For Exercises 86–89, choose your own two-component vectors $\overline{u}$ and $\overline{v}$ and repeat the given exercises.

86. Exercise 82   88. Exercise 84

87. Exercise 83   89. Exercise 85

In Exercises 90–95 determine the output of the vector field at the inputs $(-1,-1)$, $(-1,0)$, $(-1,1)$, $(0,-1)$, $(0,0)$, $(0,1)$, $(1,-1)$, $(1,0)$, and $(1,1)$ and then sketch the vector field.

90. $\overline{F}(x,y) = \begin{pmatrix} x \\ y \end{pmatrix}$

91. $\overline{F}(x,y) = \begin{pmatrix} y \\ 1 \end{pmatrix}$

92. $\overline{F}(x,y) = \begin{pmatrix} 0 \\ x \end{pmatrix}$

93. $\overline{F}(x,y) = \begin{pmatrix} y \\ -x \end{pmatrix}$

94. $\overline{F}(x,y) = \begin{pmatrix} 2y \\ x+1 \end{pmatrix}$

95. $\overline{F}(x,y) = \begin{pmatrix} x+y \\ x-y \end{pmatrix}$

In Your Own Words. In Exercises 96–105, explain the following.

96. Vector

97. Component

98. Computing vector length

99. Vector addition

100. Vector subtraction

101. Scalar multiplication

102. Geometry of scalar multiplication

103. Geometry of vector addition

104. Commutativity

105. Vector field

3.2   Linear Combinations of Vectors

This chapter works toward developing an understanding of the method of least squares, which is used in a host of different disciplines to create models of physical and social phenomena. This method is also coded into the RStudio command `fitModel`. The goal of the method of least squares is to find the best possible mathematical model for a given data set in the sense that the line or curve is as close as possible to every point in the data set simultaneously. When given exactly two data points, a curve that passes through both points can be readily found so long as the curve has two distinct parameters. For example, any two points uniquely determine a line $y = mx + b$ with its two parameters of slope m and vertical intercept b. This line is a close as possible to these points, because it actually passes through them.

However, for three or more data points, it may be (and often is) impossible to find a curve that passes through all of the given points. Even so, the best possible curve that is as close as possible to all the given data points can be found by means of the method of least squares. *Linear combinations* of vectors lie at the heart of this method and so are essential to this study. As an introduction, linear combinations are considered in an example of trying to find a line that passes through every point in a given data set for which a linear model is reasonable.

◆ **EXAMPLE 1** The table below presents how many million people used Twitter during each quarter of the given year since 2000, where 11 represents the first quarter of 2011 (January – March 2011), 11.25 represents the second quarter of 2011 (April – June 2011), and so on.

Year (Y)	11	11.25	11.5	11.75	12.75
Twitter users (U)	68	85	101	117	185

This data set is approximately linear (see Example 1 in Section 2.1) and can be represented by a model of the form $U = mY + b$.

(a) Substitute each data point into the model $U = mY + b$ to find a **system of linear equations**, which consists of the corresponding collection of linear equations involving the quantities m and b.

(b) Write the system of linear equations from (a) as a **vector equation**, which is an equality between arithmetic combinations of vectors.

Solution.

(a) While this data set is approximately linear, it is not exactly so, as is demonstrated by Example 2 in Section 2.1. However, supposing that a perfect linear relationship exists, substitute each data point into the equation $U = mY + b$ to obtain the following system of linear equations:

$$68 = 11m + b$$
$$85 = 11.25m + b$$
$$101 = 11.5m + b$$
$$117 = 11.75m + b$$
$$185 = 12.75m + b$$

(b) Now write this system of linear equations as a vector equation. Recall that scalar multiplication multiplies each component by the same real number, and vector addition adds the corresponding components of the two vectors. Reversing these operations in the following three steps provides the sought for vector equation:

$$\begin{pmatrix} 68 \\ 85 \\ 101 \\ 117 \\ 185 \end{pmatrix} = \begin{pmatrix} 11m + b \\ 11.25m + b \\ 11.5m + b \\ 11.75m + b \\ 12.75m + b \end{pmatrix} \qquad \text{Put the entries on each side into a vector}$$

$$
\begin{pmatrix} 68 \\ 85 \\ 101 \\ 117 \\ 185 \end{pmatrix} = \begin{pmatrix} 11m \\ 11.25m \\ 11.5m \\ 11.75m \\ 12.75m \end{pmatrix} + \begin{pmatrix} b \\ b \\ b \\ b \\ b \end{pmatrix} \qquad \text{Split the right side into a sum of two vectors}
$$

$$
\begin{pmatrix} 68 \\ 85 \\ 101 \\ 117 \\ 185 \end{pmatrix} = m \begin{pmatrix} 11 \\ 11.25 \\ 11.5 \\ 11.75 \\ 12.75 \end{pmatrix} + b \begin{pmatrix} 1 \\ 1 \\ 1 \\ 1 \\ 1 \end{pmatrix} \qquad \text{Factor out the scalar multiples } m \text{ and } b
$$

Such a vector equation is often expressed in the form $\overline{U} = m\overline{Y} + b\overline{1}_5$, where

$$
\overline{U} = \begin{pmatrix} 68 \\ 85 \\ 101 \\ 117 \\ 185 \end{pmatrix} \qquad \overline{Y} = \begin{pmatrix} 11 \\ 11.25 \\ 11.5 \\ 11.75 \\ 12.75 \end{pmatrix} \qquad \overline{1}_5 = \begin{pmatrix} 1 \\ 1 \\ 1 \\ 1 \\ 1 \end{pmatrix}.
$$

This vector equation is an example of what is known as a linear combination of vectors. In the vector equation $\overline{U} = m\overline{Y} + b\overline{1}_5$ from Example 1(b), the notation $\overline{1}_5$ is used to represent a vector that has an entry of 1 for all five components. This vector $\overline{1}_5$ is called the *intercept vector* because of its connection with the vertical intercept b in the linear equation $U = mY + b$. The notation $\overline{1}_5$ is used throughout this chapter and can be extended to an arbitrary number of entries.

Definition. The **intercept vector** $\overline{1}_n$ is a vector whose n components are all 1.

▶ **QUESTION 1** The following table presents the average debt load in thousands of dollars at the end of each year's spring term for bachelor's degree recipients who attended public four-year colleges and universities and borrowed money to finance their education:

Academic year (Y)	2001	2003	2005	2006
Average debt (D)	\$20,400	\$20,900	\$21,500	\$21,800

This data set is approximately linear (see Question 1 in Section 2.1) and can be represented by a model of the form $D = mY + b$.

(a) Substitute each data point into the model $D = mY + b$ to find its corresponding system of linear equations.

(b) Write the system of linear equations from (a) as a vector equation.

As suggested by Example 1 and Question 1, linear combinations of vectors can arise when developing mathematical models of data sets, and, as such, they are a key aspect of studying the method of least squares. Mathematicians have developed powerful techniques for manipulating linear combinations of vectors as well as for finding exact solutions to such vector equations when they exist. Even more, approximate solutions can be found when exact solutions do not exist. The idea of a linear combination is formally defined as follows:

Definition.

- If $\overline{u}$ and $\overline{v}$ are vectors, and x and y are scalars, then $x\overline{u} + y\overline{v}$ is a **linear combination** of $\overline{u}$ and $\overline{v}$.

- If $\overline{u}_1, \overline{u}_2, \ldots, \overline{u}_n$ are vectors and $x_1, x_2, \ldots, x_n$ are scalars, then $x_1\overline{u}_1 + x_2\overline{u}_2 + \cdots + x_n\overline{u}_n$ is a **linear combination** of $\overline{u}_1, \overline{u}_2, \ldots, \overline{u}_n$.

In this book, linear combinations were first encountered in Section 3.1 while using the definitions of scalar multiplication and vector addition to compute such quantities as $3\overline{u} - 2\overline{v}$. Further examples of these computations can help solidify an understanding of this definition.

◆ **EXAMPLE 2** Let $\overline{u} = \begin{pmatrix} 2 \\ 1 \\ 3 \\ -1 \end{pmatrix}$, $\overline{v} = \begin{pmatrix} -1 \\ 3 \\ 5 \\ -4 \end{pmatrix}$, and $\overline{w} = \begin{pmatrix} 0 \\ 2 \\ -3 \\ 1 \end{pmatrix}$, and compute each vector:

(a) $\overline{u} + 3\overline{v}$

(b) $3\overline{u} - 2\overline{v} + \overline{w}$

Solution.

(a) $\overline{u} + 3\overline{v} = \begin{pmatrix} 2 \\ 1 \\ 3 \\ -1 \end{pmatrix} + 3\begin{pmatrix} -1 \\ 3 \\ 5 \\ -4 \end{pmatrix} = \begin{pmatrix} 2 \\ 1 \\ 3 \\ -1 \end{pmatrix} + \begin{pmatrix} -3 \\ 9 \\ 15 \\ -12 \end{pmatrix} = \begin{pmatrix} 2 + (-3) \\ 1 + 9 \\ 3 + 15 \\ -1 + (-12) \end{pmatrix} = \begin{pmatrix} -1 \\ 10 \\ 18 \\ -13 \end{pmatrix}$

(b) $3\overline{u} - 2\overline{v} + \overline{w} = 3\begin{pmatrix} 2 \\ 1 \\ 3 \\ -1 \end{pmatrix} - 2\begin{pmatrix} -1 \\ 3 \\ 5 \\ -4 \end{pmatrix} + \begin{pmatrix} 0 \\ 2 \\ -3 \\ 1 \end{pmatrix} = \begin{pmatrix} 6 \\ 3 \\ 9 \\ -3 \end{pmatrix} - \begin{pmatrix} -2 \\ 6 \\ 10 \\ -8 \end{pmatrix} + \begin{pmatrix} 0 \\ 2 \\ -3 \\ 1 \end{pmatrix} = \begin{pmatrix} 8 \\ -1 \\ -4 \\ 6 \end{pmatrix}$

∎

➤ **QUESTION 2** Let $\overline{u} = \begin{pmatrix} 3 \\ 4 \\ 1 \\ 0 \end{pmatrix}$, $\overline{v} = \begin{pmatrix} -1 \\ 5 \\ 5 \\ 5 \end{pmatrix}$, and $\overline{w} = \begin{pmatrix} 6 \\ 1 \\ 2 \\ 1 \end{pmatrix}$, and compute each vector:

(a) $2\overline{u} - 2\overline{v}$

(b) $-3\overline{u} - \overline{v} + 2\overline{w}$

Finding Desired Linear Combinations

As illustrated in Example 2, the process of computing a linear combination of two vectors is a relatively straightforward application of scalar multiplication and vector addition. Moving forward, the primary focus of discussion shifts to a variation of this idea. Namely, the goal is to answer the following question, which is asked in three different, but equivalent, forms:

Given vectors $\bar{u}$, $\bar{v}$, and $\bar{w}$,
- Is $\bar{w}$ a linear combination of $\bar{u}$ and $\bar{v}$?
- Do there exist scalars x and y such that $\bar{w} = x\bar{u} + y\bar{v}$?
- Is there a solution of the vector equation $\bar{w} = x\bar{u} + y\bar{v}$?

Informally, we often ask when vectors $\bar{u}$ and $\bar{v}$ can be combined to hit (or attain) a target vector $\bar{w}$. Such questions arise in many different settings. Recall that Example 1 considered a setting in which $\bar{U} = m\bar{Y} + b\bar{1}_5$. The next natural step is to try to find the slope m and vertical intercept b that solve this vector equation. Sometimes such scalars can be found, but sometimes not. The rest of this section focuses on using the power of algebra and RStudio to determine when such a linear combination exists and finding these unknown scalars when they do exist.

◆ **EXAMPLE 3** Let $\bar{u} = \begin{pmatrix} 2 \\ -1 \end{pmatrix}$, $\bar{v} = \begin{pmatrix} 1 \\ 4 \end{pmatrix}$, and $\bar{w} = \begin{pmatrix} 7 \\ 10 \end{pmatrix}$, and find scalars x and y so that $\bar{w} = x\bar{u} + y\bar{v}$.

Solution. First, write the equation $\bar{w} = x\bar{u} + y\bar{v}$ in vector form and then perform the vector operations on the right-hand side of the equation to obtain a single vector:

$$
\begin{aligned}
\bar{w} &= x\bar{u} + y\bar{v} \\
\begin{pmatrix} 7 \\ 10 \end{pmatrix} &= x \begin{pmatrix} 2 \\ -1 \end{pmatrix} + y \begin{pmatrix} 1 \\ 4 \end{pmatrix} \\
\begin{pmatrix} 7 \\ 10 \end{pmatrix} &= \begin{pmatrix} 2x + y \\ -x + 4y \end{pmatrix}
\end{aligned}
$$

Two vectors are equal when their corresponding components are equal. Setting the first and second components of these vectors equal to each other gives the following two equations for the unknown scalars x and y:

$$
\begin{aligned}
\text{first component:} \quad & 2x + y = 7 \\
\text{second component:} \quad & -x + 4y = 10
\end{aligned}
$$

Solving the first component equation for y provides $y = 7 - 2x$. Now substitute this expression for y into the second component equation and solve for x:

$-x + 4y = 10$	Second component equation
$-x + 4(7 - 2x) = 10$	Substitute $y = 7 - 2x$ from first component equation
$-x + 28 - 8x = 10$	Multiply through by 4
$-9x = -18$	Subtract 28 from both sides
$x = 2$	Divide both sides by -9

In this way, $x = 2$. Substituting into the first component equation yields $y = 7 - 2(2) = 3$. Therefore, $\overline{w} = 2\overline{u} + 3\overline{v}$.

■

   Thinking in terms of the various forms of our motivating question, the scalars $x = 2$ and $y = 3$ provide a solution of the vector equation $\overline{w} = x\overline{u} + y\overline{v}$ in Example 3. As with standard algebraic equations, such specific numbers provide a solution of a given vector equation when substituting them into the equation results in the two resulting quantities being equal. The following definition expresses this intuitive idea more precisely:

Definition.

- If $\overline{u}$, $\overline{v}$, and $\overline{w}$ are vectors, x and y are variables, and a and b are scalars, then $x = a$ and $y = b$ is a **solution** of the vector equation $\overline{w} = x\overline{u} + y\overline{v}$ when $\overline{w}$ is equal to $a\overline{u} + b\overline{v}$.

- If $\overline{u}_1$, $\overline{u}_2$, ..., $\overline{u}_n$ and $\overline{v}$ are vectors, x_1, x_2, ..., x_n are variables, and a_1, a_2, ..., a_n are scalars, then $x_1 = a_1$, $x_2 = a_2$, ..., $x_n = a_n$ is a **solution** of the vector equation $\overline{v} = x_1\overline{u}_1 + x_2\overline{u}_2 + \cdots + x_n\overline{u}_n$ when $\overline{v}$ is equal to $a_1\overline{u}_1 + a_2\overline{u}_2 + \cdots + a_n\overline{u}_n$.

▶ **QUESTION 3** Let $\overline{u} = \begin{pmatrix} 3 \\ -5 \end{pmatrix}$, $\overline{v} = \begin{pmatrix} -1 \\ 2 \end{pmatrix}$, and $\overline{w} = \begin{pmatrix} 1 \\ -1 \end{pmatrix}$, and find the solution of the vector equation $\overline{w} = x\overline{u} + y\overline{v}$.

Vector Equations as Matrix Equations

Example 3 stepped through a standard algebraic approach to finding scalars x and y satisfying $\overline{w} = x\overline{u} + y\overline{v}$ for given vectors $\overline{u}$, $\overline{v}$, and $\overline{w}$. This process works quite well whenever just two scalars x and y need to be found. However, as more vectors are included and, consequently, more scalars are in the linear combination, the resulting algebra becomes more complicated and cumbersome. Fortunately, such questions can be formulated in terms of *matrices* and RStudio used to find the desired solutions.

Definition. A $m \times n$ **matrix** is a rectangular array of real numbers arranged in m rows and n columns. A **square matrix** is an $n \times n$ matrix with the same number of rows and columns. Each column in a matrix represents a vector with m components and, sometimes, an m-component vector is thought of as an $m \times 1$ matrix.

◆ **EXAMPLE 4** Identify the size and the vectors represented by each matrix:

(a) $\begin{pmatrix} 1 & 3 \\ 7 & 5 \end{pmatrix}$

(b) $\begin{pmatrix} 1 & -2 & 7 & 1 & 3 \\ 0 & 1 & 1 & 1 & 0 \\ 1 & -1 & -1 & 0 & 5 \end{pmatrix}$

(c) $\begin{pmatrix} 1 & 7 & 7 \\ 9 & 1 & 1 \\ 0 & 0 & 0.1 \\ 0.4 & -4 & 40 \end{pmatrix}$

Solution.

(a) This square matrix is 2×2 and represents $\begin{pmatrix} 1 \\ 7 \end{pmatrix}$ and $\begin{pmatrix} 3 \\ 5 \end{pmatrix}$.

(b) This matrix is 3×5 and represents $\begin{pmatrix} 1 \\ 0 \\ 1 \end{pmatrix}, \begin{pmatrix} -2 \\ 1 \\ -1 \end{pmatrix}, \begin{pmatrix} 7 \\ 1 \\ -1 \end{pmatrix}, \begin{pmatrix} 1 \\ 1 \\ 0 \end{pmatrix}$, and $\begin{pmatrix} 3 \\ 0 \\ 5 \end{pmatrix}$.

(c) This matrix is 4×3 and represents $\begin{pmatrix} 1 \\ 9 \\ 0 \\ 0.4 \end{pmatrix}, \begin{pmatrix} 7 \\ 1 \\ 0 \\ -4 \end{pmatrix}$, and $\begin{pmatrix} 7 \\ 1 \\ 0.1 \\ 40 \end{pmatrix}$.

■

► **QUESTION 4** Identify the size and the vectors represented by each matrix:

(a) $\begin{pmatrix} 4 & 1 & 8 \\ -3 & 16 & -7 \\ 2 & -1 & 3 \end{pmatrix}$

(b) $\begin{pmatrix} -1 & 2 & -3 & 4 \\ 1 & 0 & 1 & 0 \\ -1 & 5 & -2 & 4 \end{pmatrix}$

(c) $\begin{pmatrix} -1 & 2 \\ 1 & 0 \\ -1 & 8 \\ 5 & 6 \end{pmatrix}$

When solving for multiple unknown scalars $x_1, x_2, \ldots, x_n$ to state a desired linear combination $\bar{v} = x_1\bar{u}_1 + x_2\bar{u}_2 + \cdots + x_n\bar{u}_n$, the algebraic computations quickly become too complex. However, writing the vector equation as a matrix equation enables the use of technology to find the unknowns $x_1, x_2, \ldots, x_n$.

The process of converting a vector equation $\bar{v} = x_1\bar{u}_1 + x_2\bar{u}_2 + \cdots + x_n\bar{u}_n$ into a matrix equation has three steps. First, define a matrix with n columns consisting of the vectors $\bar{u}_1, \bar{u}_2, \ldots, \bar{u}_n$. Then insert the unknown scalars $x_1, x_2, \ldots, x_n$ into a vector. Finally, state the corresponding matrix–vector product shown below.

CONVERTING BETWEEN VECTOR AND MATRIX EQUATIONS.
The vector equation $\bar{v} = x_1\bar{u}_1 + x_2\bar{u}_2 + \cdots + x_n\bar{u}_n$ is equivalent to the matrix equation

$$\begin{pmatrix} v_1 \\ v_2 \\ \vdots \\ v_n \end{pmatrix} = \begin{pmatrix} (\bar{u}_1)_1 & (\bar{u}_2)_1 & \cdots & (\bar{u}_n)_1 \\ (\bar{u}_1)_2 & (\bar{u}_2)_2 & \cdots & (\bar{u}_n)_2 \\ \vdots & \vdots & & \vdots \\ (\bar{u}_1)_n & (\bar{u}_2)_n & \cdots & (\bar{u}_n)_n \end{pmatrix} \begin{pmatrix} x_1 \\ x_2 \\ \vdots \\ x_n \end{pmatrix}$$

where $(\bar{u}_1)_1$ is the first component of $\bar{u}_1$, $(\bar{u}_1)_2$ is the second component of $\bar{u}_1$, and so on.

Most often, this conversion process is used when the n vectors $\bar{u}_1, \bar{u}_2, \ldots, \bar{u}_n$ all have n components, which results in a corresponding square $n \times n$ matrix. Among other nice features, matrix equations almost always have unique solutions when the matrix is square.

◆ **EXAMPLE 5** Write the following vector equation as a matrix equation:

$$\begin{pmatrix} 4 \\ -1 \\ 3 \end{pmatrix} = x_1 \begin{pmatrix} 1 \\ -2 \\ 1 \end{pmatrix} + x_2 \begin{pmatrix} 4 \\ 8 \\ -7 \end{pmatrix} + x_3 \begin{pmatrix} 5 \\ 1 \\ 0 \end{pmatrix}$$

Solution. The columns of the requested matrix are the three vectors $\bar{u}_1, \bar{u}_2$, and $\bar{u}_3$, and the unknown scalars x_1, x_2, and x_3 form the components of a vector. The corresponding matrix equation is

$$\begin{pmatrix} 4 \\ -1 \\ 3 \end{pmatrix} = \begin{pmatrix} 1 & 4 & 5 \\ -2 & 8 & 1 \\ 1 & -7 & 0 \end{pmatrix} \begin{pmatrix} x_1 \\ x_2 \\ x_3 \end{pmatrix}$$

■

➤ **QUESTION 5** Write the following vector equation as a matrix equation:

$$\begin{pmatrix} 1 \\ 4 \\ 5 \\ 7 \end{pmatrix} = x_1 \begin{pmatrix} -1 \\ 0 \\ 2 \\ 7 \end{pmatrix} + x_2 \begin{pmatrix} 1 \\ 1 \\ 1 \\ 1 \end{pmatrix} + x_3 \begin{pmatrix} -5 \\ 7 \\ 3 \\ 1 \end{pmatrix} + x_4 \begin{pmatrix} 2 \\ -1 \\ -2 \\ 0 \end{pmatrix}$$

At this point, three different types of equations have been discussed: systems of linear equations, vector equations, and matrix equations. As you may have observed, these different types of equations all express the same information, in much the same way that "hello," "hola," "caio," and "nín hǎo" are equivalent ways to greet each other. For example, the following three equations are all equivalent to each other and can be readily interconverted:

$$\begin{array}{ccc} 7 = 3x + 4y & \begin{pmatrix} 7 \\ 2 \end{pmatrix} = x \begin{pmatrix} 3 \\ 1 \end{pmatrix} + y \begin{pmatrix} 4 \\ -5 \end{pmatrix} & \begin{pmatrix} 7 \\ 2 \end{pmatrix} = \begin{pmatrix} 3 & 4 \\ 1 & -5 \end{pmatrix} \begin{pmatrix} x \\ y \end{pmatrix} \\ 2 = x - 5y & & \end{array}$$

Matrix–Vector Multiplication

Multiplication of a matrix U by a vector $\bar{x}$ is performed by rewriting the matrix–vector multiplication as a linear combination of vectors. In other words, matrix–vector multiplication is expressed in terms of scalar multiplication and vector addition, which are already familiar (as in Example 2 and Question 2). In a matrix–vector product $U\bar{x}$, the columns of the matrix U represent different vectors, and the vector $\bar{x}$ represents the scalars used in the linear combination of the column vectors from U. This approach to computing a matrix–vector product is summarized in the following definition:

MATRIX–VECTOR MULTIPLICATION.
The product of an $n \times n$ matrix with an n-component vector is

$$\begin{pmatrix} (\overline{u}_1)_1 & (\overline{u}_2)_1 & \cdots & (\overline{u}_n)_1 \\ (\overline{u}_1)_2 & (\overline{u}_2)_2 & \cdots & (\overline{u}_n)_n \\ \vdots & \vdots & & \vdots \\ (\overline{u}_1)_n & (\overline{u}_2)_n & \cdots & (\overline{u}_n)_n \end{pmatrix} \begin{pmatrix} x_1 \\ x_2 \\ \vdots \\ x_n \end{pmatrix} = x_1 \begin{pmatrix} (\overline{u}_1)_1 \\ (\overline{u}_1)_2 \\ \vdots \\ (\overline{u}_1)_n \end{pmatrix} + x_2 \begin{pmatrix} (\overline{u}_2)_1 \\ (\overline{u}_2)_2 \\ \vdots \\ (\overline{u}_2)_n \end{pmatrix} + \cdots + x_n \begin{pmatrix} (\overline{u}_n)_1 \\ (\overline{u}_n)_2 \\ \vdots \\ (\overline{u}_n)_n \end{pmatrix}$$

In every matrix–vector multiplication, the number of columns in the matrix must be the same as the number of components in the vector. Also, this book only considers such products that involve a square matrix. The definition of matrix–vector multiplication can be extended to matrices with different numbers of rows and columns; an introductory course in linear algebra explores such ideas further. Finally, observe that the result of a matrix–vector multiplication is always a vector.

◆ **EXAMPLE 6** Compute each product:

(a) $\begin{pmatrix} 1 & 3 \\ 7 & 5 \end{pmatrix} \begin{pmatrix} 2 \\ 3 \end{pmatrix}$

(b) $\begin{pmatrix} 1 & -2 & 7 & 1 \\ 0 & 1 & 1 & 1 \\ 1 & -1 & -1 & 0 \\ 3 & 0 & 5 & 4 \end{pmatrix} \begin{pmatrix} -1 \\ 2 \\ 1 \\ 3 \end{pmatrix}$

Solution.

(a) $\begin{pmatrix} 1 & 3 \\ 7 & 5 \end{pmatrix} \begin{pmatrix} 2 \\ 3 \end{pmatrix} = 2 \begin{pmatrix} 1 \\ 7 \end{pmatrix} + 3 \begin{pmatrix} 3 \\ 5 \end{pmatrix} = \begin{pmatrix} 11 \\ 29 \end{pmatrix}$

(b) $\begin{pmatrix} 1 & -2 & 7 & 1 \\ 0 & 1 & 1 & 1 \\ 1 & -1 & -1 & 0 \\ 3 & 0 & 5 & 4 \end{pmatrix} \begin{pmatrix} -1 \\ 2 \\ 1 \\ 3 \end{pmatrix} = (-1) \begin{pmatrix} 1 \\ 0 \\ 1 \\ 3 \end{pmatrix} + 2 \begin{pmatrix} -2 \\ 1 \\ -1 \\ 0 \end{pmatrix} + 1 \begin{pmatrix} 7 \\ 1 \\ -1 \\ 5 \end{pmatrix} + 3 \begin{pmatrix} 1 \\ 1 \\ 0 \\ 4 \end{pmatrix} = \begin{pmatrix} 5 \\ 6 \\ -4 \\ 14 \end{pmatrix}$

∎

▶ **QUESTION 6** Compute each product:

(a) $\begin{pmatrix} 7 & 6 \\ -1 & 4 \end{pmatrix} \begin{pmatrix} 1 \\ 3 \end{pmatrix}$

(b) $\begin{pmatrix} 1 & 1 & 1 \\ 0 & 4 & 1 \\ 1 & 0 & 2 \end{pmatrix} \begin{pmatrix} 2 \\ 3 \\ 4 \end{pmatrix}$

Solving Matrix Equations

One motivation for looking at matrix equations is to enable a more efficient solution process for larger vector equations, because matrix equations provide a concise way to write a vector equation. Beginning with matrix equations involving 2×2 matrices, methods for solving matrix equations are now developed.

SOLVING 2 × 2 MATRIX EQUATIONS.

The solution $\begin{pmatrix} x \\ y \end{pmatrix}$ of the 2×2 matrix equation $\begin{pmatrix} r \\ s \end{pmatrix} = \begin{pmatrix} a & b \\ c & d \end{pmatrix} \begin{pmatrix} x \\ y \end{pmatrix}$ with $ad - bc \neq 0$ is

$$\begin{pmatrix} x \\ y \end{pmatrix} = \frac{1}{ad - bc} \begin{pmatrix} d & -b \\ -c & a \end{pmatrix} \begin{pmatrix} r \\ s \end{pmatrix}$$

If $ad - bc = 0$, then the matrix equation has either no solution or infinitely many solutions.

The following example illustrates a systematic process for computing such solutions. Usually, the matrix–vector product is first computed to produce a two component vector, followed by taking the scalar multiple of this resulting vector by $1/(ad - bc)$.

◆ **EXAMPLE 7** Solve each matrix equation:

(a) $\begin{pmatrix} 7 \\ 10 \end{pmatrix} = \begin{pmatrix} 2 & 1 \\ -1 & 4 \end{pmatrix} \begin{pmatrix} x \\ y \end{pmatrix}$
(b) $\begin{pmatrix} 2 \\ 4 \end{pmatrix} = \begin{pmatrix} 1 & 1 \\ 2 & 4 \end{pmatrix} \begin{pmatrix} x \\ y \end{pmatrix}$

Solution.

(a)
$$\begin{pmatrix} x \\ y \end{pmatrix} = \frac{1}{(2)(4) - (1)(-1)} \begin{pmatrix} 4 & -1 \\ 1 & 2 \end{pmatrix} \begin{pmatrix} 7 \\ 10 \end{pmatrix}$$ Formula for solving 2 × 2 matrix equations

$$= \frac{1}{9} \left[7 \begin{pmatrix} 4 \\ 1 \end{pmatrix} + 10 \begin{pmatrix} -1 \\ 2 \end{pmatrix} \right]$$ matrix–vector multiplication

$$= \frac{1}{9} \left[\begin{pmatrix} 28 \\ 7 \end{pmatrix} + \begin{pmatrix} -10 \\ 20 \end{pmatrix} \right]$$ Scalar multiply by 7 and 10

$$= \frac{1}{9} \begin{pmatrix} 18 \\ 27 \end{pmatrix}$$ Vector addition

$$= \begin{pmatrix} 2 \\ 3 \end{pmatrix}$$ Scalar multiply by $\frac{1}{9}$

(b)
$$\begin{pmatrix} x \\ y \end{pmatrix} = \frac{1}{(1)(4) - (2)(1)} \begin{pmatrix} 4 & -1 \\ -2 & 1 \end{pmatrix} \begin{pmatrix} 2 \\ 4 \end{pmatrix}$$ Formula for solving 2 × 2 matrix equations

$$= \frac{1}{2} \left[2 \begin{pmatrix} 4 \\ -2 \end{pmatrix} + 4 \begin{pmatrix} -1 \\ 1 \end{pmatrix} \right]$$ matrix–vector multiplication

$$= \frac{1}{2} \left[\begin{pmatrix} 8 \\ -4 \end{pmatrix} + \begin{pmatrix} -4 \\ 4 \end{pmatrix} \right]$$ Scalar multiply by 2 and 4

$$= \frac{1}{2} \begin{pmatrix} 4 \\ 0 \end{pmatrix}$$ Vector addition

$$= \begin{pmatrix} 2 \\ 0 \end{pmatrix}$$ Scalar multiply by $\frac{1}{2}$

■

➤ **QUESTION 7** Solve each matrix equation:

(a) $\begin{pmatrix} 5 \\ 10 \end{pmatrix} = \begin{pmatrix} 2 & 3 \\ 1 & 4 \end{pmatrix} \begin{pmatrix} x \\ y \end{pmatrix}$
(b) $\begin{pmatrix} -3 \\ 5 \end{pmatrix} = \begin{pmatrix} -1 & 1 \\ -2 & 1 \end{pmatrix} \begin{pmatrix} x \\ y \end{pmatrix}$

This approach to solving a 2×2 matrix equation extends to larger matrix equations, but further details lie beyond the scope of this book. Instead, technology is used to solve larger matrix equations that involve 3×3, 4×4, and larger matrices. The RStudio commands for carrying out these computations are detailed below.

Working in RStudio

In order to define and store a matrix in RStudio, first create vectors representing the columns of the matrix using the command `c()` as discussed in Section 3.1. The matrix consisting of these vectors is defined with the `matrix` command, which requires three inputs: the columns of the matrix, the number of rows, and the number of columns. A matrix whose columns are the vectors $\overline{u}_1$, $\overline{u}_2$, and $\overline{u}_3$ and that has three rows and three columns is stored by entering the command `matrix(c(u1,u2,u3),nrow=3,ncol=3)`. As a more specific example, to define and store the matrix

$$U = \begin{pmatrix} 3 & 4 & 1 \\ -6 & 1 & -2 \\ 5 & -8 & 3 \end{pmatrix}$$

first store the vectors in RStudio by entering `u1=c(3,-6,5)`, `u2=c(4,1,-8)`, and `u3=c(4,1,-8)`, and then the matrix by entering `U=matrix(c(u1,u2,u3),nrow=3, ncol=3)`.

The `solve(U,v)` command is used to solve a matrix equation of the form $\overline{v} = U\overline{x}$, where U is the matrix and $\overline{x}$ is the vector of unknown scalars. The `solve` command requires two inputs: the matrix U and the vector $\overline{v}$ from the left-hand side of the matrix equation. For example, to solve the equation

$$\begin{pmatrix} 5 \\ -19 \\ 27 \end{pmatrix} = \begin{pmatrix} 3 & 4 & 1 \\ -6 & 1 & -2 \\ 5 & -8 & 3 \end{pmatrix} \begin{pmatrix} x_1 \\ x_2 \\ x_3 \end{pmatrix}$$

first define and store the matrix U as above and then the vector by entering `v=c(5,-19,27)`. Finally, obtain the desired solution of the matrix equation by entering the command `solve(U,v)`, which outputs [1] 2 -1 3 to identify the unknown scalars as $x_1 = 2$, $x_2 = -1$, and $x_3 = 3$.

All of these commands work for any number of components, from 2×2 matrix equations to 4×4 matrix equations, and more. The only requirements, in all cases for RStudio, are that the matrix U must be a square $n \times n$ matrix and that the vector $\overline{v}$ must have the corresponding number of n components.

Examples of Commands

- u1=c(1,2)   • v=c(1,5)

- u2=c(4,-3)   • solve(U,v)

- U=matrix(c(u1,u2),nrow=2,ncol=2)

◆ **EXAMPLE 8** Solve $\begin{pmatrix} 4 \\ -1 \\ 3 \end{pmatrix} = \begin{pmatrix} 1 & 4 & 5 \\ -2 & 8 & 1 \\ 1 & -7 & 0 \end{pmatrix} \begin{pmatrix} x_1 \\ x_2 \\ x_3 \end{pmatrix}$ using RStudio.

Solution.

```
u1=c(1, -2, 1)                        v=c(4,-1,3)
u2=c(4,8,-7)                          solve(U,v)
u3=c(5,1,0)
U=matrix(c(u1,u2,u3),nrow=3,          [1] -1.097561 -0.585366  1.487805
       ncol=3)
```

Based on this RStudio output, $\begin{pmatrix} x_1 \\ x_2 \\ x_3 \end{pmatrix} = \begin{pmatrix} -1.0976 \\ -0.5854 \\ 1.4878 \end{pmatrix}$. ∎

➤ **QUESTION 8** Solve $\begin{pmatrix} 1 \\ 4 \\ 5 \\ 7 \end{pmatrix} = \begin{pmatrix} -1 & 1 & -5 & 2 \\ 0 & 1 & 7 & -1 \\ 2 & 1 & 3 & -2 \\ 7 & 1 & 1 & 0 \end{pmatrix} \begin{pmatrix} x_1 \\ x_2 \\ x_3 \\ x_4 \end{pmatrix}$ using RStudio.

As you might guess, this book only touches on certain aspects of finding scalars to achieve a target linear combination of vectors and certain aspects of solving matrix equations. Further details lie beyond the scope of this book and can be studied in a linear algebra course.

To wrap up this section, we highlight the fact that not every vector equation has a solution. In Example 1 regarding the number of Twitter users over time, the given data set is not exactly linear. As a result, the vector equation $\overline{U} = m\overline{Y} + b\overline{1}_5$ and its corresponding matrix equation do not have solutions. Section 3.3 considers the question of when such equations do not have solutions more carefully, and Section 3.4 is an introduction to finding the linear combination that is as close as possible to satisfying a vector equation via vector projection.

Summary

- A *system of linear equations* consists of a collection of linear equations. A *vector equation* is an equality between arithmetic combinations of vectors.

- The *intercept vector* $\overline{1}_n$ is a vector with n components that are all 1.

- If $\overline{u}_1$, $\overline{u}_2$, ..., $\overline{u}_n$ are vectors and x_1, x_2, ..., x_n are scalars, then $x_1\overline{u}_1 + x_2\overline{u}_2 + \cdots + x_n\overline{u}_n$ is a *linear combination* of $\overline{u}_1$, $\overline{u}_2$, ..., $\overline{u}_n$.

- If $\overline{u}_1$, $\overline{u}_2$, ..., $\overline{u}_n$ and $\overline{v}$ are vectors, x_1, x_2, ..., x_n are variables, and a_1, a_2, ..., a_n are scalars, then $x_1 = a_1$, $x_2 = a_2$, ..., $x_n = a_n$ is a *solution* of the vector equation $\overline{v} = x_1\overline{u}_1 + x_2\overline{u}_2 + \cdots + x_n\overline{u}_n$ when $\overline{v}$ is equal to $a_1\overline{u}_1 + a_2\overline{u}_2 + \cdots + a_n\overline{u}_n$.

- A *$m \times n$ matrix* is a rectangular array of real numbers arranged in m rows and n columns. A *square matrix* is an $n \times n$ matrix with the same number of rows and columns. Each column in a matrix represents a vector with m components and, sometimes, an m-component vector is thought of as an $m \times 1$ matrix.

- The *vector equation* $\overline{v} = x_1\overline{u_1} + x_2\overline{u_2} + \cdots + x_n\overline{u_n}$ is equivalent to the *matrix equation*

$$\begin{pmatrix} v_1 \\ v_2 \\ \vdots \\ v_n \end{pmatrix} = \begin{pmatrix} (\overline{u}_1)_1 & (\overline{u}_2)_1 & \dots & (\overline{u}_n)_1 \\ (\overline{u}_1)_2 & (\overline{u}_2)_2 & \dots & (\overline{u}_n)_2 \\ \vdots & \vdots & & \vdots \\ (\overline{u}_1)_n & (\overline{u}_2)_n & \dots & (\overline{u}_n)_n \end{pmatrix} \begin{pmatrix} x_1 \\ x_2 \\ \vdots \\ x_n \end{pmatrix}$$

- The *product* of an $n \times n$ matrix with an n-component vector is

$$\begin{pmatrix} (\overline{u}_1)_1 & (\overline{u}_2)_1 & \dots & (\overline{u}_n)_1 \\ (\overline{u}_1)_2 & (\overline{u}_2)_2 & \dots & (\overline{u}_n)_2 \\ \vdots & \vdots & & \vdots \\ (\overline{u}_1)_n & (\overline{u}_2)_n & \dots & (\overline{u}_n)_n \end{pmatrix} \begin{pmatrix} x_1 \\ x_2 \\ \vdots \\ x_n \end{pmatrix} = x_1 \begin{pmatrix} (\overline{u}_1)_1 \\ (\overline{u}_1)_2 \\ \vdots \\ (\overline{u}_1)_n \end{pmatrix} + \cdots + x_n \begin{pmatrix} (\overline{u}_n)_1 \\ (\overline{u}_n)_2 \\ \vdots \\ (\overline{u}_n)_n \end{pmatrix}$$

- The solution $\begin{pmatrix} x \\ y \end{pmatrix}$ of the 2×2 matrix equation $\begin{pmatrix} r \\ s \end{pmatrix} = \begin{pmatrix} a & b \\ c & d \end{pmatrix} \begin{pmatrix} x \\ y \end{pmatrix}$ with $ad - bc \neq 0$ is

$$\begin{pmatrix} x \\ y \end{pmatrix} = \frac{1}{ad - bc} \begin{pmatrix} d & -b \\ -c & a \end{pmatrix} \begin{pmatrix} r \\ s \end{pmatrix}$$

If $ad - bc = 0$, then the matrix equation has no solution or infinitely many solutions.

Exercises

In Exercises 1–4, let

$$\bar{u} = \begin{pmatrix} 1 \\ 0 \end{pmatrix} \text{ and } \bar{v} = \begin{pmatrix} 0 \\ 1 \end{pmatrix}$$

and compute the linear combination.

1. $2\bar{u} - 3\bar{v}$

2. $-4\bar{u} + 6\bar{v}$

3. $-3.4\bar{u} - 7.7\bar{v}$

4. $5.6\bar{u} - 9.1\bar{v}$

In Exercises 5–8, let

$$\bar{u} = \begin{pmatrix} 2 \\ -3 \end{pmatrix} \text{ and } \bar{v} = \begin{pmatrix} 4 \\ 1 \end{pmatrix}$$

and compute the linear combination.

5. $7\bar{u} + 5\bar{v}$

6. $-\bar{u} - 8\bar{v}$

7. $2.1\bar{u} - 1.3\bar{v}$

8. $-0.5\bar{u} + 3.2\bar{v}$

Your Turn. In Exercises 9–12, choose your own vectors $\bar{u}$ and $\bar{v}$ with the given number of components, your own scalars x and y, and compute the linear combination $x\bar{u} + y\bar{v}$.

9. One component

10. Two components

11. Three components

12. Four components

In Exercises 13–16, compute the linear combination or explain why the computation is not possible.

13. $\begin{pmatrix} 1 \\ 2 \end{pmatrix} + 2 \begin{pmatrix} 1 \\ 1 \end{pmatrix} - 4 \begin{pmatrix} -1 \\ 1 \end{pmatrix}$

14. $5 \begin{pmatrix} 1 \\ 0 \\ 0 \end{pmatrix} + 3 \begin{pmatrix} 1 \\ 1 \\ 0 \end{pmatrix} + \begin{pmatrix} 1 \\ 1 \\ 1 \end{pmatrix}$

15. $3 \begin{pmatrix} 1 \\ 2 \end{pmatrix} + 4 \begin{pmatrix} 2 \\ 3 \\ 0 \end{pmatrix} - 5 \begin{pmatrix} 3 \\ 6 \end{pmatrix}$

16. $4 \begin{pmatrix} 1 \\ 0 \\ 1 \\ 0 \end{pmatrix} - 2 \begin{pmatrix} 1 \\ 1 \\ 2 \\ 2 \end{pmatrix}$

In Exercises 17–22, identify the size and the vectors represented by the matrix.

17. $\begin{pmatrix} -2 & 7 \\ 3 & 2 \end{pmatrix}$

20. $\begin{pmatrix} 2 & 1 & 4 \\ -1 & 7 & 5 \end{pmatrix}$

18. $\begin{pmatrix} -1 & 7 & 0 \\ 0 & 1 & -1 \\ -2 & 4 & 2 \end{pmatrix}$

21. $\begin{pmatrix} 7 & 1 & 2 & 2 \\ 0 & 3 & 4 & 3 \\ 1 & 2 & 1 & 4 \end{pmatrix}$

19. $\begin{pmatrix} 7 & 1 & 2 \\ 0 & 3 & 4 \\ -1 & -2 & 1 \\ 2 & 3 & 4 \end{pmatrix}$

22. $\begin{pmatrix} -0.7 & 1.2 \\ 4.1 & 3.1 \\ -2.0 & 0.9 \\ 0.1 & 0 \\ -0.7 & 0.8 \end{pmatrix}$

In Exercises 23–26, compute the product.

23. $\begin{pmatrix} 1 & 7 \\ -2 & 4 \end{pmatrix} \begin{pmatrix} 1 \\ 2 \end{pmatrix}$

25. $\begin{pmatrix} 1 & 0 \\ 0 & 1 \end{pmatrix} \begin{pmatrix} 5 \\ 5 \end{pmatrix}$

24. $\begin{pmatrix} -2 & 5 \\ -4 & 3 \end{pmatrix} \begin{pmatrix} 0 \\ 3 \end{pmatrix}$

26. $\begin{pmatrix} -7 & 1 \\ -2 & 3 \end{pmatrix} \begin{pmatrix} 8 \\ 2 \end{pmatrix}$

In Exercises 27–30, compute the product.

27. $\begin{pmatrix} 1 & 0 & 0 \\ 0 & 1 & 0 \\ 0 & 0 & 1 \end{pmatrix} \begin{pmatrix} -7 \\ -3 \\ 5 \end{pmatrix}$

28. $\begin{pmatrix} -2 & -1 & 3 \\ 0 & 1 & 4 \\ -2 & 3 & 7 \end{pmatrix} \begin{pmatrix} 1 \\ 2 \\ -1 \end{pmatrix}$

29. $\begin{pmatrix} 1 & 7 & -2 & 1 \\ 0 & 2 & -2 & 1 \\ 3 & 4 & -2 & -2 \\ 0 & 1 & 3 & -3 \end{pmatrix} \begin{pmatrix} 1 \\ 5 \\ -2 \\ -1 \end{pmatrix}$

30. $\begin{pmatrix} 0.1 & 0.2 & -0.5 & 0.7 \\ 1.2 & -2.1 & -0.4 & 3.4 \\ 0.3 & 0.6 & -0.7 & 1.2 \\ 7.1 & -0.6 & 4.2 & 3.2 \end{pmatrix} \begin{pmatrix} 1.0 \\ -0.7 \\ 0.8 \\ 0.7 \end{pmatrix}$

In Exercises 31–34, write the system of linear equations as both a vector equation and a matrix equation.

31. $5 = 3x + 2y$
    $7 = x - 4y$

32. $2 = -x + 5y$
    $3 = x + 2y$
    $4 = x + y$

33. $0 = 2x + 3y$
    $-1 = x + y$

34. $1 = x - 4y$
    $8 = 4x - 7y$
    $4 = y$

In Exercises 35–40, write the vector equation as both a matrix equation and a system of linear equations.

35. $\begin{pmatrix} 4 \\ 5 \end{pmatrix} = x \begin{pmatrix} 1 \\ 0 \end{pmatrix} + y \begin{pmatrix} -3 \\ 4 \end{pmatrix}$

36. $\begin{pmatrix} 0 \\ -1 \end{pmatrix} = x \begin{pmatrix} 2 \\ -3 \end{pmatrix} + y \begin{pmatrix} 3 \\ 1 \end{pmatrix}$

37. $\begin{pmatrix} -6 \\ 11 \end{pmatrix} = x \begin{pmatrix} 7 \\ -9 \end{pmatrix} + y \begin{pmatrix} -2 \\ 5 \end{pmatrix}$

38. $\begin{pmatrix} 1 \\ 1 \\ 1 \end{pmatrix} = x \begin{pmatrix} 2 \\ -3 \\ 4 \end{pmatrix} + y \begin{pmatrix} 0 \\ 9 \\ -5 \end{pmatrix} + z \begin{pmatrix} -1 \\ 2 \\ -7 \end{pmatrix}$

39. $\begin{pmatrix} 4 \\ 1 \\ -3 \end{pmatrix} = x \begin{pmatrix} 2 \\ 7 \\ -1 \end{pmatrix} + y \begin{pmatrix} 3 \\ 0 \\ 0 \end{pmatrix} + z \begin{pmatrix} -2 \\ 0 \\ 9 \end{pmatrix}$

40. $\begin{pmatrix} 3.1 \\ 1.4 \\ 2.6 \end{pmatrix} = x \begin{pmatrix} 6.1 \\ 3.2 \\ 5.8 \end{pmatrix} + y \begin{pmatrix} 1.2 \\ 1.3 \\ 4.9 \end{pmatrix} + z \begin{pmatrix} 0.9 \\ 1.7 \\ 2.1 \end{pmatrix}$

In Exercises 41–44, write the matrix equation as both a vector equation and a system of linear equations.

41. $\begin{pmatrix} 1 \\ 1 \end{pmatrix} = \begin{pmatrix} 3 & 8 \\ 2 & 1 \end{pmatrix} \begin{pmatrix} x \\ y \end{pmatrix}$

42. $\begin{pmatrix} 2 \\ 4 \end{pmatrix} = \begin{pmatrix} -1 & 5 \\ 0 & 2 \end{pmatrix} \begin{pmatrix} x \\ y \end{pmatrix}$

43. $\begin{pmatrix} 0 \\ 1 \end{pmatrix} = \begin{pmatrix} 7 & 3 \\ -2 & 5 \end{pmatrix} \begin{pmatrix} x \\ y \end{pmatrix}$

44. $\begin{pmatrix} 2 \\ 1 \\ -2 \end{pmatrix} = \begin{pmatrix} 1 & 2 & 3 \\ 1 & 0 & 2 \\ -4 & 1 & 3 \end{pmatrix} \begin{pmatrix} x \\ y \\ z \end{pmatrix}$

Your Turn. In Exercises 45 and 46, choose your own vectors $\overline{u}$, $\overline{v}$, and $\overline{w}$.

45. Write the vector equation $\overline{w} = x\overline{u} + y\overline{v}$ as a system of linear equations.

46. Write the vector equation $\overline{w} = x\overline{u} + y\overline{v}$ as a matrix equation.

In Exercises 47–52, state the system of linear equations and the vector equation for the data set based on a linear model $y = mx + b$.

47. Annual e-commerce sales (S) in the United States in billions of dollars for each year (Y)

Y	2009	2010	2011	2012
S	121	143	168	192

48. High value (H) of Facebook stock each month (M) in 2014

M	1	2	3	4	5
H	63.4	71.4	72.6	63.9	64.3

49. World population growth rate (R) in year (Y)

Y	1995	2000	2005	2010
R	1.413	1.26	1.203	1.127

50. Average number of goals (G) per game in World Cup tournaments in years (Y) since 1994

Y	0	4	8	12	16
G	2.7	2.7	2.5	2.3	2.3

51. Percent (P) of each year's (Y) high school graduates enrolling in a two-year or four-year college

Y	2006	2009	2010	2012
P	66.0	70.1	68.1	66.2

52. Average maximum temperature (T) in degrees Fahrenheit each month (M) in New York City

M	2	4	7	10	12
T	39.4	61.1	88.3	67.8	43.4

In Exercises 53–56, let

$$\overline{u} = \begin{pmatrix} 1 \\ 0 \end{pmatrix} \text{ and } \overline{v} = \begin{pmatrix} 0 \\ 1 \end{pmatrix}$$

and find scalars x and y such that $\overline{w} = x\overline{u} + y\overline{v}$ algebraically.

53. $\overline{w} = \begin{pmatrix} 5 \\ 3 \end{pmatrix}$

54. $\overline{w} = \begin{pmatrix} -2.7 \\ 0 \end{pmatrix}$

55. $\overline{w} = \begin{pmatrix} 4.3 \\ -6 \end{pmatrix}$

56. $\overline{w} = \begin{pmatrix} -1 \\ 2.8 \end{pmatrix}$

In Exercises 57–60, let

$$\overline{u} = \begin{pmatrix} 1 \\ 2 \end{pmatrix} \text{ and } \overline{v} = \begin{pmatrix} -2 \\ 3 \end{pmatrix}$$

and find the solution of the vector equation $\overline{w} = x\overline{u} + y\overline{v}$ algebraically.

57. $\overline{w} = \begin{pmatrix} -1 \\ 5 \end{pmatrix}$

58. $\overline{w} = \begin{pmatrix} 0 \\ 7 \end{pmatrix}$

59. $\overline{w} = \begin{pmatrix} 5 \\ -3 \end{pmatrix}$

60. $\overline{w} = \begin{pmatrix} -2 \\ -1 \end{pmatrix}$

In Exercises 61–64, explain why the vector equation does not have a solution.

61. $\begin{pmatrix} 2 \\ 5 \end{pmatrix} = x\begin{pmatrix} -1 \\ -1 \end{pmatrix} + y\begin{pmatrix} 3 \\ 3 \end{pmatrix}$

62. $\begin{pmatrix} -1 \\ 4 \end{pmatrix} = x\begin{pmatrix} 3 \\ 1 \end{pmatrix} + y\begin{pmatrix} 6 \\ 2 \end{pmatrix}$

63. $\begin{pmatrix} 2 \\ 3 \\ 4 \end{pmatrix} = x\begin{pmatrix} 1 \\ 1 \end{pmatrix} + y\begin{pmatrix} -3 \\ 4 \end{pmatrix}$

64. $\begin{pmatrix} -1 \\ 2 \end{pmatrix} = x\begin{pmatrix} 0 \\ 0 \end{pmatrix} + y\begin{pmatrix} -3 \\ 8 \end{pmatrix}$

In Exercises 65–70, solve the matrix equation using the formula for solving 2×2 matrix equations.

65. $\begin{pmatrix} 1 \\ 2 \end{pmatrix} = \begin{pmatrix} -1 & 2 \\ -1 & 1 \end{pmatrix}\begin{pmatrix} x \\ y \end{pmatrix}$

66. $\begin{pmatrix} 0 \\ -3 \end{pmatrix} = \begin{pmatrix} 1 & 4 \\ -1 & -3 \end{pmatrix}\begin{pmatrix} x \\ y \end{pmatrix}$

67. $\begin{pmatrix} -5 \\ 2 \end{pmatrix} = \begin{pmatrix} 1 & 0 \\ 0 & 1 \end{pmatrix}\begin{pmatrix} x \\ y \end{pmatrix}$

68. $\begin{pmatrix} 4 \\ 3 \end{pmatrix} = \begin{pmatrix} 1 & -1 \\ 1 & 1 \end{pmatrix}\begin{pmatrix} x \\ y \end{pmatrix}$

69. $\begin{pmatrix} 2 \\ 2 \end{pmatrix} = \begin{pmatrix} 4 & 2 \\ 1 & 1 \end{pmatrix}\begin{pmatrix} x \\ y \end{pmatrix}$

70. $\begin{pmatrix} 2 \\ 3 \end{pmatrix} = \begin{pmatrix} 1 & 2 \\ 2 & 4 \end{pmatrix}\begin{pmatrix} x \\ y \end{pmatrix}$

In Exercises 71–74, write the system of equations as a matrix equation and find its solution using the formula for solving 2×2 matrix equations.

71. $x - 2y = -4$
    $x + y = -1$

72. $x + y = 4$
    $x + y = -1$

73. $2x + 3y = 0$
    $x + y = -1$

74. $x - 2y = 4$
    $-3x + y = 0$

In Exercises 75 and 76, let

$$\bar{u} = \begin{pmatrix} 1 \\ 0 \\ 0 \end{pmatrix}, \bar{v} = \begin{pmatrix} 1 \\ 1 \\ 1 \end{pmatrix}, \text{ and } \bar{w} = \begin{pmatrix} 2 \\ 4 \\ 4 \end{pmatrix}$$

75. Write the vector equation $\bar{w} = x\bar{u} + y\bar{v}$ as a system of three linear equations.

76. Find scalars x and y that satisfy the system of linear equations from Exercise 75.

RStudio. In Exercises 77–86, use RStudio to solve the matrix equations given here or in the earlier exercises.

77. Exercise 65   80. Exercise 68

78. Exercise 66   81. Exercise 69

79. Exercise 67   82. Exercise 70

83. $\begin{pmatrix} 1 \\ 1 \\ 2 \end{pmatrix} = \begin{pmatrix} 3 & 8 & 2 \\ 2 & 1 & 4 \\ 3 & 2 & 9 \end{pmatrix} \begin{pmatrix} x \\ y \\ z \end{pmatrix}$

84. $\begin{pmatrix} 2 \\ 4 \\ 5 \end{pmatrix} = \begin{pmatrix} -1 & 5 & -2 \\ 0 & 2 & 1 \\ 2 & 5 & -6 \end{pmatrix} \begin{pmatrix} x \\ y \\ z \end{pmatrix}$

85. $\begin{pmatrix} 0 \\ -1 \\ 3 \end{pmatrix} = \begin{pmatrix} 7 & 3 & -5 \\ -2 & 5 & -4 \\ 1 & 2 & 3 \end{pmatrix} \begin{pmatrix} x \\ y \\ z \end{pmatrix}$

86. $\begin{pmatrix} 2 \\ 1 \\ -2 \end{pmatrix} = \begin{pmatrix} 1 & 2 & 3 \\ 1 & 0 & 2 \\ -4 & 1 & 3 \end{pmatrix} \begin{pmatrix} x \\ y \\ z \end{pmatrix}$

RStudio. In Exercises 87–92, use RStudio to solve the vector equations from Exercises 35–40.

87. Exercise 35   90. Exercise 38

88. Exercise 36   91. Exercise 39

89. Exercise 37   92. Exercise 40

In Your Own Words. In Exercises 93–101, explain the following.

93. System of linear equations

94. Vector equation

95. Linear combination

96. Intercept vector

97. Solution of a vector equation

98. Matrix

99. Square matrix

100. Matrix equation

101. Solving 2×2 matrix equations

3.3 Existence of Linear Combinations

Section 3.2 introduced the idea of a linear combination and we learned how to find the scalars needed to obtain a target linear combination of vectors, in response to the following three equivalent questions:

Given vectors $\bar{u}$, $\bar{v}$, and $\bar{w}$,

- Is $\bar{w}$ a linear combination of $\bar{u}$ and $\bar{v}$?

- Do there exist scalars x and y such that $\bar{w} = x\bar{u} + y\bar{v}$?

- Is there a solution of the vector equation $\bar{w} = x\bar{u} + y\bar{v}$?

As it turns out, we cannot always find the scalars that give a desired linear combination. In fact, the scalars needed for such a linear combination usually do not exist, regardless of the number of vectors. This section discusses ways to demonstrate that a target vector cannot be expressed by a linear combination of a given set of vectors.

The simplest case to consider is the linear combination of a single vector. Given only one vector $\bar{u}$, every linear combination of this vector must have exactly the form $x\bar{u}$. In other words, every linear combination of $\bar{u}$ is a scalar multiple of $\bar{u}$. Recall that scalar multiplication of a vector can change its length and either preserve or reverse its direction. These properties mean that all scalar multiples fall on the same line as the original vector $\bar{u}$. Determining if $\bar{v}$ is a linear combination of $\bar{u}$ is accomplished graphically by checking whether they lie on the same line and is accomplished algebraically by checking whether $\bar{v}$ is a scalar multiple of $\bar{u}$.

◆ **EXAMPLE 1** Let $\bar{u} = \begin{pmatrix} 3 \\ 1 \end{pmatrix}$ and $\bar{v} = \begin{pmatrix} 6 \\ 3 \end{pmatrix}$, and determine if $\bar{v}$ is a linear combination of $\bar{u}$:

(a) Graphically   (b) Algebraically

Solution.

(a) The graph in Figure 1 shows that $\bar{u}$ and $\bar{v}$ do not lie on the same line. Thus, $\bar{v}$ is not a scalar multiple of $\bar{u}$, which means that $\bar{v}$ is not a linear combination of $\bar{u}$.

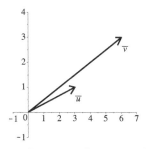

Figure 1: $\bar{v}$ is not a linear combination of $\bar{u}$ for Example 1

(b) The algebraic approach to determining whether $\bar{v}$ is a linear combination of $\bar{u}$ is based on trying to find a scalar x such that $\bar{v} = x\bar{u}$:

$$\bar{v} = x\bar{u} \qquad \text{Linear combination equation}$$

$$\begin{pmatrix} 6 \\ 3 \end{pmatrix} = x \begin{pmatrix} 3 \\ 1 \end{pmatrix} \qquad \text{Substitute the vectors}$$

$$\begin{pmatrix} 6 \\ 3 \end{pmatrix} = \begin{pmatrix} 3x \\ x \end{pmatrix} \qquad \text{Scalar multiply}$$

Two vectors are equal when their corresponding components are equal. If the first components were equal, then $3x = 6$, which gives $x = 2$. If the second components were equal, then $x = 3$. As a result, no single value of x can provide $\bar{v} = x\bar{u}$, which means that $\bar{v}$ is not a linear combination of $\bar{u}$.   ■

DETERMINING LINEAR COMBINATIONS OF A SINGLE VECTOR.
Graphically, a vector $\bar{v}$ is a linear combination of vector $\bar{u}$ if they lie on the same line. Algebraically, $\bar{v}$ must equal $x\bar{u}$ for some scalar x.

➤ **QUESTION 1** Let $\bar{u} = \begin{pmatrix} -2 \\ 5 \end{pmatrix}$ and $\bar{v} = \begin{pmatrix} 1 \\ -2 \end{pmatrix}$, and determine if $\bar{v}$ is a linear combination of $\bar{u}$:

  (a) Graphically   (b) Algebraically

➤ **QUESTION 2** Let $\bar{u} = \begin{pmatrix} 3 \\ -9 \end{pmatrix}$ and $\bar{v} = \begin{pmatrix} -1 \\ 3 \end{pmatrix}$, and determine if $\bar{v}$ is a linear combination of $\bar{u}$:

  (a) Graphically   (b) Algebraically

Linear Combinations of Two Vectors

Moving on to considering linear combinations of two vectors $\bar{u}$ and $\bar{v}$, the graphical process for determining if a vector $\bar{w}$ is a linear combination of $\bar{u}$ and $\bar{v}$ is cumbersome and not very precise. Instead, algebra and technology are used to recognize linear combinations.

◆ **EXAMPLE 2** Let $\bar{u} = \begin{pmatrix} 3 \\ 1 \end{pmatrix}$, $\bar{v} = \begin{pmatrix} -1 \\ 2 \end{pmatrix}$, and $\bar{w} = \begin{pmatrix} 3 \\ 8 \end{pmatrix}$, and determine if $\bar{w}$ is a linear combination of $\bar{u}$ and $\bar{v}$:

  (a) Algebraically   (b) Using RStudio

Solution.

  (a) The goal is to decide whether or not there exist scalars x and y such that $\bar{w} = x\bar{u} + y\bar{v}$. Using the tools developed in Section 3.2, express this vector equation as its associated matrix equation:

$$\bar{w} = x\bar{u} + y\bar{v} \qquad \text{Linear combination equation}$$

$$\begin{pmatrix} 3 \\ 8 \end{pmatrix} = x \begin{pmatrix} 3 \\ 1 \end{pmatrix} + y \begin{pmatrix} -1 \\ 2 \end{pmatrix} \qquad \text{Substitute the vectors}$$

$$\begin{pmatrix} 3 \\ 8 \end{pmatrix} = \begin{pmatrix} 3 & -1 \\ 1 & 2 \end{pmatrix} \begin{pmatrix} x \\ y \end{pmatrix} \qquad \text{Express as a matrix equation}$$

Applying the formula for solving 2×2 matrix equations gives

$$\begin{pmatrix} x \\ y \end{pmatrix} = \frac{1}{(3)(2) - (1)(-1)} \begin{pmatrix} 2 & 1 \\ -1 & 3 \end{pmatrix} \begin{pmatrix} 3 \\ 8 \end{pmatrix} = \frac{1}{7} \begin{pmatrix} 6 + 8 \\ -3 + 24 \end{pmatrix} = \begin{pmatrix} 2 \\ 3 \end{pmatrix}$$

Therefore, $\bar{w} = 2\bar{u} + 3\bar{v}$, which means that $\bar{w}$ is a linear combination of $\bar{u}$ and $\bar{v}$.

  (b) Use the `solve` command in RStudio to look for the solution of the matrix equation from part (a). See the end of Section 3.2 for a refresher on these commands.

```
u = c(3, 1)                          w = c(3,8)
v = c(-1, 2)                         solve(U,w)
U = matrix(c(u,v),nrow=2,
           ncol=2)                   [1] 2 3
```

From this RStudio output

$$\begin{pmatrix} x \\ y \end{pmatrix} = \begin{pmatrix} 2 \\ 3 \end{pmatrix},$$

which provides the same result as in part (a). Namely, $x = 2$ and $y = 3$, which gives $\overline{w} = 2\overline{u} + 3\overline{v}$ and means that $\overline{w}$ is a linear combination of $\overline{u}$ and $\overline{v}$. ■

While using algebra and technology to determine linear combinations, keep in mind the graphical interpretation of these results. In Example 2, finding $x = 2$ and $y = 3$ corresponds to knowing how to scale the vectors $\overline{u}$ and $\overline{v}$ so that they sum to $\overline{w}$. The graph in Figure 2 depicts this scaling of $\overline{u}$ and $\overline{v}$ as well as the resulting sum $\overline{w} = 2\overline{u}+3\overline{v}$.

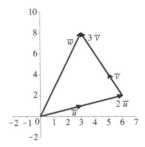

Figure 2: Graph of $\overline{w} = 2\overline{u} + 3\overline{v}$ for Example 2

➤ **QUESTION 3** Let $\overline{u} = \begin{pmatrix} 4 \\ -1 \end{pmatrix}$, $\overline{v} = \begin{pmatrix} -2 \\ 3 \end{pmatrix}$, and $\overline{w} = \begin{pmatrix} -8 \\ 2 \end{pmatrix}$, and determine if $\overline{w}$ is a linear combination of $\overline{u}$ and $\overline{v}$:

(a) Algebraically   (b) Using RStudio

Linear Combinations of Three or More Vectors

In some settings, linear combinations with three or more vectors need to be found, at which point RStudio or some other software package becomes the appropriate tool (rather than algebraic computations by hand).

◆ **EXAMPLE 3** Let $\overline{u_1} = \begin{pmatrix} 0 \\ 1 \\ 2 \end{pmatrix}$, $\overline{u_2} = \begin{pmatrix} 1 \\ 1 \\ 0 \end{pmatrix}$, $\overline{u_3} = \begin{pmatrix} 2 \\ 2 \\ 0 \end{pmatrix}$, and $\overline{v} = \begin{pmatrix} 2 \\ 3 \\ 1 \end{pmatrix}$.

Use RStudio to determine if $\overline{v}$ is a linear combination of $\overline{u_1}$, $\overline{u_2}$, and $\overline{u_3}$.

Solution. First, express the vector equation as its associated matrix equation:

$$\bar{v} = x_1\bar{u}_1 + x_2\bar{u}_2 + x_3\bar{u}_3 \qquad \text{Linear combination equation}$$

$$\begin{pmatrix} 2 \\ 3 \\ 1 \end{pmatrix} = x_1 \begin{pmatrix} 0 \\ 1 \\ 2 \end{pmatrix} + x_2 \begin{pmatrix} 1 \\ 1 \\ 0 \end{pmatrix} + x_3 \begin{pmatrix} 2 \\ 2 \\ 0 \end{pmatrix} \qquad \text{Substitute the vectors}$$

$$\begin{pmatrix} 2 \\ 3 \\ 1 \end{pmatrix} = \begin{pmatrix} 0 & 1 & 2 \\ 1 & 1 & 2 \\ 2 & 0 & 0 \end{pmatrix} \begin{pmatrix} x_1 \\ x_2 \\ x_3 \end{pmatrix} \qquad \text{Express as a matrix equation}$$

Now use the `solve` command in RStudio to seek the solution of this matrix equation.

```
u1 = c(0,1,2)
u2 = c(1,1,0)
u3 = c(2,2,0)
U = matrix(c(u1,u2,u3),
           nrow=3,ncol=3)
```

```
v = c(2, 3, 1)
solve(U,v)
```

```
Error in solve.default(U, v):
Lapack routine dgesv:  system is
exactly singular:  U[3,3] = 0
```

The RStudio output of `solve(U,v)` indicates an **Error** and states that the `system is exactly singular`. Unlike many other RStudio errors, this particular message does not result from a typing or syntax mistake. Rather, this `system is exactly singular` error indicates that the matrix equation has no solution and means that $\bar{v}$ is not a linear combination of $\bar{u}_1$, $\bar{u}_2$, and $\bar{u}_3$. ■

The phrase `system is exactly singular` encountered in Example 3 is RStudio's way of indicating that a matrix equation has either no solution or infinitely many solutions. In this book, the questions and exercises are set up to ensure that whenever RStudio reports such an error, the matrix equation has no solutions (rather than infinitely many).

RStudio uses various synonymous expressions to indicate that a matrix equation has no solution. RStudio usually reports `Error in solve.default(U,v):`, but sometimes just `Error`. Likewise, RStudio usually reports `system is exactly singular`, but sometimes uses the phrase `system is computationally singular`. All of these various expressions indicate that a matrix equation has no solution for the questions and exercises in this book.

In the case of 2×2 matrix equations, such an error is equivalent to the case when $ad - bc = 0$ in the formula for solving 2×2 matrix equations. The terminology "exactly singular" comes from the field of linear algebra, and further study lies beyond the scope of this book. However, remember these error messages so that appropriate conclusions can be drawn about the existence and non-existence of solutions of matrix equations.

SYSTEM IS SINGULAR ERROR.

RStudio error messages indicating that the corresponding matrix equation has no solution include

- `Error in solve.default(U,v): Lapack routine dgesv: system is exactly singular:`

- `Error in solve.default(U,v): system is computationally singular:`

▶ **QUESTION 4** Determine if $\overline{v}$ is a linear combination of $\overline{u}_1$, $\overline{u}_2$, $\overline{u}_3$, and $\overline{u}_4$ using RStudio:

$$\overline{u}_1 = \begin{pmatrix} -2 \\ 5 \\ 2 \\ 1 \end{pmatrix} \quad \overline{u}_2 = \begin{pmatrix} 0 \\ 7 \\ 3 \\ -2 \end{pmatrix} \quad \overline{u}_3 = \begin{pmatrix} 1 \\ 0 \\ -8 \\ 3 \end{pmatrix} \quad \overline{u}_4 = \begin{pmatrix} -1 \\ 12 \\ -3 \\ 2 \end{pmatrix} \quad \overline{v} = \begin{pmatrix} 1 \\ 2 \\ 3 \\ 4 \end{pmatrix}$$

Linear Combinations and Data Sets

In Examples 2 and 3, the matrix equations contained a square matrix with the same number of rows and columns. However, when working in real-life, data-focused settings, the resulting linear combinations almost universally involve vectors with more components than unknown scalars. Example 1 and Question 1 in Section 3.2 illustrate this fact, and the study of these settings continues in Example 4 and Question 5 below.

When working with vectors that have more components than unknown scalars, the resulting matrix equation has a nonsquare matrix, which the methods studied thus far are not equipped to handle. In such cases, the standard approach is to consider a small enough initial segment of the data to produce a square matrix. A solution to the corresponding square matrix equation is sought and then this solution is checked to see if it works for all the data. This process for determining the existence of such linear combinations is summarized as follows:

EXISTENCE OF LINEAR COMBINATIONS.

If the vectors $\overline{u}_1$, $\overline{u}_2$, ..., $\overline{u}_n$ in the vector equation $\overline{v} = x_1\overline{u}_1 + x_2\overline{u}_2 + \cdots + x_n\overline{u}_n$ have more than n components, then determine if a solution exists via the following three steps:

(1) Form a new vector equation using truncated versions of $\overline{u}_1$, $\overline{u}_2$, ..., $\overline{u}_n$ that contain just the first n components of the original vectors.

(2) Find the solution x_1, x_2, ..., x_n of the vector equation from step (1), if possible.

EXISTENCE OF LINEAR COMBINATIONS. (CONTINUED)

(3) Check whether the solution x_1, x_2, ..., x_n from step (2) satisfies the original vector equation by computing $x_1\overline{u}_1 + x_2\overline{u}_2 + \cdots + x_n\overline{u}_n$ and comparing the result with $\overline{v}$.

This process applies not only to data sets for which linear models are reasonable, but also to exponential and quadratic data sets as discussed later in this section. In addition, the first example uses algebra to check solutions, while RStudio is used in the subsequent examples.

◆ **EXAMPLE 4** The following table presents how many million people used Twitter during each quarter of the given year since 2000, where 11 represents the first quarter of 2011 (January – March 2011), 11.25 represents the second quarter of 2011 (April – June 2011), and so on:

Year (Y)	11	11.25	11.5	11.75	12.75
Twitter users (U)	68	85	101	117	185

This data set is approximately linear (see Example 1 in Section 2.1) and can be represented by a model of the form $U = mY + b$, as in Example 1 in Section 3.2. Substituting each data point into this model yields the associated vector equation $\overline{U} = m\overline{Y} + b\overline{1}_5$ stated below. Show that this vector equation has no solution algebraically:

$$\begin{pmatrix} 68 \\ 85 \\ 101 \\ 117 \\ 185 \end{pmatrix} = m \begin{pmatrix} 11 \\ 11.25 \\ 11.5 \\ 11.75 \\ 12.75 \end{pmatrix} + b \begin{pmatrix} 1 \\ 1 \\ 1 \\ 1 \\ 1 \end{pmatrix}$$

Solution.
Step (1): The vectors in this equation have five components, but there are only two unknown scalars m and b. Thus, a new vector equation is formed using just the first two components of the original vectors, because there are two unknown scalars. This new vector equation and its corresponding matrix equation are

$$\begin{pmatrix} 68 \\ 85 \end{pmatrix} = m \begin{pmatrix} 11 \\ 11.25 \end{pmatrix} + b \begin{pmatrix} 1 \\ 1 \end{pmatrix} \qquad \qquad \begin{pmatrix} 68 \\ 85 \end{pmatrix} = \begin{pmatrix} 11 & 1 \\ 11.25 & 1 \end{pmatrix} \begin{pmatrix} m \\ b \end{pmatrix}$$

Step (2): Apply the formula for solving 2×2 matrix equations:

$$\begin{pmatrix} m \\ b \end{pmatrix} = \frac{1}{(11)(1) - (11.25)(1)} \begin{pmatrix} 1 & -1 \\ -11.25 & 11 \end{pmatrix} \begin{pmatrix} 68 \\ 85 \end{pmatrix} \qquad \text{Formula for solving matrix equations}$$

$$= \frac{1}{(-0.25)} \left[68 \begin{pmatrix} 1 \\ -11.25 \end{pmatrix} + 85 \begin{pmatrix} -1 \\ 11 \end{pmatrix} \right] \qquad \text{Matrix–vector multiplication}$$

$$= (-4) \begin{pmatrix} -17 \\ 170 \end{pmatrix} \qquad \text{Compute the linear combination}$$

$$= \begin{pmatrix} 68 \\ -680 \end{pmatrix} \qquad \text{Scalar multiply by } -4$$

Thus, $m = 68$ and $b = -680$ is the conjectured solution of the original vector equation.

Step (3): Determine if the m and b from step (2) satisfy every component of the original vector equation by substituting $m = 68$ and $b = -680$ into the right-hand side of the complete vector equation $\overline{U} = m\overline{Y} + b\overline{1}_5$ and comparing the result with the complete vector $\overline{U}$:

$$m\overline{Y} + b\overline{1}_5 = 68 \begin{pmatrix} 11 \\ 11.25 \\ 11.5 \\ 11.75 \\ 12.75 \end{pmatrix} + (-680) \begin{pmatrix} 1 \\ 1 \\ 1 \\ 1 \\ 1 \end{pmatrix} = \begin{pmatrix} 68 \\ 85 \\ 102 \\ 119 \\ 187 \end{pmatrix} \quad \text{compared with} \quad \overline{U} = \begin{pmatrix} 68 \\ 85 \\ 101 \\ 117 \\ 185 \end{pmatrix}$$

Note that only the first two components of $68\overline{Y} + (-680)\overline{1}_5$ and $\overline{U}$ are the same, which means that these two vectors are not equal. Therefore, the vector equation has no solution. ∎

In Example 4, only the first two components of $68\overline{Y} + (-680)\overline{1}_5$ were equal to the corresponding components of $\overline{U}$, and the remaining three components were all different. While this outcome with *all* the other components being unequal is not uncommon, note that to demonstrate that a solution does not exist, it is only necessary to show that just one of the subsequent pairs of components is unequal.

▶ **QUESTION 5** The following table presents the average debt load in thousands of dollars at the end of each year's spring term for bachelor's degree recipients who attended public four-year colleges and universities and borrowed money to finance their education:

Academic year (Y)	2001	2003	2005	2006
Average debt (D)	$20,400	$20,900	$21,500	$21,800

This data set is approximately linear (see Question 1 in Section 2.1) and can be represented by a model of the form Debt $= m \cdot$ Year $+ b$, as in Question 1 in Section 3.2. Substituting each data point into this model yields the associated vector equation $\overline{\text{Debt}} = m \cdot \overline{\text{Year}} + b\overline{1}_4$ as follows:

$$\begin{pmatrix} 20{,}400 \\ 20{,}900 \\ 21{,}500 \\ 21{,}800 \end{pmatrix} = m \begin{pmatrix} 2001 \\ 2003 \\ 2005 \\ 2006 \end{pmatrix} + b \begin{pmatrix} 1 \\ 1 \\ 1 \\ 1 \end{pmatrix}$$

Show that this vector equation has no solution algebraically.

The data sets given in Example 4 and Question 5 are well represented by linear models, which is a natural starting place in light of considering systems of linear equations. However, as introduced in Chapter 2, data sets vary widely, and many different modeling functions are used to describe our diverse reality. As it turns out, this method for working with systems of linear equations extends to other systems of equations, because other modeling functions correspond to lines once the data set has been transformed appropriately.

Working in this direction, consider the case of a data set for which an exponential model $y = Ce^{k(x-h)}$ is reasonable. Recall from Section 2.2 that the natural logarithms of the outputs are linear with respect to the original inputs, which is graphically apparent in the corresponding semi-log plot of points $(x, \ln(y))$ being linear. Similarly, transforming the data set with the natural logarithm allows an exponential data set to be expressed in terms of an associated system of linear equations and enables an analysis with the methods of this chapter. The next example illustrates this approach.

◆ **EXAMPLE 5** The following table presents U.S. coal prices in dollars per short ton in the given year:

Year	1989	1990	1992	1996	1998
Price	$31.36	$30.11	$27.46	$22.25	$20.65

An exponential model Price $= Ce^{k(\text{Year}-1989)}$ is reasonable for this data set (see Example 3 of Section 2.2). Taking the natural logarithm of the outputs, substitute each data point into the corresponding linear equation $\ln(\text{Price}) = k(\text{Year} - 1989) + b$ to obtain the associated vector equation $\overline{\ln(\text{Price})} = k\,\overline{(\text{Year} - 1989)} + b\bar{1}_5$ as follows:

$$\begin{pmatrix} 3.446 \\ 3.405 \\ 3.313 \\ 3.102 \\ 3.028 \end{pmatrix} = k \begin{pmatrix} 0 \\ 1 \\ 3 \\ 7 \\ 9 \end{pmatrix} + b \begin{pmatrix} 1 \\ 1 \\ 1 \\ 1 \\ 1 \end{pmatrix}$$

Show that this vector equation has no solution using RStudio.

Solution.
Step (1): The vectors in this equation have five components, but there are only two unknown scalars k and b. Thus, a new vector equation is formed using just the first two components of the original vectors because there are two unknown scalars. This new vector equation and its corresponding matrix equation are.

$$\begin{pmatrix} 3.446 \\ 3.405 \end{pmatrix} = k \begin{pmatrix} 0 \\ 1 \end{pmatrix} + b \begin{pmatrix} 1 \\ 1 \end{pmatrix} \qquad\qquad \begin{pmatrix} 3.446 \\ 3.405 \end{pmatrix} = \begin{pmatrix} 0 & 1 \\ 1 & 1 \end{pmatrix} \begin{pmatrix} k \\ b \end{pmatrix}$$

Step (2): Use RStudio to solve this matrix equation:

```
u = c(0,1)                              solve(U,w)
v = c(1,1)
U = matrix(c(u,v),nrow=2,ncol=2)       [1] -0.041  3.446
w = c(3.446,3.405)
```

Thus, $k = -0.041$ and $b = 3.446$ is the conjectured solution of the original vector equation.

Step (3): Determine if the k and b from step (2) satisfy every component of the original vector equation by substituting $k = -0.041$ and $b = 3.446$ into the right-hand side of the vector equation $k \cdot \overline{(\text{Year} - 1989)} + b\bar{1}_5$ and comparing the result with the complete vector $\overline{\ln(\text{Price})}$:

```
Year=c(0,1,3,7,9)                  -0.041*Year + 3.446*Int
Int=c(1,1,1,1,1)
                                   [1] 3.446 3.405 3.323 3.159 3.077
```

This RStudio output provides the following result and enables the appropriate comparison:

$$(-0.041)\overline{(\text{Year} - 1989)} + 3.446\bar{1}_5 = \begin{pmatrix} 3.446 \\ 3.405 \\ 3.323 \\ 3.159 \\ 3.077 \end{pmatrix} \quad \text{compared with} \quad \overline{\ln(\text{Price})} = \begin{pmatrix} 3.446 \\ 3.405 \\ 3.313 \\ 3.102 \\ 3.028 \end{pmatrix}$$

Note that only the first two components of these vectors are the same, which means these two vectors are not equal. Therefore, the vector equation has no solution. ∎

▶ **QUESTION 6** The following table presents the average weight of male fetuses in grams during the given gestational week:

Week	25	26	30	31	33
Weight	777	888	1435	1633	2058

An exponential model $\text{Weight} = Ce^{k(\text{Week}-25)}$ is reasonable for this data set (see Example 1 of Section 2.2). Taking the natural logarithm of the outputs, substitute each data point into the corresponding linear equation $\ln(\text{Weight}) = k(\text{Week} - 25) + b$ to obtain the associated vector equation $\overline{\ln(\text{Weight})} = k\,\overline{(\text{Week} - 25)} + b\bar{1}_5$ as follows:

$$\begin{pmatrix} 6.655 \\ 6.789 \\ 7.269 \\ 7.398 \\ 7.629 \end{pmatrix} = k \begin{pmatrix} 0 \\ 1 \\ 5 \\ 6 \\ 8 \end{pmatrix} + b \begin{pmatrix} 1 \\ 1 \\ 1 \\ 1 \\ 1 \end{pmatrix}$$

Show that this vector equation has no solution using RStudio.

These techniques can also be applied to data sets that are well represented by models with three or more parameters in certain cases. In particular, such models must be **linear in their parameters**, which means that the parameters can only be multiplied by the variables or by functions of the variables in the modeling equations. For example, $y = mx + b$ and $y = ax^2 + be^x$ are linear in their parameters. In contrast, when the parameters appear in the exponent or inside a transcendental function, such as $y = e^{kx+b}$ and $y = \sin(ax)$, the model is not linear in its parameters. The following example investigates a data set that is well represented by a model with three parameters:

◆ **EXAMPLE 6** The following table presents U.S. Census data for Austin, Texas in thousands of people for each year:

Year	1950	1960	1970	1980	1990
Population	132	187	252	346	466

A quadratic model of the form Population $= A(\text{Year} - 1950)^2 + B(\text{Year} - 1950) + C$ is reasonable for this data set. Substituting each data point into this model yields the associated vector equation as follows:

$$\begin{pmatrix} 132 \\ 187 \\ 252 \\ 346 \\ 466 \end{pmatrix} = A \begin{pmatrix} 0 \\ 100 \\ 400 \\ 900 \\ 1600 \end{pmatrix} + B \begin{pmatrix} 0 \\ 10 \\ 20 \\ 30 \\ 40 \end{pmatrix} + C \begin{pmatrix} 1 \\ 1 \\ 1 \\ 1 \\ 1 \end{pmatrix}$$

Show that this vector equation has no solution using RStudio.

Solution.

Step (1): The vectors in this equation have five components, but there are only three unknown scalars A, B, and C. Thus, a new vector equation is formed using just the first three components of the original vectors, because there are three unknown scalars. This new vector equation and its corresponding matrix equation are

$$\begin{pmatrix} 132 \\ 187 \\ 252 \end{pmatrix} = A \begin{pmatrix} 0 \\ 100 \\ 400 \end{pmatrix} + B \begin{pmatrix} 0 \\ 10 \\ 20 \end{pmatrix} + C \begin{pmatrix} 1 \\ 1 \\ 1 \end{pmatrix} \qquad \begin{pmatrix} 132 \\ 187 \\ 252 \end{pmatrix} = \begin{pmatrix} 0 & 0 & 1 \\ 100 & 10 & 1 \\ 400 & 20 & 1 \end{pmatrix} \begin{pmatrix} A \\ B \\ C \end{pmatrix}$$

Step (2): Use RStudio to solve this matrix equation:

```
u1 = c(0,100,400)                          v = c(132,187,252)
u2 = c(0,10,20)                            solve(U,v)
u3 = c(1,1,1)
U = matrix(c(u1,u2,u3),nrow=3,      [1]    0.05    5.00 132.00
          ncol=3)
```

Thus, $A = 0.05$, $B = 5$, and $C = 132$ is the conjectured solution of the original vector equation. Now, determine if this A, B, and C satisfy every component of the original vector equation by substituting $A = 0.05$, $B = 5.00$, $C = 132.00$ into the right-hand side of the complete vector equation and comparing the result with the complete output vector on its left-hand side:

```
Year1=c(0,100,400,900,1600)             0.05*Year1+5*Year2+132*Int
Year2=c(0,10,20,30,40)
Int=c(1,1,1,1,1)                        [1] 132 187 252 327 412
```

This RStudio output provides the following result and enables the appropriate comparison:

$$0.05 \begin{pmatrix} 0 \\ 100 \\ 400 \\ 900 \\ 1600 \end{pmatrix} + 5.00 \begin{pmatrix} 0 \\ 10 \\ 20 \\ 30 \\ 40 \end{pmatrix} + 132 \begin{pmatrix} 1 \\ 1 \\ 1 \\ 1 \\ 1 \end{pmatrix} = \begin{pmatrix} 132 \\ 187 \\ 252 \\ 327 \\ 412 \end{pmatrix} \quad \text{compared with} \quad \begin{pmatrix} 132 \\ 187 \\ 252 \\ 346 \\ 466 \end{pmatrix}$$

Because only the first three components of these vectors are the same, the two vectors are not equal. Therefore, the vector equation has no solution. ∎

Summary

- Often a vector $\overline{v}$ *cannot* be written as a linear combination of a given collection of vectors $\overline{u}_1$, $\overline{u}_2$, ..., $\overline{u}_n$.

- Graphically, a vector $\overline{v}$ is a linear combination of vector $\overline{u}$ if they lie on the same line. Algebraically, $\overline{v}$ must equal $x\overline{u}$ for some scalar x.

- The RStudio messages `Error in solve.default(U,v): Lapack routine dgesv:system is exactly singular` and `Error in solve.default(U,v): system is computationally singular` indicate that the corresponding matrix equation has no solution.

- If the vectors $\overline{u}_1$, $\overline{u}_2$, ..., $\overline{u}_n$ in the vector equation $\overline{v} = x_1\overline{u}_1 + x_2\overline{u}_2 + \cdots + x_n\overline{u}_n$ have more than n components, then determine if a solution exists via the following three steps:

  (1) Form a new vector equation using truncated versions of $\overline{u}_1$, $\overline{u}_2$, ..., $\overline{u}_n$ that contain just the first n components of the original vectors.

  (2) Find the solution x_1, x_2, ..., x_n of the vector equation from step (1), if possible.

  (3) Check whether the solution x_1, x_2, ..., x_n from step (2) satisfies the original vector equation by computing $x_1\overline{u}_1 + x_2\overline{u}_2 + \cdots + x_n\overline{u}_n$ and comparing the result with $\overline{v}$.

- A model is *linear in its parameters* if the parameters are only multiplied by the variables or by functions of the variables in the modeling equation.

Exercises

In Exercises 1–6, let $\overline{u} = \begin{pmatrix} 1 \\ 2 \end{pmatrix}$. Graphically determine if $\overline{v}$ is a linear combination of $\overline{u}$.

In Exercises 7–12, let $\overline{u} = \begin{pmatrix} -3 \\ 4 \end{pmatrix}$. Graphically determine if $\overline{v}$ is a linear combination of $\overline{u}$.

1. $\overline{v} = \begin{pmatrix} 5 \\ 10 \end{pmatrix}$ 4. $\overline{v} = \begin{pmatrix} 0 \\ 0 \end{pmatrix}$ 7. $\overline{v} = \begin{pmatrix} 12 \\ 16 \end{pmatrix}$ 10. $\overline{v} = \begin{pmatrix} -24 \\ 32 \end{pmatrix}$

2. $\overline{v} = \begin{pmatrix} -1 \\ 2 \end{pmatrix}$ 5. $\overline{v} = \begin{pmatrix} 3 \\ 4 \end{pmatrix}$ 8. $\overline{v} = \begin{pmatrix} 0.3 \\ 0.4 \end{pmatrix}$ 11. $\overline{v} = \begin{pmatrix} 6 \\ -6 \end{pmatrix}$

3. $\overline{v} = \begin{pmatrix} 1 \\ 2 \end{pmatrix}$ 6. $\overline{v} = \begin{pmatrix} -0.25 \\ -0.5 \end{pmatrix}$ 9. $\overline{v} = \begin{pmatrix} -6 \\ 8 \end{pmatrix}$ 12. $\overline{v} = \begin{pmatrix} -1 \\ 4 \end{pmatrix}$

In Exercises 13–18, let $\overline{u} = \begin{pmatrix} 1 \\ 2 \end{pmatrix}$. Algebraically determine if $\overline{v}$ is a linear combination of $\overline{u}$. If so, find the scalar x such that $\overline{v} = x\overline{u}$.

13. $\overline{v} = \begin{pmatrix} 5 \\ 10 \end{pmatrix}$ 16. $\overline{v} = \begin{pmatrix} 0 \\ 0 \end{pmatrix}$

14. $\overline{v} = \begin{pmatrix} -1 \\ 2 \end{pmatrix}$ 17. $\overline{v} = \begin{pmatrix} 3 \\ 4 \end{pmatrix}$

15. $\overline{v} = \begin{pmatrix} 1 \\ 2 \end{pmatrix}$ 18. $\overline{v} = \begin{pmatrix} -0.25 \\ -0.5 \end{pmatrix}$

In Exercises 19–22, let $\overline{u} = \begin{pmatrix} -3 \\ 4 \end{pmatrix}$. Algebraically determine if $\overline{v}$ is a linear combination of $\overline{u}$. If so, find the scalar x such that $\overline{v} = x\overline{u}$.

19. $\overline{v} = \begin{pmatrix} 12 \\ 16 \end{pmatrix}$ 21. $\overline{v} = \begin{pmatrix} -6 \\ 8 \end{pmatrix}$

20. $\overline{v} = \begin{pmatrix} 0.3 \\ 0.4 \end{pmatrix}$ 22. $\overline{v} = \begin{pmatrix} -24 \\ 32 \end{pmatrix}$

In Exercises 23–28, let

$$\overline{u} = \begin{pmatrix} 5 \\ 3 \\ 1 \end{pmatrix}$$

Algebraically determine if $\overline{v}$ is a linear combination of $\overline{u}$. If so, find the scalar x such that $\overline{v} = x\overline{u}$.

23. $\overline{v} = \begin{pmatrix} 0.5 \\ 0.3 \\ 0.1 \end{pmatrix}$ 26. $\overline{v} = \begin{pmatrix} 0 \\ 0 \\ 0 \end{pmatrix}$

24. $\overline{v} = \begin{pmatrix} 5 \\ 3 \\ 0 \end{pmatrix}$ 27. $\overline{v} = \begin{pmatrix} 1 \\ 2 \\ 4 \end{pmatrix}$

25. $\overline{v} = \begin{pmatrix} 25 \\ 15 \\ 5 \end{pmatrix}$ 28. $\overline{v} = \begin{pmatrix} 75 \\ 40 \\ 15 \end{pmatrix}$

In Exercises 29–32, let

$$\overline{u} = \begin{pmatrix} 1 \\ 2 \end{pmatrix} \text{ and } \overline{v} = \begin{pmatrix} 2 \\ 1 \end{pmatrix}$$

Algebraically determine if $\overline{w}$ is a linear combination of $\overline{u}$ and $\overline{v}$. If so, find scalars x and y such that $\overline{w} = x\overline{u} + y\overline{u}$.

29. $\overline{w} = \begin{pmatrix} 5 \\ 10 \end{pmatrix}$ 31. $\overline{w} = \begin{pmatrix} 0 \\ 0 \end{pmatrix}$

30. $\overline{w} = \begin{pmatrix} 2 \\ 3 \end{pmatrix}$ 32. $\overline{w} = \begin{pmatrix} -0.5 \\ -0.25 \end{pmatrix}$

In Exercises 33–36, let

$$\overline{u} = \begin{pmatrix} 1 \\ 2 \end{pmatrix} \text{ and } \overline{v} = \begin{pmatrix} -3 \\ -6 \end{pmatrix}$$

Algebraically determine if $\overline{w}$ is a linear combination of $\overline{u}$ and $\overline{v}$. If so, find scalars x and y such that $\overline{w} = x\overline{u} + y\overline{v}$.

33. $\overline{w} = \begin{pmatrix} 5 \\ 10 \end{pmatrix}$ 35. $\overline{w} = \begin{pmatrix} 1 \\ -3 \end{pmatrix}$

34. $\overline{w} = \begin{pmatrix} 3 \\ 1 \end{pmatrix}$ 36. $\overline{w} = \begin{pmatrix} -0.5 \\ -0.25 \end{pmatrix}$

RStudio. In Exercises 37–40, use RStudio to determine if $\overline{w}$ is a linear combination of $\overline{u}$ and $\overline{v}$, where

$$\overline{u} = \begin{pmatrix} 1 \\ 2 \end{pmatrix} \text{ and } \overline{v} = \begin{pmatrix} 2 \\ 1 \end{pmatrix}$$

If so, find scalars x and y such that $\overline{w} = x\overline{u} + y\overline{v}$.

37. $\overline{w} = \begin{pmatrix} 5 \\ 10 \end{pmatrix}$ 39. $\overline{w} = \begin{pmatrix} -1.7 \\ 5.8 \end{pmatrix}$

38. $\overline{w} = \begin{pmatrix} 3 \\ -4 \end{pmatrix}$ 40. $\overline{w} = \begin{pmatrix} -0.5 \\ -0.25 \end{pmatrix}$

RStudio. In Exercises 41–44, use RStudio to determine if $\overline{w}$ is a linear combination of $\overline{u}$ and $\overline{v}$, where

$$\overline{u} = \begin{pmatrix} 1 \\ 2 \end{pmatrix} \text{ and } \overline{v} = \begin{pmatrix} -6 \\ -3 \end{pmatrix}$$

If so, find scalars x and y such that $\overline{w} = x\overline{u} + y\overline{v}$.

41. $\overline{w} = \begin{pmatrix} 5 \\ 10 \end{pmatrix}$ 43. $\overline{w} = \begin{pmatrix} -1.3 \\ 2.6 \end{pmatrix}$

42. $\overline{w} = \begin{pmatrix} 0 \\ 0 \end{pmatrix}$ 44. $\overline{w} = \begin{pmatrix} 1 \\ -3 \end{pmatrix}$

In Exercises 45–48, let

$$\overline{u}_1 = \begin{pmatrix} 5 \\ 3 \\ 1 \end{pmatrix}, \overline{u}_2 = \begin{pmatrix} 4 \\ 1 \\ 0 \end{pmatrix}, \text{ and } \overline{u}_3 = \begin{pmatrix} 8 \\ 2 \\ 0 \end{pmatrix}$$

and determine if $\overline{v}$ is a linear combination of $\overline{u}_1$, $\overline{u}_2$, and $\overline{u}_3$. If so, find scalars x_1, x_2, and x_3 such that $\overline{v} = x_1\overline{u}_1 + x_2\overline{u}_2 + x_3\overline{u}_3$.

45. $\overline{v} = \begin{pmatrix} 0.5 \\ 0.3 \\ 0.1 \end{pmatrix}$ 47. $\overline{v} = \begin{pmatrix} 1 \\ 3 \\ 5 \end{pmatrix}$

46. $\overline{v} = \begin{pmatrix} 0 \\ 2 \\ 4 \end{pmatrix}$ 48. $\overline{v} = \begin{pmatrix} 25 \\ 15 \\ 5 \end{pmatrix}$

In Exercises 49–52, let

$$\overline{u}_1 = \begin{pmatrix} 1 \\ 1 \\ 1 \end{pmatrix}, \overline{u}_2 = \begin{pmatrix} 1 \\ 1 \\ 0 \end{pmatrix}, \text{ and } \overline{u}_3 = \begin{pmatrix} 1 \\ 0 \\ 0 \end{pmatrix}$$

and determine if $\overline{v}$ is a linear combination of $\overline{u}_1$, $\overline{u}_2$, and $\overline{u}_3$. If so, find scalars x_1, x_2, and x_3 such that $\overline{v} = x_1\overline{u}_1 + x_2\overline{u}_2 + x_3\overline{u}_3$.

49. $\overline{v} = \begin{pmatrix} 0.5 \\ 0.3 \\ 0.1 \end{pmatrix}$ 51. $\overline{v} = \begin{pmatrix} 1 \\ 3 \\ 5 \end{pmatrix}$

50. $\overline{v} = \begin{pmatrix} 0 \\ 2 \\ 4 \end{pmatrix}$ 52. $\overline{v} = \begin{pmatrix} 25 \\ 15 \\ 5 \end{pmatrix}$

In Exercises 53–60, assume the data is linear and modeled by $y = mx + b$. State the corresponding vector equation in terms of the scalars m and b, and determine if this equation has a solution algebraically.

53.

x	1	2	3	4	5
y	111	69	26	−15	−57

54.

x	8	10	12	14	16
y	−2.5	4	11.5	18.5	25.5

55.

x	1	2	4	5	6
y	77.4	81	84.1	87.5	92

56.

x	1	2	4	5
y	−23.7	−36.8	−49.1	−59.8

57.

x	2	8	12	16	18
y	33.7	1.9	−19.3	−40.5	−51.1

58.

x	3	4	7	8	10
y	7.1	11.9	18.2	20.3	24.5

59.

x	0	5	10	15	20
y	5.0	4.5	4.0	3.5	3.0
x	30	35	40	45	50
y	2.0	1.5	1.0	0.5	0

60.

x	0	1	2	3	4
y	8	10	12	14	16
x	5	6	7	8	9
y	18	20	22	24	26

RStudio. In Exercises 61–66, use RStudio to determine if the vector equations from the given exercises have solutions.

61. Exercise 53   64. Exercise 56

62. Exercise 54   65. Exercise 59

63. Exercise 55   66. Exercise 60

In Exercises 67 and 68, consider the length (L) of the tornado season in days between the first and last tornado for each year (Y) in the warning area of the National Weather Service Office in Goodland, Kansas.

Y	1990	2000	2005	2010
L	120	142	149	155

The linear model $L = mY + b$ yields the following associated vector equation:

$$\begin{pmatrix} 120 \\ 142 \\ 149 \\ 155 \end{pmatrix} = m \begin{pmatrix} 1990 \\ 2000 \\ 2005 \\ 2010 \end{pmatrix} + b \begin{pmatrix} 1 \\ 1 \\ 1 \\ 1 \end{pmatrix}$$

67. Show that the associated vector equation has no solution algebraically.

68. Show that the associated vector equation has no solution using RStudio.

In Exercises 69 and 70, consider the U.S. retail prescription drug sales (S) in billions of dollars for each year (Y); that is, for example, in 2006, \$262.7 billion was spent in prescriptions in the United States.

Y	2006	2007	2009	2010
S	\$262.7	\$261.6	\$266.8	\$266.4

The linear model $S = mY + b$ yields the following associated vector equation:

$$\begin{pmatrix} 262.7 \\ 261.6 \\ 266.8 \\ 266.4 \end{pmatrix} = m \begin{pmatrix} 2006 \\ 2007 \\ 2009 \\ 2010 \end{pmatrix} + b \begin{pmatrix} 1 \\ 1 \\ 1 \\ 1 \end{pmatrix}$$

69. Show that the associated vector equation has no solution algebraically.

70. Show that the associated vector equation has no solution using RStudio.

In Exercises 71 and 72, consider the number of red cards (R) given per men's World Cup tournament in each year (Y).

Y	1978	1982	1986	1994	1998
R	3	5	8	15	22

For the exponential model

$$R = Ce^{k(Y-1978)}$$

the equation $\ln(R) = k(Y - 1978) + b$ yields the following associated vector equation:

$$\begin{pmatrix} 1.099 \\ 1.609 \\ 2.079 \\ 3.091 \end{pmatrix} = k \begin{pmatrix} 0 \\ 4 \\ 8 \\ 16 \\ 20 \end{pmatrix} + b \begin{pmatrix} 1 \\ 1 \\ 1 \\ 1 \\ 1 \end{pmatrix}$$

71. Show that the associated vector equation has no solution algebraically.

72. Show that the associated vector equation has no solution using RStudio.

In Exercises 73 and 74, consider the average debt load (D) in thousands of dollars at the end of each year's (Y) spring term for bachelor's degree recipients who attended public four-year colleges and universities and borrowed money to finance their education.

Y	2008	2010	2011	2012
D	\$22	\$23.6	\$24.6	\$25.3

For the exponential model

$$D = Ce^{k(Y-2008)}$$

the equation $\ln(D) = k(Y - 2008) + b$ yields the following associated vector equation:

$$\begin{pmatrix} 3.091 \\ 3.161 \\ 3.203 \\ 3.231 \end{pmatrix} = k \begin{pmatrix} 0 \\ 2 \\ 3 \\ 4 \end{pmatrix} + b \begin{pmatrix} 1 \\ 1 \\ 1 \\ 1 \end{pmatrix}$$

73. Show that the associated vector equation has no solution algebraically.

74. Show that the associated vector equation has no solution using RStudio.

In Exercises 75 – 77, consider the total number of prescription drugs (P) sold in the United States in millions per year (Y); that is, for example, in 2006, 3, 530, 000, 000 prescriptions were sold.

Y	2006	2007	2009	2010
P	3419	3530	3633	3676

75. For the linear model $P = mY + b$, state the associated vector equation.

76. Determine if the vector equation in Exercise 75 has a solution algebraically.

77. Determine if the vector equation in Exercise 75 has a solution using RStudio.

In Exercises 78–80, consider the total number of burgers (B) sold by McDonald's in billions as of each year (Y).

Y	1976	1987	1990	1994
B	20	65	80	100

78. For the linear model $B = mY + b$, state the associated vector equation.

79. Determine if the vector equation in Exercise 78 has a solution algebraically.

80. Determine if the vector equation in Exercise 78 has a solution using RStudio.

In Exercises 81–83, consider the global gender ratio based on the number of males per 100 females (M) in each year (Y).

Y	1990	1995	2005	2010
M	101.5	101.5	101.6	101.6

81. For the linear model M $= mY + b$, state the associated vector equation.

82. Determine if the vector equation in Exercise 81 has a solution algebraically.

83. Determine if the vector equation in Exercise 81 has a solution using RStudio.

In Exercises 84–86, consider the world population growth rate (P) by year (Y).

Y	1995	2000	2005	2010	2015
P	1.413	1.26	1.203	1.127	1.06

84. For the linear model P $= mY + b$, state the associated vector equation.

85. Determine if the vector equation in Exercise 84 has a solution algebraically.

86. Determine if the vector equation in Exercise 84 has a solution using RStudio.

In Exercises 87–89, consider the total number of tornadoes (T) each year (Y) in the warning area of the National Weather Service Office in Goodland, Kansas.

Y	1975	1985	1995	2000	2005
T	5	8	18	28	33

87. For the exponential model

$$T = Ce^{k(Y-1975)}$$

state the associated vector equation of $\ln(T) = k(Y - 1975) + b$.

88. Determine if the vector equation in Exercise 87 has a solution algebraically.

89. Determine if the vector equation in Exercise 87 has a solution using RStudio.

In Exercises 90–92, consider the population of Ireland in millions of people (P) each year (Y).

Y	1834	1847	1854	1867	1881
P	7.89	8.02	6.08	5.64	3.87

90. For the exponential model

$$P = Ce^{k(Y-1834)}$$

state the associated vector equation of $\ln(P) = k(Y - 1834) + b$.

91. Determine if the vector equation in Exercise 90 has a solution algebraically.

92. Determine if the vector equation in Exercise 90 has a solution using RStudio.

In Exercises 93–95, consider the highest value (V) of Facebook stock each month (M) in 2013, where the value of M identifies the month; that is, for example, 8 corresponds to August, the 8th month of the year.

M	5	7	8	9	10
V	29.1	38.3	42.5	51.6	54.8

93. For the exponential model

$$V = Ce^{k(M-5)}$$

state the associated vector equation of $\ln(V) = k(M-5) + b$.

94. Determine if the vector equation in Exercise 93 has a solution algebraically.

95. Determine if the vector equation in Exercise 93 has a solution using RStudio.

In Exercises 96–98, consider the S&P 500 stock market closing value (V) on June 1 of each year (Y) in U.S. dollars.

Y	1975	1985	1995	2005	2015
V	95	192	545	1191	2077

96. For the exponential model

$$V = Ce^{k(Y-1975)}$$

state the associated vector equation of $\ln(V) = k(Y - 1975) + b$.

97. Determine if the vector equation in Exercise 96 has a solution algebraically.

98. Determine if the vector equation in Exercise 96 has a solution using RStudio.

In Your Own Words. In Exercises 99–103, explain the following.

99. The process for determining if $\overline{v}$ is a linear combination of $\overline{u}$ graphically.

100. The process for determining if $\overline{v}$ is a linear combination of $\overline{u}$ algebraically.

101. The process for determining if $\overline{v}$ is a linear combination of $\overline{u}_1, \ldots, \overline{u}_n$ by hand algebraically.

102. The process for determining if $\overline{v}$ is a linear combination of $\overline{u}_1, \ldots, \overline{u}_n$ using RStudio.

103. A model is linear in its parameters.

3.4   Vector Projection

Section 3.3 introduced a process for determining whether or not a target vector $\overline{v}$ is a linear combination of the vectors $\overline{u}_1, \overline{u}_2, \ldots, \overline{u}_n$. In most real-life settings, the target vector $\overline{v}$ is *not* a linear combination of the given vectors, meaning that there does not exist a solution set of scalars $x_1, x_2, \ldots, x_n$ so that $\overline{v} = x_1\overline{u}_1 + x_2\overline{u}_2 + \cdots + x_n\overline{u}_n$. When an exact linear combination does not exist, we adjust our approach and try to find the linear combination $x_1\overline{u}_1 + x_2\overline{u}_2 + \cdots + x_n\overline{u}_n$ that is "as close as possible" to the target vector $\overline{v}$. Such a linear combination is produced using the method of least squares, which relies on the idea of *vector projection*. This study of vector projection begins by discussing a type of vector multiplication known as the *dot product*.

The Dot Product

Section 3.1 highlighted three basic arithmetic vector operations: vector addition, vector subtraction, and scalar multiplication. Recall that scalar multiplication is the product of a number and a vector, and has many nice properties. A natural next question to consider is: what about the product of two vectors with each other? As mentioned in Section 3.1, in addition to scalar multiplication, there are two other approaches to multiplication of vectors, neither of which preserves all of the traditional, desirable properties of standard multiplication of numbers (e.g., commutativity, associativity, distributivity, etc.). This section discusses the *dot product*, which multiplies two vectors together to produce a number (or scalar), and examines a few of its properties. (The other approach, the cross product, which we shall not consider in this book, also multiplies two vectors together, but this time to produce another vector.)

Definition. The **dot product** of two 2-component vectors is

$$\overline{u} \cdot \overline{v} = \begin{pmatrix} u_1 \\ u_2 \end{pmatrix} \cdot \begin{pmatrix} v_1 \\ v_2 \end{pmatrix} = u_1 v_1 + u_2 v_2$$

Namely, multiply each component of $\overline{u}$ by the corresponding component of $\overline{v}$ and sum the resulting products together to obtain a scalar. More generally, the **dot product** of two vectors is

$$\overline{u} \cdot \overline{v} = \begin{pmatrix} u_1 \\ u_2 \\ \vdots \\ u_n \end{pmatrix} \cdot \begin{pmatrix} v_1 \\ v_2 \\ \vdots \\ v_n \end{pmatrix} = u_1 v_1 + u_2 v_2 + \cdots + u_n v_n$$

Sometimes, the dot product is called the **scalar product** of two vectors, because the output of this vector multiplication is a scalar. Always keep in mind that the dot product takes two *vectors* and produces a *scalar*.

◆ **EXAMPLE 1** Compute each dot product for $\overline{u} = \begin{pmatrix} 3 \\ 1 \end{pmatrix}$, $\overline{v} = \begin{pmatrix} -2 \\ 3 \end{pmatrix}$, and $\overline{w} = \begin{pmatrix} 1 \\ 1 \end{pmatrix}$:

(a) $\overline{u} \cdot \overline{v}$ (b) $\overline{v} \cdot \overline{u}$ (c) $\overline{v} \cdot \overline{w}$ (d) $(3\overline{u}) \cdot (-\overline{w})$

Solution.

(a) $\overline{u} \cdot \overline{v} = \begin{pmatrix} 3 \\ 1 \end{pmatrix} \cdot \begin{pmatrix} -2 \\ 3 \end{pmatrix} = (3)(-2) + (1)(3) = (-6) + 3 = -3$

(b) $\overline{v} \cdot \overline{u} = \begin{pmatrix} -2 \\ 3 \end{pmatrix} \cdot \begin{pmatrix} 3 \\ 1 \end{pmatrix} = (-2)(3) + (3)(1) = (-6) + 3 = -3$

(c) $\overline{v} \cdot \overline{w} = \begin{pmatrix} -2 \\ 3 \end{pmatrix} \cdot \begin{pmatrix} 1 \\ 1 \end{pmatrix} = (-2)(1) + (3)(1) = (-2) + 3 = 1$

(d) $(3\overline{u}) \cdot (-\overline{w}) = \begin{pmatrix} 9 \\ 3 \end{pmatrix} \cdot \begin{pmatrix} -1 \\ -1 \end{pmatrix} = (9)(-1) + (3)(-1) = (-9) + (-3) = -12$

■

▶ **QUESTION 1** Compute each dot product for $\overline{u} = \begin{pmatrix} 1 \\ 4 \end{pmatrix}$, $\overline{v} = \begin{pmatrix} -5 \\ 2 \end{pmatrix}$, and $\overline{w} = \begin{pmatrix} -8 \\ -2 \end{pmatrix}$:

(a) $\overline{u} \cdot \overline{v}$ (b) $\overline{v} \cdot \overline{u}$ (c) $(2\overline{u}) \cdot \overline{w}$ (d) $\overline{u} \cdot (2\overline{w})$

Example 1 and Question 1 illustrate two important properties of the dot product that can be summarized as follows:

- The dot product is **commutative** because $\overline{u} \cdot \overline{v} = \overline{v} \cdot \overline{u}$, for every possible choice of vectors $\overline{u}$ and $\overline{v}$ with the same number of components. In other words, the order of the vectors in a dot product does not matter.

- Scalar multiplication **commutes** with the dot product because, if the real number m is a scalar, then $(m\overline{u}) \cdot \overline{v} = \overline{u} \cdot (m\overline{v}) = m(\overline{u} \cdot \overline{v})$, for every possible choice of vectors $\overline{u}$ and $\overline{v}$ with the same number of components.

While Example 1 and Question 1 provide evidence that these properties are true, mathematicians have also created a general argument, or proof, to show that they hold for every possible choice of vectors and scalars. In other words, for commutativity, the equality of Examples 1(a) and 1(b), and of Questions 1(a) and 1(b) suggest that this property holds for all vectors $\overline{u}$ and $\overline{v}$. Mathematicians know that commutativity holds for all vectors by computing $\overline{u} \cdot \overline{v}$ and $\overline{v} \cdot \overline{u}$ separately for arbitrary vectors $\overline{u}$ and $\overline{v}$, and showing they are equal. Similarly, while Questions 1(c) and 1(d) suggest that scalars commute with the dot product, a general argument computes $(m\overline{u}) \cdot \overline{v}$, $\overline{u} \cdot (m\overline{v})$, and $m(\overline{u} \cdot \overline{v})$ separately for arbitrary vectors $\overline{u}$, $\overline{v}$ and an arbitrary scalar m, and shows that they are all equal.

While Example 1 and Question 1 both work with two-component vectors, the natural extension of the dot product to vectors with three, four, and more components is illustrated in the next example.

◆ **EXAMPLE 2** Compute each dot product:

(a) $\begin{pmatrix} 6 \\ 2 \\ -5 \end{pmatrix} \cdot \begin{pmatrix} 7 \\ 3 \\ 1 \end{pmatrix}$ (b) $\begin{pmatrix} 3 \\ 1 \\ -2 \\ 0 \end{pmatrix} \cdot \begin{pmatrix} -2 \\ 4 \\ 5 \\ 3 \end{pmatrix}$

Solution.

(a) $\begin{pmatrix} 6 \\ 2 \\ -5 \end{pmatrix} \cdot \begin{pmatrix} 7 \\ 3 \\ 1 \end{pmatrix} = (6)(7) + (2)(3) + (-5)(1) = 42 + 6 + (-5) = 43$

(b) $\begin{pmatrix} 3 \\ 1 \\ -2 \\ 0 \end{pmatrix} \cdot \begin{pmatrix} -2 \\ 4 \\ 5 \\ 3 \end{pmatrix} = (3)(-2) + (1)(4) + (-2)(5) + (0)(3) = -12$

■

▶ **QUESTION 2** Compute each dot product:

(a) $\begin{pmatrix} -2 \\ 4 \\ 6 \\ 2 \end{pmatrix} \cdot \begin{pmatrix} 3 \\ 5 \\ -1 \\ 7 \end{pmatrix}$
(b) $\begin{pmatrix} 1 \\ 2 \\ 3 \\ 4 \\ 5 \end{pmatrix} \cdot \begin{pmatrix} -3 \\ 2 \\ 3 \\ 0 \\ -2 \end{pmatrix}$

Geometry of the Dot Product

The dot product is relatively easy to compute and two important algebraic properties of this vector multiplication were identified above. The dot product also has important geometric properties, including a connection between the dot product and the length of a vector.

LENGTH OF A VECTOR. The length of a vector $\bar{u}$ is $\|\bar{u}\| = \sqrt{\bar{u} \cdot \bar{u}}$. Alternatively, $\|\bar{u}\| = \sqrt{\bar{u}^2}$, provided $\bar{u}^2$ is interpreted as meaning $\bar{u} \cdot \bar{u}$.

This relationship between the length $\|\bar{u}\|$ of a vector $\bar{u}$ and the dot product $\bar{u} \cdot \bar{u}$ with itself is a consequence of the Pythagorean theorem. The number of components in a vector $\bar{u}$ does not impact this equality; rather, $\|\bar{u}\| = \sqrt{\bar{u} \cdot \bar{u}}$ is always true. While humans find it impossible to visualize a vector with four or more components, the length of such vectors can still be discussed and this length calculated using the dot product.

In addition to information about vector length, the dot product also provides information about the angle between two vectors. Thinking about two vectors in the plane, you can visualize two angles between the vectors that depend on the direction of travel from one vector to the other. The angle between two vectors is defined to be the smaller of the resulting two angles, which will always have a measure between 0° and 180° (i.e., between 0 and π radians). In two dimensions, two lines are perpendicular when the angle between them is 90° or $\pi/2$ radians. This idea extends to vectors: two vectors are perpendicular or, more formally, *orthogonal*, if the angle between them is 90° or $\pi/2$ radians.

Definition. Two vectors $\bar{u}$ and $\bar{v}$ are **orthogonal** when $\bar{u} \cdot \bar{v} = 0$, where "orthogonal" is an extension of the two-dimensional idea of "perpendicular" to any number of dimensions.

The preceding definition of orthogonal asserts that only orthogonal vectors have the geometric relationship of being perpendicular. Also, note that only one of $\bar{u} \cdot \bar{v}$ or $\bar{v} \cdot \bar{u}$

needs to be computed in order to determine orthogonality, because the dot product is a commutative operation on vectors.

◆ **EXAMPLE 3** Determine if each pair of vectors is orthogonal:

(a) $\bar{u} = \begin{pmatrix} 3 \\ 1 \end{pmatrix}$ and $\bar{v} = \begin{pmatrix} -1 \\ 3 \end{pmatrix}$ (b) $\bar{u} = \begin{pmatrix} 3 \\ 1 \end{pmatrix}$ and $\bar{w} = \begin{pmatrix} 2 \\ 2 \end{pmatrix}$

Solution.

(a) Compute $\bar{u} \cdot \bar{v}$ to determine if these vectors are orthogonal. In particular, $\bar{u} \cdot \bar{v} = (3)(-1) + (1)(3) = -3 + 3 = 0$. Therefore, $\bar{u}$ and $\bar{v}$ are orthogonal, and the angle between them is $90° = \pi/2$ radians, as illustrated in Figure 4(a).

(b) Computing $\bar{u} \cdot \bar{w}$ gives $\bar{u} \cdot \bar{w} = (3)(2) + (1)(2) = 6 + 2 = 8$. Because $\bar{u} \cdot \bar{w} = 8 \neq 0$, $\bar{u}$ and $\bar{w}$ are not orthogonal, as illustrated in Figure 4(b).

(a)   (b)

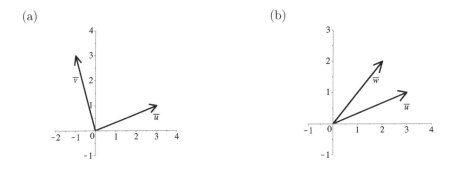

Figure 1: Graphs of vectors for Example 3

➤ **QUESTION 3**

(a) Determine if the vectors $\bar{u} = \begin{pmatrix} -2 \\ 5 \end{pmatrix}$ and $\bar{v} = \begin{pmatrix} -10 \\ 5 \end{pmatrix}$ are orthogonal.

(b) Determine if the vectors $\bar{u}$ and $\bar{v}$ graphed in Figure 2 are or are not orthogonal.

➤ **QUESTION 4** State three different vectors that are orthogonal to $\bar{u} = \begin{pmatrix} 3 \\ 1 \end{pmatrix}$.

◆ **EXAMPLE 4** Determine which pairs of the following vectors are orthogonal:

$$\bar{u} = \begin{pmatrix} 2 \\ -1 \\ 3 \\ 1 \end{pmatrix} \qquad \bar{v} = \begin{pmatrix} 1 \\ 1 \\ 1 \\ 1 \end{pmatrix} \qquad \bar{w} = \begin{pmatrix} 2 \\ 2 \\ 1 \\ -5 \end{pmatrix}$$

Solution. The pairs of vectors that are orthogonal is determined by computing all possible dot products, keeping in mind the commutativity of the dot product. This property allows computation of just one of $\bar{u} \cdot \bar{v}$ and $\bar{v} \cdot \bar{u}$, rather than both.

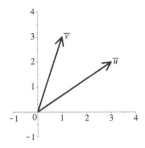

Figure 2: Vectors for Question 3(b)

- $\overline{u} \cdot \overline{v} = \begin{pmatrix} 2 \\ -1 \\ 3 \\ 1 \end{pmatrix} \cdot \begin{pmatrix} 1 \\ 1 \\ 1 \\ 1 \end{pmatrix} = (2)(1) + (-1)(1) + (3)(1) + (1)(1) = 2 + (-1) + 3 + 1 = 5$

- $\overline{u} \cdot \overline{w} = \begin{pmatrix} 2 \\ -1 \\ 3 \\ 1 \end{pmatrix} \cdot \begin{pmatrix} 2 \\ 2 \\ 1 \\ -5 \end{pmatrix} = (2)(2) + (-1)(2) + (3)(1) + (1)(-5) = 4 + (-2) + 3 + (-5) = 0$

- $\overline{v} \cdot \overline{w} = \begin{pmatrix} 1 \\ 1 \\ 1 \\ 1 \end{pmatrix} \cdot \begin{pmatrix} 2 \\ 2 \\ 1 \\ -5 \end{pmatrix} = (1)(2) + (1)(2) + (1)(1) + (1)(-5) = 2 + 2 + 1 + (-5) = 0$

Because $\overline{u} \cdot \overline{w} = 0$, the vectors $\overline{u}$ and $\overline{w}$ are orthogonal. Similarly, because $\overline{v} \cdot \overline{w} = 0$, the vectors $\overline{v}$ and $\overline{w}$ are orthogonal. However, $\overline{u} \cdot \overline{v} = 5 \neq 0$, which means that $\overline{u}$ and $\overline{v}$ are not orthogonal. ∎

➤ **QUESTION 5** Determine which pairs of the following vectors are orthogonal:

$$\overline{u} = \begin{pmatrix} 1 \\ 2 \\ 3 \\ 4 \\ 5 \end{pmatrix} \qquad \overline{v} = \begin{pmatrix} -3 \\ 2 \\ 3 \\ 0 \\ -2 \end{pmatrix} \qquad \overline{w} = \begin{pmatrix} 2 \\ 1 \\ -3 \\ 2 \\ 0 \end{pmatrix}$$

Residual Vectors

Working in the direction of motivating the definition of vector projection, the notion of a *residual vector* is first introduced. A residual vector measures how close a linear combination $x\overline{u}$ of the vector $\overline{u}$ is to a target vector $\overline{v}$ as defined next.

Definition. For vectors $\overline{v}$ and $\overline{u}$ and a scalar x, the **residual vector** for the linear combination $x\overline{u}$ and target vector $\overline{v}$ is $\overline{r} = \overline{v} - x\overline{u}$.

◆ **EXAMPLE 5** Let

$$\overline{u} = \begin{pmatrix} 4 \\ 1 \\ 2 \end{pmatrix} \text{ and } \overline{v} = \begin{pmatrix} -2 \\ 3 \\ 4 \end{pmatrix}$$

and compute the residual vector $\overline{r}$ for the linear combination $2\overline{u}$ and target vector $\overline{v}$.

Solution. First, calculate the linear combination

$$2\overline{u} = 2\begin{pmatrix} 4 \\ 1 \\ 2 \end{pmatrix} = \begin{pmatrix} 2 \cdot 4 \\ 2 \cdot 1 \\ 2 \cdot 2 \end{pmatrix} = \begin{pmatrix} 8 \\ 2 \\ 4 \end{pmatrix}$$

and then compute the corresponding residual vector $\overline{r}$:

$$\overline{r} = \overline{v} - 2\overline{u} \qquad\qquad \text{Definition of residual}$$

$$= \begin{pmatrix} -2 \\ 3 \\ 4 \end{pmatrix} - \begin{pmatrix} 8 \\ 2 \\ 4 \end{pmatrix} \qquad\qquad \text{Substitute known vectors}$$

$$= \begin{pmatrix} -10 \\ 1 \\ 0 \end{pmatrix} \qquad\qquad \text{Simplify}$$

∎

➤ **QUESTION 6** Let

$$\overline{u} = \begin{pmatrix} 5 \\ -1 \\ 3 \\ 2 \end{pmatrix} \text{ and } \overline{v} = \begin{pmatrix} 6 \\ -3 \\ 5 \\ 3 \end{pmatrix}$$

and compute the residual vector $\overline{r}$ for the linear combination $1.5\overline{u}$ and target vector $\overline{v}$.

◆ **EXAMPLE 6** Let $\overline{v} = \begin{pmatrix} 2 \\ 4 \end{pmatrix}$ and $\overline{u} = \begin{pmatrix} 4 \\ 2 \end{pmatrix}$.

(a) Compute the residual vectors for the linear combinations $\frac{1}{2}\overline{u}$, $\frac{4}{5}\overline{u}$, and $\frac{5}{4}\overline{u}$, and target vector $\overline{v}$.

(b) Compute the lengths of the residual vectors for $\frac{1}{2}\overline{u}$, $\frac{4}{5}\overline{u}$, and $\frac{5}{4}\overline{u}$.

(c) For each linear combination $x\overline{u}$, sketch the vectors $\overline{u}$, $x\overline{u}$, and $\overline{v}$ in standard position, and the corresponding residual vector $\overline{r}$ starting at the tip of $x\overline{u}$.

Solution.

(a) Use the formula $\bar{r} = \bar{v} - x\bar{u}$ to compute each residual vector. Note that the scalar x is used as a subscript on $\bar{r}$ to identify each residual with its linear combination.

$$\bar{r}_{\frac{1}{2}} = \begin{pmatrix} 2 \\ 4 \end{pmatrix} - \frac{1}{2}\begin{pmatrix} 4 \\ 2 \end{pmatrix} = \begin{pmatrix} 2 \\ 4 \end{pmatrix} - \begin{pmatrix} 2 \\ 1 \end{pmatrix} = \begin{pmatrix} 0 \\ 3 \end{pmatrix}$$

$$\bar{r}_{\frac{4}{5}} = \begin{pmatrix} 2 \\ 4 \end{pmatrix} - \frac{4}{5}\begin{pmatrix} 4 \\ 2 \end{pmatrix} = \begin{pmatrix} 2 \\ 4 \end{pmatrix} - \begin{pmatrix} 3.2 \\ 1.6 \end{pmatrix} = \begin{pmatrix} -1.2 \\ 2.4 \end{pmatrix}$$

$$\bar{r}_{\frac{5}{4}} = \begin{pmatrix} 2 \\ 4 \end{pmatrix} - \frac{5}{4}\begin{pmatrix} 4 \\ 2 \end{pmatrix} = \begin{pmatrix} 2 \\ 4 \end{pmatrix} - \begin{pmatrix} 5 \\ 2.5 \end{pmatrix} = \begin{pmatrix} -3 \\ 1.5 \end{pmatrix}$$

(b) Calculate the length of each residual vector as follows. Note that $\bar{r}_{\frac{4}{5}}$ has the shortest length among these residuals.

$$\bar{r}_{\frac{1}{2}} = \sqrt{0^2 + 3^2} = \sqrt{9} = 3$$

$$\bar{r}_{\frac{4}{5}} = \sqrt{(-1.2)^2 + (2.4)^2} = \sqrt{1.44 + 5.76} = \sqrt{7.2} \approx 2.68$$

$$\bar{r}_{\frac{5}{4}} = \sqrt{(-3)^2 + (1.5)^2} = \sqrt{9 + 2.25} = \sqrt{11.25} \approx 3.35$$

(c) Figure 3 presents the plots of the vectors $\bar{u}$, $x\bar{u}$, $\bar{v}$, and $\bar{r}$ for each linear combination, where Figure 3(i) is the graph for $x = \dfrac{1}{2}$, Figure 3(ii) is the graph for $x = \dfrac{4}{5}$, and Figure 3(iii) is the graph for $x = \dfrac{5}{4}$. ∎

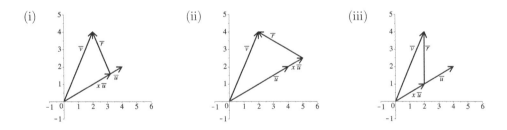

Figure 3: Graphs of linear combinations and residuals for Example 6

▶ **QUESTION 7** Let $\bar{v} = \begin{pmatrix} 1 \\ 6 \end{pmatrix}$ and $\bar{u} = \begin{pmatrix} 4 \\ 4 \end{pmatrix}$.

(a) Compute the residual vectors for the linear combinations $\dfrac{1}{2}\bar{u}$, $\dfrac{7}{8}\bar{u}$, and $2\bar{u}$, and target vector $\bar{v}$.

(b) Compute the length of the residuals for $\dfrac{1}{2}\bar{u}$, $\dfrac{7}{8}\bar{u}$, and $2\bar{u}$.

(c) For each linear combination $x\overline{u}$, sketch the vectors $\overline{u}$, $x\overline{u}$, and $\overline{v}$ in standard position, and the corresponding residual vector $\overline{r}$ starting at the tip of $x\overline{u}$.

Observe that each linear combination in Example 6 and Question 7 has a different residual vector. As mentioned earlier, the residual measures how close a linear combination $x\overline{u}$ is to a target vector $\overline{v}$. In more detail, the length of the residual vector provides a precise measure of "closeness," where a shorter residual vector $\overline{r} = \overline{v} - x\overline{u}$ indicates that the linear combination $x\overline{u}$ is closer to the target vector $\overline{v}$ and a longer $\overline{r}$ indicates that $x\overline{u}$ is farther from $\overline{v}$. Example 6 provides a specific illustration of this behavior. Namely, both the computation of length in part (b) and the vector plots in part (c) indicate that the shortest residual vector is $\overline{r}_{\frac{4}{5}}$ and so the linear combination $\frac{4}{5}\overline{u}$ is closest to the target vector $\overline{v}$ among these three vectors and, in fact, among all possible sets of vectors.

A closer inspection of the three plots in Figure 3 reveals that this residual vector $\overline{r}_{\frac{4}{5}}$ is also perpendicular to linear combination $\frac{4}{5}\overline{u}$. This observation that the shortest residual occurs when the linear combination $x\overline{u}$ is orthogonal to the residual vector $\overline{r}$ provides the foundation for the concept of *vector projection*, as discussed in more detail next.

Vector Projection

This section began with the observation that in most real-life settings a target vector $\overline{v}$ is *not* a linear combination of a given vector $\overline{u}$. In response, the focus turned toward finding the linear combination $x\overline{u}$ that is as close as possible to $\overline{v}$. As mentioned above, this closest possible linear combination $x\overline{u}$ occurs when the corresponding residual vector $\overline{r}$ is orthogonal to $x\overline{u}$, and is called the *projection of $\overline{v}$ onto $\overline{u}$*.

Definition. For a vector $\overline{v}$ and a nonzero vector $\overline{u}$, the **projection of $\overline{v}$ onto $\overline{u}$** is the vector $x\overline{u}$ where

$$x = \frac{\overline{v} \cdot \overline{u}}{\overline{u} \cdot \overline{u}}$$

When $\overline{u} = \overline{0}$ is the zero vector, the projection of $\overline{v}$ onto $\overline{u}$ is $\overline{0}$.

While the definition of the projection $x\overline{u}$ of $\overline{v}$ onto $\overline{u}$ includes a formula for the scalar x, this number is not the projection. Rather, the projection is a vector, about which a great deal is known, including the formula given in the definition.

◆ **EXAMPLE 7** Find the projection of $\overline{v}$ onto $\overline{u}$ for each pair of vectors $\overline{v}$ and $\overline{u}$:

(a) $\overline{v} = \begin{pmatrix} 1 \\ 5 \end{pmatrix}$ and $\overline{u} = \begin{pmatrix} 1 \\ 3 \end{pmatrix}$ (b) $\overline{v} = \begin{pmatrix} 1 \\ 5 \\ -2 \end{pmatrix}$ and $\overline{u} = \begin{pmatrix} 2 \\ 3 \\ 1 \end{pmatrix}$

Solution.

(a) Find x using the formula $x = \dfrac{\overline{v} \cdot \overline{u}}{\overline{u} \cdot \overline{u}}$ from the definition of projection:

$$x = \frac{\overline{v} \cdot \overline{u}}{\overline{u} \cdot \overline{u}} = \frac{(1)(1) + (5)(3)}{(1)(1) + (3)(3)} = \frac{16}{10} = \frac{8}{5} = 1.6$$

Thus, the projection of $\overline{v}$ onto $\overline{u}$ is given by $x\overline{u} = \dfrac{8}{5}\begin{pmatrix} 1 \\ 3 \end{pmatrix} = \begin{pmatrix} 8/5 \\ 24/5 \end{pmatrix} = \begin{pmatrix} 1.6 \\ 4.8 \end{pmatrix}$.

(b) Find x using the formula $x = \dfrac{\overline{v} \cdot \overline{u}}{\overline{u} \cdot \overline{u}}$ from the definition of projection:

$$x = \frac{\overline{v} \cdot \overline{u}}{\overline{u} \cdot \overline{u}} = \frac{(1)(2) + (5)(3) + (-2)(1)}{(2)(2) + (3)(3) + (1)(1)} = \frac{15}{14}$$

Thus, the projection of $\overline{v}$ onto $\overline{u}$ is given by $x\overline{u} = \dfrac{15}{14}\begin{pmatrix} 2 \\ 3 \\ 1 \end{pmatrix} = \begin{pmatrix} 30/14 \\ 45/14 \\ 15/14 \end{pmatrix} \approx \begin{pmatrix} 2.14 \\ 3.21 \\ 1.07 \end{pmatrix}$.

∎

▶ **QUESTION 8** Find the projection of $\overline{v}$ onto $\overline{u}$ for each pair of vectors $\overline{v}$ and $\overline{u}$:

(a) $\overline{v} = \begin{pmatrix} -7 \\ 3 \end{pmatrix}$ and $\overline{u} = \begin{pmatrix} 5 \\ -2 \end{pmatrix}$

(b) $\overline{v} = \begin{pmatrix} 2 \\ 3 \\ 2 \\ 4 \end{pmatrix}$ and $\overline{u} = \begin{pmatrix} 1 \\ 5 \\ -7 \\ 4 \end{pmatrix}$

Understanding Vector Projection

The task of finding $x\overline{u}$ as close as possible to a vector $\overline{v}$ is accomplished by making the residual vector $\overline{r} = \overline{v} - x\overline{u}$ as small as possible. The precise formula for the scalar x when $\overline{u}$ is not the zero vector from the definition of vector projection is obtained from an analysis of the geometric setting illustrated in Figure 4.

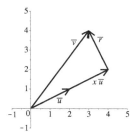

Figure 4: Finding $x\overline{u}$ closest to $\overline{v}$

For the moment, assume that $\overline{v}$ and $\overline{u}$ are not orthogonal, which corresponds exactly to the case when x is nonzero. Note that the sum of $x\overline{u}$ and the residual vector $\overline{r}$ is the vector $\overline{v}$. Algebraically, this fact is expressed as follows:

$$x\overline{u} + \overline{r} = \overline{v}$$

In addition, the residual vector $\overline{r}$ needs to be as small as possible, which occurs when $\overline{r}$ is orthogonal to the linear combination $x\overline{u}$. Expressing this orthogonality in terms of the dot product provides $\overline{r} \cdot (x\overline{u}) = 0$, which can be rewritten as $x(\overline{r} \cdot \overline{u}) = 0$ because scalar multiplication commutes with the dot product. Dividing both sides by x (which is nonzero because $\overline{v}$ and $\overline{u}$ are not orthogonal) gives $\overline{r} \cdot \overline{u} = 0$. Therefore, the desired scalar x must satisfy both of the following two equations:

$$x\overline{u} + \overline{r} = \overline{v}$$
$$\overline{r} \cdot \overline{u} = 0$$

The next step is to solve for x by computing the dot product of the vector $\overline{u}$ with both sides of the first equation $x\overline{u} + \overline{r} = \overline{v}$, which gives the following:

$$(x\overline{u} + \overline{r}) \cdot \overline{u} = \overline{v} \cdot \overline{u} \qquad \text{Dot product of both sides of } x\overline{u} + \overline{r} = \overline{v} \text{ by } \overline{u}$$
$$(x\overline{u}) \cdot \overline{u} + \overline{r} \cdot \overline{u} = \overline{v} \cdot \overline{u} \qquad \text{Distribute } \overline{u}$$
$$x(\overline{u} \cdot \overline{u}) + \overline{r} \cdot \overline{u} = \overline{v} \cdot \overline{u} \qquad \text{Scalar multiplication commutes}$$
$$x(\overline{u} \cdot \overline{u}) + 0 = \overline{v} \cdot \overline{u} \qquad \text{Substitute } \overline{r} \cdot \overline{u} = 0$$
$$x(\overline{u} \cdot \overline{u}) = \overline{v} \cdot \overline{u} \qquad \text{Simplify}$$

The dot product $\overline{u} \cdot \overline{u}$ is a nonzero scalar because of the assumption that $\overline{u}$ is not the zero vector and the definition of the dot product. Dividing both sides of the last equation above, $x(\overline{u} \cdot \overline{u}) = \overline{u} \cdot \overline{v}$, by this nonzero scalar $\overline{u} \cdot \overline{u}$ provides the following formula for x that is stated in the definition of the projection of $\overline{v}$ onto $\overline{u}$:

$$x = \frac{\overline{v} \cdot \overline{u}}{\overline{u} \cdot \overline{u}}$$

For settings where $\overline{v}$ and $\overline{u}$ are orthogonal, the projection vector happens to always be the zero vector $\overline{0}$. In such a case, the projection $x\overline{u}$ has $x = 0$, and the corresponding residual vector is $\overline{r} = \overline{v}$ because

$$\overline{v} = x\overline{u} + \overline{r} = 0\overline{u} + \overline{r} = \overline{0} + \overline{r} = \overline{r}$$

By definition, when $\overline{v}$ and $\overline{u}$ are orthogonal, $\overline{v} \cdot \overline{u} = 0$, allowing the application of the formula $x = \dfrac{\overline{v} \cdot \overline{u}}{\overline{u} \cdot \overline{u}}$ in settings where $\overline{v}$ and $\overline{u}$ are orthogonal as well.

Working in RStudio

The dot product of two vectors is computed in RStudio using the `dot` command, which accepts two vectors separated by a comma as inputs. For example, compute the dot product

$$\begin{pmatrix} 5 \\ -2 \end{pmatrix} \cdot \begin{pmatrix} -7 \\ 3 \end{pmatrix}$$

by first defining the vectors with the commands u=c(5,-2) and v=c(-7,3), and then entering the command dot(u,v), which outputs [1] -41 to indicate the value of the preceding dot product is -41. The dot command is robust enough to also accept inputs involving scalar multiplication. Continuing the preceding example, the dot product

$$4 \begin{pmatrix} 5 \\ -2 \end{pmatrix} \cdot \begin{pmatrix} -7 \\ 3 \end{pmatrix}$$

is computed by entering dot(4*u,v).

Examples of Commands

- u=c(4,1,2)
- v=c(1,7,-2)

- dot(u,v)
- dot(u,2*v)

◆ **EXAMPLE 8** Use RStudio to compute each dot product for $\overline{u} = \begin{pmatrix} 1 \\ 2 \\ 3 \end{pmatrix}$ and $\overline{v} = \begin{pmatrix} -4 \\ 7 \\ -2 \end{pmatrix}$:

(a) $\overline{u} \cdot \overline{v}$

(b) $\overline{v} \cdot (2.5\overline{u})$

Solution.

(a)   u = c(1,2,3)
       v = c(-4,7,-2)
       dot(u,v)

    [1] 4

(b)   dot(v,2.5*u)

    [1] 10

This RStudio output shows that $\overline{u} \cdot \overline{v} = 4$ and that $\overline{v} \cdot (2.5\overline{u}) = 10$. ■

➤ **QUESTION 9** Compute each dot product for

$$\overline{u} = \begin{pmatrix} 1 \\ 8 \\ 2 \\ 5 \end{pmatrix} \text{ and } \overline{v} = \begin{pmatrix} 1 \\ 0 \\ 9 \\ 11 \end{pmatrix}$$

using RStudio:

(a) $\overline{u} \cdot \overline{v}$

(b) $\overline{u} \cdot (3\overline{v})$

(c) $(-1.1\overline{v}) \cdot (4.3\overline{u})$

Working in RStudio

Vector projections are computed in RStudio using the `project` command. The inputs to `project` are the vectors $\overline{v}$ and $\overline{u}$ separated by the tilde symbol "$\sim$". For example, compute the projection of $\overline{v} = \begin{pmatrix} 1 \\ 5 \end{pmatrix}$ onto $\overline{u} = \begin{pmatrix} 1 \\ 3 \end{pmatrix}$, by first defining the two vectors with the commands `v=c(1,5)` and `u=c(1,3)`. Then enter the command `project(v~u)` to obtain an output of 1.6, which is the value of the scalar x for the projection $x\overline{u}$. This value can be stored in RStudio with the name x for future reference by entering `x=project(v~u)`, which then allows computation of the projection vector with the command `x*u`. The residual vector $\overline{r} = \overline{v} - x\overline{u}$ is obtained by entering the command `v-x*u`, provided x has already been computed and stored as above.

Examples of Commands

- `u=c(4,1,2)`

- `v=c(1,7,-2)`

- `x=project(v~u)`

- `x*u`

- `v-x*u`

◆ **EXAMPLE 9** Consider the vectors

$$\overline{v} = \begin{pmatrix} 1 \\ -3 \\ 5 \end{pmatrix} \text{ and } \overline{u} = \begin{pmatrix} -7 \\ 4 \\ -2 \end{pmatrix}$$

Compute both the projection of $\overline{v}$ onto $\overline{u}$ and the corresponding residual vector $\overline{r} = \overline{v} - x\overline{u}$ using RStudio.

Solution.

```
u = c(-7,4,-2)
v = c(1, -3, 5)
project(v~u)

        u
-0.42029

x = project(v~u)
```

```
x*u

[1]   2.94203 -1.68116  0.84058

v-x*u

[1]  -1.94203 -1.31884  4.15942
```

This RStudio output provides the projection $x\overline{u}$ of $\overline{v}$ onto $\overline{u}$ and its residual vector $\overline{r}$:

$$x\overline{u} = \begin{pmatrix} 2.9420 \\ -1.6812 \\ 0.8406 \end{pmatrix} \qquad \overline{r} = \begin{pmatrix} -1.9420 \\ -1.3188 \\ 4.1594 \end{pmatrix}$$

■

➤ **QUESTION 10** Consider the vectors

$$\bar{v} = \begin{pmatrix} 0 \\ 7 \\ 8 \end{pmatrix} \text{ and } \bar{u} = \begin{pmatrix} 1 \\ 2 \\ 4 \end{pmatrix}$$

Use RStudio to compute both the projection of $\bar{v}$ onto $\bar{u}$ and the residual $\bar{r} = \bar{v} - x\bar{u}$.

Summary

- The *dot product* of two vectors $\bar{u}$ and $\bar{v}$ is $\bar{u} \cdot \bar{v} = u_1 v_1 + u_2 v_2 + \cdots + u_n v_n$.

- The dot product is *commutative* because $\bar{u} \cdot \bar{v} = \bar{v} \cdot \bar{u}$ for all vectors $\bar{u}$ and $\bar{v}$.

- Scalar multiplication *commutes* with the dot product because

$$(m\bar{u}) \cdot \bar{v} = \bar{u} \cdot (m\bar{v}) = m(\bar{u} \cdot \bar{v})$$

  for all vectors $\bar{u}$ and $\bar{v}$, and scalars m.

- The *length* of a vector $\bar{u}$ is $\|\bar{u}\| = \sqrt{\bar{u} \cdot \bar{u}} = \sqrt{\bar{u}^2}$, where $\bar{u}^2 = \bar{u} \cdot \bar{u}$.

- Two vectors $\bar{u}$ and $\bar{v}$ are *orthogonal* when $\bar{u} \cdot \bar{v} = 0$. "Orthogonal" is interpreted as an extension of the two-dimensional idea of "perpendicular" to any number of dimensions.

- For vectors $\bar{v}$ and $\bar{u}$, the *residual vector* corresponding to the linear combination $x\bar{u}$ and target vector $\bar{v}$ is $\bar{r} = \bar{v} - x\bar{u}$.

- The *projection* of a vector $\bar{v}$ onto a nonzero vector $\bar{u}$ is the vector $x\bar{u}$ where

$$x = \frac{\bar{v} \cdot \bar{u}}{\bar{u} \cdot \bar{u}}$$

  When $\bar{u}$ is the zero vector, the projection is $\bar{0}$.

Exercises

In Exercises 1–6, compute the dot product of the pair of vectors, and state whether they are or are not orthogonal.

1. $\begin{pmatrix} 1 \\ 0 \end{pmatrix}, \begin{pmatrix} 0 \\ -1 \end{pmatrix}$

2. $\begin{pmatrix} 3 \\ 2 \end{pmatrix}, \begin{pmatrix} -2 \\ 5 \end{pmatrix}$

3. $\begin{pmatrix} 4 \\ 1 \end{pmatrix}, \begin{pmatrix} 3 \\ 12 \end{pmatrix}$

4. $\begin{pmatrix} -1 \\ -1 \end{pmatrix}, \begin{pmatrix} 4 \\ 5 \end{pmatrix}$

5. $\begin{pmatrix} -1 \\ 1 \end{pmatrix}, \begin{pmatrix} 5 \\ 5 \end{pmatrix}$

6. $\begin{pmatrix} 6 \\ 3 \end{pmatrix}, \begin{pmatrix} 2 \\ 4 \end{pmatrix}$

In Exercises 7–12, compute the dot product of the pair of vectors, and state whether they are or are not orthogonal.

7. $\begin{pmatrix} 3 \\ 2 \\ 1 \end{pmatrix}, \begin{pmatrix} -2 \\ 5 \\ -4 \end{pmatrix}$

8. $\begin{pmatrix} 4 \\ 1 \\ 5 \end{pmatrix}, \begin{pmatrix} 3 \\ 12 \\ -7 \end{pmatrix}$

9. $\begin{pmatrix} -1 \\ -1 \\ -1 \end{pmatrix}, \begin{pmatrix} 4 \\ 5 \\ 9 \end{pmatrix}$

10. $\begin{pmatrix} -1 \\ 1 \\ -1 \end{pmatrix}, \begin{pmatrix} 5 \\ 5 \\ 0 \end{pmatrix}$

11. $\begin{pmatrix} 1 \\ 0 \\ 6 \\ 7 \end{pmatrix}, \begin{pmatrix} 0 \\ -1 \\ -7 \\ 6 \end{pmatrix}$

12. $\begin{pmatrix} 6 \\ 3 \\ 1 \\ 9 \end{pmatrix}, \begin{pmatrix} -2 \\ 4 \\ -9 \\ 1 \end{pmatrix}$

In Exercises 13–18, use the dot product to compute the length of the vector.

13. $\begin{pmatrix} 3 \\ -4 \end{pmatrix}$ 16. $\begin{pmatrix} -1 \\ -1 \end{pmatrix}$

14. $\begin{pmatrix} -2 \\ 5 \end{pmatrix}$ 17. $\begin{pmatrix} 0 \\ 4 \end{pmatrix}$

15. $\begin{pmatrix} 3 \\ 12 \end{pmatrix}$ 18. $\begin{pmatrix} 1 \\ 1 \end{pmatrix}$

In Exercises 19–24, use the dot product to compute the length of the vector.

19. $\begin{pmatrix} 3 \\ 2 \\ 1 \end{pmatrix}$ 22. $\begin{pmatrix} 3 \\ 12 \\ -7 \end{pmatrix}$

20. $\begin{pmatrix} -2 \\ 5 \\ -4 \end{pmatrix}$ 23. $\begin{pmatrix} 1 \\ 0 \\ 6 \\ 2 \end{pmatrix}$

21. $\begin{pmatrix} 4 \\ 1 \\ 5 \end{pmatrix}$ 24. $\begin{pmatrix} 0 \\ -1 \\ -7 \\ 3 \end{pmatrix}$

In Exercises 25–30, graphically determine if the pair of vectors is or is not orthogonal.

25. $\begin{pmatrix} 1 \\ 0 \end{pmatrix}, \begin{pmatrix} 0 \\ -1 \end{pmatrix}$ 28. $\begin{pmatrix} -1 \\ -1 \end{pmatrix}, \begin{pmatrix} 4 \\ 5 \end{pmatrix}$

26. $\begin{pmatrix} 3 \\ 2 \end{pmatrix}, \begin{pmatrix} -2 \\ 5 \end{pmatrix}$ 29. $\begin{pmatrix} -1 \\ 1 \end{pmatrix}, \begin{pmatrix} 5 \\ 5 \end{pmatrix}$

27. $\begin{pmatrix} 4 \\ 1 \end{pmatrix}, \begin{pmatrix} 3 \\ 12 \end{pmatrix}$ 30. $\begin{pmatrix} 6 \\ 3 \end{pmatrix}, \begin{pmatrix} 2 \\ 4 \end{pmatrix}$

In Exercises 31–38, compute the residual $\bar{r} = \bar{v} - x\bar{u}$ for vectors $\bar{v}$ and $\bar{u}$, and scalar x.

31. $\bar{v} = \begin{pmatrix} 6 \\ 9 \end{pmatrix}, \bar{u} = \begin{pmatrix} 4 \\ 5 \end{pmatrix}$, and $x = 2$

32. $\bar{v} = \begin{pmatrix} 2 \\ 1 \end{pmatrix}, \bar{u} = \begin{pmatrix} 3 \\ 4 \end{pmatrix}$, and $x = \dfrac{1}{2}$

33. $\bar{v} = \begin{pmatrix} 4 \\ 2 \end{pmatrix}, \bar{u} = \begin{pmatrix} 2 \\ 4 \end{pmatrix}$, and $x = 2$

34. $\bar{v} = \begin{pmatrix} 0 \\ 4 \end{pmatrix}, \bar{u} = \begin{pmatrix} 2 \\ 2 \end{pmatrix}$, and $x = 2$

35. $\bar{v} = \begin{pmatrix} 1 \\ 1 \\ 1 \end{pmatrix}, \bar{u} = \begin{pmatrix} 0.5 \\ 0.5 \\ 0 \end{pmatrix}$, and $x = 3$

36. $\bar{v} = \begin{pmatrix} -3 \\ 4 \\ 1 \end{pmatrix}, \bar{u} = \begin{pmatrix} 6 \\ -2 \\ 4 \end{pmatrix}$, and $x = \dfrac{1}{2}$

37. $\bar{v} = \begin{pmatrix} 1 \\ 2 \\ -3 \\ 0 \end{pmatrix}, \bar{u} = \begin{pmatrix} 3 \\ 6 \\ -9 \\ 0 \end{pmatrix}$, and $x = \dfrac{1}{3}$

38. $\bar{v} = \begin{pmatrix} -2 \\ 3 \\ 4 \\ 1 \end{pmatrix}, \bar{u} = \begin{pmatrix} 0.5 \\ 1.5 \\ 1 \\ 0.1 \end{pmatrix}$, and $x = 4$

In Exercises 39–44, let $\bar{v} = \begin{pmatrix} 7 \\ 5 \end{pmatrix}$ and compute both the projection $x\bar{u}$ of $\bar{v}$ onto $\bar{u}$ and the corresponding residual vector $\bar{r} = \bar{v} - x\bar{u}$.

39. $\bar{u} = \begin{pmatrix} 1 \\ 0 \end{pmatrix}$

42. $\bar{u} = \begin{pmatrix} -1 \\ 0 \end{pmatrix}$

40. $\bar{u} = \begin{pmatrix} 0 \\ 1 \end{pmatrix}$

43. $\bar{u} = \begin{pmatrix} 3 \\ 2 \end{pmatrix}$

41. $\bar{u} = \begin{pmatrix} 1 \\ 1 \end{pmatrix}$

44. $\bar{u} = \begin{pmatrix} -2 \\ 5 \end{pmatrix}$

In Exercises 45–50, let

$$\bar{v} = \begin{pmatrix} -3 \\ 5 \\ 4 \end{pmatrix}$$

and compute both the projection $x\bar{u}$ of $\bar{v}$ onto $\bar{u}$ and the corresponding residual vector $\bar{r} = \bar{v} - x\bar{u}$.

45. $\bar{u} = \begin{pmatrix} 1 \\ 0 \\ 0 \end{pmatrix}$

48. $\bar{u} = \begin{pmatrix} -1 \\ 1 \\ 1 \end{pmatrix}$

46. $\bar{u} = \begin{pmatrix} 0 \\ 0 \\ 1 \end{pmatrix}$

49. $\bar{u} = \begin{pmatrix} 1 \\ 3 \\ 5 \end{pmatrix}$

47. $\bar{u} = \begin{pmatrix} 1 \\ 1 \\ 1 \end{pmatrix}$

50. $\bar{u} = \begin{pmatrix} -2 \\ 5 \\ -3 \end{pmatrix}$

In Exercises 51–54, and compute both the projection $x\bar{u}$ of $\bar{v}$ onto $\bar{u}$ and the corresponding residual vector $\bar{r} = \bar{v} - x\bar{u}$.

51. $\bar{v} = \begin{pmatrix} 4 \\ -5 \\ 2 \end{pmatrix}$ and $\bar{u} = \begin{pmatrix} 0 \\ 1 \\ 1 \end{pmatrix}$

52. $\bar{v} = \begin{pmatrix} 0.5 \\ -0.5 \\ -0.5 \end{pmatrix}$ and $\bar{u} = \begin{pmatrix} 4 \\ -6 \\ 8 \end{pmatrix}$

53. $\bar{v} = \begin{pmatrix} 3 \\ 1 \\ 5 \\ 3 \end{pmatrix}$ and $\bar{u} = \begin{pmatrix} 3 \\ 1 \\ 2 \\ 1 \end{pmatrix}$

54. $\bar{v} = \begin{pmatrix} 5 \\ -3 \\ 1 \\ 0 \end{pmatrix}$ and $\bar{u} = \begin{pmatrix} -2 \\ 5 \\ -3 \\ 1 \end{pmatrix}$

In Exercises 55–58, and compute both the projection $x\bar{u}$ of $\bar{v}$ onto $\bar{u}$ and the corresponding residual vector $\bar{r} = \bar{v} - x\bar{u}$. Also, graph $\bar{u}$, $\bar{v}$, $x\bar{u}$, and $\bar{r}$, with $\bar{r}$ starting from the tip of $x\bar{u}$.

55. $\bar{v} = \begin{pmatrix} 1 \\ 2 \end{pmatrix}$ and $\bar{u} = \begin{pmatrix} 3 \\ 1 \end{pmatrix}$

56. $\bar{v} = \begin{pmatrix} 4 \\ 6 \end{pmatrix}$ and $\bar{u} = \begin{pmatrix} 2 \\ 2 \end{pmatrix}$

57. $\bar{v} = \begin{pmatrix} 1 \\ 3 \end{pmatrix}$ and $\bar{u} = \begin{pmatrix} -2 \\ 5 \end{pmatrix}$

58. $\bar{v} = \begin{pmatrix} 1 \\ 1 \end{pmatrix}$ and $\bar{u} = \begin{pmatrix} -1 \\ 1 \end{pmatrix}$

Your Turn. In Exercises 59–62, choose your own two nonzero vectors $\bar{v}$ and $\bar{u}$ with the given number of components and, if possible, compute both the projection $x\bar{u}$ of $\bar{v}$ onto $\bar{u}$ and the corresponding residual vector $\bar{r} = \bar{v} - x\bar{u}$.

59. $\bar{v}$ that is 2×1 and $\bar{u}$ that is 2×1

60. $\bar{v}$ that is 1×2 and $\bar{u}$ that is 2×1

61. $\bar{v}$ that is 3×1 and $\bar{u}$ that is 3×1

62. $\bar{v}$ that is 2×1 and $\bar{u}$ that is 3×1

In Exercises 63–68, let

$$\bar{v} = \begin{pmatrix} 1 \\ 3 \end{pmatrix} \text{ and } \bar{u} = \begin{pmatrix} 2 \\ 2 \end{pmatrix}$$

Compute the residual vector $\bar{r} = \bar{v} - x\bar{u}$ and the length of $\bar{r}$.

63. $x = -1$

66. $x = 1$

64. $x = 0$

67. $x = 2$

65. $x = 0.5$

68. $x = 4$

69. Based on Exercises 63–68, explain why $x = 0.5$ and $x = 2$ cannot provide the projection $x\bar{u}$ of $\bar{v}$ onto $\bar{u}$.

In Exercises 70–76, let

$$\overline{v} = \begin{pmatrix} 1 \\ 1 \end{pmatrix} \text{ and } \overline{u} = \begin{pmatrix} 2 \\ 6 \end{pmatrix}$$

70. Find the projection $x\overline{u}$ of $\overline{v}$ onto $\overline{u}$.

71. Find the corresponding residual vector $\overline{r}$.

72. Find the length of the residual vector $\overline{r}$.

73. Compute $\overline{r}_{-2} = \overline{v} - (-2)\overline{u}$ and $\|\overline{r}_{-2}\|$.

74. Compute $\overline{r}_1 = \overline{v} - \overline{u}$ and $\|\overline{r}_1\|$.

75. Compute $\overline{r}_4 = \overline{v} - 4\overline{u}$ and $\|\overline{r}_4\|$.

76. Based on Exercises 72–75, explain why $x = -2$, $x = 1$, and $x = 4$ cannot provide the projection $x\overline{u}$ of $\overline{v}$ onto $\overline{u}$.

In Exercises 77–83, let

$$\overline{v} = \begin{pmatrix} 2 \\ 1 \\ 1 \end{pmatrix} \text{ and } \overline{u} = \begin{pmatrix} 3 \\ 4 \\ 5 \end{pmatrix}$$

77. Find the projection $x\overline{u}$ of $\overline{v}$ onto $\overline{u}$.

78. Find the corresponding residual vector $\overline{r}$.

79. Find the length of the residual vector $\overline{r}$.

80. Compute $\overline{r}_{-2} = \overline{v} - (-2)\overline{u}$ and $\|\overline{r}_{-2}\|$.

81. Compute $\overline{r}_1 = \overline{v} - \overline{u}$ and $\|\overline{r}_1\|$.

82. Compute $\overline{r}_4 = \overline{v} - 4\overline{u}$ and $\|\overline{r}_4\|$.

83. Based on Exercises 79–82, explain why $x = -2$, $x = 1$, and $x = 4$ cannot provide the projection $x\overline{u}$ of $\overline{v}$ onto $\overline{u}$.

In Exercises 84–90, let

$$\overline{v} = \begin{pmatrix} -2 \\ 4 \\ 10 \\ 5 \end{pmatrix} \text{ and } \overline{u} = \begin{pmatrix} -2 \\ 1 \\ -1 \\ 2 \end{pmatrix}$$

84. Find the projection $x\overline{u}$ of $\overline{v}$ onto $\overline{u}$.

85. Find the corresponding residual vector $\overline{r}$.

86. Find the length of the residual vector $\overline{r}$.

87. Compute $\overline{r}_{-2} = \overline{v} - (-2)\overline{u}$ and $\|\overline{r}_{-2}\|$.

88. Compute $\overline{r}_1 = \overline{v} - \overline{u}$ and $\|\overline{r}_1\|$.

89. Compute $\overline{r}_4 = \overline{v} - 4\overline{u}$ and $\|\overline{r}_4\|$.

90. Based on Exercises 86–89, explain why $x = -2$, $x = 1$, and $x = 4$ cannot provide the projection $x\overline{u}$ of $\overline{v}$ onto $\overline{u}$.

In Exercises 91–96, verify that every pair (x, y) in the data set lies on the same line through the origin. Also, confirm that the slope of this line is equal to the scalar $\dfrac{\overline{y} \cdot \overline{x}}{\overline{x} \cdot \overline{x}}$.

91.

x	1	2
y	3	6

92.

x	2	4
y	-1	-2

93.

x	1	3	5
y	5	15	25

94.

x	-5	0	2
y	10	0	-4

95.

x	-3	2	3	5
y	-1.5	1	1.5	2.5

96.

x	-4	-2	5	8
y	-2.4	-1.2	3	4.8

RStudio. In Exercises 97–102, use RStudio to compute the dot product of the pair of vectors.

97. $\overline{u} = \begin{pmatrix} 7 \\ 6 \end{pmatrix}$ and $\overline{v} = \begin{pmatrix} 1 \\ -4 \end{pmatrix}$

98. $\overline{u} = \begin{pmatrix} 1 \\ 2 \end{pmatrix}$ and $\overline{v} = \begin{pmatrix} -6 \\ -12 \end{pmatrix}$

99. $\overline{u} = \begin{pmatrix} 2 \\ 3 \\ 4 \end{pmatrix}$ and $\overline{v} = \begin{pmatrix} -2 \\ 2 \\ -0.5 \end{pmatrix}$

100. $\overline{u} = \begin{pmatrix} 6 \\ -2 \\ 1 \end{pmatrix}$ and $\overline{v} = \begin{pmatrix} 9 \\ 8 \\ 0 \end{pmatrix}$

101. $\overline{u} = \begin{pmatrix} 1 \\ 2 \\ 3 \\ 4 \end{pmatrix}$ and $\overline{v} = \begin{pmatrix} -2 \\ 0 \\ 1 \\ 5 \end{pmatrix}$

102. $\overline{u} = \begin{pmatrix} -8 \\ 7 \\ 1 \\ 0 \\ 2 \end{pmatrix}$ and $\overline{v} = \begin{pmatrix} -2 \\ 0 \\ 2 \\ 5 \\ -9 \end{pmatrix}$

RStudio. In Exercises 103–108, use RStudio to compute the projection $x\overline{u}$ of $\overline{v}$ onto $\overline{u}$ and the corresponding residual vector $\overline{r} = \overline{v} - x\overline{u}$.

103. $\overline{u} = \begin{pmatrix} 3 \\ 4 \end{pmatrix}$ and $\overline{v} = \begin{pmatrix} -2 \\ 5 \end{pmatrix}$

104. $\overline{u} = \begin{pmatrix} 0 \\ 7 \end{pmatrix}$ and $\overline{v} = \begin{pmatrix} 1 \\ -1 \end{pmatrix}$

105. $\overline{u} = \begin{pmatrix} 0 \\ 3 \\ -5 \end{pmatrix}$ and $\overline{v} = \begin{pmatrix} 12 \\ 1 \\ 3 \end{pmatrix}$

106. $\overline{u} = \begin{pmatrix} 1 \\ -2 \\ 3 \end{pmatrix}$ and $\overline{v} = \begin{pmatrix} -7 \\ 5 \\ 1 \end{pmatrix}$

107. $\overline{u} = \begin{pmatrix} -1 \\ 5 \\ -2 \\ 3 \end{pmatrix}$ and $\overline{v} = \begin{pmatrix} 5 \\ 1 \\ 1 \\ -3 \end{pmatrix}$

108. $\overline{u} = \begin{pmatrix} -4 \\ 1 \\ 6 \\ 1 \\ 2 \end{pmatrix}$ and $\overline{v} = \begin{pmatrix} 7 \\ -2 \\ 9 \\ 0 \\ -1 \end{pmatrix}$

In Your Own Words. In Exercises 109–115, explain the following.

109. The dot product of two vectors $\overline{u}$ and $\overline{v}$

110. The connection between the length of a vector and the dot product

111. Vectors $\overline{u}$ and $\overline{v}$ are orthogonal

112. Residual vectors

113. Geometry of vector projection

114. Computing vector projection

115. The key property of the residual vector for vector projections

3.5   The Method of Least Squares

This chapter develops the mathematical framework needed for the *method of least squares*. The goal of this method is to determine the parameters m and b of the best linear model $y = mx + b$ for a given data set. In addition, least squares can be applied to many nonlinear models as well as to multivariable data sets. This section focuses on applying the method of least squares to data sets for which a linear model is reasonable.

Example 1 in Section 3.2 and Example 4 in Section 3.3 examined the following data set, which presents how many millions of people used Twitter during each quarter of a year since 2000, where 11 represents the first quarter of 2011 (January – March 2011), 11.25 represents the second quarter of 2011 (April – June 2011), and so on.

Year (Y)	11	11.25	11.5	11.75	12.75
Twitter users (U)	68	85	101	117	185

If this data set were perfectly linear, then the following vector equation $\overline{U} = m\overline{Y} + b\overline{1}_5$ would have a unique solution:

$$\begin{pmatrix} 68 \\ 85 \\ 101 \\ 117 \\ 185 \end{pmatrix} = m \begin{pmatrix} 11 \\ 11.25 \\ 11.5 \\ 11.75 \\ 12.75 \end{pmatrix} + b \begin{pmatrix} 1 \\ 1 \\ 1 \\ 1 \\ 1 \end{pmatrix}$$

However, Example 4 of Section 3.3 showed that this vector equation has no solution. In more detail, the first two data points have a linear model of $U = 68 \cdot Y - 680$. However, substituting the third data point $(11.5, 101)$ gives $U = 68 \cdot 11.5 - 680 = 102$, which is not equal to 101. This lack of equality between the value from the data set and the modeled value means that there does not exist an exact linear model of Twitter users U as a function of the quarter of the year Y. Therefore, adopting the approach of vector projections from Section 3.4, our focus shifts to finding the linear combination

$$m\overline{Y} + b\overline{1}_5 = m \begin{pmatrix} 11 \\ 11.25 \\ 11.5 \\ 11.75 \\ 12.75 \end{pmatrix} + b \begin{pmatrix} 1 \\ 1 \\ 1 \\ 1 \\ 1 \end{pmatrix}$$

that is as close as possible to the target vector

$$\overline{U} = \begin{pmatrix} 68 \\ 85 \\ 101 \\ 117 \\ 185 \end{pmatrix}$$

Applying the Method of Least Squares

Suppose that for a given set of n data points $(x_1, y_1), \dots, (x_n, y_n)$, a linear model $y = mx + b$ is known to be reasonable. If every data point lies on the same line $y = mx + b$, then the corresponding vector equation $\overline{y} = m\overline{x} + b\overline{1}_n$ is valid. However, in most settings, not every given data point will lie exactly on any line $y = mx + b$, and so the modeling goal is adjusted to finding the values of m and b for the line $y = mx + b$ that is "as close as possible" to the given data points.

Expressing these ideas in terms of vectors, a data set that does not lie perfectly on any line $y = mx + b$ corresponds to the fact that no linear combination $m\overline{x} + b\overline{1}_n$ is equal to the target vector $\overline{y}$. Consequently, our goal becomes that of finding the linear combination $m\overline{x} + b\overline{1}_n$ that is as close as possible to $\overline{y}$. The measure of closeness is the length of the residual vector $\bar{r} = \bar{y} - (m\bar{x} + b\bar{1}_n)$, which means that the goal is for the length of the residual vector to be as short as possible. The method of least squares enables us to find m and b for such a best possible linear combination $m\bar{x} + b\bar{1}_n$. As discussed in more detail soon, least squares is an extension of the notion of vector projection to linear combinations involving more than one vector.

METHOD OF LEAST SQUARES.
The best linear model $y = mx + b$ for a data set with n points $(x_1, y_1), \ldots, (x_n, y_n)$, stored in a vector $\bar{x}$ of input values and a vector $\bar{y}$ of output values, is given by the solution to the following matrix equation with unknowns m and b:

$$\begin{pmatrix} \bar{y} \cdot \bar{x} \\ \bar{y} \cdot \bar{1}_n \end{pmatrix} = \begin{pmatrix} \bar{x} \cdot \bar{x} & \bar{1}_n \cdot \bar{x} \\ \bar{x} \cdot \bar{1}_n & \bar{1}_n \cdot \bar{1}_n \end{pmatrix} \begin{pmatrix} m \\ b \end{pmatrix}$$

Recall that the dot product operation always outputs a scalar, providing a 2×1 vector of numbers on the left side of the least squares matrix equation and a 2×2 matrix of numbers on the right side. This section first examines applications of the method of least squares to some specific data sets and then explores the reasons why least squares works.

◆ **EXAMPLE 1** Use the method of least squares to find the best possible linear model $y = mx + b$ for the data set below. Also, graph this data set and its least squares model on the same axes, and discuss the model's goodness of fit.

x	-1	2	3
y	-1	9	11

Solution. First, identify the specific vectors $\bar{x}$, $\bar{y}$, and $\bar{1}_3$ used by the method of least squares in this setting. The data set contains three points, which gives $n = 3$ and the following three vectors:

$$\bar{x} = \begin{pmatrix} -1 \\ 2 \\ 3 \end{pmatrix} \qquad \bar{y} = \begin{pmatrix} -1 \\ 9 \\ 11 \end{pmatrix} \qquad \bar{1}_3 = \begin{pmatrix} 1 \\ 1 \\ 1 \end{pmatrix}$$

Next, compute the values of the five dot products that appear in the least squares matrix equation. In this setting, the commutativity of the dot products $\bar{x} \cdot \bar{1}_3 = \bar{1}_3 \cdot \bar{x}$ is used to determine the numeric value in the first component of the second column of the matrix.

$$\bar{y} \cdot \bar{x} = \begin{pmatrix} -1 \\ 9 \\ 11 \end{pmatrix} \cdot \begin{pmatrix} -1 \\ 2 \\ 3 \end{pmatrix} = 1 + 18 + 33 = 52 \qquad \bar{x} \cdot \bar{1}_3 = \begin{pmatrix} -1 \\ 2 \\ 3 \end{pmatrix} \cdot \begin{pmatrix} 1 \\ 1 \\ 1 \end{pmatrix} = (-1) + 2 + 3 = 4$$

$$\bar{y} \cdot \bar{1}_3 = \begin{pmatrix} -1 \\ 9 \\ 11 \end{pmatrix} \cdot \begin{pmatrix} 1 \\ 1 \\ 1 \end{pmatrix} = (-1) + 9 + 11 = 19 \qquad \bar{1}_3 \cdot \bar{1}_3 = \begin{pmatrix} 1 \\ 1 \\ 1 \end{pmatrix} \cdot \begin{pmatrix} 1 \\ 1 \\ 1 \end{pmatrix} = 1 + 1 + 1 = 3$$

$$\bar{x} \cdot \bar{x} = \begin{pmatrix} -1 \\ 2 \\ 3 \end{pmatrix} \cdot \begin{pmatrix} -1 \\ 2 \\ 3 \end{pmatrix} = 1 + 4 + 9 = 14$$

Substitute these values into the least squares matrix equation as follows.

$$\begin{pmatrix} 52 \\ 19 \end{pmatrix} = \begin{pmatrix} 14 & 4 \\ 4 & 3 \end{pmatrix} \begin{pmatrix} m \\ b \end{pmatrix}$$

Now use the formula for solving 2×2 matrix equations:

$$\begin{aligned}
\begin{pmatrix} m \\ b \end{pmatrix} &= \frac{1}{(14)(3) - (4)(4)} \begin{pmatrix} 3 & -4 \\ -4 & 14 \end{pmatrix} \begin{pmatrix} 52 \\ 19 \end{pmatrix} && \text{Formula for solving } 2 \times 2 \text{ matrix equations} \\
&= \frac{1}{26} \left[52 \begin{pmatrix} 3 \\ -4 \end{pmatrix} + 19 \begin{pmatrix} -4 \\ 14 \end{pmatrix} \right] && \text{Matrix–vector multiplication} \\
&= \frac{1}{26} \left[\begin{pmatrix} 156 \\ -208 \end{pmatrix} + \begin{pmatrix} -76 \\ 266 \end{pmatrix} \right] && \text{Scalar multiply by 52 and 19} \\
&= \frac{1}{26} \begin{pmatrix} 80 \\ 58 \end{pmatrix} && \text{Vector addition} \\
&= \begin{pmatrix} 80/26 \\ 58/26 \end{pmatrix} \approx \begin{pmatrix} 3.077 \\ 2.231 \end{pmatrix} && \text{Scalar multiply by } \frac{1}{26}
\end{aligned}$$

Thus, the least squares solution is $m = 80/26 \approx 3.077$ and $b = 58/26 \approx 2.231$, which means that the best linear model for this data set is approximately $y = 3.077x + 2.231$. Figure 1 provides the graph of this least squares linear model on a plot of the data and affirms its quality of fit. Namely, the model expresses both the general trend of the data and pretty closely approximates the actual values of the given data points.

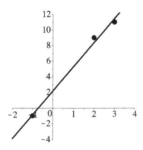

Figure 1: Data set and least squares model for Example 1

◆ **EXAMPLE 2** The table below presents how many millions of people used Twitter during each quarter of a year since 2000, where 11 represents the first quarter of 2011 (January – March 2011), 11.25 represents the second quarter of 2011 (April – June 2011), and so on. Use the method of least squares to find the best possible linear model $U = mY + b$ for this data set. Also, graph the data set and its least squares model on the same axes, and discuss the model's goodness of fit.

Year (Y)	11	11.25	11.5	11.75	12.75
Twitter users (U)	68	85	101	117	185

Solution. First, identify the corresponding vectors used by the method of least squares in this setting. The data set contains five points, which gives $n = 5$ and the following three vectors:

$$\overline{Y} = \begin{pmatrix} 11 \\ 11.25 \\ 11.5 \\ 11.75 \\ 12.75 \end{pmatrix} \qquad \overline{U} = \begin{pmatrix} 68 \\ 85 \\ 101 \\ 117 \\ 185 \end{pmatrix} \qquad \overline{1}_5 = \begin{pmatrix} 1 \\ 1 \\ 1 \\ 1 \\ 1 \end{pmatrix}$$

Next, compute the values of the five dot products that appear in the least squares matrix equation, again, based on the commutativity of the dot products $\overline{Y} \cdot \overline{1}_5 = \overline{1}_5 \cdot \overline{Y}$:

$$\overline{U} \cdot \overline{Y} = \begin{pmatrix} 11 \\ 11.25 \\ 11.5 \\ 11.75 \\ 12.75 \end{pmatrix} \cdot \begin{pmatrix} 68 \\ 85 \\ 101 \\ 117 \\ 185 \end{pmatrix} = 6599.25 \qquad \overline{Y} \cdot \overline{1}_5 = \begin{pmatrix} 11 \\ 11.25 \\ 11.5 \\ 11.75 \\ 12.75 \end{pmatrix} \cdot \begin{pmatrix} 1 \\ 1 \\ 1 \\ 1 \\ 1 \end{pmatrix} = 58.25$$

$$\overline{U} \cdot \overline{1}_5 = \begin{pmatrix} 68 \\ 85 \\ 101 \\ 117 \\ 185 \end{pmatrix} \cdot \begin{pmatrix} 1 \\ 1 \\ 1 \\ 1 \\ 1 \end{pmatrix} = 556 \qquad \overline{1}_5 \cdot \overline{1}_5 = \begin{pmatrix} 1 \\ 1 \\ 1 \\ 1 \\ 1 \end{pmatrix} \cdot \begin{pmatrix} 1 \\ 1 \\ 1 \\ 1 \\ 1 \end{pmatrix} = 5$$

$$\overline{Y} \cdot \overline{Y} = \begin{pmatrix} 11 \\ 11.25 \\ 11.5 \\ 11.75 \\ 12.75 \end{pmatrix} \cdot \begin{pmatrix} 11 \\ 11.25 \\ 11.5 \\ 11.75 \\ 12.75 \end{pmatrix} = 680.4375$$

Substitute these values into the least squares matrix equation as follows:

$$\begin{pmatrix} 6599.25 \\ 556 \end{pmatrix} = \begin{pmatrix} 680.4375 & 58.25 \\ 58.25 & 5 \end{pmatrix} \begin{pmatrix} m \\ b \end{pmatrix}$$

Now use the formula for solving 2×2 matrix equations, beginning with the leading scalar:

$$\frac{1}{ad - bc} = \frac{1}{(680.4375)(5) - (58.25)(58.25)} = \frac{1}{9.125}$$

Substituting into the complete formula gives the following:

$$\begin{pmatrix} m \\ b \end{pmatrix} = \frac{1}{9.125} \begin{pmatrix} 5 & -58.25 \\ -58.25 & 680.4375 \end{pmatrix} \begin{pmatrix} 6599.25 \\ 556 \end{pmatrix}$$ Formula for solving matrix equation

$$= \frac{1}{9.125} \left[6599.25 \begin{pmatrix} 5 \\ -58.25 \end{pmatrix} + 556 \begin{pmatrix} -58.25 \\ 680.4375 \end{pmatrix} \right]$$ Matrix–vector multiplication

$$= \frac{1}{9.125} \left[\begin{pmatrix} 32996.25 \\ -384406.3 \end{pmatrix} + \begin{pmatrix} -32387 \\ 378323.2 \end{pmatrix} \right]$$ Scalar multiply by 6599.25 and 556

$$= \frac{1}{9.125} \begin{pmatrix} 609.25 \\ -6083.1 \end{pmatrix}$$ Vector addition

$$= \begin{pmatrix} 66.767 \\ 666.641 \end{pmatrix}$$ Scalar multiply by $\frac{1}{9.125}$

Thus, the least squares solution is $m = 66.767$ and $b = 666.641$, which means that the best linear model for this data set is $U = 66.767Y - 666.641$. Figure 2 shows the graph of this least squares linear model on a plot of the data and affirms its quality of fit. Namely, the model expresses both the general trend of the data and very closely approximates all the actual values of the given data points.

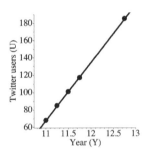

Figure 2: Data set and least squares model for Example 2

► **QUESTION 1** The table below presents the average debt load in thousands of dollars at the end of each year's spring term for bachelor's degree recipients who attended public four-year colleges and universities and borrowed money to finance their education. Use the method of least squares to find the best possible linear model $D = mY + b$ for this data set. Also, graph the data set and its least squares model on the same axes, and discuss the model's goodness of fit.

Academic year (Y)	2001	2003	2005	2006
Average debt (D)	$20,400	$20,900	$21,500	$21,800

In general, the method of least squares produces a line $y = mx + b$ that is as close as possible to all of the given data points. This fact holds independently of how good the fit actually is between the data and the linear model. In some cases, the fit may appear to be quite poor, but, even so, the least squares linear model remains the best possible line that can be used to model the given data set. Sometimes linear data sets are scattered, in which case, the fit might not be expected to be very good. In other cases, a poor fit could indicate that a linear model is not reasonable and that a different type of model should be considered.

The Residual Vector of Minimal Length

The method of least squares selects the linear combination $m\overline{x} + b\overline{1}_n$ that is closest to the target vector $\overline{y}$ in the sense that the corresponding residual vector $\overline{r} = \overline{y} - (m\overline{x} + b\overline{1}_n)$ has the shortest length possible. While the length of a residual is rarely calculated for its own sake, keep in mind this fact that the shortest possible length of a residual vector among all possible residuals of lines occurs for the least squares linear model. This idea is explored further in the context of our model for the number of Twitter users from Example 2, comparing the residual of the least squares model with the residual of the model identified in Chapter 2.

◆ **EXAMPLE 3** Example 2 used the method of least squares to identify the linear model $U_1 = 66.767 \cdot Y - 666.641$ for the number of Twitter users in a given year. Example 3 of Section 2.1 conjectured the model $U_2 = 66 \cdot Y - 657.7$ for this same data set. Compare the lengths of the residual vectors associated with these models.

Solution. First, the residual vectors are calculated for these two models, where $\overline{r}_1$ denotes the least squares residual and $\overline{r}_2$ denotes the residual for the linear model from Section 2.1:

$$\overline{r}_1 = \overline{U} - \overline{U}_1 = \overline{U} - (66.767 \cdot \overline{Y} - 666.641 \cdot \overline{1}_5)$$

$$= \begin{pmatrix} 68 \\ 85 \\ 101 \\ 117 \\ 185 \end{pmatrix} - \left[66.767 \begin{pmatrix} 11 \\ 11.25 \\ 11.5 \\ 11.75 \\ 12.75 \end{pmatrix} - 666.641 \begin{pmatrix} 1 \\ 1 \\ 1 \\ 1 \\ 1 \end{pmatrix} \right] = \begin{pmatrix} 0.204 \\ 0.512 \\ -0.180 \\ -0.871 \\ 0.362 \end{pmatrix}$$

$$\overline{r}_2 = \overline{U} - \overline{U}_2 = \overline{U} - (66 \cdot \overline{Y} - 657.7 \cdot \overline{1}_5)$$

$$= \begin{pmatrix} 68 \\ 85 \\ 101 \\ 117 \\ 185 \end{pmatrix} - \left[66 \begin{pmatrix} 11 \\ 11.25 \\ 11.5 \\ 11.75 \\ 12.75 \end{pmatrix} - 657.7 \begin{pmatrix} 1 \\ 1 \\ 1 \\ 1 \\ 1 \end{pmatrix} \right] = \begin{pmatrix} -0.3 \\ 0.2 \\ -0.3 \\ -0.8 \\ 1.2 \end{pmatrix}$$

Next, the lengths of these residual vectors are computed as follows:

$$\|\overline{r}_1\| = \sqrt{(0.204)^2 + (0.512)^2 + (-0.180)^2 + (-0.871)^2 + (0.362)^2} \approx 1.107$$
$$\|\overline{r}_2\| = \sqrt{(-0.3)^2 + (0.2)^2 + (-0.3)^2 + (-0.8)^2 + (1.2)^2} \approx 1.517$$

As can be seen, the length of the residual $\overline{r}_1$ associated with the method of least squares model $U_1 = 66.767 \cdot Y - 666.641$ is less than the length of the residual $\overline{r}_2$ associated with the model $U_2 = 66 \cdot Y - 657.7$ obtained using the method of Section 2.1. ■

Example 3 compared the least squares model with one particular linear model and confirmed the relatively smaller length of the least squares residual. The power of the method of least squares is that every such comparison has the exact same outcome for all possible linear combinations $U = m \cdot Y + b$

➤ **QUESTION 2** In Question 1, the method of least squares produces the linear model $D_1 = 281.36Y - 542616.95$ for the average debt load of certain bachelor's degree recipients. Question 3 of Section 2.1 conjectures a linear model for this same data set of $D_2 = 283.33Y - 546581.67$. Compare the lengths of the residual vectors associated with these models.

Understanding the Method of Least Squares

As mentioned earlier, the method of least squares extends the notion of vector projection to settings where the corresponding linear combinations involve more than one vector. The goal is to find the linear combination $m\bar{x} + b\bar{1}_n$ of the vectors $\bar{x}$ and $\bar{1}_n$ that is as close as possible to the target vector $\bar{y}$. With some computational facility using the method of least squares in hand, this discussion turns to examining why the method of least squares works.

Figure 3 illustrates the relationships among the various vectors $\bar{x}$, $\bar{1}_n$, $\bar{y}$, $m\bar{x} + b\bar{1}_n$, and $\bar{r}$. The rectangular box represents all possible linear combinations of the vectors $\bar{x}$ and $\bar{1}_n$. Note that both $\bar{x}$ and $\bar{1}_n$ are in this set of linear combinations, because $\bar{x} = 1 \cdot \bar{x} + 0 \cdot \bar{1}_n$ and $\bar{1}_n = 0 \cdot \bar{x} + 1 \cdot \bar{1}_n$.

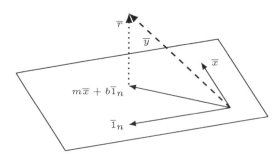

Figure 3: Graphical representation of the method of least squares

The method of least squares is applied when the given data points do not lie perfectly on any line $y = mx + b$. In terms of vectors, this means that the target output vector $\bar{y}$ cannot be written as a linear combination of the input vector $\bar{x}$ and the intercept vector $\bar{1}_n$. Notice that in Figure 3, the vector $\bar{y}$ lies outside the rectangular box, indicating that $\bar{y}$ is not a linear combination of $\bar{x}$ and $\bar{1}_n$.

As with vector projections, the linear combination $m\bar{x} + b\bar{1}_n$ closest to $\bar{y}$ occurs when the residual vector $\bar{r}$ is orthogonal to all linear combinations of $\bar{x}$ and $\bar{1}_n$. Figure 3 illustrates this orthogonal relationship between $\bar{r} = \bar{y} - (m\bar{x} + b\bar{1}_n)$ and $m\bar{x} + b\bar{1}_n$, where the residual vector $\bar{r}$ is depicted with a dotted line to indicate that $\bar{r}$ is not a linear combination of $\bar{x}$ and $\bar{1}_n$.

Translating this graphical representation of the method of least squares into its corresponding algebraic formulation provides the least squares matrix equation. First, solving the residual vector formula $\bar{r} = \bar{y} - (m\bar{x} + b\bar{1}_n)$ for $\bar{y}$ yields the following vector equation:

$$\bar{y} = m\bar{x} + b\bar{1}_n + \bar{r}$$

The goal of finding the values of the scalars m and b is accomplished by identifying two equations in these two unknowns. Working in this direction, take the dot product of both sides of this equation $\bar{y} = m\bar{x} + b\bar{1}_n + \bar{r}$ first with $\bar{x}$ and then with $\bar{1}_n$. Distributing these dot products across the vector additions and scalar multiplications on the left-hand side of the resulting equations to obtain the following:

$$\bar{y} \cdot \bar{x} = m(\bar{x} \cdot \bar{x}) + b(\bar{1}_n \cdot \bar{x}) + \bar{r} \cdot \bar{x}$$
$$\bar{y} \cdot \bar{1}_n = m(\bar{x} \cdot \bar{1}_n) + b(\bar{1}_n \cdot \bar{1}_n) + \bar{r} \cdot \bar{1}_n$$

As discussed with regard to the preceding graphical description of the method of least squares, the residual vector $\bar{r}$ is orthogonal to all linear combinations of $\bar{x}$ and $\bar{1}_n$, including these two vectors by themselves. In other words, $\bar{r}$ is orthogonal to both $\bar{x}$ and $\bar{1}_n$, which is expressed algebraically as $\bar{1}_n \cdot \bar{r} = 0$ and $\bar{x} \cdot \bar{r} = 0$. Substituting these zeros into the previous two equations results in the following system of linear equations for the method of least squares:

$$\bar{y} \cdot \bar{x} = m(\bar{x} \cdot \bar{x}) + b(\bar{1}_n \cdot \bar{x})$$
$$\bar{y} \cdot \bar{1}_n = m(\bar{x} \cdot \bar{1}_n) + b(\bar{1}_n \cdot \bar{1}_n)$$

Finally, this system of linear equations can be written as the least squares matrix equation

$$\begin{pmatrix} \bar{y} \cdot \bar{x} \\ \bar{y} \cdot \bar{1}_n \end{pmatrix} = \begin{pmatrix} \bar{x} \cdot \bar{x} & \bar{1}_n \cdot \bar{x} \\ \bar{x} \cdot \bar{1}_n & \bar{1}_n \cdot \bar{1}_n \end{pmatrix} \begin{pmatrix} m \\ b \end{pmatrix}$$

When applying the method of least squares, the vectors $\bar{x}$, $\bar{y}$, and $\bar{1}_n$ are known, which enables computation of all the various dot products appearing in this matrix equation. Section 3.2 explained how to solve such 2×2 matrix equations by hand or by using RStudio. Either approach produces the values of the two unknowns m and b and provides the best possible linear model $y = mx + b$ for the given data set.

Working in RStudio

Recall from Section 3.4 that the projection of a vector $\bar{v}$ onto a vector $\bar{u}$ is computed in RStudio using the command x=project(v~u). The output of this command is the scalar x with the property that the linear combination $x\bar{u}$ is as close as possible to $\bar{v}$, providing the best possible model of $\bar{v}$. The project command finds the best linear model for a given data set because the method of least squares is an extension of vector projection.

Least squares essentially seeks the projection of a target vector $\bar{y}$ onto the set of all linear combinations $m\bar{x} + b\bar{1}_n$ of the vectors $\bar{x}$ and $\bar{1}_n$. This task is accomplished by first storing the input vector $\bar{x}$ in RStudio by entering the command x=c(,) and the output vector $\bar{y}$ by entering y=c(,), both of which contain a list of the components of each vector separated by commas. For the intercept vector $\overline{1}_n$, RStudio accepts 1 in the `project` command without requiring explicit identification of the number of components; RStudio is programmed to automatically determine the appropriate number of components based on input and output vectors because the intercept is used so commonly in this command.

Once the input and output vectors have been defined, the desired least squares solution is obtained by entering the command `project(y~x+1)`. Notice that the two vectors x and 1 used in the linear combination appear to the right of the tilde "~", separated by the addition symbol "+".

Examples of Commands

- `x=c(-1,2,3)`

- `y=c(-1,9,11)`

- `project(y~x+1)`

◆ **EXAMPLE 4** The table below from Example 2 presents how many millions of people used Twitter during each quarter of a year since 2000, where 11 represents the first quarter of 2011 (January – March 2011), 11.25 represents the second quarter of 2011 (April – June 2011), and so on. Use RStudio to find the least squares model $U = mY + b$ for this data set. Also, graph the data set and its least squares model on the same axes, and discuss the model's goodness of fit.

Year (Y)	11	11.25	11.5	11.75	12.75
Twitter users (U)	68	85	101	117	185

Solution. First, enter the data into RStudio using `c(,)` and then apply the `project` command to find the unknown coefficients m and b of the least squares solution:

```
Y=c(11, 11.25, 11.5, 11.75, 12.75)        project(U~Y+1)
U=c(68, 85, 101, 117, 185)
                                          (Intercept)           Y
                                            -666.6370      66.7671
```

The number listed under (Intercept) identifies the value of $b = -666.637$ and the number listed under Y identifies the value of $m = 66.7671$ in the model $U = mY + b$. Therefore, the best linear model for this data set is $U = 66.7671Y - 666.637$. The slight difference between this RStudio least squares model and the least squares model $U = 66.767Y - 666.641$ from Example 2 is the result of rounding. Figure 4 presents the plot of the data set and the graph of the least squares model, and affirms the quality of fit of the latter. Namely, the model expresses both the general trend of the data and very closely approximates all the actual values of the given data points. ■

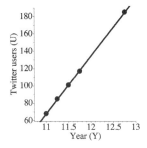

Figure 4: Data set and least squares model for Example 4

➤ **QUESTION 3** The table below from Question 1 presents the average debt load in thousands of dollars at the end of each year's spring term for bachelor's degree recipients who attended public four-year colleges and universities and borrowed money to finance their education. Use RStudio to find the least squares model $D = mY + b$ for this data set. Also, graph the data set and its least squares model on the same axes, and discuss the model's goodness of fit.

Academic year (Y)	2001	2003	2005	2006
Average debt (D)	$20,400	$20,900	$21,500	$21,800

Working in RStudio

Recall from Section 2.1 that data sets are plotted in RStudio by using the command **plotPoints** with required arguments of the name of the output variable, followed by the tilde symbol "~", and then the name of the input variable. When working with a larger data set "**data=**", the name of the data set" must also be included. The graph of a function is added to the existing plot of a data set with a **plotFun** command that includes the argument **add=TRUE**.

Examples of Commands

- `plotPoints(U~Y)`

- `plotFun(66.7671*Y-666.637~Y,add=TRUE)`

- `plotPoints(PercentGDP~Year,data=HealthExpenditure)`

An important advantage of using RStudio and the **project** command to determine the best linear model for a given data set is that this command works equally well for both small and large sets of data. The relative ease of working with a larger data set is illustrated in Example 5. If needed, the reader is referred to Section 2.1 for a review of the RStudio commands used to load and work with large data sets.

◆ **EXAMPLE 5** The data set stored in `HealthExpenditure` contains the World Bank's data for total U.S. health expenditures as a percentage of U.S. gross domestic product (or GDP) from 1995 to 2012. Use RStudio to find the best linear model for this data set. Also, graph the data set and this model on the same axes, and discuss the model's goodness of fit.

Solution. First, determine the names of the variables using `names` (or alternatively `head`). Next, use the `project` command to find the best linear model for this data set.

```
names(HealthExpenditure)                    project(PercentGDP~Year+1,
                                                    data=HealthExpenditure)
[1] "Year"          "PercentGDP"
                                            (Intercept)            Year
                                            -585.21396          0.29978
```

This RStudio output provides $b = -585.21396$ and $m = 0.29978$, which means that the best linear model for this data set is PercentGDP $= 0.29978 \cdot$ Year $- 585.21396$. Now use RStudio to plot the data set and this linear model on the same axes.

```
plotPoints(PercentGDP~Year,data=HealthExpenditure)
plotFun(0.29978*Year-585.21396~Year,add=TRUE)
```

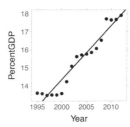

The graph of the model on a plot of the data indicates its relative goodness of fit in this setting where the data is somewhat linear. Because the data is not perfectly linear, a line cannot provide a perfect approximation of all the given data points. However, the model does provide a pretty reasonable approximation of the overall behavior of the data set.   ■

➤ **QUESTION 4** The data stored in `WeightChange` contains the change in a woman's weight during her pregnancy, from the beginning of her second trimester until birth. Use RStudio to find the best linear model for this set of data. Also, graph the data set and this model on the same axes, and discuss the model's goodness of fit.

On a final note, Example 5 in Section 2.1 used the RStudio command `fitModel` to identify the best linear model for the data set stored in `HealthExpenditure` as follows:

```
healthExpendModel=fitModel(PercentGDP~m*Year+b,data=HealthExpenditure)
coef(healthExpendModel)
```

```
       m          b
  0.29978 -585.21396
```

This RStudio output provides the linear model $\text{PercentGDP} = 0.29978 \cdot \text{Year} - 585.21396$, which is exactly the same as the model produced by the project command in Example 5. On the one hand, this agreement between our two methods for finding the best linear model should be encouraging. On the other hand, one might wonder why we study two different methods. As it turns out, these two methods can be applied to different types of data sets, but both happen to work quite well for linear data sets. For the exercises in this section, you are encouraged to use the project command to develop a facility with this particular tool from RStudio.

Summary

- *Method of least squares*: The best linear model $y = mx + b$ for a data set with n data points $(x_1, y_1), \ldots, (x_n, y_n)$, stored in an input vector $\overline{x}$ and an output vector $\overline{y}$, is given by the solution to the following matrix equation with unknowns m and b:

$$\begin{pmatrix} \overline{y} \cdot \overline{x} \\ \overline{y} \cdot \overline{1}_n \end{pmatrix} = \begin{pmatrix} \overline{x} \cdot \overline{x} & \overline{1}_n \cdot \overline{x} \\ \overline{x} \cdot \overline{1}_n & \overline{1}_n \cdot \overline{1}_n \end{pmatrix} \begin{pmatrix} m \\ b \end{pmatrix}$$

- The method of least squares produces the best possible linear model $y = mx + b$ of a given set of n data points in the sense that the corresponding residual vector $\overline{r} = \overline{y} - (m\overline{x} + b\overline{1})$ has the smallest possible length $\|\overline{r}\|$ among the residuals of all possible linear combinations of $\overline{x}$ and $\overline{1}_n$.

Exercises

In Exercises 1–14, use the method of least squares to find the best linear model for the data set.

1.
x	2	5	7	8
y	1	2	3	3

2.
x	−1	0	1	2
y	0	1	2	4

3.
x	0	1	2	3
y	1	1	2	2

4.
x	1	4	8	11
y	1	2	4	5

5.
x	2	3	5	6
y	3	2	1	0

6.
x	−2	−1	0	−1
y	2	1	0	−1

7.
x	−3	−2	−1	0	1
y	4.2	2.4	−0.2	−1.7	−4.3

8.

x	-4	-2	0	2	4
y	19.3	9.2	1.1	-7.4	-13.4

9.

x	1	2	3	4	5
y	5.42	7.51	11.48	14.40	16.48

10.

x	-2	-1	1	2	4
y	10.85	6.84	-1.32	-4.45	-12.55

11.

x	-5	-4	-3	-2	-1
y	-14.16	-10.88	-8.37	-5.93	-2.85

12.

x	-2	2	6	10	14
y	3.55	1.30	-1.79	-4.44	-7.38

13.

x	-5	-4	-3	-2	-1
y	-28.0	-20.7	-16.2	-8.4	-3.4
x	0	1	2	3	4
y	3.5	10.3	16.5	21.7	27.3

14.

x	-10	-5	0	5
y	65.9	23.4	-12.4	-49.2
x	15	20	25	35
y	113.9	-154.8	-185.8	-270.5

In Exercises 15–22, use the method of least squares to find the best linear model for the data set.

15. Length of the tornado season (L) in days between the first and last tornado for each year (Y) in the warning area of the National Weather Service Office in Goodland, Kansas

Y	1990	2000	2005	2010
L	120	142	149	155

16. U.S. retail prescription drug sales (S) in billions of dollars for each year (Y); that is, for example, in 2006, $262.7 billion was spent in prescriptions in the United States

Y	2006	2007	2009	2010
S	262.7	261.6	266.8	266.4

17. Total number of prescription drugs sold (S) in the United States in billions per year (Y); that is, for example, in 2006, $3,420,000,000 prescriptions were sold in the United States

Y	2006	2007	2009	2010
S	3.42	3.53	3.63	3.68

18. Total number of burgers (B) sold by McDonald's in billions as of each year (Y)

Y	1976	1987	1990	1994
B	20	65	80	100

19. Annual e-commerce sales (S) in the United States in billions of dollars for each year (Y)

Y	2009	2010	2011	2012
S	121	143	168	192

20. Global gender ratio based on the number of males per 100 females (M) in each year (Y)

Y	1990	1995	2005	2010
M	101.5	101.5	101.6	101.6

21. World population growth rate (R) in each year (Y) since 1995

Y	0	5	10	15	20
R	1.41	1.26	1.20	1.13	1.06

22. Average number of goals (G) scored per game in World Cup tournaments in each year (Y) since 1994

Y	0	4	8	12	16
G	2.7	2.7	2.5	2.3	2.3

Your Turn. In Exercises 23–28, identify a real-life data set in the area with at least four points for which a linear model is reasonable and use the method of least squares to find its best linear model.

23. Athletics

24. Economics

25. Engineering

26. Environment

27. Psychology

28. Sociology

In Exercises 29–31, consider the data set from Exercise 15, which presents the length of the tornado season in days.

29. Compute the length of the residual vector associated with the least squares linear model Length = 1.77 · Year − 3403.57.

30. Compute the length of the residual vector associated with the Section 2.1 linear model Length = 1.6 · Year − 3060.5.

31. Compare the lengths of the residuals in Exercises 29 and 30.

In Exercises 32–34, consider the data set from Exercise 16, which presents U.S. retail prescription drug sales in billions of dollars for each year.

32. Compute the length of the residual vector associated with the least squares linear model Sales = 1.26 · Year − 2265.71.

33. Compute the length of the residual vector associated with the Section 2.1 linear model Sales = 0.37·Year− 478.58.

34. Compare the lengths of the residuals in Exercises 32 and 33.

In Exercises 35–37, consider the data set from Exercise 17, which presents the total number of prescription drugs sold in the United States in billions per year.

35. Compute the length of the residual vector associated with the least squares linear model Prescriptions = 0.062 · Year − 120.931.

36. Compute the length of the residual vector associated with the Section 2.1 linear model Prescriptions = 0.0685 · Year − 133.984.

37. Compare the lengths of the residuals in Exercises 35 and 36.

In Exercises 38–40, consider the data set from Exercise 18, which presents the total number of burgers sold by McDonald's in billions as of each year.

38. Compute the length of the residual vector associated with the least squares linear model Burgers = 4.4 · Year − 8672.7.

39. Compute the length of the residual vector associated with the Section 2.1 linear model Burgers = 4.7 · Year − 9266.8.

40. Compare the lengths of the residuals in Exercises 38 and 39.

In Exercises 41–43, consider the data set from Exercise 19, which presents the annual e-commerce sales in the United States in billions of dollars for each year.

41. Compute the length of the residual vector associated with the least squares linear model Sales = 23.8 · Year − 47,693.9.

42. Compute the length of the residual vector associated with the Section 2.1 linear model Sales = 23.67 · Year − 47,432.54.

43. Compare the lengths of the residuals in Exercises 41 and 42.

In Exercises 44–46, consider the data set from Exercise 20, which presents the global gender ratio based on the number of males per 100 females in each year.

44. Compute the length of the residual vector associated with the least squares linear model Males = 0.006· Year + 89.55.

45. Compute the length of the residual vector associated with the Section 2.1 linear model Males $= 0.0067 \cdot$ Year $+ 88.145$.

46. Compare the lengths of the residuals in Exercises 44 and 45.

RStudio. In Exercises $47 - 60$, use RStudio to find the best linear model for each of the data sets from Exercises $1 - 14$ using the `project` command.

47. Exercise 1

48. Exercise 2

49. Exercise 3

50. Exercise 4

51. Exercise 5

52. Exercise 6

53. Exercise 7

54. Exercise 8

55. Exercise 9

56. Exercise 10

57. Exercise 11

58. Exercise 12

59. Exercise 13

60. Exercise 14

RStudio. In Exercises $61 - 68$, use RStudio to find the best linear model for each of the data sets from Exercises $15 - 22$ using the `project` command.

61. Exercise 15

62. Exercise 16

63. Exercise 17

64. Exercise 18

65. Exercise 19

66. Exercise 20

67. Exercise 21

68. Exercise 22

RStudio. In Exercises $69 - 72$, use RStudio to study the U.S. monthly unemployment rate from January 2010 to December 2014 at `MonthlyUnemployment`.

69. Use the `project` command to find the best linear model for this data set.

70. Plot the data set and the linear model from Exercise 69 on the same axes, and discuss the model's goodness of fit.

71. Evaluate the model from Exercise 69 at $x = 36$. Explain the answer's meaning.

72. Evaluate the model from Exercise 69 at $x = 65$. Explain the answer's meaning.

RStudio. In Exercises $73 - 76$, use RStudio to study the total midyear population for the world from 1950 to 2015 at `WorldPopulation`.

73. Use the `project` command to find the best linear model for this data set.

74. Plot the data set and the linear model from Exercise 73 on the same axes, and discuss the model's goodness of fit.

75. Evaluate the model from Exercise 73 at $x = 1988$. Explain the answer's meaning.

76. Evaluate the model from Exercise 73 at $x = 2020$. Explain the answer's meaning.

RStudio. In Exercises $77 - 80$, use RStudio to study the interest rates on 15-year, fixed-rate conventional home mortgages annually from 1992 to 2014 at `Mortgage15YrAnnual`.

77. Use the `project` command to find the best linear model for this data set.

78. Plot the data set and the linear model from Exercise 77 on the same axes, and discuss the model's goodness of fit.

79. Evaluate the model from Exercise 77 at $x = 2007$. Explain the answer's meaning.

80. Evaluate the model from Exercise 77 at $x = 2015$. Explain the answer's meaning.

RStudio. In Exercises 81–84, use RStudio to study the number of Facebook users in millions from 2009 through 2012 at `FacebookUsers`.

81. Use the `project` command to find the best linear model for this data set.

82. Plot the data set and the linear model from Exercise 81 on the same axes, and discuss the model's goodness of fit.

83. Evaluate the model from Exercise 77 at $x = 30$. Explain the answer's meaning.

84. Evaluate the model from Exercise 77 at $x = 60$. Explain the answer's meaning.

RStudio. In Exercises 85–88, use RStudio to study the high school dropout rate in the United States from 1970 through 2012 at `HSDropoutRate`.

85. Use the `project` command to find the best linear model for this data set.

86. Plot the data set and the linear model from Exercise 85 on the same axes, and discuss the model's goodness of fit.

87. Evaluate the model from Exercise 81 at $x = 2005$. Explain the answer's meaning.

88. Evaluate the model from Exercise 81 at $x = 2015$. Explain the answer's meaning.

RStudio. In Exercises 89–92, use RStudio to study U.S. carbon dioxide emissions in kT annually from 1980 to 2008 according to the World Bank at `USCO2Emissions`.

89. Use the `project` command to find the best linear model for this data set.

90. Plot the data set and the linear model from Exercise 89 on the same axes, and discuss the model's goodness of fit.

91. Evaluate the model from Exercise 89 at $x = 2000$. Explain the answer's meaning.

92. Evaluate the model from Exercise 89 at $x = 2020$. Explain the answer's meaning.

RStudio. In Exercises 93–96, use RStudio to study the average interest rate for conventional 30-year mortgages each year from 1981 to 2012 at `Mortgage30YrAnnual`.

93. Use the `project` command to find the best linear model for this data set.

94. Plot the data set and the linear model from Exercise 93 on the same axes, and discuss the model's goodness of fit.

95. Evaluate the model from Exercise 93 at $x = 2000$. Explain the answer's meaning.

96. Evaluate the model from Exercise 93 at $x = 2015$. Explain the answer's meaning.

Your Turn. In Exercises 97 and 98 find your own real-life data set with a minimum of ten data points for which a linear model is reasonable.

97. Use the `project` command to find the best linear model for this data set.

98. Plot the data set and the linear model from Exercise 97 on the same axes, and discuss the model's goodness of fit.

In Your Own Words. In Exercises 99 – 101, explain the following.

99. Method of least squares

100. Least squares residual vector

101. Figure 3

Chapter 4

Derivatives

Our world is constantly changing. In our physical world, temperatures and weather vary over the course of the day and from season to season. In our economic world, prices of consumer goods depend on supply and demand. In our social world, technologies have enabled greater connectivity and real-time relationships across continents. In our medical world, ongoing research provides a much greater understanding of the microscopic processes that determine the health of individuals and of societies. The mathematical tool for analyzing and understanding such changes is called the *derivative*.

Adopting a big picture view, this book's quest is to enable a better understanding of reality via the perspectives of mathematics. Chapters 1 and 2 studied a variety of functions that provide models for sets of data, including linear, exponential, power, sine, and sigmoidal functions. These models enable us to describe, understand, and, often, make predictions about diverse phenomenon. Chapter 3 developed various additional modeling techniques, most importantly, the method of least squares as a means of obtaining best possible models when exact models do not exist.

This chapter and the next study functions in a more dynamic way, focusing on how a function's values change as its inputs increase. Particularly important features of functions include how quickly or slowly they are changing, where they are increasing, constant, or decreasing, and their maximum and minimum values. Among other things, such information facilitates a more thorough analysis of the common modeling functions and so a better understanding of the phenomena they model.

These ideas are explored in terms of the *rate of change* of a function, which is called its derivative, and are pivotal to understanding reality from a mathematical point of view. This chapter develops a strong understanding of the idea of the derivative and studies methods for computing derivatives. These "calculating" tools are part of what earned "calculus" its name.

4.1 Rates of Change

Rates of change are common in many aspects of our lives. What is our average speed in miles per hour while traveling home from school? How many characters can we text per second? How does our number of heartbeats per minute change during a workout?

This section studies two methods for answering such questions. First, the *average rate of change* of a function expresses how the function changes over an interval of input values. Alternatively, the *instantaneous rate of change* of a function expresses how the function changes at an individual input value. As might be expected, these two different understandings of how a function changes are interconnected, and studying them together enables a better understanding of both.

Average Rate of Change

While the term may be new, the idea of an *average rate of change* is probably already familiar. Consider the following example:

◆ **EXAMPLE 1**   At the end of the Fall Term, you make the 293-mile trip home for break. After you arrive home, your parents ask, "How was the trip?" "Fine," you respond. "How long did it take you?" your mom asks. "I was able to drive straight through. It only took four and a half hours." During your trip home, what was your average velocity (or, more colloquially, your average speed)? Also, what is the dimension of your average velocity?

Solution. The average velocity of such a trip is calculated by dividing the total number of miles traveled by the total amount of time it took to complete the trip:

$$\text{average velocity} = \frac{293 \text{ miles}}{4.5 \text{ hours}} = 65.1 \text{ miles per hour}$$

The dimension of average velocity is length divided by time, which is expressed symbolically as LT^{-1}. As indicated above, an appropriate choice of units is miles per hour. ■

This scenario and many other such real-life events provide the inspiration for the definition of average rate of change. Namely, the average rate of change of a function over a given input interval is equal to the total change in the output values of the function divided by the total change in its input values. Suppose that a function $f(x)$ over an input interval $[a, b]$ is given. In order to calculate the total change in the output of the function, take the final value $f(b)$ on the interval $[a, b]$ and subtract the initial value $f(a)$; that is, compute $f(b) - f(a)$. Similarly, calculate the change in its inputs on the interval by taking the final input b and subtracting the initial input a; that is, compute $b - a$. Finally, divide these two values as indicated in the following definition:

Definition. The **average rate of change** of a function $f(x)$ over the interval $[a, b]$ is the total change in output values of $f(x)$ divided by the total change in input values:

$$\text{average rate of change of } f(x) \text{ over } [a, b] = \frac{\text{change in outputs}}{\text{change in inputs}} = \frac{f(b) - f(a)}{b - a}$$

Example 1 presented the distance traveled from school to home as a function of how much time elapsed while driving, which means that this function's input value was time and its output value was distance. Therefore, when computing average speed, the total change in distance (i.e., the difference in the output values of the function) is divided by the total change in time (i.e., the difference in the input values to the function).

◆ **EXAMPLE 2** One rainy summer day, Morgan checked a rain gauge every hour from 8 a.m. to 6 p.m. and recorded the measurements in inches:

Time	8 a.m.	9	10	11	noon	1 p.m.	2	3	4	5
Amount	0.15	0.17	0.20	0.45	0.48	0.75	1.03	1.20	1.45	1.60

(a) How much rain fell during the entire time period from 8 a.m. to 5 p.m.?

(b) What was the average rate of rainfall from 8 a.m. to 5 p.m.?

(c) How much rain fell between 10 a.m. and 11 a.m.?

(d) How much rain fell between 11 a.m. and noon?

(e) What was the average rate of rainfall from 10 a.m. to 12 p.m.?

Solution.

(a) Total rainfall during any time period is found by subtracting the amount of water in the rain gauge at the beginning of the time period from the amount at the end. In this case, the total rainfall from 8 a.m. to 5 p.m. was $1.60 - 0.15 = 1.45$ inches.

(b) The average rate of rainfall is computed by dividing the total rainfall during the time period by the length of the time period. From part (a), the total rainfall from 8 a.m. to 5 p.m. was 1.45 inches, and the the total time elapsed from 8 a.m. to 5 p.m. was 9 hours. Now apply the formula for the average rate of change:

$$\text{average rate of rainfall} \;=\; \frac{1.45 \text{ inches}}{9 \text{ hours}} \;=\; 0.16 \text{ inches of rain per hour}$$

(c) Between 10 a.m. and 11 a.m., the total rainfall was $0.45 - 0.20 = 0.25$ inches.

(d) Between 11 a.m. and noon, the total rainfall was $0.48 - 0.45 = 0.03$ inches.

(e) From 10 a.m. to noon, the total rainfall was $0.48 - 0.20 = 0.28$ inches. Alternatively, the results of parts (c) and (d) can be used to determine this total rainfall of $0.25 + 0.03 = 0.28$ inches. The total time elapsed was 2 hours. Applying the formula for the average rate of change gives:

$$\text{average rate of rainfall} \;=\; \frac{0.28 \text{ inches}}{2 \text{ hours}} \;=\; 0.14 \text{ inches of rain per hour}$$

■

In light of the average rate of change being defined in terms of a quotient, the dimensions of any such rate are the dimensions of the output values divided by the dimensions of the input values. Similarly, the units for an average rate of change are the units of the output values divided by the units of the input values. These units are usually expressed using the word "per," as in miles per hour or inches of rain per hour.

Example 1 determined the average rate of change in position (i.e., average velocity) as the change in position divided by the change in time. In such a case, the dimensions are LT^{-1}, with possible units including miles per hour and meters per second. In Example 2, the average rate of rainfall has dimensions of LT^{-1}, with units of inches per hour. Additional examples of units for rates of change include dollars per item manufactured, with dimensions BM^{-1}, and atmospheres of pressure per meter of depth in the ocean, with dimensions of $(ML^{-1}T^{-2})L^{-1}$. Dimensions and units for rates of change are discussed in greater detail in Section 4.2.

➤ **QUESTION 1** The following table presents median home prices in thousands of dollars from 2001 to 2010 according to the U.S. Census Bureau:

Year	2001	2002	2003	2004	2005	2006	2007	2008	2009	2010
Price	175.2	187.6	195.0	221.0	240.9	246.5	247.9	232.1	216.7	221.8

Compute the average rate of change in median home price over each time period. Justify the answers and use appropriate units. Also, what is the dimension of these average rates of change?

(a) 2004 to 2007   (b) 2004 to 2009   (c) 2001 to 2010

Note the similarity between calculating the average rate of change of a function over an interval and calculating the slope of the line joining two points on the graph of a function. Both of these values are found by dividing the difference between the output values of the function by the difference between the corresponding input values. A line connecting two points on the graph of a function is referred to as a **secant line**. Figure 1 presents two different functions and their respective secant lines on the input interval $[0, 4]$.

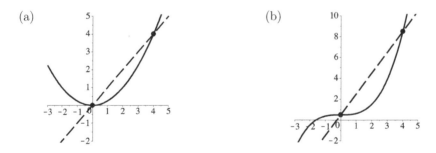

Figure 1: Secant lines over $[0, 4]$ for two different functions

In light of the equivalence of these two processes, the definition of average rate of change can be extended as follows:

Definition. The **average rate of change** of a function $f(x)$ over the interval $[a, b]$ is the slope of the line through $(a, f(a))$ and $(b, f(b))$.

This connection between the average rate of change of $f(x)$ over the interval $[a, b]$ and the slope of the secant line through the point $(a, f(a))$ and $(b, f(b))$ enables a graphical interpretation of the average rate of change of a function.

In Figure 1, both graphs show the secant line that connects the output for the input $x = 0$ to the output for the input $x = 4$, and the slopes of these secant lines provide the average rates of change of the corresponding functions on the interval $[0, 4]$. Notice that the slope of the secant line in Figure 1(a) is less than the slope of the secant line in Figure 1(b), as indicated by the relative steepness of these two lines. This graphical observation matches the definition of average rate of change. The total change in the input values is the same for both secant lines, but the change in the output values of Figure 1(a) is less than the change in the output values of Figure 1(b). Consequently, the average rate of change of Figure 1(a) is less than the average rate of change of Figure 1(b).

Alternatively, this phenomenon can be thought about in terms of the total change in function values corresponding to a uniform, average change. The total change in function values in Figure 1(a) is smaller than in Figure 1(b). Therefore, the average rate at which Figure 1(a) grows must be smaller in order to achieve the overall smaller change in output values over the same change in input values.

▶ **QUESTION 2** Compute the average rate of change of each function over the given interval and sketch the corresponding secant line:

(a) over $[0, 4]$

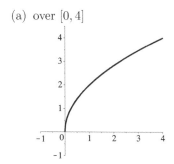

(b) over $[-2, 1]$

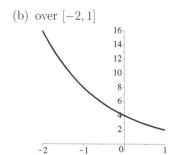

Instantaneous Rate of Change

The idea of the average rate of change of a function over an interval can be extended to the idea of the *instantaneous rate of change* of a function at a point, which measures how a function changes at one specific point.

Imagine that you are driving down the road and that you see a police officer with a radar gun pointed at your car. You immediately check the speedometer and see that you are going 63 mph. Fortunately, they let you drive on past, even though you are driving through a 55 mph zone! The moment that you checked the speedometer, you determined the instantaneous rate of change of your car (or its instantaneous velocity) as you measured its change in distance with respect to time. In other words, the instantaneous rate of change of a function measures the rate at which the function is changing at a single, specific input value.

The instantaneous rate of change of a function at input $x = a$ is usually very close to the average rates of change of the function over very short input intervals containing $x = a$. For many functions $f(x)$, the various average rates of change of $f(x)$ on such smaller and smaller intervals become really close to a unique number, which is called the instantaneous rate of change of the function at the given point. As such, a first strategy for finding an instantaneous rate of change is to compute the average rate of change over smaller and smaller intervals containing the input value for which the instantaneous rate is being sought.

Definition. The **instantaneous rate of change** of a function $f(x)$ when $x = a$ is the number approached by the average rates of change of $f(x)$ when these rates are computed over smaller and smaller intervals containing $x = a$. If the average rates of change do not approach a unique, fixed number, the instantaneous rate of change is undefined when $x = a$.

A more precise definition of instantaneous rate of change can be given using the mathematical idea of a *limit*, which carefully describes the idea of "approaching a unique, fixed number." The relationships among limits, instantaneous rates of change, and derivatives are discussed in Section 4.7.

The connection between instantaneous rates of change and average rates of change enables a process of making informed conjectures about the instantaneous rate of change of a function when $x = a$ by computing a sequence of average rates of change over smaller and smaller intervals. The next example illustrates this process.

◆ **EXAMPLE 3** Compute the average rate of change of the function $f(x) = x^3 - 7x$ over each input interval of x-values and use these results to conjecture the instantaneous rate of change of $f(x)$ when $x = 2$:

(a) $[2, 2.1]$ (b) $[2, 2.01]$ (c) $[2, 2.001]$ (d) $[1.99999, 2]$

Solution. Compute the average rate of change of $f(x)$ over each interval as follows:

(a) $\dfrac{f(2.1) - f(2)}{2.1 - 2} = \dfrac{[(2.1)^3 - 7 \cdot 2.1] - [2^3 - 7 \cdot 2]}{0.1} \approx 5.61$

(b) $\dfrac{f(2.01) - f(2)}{2.01 - 2} = \dfrac{[(2.01)^3 - 7 \cdot 2.01] - [2^3 - 7 \cdot 2]}{0.01} \approx 5.0601$

(c) $\dfrac{f(2.001) - f(2)}{2.001 - 2} = \dfrac{[(2.001)^3 - 7 \cdot 2.001] - [2^3 - 7 \cdot 2]}{0.001} \approx 5.006001$

(d) $\dfrac{f(1.99999) - f(2)}{1.99999 - 2} = \dfrac{[(1.99999)^3 - 7 \cdot 1.99999] - [2^3 - 7 \cdot 2]}{-0.00001} \approx 4.99994$

From this sequence of computations, observe that the average rate of change of $f(x)$ appears to become closer and closer to the number 5 as the intervals containing 2 become smaller and smaller. Based on this observation, the instantaneous rate of change of $f(x) = x^3 - 7x$ when $x = 2$ is conjectured to be 5. Symbolically, this answer would be presented as $f'(2) = 5$. ∎

The notation "$f'(2) = 7$" is discussed in greater detail later in this section. For now, observe that the value of $f'(2)$ in this setting measures the rate of change of $f(x)$ at the point $(2, f(2)) = (2, -2)$ in essentially the same way that the slope of a line measures the rate of change of a line.

➤ **QUESTION 3** Compute the average rate of change of the function $f(x) = x^2 + 3x$ over each input interval of x-values, and use these results to conjecture the instantaneous rate of change of $f(x)$ when $x = 3$.

   (a) $[3, 3.1]$ (b) $[3, 3.01]$ (c) $[3, 3.0001]$ (d) $[2.9999, 3]$

➤ **QUESTION 4** The following table from Question 1 presents median home prices P in thousands of dollars each year Y from 2001 to 2010 according to the U.S. Census Bureau:

Year	2001	2002	2003	2004	2005	2006	2007	2008	2009	2010
Price	175.2	187.6	195.0	221.0	240.9	246.5	247.9	232.1	216.7	221.8

Applying `fitModel` in RStudio produces the following model for this data set:

$$P(Y) = 0.17(Y-2001)^4 - 3.2(Y-2001)^3 + 16.7(Y-2001)^2 - 10.9(Y-2001) + 177.7$$

A graph of the data with this model is shown in Figure 2. Compute the average rate of change of this model $P(Y)$ over each of the following intervals, and use these results to conjecture the instantaneous rate of change of median home prices in 2005:

   (a) $[2005, 2006]$ (b) $[2005, 2005.01]$ (c) $[2005, 2005.0001]$

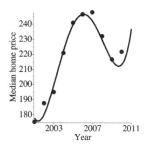

Figure 2: Median home prices from 2001 to 2010 for Question 4

For many functions, this method of computing the average rates of change of a given function over smaller and smaller intervals provides an accurate conjecture for the value of the instantaneous rate of change of the given function for a given input. However, as with all conjectures, any conclusions must be taken with a grain of salt, because additional data may require modifying an initial conjecture. The study of limits in Section 4.7 explores some of the subtleties that arise when following this numerical approach to computing instantaneous rates of change. Fortunately, analytic techniques have been developed that enable the computation of exact values of instantaneous rates of change for many functions, including the common modeling functions. Sections 4.3 – 4.6 present the most important of these analytic techniques.

Tangent Line Question and Fermat's Solution

Just as secant lines provide a graphical interpretation of average rates of changes, tangent lines provide a graphical interpretation for instantaneous rates of change. As will become apparent, tangent lines are key to analyzing a diverse collection of functional behaviors. As such, in the early 1600s, mathematicians focused on techniques for answering what this book will refer to as the *tangent line question*. Intuitively, the **tangent line** to the graph of a function $f(x)$ at a point $(a, f(a))$ is the line that intersects the graph of the function at exactly this one point, at least within the immediate, local vicinity of the point.

TANGENT LINE QUESTION.
Given the graph of a function $f(x)$ and the point $(a, f(a))$ on this graph, what is the equation of the tangent line to $f(x)$ when $x = a$?

The standard approach to determining the equation of a line $y = mx + b$ begins with finding two points on the line and then computing the slope. Section 1.3 described various approaches to computing the slope m of the line through the points (x_1, y_1) and (x_2, y_2), including the following when $x_1 \neq x_2$:

$$m = \frac{y_2 - y_1}{x_2 - x_1} = \frac{y_1 - y_2}{x_1 - x_2} = \frac{\Delta y}{\Delta x} = \frac{\text{rise}}{\text{run}}$$

Unfortunately, when determining the equation of a tangent line to $f(x)$ when $x = a$, only one point $(a, f(a))$ on the tangent line is known (rather than two), so the above pre-calculus approach to finding the slope of the line cannot be used. The key to overcoming this obstacle is the instantaneous rate of change.

Recall that the average rate of change over an input interval is represented graphically as the slope of the secant line passing through the two points on the graph corresponding to the endpoints of the interval. As a specific graphical example, Figure 3 presents three secant lines and a horizontal tangent line to a function $f(x)$. The slope of the secant line through points P and Q is the same as the average rate of change of the function over the input interval $[1, 2.3]$. Similarly, the slope of the secant line through points P and R is the same as the average rate of change of the function over the input interval $[1, 2]$, and for P and S over the input interval $[1, 1.5]$.

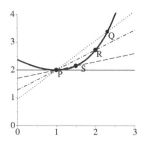

Figure 3: Secant lines approaching the tangent line

Observe in Figure 3 that as the input intervals become smaller and smaller, the corresponding secant lines approach the tangent line to the graph of $f(x)$ at the point $(1, 2)$. In particular, the secant line through PR for the input interval $[1, 2]$ is closer to the tangent line than the secant line through PQ for $[1, 2.3]$, and the secant line through PS for $[1, 1.5]$ is even closer. For these reasons, the slope of this tangent line is said to be equal to the instantaneous rate of change of $f(x)$ when $x = 1$.

Figure 3 illustrates the idea of the tangent line as a *limit* of secant lines. Namely, the secant lines approach closer and closer to the tangent line. The study of limits is subtle, but it is essential to a careful explanation of why calculus works, and Section 4.7 takes a more careful look at the idea of limits.

➤ **QUESTION 5** On a graph of $y = \sqrt{x}$, sketch secant lines for the input intervals $[1, 16]$, $[1, 9]$, and $[1, 4]$, as well as the tangent line to $y = \sqrt{x}$ when $x = 1$.

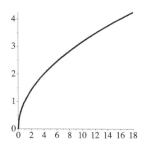

Because the instantaneous rate of change of a function $f(x)$ when $x = a$ is fundamentally important, this idea has a number of different names depending on the context. The next definition highlights various other phrases that are used synonymously for instantaneous rate of change.

Definition. The instantaneous rate of change of a function $f(x)$ when $x = a$ is called the **derivative** of $f(x)$ when $x = a$ and is denoted by $f'(a)$. The following names are used interchangably to identify this quantity:

- the instantaneous rate of change of $f(x)$ when $x = a$

- the derivative of $f(x)$ when $x = a$, which is denoted by $f'(a)$ and by $\left.\dfrac{df}{dx}\right|_{x=a}$

- the rate of change of $f(x)$ when $x = a$

- the slope of the tangent line to $f(x)$ when $x = a$

- the slope of the graph of $f(x)$ when $x = a$

The most commonly used of these various names is the derivative of $f(x)$ when $x = a$. Moving forward, you will want to be able to work flexibly with these names, shifting seamlessly among them to compute the instantaneous rate of change.

Because $f'(a)$ provides the slope of the tangent line to $f(x)$ when $x = a$, the tangent line question can now be answered. Recall that the point–slope form of a line with slope $m = f'(a)$ through the point $(a, f(a))$ is $y - f(a) = f'(a) \cdot (x - a)$. Rearranging by adding $f(a)$ to both sides yields $y = f(a) + f'(a) \cdot (x - a)$, which is the standard form of the equation of the tangent line to $f(x)$ when $x = a$.

Definition. The **equation of the tangent line** to $f(x)$ at $(a, f(a))$ is given by
$$y = f(a) + f'(a)(x - a).$$

In general, the equation of a tangent line is found by completing the following three steps:

(1) Calculate $f(a)$.

(2) Calculate or approximate $f'(a)$.

(3) Substitute these values into the equation of the tangent line $y = f(a) + f'(a)(x - a)$.

The next example illustrates this three-step process.

◆ **EXAMPLE 4** Find the equation of the tangent line to $f(x) = x^3 - 7x$ when $x = 2$, using the answer from Example 3.

Solution. First, calculate $f(2) = 2^3 - 7 \cdot 2 = 8 - 14 = -6$. Example 3 found that $f'(2) = 5$. Thus, the equation of the tangent line to $f(x) = x^3 - 7x$ when $x = 2$ is $y = -6 + 5 \cdot (x - 2)$ or $y = 5x - 16$. ∎

➤ **QUESTION 6** Find the equation of the tangent line to $f(x) = x^2 + 3x$ when $x = 3$, using the answer to Question 3.

This approach to answering the tangent line question was first developed by the French mathematician Pierre de Fermat. While working as a lawyer and government official in Toulouse in the mid-1600s, Fermat spent his free time studying mathematical questions. And not just any questions, but some of the most challenging research questions of his time. Fermat made fundamental contributions to diverse areas of mathematics, including this work with calculus, the development of the foundational ideas of probability and statistics with Blaise Pascal, and important questions in number theory.

NOTATION. The ideas of calculus were developed independently by the English mathematician–physicist Sir Isaac Newton and the German mathematician–philosopher Gottfried Leibniz. As might be expected when people work separately from one another, they used different notations for the derivative of a function at a point. The symbolism $f'(a)$ for the derivative of $f(x)$ when $x = a$ is a modification of Newton's notation, while Leibniz used the notation $\left. \dfrac{df}{dx} \right|_{x=a}$. These notations have their respective strengths and will be used interchangably throughout this book.

RStudio can be used to compute average rates of change, either by working with RStudio as a fancy calculator for individual computations or by using the `makeFun` command for computing multiple average rates of change.

Working in RStudio

The RStudio command `makeFun` can be used to compute average rates of change. Recall the definition:

$$\text{average rate of change of } f(x) \text{ over } [a, b] \ = \ \frac{f(b) - f(a)}{b - a}.$$

First, use `makeFun` to define and store the function $f(x)$ in RStudio. If $f(x) = 6x^3 - 3x^2$, enter the command `f=makeFun(6*x^3-3*x^2~x)`. Then, use `makeFun` again to define the multivariable average rate of change function in RStudio with input variables a and b (as in the above definition):

    AROC=makeFun((f(b)-f(a))/(b-a)~a&b)

Typically, the first variable a corresponds to the left endpoint of the input interval and the second variable b is its right endpoint, although the order can be swapped as in any standard slope computation. The average rate of change of $f(x)$ over $[a, b]$ is obtained by substituting the left and right endpoints of the given input interval into this multivariable function `AROC(a,b)`. Continuing the example, the average rate of change of $f(x) = 6x^3 - 3x^2$ over $[2, 3]$ is found by entering the command `AROC(a=2,b=3)`.

Examples of Commands:

- `f=makeFun(6*x^3-3*x^2~x)`

- `AROC=makeFun((f(b)-f(a))/(b-a)~a&b)`

- `AROC(a=2,b=3)`

◆ **EXAMPLE 5** Use RStudio to solve Example 3. Compute the average rate of change of the function $f(x) = x^3 - 5x$ over each input interval. Based on the results, conjecture the instantaneous rate of change of $f(x)$ when $x = 2$.

(a) $[2, 2.1]$ (b) $[2, 2.01]$ (c) $[2, 2.001]$ (d) $[1.99999, 2]$

Solution. First, define the function $f(x) = x^3 - 5x$ and its corresponding average rate of change function, and store them in RStudio:

```
func=makeFun(x^3 - 5*x ~ x)
AROC=makeFun( (func(b) - func(a))/(b-a) ~ a & b)
```

Now, substitute the endpoints of each interval:

`AROC(a=2,b=2.1)`   `AROC(a=2,b=2.001)`

`[1] 7.61`   `[1] 7.006`

`AROC(a=2,b=2.01)`   `AROC(a=1.99999,b=2)`

`[1] 7.0601`   `[1] 6.99994`

This RStudio output indicates that

(a) the average rate of change of $f(x)$ over $[2, 2.1]$ is 7.61,

(b) the average rate of change of $f(x)$ over $[2, 2.01]$ is 7.0601,

(c) the average rate of change of $f(x)$ over $[2, 2.001]$ is 7.006, and

(d) the average rate of change of $f(x)$ over $[1.99999, 2]$ is 6.99994.

As in Example 3, observe that the average rates of changes approach the value 7 as the intervals containing $x = 2$ become smaller and smaller. Therefore, the instantaneous rate of change of $f(x) = x^3 - 5x$ when $x = 2$ is conjectured to be 7 or, symbolically, $f'(2) = 7$.   ■

➤ **QUESTION 7** Use RStudio to solve Question 3. Compute the average rate of change of the function $f(x) = x^2 + 3x$ over each input interval, and use these results to conjecture the instantaneous rate of change of $f(x)$ when $x = 3$:

(a) $[3, 3.1]$ (b) $[3, 3.01]$ (c) $[3, 3.0001]$ (d) $[2.99999, 3]$

➤ **QUESTION 8** Question 4 introduced the following model P(Y) for median home prices P in thousands of dollars each year Y from 2001 to 2010 based on data from the U.S. Census Bureau:

$$P(Y) = 0.17(Y - 2001)^4 - 3.2(Y - 2001)^3 + 16.7(Y - 2001)^2 - 10.9(Y - 2001) + 177.7$$

The following table presents median home prices P in thousands of dollars each year Y from 2001 to 2010 according to the U.S. Census Bureau:

Year	2001	2002	2003	2004	2005	2006	2007	2008	2009	2010
Price	175.2	187.6	195.0	221.0	240.9	246.5	247.9	232.1	216.7	221.8

Applying `fitModel` in RStudio produces the following model for this data set:

$$P(Y) = 0.17(Y - 2001)^4 - 3.2(Y - 2001)^3 + 16.7(Y - 2001)^2 - 10.9(Y - 2001) + 177.7$$

Use RStudio to compute the average rate of change in median home prices over each input interval and then conjecture the instantaneous rate of change in 2007:

(a) $[2007, 2007.1]$ (c) $[2006.9, 2007]$

(b) $[2007, 2007.00001]$ (d) $[2006.99999, 2007]$

On a final note, Section 4.3 introduces the RStudio approach to computing instantaneous rates of change, or derivatives.

Summary

- The *average rate of change* of a function $f(x)$ over an interval $[a, b]$ is the same as the slope of the *secant line* joining the points $(a, f(a))$ and $(b, f(b))$:

$$\text{average rate of change} = \frac{\text{change in outputs}}{\text{change in inputs}} = \frac{f(b) - f(a)}{b - a}$$

- The *instantaneous rate of change* of a function $f(x)$ when $x = a$ is the number approached by the average rates of change of $f(x)$ on smaller and smaller intervals containing $x = a$.

Summary (continued)

- The *derivative* of a function $f(x)$ when $x = a$ is the instantaneous rate of change of $f(x)$ when $x = a$ and is denoted by $f'(a)$ or by $\left.\dfrac{df}{dx}\right|_{x=a}$.

- The derivative of $f(x)$ when $x = a$ is sometimes referred to as the *rate of change of $f(x)$ when $x = a$*, the *slope of the tangent line to $f(x)$ when $x = a$*, and the *slope of the graph of $f(x)$ when $x = a$*.

- *Tangent line question*: Given the graph of a function $f(x)$ and the point $(a, f(a))$ on this graph, what is the equation of the tangent line to $f(x)$ when $x = a$?

- The *equation of the tangent line* to $f(x)$ at $(a, f(a))$ is $y = f(a) + f'(a)(x - a)$.

Exercises

In Exercises 1–16, compute the average rate of change of the function over the given interval.

1. $f(x) = x + 5$ over $[1, 3]$

2. $f(x) = x + 5$ over $[3, 5]$

3. $f(x) = 3 - 2x$ over $[-4, -2]$

4. $f(x) = 3 - 2x$ over $[-2, 3]$

5. $f(x) = x^2 + 4$ over $[0, 5]$

6. $f(x) = x^2 + 4$ over $[1, 3]$

7. $f(x) = x^2 + 4$ over $[-1, 3]$

8. $f(x) = x^2 + 4$ over $[-2, 2]$

9. $f(x) = -3x^2 + 6x$ over $[-1, 1]$

10. $f(x) = -3x^2 + 6x$ over $[0, 4]$

11. $f(x) = x^3 - 2x + 1$ over $[0, 2]$

12. $f(x) = 2 - 4x^2 - x^4$ over $[0, 1]$

13. $f(x) = mx + b$ over $[2, 5]$

14. $f(x) = mx + b$, $[c, d]$

15. $f(x) = x^2$ over $[a, b]$

16. $f(x) = cx^2$ over $[a, b]$

In Exercises 17–22, compute the average rate of change of the annual total retail sales taxes collected in the United States in billions of dollars each year (for example, during 2006, $141,100,000,000 in taxes was collected) over the given input interval.

Year	2006	2008	2010	2012
Tax	$141.1	$141.4	$139.4	$150.4

17. $[2006, 2008]$ 20. $[2008, 2010]$

18. $[2006, 2010]$ 21. $[2008, 2012]$

19. $[2006, 2012]$ 22. $[2010, 2012]$

In Exercises 23–28, compute the average rate of change of the plasma concentration of Prozac in nanograms per milliliter (ng/mL) for a person who has taken 20 mg of Prozac for 30 days, but stops cold turkey, over the given input interval. (Note that discontinuing any medication should always be done in consultation with a healthcare provider.)

Day	0	5	10	22	27
Conc.	79	40	19.6	4.3	2.5

23. $[0, 5]$ 26. $[5, 22]$

24. $[0, 27]$ 27. $[10, 22]$

25. $[5, 10]$ 28. $[10, 27]$

In Exercises 29–34, compute the average rate of change of the number of Facebook users in millions of people per month since December 2004 over the given input interval.

Time	24	28	34	44	49
Users	12	20	50	100	150
Time	50	52	55	57	60
Users	175	200	300	350	400

29. $[24, 34]$ 32. $[49, 55]$

30. $[34, 49]$ 33. $[49, 60]$

31. $[49, 50]$ 34. $[52, 60]$

In Exercises 35–38, estimate the average rate of change of the graphed function over the given interval.

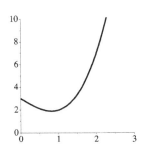

35. $[0, 1]$ 37. $[0.5, 2]$

36. $[0, 2]$ 38. $[1, 2]$

In Exercises 39–44, estimate the average rate of change of the graphed function over the given interval.

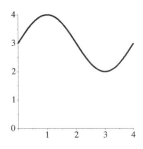

39. $[1, 4]$ 42. $[2, 4]$

40. $[1, 3]$ 43. $[2, 3]$

41. $[1, 2]$ 44. $[3, 4]$

RStudio. In Exercises 45–56, use RStudio to compute the average rate of change of the function from Exercises 1–12 over the given interval.

45. Exercise 1   51. Exercise 7

46. Exercise 2   52. Exercise 8

47. Exercise 3   53. Exercise 9

48. Exercise 4   54. Exercise 10

49. Exercise 5   55. Exercise 11

50. Exercise 6   56. Exercise 12

RStudio. In Exercises 57–64, use RStudio to compute the average rate of change of the function over the given interval.

57. $f(x) = e^x$ over $[0, 4]$

58. $f(x) = 2e^{x+1}$ over $[-2, 2]$

59. $f(x) = 3e^{-x}$ over $[-1, 1.5]$

60. $f(x) = xe^x$ over $[0, 1]$

61. $f(x) = e^{x^2}$ over $[0, 0.1]$

62. $f(x) = \ln(x)$ over $[2, 3]$

63. $f(x) = \sin(x)$ over $[0, 0.5]$

64. $f(x) = \cos(2x)$ over $[-1, 1]$

Your Turn. In Exercises 65–68, define your own function and compute the average rate of change over the given intervals. Use RStudio if you like.

65. $[1, 5]$ 67. $[1, 1.5]$

66. $[1, 2]$ 68. $[1, 1.01]$

In Exercises 69–72, compute the average rate of change of $f(x) = x^2$ over the given sequence of intervals. Use these results to conjecture the instantaneous rate of change $f'(a)$ and state the equation of the tangent line at $(a, f(a))$.

69. For $a = 3$:

    (a) $[3, 3.1]$

    (b) $[3, 3.01]$

    (c) $[3, 3.001]$

    (d) $f'(3)$

    (e) Equation of the tangent line

70. For $a = 5$:

    (a) $[4.9, 5]$

    (b) $[4.99, 5]$

    (c) $[4.999, 5]$

    (d) $f'(5)$

    (e) Equation of the tangent line

71. For $a = -2$:

    (a) $[-2, -1.9]$

    (b) $[-2, -1.99]$

    (c) $[-2, -1.999]$

    (d) $f'(-2)$

    (e) Equation of the tangent line

72. For $a = -4$:

    (a) $[-4.1, -4]$

    (b) $[-4.01, -4]$

    (c) $[-4.001, -4]$

    (d) $f'(-4)$

    (e) Equation of the tangent line

In Exercise 73, compute the average rates of change of the function $f(x) = x^2$ over the given intervals and use the results to conjecture the value of the instantaneous rate of change $f'(c)$ for x^2 for an arbitrary input $x = c$. Use the identity $(c \pm d)^2 - c^2 = \pm 2cd + d^2$.

73. (a) $[c, c + 0.1]$

    (b) $[c, c + 0.001]$

    (c) $[c - 0.1, c]$

    (d) $[c - 0.001, c]$

    (e) $f'(c)$

In Exercises 74 and 75, consider the linear model $\text{GDP}(Y) = 0.29978 \cdot Y - 585.21396$ of the World Bank's data for the total U.S. health expenditures as a percentage of GDP from 1995 to 2012. Compute the average rate of change of GDP(Year) over the sequence of intervals. Use these results to conjecture the instantaneous rate of change $\text{GDP}'(Y)$ and state the equation of the tangent line at $(Y, \text{GDP}(Y))$.

74. For Year $= 2000$:

    (a) $[2000, 2001]$

    (b) $[2000, 2000.0001]$

    (c) $[1999.99, 2000]$

    (d) $\text{GDP}'(2000)$

    (e) Equation of the tangent line

75. For Year $= 2010$:

    (a) $[2010, 2011]$

    (b) $[2010, 2010.0001]$

    (c) $[2009.99, 2010]$

    (d) $\text{GDP}'(2010)$

    (e) Equation of the tangent line

In Exercises 76 and 77, consider the exponential model Weight $= 781.9e^{0.1216(\text{Week}-25)}$ for the average weight of male fetuses in grams in the given gestational week. Compute the average rate of change of Weight(Week) over the sequence of intervals. Use these results to conjecture the instantaneous rate of change Weight$'$(Week) and state the equation of the tangent line at (Week, Weight(Week)).

76. For Week $= 26$:

   (a) $[26, 27]$

   (b) $[26, 26.0001]$

   (c) $[25.9999, 26]$

   (d) Weight$'(26)$

   (e) Equation of the tangent line

77. For Week $= 33$:

   (a) $[33, 35]$

   (b) $[33, 33.0001]$

   (c) $[32.99999, 33]$

   (d) Weight$'(33)$

   (e) Equation of the tangent line

In Exercises 78 and 79, consider the power function model

$$\text{RPM(Mass)} = 13545.5 \cdot \text{Mass}^{-0.339622}$$

for revolutions per minute (RPM) of an engine as a function of the mass of the engine in kilograms. Compute the average rate of change of RPM(Mass) over the sequence of intervals. Use these results to conjecture the instantaneous rate of change RPM$'$(Mass) and state the equation of the tangent line at (Mass, RPM(Mass)).

78. For Mass $= 2$:

   (a) $[2, 3]$

   (b) $[2, 2.001]$

   (c) $[1.999999, 2]$

   (d) RPM$'(2)$

   (e) Equation of the tangent line

79. For Mass $= 300$:

   (a) $[300, 301]$

   (b) $[300, 300.1]$

   (c) $[299.999, 300]$

   (d) RPM$'(300)$

   (e) Equation of the tangent line

In Exercises 80–83, sketch secant lines on the given intervals and the tangent line to $f(x)$ for the given input.

80. Secant lines on $[1, 3]$, $[1, 2]$, $[1, 1.5]$
    Tangent line when $x = 1$

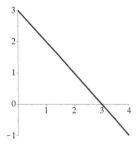

81. Secant lines on $[2, 4]$, $[2, 3]$, $[2, 2.5]$
    Tangent line when $x = 2$

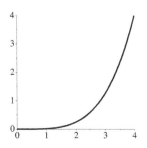

82. Secant lines on $[-2, 2]$, $[-1, 1]$, $[0.5, 0.5]$
    Tangent line when $x = 0$

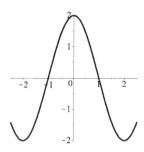

83. Secant lines on $[-3, 0]$, $[-2, 0]$, $[-1, 0]$
    Tangent line when $x = 0$

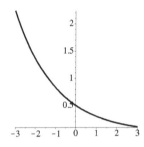

In Exercises 84–89, use the following graph to estimate the instantaneous rate of change $f'(a)$ at the point $(a, f(a))$:

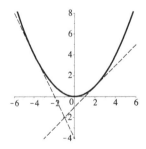

84. $f'(-4)$ at the point $(-4, 4)$

85. $f'(0)$ at the point $(0, 0)$

86. $f'(2)$ at the point $(2, 1)$

87. $f'(-2)$ at the point $(-2, 1)$

88. $f'(4)$ at the point $(4, 4)$

89. $f'(5)$ at the point $(5, 6)$

In Exercises 90–92, use the following graph to estimate the instantaneous rate of change $f'(a)$ at the point $(a, f(a))$:

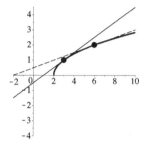

90. $f'(3)$ at the point $(3, 1)$

91. $f'(6)$ at the point $(6, 2)$

92. $f'(8)$ at the point $(8, 2.5)$

In Your Own Words. In Exercises 93–100, explain the following.

93. Average rate of change of $f(x)$ over $[a, b]$

94. Average rate of change and slopes of lines

95. Instantaneous rate of change of $f(x)$ when $x = a$

96. Derivative of $f(x)$ when $x = a$

97. Secant lines versus tangent lines

98. Tangent line question

99. Fermat's solution to the tangent line question

100. The equation of the tangent line to $f(x)$ at $(a, f(a))$

4.2   The Derivative as a Function

This section studies a functional approach to the instantaneous rate of change of a given function. Recall that a function provides unambiguous information by mapping each input value to exactly one output value. For example, a function might provide the number of textbooks sold at a given price point, the distance traveled by a car during a given time period, or the average GPA of an individual student or a particular group of students.

Section 4.1 defined the derivative $f'(a)$ of a function $f(x)$ for a particular input $x = a$. Recall that the number $f'(a)$ is equal to the instantaneous rate of change of the function $f(x)$ when $x = a$. Graphically, $f'(a)$ is interpreted as the slope of the tangent line to $f(x)$ at the point $(a, f(a))$. This section shifts from thinking about the value of the derivative for a single input to studying the value of the derivative over a domain of inputs. In other words, we focus on the function $f'(x)$ that provides the instantaneous rate of change of $f(x)$ for multiple input values x. By way of introducing these ideas, graphical examples of derivative functions are examined and then these ideas are applied to data sets.

◆ **EXAMPLE 1** Consider the graph of the function $f(x)$ given in Figure 1.

(a) Approximate the derivative of $f(x)$ for input values $x = -3, -2, 0, 1, 2, 3$, and 4.

(b) Create a rough sketch of the graph of $f'(x)$ using the values from part (a).

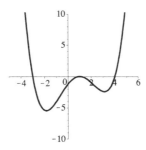

Figure 1: Graph of $f(x)$ for Example 1

Solution.

(a) The input values $x = -2$, $x = 1$, and $x = 3$ all correspond to points where $f(x)$ has a horizontal tangent line. Because the slope of a horizontal line is zero, the value of the derivative is equal to zero for all three of these inputs:

$$f'(-2) = 0 \qquad\qquad f'(1) = 0 \qquad\qquad f'(3) = 0$$

The other four input values have nonhorizontal tangent lines, so the derivative of $f(x)$ (i.e., its instantaneous rate of change) is approximated by computing the average rate of change of $f(x)$ over a small interval containing the input. Recall the formula introduced in Section 4.1 for computing an average rate of change:

$$\text{average rate of change of } f(x) \text{ over } [a, b] \;=\; \frac{f(b) - f(a)}{b - a}$$

This example mostly uses input intervals that are symmetric around the given input and that have integer outputs at their endpoints in an effort to simplify the computations (other choices could certainly be made):

$$f'(-3) \approx \frac{-1 - 1}{-2.9 - (-3.1)} = -10 \qquad\qquad f'(2) \approx \frac{-2 - 0.5}{2.5 - 1.5} = -1.5$$

$$f'(0) \approx \frac{-0.5 - (-2)}{0.5 - (-0.5)} = 1.5 \qquad\qquad f'(4) \approx \frac{1 - (-1)}{4.1 - 3.8} \approx 6.67$$

(b) The rough sketch of the graph of $f'(x)$ in Figure 2 is created by plotting the points from part (a) and then connecting them with straight lines. The resulting graph is not exactly $f'(x)$, but does provide a rough illustration of $f'(x)$. ∎

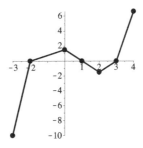

Figure 2: Rough sketch of $f'(x)$ for Example 1

As discussed in Section 4.1, more accurate approximations of an instantaneous rate of change are obtained by computing average rates of change over smaller and smaller intervals. In Example 1, other choices of input intervals could be made to achieve better approximations, particularly if some motivating question requires a more precise answer. A couple more examples are presented, after which the process of selecting input intervals to approximate the derivative with average rates of change is discussed more fully.

➤ **QUESTION 1** Consider the graph of the function $f(x)$ given in Figure 3.

(a) Approximate the derivative of $f(x)$ for input values $x = -2, -1, 0, 2$, and 3.25.

(b) Create a rough sketch of the graph of $f'(x)$ using the values from part (a).

Example 1 and Question 1 suggest that the value of the derivative $f'(x)$ can be found for any function $f(x)$ and any input x. As it turns out, exceptions to this claim do exist, because the derivative cannot be computed for every function at every input and, in fact, some functions are nowhere differentiable. However, many functions (including the common modeling functions) do have a derivative at every point in their domain, enabling the definition of the derivative $f'(x)$ as a function.

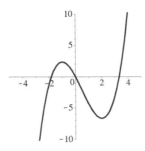

Figure 3: Graph of $f(x)$ for Question 1

Definition. Let $f(x)$ be a function. The **derivative function** $f'(x)$ outputs the instantaneous rate of change of $f(x)$ for each input x. A function is called **differentiable** if and only if its derivative exists.

Section 4.1 highlighted several different phrases that synonymously reference the derivative of $f(x)$ for an input $x = a$, including the **slope of the tangent line** to $f(x)$ when $x = a$ and the **slope** of $f(x)$ when $x = a$. These same identifiers and interpretations apply when thinking of the derivative as a function. Thus, $f'(x)$ is referred to as the **slope of the tangent line** to $f(x)$.

◆ **EXAMPLE 2** The table below presents the average SAT Math Score in the state of Kentucky for each year. Let $f(x)$ denote the function that outputs the average SAT Math Score for year x. Approximate the value of $f'(x)$ for each input and create a rough sketch of its graph.

Year	1975	1980	1985	1990	1995	2000	2005	2010
SAT Math Score	498	492	500	501	506	514	520	515

Solution. Approximate the value of $f'(x)$ by calculating the average rate of change over intervals containing each input. Working from left to right, the shortest available input interval is chosen in each case. For example, $f'(1975)$ is approximated by calculating the average rate of change over the input interval $[1975, 1980]$. Similarly, $f'(1980)$ is approximated based on the input interval $[1980, 1985]$, $f'(1985)$ based on $[1985, 1990]$, and so on. Applying the formula for the average rate of change, approximate $f'(x)$ when $x = 1975$ and $x = 1980$ as follows:

$$f'(1975) \approx \frac{492 - 498}{1980 - 1975} = -1.2 \qquad\qquad f'(1980) \approx \frac{500 - 492}{1985 - 1980} = 1.6$$

Continuing in this fashion for the remaining years provides the following approximations:

Year	1975	1980	1985	1990	1995	2000	2005
f'(Year)	−1.2	1.6	0.2	1	1.6	1.2	−1

A rough sketch of the graph of f'(Year) is given in Figure 4. ∎

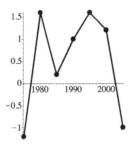

Figure 4: Rough sketch of $f'(\text{Year})$ for Example 2

Example 2 approximated $f'(a)$ using an input interval for which the input $x = a$ was the left endpoint. Alternatively, one could have used an input interval for which $x = a$ was the right endpoint to obtain different, but still reasonable, approximations of $f'(a)$.

▶ **QUESTION 2** The table below presents the U.S. coal prices in dollars per short ton for each year. Let $f(x)$ denote the function that outputs the price of coal per short ton in year x. Approximate the value of $f'(x)$ for each input and create a rough sketch of its graph.

Year	1989	1990	1992	1996	1998
Coal price	$31.36	$30.11	$27.46	$22.25	$20.65

When using an average rate of change to approximate the derivative $f'(x)$ for specific inputs $x = a$, the choice of input intervals plays a significant role in determining the results. Thus far, three different approaches have been noted for selecting input intervals. Example 1 chose intervals with endpoints symmetrically about each input and with integer outputs to simplify the computations. Example 2 made each input the left endpoint of the interval and used the next data point as its right endpoint. Alternatively, the right endpoint of the input interval could be used when approximating $f'(x)$.

All of these approaches provide a reasonable approximation of the derivative and any of these (and other) choices of input intervals can be used when computing average rates of change. As might be expected, different choices for input intervals will usually produce different (but still reasonable) approximations. With so many choices available, two guiding principles should be kept in mind when selecting input intervals to approximate the derivative:

INTERVALS FOR AVERAGE RATES OF CHANGE. When using average rate of change to approximate the derivative, input intervals should

- be as small as the available information allows, and

- contain the input x for which $f'(x)$ is being approximated.

Existence of Derivatives

The derivative function $f'(x)$ is not defined at every input for every function $f(x)$. In fact, the derivative is not defined for any input in the domain of some functions. The common modeling functions turn out to be differentiable on their domains, but piecewise functions defined in terms of the modeling functions are sometimes not differentiable at the boundaries between pieces. Besides being of interest in its own right, recognizing where a function is differentiable and where it is not is a key step in the study of optimization in Chapter 5.

Graphical features that correspond to a function being nondifferentiable are sharp points, including corners and cusps, as well as breaks in the graph, including holes, jumps, and vertical asymptotes. For the following example illustrating where a derivative function does not exist, keep in mind that $f'(x)$ provides the slope of the tangent line to $f(x)$ at each point and that the value of $f'(x)$ is the number approached by average rates of change computed on smaller and smaller input intervals.

◆ **EXAMPLE 3** Verify that the functions shown in Figure 5 are not differentiable when $x = 0$.

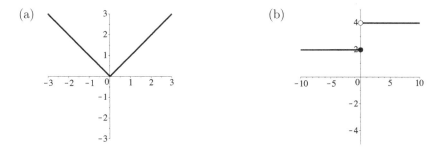

Figure 5: Nondifferentiable functions when $x = 0$ for Example 3

Solution.

(a) On input intervals $[h, 0]$ with h negative, the average rate of change (or the slope of the line) is always -1, while on input intervals $[0, h]$ with h positive, the average rate of change (or slope of the line) is always $+1$. Because these values are different, the function in Figure 3(a) is not differentiable when $x = 0$.

(b) On input intervals $[h, 0]$ with h negative, the average rate of change (or the slope of the line) is always 0, because the function is a horizontal line for inputs less than or equal to zero. However, on input intervals $[0, h]$ with h positive, the average rate of change (or slope of the line) is always positive and, as h becomes smaller and smaller, the slopes become larger and larger, as illustrated in Figure 6. Therefore, the function in Figure 3(b) is not differentiable when $x = 0$, because the average rates of change on $[0, h]$ as h becomes close to zero do not approach a single value but become larger without bound.

■

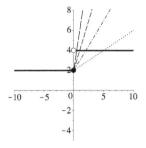

Figure 6: Secant lines for Example 3(b)

◆ **EXAMPLE 4** Verify that the derivative of the function $f(x)$ given in Figure 7 does not exist when $x = 0$ and $x = 2$.

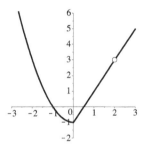

Figure 7: Graph of $f(x)$ for Example 4

Solution. When $x = 0$, the average rates of change approach zero when computed on smaller and smaller intervals $[h, 0]$ with h negative, while the average rates of change are always 2 when computed on intervals $[0, h]$ with h positive. Because these values are different, $f(x)$ is not differentiable when $x = 0$, or, symbolically, $f'(0)$ does not exist.

When $x = 2$, average rates of change cannot be computed for $f(x)$ on any input interval. Namely, the hole in the graph of $f(x)$ when $x = 2$ indicates that $f(2)$ is undefined, which means that the difference quotient $[f(2) - f(h)]/(2 - h)$ is also undefined for every possible h. Therefore, $f'(2)$ does not exist.   ■

➤ **QUESTION 3** Identify the inputs where the function $f(x)$ given in Figure 8 is not differentiable.

Dimensions and Derivatives

Working with real-life data sets (as in Example 2 and Question 2) raises the questions of dimensions and units for derivatives. Before looking at specific examples, these questions are considered from a general perspective. The average rate of change of $f(x)$ is

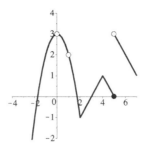

Figure 8: Graph of $f(x)$ for Question 3

computed over small input intervals to determine the instantaneous rate of change of a function. As such, the dimension and units for average and instantaneous rates of change are identical, and so this analysis begins with average rates of change.

Recall that the average rate of change of a function $f(x)$ over $[a, b]$ is computed via the following formula:

$$\text{average rate of change of } f(x) \text{ over } [a, b] \;=\; \frac{f(b) - f(a)}{b - a}$$

Applying rule 1 of dimensional arithmetic from Section 2.7, $f(a)$ and $f(b)$ must have the same dimension and $f(b) - f(a)$ has the same dimension as $f(a)$ and $f(b)$. Similarly, the quantities a, b, and $b - a$ must all have the same dimension. Now, apply rule 2 of dimensional arithmetic, which states that the dimension of a quotient is equal to the quotient of the dimensions of the quantities being divided.

In this setting, suppose that the dimension of the function $f(x)$ is denoted by $[f(x)] = $ [output] and that the dimension of the input x is denoted by $[x] = $ [input]. Using this symbolism, the dimension of the average rate of change of $f(x)$ over $[a, b]$ is given by

$$\left[\frac{f(b) - f(a)}{b - a}\right] \;=\; \frac{[f(b) - f(a)]}{[b - a]} \;=\; \frac{[\text{output}]}{[\text{input}]} \;=\; [\text{output}] \cdot [\text{input}]^{-1}$$

Because the dimensions and units for average rate of change and instantaneous rate of change are identical, the dimension of the derivative $f'(x)$ is the same:

$$[f'(x)] = \frac{[\text{output}]}{[\text{input}]} = [\text{output}] \cdot [\text{input}]^{-1}$$

The following examples determine the dimensions and units of functions and their derivatives in specific settings and also solidify an understanding of the information provided by a function and its derivative:

◆ **EXAMPLE 5** Let $h(t)$ be a function providing the total number of heartbeats that have occurred t minutes since an athlete began a 60-minute workout.

(a) Interpret $h(15) = 2760$ and state both the dimensions and units of t and $h(t)$.

(b) Interpret $h'(62) = -15$ and state both the dimension and units of $h'(t)$.

Solution.

(a) The expression $h(15) = 2760$ means that 15 minutes into the workout the athlete has had 2760 heartbeats. The dimension of the input value t is time, which is expressed symbolically as $[t] = T$, with units of minutes. The dimension of the output value $h(t)$ is an amount, or $[h(t)] = N$, with units of "number of heartbeats."

(b) The derivative $h'(t)$ provides the rate of change in the number of heartbeats per minute. Therefore, the expression $h'(62) = -15$ means that the athlete's number of heartbeats is decreasing at a rate of 15 heartbeats per minute 62 minutes after the beginning of the 60-minute workout, or 2 minutes after the workout is over. In order to find the dimension of $h'(t)$, divide the dimension of the output $[h(t)] = N$ by the dimension of the input $[t] = T$, which results in $[h'(t)] = NT^{-1}$. Similarly, the units of $h'(t)$ are found via division, which gives them as heartbeats per minute.

◆ **EXAMPLE 6** Let $U(t)$ represent the number of active Twitter users worldwide in millions of people as a function of time measured in years.

(a) Interpret $U(12) = 133$ and state both the dimensions and units of t and $U(t)$.

(b) Interpret $U'(12) = 68$ and state both the dimension and units of $U'(t)$.

Solution.

(a) The expression $U(12) = 133$ means that there were 133 million Twitter users worldwide at the beginning of 2012. The dimension of the input value t is time, which is expressed symbolically as $[t] = T$, with units of years. The dimension of the output value $U(t)$ is an amount, or $[U(t)] = N$, with units of "number of Twitter users."

(b) The expression $U'(12) = 68$ means that 68 million Twitter users were being added at the beginning of 2012. In order to find the dimension of $U'(t)$, divide the dimension of the output $[U(t)] = N$ by the dimension of the input $[t] = T$, which results in $[U'(t)] = NT^{-1}$. Similarly, the units of $U'(t)$ are found via division, which gives them as millions of Twitter users per year.

➤ **QUESTION 4** Let $s(t)$ be the function providing the total miles driven t hours since the beginning of a student's Spring Break trip.

(a) Interpret $s(8) = 493$ and state both the dimensions and units of t and $s(t)$.

(b) Interpret $s'(7.45) = 0$ and state both the dimension and units of $s'(t)$.

Higher-Order Derivatives

Differentiable functions (i.e., functions whose derivatives can be computed) are special, with many nice properties. Researchers are also often interested in the **second derivative** $f''(x)$ of a given function $f(x)$, which is obtained by computing the derivative of

$f'(x)$, so that $f''(x) = [f'(x)]'$. Similarly, the following expressions provide the **third derivative** and the **fourth derivative** of $f(x)$:

$$f'''(x) = f^{(3)}(x) = [f''(x)]' \qquad f''''(x) = f^{(4)}(x) = \left[f^{(3)}(x)\right]'$$

Higher-order derivatives are of the form $f^{(n)}(x) = \left[f^{(n-1)}(x)\right]'$.

As mentioned when defining the derivative function, the first derivative of any given function cannot always be computed. Similarly, the second derivative cannot always be computed, nor the third derivative, nor higher-order derivatives of every function. Consequently, functions for which the first, second, third, and more derivatives can be found are distinctive and important. Among other things, the calculus results used to analyze mathematical models hold for such functions. Therefore, these special functions are identified by the name *smooth*, which is defined formally as follows:

Definition. A function $f(x)$ is **smooth** if and only if $f(x)$ has derivatives of all orders.

More symbolically, a function $f(x)$ is smooth when all its derivatives $f'(x)$, $f''(x)$, $f^{(3)}(x)$, and so on exist and are well-defined. As examples, all common modeling functions are smooth, because their derivatives of all orders can be computed on their domains. Functions that are not smooth are discussed in Section 4.7.

Determining the dimensions of higher-order derivatives closely parallels the approach to finding the dimension of the first derivative of a function. Namely, in order to find the dimension of the second derivative $f''(x)$ of a given function $f(x)$, divide the dimension of its first derivative $[f'(x)] = [\text{output}] \cdot [\text{input}]^{-1}$ by the dimension of the input variable $[x] = [\text{input}]$ to obtain the dimension $[f''(x)] = [\text{output}] \cdot [\text{input}]^{-2}$. Continuing in this fashion provides the following relationships among the dimensions of a functions and its first, second, third, and higher-order derivatives:

$$[f(x)] = [\text{output}] \quad \Rightarrow \quad [f'(x)] = [\text{output}] \cdot [\text{input}]^{-1}$$
$$\Rightarrow \quad [f''(x)] = [\text{output}] \cdot [\text{input}]^{-2}$$
$$\Rightarrow \quad [f^{(3)}(x)] = [\text{output}] \cdot [\text{input}]^{-3}$$
$$\vdots$$
$$\Rightarrow \quad [f^{(n)}(x)] = [\text{output}] \cdot [\text{input}]^{-n}$$

Monotonicity of Functions

The derivative is a powerful tool for analyzing the behavior of functions, including the determination of when a function is increasing, constant, or decreasing. In more technical language, the derivative is used to investigate the *monotonic behavior* of functions. Section 1.3 gave the following definitions for when functions are increasing, constant, and decreasing:

> **Definition.** The **monotonicity** of a function $f(x)$ on an open interval (a, b) is defined as follows.
>
> - $f(x)$ is **increasing** on (a, b) if and only if for all numbers c and d in (a, b), when $c < d$ then $f(c) < f(d)$.
>
> - $f(x)$ is **constant** on (a, b) if and only if for all numbers c and d in (a, b), $f(c) = f(d)$.
>
> - $f(x)$ is **decreasing** on (a, b) if and only if for all numbers c and d in (a, b), when $c < d$ then $f(c) > f(d)$.

These definitions provide a mathematically precise way to express functional behavior that is readily identifiable from a graph. When describing functions as increasing, constant, or decreasing, the standard choice is to think about input values becoming larger, which corresponds to reading inputs from left to right on a graph.

When reading inputs from left to right, if the output values of a function become larger, then the function is increasing. Visually, the graph of an increasing function moves upward when following along the graph from left to right. Similarly, if the output values of a function become smaller, then the function is decreasing. Visually, the graph of a decreasing function moves downward when following along the graph from left to right. Finally, a function is constant if every input produces the same output, which corresponds to a horizontal line.

◆ **EXAMPLE 7** Describe the monotonic behavior of the function $f(x)$ given in Figure 9 by identifying the intervals on which $f(x)$ is increasing, constant, and decreasing.

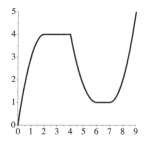

Figure 9: Monotonic behavior of $f(x)$ for Example 7

Solution. The function $f(x)$ is increasing on the input intervals $(0, 2)$ and $(7, 9)$, because the graph is rising upward over these intervals. The function $f(x)$ is constant on the input intervals $(2, 4)$ and $(6, 7)$, because the graph is horizontal over these intervals. The function $f(x)$ is decreasing on the input interval $(4, 6)$, because the graph is falling downward over this interval.   ■

➤ **QUESTION 5** Describe the monotonic behavior of the function $f(x)$ given in Figure 10 by identifying the intervals on which $f(x)$ is increasing, constant, and decreasing.

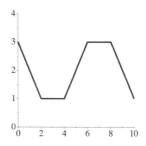

Figure 10: Monotonic behavior of $f(x)$ for Question 5

The monotonic behavior of many functions can be determined from their derivatives. Recall that the value of $f'(a)$, the derivative of $f(x)$ when $x = a$, is equal to the slope of the tangent line to $f(x)$ when $x = a$. Working graphically, one can readily identify whether the slope of the tangent line to a function is positive, zero, or negative at any point and, thus, the corresponding monotonic behavior of the function.

Example 7 found that $f(x)$ is increasing on the intervals $(0, 2)$ and $(7, 9)$. Observe that in these intervals, the tangent line to $f(x)$ at each input x has a positive slope. Therefore, $f'(x)$ is positive for all x in $(0, 2)$ and $(7, 9)$. Also, when $f(x)$ is constant on the intervals $(2, 4)$ and $(6, 7)$, the tangent lines are horizontal and have slope zero. Thus, $f'(x) = 0$ for all x in $(2, 4)$ and $(6, 7)$. Finally, $f(x)$ is decreasing on the interval $(4, 6)$. The tangent lines to $f(x)$ on this interval have negative slopes, and $f'(x)$ is negative for all x in $(4, 6)$. Such connections between the monotonic behavior of a function and the sign of its derivative are valid for many functions.

THE DERIVATIVE AND MONOTONICITY.
Let $f(x)$ be a smooth function on an interval containing $x = a$.

- $f'(a) > 0$ if and only if $f(x)$ is increasing when $x = a$.

- $f'(a) = 0$ if and only if $f(x)$ is constant when $x = a$.

- $f'(a) < 0$ if and only if $f(x)$ is decreasing when $x = a$.

◆ **EXAMPLE 8** Use the graph of $f'(x)$ given in Figure 11 to identify the intervals, if any, on which $f(x)$ exhibits each type of monotonic behavior:

(a) increasing   (b) constant   (c) decreasing

Solution.

(a) The function $f(x)$ is increasing on the input intervals where $f'(x)$ is positive or where $f'(x)$ lies above the x-axis. In Figure 11, observe that $f'(x)$ is positive on $(-3, -1)$ and $(3, \infty)$, which means that $f(x)$ is increasing on $(-3, -1)$ and $(3, \infty)$.

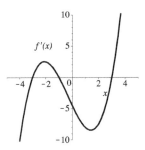

Figure 11: Graph of $f'(x)$ for Example 8

(b) The function $f(x)$ is constant on the input intervals where $f'(x)$ is zero or where $f'(x)$ lies on the x-axis. Because there are no intervals where $f'(x) = 0$, there are no intervals where $f(x)$ is constant. Note that $f'(x) = 0$ when $x = -3$, $x = -1$, and $x = 3$, all of which correspond to points where $f(x)$ has a horizontal tangent line. Such points are important for the study of optimization in Chapter 5.

(c) The function $f(x)$ is decreasing on the input intervals where $f'(x)$ is negative, or where $f'(x)$ is below the x-axis. From Figure 11, observe that $f'(x)$ is negative on $(-\infty, -3)$ and $(-1, 3)$, which means that $f(x)$ is decreasing on $(-\infty, -3)$ and $(-1, 3)$.   ∎

➤ **QUESTION 6** Use the graph of $f'(x)$ in Figure 12 to identify the intervals, if any, on which $f(x)$ is increasing, constant, or decreasing.

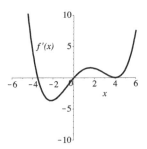

Figure 12: Graph of $f'(x)$ for Question 6

◆ **EXAMPLE 9** Use the graph of $f(x)$ given in Figure 13 to identify the graph of its derivative $f'(x)$ in Figure 14.

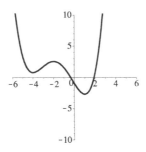

Figure 13: Graph of $f(x)$ for Example 9

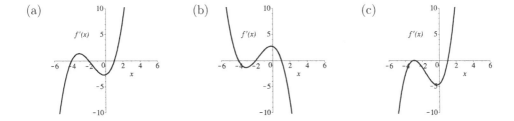

Figure 14: Possible graphs of $f'(x)$ for Example 9

Solution. The monotonicity of $f(x)$ and the corresponding signs of $f'(x)$ are used to decide which graph is $f'(x)$. Working from left to right, observe that $f(x)$ is decreasing on the interval $(-\infty, -4)$, which means that $f'(x)$ is negative (and lies below the x-axis) on this interval. Similarly, on the interval $(-2, 1)$, $f(x)$ is decreasing and so $f'(x)$ is also negative on $(-2, 1)$. On the intervals $(-4, -2)$ and $(1, \infty)$, $f(x)$ is increasing, which means that $f'(x)$ is positive on these intervals (and lies above the x-axis). In summary, the graph of $f'(x)$ lies below the x-axis on the intervals $(-\infty, -4)$ and $(-2, 1)$, and above the x-axis on the intervals $(-4, -2)$ and $(1, \infty)$. Figure 14(a) exhibits these behaviors and presents the graph of $f'(x)$. ◼

➤ **QUESTION 7** The function $c(x) = \dfrac{e}{3}xe^{-x/3}$ provides a reasonable model for the concentration of a medication in the bloodstream of a patient x hours after the medication is injected, in milligrams per milliliter. Use the graph of $c(x)$ given in Figure 15 to identify the graph of its derivative $c'(x)$ among those in Figure 16.

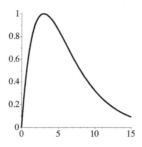

Figure 15: Graph of $c(x)$ for Question 7

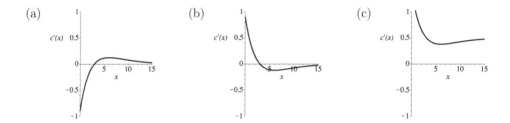

Figure 16: Possible graphs of $c'(x)$ for Question 7

Summary

- When using average rates of change to approximate the derivative, input intervals should

  - be as small as the available information allows, and

  - contain the input $x = a$ for which $f'(x)$ is being approximated.

- The average rate of change of $f(x)$ on a small enough interval containing x provides a reasonable approximation of the instantaneous rate of change $f'(x)$.

- The *derivative function* $f'(x)$ equals the instantaneous rate of change of $f(x)$ at each input x. A function is called *differentiable* if and only if its derivative exists.

- The dimension of $f'(x)$ is the dimension of the output $f(x)$ divided by the dimension of the input x. Symbolically, $[f'(x)] = [\text{output}] \cdot [\text{input}]^{-1}$ and $[f^{(n)}(x)] = [\text{output}] \cdot [\text{input}]^{-n}$.

Summary (continued)

- The *second derivative* $f''(x) = [f'(x)]'$ of a function $f(x)$ is the derivative of $f'(x)$. In general, $f^{(n)}(x) = \left[f^{(n-1)}(x)\right]'$ denotes the *nth derivative* of $f(x)$.

- A function $f(x)$ is *smooth* if and only if $f(x)$ has derivatives of all orders.

- The *monotonicity* of a function $f(x)$ on an open interval (a, b) is defined as follows:

  - $f(x)$ is *increasing* on (a, b) if and only if for all numbers c and d in (a, b), when $c < d$ then $f(c) < f(d)$.

  - $f(x)$ is *constant* on (a, b) if and only if for all numbers c and d in (a, b), $f(c) = f(d)$.

  - $f(x)$ is *decreasing* on (a, b) if and only if for all numbers c and d in (a, b), when $c < d$ then $f(c) > f(d)$.

- Let $f(x)$ be a smooth function on an interval containing $x = a$.

  - $f'(a) > 0$ if and only if $f(x)$ is increasing when $x = a$.

  - $f'(a) = 0$ if and only if $f(x)$ is constant when $x = a$.

  - $f'(a) < 0$ if and only if $f(x)$ is decreasing when $x = a$.

Exercises

In Exercises $1-4$, approximate the derivative $f'(x)$ at the given inputs using an average rate of change on appropriate input interval(s), and draw a rough sketch of $f'(x)$.

1. $x = 0, 2$

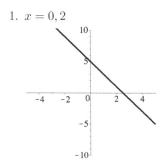

2. $x = -5, -1, 4, 7$

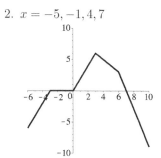

3. $x = -2, 0, 2$

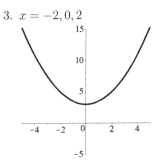

4. $x = 1, 2, 3, 4, 5$

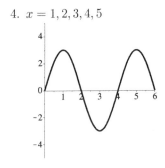

In Exercises 5–8, approximate the derivative $f'(x)$ for each input in the given data set.

5. Percent of high school graduates (P) enrolled in a two-year or four-year college as a function of the year since 2005

Y	0	1	4	5	7
P	68.6	66.0	70.1	68.1	66.2

6. Percent of Americans (P) with incomes below the poverty line each year (Y) since 1970

Y	0	10	20	30	40
P	12.6	13.0	13.5	11.3	15.1

7. S&P 500 stock market closing value (V) on June 1 of each year (Y) since 1975 in U.S. dollars

Y	0	10	20	30	40
V	95	192	545	1191	2077

8. Average monthly maximum temperature (T) in degrees Fahrenheit each month (M) in New York City

M	2	4	7	10	12
T	39.4	61.1	88.3	67.8	43.4

In Exercises 9–12, identify one of the two graphs as the *derivative* $f'(x)$ of the function $f(x)$ defined by the data in the given table using approximations to the average rate of change.

9.

x	-3	-2	0	1	3	4
$f(x)$	11	9	5	3	-1	-3

10.

x	-6	-5	-4	-3	-2
$f(x)$	2.2	2.0	1.8	1.76	1.76
x	-1	0	1	2	3
$f(x)$	1.8	2.0	2.2	2.6	3.0

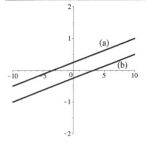

11.

x	-10	-8	-4	-2
$f(x)$	-4.0	0.8	4.2	3.6
x	4	8	10	12
$f(x)$	-2.8	-7.0	-8.0	-7.7

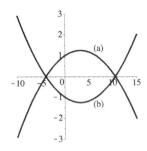

12.

x	-6	-5	-4	-3	-2
$f(x)$	-2.7	-1.2	-0.1	0.7	1.2

x	-1	0	1	2	3
$f(x)$	1.7	2.0	2.3	2.8	3.3

In Exercises 13 – 18, state the dimensions and appropriate units for the input, the output, and the rate of change of the function.

13. $c(f)$ equal to the clock speed of a computer as a function of the frequency of its crystal oscillator

14. $N(t)$ equal to the population of the Netherlands each year

15. $p(u)$ equal to the net profit from a manufactured good as a function of the number of units manufactured

16. $r(t)$ equal to the amount of material recycled from a college campus each month

17. $T(\ell)$ equal to the temperature at a given location on the day of the winter solstice

18. $w(t)$ equal to the amount of water in a leaky bucket at any time after being filled

In Exercises 19 – 22, interpret the expression using appropriate units, where $s(v)$ is the concentration of salt dissolved in a bucket of water in grams per liter as a function of the number of liters of water.

19. $s(9) = 600$ 21. $s'(5) = 50$

20. $s(3) = 1750$ 22. $s'(2) = 751$

In Exercises 23 – 28, interpret the expression using appropriate units, where $m(t)$ is the concentration of a medication in the bloodstream in nanograms per milliliter as a function of hours since the medication was administered.

23. $m(0) = 20$ 26. $m'(1) = 2$

24. $m(2) = 25$ 27. $m'(4) = -3$

25. $m(5) = 10$ 28. $m''(3) = 1$

In Exercises 29 – 42, use the given graph of $f(x)$ to identify the input intervals on which $f(x)$ is increasing, constant, and decreasing. Also, state the sign of its derivative $f'(x)$ on these intervals.

29.

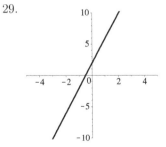

30.

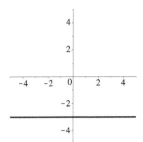

34.

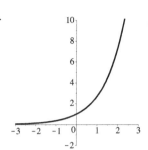

31.

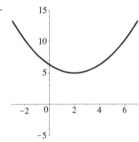

35.

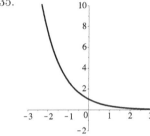

32.

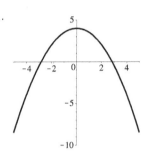

36.

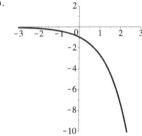

33.

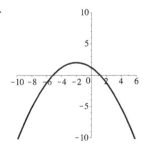

37.

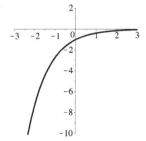

38.

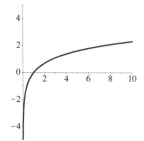

42.

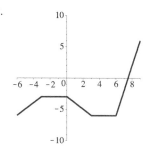

In Exercises 43 – 46, sketch the graph of a function $f(x)$ satisfying the set of conditions.

43. $f(x)$ increasing on $(-\infty, 2)$
$f(x)$ constant on $(2, \infty)$

44. $f(x)$ decreasing on $(-\infty, -1)$
$f(x)$ increasing on $(-1, \infty)$

45. $f(x)$ increasing on $(-\infty, \pi)$
$f(x)$ constant on $(\pi, 10)$
$f(x)$ decreasing on $(10, \infty)$

46. $f(x)$ increasing on $(-\infty, -2)$
$f(x)$ decreasing on $(-2, 4)$
$f(x)$ increasing on $(4, \infty)$

39.

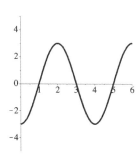

In Exercises 47 – 54, use the given graph of $f(x)$ to sketch the graph of its derivative $f'(x)$ as accurately as possible.

40.

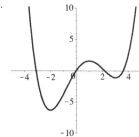

47.

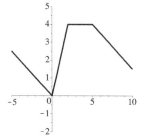

41.

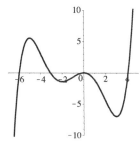

48.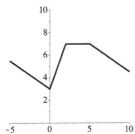

49.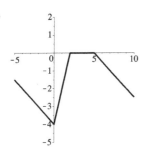

50.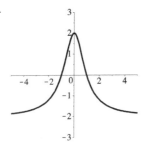

51.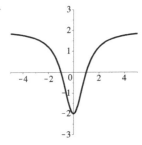

52.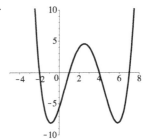

53.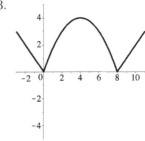

54.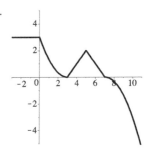

In Exercises 55–60, identify the inputs where the function given here or in the earlier exercise is not differentiable.

55. Exercise 41   57. Exercise 53

56. Exercise 42   58. Exercise 54

59.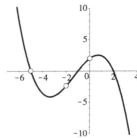

60.

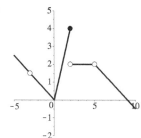

63.

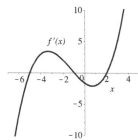

In Exercises 61–74, use the graph of the derivative $f'(x)$ to identify the intervals on which $f(x)$ is increasing, constant, and decreasing.

64.

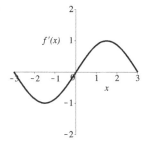

61.

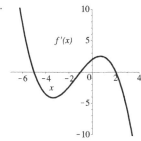

65.

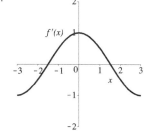

62.

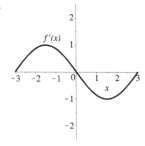

66.

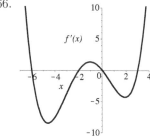

67.

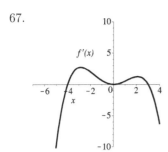

71.

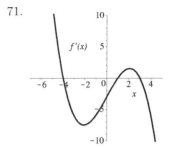

68.

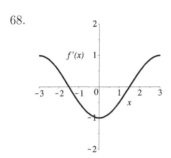

72.

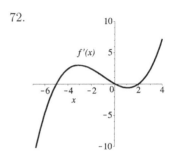

69.

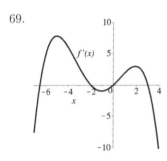

73.

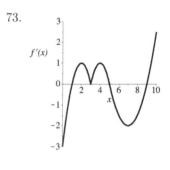

74.

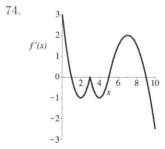

70.

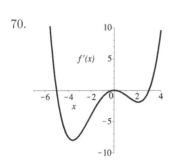

In Exercises 75–88, use the graph of each derivative $f'(x)$ from Exercises 61–74 to produce a rough sketch of a possible $f(x)$. Each exercise has multiple correct solutions.

75. Exercise 61

76. Exercise 62

77. Exercise 63

78. Exercise 64

79. Exercise 65

80. Exercise 66

81. Exercise 67

82. Exercise 68

83. Exercise 69

84. Exercise 70

85. Exercise 71

86. Exercise 72

87. Exercise 73

88. Exercise 74

In Exercises 89 – 96, sketch the graph of a function $f(x)$ whose derivative $f'(x)$ satisfies the set of conditions. Each exercise has multiple correct answers.

89. $f'(x) > 0$ on $(-\infty, \infty)$

90. $f'(x) = 0$ on $(-\infty, \infty)$

91. $f'(x) < 0$ on $(-\infty, \infty)$

92. $f'(x) > 0$ on $(-\infty, 4)$
    $f'(x) = 0$ on $(4, \infty)$

93. $f'(x) > 0$ on $(-\infty, \pi)$
    $f'(x) < 0$ on (π, ∞)

94. $f'(x) = 0$ on $(-\infty, -2)$
    $f'(x) < 0$ on $(2, \infty)$

95. $f'(x) > 0$ on $(-\infty, 2)$
    $f'(x) = 0$ on $(2, 8)$
    $f'(x) < 0$ on $(8, \infty)$

96. $f'(x) < 0$ on $(2, 8)$ and $(14, 20)$
    $f'(x) = 0$ on $(-\infty, 2)$
    $f'(x) > 0$ on $(8, 14)$ and $(20, \infty)$

In Exercises 97 – 100, determine whether $f'(x) = g(x)$ or $g'(x) = f(x)$.

97.

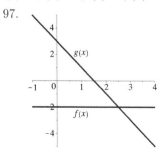

98.

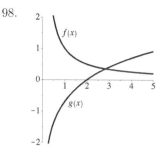

99.

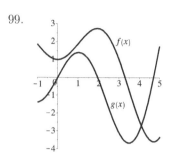

100.

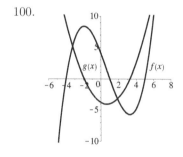

In Your Own Words. In Exercises 101 – 106, explain the following.

101. Derivative function

102. Dimension of $f'(x)$

103. Units of $f'(x)$

104. Smooth function

105. Monotonic behavior of functions

106. The relationship between the monotonocity of $f(x)$ and the sign of its derivative $f'(x)$

4.3   Derivatives of Modeling Functions

This chapter so far has focused on interpreting the information the derivative provides about functions. By definition, the **derivative** $f'(a)$ is equal to the rate at which a function $f(x)$ is changing for a specific input $x = a$, and the **derivative function** $f'(x)$ provides this instantaneous rate of change of $f(x)$ for multiple input values on some domain. From a graphical perspective, the value of $f'(a)$ is equal to the slope of the tangent line to $f(x)$ at the point $(a, f(a))$, and the derivative function $f'(x)$ provides such a slope $f'(a)$ of the tangent line to $f(x)$ depending on the input $x = a$.

This section and the next study rules for computing the derivative of many common functions. Historically, these "calculating" rules for derivatives are the reason why this field of mathematics is called "calculus." In addition to learning how to differentiate common functions, differentiation rules are introduced for various combinations of functions, including sums, differences, products, quotients, and compositions. As one motivation for the existence of such rules, consider the following example of approximations of derivatives by average rates of change:

◆ **EXAMPLE 1**   The tables below present average rates of change (AROC) of $f(x) = \frac{1}{3}x^3$ on various input intervals. Based on these average rates of change, conjecture the value of the derivative $f'(x)$ for $x = -4$, $x = 2$, and $x = 3$, and also a general formula for $f'(x)$.

Interval	AROC
$[-4, -3.9]$	15.6033
$[-4, -3.99]$	15.9600
$[-4, -3.999]$	15.9960

Interval	AROC
$[2, 2.1]$	4.2033
$[2, 2.01]$	4.0200
$[2, 2.001]$	4.0020

Interval	AROC
$[3, 3.1]$	9.3033
$[3, 3.01]$	9.0300
$[3, 3.001]$	9.0030

Solution. Based on these tables, the following provide reasonable conjectures for the values of these derivatives:

$$f'(-4) = 16 \qquad\qquad f'(2) = 4 \qquad\qquad f'(3) = 9$$

For each of these inputs, observe that $f'(x) = x^2$:

$$f'(-4) = 16 = (-4)^2 \qquad\qquad f'(2) = 4 = 2^2 \qquad\qquad f'(3) = 9 = 3^2$$

Based on this evidence, a reasonable further conjecture is that $f(x) = \frac{1}{3}x^3$ has a derivative of $f'(x) = x^2$ for all inputs.   ■

▶ **QUESTION 1**   The tables below present average rates of change (AROC) of $g(x) = x^2$ on various input intervals. Based on these average rates of change, conjecture the value of the derivative $g'(x)$ for $x = -4$, $x = -1$, and $x = 3$, and also a general formula for $g'(x)$.

Interval	AROC
$[-4, -3.9]$	-7.900
$[-4, -3.99]$	-7.990
$[-4, -3.999]$	-7.999

Interval	AROC
$[-1.1, -1]$	-2.100
$[-1.01, -1]$	-2.010
$[-1.001, -1]$	-2.001

Interval	AROC
$[3, 3.1]$	6.100
$[3, 3.01]$	6.010
$[3, 3.001]$	6.001

▶ **QUESTION 2** Calculate the average rate of change of $h(x) = 3x$ on appropriate subintervals for $x = -3$, $x = 1$, and $x = 8$. Based on these average rates of change, conjecture the value of the derivative $h'(x)$ for each of these three inputs and also a general formula for $h'(x)$.

In Example 1 and Questions 1 and 2, patterns in the average rates of change for multiple inputs enabled a reasonable conjecture of a formula for the corresponding derivative function. This section takes this pattern recognition for individual functions to the next level as formulas are articulated for whole families of functions.

As an illustration, notice that Example 1 found that when $f(x) = \frac{1}{3}x^3$, then $f'(x) = \frac{1}{3} \cdot 3 \cdot x^2$; Question 1 found that when $g(x) = x^2$, then $g'(x) = 2 \cdot x^1$; and Question 2 found that when $h(x) = 3x$, then $h'(x) = 3 \cdot 1 \cdot x^0$. In all three cases, the degree of a power function (or, informally, the highest power of x) is one more than the degree of that function's derivative. In addition, the degree of the function is a factor of the coefficient of the derivative. Symbolically, the derivative of $c \cdot x^n$ appears to be $c \cdot n \cdot x^{n-1}$. As it turns out, this calculational rule holds for all functions of the form $c \cdot x^n$ and serves as an example of the kinds of rules that are studied in this section and the next.

In the midst of this shift to learning such rules, the reader is encouraged to keep in mind the interpretation of the derivative: $f'(x)$ provides the instantaneous rate of change of $f(x)$.

Derivatives of Linear Functions

The first differentiation rules are for the simplest modeling functions: linear functions of the form $f(x) = mx + b$. As illustrated in Figure 1(a), the rate of change of a line is equal to the constant slope of the line, giving $f'(x) = m$ for all inputs x. For example, if $f(x) = 2x + 1$ with slope $m = 2$, then $f'(x) = 2$ and, if $g(x) = -0.5x + 1.3$ with slope $m = -0.5$, then $g'(x) = -0.5$.

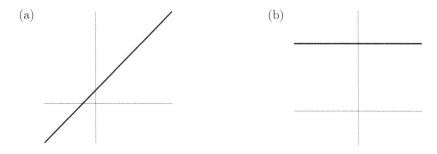

(a)   (b)

Figure 1: Graphs of linear functions

As a special case of this rule consider horizontal lines. As illustrated in Figure 1(b), a horizontal line $f(x) = b$ has a slope m of zero, giving $f'(x) = 0$. For example, if $f(x) = 1$ with slope $m = 0$, then $f'(x) = 0$ and, if $g(x) = \pi$ with slope $m = 0$, then $g'(x) = 0$. This observation corresponds to applying the rule for linear functions noted above. For horizontal lines, $f(x) = b = 0 \cdot x + b$ with slope $m = 0$ and so $f'(x) = m = 0$ for all such lines.

This discussion is summarized by stating the first two differentiation rules. Leibniz's differential notation is used when presenting these rules: recall from Section 4.1 that $\dfrac{d}{dx}[f(x)]$ denotes $f'(x)$.

DIFFERENTIATION RULES FOR LINEAR FUNCTIONS.
Let m and b be real numbers.

Line rule: $\dfrac{d}{dx}[mx + b] = m$ **Constant rule:** $\dfrac{d}{dx}[b] = 0$

An important aspect of developing a proficiency with these differentiation rules is identifying a personally effective means for recalling them. Some people can visually picture the symbolic versions of these rules for linear functions as given above. Other people better recall phrases that express these rules, such as the following.

- For the line rule $\dfrac{d}{dx}[mx + b] = m$, "the derivative of a line is its slope."

- For the constant rule $\dfrac{d}{dx}[b] = 0$, "the derivative of a constant is zero."

While working through these next examples, the reader is encouraged to identify a way of expressing these rules that seems most natural and helpful.

◆ **EXAMPLE 2** Differentiate each function using the rules for linear functions:

(a) $f(x) = 5x - 6$ (b) $g(x) = e^2 - 8$

Solution.

(a) The linear function $f(x) = 5x - 6 = mx + b$ has slope $m = 5$. Applying the line rule $\dfrac{d}{dx}[mx + b] = m$ yields $f'(x) = \dfrac{d}{dx}[5x - 6] = 5$.

(b) Applying the constant rule $\dfrac{d}{dx}[b] = 0$ to the constant function $g(x) = e^2 - 8 = b$ yields $g'(x) = \dfrac{d}{dx}[e^2 - 8] = 0$. ∎

➤ **QUESTION 3** Differentiate each function using the rules for linear functions:

(a) $f(x) = \sqrt{3} + \sqrt{5}$

(b) $g(x) = \pi x + e^2$

◆ **EXAMPLE 3** The U.S. monthly unemployment rate from January 2010 to December 2014 according to the U.S. Bureau of Labor Statistics is stored in `MonthlyUnemployment`. Applying `fitModel` in RStudio produces the following linear model for this data set:

$$\text{Unemploy}(\text{Month}) = -0.07053626 \cdot \text{Month} + 10.17802260$$

Find and explain the meaning of both $\text{Unemploy}(22)$ and $\text{Unemploy}'(22)$.

Solution. The value of $\text{Unemploy}(22)$ is found by substituting Month $= 22$ into the given function and simplifying:

$$\text{Unemploy}(22) = -0.07053626 \cdot 22 + 10.17802260 = 8.626225$$

Therefore, according to this model, U.S. unemployment was 8.63% in October 2011, which is 22 months after the first month of January 2010 in the domain of this model.

Unemploy$'(22)$ is found by differentiating $\text{Unemploy}(\text{Month})$ and substituting Month $= 22$ into the resulting derivative. In this setting, the variable "Month" is synonymous with the variable "x" in the line rule $\frac{d}{dx}[mx + b] = m$, yielding Unemploy$'(\text{Month}) = -0.07053626$. Substituting Month $= 22$ into this constant function gives Unemploy$'(22) = -0.07053626$. Thus, the U.S. unemployment rate was decreasing at a rate of 0.07% per month in October 2011. ∎

➤ **QUESTION 4** The following table presents the global gender ratio based on the number of males per 100 females in each year according to the United Nations Department of Economic and Social Affairs:

Year	1990	1995	2005	2010
Males	101.5	101.5	101.6	101.6

Applying `fitModel` in RStudio produces the linear model $\text{Gender}(\text{Year}) = 0.006 \cdot \text{Year} + 89.550$ for this data set. Find and explain the meanings of both $\text{Gender}(2008)$ and $\text{Gender}'(2008)$.

The Power Rule

With this proficiency in differentiating any linear function in hand, more complex functions can be considered. As one jumping-off point, consider quadratic functions, such as $f(x) = 3x^2 + 4x + 1$ and $g(x) = x^2$, which are commonly used as modeling functions. Earlier in this section, Question 1 conjectured that $g(x) = x^2$ has a derivative $g'(x) = 2 \cdot x^1$. The ensuing discussion suggested that this differentiation rule for x^2

extends in a very natural way to *all* power functions of the form x^n for any real number n, providing essentially the same differentiation rule for all of the following functions:

$$a(x) = \sqrt{x} = x^{\frac{1}{2}} \qquad\qquad b(x) = \frac{1}{x^8} = x^{-8} \qquad\qquad c(x) = \frac{1}{\sqrt[3]{x^7}} = x^{-\frac{7}{3}}$$

As these examples suggest, the exponent rules from Section 1.4 are sometimes needed to identify the power n for these functions. The third differentiation rule for power functions is as follows:

> **POWER RULE.** If n is a real number, then $\dfrac{d}{dx}\left[\, x^n \,\right] \;=\; n \cdot x^{n-1}$.

In the spirit of providing phrases expressing the symbolic versions of these rules, one algorithmic description of this differentiation rule is

> "Bring the power down in front and then subtract one from the power."

For the moment, the power rule is used to differentiate several power functions. After stating further differentiation rules, some graphical evidence is provided to support the validity of this power rule as well as some other differentiation rules.

◆ **EXAMPLE 4** Differentiate each function using the power rule:

(a) $a(x) = \sqrt[3]{x^2}$ $\qquad\qquad$ (b) $b(x) = \dfrac{1}{x^2}$ $\qquad\qquad$ (c) $c(x) = \dfrac{1}{\sqrt[4]{x^5}}$

Solution.

(a) The exponent rule $\sqrt[s]{x^r} = x^{\frac{r}{s}}$ gives $a(x) = \sqrt[3]{x^2} = x^{\frac{2}{3}}$ with an exponent of $n = \frac{2}{3}$. Now apply the power rule:

$$a'(x) \;=\; \frac{d}{dx}\left[\, \sqrt[3]{x^2} \,\right] \;=\; \frac{d}{dx}\left[\, x^{\frac{2}{3}} \,\right] \;=\; \frac{2}{3} x^{\frac{2}{3}-1} \;=\; \frac{2}{3} x^{-\frac{1}{3}} \;=\; \frac{2}{3x^{\frac{1}{3}}} \;=\; \frac{2}{3\sqrt[3]{x}}$$

(b) The exponent rule $1/x^s = x^{-s}$ gives $b(x) = 1/x^2 = x^{-2}$ with an exponent of $n = -2$. Now apply the power rule:

$$b'(x) \;=\; \frac{d}{dx}\left[\, \frac{1}{x^2} \,\right] \;=\; \frac{d}{dx}\left[\, x^{-2} \,\right] \;=\; (-2) \cdot x^{-2-1} \;=\; -2x^{-3} \;=\; \frac{-2}{x^3}$$

(c) The exponent rule $1/\sqrt[s]{x^r} = x^{-\frac{r}{s}}$ gives $c(x) = 1/\sqrt[4]{x^5} = x^{-\frac{5}{4}}$ with an exponent of $n = -\frac{5}{4}$. Now apply the power rule:

$$c'(x) \;=\; \frac{d}{dx}\left[\, \frac{1}{\sqrt[4]{x^5}} \,\right] \;=\; \frac{d}{dx}\left[\, x^{-\frac{5}{4}} \,\right] \;=\; -\frac{5}{4} \cdot x^{-\frac{5}{4}-1} \;=\; -\frac{5}{4} x^{-\frac{9}{4}} \;=\; \frac{-5}{4\sqrt[4]{x^9}}$$

■

➤ **QUESTION 5** Differentiate each function using the power rule:

(a) $a(x) = \dfrac{1}{\sqrt{x}}$

(b) $b(x) = \dfrac{1}{\sqrt[4]{x^3}}$

Differentiation and Basic Arithmetic

We now consider how to differentiate functions that are created by combining simpler functions using the operations of basic arithmetic. Namely, if the derivatives of $f(x)$ and $g(x)$ are known, what is the derivative of their sum $f(x) + g(x)$? And, is the derivative of this sum $f(x) + g(x)$ related to the derivatives $f'(x)$ and $g'(x)$ of the given functions $f(x)$ and $g(x)$? These same questions can be asked about a difference $f(x) - g(x)$ of two differentiable functions $f(x)$ and $g(x)$. Similarly, how does one compute the derivative of $c \cdot f(x)$, when c is a real number and $f(x)$ is a differentiable function? The following rules provide differentiation algorithms for these three arithmetic operations:

DIFFERENTIATION RULES FOR BASIC ARITHMETIC.
Let $f(x)$ and $g(x)$ be differentiable functions, and let c be a real number.

- **Sum rule:** $\dfrac{d}{dx}\left[f(x) + g(x) \right] = f'(x) + g'(x)$

- **Difference rule:** $\dfrac{d}{dx}\left[f(x) - g(x) \right] = f'(x) - g'(x)$

- **Constant multiple rule:** $\dfrac{d}{dx}\left[c \cdot f(x) \right] = c \cdot f'(x)$

Phrases expressing these rules include the following:

- For the sum rule, "the derivative of a sum is the sum of their derivatives."

- For the difference rule, "the derivative of a difference is the difference of their derivatives."

- For the constant multiple rule, "repeat the constant and differentiate the function," or "the constant is along for the ride in a derivative."

Consider the following examples of using these differentiation rules for basic arithmetic:

◆ **EXAMPLE 5** Differentiate each function:

(a) $a(x) = x^3 - 4x^2 + 5x - 6$

(b) $b(x) = \dfrac{5}{x^2} + 7 - 6x^3$

Solution.

(a) $a'(x) = \dfrac{d}{dx}\left[x^3 - 4x^2 + 5x - 6\right]$ Given function

$\qquad = \dfrac{d}{dx}\left[x^3\right] - \dfrac{d}{dx}\left[4x^2\right] + \dfrac{d}{dx}\left[5x - 6\right]$ Difference and sum rules

$\qquad = 3x^2 - 4\dfrac{d}{dx}\left[x^2\right] + 5$ Power, constant multiple, and line rules

$\qquad = 3x^2 - 4 \cdot 2x^1 + 5$ Power rule

$\qquad = 3x^2 - 8x + 5$ Simplify

(b) $b'(x) = \dfrac{d}{dx}\left[\dfrac{5}{x^2} + 7 - 6x^3\right]$ Given function

$\qquad = \dfrac{d}{dx}\left[\dfrac{5}{x^2}\right] + \dfrac{d}{dx}\left[7\right] - \dfrac{d}{dx}\left[6x^3\right]$ Sum and difference rules

$\qquad = 5\dfrac{d}{dx}\left[x^{-2}\right] + \dfrac{d}{dx}\left[7\right] - 6\dfrac{d}{dx}\left[x^3\right]$ Constant multiple and exponent rules

$\qquad = 5 \cdot (-2)x^{-3} + 0 - 6 \cdot 3x^2$ Power and constant rules

$\qquad = -\dfrac{10}{x^3} - 18x^2$ Simplify

■

▶ **QUESTION 6** Differentiate each function:

(a) $a(x) = x^8 + 5 - \dfrac{6}{x} + 3x^4$ (b) $b(x) = 4\sqrt{x^3} + 7x^6 - 8x^2 - 5$

◆ **EXAMPLE 6** The field metabolic rate of certain individual birds and mammals in kilojoules per day as a function of body mass in kilograms according to the *Journal of Ecology* is stored in `BodyMassMetabolicRate`. Applying `fitModel` in RStudio produces the power function model $\text{Metabolic(Mass)} = 911.1768 \cdot \text{Mass}^{0.6528129}$ for this data set. Find and explain the meaning of both $\text{Metabolic}(1.1)$ and $\text{Metabolic}'(1.1)$.

Solution. $\text{Metabolic}(1.1)$ is found by substituting $\text{Mass} = 1.1$ into the given model and simplifying:

$$\text{Metabolic}(1.1) = 911.1768 \cdot (1.1)^{0.6528129} = 969.6708$$

Therefore, according to this model, birds or mammals with a mass of 1.1 kilograms have a field metabolic rate of 969.6708 kilojoules per day.

$\text{Metabolic}'(1.1)$ is found by differentiating Metabolic(Mass) and substituting $\text{Mass} = 1.1$ into the resulting derivative. In this setting, the variable "Mass" is synonymous with the variable "x" in the constant multiple and power rules:

$$\text{Metabolic}'(\text{Mass}) = 911.1768 \cdot 0.6528129 \cdot \text{Mass}^{0.6528129 - 1} = 594.828 \cdot \text{Mass}^{-0.3471871}$$

Substituting yields $\text{Metabolic}'(1.1) = 594.828 \cdot (1.1)^{-0.3471871} = 575.467$. Therefore, for birds or mammals with a mass of 1.1 kilograms, the field metabolic rate is increasing at a rate of 575.467 kilojoules per day per kilogram. Symbolically, the units for $\text{Metabolic}'(\text{Mass})$ are kJ/(day·kg).

■

➤ **QUESTION 7** The running speed in centimeters per second of various animals as a function of their length in centimeters is stored in `RunningSpeed`. Applying `fitModel` in RStudio produces the power function model $\text{Running}(\text{Length}) = 21.87416 \cdot \text{Length}^{1.046521}$ for this data set. Find and explain the meaning of both $\text{Running}(30)$ and $\text{Running}'(30)$.

An economic principle known as the *law of diminishing returns* can be analyzed in terms of the monotonic behavior of functions.

◆ **EXAMPLE 7** Suppose the production function of some economic good is modeled by $p(x) = 100x^{0.35}$. Verify that production is increasing as inputs increase, but that this increase happens at a decreasing rate.

Solution. This setting asks for the verification of two functional behaviors: $p(x)$ is increasing and $p'(x)$ is decreasing. The production function is shown to be always increasing based on its derivative always being positive. Applying the constant multiple and power rules to $p(x)$ yields $p'(x) = 100(0.35)x^{-0.65} = 35x^{-0.65}$. Because x represents the input to a production process, the only situationally relevant values of x are positive, which implies that $p'(x) = 35x^{-0.65} > 0$. Therefore, $p(x)$ is an increasing function for positive inputs.

The production function is shown to be increasing at a decreasing rate by demonstrating that $p'(x)$ is decreasing. A function is decreasing when its derivative is negative, which means that $p'(x)$ is decreasing when its derivative $[p'(x)]' = p''(x)$ is negative. Aplying the constant multiple and power rules to $p'(x)$ yields $p''(x) = 35(-0.65)x^{-1.65} = -22.75x^{-1.65}$, which is always negative because only positive values of x are considered in this setting. Therefore, $p(x)$ is increasing at a decreasing rate. ∎

Differentiating Other Modeling Functions

Now consider the simplest forms of the other basic modeling functions. As with lines and power functions, mathematicians have found differentiation rules that enable the algorithmic computation of the derivatives of these functions.

DIFFERENTIATION RULES FOR OTHER MODELING FUNCTIONS.

- $\dfrac{d}{dx}\left[e^x\right] = e^x$

- $\dfrac{d}{dx}\left[a^x\right] = a^x \cdot \ln(a)$

- $\dfrac{d}{dx}\left[\sin(x)\right] = \cos(x)$

- $\dfrac{d}{dx}\left[\cos(x)\right] = -\sin(x)$

- $\dfrac{d}{dx}\left[\ln(x)\right] = \dfrac{1}{x}$ when $x > 0$

◆ **EXAMPLE 8** Differentiate each function:

(a) $a(x) = \cos(x)$ (b) $b(x) = \ln(x) + 2^x$ (c) $c(x) = 3e^x$

Solution.

(a) $a'(x) = \dfrac{d}{dx}[\cos(x)] = -\sin(x)$

(b) $b'(x) = \dfrac{d}{dx}[\ln(x) + 2^x] = \dfrac{d}{dx}[\ln(x)] + \dfrac{d}{dx}[2^x] = \dfrac{1}{x} + 2^x \cdot \ln(2)$

(c) $c'(x) = \dfrac{d}{dx}[3e^x] = 3\dfrac{d}{dx}[e^x] = 3e^x$

■

➤ **QUESTION 8** Differentiate each function:

(a) $a(x) = \sin(x)$ (b) $b(x) = 10^x - e^x$ (c) $c(x) = 5\ln(x)$

The actual application of differentiation to study and analyze mathematical models requires further extensions of the preceding differentiation formulas for the simplest forms of the basic modeling functions. Most importantly, Section 4.5 introduces the *chain rule* for computing the derivatives of compositions of differentiable functions. As a small step in this direction, the following rule provides the derivative of the natural exponential function composed with a linear function:

EXTENDED EXPONENTIAL RULE.

Let m and b be real numbers. Then $\dfrac{d}{dx}\left[e^{mx+b}\right] = e^{mx+b} \cdot m.$

When presenting a constant multiple of an exponential, mathematicians and scientists usually write the constant first, then the exponential function. For example, one typically will read about $2e^{2x+4}$ rather than $e^{2x+4}2$. Thus, this extended exponential rule might be presented with some subsequent simplification as follows:

$$\frac{d}{dx}\left[e^{mx+b}\right] = e^{mx+b} \cdot m = m \cdot e^{mx+b} = me^{mx+b}$$

The ordering given above will facilitate learning how to apply the chain rule in Section 4.5. For now, this differentiation rule is used to analyze certain exponential models of real-life phenomena.

◆ **EXAMPLE 9** The number of physicians per 1000 people as a function of average life expectancy in various countries in 2010 according to the World Bank is stored in `LifeExpectancyPhysicians`. Applying `fitModel` in RStudio produces the following exponential model for this data set:

$$\text{Physicians}(\text{Years}) = e^{0.1489836 \cdot \text{Years} - 10.6608944}$$

Find and explain the meaning of both Physicians(80) and Physicians$'$(80).

Solution. Physicians(80) is found by substituting Years = 80 into the given exponential model and simplifying:

$$\text{Physicians}(80) = e^{0.1489836 \cdot 80 - 10.6608944} = 3.517$$

Therefore, according to this model, a country where the life expectancy is 80 years has 3.517 physicians per 1000 people.

Physicians$'$(80) is found by differentiating Physicians(Years) and substituting Years = 80. In this setting, the variable "Years" is synonymous with the variable "x" in the differentiation rule for exponential functions.

$$\text{Physicians}'(\text{Years}) = 0.1489836 \cdot e^{0.1489836 \cdot \text{Years} - 10.6608944}$$

Substituting yields Physicians$'$(80) $= 0.1489836 \cdot e^{0.1489836 \cdot 80 - 10.6608944} = 0.524$. Thus, for a country where the life expectancy is 80 years, the number of physicians per 1000 people is changing at a rate of 0.524 physicians per 1000 people per year of life expectancy. ■

▶ **QUESTION 9** The closing NASDAQ stock market value in U.S. dollars at the end of each quarter from March 1938 (quarter 1) through December 2014 (quarter 308) is stored in `NASDAQQuarterly`. Applying `fitModel` in RStudio produces the following exponential model for this data set:

$$\text{Nasdaq}(\text{Quarter}) = e^{0.02258515 \cdot \text{Quarter} + 1.4908287}$$

Find and explain the meaning of both Nasdaq(300) and Nasdaq$'$(300).

Evidence for Differentiation Rules

Beyond knowing these differentiation rules and how to apply them to various functions, an important goal is to understand why they are true. What arguments support the validity of these rules holding for *all* functions with a particular form?

Complete proofs establishing differentiation rules require the notion of a limit, which is introduced in Section 4.7. In the meantime, the graphical relationship between a function $f(x)$ and its derivative $f'(x)$ is explored as one type of evidence supporting the validity of the differentiation rules for the functions x^2, $\sqrt{x}$, e^x, and $\sin(x)$.

Particular attention is paid to the connection between the monotonic behavior of functions and the signs of their derivatives. Recall from Section 4.2 that a smooth function $f(x)$ is increasing on an input interval if and only if its derivative $f'(x)$ is positive throughout the interval. Similarly, for smooth functions, $f(x)$ is constant if and only if $f'(x)$ is equal to zero, and $f(x)$ is decreasing if and only if $f'(x)$ is negative on the corresponding input intervals.

This study of such graphical evidence for differentiation rules begins with power functions.

◆ **EXAMPLE 10** Differentiate each function, graph the function and its derivative on the same axes, and then discuss the correspondence between the monotonic behavior of the function and the sign of its derivative:

(a) $a(x) = x^2$

(b) $b(x) = \sqrt{x}$

Solution.

(a) The function $a(x) = x^2$ has an exponent $n = 2$. Now apply the power rule:

$$a'(x) \;=\; \frac{d}{dx}\left[x^2\right] \;=\; 2 \cdot x^{2-1} \;=\; 2x^1 \;=\; 2x$$

Figure 2(a) presents the graphs of $a(x) = x^2$ and $a'(x) = 2x$ on the same axes. Observe that when $x = 0$, the parabola $a(x) = x^2$ has a horizontal tangent line with a slope of zero, and its derivative $a'(x) = 2x$ is also equal to zero. Furthermore, $a(x)$ is decreasing and $a'(x)$ is negative for all inputs x from the interval $(-\infty, 0)$. Similarly, $a(x)$ is increasing and $a'(x)$ is positive for all inputs x from the interval $(0, \infty)$. These appropriately corresponding functional behaviors provide graphical evidence in support of the fact that the derivative of x^2 is $2x$.

(b) The exponent rule $\sqrt[r]{x} = x^{\frac{1}{r}}$ gives $b(x) = \sqrt{x} = x^{\frac{1}{2}}$ with an exponent of $n = \frac{1}{2}$. Now apply the power rule:

$$b'(x) \;=\; \frac{d}{dx}\left[x^{\frac{1}{2}}\right] \;=\; \frac{1}{2} \cdot x^{\frac{1}{2}-1} \;=\; \frac{1}{2}x^{-\frac{1}{2}} \;=\; \frac{1}{2\sqrt{x}}$$

Figure 2(b) presents the graphs of $b(x) = \sqrt{x}$ and $b'(x) = 1/(2\sqrt{x})$ on the same axes. The square root function $b(x) = \sqrt{x}$ is always increasing and its derivative $b'(x)$ is always positive. Also, observe that $b(x)$ flattens out as the inputs increase, which means that the corresponding tangent lines become more horizontal, with slopes closer to zero. In parallel with this $b(x)$ behavior, $b'(x)$ approaches zero as the values of the inputs increase. These appropriately corresponding functional behaviors indicate that the derivative of $\sqrt{x}$ is $1/(2\sqrt{x})$.   ■

➤ **QUESTION 10** Differentiate the function $f(x) = x^3$, graph $f(x)$ and $f'(x)$ on the same axes, and then discuss the correspondence between the monotonic behavior of $f(x)$ and the sign of $f'(x)$.

Exponential functions are considered next. One defining property of an exponential function is that the value of its instantaneous rate of change is directly proportional to the value of the function. Symbolically, this property is expressed by the equation

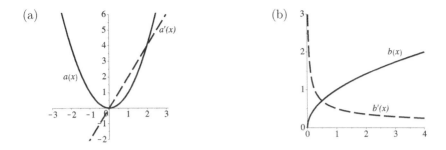

Figure 2: Graphs of power functions and their derivatives for Example 10

$f'(x) = k \cdot f(x)$, where k is a constant real number that is unique to each exponential $f(x) = a^x$. In the case of the natural exponential function, $f(x) = e^x$ satisfies the equation $f'(x) = k \cdot f(x)$ for $k = 1$, giving $f'(x) = 1 \cdot f(x) = f(x)$, or $[e^x]' = e^x$. Therefore, for every input x, the value of the derivative $f'(x) = e^x$ is the same as the function value $f(x) = e^x$. This differentiation property is the calculus-based reason for e^x to be called the "natural" exponential function.

Figure 3 indicates that the slope of e^x is equal to the height of e^x for every input x. In more detail, when the height of e^x is close to zero to the left of the y-axis, the slope of e^x is relatively flat; when the height is $e^0 = 1$ as e^x crosses the y-axis, the slope of e^x is also 1; and as the height of e^x increases to the right of the y-axis, the slope of e^x becomes steeper. These appropriately corresponding functional behaviors provide graphical evidence that the derivative of e^x is e^x.

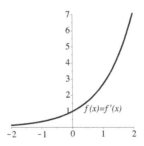

Figure 3: The graphs of $f(x) = e^x$ and its derivative $f'(x) = e^x$

Finally, consider the trigonometric function sine. As might be expected, the derivative of this periodic function is also periodic. The differentiation rules for other modeling functions assert that $\sin(x)$ and $\cos(x)$ "swap" when computing their derivatives, and the sign changes when differentiating $\cos(x)$. Figure 4 presents the graph of $f(x) = \sin(x)$ and its derivative $f'(x) = \cos(x)$ on the same axes.

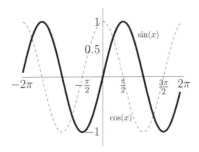

Figure 4: The graphs of $f(x) = \sin(x)$ and its derivative $f'(x) = \cos(x)$

In Figure 4, note the correspondence between the horizontal tangent lines of $f(x) = \sin(x)$ and the zeros of $f'(x) = \cos(x)$. Both of these behaviors occur for the same inputs of $x = -\pi$, $x = 0$, and $x = \pi$ in Figure 4. More generally, when $x = n\pi$ for any integer n equal to one of $\ldots, -3, -2, -1, 0, 1, 2, 3, \ldots$, the function $f(x) = \sin(x)$ has a horizontal tangent line and its derivative $f'(x) = \cos(x)$ is equal to zero. Also, observe that the monotonic behavior of $\sin(x)$ corresponds to the sign of its derivative $\cos(x)$. In more detail, $f(x) = \sin(x)$ is decreasing and $f'(x) = \cos(x)$ is negative on input intervals $(-3\pi/2, -\pi/2)$ and $(\pi/2, 3\pi/2)$; similarly, $f(x) = \sin(x)$ is increasing and $f'(x) = \cos(x)$ is positive on input interval $(-\pi/2, \pi/2)$. These matching patterns repeat throughout these periodic functions' domains of $(-\infty, \infty)$. These appropriately corresponding functional behaviors provide graphical evidence that the derivative of $\sin(x)$ is $\cos(x)$.

Tangent Lines and Linear Approximations

Section 4.1 introduced the tangent line question: Given the graph of a function $f(x)$ and the point $(a, f(a))$ on this graph, what is the equation of the tangent line to $f(x)$ when $x = a$? Recall that the slope of the tangent line to $f(x)$ when $x = a$ is given by $f'(a)$ and that the equation of the tangent line to $f(x)$ when $x = a$ is $y = f(a) + f'(a) \cdot (x - a)$. Therefore, in order to state an equation of a tangent line, finding the values of $f(a)$ and $f'(a)$ are usually the primary tasks. In Section 4.1, the only method provided for finding the value of $f'(a)$ was by means of an approximation based on average rates of change over smaller and smaller input intervals containing the given input $x = a$. This section has greatly simplified and improved the process of accurately determining $f'(a)$ by means of the differentiation rules, as illustrated next.

◆ **EXAMPLE 11**   Repeat Example 4 from Section 4.1:   find the tangent line to $f(x) = x^3 - 7x$ when $x = 2$.

Solution. Applying the difference and power rules to $f(x) = x^3 - x^2 - 1$ gives $f'(x) = 3x^2 - 7$. Now substitute $x = 2$ to find both the value of the function and the slope of its tangent line:

$$f(2) = (2)^3 - 7 \cdot 2 = 8 - 14 = -6 \qquad\qquad f'(2) = 3(2)^2 - 7 = 12 - 7 = 5$$

Thus, the point $(2, f(2)) = (2, -6)$ lies on both the graph of $f(x)$ and the tangent line to $f(x)$ when $x = 2$. Substituting these values into the formula for the equation of a tangent line gives $y = -6 + 5(x - 2)$ or, simplifying, $y = 5x - 16$. ∎

➤ **QUESTION 11** Repeat Question 6 from Section 4.1: find the equation of the tangent line to $g(x) = x^2 + 3x$ when $x = 3$.

➤ **QUESTION 12** Find the tangent line to $h(x) = \ln(x) + x$ when $x = 5$.

Among their many uses, tangent lines happen to provide a relatively simple way to approximate more complicated functions. Linear functions are the simplest, best understood functions, making them an excellent choice for approximating more complicated functions. In addition, for many functions $f(x)$, the tangent line to $f(x)$ when $x = a$ outputs values close to $f(x)$ and behaves similarly to $f(x)$, at least for inputs near (or local to) $x = a$. As an example, consider Figure 5, which presents the tangent lines to $f(x) = \frac{1}{4}x^2$ when $x = -4$ and when $x = 2$. Notice that for inputs in the local vicinity of $x = -4$ and $x = 2$, the output values of the function and the corresponding tangent lines are very close together, and they exhibit similar behaviors.

Such close interconnections occur for many functions, which enables the use of the tangent line to obtain a good, local approximation of a function.

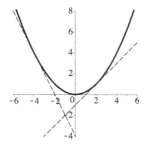

Figure 5: Tangent line approximation of a function

LINEAR APPROXIMATIONS.

For a smooth function $f(x)$, the tangent line to $f(x)$ when $x = a$ given by $L(x) = f(a) + f'(a) \cdot (x - a)$ provides a **linear approximation** of $f(x)$ when $x = a$ in the sense that for inputs near $x = a$, both

- the output values of $f(x)$ and $L(x)$ are approximately equal, and

- $f(x)$ and $L(x)$ exhibit similar monotonic behavior.

◆ **EXAMPLE 12** The following table presents median home prices P in thousands of dollars each year Y from 2001 to 2010 according to the U.S. Census Bureau:

Year	2001	2002	2003	2004	2005	2006	2007	2008	2009	2010
Price	175.2	187.6	195.0	221.0	240.9	246.5	247.9	232.1	216.7	221.8

Applying `fitModel` in RStudio produces the following model for this data set:

$$P(Y) = 0.17(Y-2001)^4 - 3.2(Y-2001)^3 + 16.7(Y-2001)^2 - 10.9(Y-2001) + 177.7$$

(a) Find a linear approximation of P(Y) when Y = 2002.

(b) Compare the value of this linear approximation and the given model when Y = 2002.5.

Solution.

(a) P(2002) is found by substituting Y = 2002 into the given function P(Y) to obtain P(2002) = 180.47. The slope P′(2002) of the approximating line is found by differentiating P(Y) and substituting Y = 2002. In this setting, the variable "Y" is synonymous with the variable "x" in the differentiation rules. Differentiating P(Y) requires a slight extension of the power rule: $[(x-a)^n]' = n \cdot (x-a)^{n-1}$ for all real numbers a, including $a = 2001$; this extension follows from the chain rule that will be introduced in Section 4.5. Applying the differentiation rules gives the following:

$$P'(Y) = 0.68(Y-2001)^3 - 9.6(Y-2001)^2 + 33.4(Y-2001) - 10.9.$$

Substituting Y = 2002 yields P′(2002) = 13.58. Therefore, the linear approximation of P(Y) when Y = 2002 is given by $L(Y) = 13.58(Y-2002) + 180.47$.

(b) The value of the function is P(2002.5) = 188.99, and the value given by the linear approximation is L(2002.5) = 187.26. While these values are not exactly equal, they are approximately equal, and working with the linear approximation $L(Y) = 13.58(Y-2002)+180.47$ is much easier than the original, more complicated function P(Y) given above. ■

➤ **QUESTION 13** Example 9 introduced $Physicians(Years) = e^{0.1489836 \cdot Years - 10.6608944}$ as a reasonable exponential model of the number of physicians per 1000 people as a function of average life expectancy in various countries in 2010.

(a) Find a linear approximation of Physicians(Years) when Years = 60.

(b) Compare the value of this linear approximation and the model when Years = 62.

Working in RStudio

Derivatives are computed in RStudio using the "D" command inside the `mosaicCalc` package, which can be accessed after entering the command `require(mosaicCalc)`.

As with `makeFun` and `plotFun`, the command D is followed by a function, a tilde "~", and then the variable of differentiation. For example, $f(x) = x^2 + e^{3x-1} - 4\sin(x)$ is differentiated by entering

    D(x^2+exp(3*x-1)-4*sin(x)~x)

As with `makeFun`, the resulting derivative function can be stored and named by beginning this command with `df=` or some other appropriate name based on the context. This name `df` can then be used to evaluate the derivative on various inputs.

The derivatives of functions previously defined and stored in RStudio can also be computed using the D command. For example, suppose $f(x) = x^3 + 2$ has been defined via the command `f=makeFun(x^3+2~x)`. RStudio computes $f'(x)$ and simultaneously names this derivative function based on the command `df=D(f(x)~x)`. When using this approach to compute a derivative, the "`(x)`" must be included with the name "`f`".

Examples of Commands:

- `D(6*x^3-3*x^2~x)`   - `df=D(f(x)~x)`

- `f=makeFun(exp(6*x^3-3)+x^2~x)`   - `df(x=3)`

◆ **EXAMPLE 13** Find the tangent line to $f(x) = x^5 + 3x^2 + 4x$ when $x = -3$ using RStudio.

Solution. The following commands define $f(x) = x^5 + 3x^2 + 4x$ in RStudio, compute its derivative $f'(x)$, and evaluate both for $x = -3$:

```
f = makeFun(x^5+3*x^2+4*x~x)          df = D(f(x)~x)
f(x=-3)                               df(x=-3)

[1] -228                              [1] 391
```

Based on this output, $f(-3) = -228$ and $f'(-3) = 391$. Substituting into the formula $y = f(a) + f'(a) \cdot (x - a)$ for the equation of a tangent line yields $y = -228 + 391(x + 3)$ or, simplifying, $y = 391x + 945$.   ■

➤ **QUESTION 14** Find the tangent line to $g(x) = \dfrac{3}{\pi}x - \cos(x)$ when $x = \pi$ using RStudio.

Summary

- Let b, m, n, and c be real numbers, and $f(x)$ and $g(x)$ be differentiable functions.

  - $\dfrac{d}{dx}[b] = 0$
  - $\dfrac{d}{dx}[mx + b] = m$
  - $\dfrac{d}{dx}[x^n] = n \cdot x^{n-1}$
  - $\dfrac{d}{dx}[e^x] = e^x$
  - $\dfrac{d}{dx}[a^x] = a^x \cdot \ln(a)$
  - $\dfrac{d}{dx}[\ln(x)] = \dfrac{1}{x}$, when $x > 0$

  - $\dfrac{d}{dx}[\sin(x)] = \cos(x)$
  - $\dfrac{d}{dx}[\cos(x)] = -\sin(x)$
  - $\dfrac{d}{dx}[f(x) + g(x)] = f'(x) + g'(x)$
  - $\dfrac{d}{dx}[f(x) - g(x)] = f'(x) - g'(x)$
  - $\dfrac{d}{dx}[c \cdot f(x)] = c \cdot f'(x)$

- *Extended Differentiation Rules*: Let m, b, and a be real numbers.

  - $\dfrac{d}{dx}[e^{mx+b}] = m \cdot e^{mx+b}$
  - $\dfrac{d}{dx}[(x-a)^n] = n \cdot (x-a)^{n-1}$

- For a smooth function $f(x)$, the tangent line $L(x) = f(a) + f'(a) \cdot (x - a)$ provides a *linear approximation* of $f(x)$ when $x = a$ in the sense that for inputs near $x = a$, both

  - the output values of $f(x)$ and $L(x)$ are approximately equal, and
  - $f(x)$ and $L(x)$ exhibit similar monotonic behavior.

Exercises

In Exercises 1–8, differentiate the function using the rules for linear functions.

1. 27

2. $\ln(2)$

3. $e^3 - 4$

4. $\sin(4)$

5. $7x - \pi$

6. $4 - \pi x$

7. $ex + e$

8. $-8x + 2$

In Exercises 9–24, differentiate the function using the power rule and exponent rules.

9. x^8

10. x^{22}

11. x^{-6}

12. $\sqrt{x}$

13. $\sqrt[4]{x}$

14. $\sqrt[3]{x^5}$

15. $\sqrt{x^3}$

16. $x\sqrt{x}$

17. $x^2 \sqrt[3]{x}$

18. $\dfrac{1}{x^3}$

19. $\dfrac{x^4}{\sqrt{x}}$

20. $\dfrac{1}{\sqrt[5]{x^3}}$

21. $\dfrac{x^3 \cdot x^2}{\sqrt{x}}$

22. $\dfrac{1}{x\sqrt[5]{x}}$

23. $\dfrac{x^6}{\sqrt{x}}$

24. $\dfrac{x^2}{x\sqrt[4]{x^8}}$

48. $e^{\pi x - 8} - 5\sin(9x - 8x)$

49. $\ln(\sqrt[3]{x^4}) - \sin(\pi) + 7^x$

50. $\ln(x) - 8x + 2 - e^x$

51. $\ln\left[\dfrac{1}{x}\right] + \ln\left[\dfrac{1}{\sqrt{x}}\right]$

52. $17\sin(x) + \pi - \pi x$

In Exercises 25–32, differentiate the function using the rules for exponential functions.

25. e^{4x+8}

29. e^{2x+6}

26. e^{-2x+3}

30. e^{6-2x}

27. $e^{\pi x + 7}$

31. $e^{-\pi x + 8}$

28. e^{13-ex}

32. e^{ex-3}

In Exercises 33–52, differentiate the function.

33. $\ln(6) + x^4 - e^x$

34. $\dfrac{1}{x^6} - \pi^3 + 4x^3 + e$

35. $\sqrt[4]{x^7} + e^7 - 4 - \dfrac{3x^6}{x^2}$

36. $5\ln(x) + 3\cos(\pi + 2)$

37. $ex + 7 - \dfrac{1}{x\sqrt[5]{x^9}}$

38. $6e^{\pi x + 7} - 7x + 9\cos(x)$

39. $3e^x - 4 - 2\sin(x)$

40. $5e^{8x-4} - 3e^{9-6x}$

41. $\sqrt[3]{x} - 10\ln(x) + 14$

42. $\sqrt{x^5} + 2\cos(4x - 3x) - 6e^{6-3x}$

43. $x\sqrt{x^3} - \pi\ln(x) + 22 \cdot e^{-x+13}$

44. $x^3\sqrt{x} - 9x^{10} + \sin(x)$

45. $\dfrac{x^3}{x^8} - \dfrac{1}{\sqrt[5]{x^3}} - 6\ln(x)$

46. $\dfrac{x^2 \cdot x^7}{\sqrt{x}} + \dfrac{17}{x\sqrt[5]{x}}$

47. $\dfrac{x^4}{\sqrt[7]{x}} + 3\cos(x) - 3e^x$

In Exercises 53–56, consider the linear model $\text{HE(Year)} = 0.2998 \cdot \text{Year} - 585.214$ of total U.S. health expenditures as a percentage of GDP from 1995 to 2012. Find and explain the meaning of the quantity.

53. (a) HE(1995)   (b) HE$'$(1995)

54. (a) HE(2000)   (b) HE$'$(2000)

55. (a) HE(2005)   (b) HE$'$(2005)

56. (a) HE(2010)   (b) HE$'$(2010)

In Exercises 57–60, consider the power function model

$$\text{RPM(Mass)} = 13545.5 \cdot \text{Mass}^{-0.339622}$$

of the revolutions per minute of an engine as a function of its mass in kilograms. Find and explain the meaning of the quantity.

57. (a) RPM(0.4)   (b) RPM$'$(0.4)

58. (a) RPM(50)   (b) RPM$'$(50)

59. (a) RPM(200)   (b) RPM$'$(200)

60. (a) RPM(1000)   (b) RPM$'$(1000)

In Exercises 61–64, consider the exponential model

$$\text{Wt(Week)} = 781.9e^{0.1216(\text{Week}-25)}$$

of the average weight of male fetuses in grams as a function of gestational week. Find and explain the meaning of the quantity.

61. (a) Wt(20) (b) Wt$'(20)$

62. (a) Wt(25) (b) Wt$'(25)$

63. (a) Wt(30) (b) Wt$'(30)$

64. (a) Wt(35) (b) Wt$'(35)$

In Exercises 65–72, find the linear approximation $L(x)$ of the function for the given input.

65. $f(x) = 3x - 4$ for $x = 2$

66. $f(x) = 3x^2 + 3x$ for $x = 1$

67. $f(x) = x^4 + 7$ for $x = -1$

68. $f(x) = e^x$ for $x = 2$

69. $f(x) = e^{2x-1}$ for $x = 1.5$

70. $f(x) = \ln(x)$ for $x = 2$

71. $f(x) = \sin(x)$ for $x = 0$

72. $f(x) = \cos(x)$ for $x = \pi$

In Exercises 73–76, find the linear approximation of the model for the given input.

73. Weight $= 0.0632 \cdot$ Week $+ 1.9034$ models the change in a woman's weight during a pregnancy from the beginning of her second trimester until birth, for Week $= 35$.

74. Flying $= 636.4227 \cdot$ Length$^{0.255245}$ models the maximum flying speed of animals in centimeters per second as function of their length in centimeters, for Length $= 20$.

75. Average $= e^{-56.461488} \cdot$ Year$^{8.206029}$ models the atmospheric carbon dioxide from Mauna Loa in ppmv as a function of years from 1958 to 2008, for Year $= 2005$.

76. WPop $= 2.633746e^{0.016511(\text{Year}-1950)}$ models the U.S. Census Bureau's estimates of world population in billions of people since 1950, for Year $= 2000$.

In Exercises 77–80, differentiate the function, graph the function and its derivative on the same axes, and discuss the correspondence between the monotonic behavior of the function and the sign of its derivative.

77. $f(x) = -x^3$

78. $f(x) = 4x^{1/3}$

79. $f(x) = \ln(x)$ for $x > 0$

80. $f(x) = \cos(x)$

RStudio. In Exercises 81–88, use RStudio to differentiate the function.

81. $a(x) = x^3 - 4x^2 + 5x - 6$

82. $b(x) = 7\sin(x + 8) + \pi x^{100}$

83. $c(x) = 3 \cdot 2^{x+8} + e^x$

84. $d(x) = 5\cos(\pi x + 3) + 12.1e^{3x-14}$

85. $e(x) = \dfrac{x^2 + 2x}{\sqrt{x}} + 2^{3(x-45)+13}$

86. $f(x) = \dfrac{e^{3x-2}}{e^{5x+3}} - 14.5\cos(ex + \pi)$

87. $g(x) = \ln((5x)^4 + 9) - \ln(\sqrt{x})$

88. $h(x) = 4\ln(3x + 7) + 5\cos(2x - 9)$

RStudio. In Exercises 89–94, use RStudio to find the linear approximation of the function for the given input.

89. $a(x) = x^2 - 3x + 4$ for $x = 2$

90. $b(x) = 4e^{2x-3}$ for $x = 1/2$

91. $c(x) = 2^{3x} + 3^{4x}$ for $x = -1$

92. $d(x) = 2\ln(4x - 9)$ for $x = 3$

93. $b(x) = 5\cos(x + 3)$ for $x = 1$

94. $d(x) = 4\sin(3x + 7)$ for $x = \pi$

4.4 Product and Quotient Rules

Section 4.3 provided differentiation rules for the simplest forms of the common modeling functions and for basic arithmetic operations. This section continues the development of these calculational rules by studying the differentiation rules for the remaining two arithmetic operations: multiplication and division. For both of these operations, the derivative of the function created by combining two functions $f(x)$ and $g(x)$ depends on their derivatives $f'(x)$ and $g'(x)$. The most straightforward of such connections occur for sums, differences, and constant multiplication: the derivative of a sum of differentiable functions is the sum of the derivatives of the functions, and so on. The formulas for products and quotients are more elaborate, but they still provide a definitive algorithmic process for computing derivatives.

Many different reasons motivate an interest and need for studying these rules. Some people appreciate a certain sense of completeness that arises from knowing how to differentiate all arithmetic combinations of functions. Others enjoy the increasing power and competency that comes from expanding the collection of readily differentiable functions. Alternatively, some people are more interested in using the derivative as a tool to analyze and understand our physical and social world. All of these perspectives provide strong reasons for studying these rules.

While Section 4.3 introduced differentiation rules for the simplest forms of the common modeling functions, many phenomena exhibit more subtle behaviors, and differentiating these more sophisticated models requires rules for products and quotients of functions. For example, the concentration of a medication in the bloodstream as a function of the time t elapsed since taking the medication is often modeled by the product $(mt + b)e^{-at}$ of a linear function $mt + b$ and a decaying exponential e^{-at}, where m, b, and a are context-dependent constants. Another such example arises from the study of long-term population trends of humans and other living beings that are often modeled by sigmoidal functions, which are typically presented as quotients:

$$\text{Population(Year)} = \frac{L}{1 + Ce^{-k(\text{Year}-h)}} + v$$

In order to study these and other similarly more sophisticated models, differentiation rules are needed for products and quotients of functions.

While learning these new differentiation rules, the reader will want to keep in mind the meaning and interpretation of the derivative. Recall that the derivative $f'(x)$ provides the instantaneous rate of change of a function $f(x)$ and enables the analysis of the monotonic behavior of $f(x)$. The derivative is also used to obtain linear approximations of differentiable functions that provide a simple means to approximate functional values for inputs near (or local to) where the corresponding tangent line is based. Further applications of the derivative are explored in Chapter 5.

The Product Rule

Section 4.3 introduced the following differentiation rules for addition and subtraction:

$$\frac{d}{dx}\left[f(x) + g(x)\right] = f'(x) + g'(x) \qquad \frac{d}{dx}\left[f(x) - g(x)\right] = f'(x) - g'(x)$$

In other words, the derivative of a sum is the sum of the derivatives, and the derivative of a difference is the difference of the derivatives. Based on the pattern of these two rules, a reasonable conjecture would be that the derivative of a product $f(x) \cdot g(x)$ is equal to the product $f'(x) \cdot g'(x)$ of the derivatives of the given functions. However, this pattern does not hold for all products, as demonstrated by the following examples:

◆ **EXAMPLE 1** Show $\dfrac{d}{dx}\left[f(x) \cdot g(x)\right] \neq f'(x) \cdot g'(x)$ when $f(x) = x^2$ and $g(x) = \dfrac{1}{x}$.

Solution. First compute $f(x) \cdot g(x)$ and then simplify using the exponent rule $a^r/a^s = a^{r-s}$ as follows:

$$f(x) \cdot g(x) = x^2 \cdot \frac{1}{x} = \frac{x^2}{x^1} = x^{2-1} = x^1 = x$$

Applying the differentiation rule for linear functions $\dfrac{d}{dx}(mx + b) = m$ to $x = 1 \cdot x$ with slope $m = 1$ gives $\dfrac{d}{dx}[f(x) \cdot g(x)] = \dfrac{d}{dx}[x] = 1$.

Next find $f'(x) \cdot g'(x)$ by differentiating both factors with the power rule and then multiplying the results together. The function $f(x) = x^2$ with exponent $n = 2$ has derivative $f'(x) = 2x$, and $g(x) = x^{-1}$ with exponent $n = -1$ has derivative $g'(x) = (-1)x^{-2} = -x^{-2}$. Now multiply $f'(x) \cdot g'(x)$ and simplify using the exponent rule $a^r \cdot a^s = a^{r+s}$ as follows:

$$f'(x) \cdot g'(x) = (2x^1)(-x^{-2}) = (-2)x^{1-2} = -2x^{-1} = \frac{-2}{x}$$

As can be seen, $\dfrac{d}{dx}\left[f(x) \cdot g(x)\right] \neq f'(x) \cdot g'(x)$ when $f(x) = x^2$ and $g(x) = \dfrac{1}{x}$. ■

▶ **QUESTION 1** Show that $\dfrac{d}{dx}\left[f(x) \cdot g(x)\right] \neq f'(x) \cdot g'(x)$ when $f(x) = x$ and $g(x) = x^4$.

While exceptions do exist (such as when $f(x)$ and $g(x)$ are both constant functions), it turns out that $[f(x) \cdot g(x)]'$ is not equal to $f'(x) \cdot g'(x)$ for almost all differentiable functions $f(x)$ and $g(x)$; in other words, the derivative of a product is almost always not equal to the product of the derivatives. While this most obvious conjecture does not hold as a general rule, mathematicians have identified the correct product rule that relates the derivative of $f(x) \cdot g(x)$ to the derivatives $f'(x)$ and $g'(x)$. This differentiation rule is more complicated than what might be desired, but still provides a clear algorithm for computing the derivative of a product.

PRODUCT RULE.
Let $f(x)$ and $g(x)$ be differentiable functions.

$$\frac{d}{dx}[f(x) \cdot g(x)] = g(x) \cdot f'(x) + f(x) \cdot g'(x)$$

In addition to this symbolic version of the product rule, some people best remember this rule by learning one of the following phrases:

- "take turns differentiating and then add," or

- "the second times the derivative of the first plus the first times the derivative of the second," where "second" refers to $g(x)$ and "first" refers to $f(x)$.

Both addition and multiplication are commutative, so alternative orderings can be used when stating or applying the product rule. The particular ordering of the functions and their derivatives given in the above statement of the product rule was intentionally chosen to facilitate upcoming work with the quotient rule.

For almost everyone, the product rule at first seems a bit surprising and even mysterious. Why does this particular combination of the functions $f(x)$ and $g(x)$ and their derivatives $f'(x)$ and $g'(x)$ correspond to the derivative of the product $f(x) \cdot g(x)$? As it turns out, sometimes reality just happens to be more subtle than might be desired or expected. Section 4.7 discusses limits and the definition of the derivative, and the product rule results from applying this definition. A complete argument for the validity of the product rule lies beyond the scope of this book, which instead focuses on developing a proficiency with using this rule.

◆ **EXAMPLE 2** Differentiate each function using the product rule:

(a) $\left(x^2 + 3x\right)\sin(x)$
(b) $\left(\dfrac{1}{x^2} + x^7\right)e^x$

Solution. First identify the two factors $f(x)$ and $g(x)$ in each product and then differentiate.

(a) For $\left(x^2 + 3x\right) \cdot \sin(x)$, let $f(x) = x^2 + 3x$ with $f'(x) = 2x + 3$ and let $g(x) = \sin(x)$ with $g'(x) = \cos(x)$. Now apply the product rule:

$$\frac{d}{dx}\left[(x^2 + 3x)\sin(x)\right] = \sin(x) \cdot \frac{d}{dx}\left[x^2 + 3x\right] + (x^2 + 3x) \cdot \frac{d}{dx}[\sin(x)]$$
$$= \sin(x) \cdot (2x + 3) + (x^2 + 3x) \cdot \cos(x)$$
$$= (2x + 3)\sin(x) + (x^2 + 3x)\cos(x)$$

(b) For $\left(\dfrac{1}{x^2} + x^7\right) \cdot e^x$, let $f(x) = \dfrac{1}{x^2} + x^7 = x^{-2} + x^7$ with $f'(x) = -2x^{-3} + 7x^6$ and let $g(x) = e^x$ with $g'(x) = e^x$. Now apply the product rule:

$$\frac{d}{dx}\left[\left(\frac{1}{x^2} + x^7\right)e^x\right] = e^x \cdot \frac{d}{dx}\left[\frac{1}{x^2} + x^7\right] + \left(\frac{1}{x^2} + x^7\right) \cdot \frac{d}{dx}\left[e^x\right]$$

$$= e^x \cdot \left(-2x^{-3} + 7x^6\right) + \left(x^{-2} + x^7\right) \cdot e^x$$

$$= \left(-2x^{-3} + 7x^6\right)e^x + \left(x^{-2} + x^7\right)e^x$$

$$= \left(-2x^{-3} + 7x^6 + x^{-2} + x^7\right)e^x$$

■

◆ **EXAMPLE 3** Differentiate each function using the product rule:

(a) $\ln(x)\cos(x)$ (b) $(3x + 5)4^x$

Solution. First identify the two factors $f(x)$ and $g(x)$ in each product and then differentiate.

(a) For $\ln(x)\cos(x)$, let $f(x) = \ln(x)$ with $f'(x) = \dfrac{1}{x}$ and $g(x) = \cos(x)$ with $g'(x) = -\sin(x)$, and apply the product rule:

$$\frac{d}{dx}\left[\ln(x)\cos(x)\right] = \cos(x) \cdot \frac{d}{dx}\left[\ln(x)\right] + \ln(x) \cdot \frac{d}{dx}\left[\cos(x)\right]$$

$$= \cos(x) \cdot \frac{1}{x} + \ln(x) \cdot \left[-\sin(x)\right]$$

$$= \frac{\cos(x)}{x} - \ln(x)\sin(x)$$

(b) For $(3x + 5)4^x$, let $f(x) = 3x + 5$ with $f'(x) = 3$ and $g(x) = 4^x$ with $g'(x) = \ln(4) \cdot 4^x$, and apply the product rule:

$$\frac{d}{dx}\left[(3x + 5)4^x\right] = 4^x \cdot \frac{d}{dx}\left[3x + 5\right] + (3x + 5) \cdot \frac{d}{dx}\left[4^x\right]$$

$$= 4^x \cdot 3 + (3x + 5) \cdot \left[\ln(4) \cdot 4^x\right]$$

$$= 3 \cdot 4^x + \ln(4)(3x + 5)4^x$$

■

▶ **QUESTION 2** Differentiate each function using the product rule:

(a) $(x^3 + 9x^2)\sin(x)$ (b) $e^x\cos(x)$ (c) $x^2\ln(x)$

◆ **EXAMPLE 4** The function $\text{Conc}(\text{Hr}) = 16 \cdot \text{Hr} \cdot e^{-\text{Hr}}$ graphed in Figure 1 provides a reasonable model for a patient's bloodstream concentration of a certain immediate-release medication in nanomoles per liter as a function of the number of hours Hr that have elapsed since the medication was taken.

(a) Find and explain the meaning of $\text{Conc}(3)$ and $\text{Conc}'(3)$.

(b) Find and explain the meaning of $\text{Conc}(1)$ and $\text{Conc}'(1)$ using RStudio.

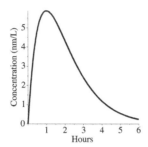

Figure 1: Model for the bloodstream concentration of a medication

Solution.

(a) Find $\text{Conc}(3)$ by substituting $\text{Hr} = 3$ into the given function $\text{Conc}(\text{Hr})$ to obtain $\text{Conc}(3) = 2.389779$. Thus, based on the model, the patient's bloodstream concentration of the medication is 2.390 nanomoles per liter three hours after the patient takes the medication.

$\text{Conc}'(3)$ is the rate of change of the bloodstream concentration of the medication three hours after taking the medication. First, compute the derivative using the product and line rules, as well as the extended exponent rule $[\, e^{mx+b}\,]' = m \cdot e^{mx+b}$ in the second line below. In this setting, the variable "Hr" is synonymous with the variable "x" in the standard presentation of the differentiation rules and when using differential notation.

$$\begin{aligned}
\text{Conc}'(\text{Hr}) &= e^{-\text{Hr}} \cdot \frac{d}{d\text{Hr}}\,[16\,\text{Hr}] + (16\,\text{Hr}) \cdot \frac{d}{d\text{Hr}}\,[e^{-\text{Hr}}] & \text{Product rule} \\
&= e^{-\text{Hr}} \cdot 16 + (16\,\text{Hr}) \cdot (-e^{-\text{Hr}}) & \text{Line and exponent rules} \\
&= 16\,e^{-\text{Hr}} - 16\,\text{Hr}\,e^{-\text{Hr}} & \text{Simplify} \\
&= (16 - 16\,\text{Hr})\,e^{-\text{Hr}} & \text{Factor}
\end{aligned}$$

Substituting $\text{Hr} = 3$ into this derivative gives $\text{Conc}'(3) = -1.593186$. Therefore, based on the model, three hours after taking the medication, the patient's bloodstream concentration of the medication is decreasing at rate of 1.593 nanomoles per liter per hour.

(b) First define $\text{Conc}(\text{Hr})$ in RStudio and evaluate the stored function at $\text{Hr} = 1$. Then differentiate this function and evaluate the resulting derivative at $\text{Hr} = 1$.

```
Conc=makeFun(16*Hr*exp(-Hr)~Hr)        dConc=D(Conc(Hr)~Hr)
Conc(Hr=1)                             dConc(Hr=1)
```

```
[1] 5.88607                            [1] 0
```

Thus, based on the model, one hour after the patient takes the medication, the patient's bloodstream concentration of the medication is 5.886 nanomoles per liter and this concentration is not changing with respect to time. In Figure 1, notice that the peak bloodstream concentration occurs exactly one hour after taking the medication, and so the instantaneous rate of change, or the slope of the corresponding horizontal tangent line, is zero.

■

➤ **QUESTION 3** When a car runs over a pothole, its suspension system absorbs the "shock" of hitting such an irregularity in the road. The change of height in inches of the passenger compartment relative to the road in seconds after striking a pothole is modeled by the function Amplitude(Second) $= 3\,e^{-0.75\text{Second}}\cos(4\cdot\text{Second})$, graphed in Figure 2. Find and explain the meaning of Amplitude(1.5) and Amplitude$'$(1.5). Answering this question requires the use of the extended exponent rule $\left[e^{mx+b}\right]' = m\cdot e^{mx+b}$, as well as a similar extension of the differentiation rule for the cosine:

$$\left[\cos(mx+b)\right]' = m\cdot\cos(mx+b)$$

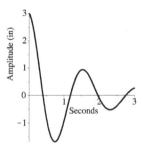

Figure 2: Model for the suspension system of a car after it runs over a pothole

➤ **QUESTION 4** Continuing to work with the following model for Amplitude(Second) from Question 3, find and explain the meaning of Amplitude(3.5) and Amplitude$'$(3.5) using RStudio:

$$\text{Amplitude(Second)} = 3\,e^{-0.75\text{Second}}\cos(4\,\text{Second})$$

Finally, observe that the constant multiple rule can be thought of as a special case of the product rule. Recall from Section 4.3 that if $h(x)$ is any differentiable function and c is any real number, then $[c \cdot h(x)]' = c \cdot h'(x)$. For the product $c \cdot h(x)$, let $f(x) = c$ with derivative $f'(x) = 0$ and $g(x) = h(x)$ with derivative $g'(x) = h'(x)$. Now apply the product rule $[f(x) \cdot g(x)]' = g(x) \cdot f'(x) + f(x) \cdot g'(x)$ to obtain the constant multiple rule:

$$\frac{d}{dx}[c \cdot h(x)] = h(x) \cdot 0 + c \cdot h'(x) = 0 + c \cdot h'(x) = c \cdot h'(x)$$

The process of deciding which differentiation rules to study and learn involves finding a balance between the number of rules and the relative ease of using them. While the constant multiple rule is not necessary as a distinct rule, because it is a special case of the product rule, constant multiples of functions arise so often in practice that having a specific rule for them is worth the investment.

The Quotient Rule

As with the product rule, one might reasonably conjecture that the derivative of a quotient $f(x)/g(x)$ would be equal to the quotient $f'(x)/g'(x)$ of the derivatives of the given functions $f(x)$ and $g(x)$. As illustrated in Exercises 53 and 54 at the end of this section, $\frac{d}{dx}\left[\frac{f(x)}{g(x)}\right]$ is not equal to $f'(x)/g'(x)$ for almost all differentiable functions $f(x)$ and $g(x)$, which means that the derivative of a quotient is almost always not equal to the quotient of the derivatives. While this most obvious conjecture does not hold, mathematicians have identified the correct quotient rule that relates the derivative of $f(x)/g(x)$ to the derivatives $f'(x)$ and $g'(x)$. As with the product rule, this differentiation rule is more elaborate than might be desired, but still provides a clear algorithm for computing the derivative of a quotient.

QUOTIENT RULE.
Let $f(x)$ and $g(x)$ be differentiable functions. When $g(x) \neq 0$, the following holds:

$$\frac{d}{dx}\left[\frac{f(x)}{g(x)}\right] = \frac{g(x) \cdot f'(x) - f(x) \cdot g'(x)}{[g(x)]^2}$$

Various phrases describing this differentiation rule include

- "the bottom times the derivative of the top minus the top times the derivative of the bottom all over the bottom squared," where "bottom" refers to $g(x)$ and "top" refers to $f(x)$, and

- "lo dee hi minus hi dee lo all over lo lo," where "lo" refers to $g(x)$ and "hi" refers to $f(x)$.

Because of the subtraction in the numerator of the quotient rule, the order in which the functions $f(x)$ and $g(x)$ appear in this rule is stricter than in the product rule.

One common mistake when using the quotient rule is to switch the roles of $f(x)$ and $g(x)$, particularly if a different ordering for the product rule has been learned. However, recalling the product rule as $[f(x) \cdot g(x)]' = g(x) \cdot f'(x) + f(x) \cdot g'(x)$ means that the numerator of the quotient rule is almost the same, with the exception of being a difference rather than a sum.

While perhaps even more surprising than the product rule, the quotient rule also follows directly from the definition of the derivative that is discussed in Section 4.7. A complete argument for the validity of the quotient rule lies beyond the scope of this book, which instead focuses on developing a proficiency with using this rule.

◆ **EXAMPLE 5** Differentiate each function using the quotient rule:

(a) $\dfrac{x^5 + 3x^2}{x^7 + 4x}$

(b) $\dfrac{x^2 + 9x}{\sin(x)}$

Solution. First identify the numerator $f(x)$ and the denominator $g(x)$ of each quotient, and then differentiate.

(a) The numerator is $f(x) = x^5 + 3x^2$ with $f'(x) = 5x^4 + 6x$ and the denominator is $g(x) = x^7 + 4x$ with $g'(x) = 7x^6 + 4$. Now apply the quotient rule:

$$\frac{d}{dx}\left[\frac{x^5 + 3x^2}{x^7 + 4x}\right] = \frac{(x^7 + 4x) \cdot \dfrac{d}{dx}\left[x^5 + 3x^2\right] - (x^5 + 3x^2) \cdot \dfrac{d}{dx}\left[x^7 + 4x\right]}{(x^7 + 4x)^2}$$

$$= \frac{(x^7 + 4x)(5x^4 + 6x) - (x^5 + 3x^2)(7x^6 + 4)}{(x^7 + 4x)^2}$$

(b) The numerator is $f(x) = x^2 + 9x$ with $f'(x) = 2x + 9$ and the denominator is $g(x) = \sin(x)$ with $g'(x) = \cos(x)$. Now apply the quotient rule:

$$\frac{d}{dx}\left[\frac{x^2 + 9x}{\sin(x)}\right] = \frac{\sin(x) \cdot \dfrac{d}{dx}\left[x^2 + 9x\right] - (x^2 + 9x) \cdot \dfrac{d}{dx}\left[\sin(x)\right]}{[\sin(x)]^2}$$

$$= \frac{\sin(x) \cdot (2x + 9) - (x^2 + 9x) \cdot \cos(x)}{\sin^2(x)}$$

$$= \frac{(2x + 9)\sin(x) - (x^2 + 9x)\cos(x)}{\sin^2(x)}$$

∎

Before considering the next example, the notation used for exponents of trigonometric functions is discussed further. As suggested by the denominator in the solution of Example 5(b), $[\sin(x)]^2$ is identified with $\sin^2(x)$ and, in general, $[\sin(x)]^n = \sin^n(x)$ for any real number n except $n = -1$. The exponent $n = -1$ in $\sin^{-1}(x)$ is often reserved to identify the inverse sine function rather than the reciprocal of $\sin(x)$. Similarly, for all trigonometric functions and for all real numbers except $n = -1$, the notation $\mathrm{trig}^n(x)$ denotes $[\,\mathrm{trig}(x)\,]^n$, where "trig" is any of sin, cos, tan, csc, sec, or cot.

◆ **EXAMPLE 6** Differentiate each function using the quotient rule:

(a) $\dfrac{2^x}{\sin(x) + \ln(x)}$

(b) $\dfrac{\sin(x)}{\cos(x)}$

Solution. First identify the numerator $f(x)$ and the denominator $g(x)$ of each quotient, and then differentiate.

(a) The numerator is $f(x) = 2^x$ with $f'(x) = \ln(2) \cdot 2^x$, and the denominator is $g(x) = \sin(x) + \ln(x)$ with $g'(x) = \cos(x) + 1/x$. Now apply the quotient rule:

$$\frac{d}{dx}\left[\frac{2^x}{\sin(x) + \ln(x)}\right] = \frac{[\sin(x) + \ln(x)] \cdot \dfrac{d}{dx}[2^x] - 2^x \cdot \dfrac{d}{dx}[\sin(x) + \ln(x)]}{[\sin(x) + \ln(x)]^2}$$

$$= \frac{[\sin(x) + \ln(x)] \cdot \ln(2) \cdot 2^x - 2^x \cdot \left[\cos(x) + \dfrac{1}{x}\right]}{[\sin(x) + \ln(x)]^2}$$

(b) The numerator is $f(x) = \sin(x)$ with $f'(x) = \cos(x)$, and the denominator is $g(x) = \cos(x)$ with $g'(x) = -\sin(x)$. Now apply the quotient rule:

$$\frac{d}{dx}\left[\frac{\sin(x)}{\cos(x)}\right] = \frac{\cos(x) \cdot \dfrac{d}{dx}[\sin(x)] - \sin(x) \cdot \dfrac{d}{dx}[\cos(x)]}{[\cos(x)]^2}$$

$$= \frac{\cos(x) \cdot \cos(x) - \sin(x) \cdot [-\sin(x)]}{\cos^2(x)}$$

$$= \frac{\cos^2(x) + \sin^2(x)}{\cos^2(x)}$$

■

▶ **QUESTION 5** Differentiate each function using the quotient rule:

(a) $\dfrac{4x + 5}{7x - 3}$

(b) $\dfrac{\cos(x)}{\ln(x)}$

(c) $\dfrac{e^x}{x^2 + \sin(x)}$

◆ **EXAMPLE 7** The population of the Netherlands in millions of people at the beginning of each decade since 1700 is stored in `NetherlandsPopulations` and graphed in Figure 3. Applying `fitModel` in RStudio produces the following sigmoidal model for this data set:

$$\text{Population(Year)} = \frac{18}{1 + 1.25e^{-0.031(\text{Year}-1950)}} + 2$$

(a) Find and explain the meaning of Population(1970) and Population′(1970).

(b) Find and explain the meaning of Population(2010) and Population′(2010) using RStudio.

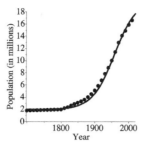

Figure 3: Population of the Netherlands in millions of people

Solution.

(a) Find Population(1970) by substituting Year = 1970 into the given function to obtain Population(1970) = 12.76278. Thus, based on the model, 12,762,780 people lived in the Netherlands in 1970; the data set provides an actual population of 12,957,600 people.

Population$'$(1970) is the rate of change of the population of the Netherlands in 1970. First, compute the derivative using the sum and quotient rules, as well as the extended exponent rule $\left[e^{mx+b}\right]' = m \cdot e^{mx+b}$ in the third line below. In this setting, the variable "Year" is synonymous with the variable "x" in the standard presentation of the differentiation rules and when using differential notation. Because $\dfrac{d}{d\text{Year}}[18] = 0$, the first half of the numerator in the quotient rule is zeroed out.

$$\text{Population}'(\text{Year}) = \frac{d}{d\text{Year}}\left[\frac{18}{1+1.25e^{-0.031(\text{Year}-1950)}}\right] + \frac{d}{d\text{Year}}[2]$$

$$= \frac{0 - 18 \cdot \dfrac{d}{d\text{Year}}\left[1+1.25e^{-0.031(\text{Year}-1950)}\right]}{\left[1+1.25e^{-0.031(\text{Year}-1950)}\right]^2}$$

$$= \frac{-18 \cdot \left[0 + 1.25 \cdot (-0.031) \cdot e^{-0.031(\text{Year}-1950)}\right]}{\left[1+1.25e^{-0.031(\text{Year}-1950)}\right]^2}$$

$$= \frac{0.6975 e^{-0.031(\text{Year}-1950)}}{\left[1+1.25e^{-0.031(\text{Year}-1950)}\right]^2}$$

Substituting Year = 1970 into this derivative gives Population$'$(1970) = 0.1341484. Therefore, based on the model, the population of the Netherlands was increasing at a rate of 134,148 people per year in 1970.

(b) First define Population(Year) in RStudio and evaluate the stored function at Year = 2010. Then differentiate this function and evaluate the resulting derivative at Year = 2010.

```
Population=makeFun(18/(1+1.25*exp(-0.031*(Year-1950)))+2~Year)
Population(Year=2010)
```

```
[1] 17.0679
```

```
dPop=D(Population(Year)~Year)
dPop(Year=2010)
```

```
[1] 0.0760883
```

Thus, based on the model, the population of the Netherlands was 17,067,920 and it was increasing at a rate of 76,088 people per year in 2010. ∎

➤ **QUESTION 6** The cumulative number of Ebola cases in Sierra Leone from May 1, 2014 to December 16, 2015 is stored in **EbolaSierraLeone** and graphed in Figure 4. Applying **fitModel** in RStudio produces the following sigmoidal model for this data set:

$$\text{Cases(Day)} \ = \ \frac{13420}{1 + 1.016e^{-0.0233(\text{Day}-200)}}$$

Find and explain the meaning of Cases(60) and Cases$'$(60).

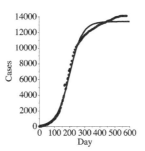

Figure 4: Cumulative number of Ebola cases in Sierra Leone in 2014 and 2015

➤ **QUESTION 7** Continuing to work with the model Cases(Day) from Question 6, find and explain the meaning of Cases(100) and Cases$'$(100) using RStudio.

More Trigonometric Derivatives

As you may have recognized, the solution to Example 6(b) provides the derivative of the tangent function. The most common presentation of the derivative of $\tan(x)$ is obtained after some additional simplification using trigonometric definitions and identities from Section 1.7:

$$\frac{d}{dx}\left[\tan(x)\right] = \frac{d}{dx}\left[\frac{\sin(x)}{\cos(x)}\right] \qquad \text{Definition of tangent}$$

$$= \frac{\cos^2(x) + \sin^2(x)}{\cos^2(x)} \qquad \text{Quotient rule from Example 6(b)}$$

$$= \frac{1}{\cos^2(x)} \qquad \text{Pythagorean identity } \cos^2(x) + \sin^2(x) = 1$$

$$= \left[\frac{1}{\cos(x)}\right]^2 \qquad \text{Exponent rule } \frac{a^r}{b^r} = \left(\frac{a}{b}\right)^r \text{ and } 1^2 = 1$$

$$= \left[\sec(x)\right]^2 \qquad \text{Definition of secant}$$

$$= \sec^2(x) \qquad \text{Exponent notation for trigonometric functions}$$

In much this same way, the quotient rule can be used to differentiate the cotangent, secant, and cosecant functions, because they are all equal to ratios (or quotients) of sine and cosine. Applying the appropriate trigonometric definitions and identities, as has just been done with the derivative of $\tan(x)$, results in the following standard forms for the derivatives of these trigonometric functions:

DERIVATIVES OF OTHER TRIGONOMETRIC FUNCTIONS.
For all real numbers x in the domain of each function, the following hold:

- $\dfrac{d}{dx}\left[\tan(x)\right] = \sec^2(x)$ - $\dfrac{d}{dx}\left[\sec(x)\right] = \sec(x)\tan(x)$

- $\dfrac{d}{dx}\left[\cot(x)\right] = -\csc^2(x)$ - $\dfrac{d}{dx}\left[\csc(x)\right] = -\csc(x)\cot(x)$

A couple of patterns arise among the derivatives of the trigonometric functions. First, recall from Section 4.3 that $\left[\sin(x)\right]' = \cos(x)$ and that $\left[\cos(x)\right]' = -\sin(x)$. As can be seen, the derivatives of all the "co" trigonometric functions (i.e., $\cos(x)$, $\csc(x)$, and $\cot(x)$) include a negative sign, while the others do not. In addition, these differentiation rules come in pairs of parallel results, as indicated by the vertical listing of the rules above.

▶ **QUESTION 8** Show that $\dfrac{d}{dx}\left[\csc(x)\right] = -\csc(x)\cot(x)$ by applying the quotient rule and simplifying the resulting ratio.

These differentiation rules for the other trigonometric functions are used in the exact same way as the rules for the simplest forms of the common modeling functions.

◆ **EXAMPLE 8** Differentiate each function:

(a) $(x^4 + x^2)\tan(x)$ (b) $\dfrac{\csc(x)}{e^x + 5}$

Solution.

(a) For the given product $(x^4 + x^2)\tan(x)$, let $f(x) = x^4 + x^2$ with $f'(x) = 4x^3 + 2x$ and let $g(x) = \tan(x)$ with $g'(x) = \sec^2(x)$. Now apply the product rule $[f(x) \cdot g(x)]' = g(x) \cdot f'(x) + f(x) \cdot g'(x)$:

$$\frac{d}{dx}\left[(x^4 + x^2)\tan(x)\right] = \tan(x) \cdot \frac{d}{dx}\left[x^4 + x^2\right] + (x^4 + x^2) \cdot \frac{d}{dx}[\tan(x)]$$
$$= \tan(x) \cdot (4x^3 + 2x) + (x^4 + x^2) \cdot \sec^2(x)$$
$$= (4x^3 + 2x)\tan(x) + (x^4 + x^2)\sec^2(x)$$

(b) For the given quotient, the numerator is $f(x) = \csc(x)$ with $f'(x) = -\csc(x)\cot(x)$, and the denominator is $g(x) = e^x + 5$ with $g'(x) = e^x$. Now apply the quotient rule:

$$\frac{d}{dx}\left[\frac{\csc(x)}{e^x + 5}\right] = \frac{(e^x + 5) \cdot \dfrac{d}{dx}[\csc(x)] - \csc(x) \cdot \dfrac{d}{dx}[e^x + 5]}{(e^x + 5)^2}$$
$$= \frac{(e^x + 5) \cdot [-\csc(x)\cot(x)] - \csc(x) \cdot e^x}{(e^x + 5)^2}$$
$$= \frac{-(e^x + 5)\csc(x)\cot(x) - e^x\csc(x)}{(e^x + 5)^2}$$

■

▶ **QUESTION 9** Differentiate each function:

(a) $2^x \cot(x)$

(b) $\dfrac{3x + 7}{\sec(x)}$

Differentiating with Multiple Rules

When computing the derivative of a given function, more than one differentiation rule often needs to be applied. The key idea is to follow the standard order of operations backwards, applying the differentiation rules beginning with the outermost operation (which would apply to an input last) and working toward the innermost operation (which would apply to an input first). This process of differentiating in the reverse of the standard order of operations is illustrated in the next example.

◆ **EXAMPLE 9** Differentiate $\dfrac{e^x}{x^3 \sin(x)}$.

Solution. The last, outermost operation is division, which indicates that the quotient rule is applied first when differentiating this function. For the numerator, recall that e^x is its own derivative. The denominator $x^3 \sin(x)$ is a product of $f(x) = x^3$ with $f'(x) = 3x^2$ and $g(x) = \sin(x)$ with $g'(x) = \cos(x)$; therefore, the product rule is used to obtain the denominator's derivative of $\sin(x) \cdot (3x^2) + x^3 \cdot \cos(x)$ for that portion of the quotient rule (on the second line below). Combining this work together provides the derivative of the given function.

$$\frac{d}{dx}\left[\frac{e^x}{x^3\sin(x)}\right] = \frac{x^3\sin(x)\cdot\dfrac{d}{dx}\left[e^x\right] - e^x\cdot\dfrac{d}{dx}\left[x^3\sin(x)\right]}{\left[x^3\sin(x)\right]^2}$$

$$= \frac{\left[x^3\sin(x)\right]\cdot e^x - e^x\cdot\left[3x^2\sin(x) + x^3\cos(x)\right]}{(x^3)^2\sin^2(x)}$$

$$= \frac{x^3 e^x\sin(x) - 3x^2 e^x\sin(x) - x^3 e^x\cos(x)}{x^6\sin^2(x)}$$

◼

◆ EXAMPLE 10 Find the linear approximation of $f(x) = \dfrac{e^x}{x^3\sin(x)}$ when $x = \dfrac{\pi}{2}$.

Solution. Recall from Section 4.3 that the tangent line $L(x) = f(a) + f'(a)\cdot(x - a)$ provides the linear approximation of $f(x)$ when $x = a$. Substituting $x = \pi/2$ into the given expression for $f(x)$ and into the expression for $f'(x)$ from Example 9 provides the necessary numeric values:

- $f\left(\dfrac{\pi}{2}\right) = \dfrac{e^{\pi/2}}{\left(\dfrac{\pi}{2}\right)^3\sin\left(\dfrac{\pi}{2}\right)} \approx 1.241$

- $f'\left(\dfrac{\pi}{2}\right) = \dfrac{\left(\dfrac{\pi}{2}\right)^3 e^{\pi/2}\sin\left(\dfrac{\pi}{2}\right) - 3\left(\dfrac{\pi}{2}\right)^2 e^{\pi/2}\sin\left(\dfrac{\pi}{2}\right) - \left(\dfrac{\pi}{2}\right)^3 e^{\pi/2}\cos\left(\dfrac{\pi}{2}\right)}{\left(\dfrac{\pi}{2}\right)^6\sin^2\left(\dfrac{\pi}{2}\right)} \approx -1.129$

Substituting these values provides the sought-for linear approximation:

$$L(x) = f\left(\frac{\pi}{2}\right) + f'\left(\frac{\pi}{2}\right)\cdot\left(x - \frac{\pi}{2}\right) \approx 1.241 - 1.129\,(x - 1.571)$$

◼

➤ QUESTION 10 Differentiate $\dfrac{x\cos(x)}{x + 4}$.

➤ QUESTION 11 Find the linear approximation of $g(x) = \dfrac{x\cos(x)}{x + 4}$ when $x = 0$.

Tabular Functions

Sometimes limited information is available about the values of functions and their derivatives, particularly when working with data sets. Even so, the differentiation rules can still be used to determine rates of change. In such settings, first compute the derivative with the standard differentiation rules and then substitute known functional values to find the instantaneous rate of change for the given inputs. This process is illustrated in the next example.

◆ **EXAMPLE 11** Find the value of $h'(a)$ for each function $h(x)$ and each input $x = a$ using the information in the input–output table below.

(a) $h(x) = x^2 + f(x)$ for $x = 2$

(b) $h(x) = f(x) \cdot g(x)$ for $x = 4$

(c) $h(x) = (3x) \cdot g(x)$ for $x = 4$

(d) $h(x) = \dfrac{2f(x)}{g(x)}$ for $x = 16$

x	2	4	16
$f(x)$	3	1	-2
$g(x)$	-7	7	5
$f'(x)$	-3	6	2
$g'(x)$	8	9	3

Solution. First, compute $h'(x)$ using the differentiation rules and then substitute $x = a$.

(a) For $h(x) = x^2 + f(x)$, differentiate to obtain $h'(x) = 2x + f'(x)$ based on the sum and power rules. Now substitute $x = 2$ to find $h'(2)$:

$$h'(2) \ = \ 2 \cdot 2 + f'(2) \ = \ 4 + (-3) \ = \ 1$$

(b) For $h(x) = f(x) \cdot g(x)$, differentiate to obtain $h'(x) = g(x) \cdot f'(x) + f(x) \cdot g'(x)$ based on the product rule. Now substitute $x = 4$ to find $h'(4)$:

$$h'(4) \ = \ g(4) \cdot f'(4) + f(4) \cdot g'(4) \ = \ 7 \cdot 6 + 1 \cdot 9 \ = \ 42 + 9 \ = \ 51$$

(c) For $h(x) = (3x) \cdot g(x)$, differentiate to obtain $h'(x) = g(x) \cdot 3 + (3x) \cdot g'(x)$ based on the product rule. Now substitute $x = 4$ to find $h'(4)$:

$$h'(4) \ = \ g(4) \cdot 3 + (3 \cdot 4) \cdot g'(4) \ = \ 7 \cdot 3 + 12 \cdot 9 \ = \ 21 + 108 \ = \ 129$$

(d) For $h(x) = \dfrac{2f(x)}{g(x)}$, differentiate to obtain $h'(x) = \dfrac{g(x) \cdot 2f'(x) - 2f(x) \cdot g'(x)}{[g(x)]^2}$ based on the quotient and constant multiple rules. Now substitute $x = 16$ to find $h'(16)$:

$$h'(16) \ = \ \frac{g(16) \cdot 2f'(16) - 2f(16) \cdot g'(16)}{[g(16)]^2}$$

$$= \ \frac{5 \cdot 2 \cdot 2 - 2 \cdot (-2) \cdot 3}{[5]^2} \ = \ \frac{32}{25} \ = \ 1.28$$

■

▶ **QUESTION 12** Find the value of $h'(a)$ for each function $h(x)$ and each input $x = a$ using the information in the input–output table below.

(a) $h(x) = e^x + g(x)$ for $x = 0$

(b) $h(x) = f(x) \cdot [5g(x) + 1]$ for $x = 1$

(c) $h(x) = \dfrac{f(x)}{g(x) + 4}$ for $x = 2$

x	0	1	2
$f(x)$	4	5	-1
$g(x)$	2	9	6
$f'(x)$	-3	-1	4
$g'(x)$	3	7	1

Summary

- For all differentiable functions $f(x)$ and $g(x)$ for which these expressions are defined, the following differentiation rules hold:

  ○ *Product rule:* $\dfrac{d}{dx}\left[f(x)g(x)\right] = g(x) \cdot f'(x) + f(x) \cdot g'(x)$

  ○ *Quotient rule:* $\dfrac{d}{dx}\left[\dfrac{f(x)}{g(x)}\right] = \dfrac{g(x) \cdot f'(x) - f(x) \cdot g'(x)}{[g(x)]^2}$

- For all trigonometric functions (sin, cos, tan, csc, sec, and cot) and for all real numbers except $n = -1$, the notation $\text{trig}^n(x)$ denotes $\left[\text{trig}(x)\right]^n$.

- For all real numbers x in the domain of each function, the following hold:

  ○ $\dfrac{d}{dx}\left[\tan(x)\right] = \sec^2(x)$ ○ $\dfrac{d}{dx}\left[\sec(x)\right] = \sec(x)\tan(x)$

  ○ $\dfrac{d}{dx}\left[\cot(x)\right] = -\csc^2(x)$ ○ $\dfrac{d}{dx}\left[\csc(x)\right] = -\csc(x)\cot(x)$

Exercises

In Exercises 1–10, differentiate the function using the product rule.

1. $5x^9(x^3 + x^2)$

2. $(x^5 + x)(3x + 1)$

3. $(x^5 - x)\cos(x)$

4. $(3x + 4)\sin(x)$

5. $(x^2 + x^{-3})\,e^x$

6. $e^x \sin(x)$

7. $3^x \cos(x)$

8. $\sin(x)\cos(x)$

9. $8x^6 \ln(x)$

10. $4^x \ln(x)$

In Exercises 11–20, differentiate the function using the quotient rule.

11. $\dfrac{x^9 + 2}{x^3 + 2}$

12. $\dfrac{2x^5 + 4x}{3x + 1}$

13. $\dfrac{\sin(x)}{x + \cos(x)}$

14. $\dfrac{x^2}{e^x + \sin(x)}$

15. $\dfrac{\cos(x)}{3x^2 + 5x}$

16. $\dfrac{e^x + 5x}{x^2 + 7x}$

17. $\dfrac{\ln(x)}{x^4 + x^2}$

18. $\dfrac{\cos(x) - 4x}{\sin(x) + x^2}$

19. $\dfrac{x^2 - e^x}{2x^2 + \cos(x)}$

20. $\dfrac{3^x + x}{2^x + x^8}$

In Exercises 21–44, differentiate the function.

21. $x^2 + \sin(x)$

22. $2x + \cos(x)$

23. $\cos(x) - 7x^4$

24. $\cos(x)\ln(x)$

25. $\sin(x) - e^{2x}$

26. $\sin(x) + xe^x$

27. $\ln(x)\sin(x) + x^5$

28. $(x^3 + x)e^x$

29. $[x + \sin(x)]\cos(x)$

30. $e^x \ln(x) - x^6$

31. $e^x \sin(x)$

32. $\cos(x)\ln(x)$

33. $6e^x - e6^x$

34. $3x^2 \ln(x)$

35. $7^x + \dfrac{\ln(x)}{x}$

36. $\dfrac{3^x}{e^x - 2x}$

37. $\dfrac{x + \sin(x)}{\ln(x)}$

38. $\dfrac{x + 5\cos(x)}{x^2 - 5x}$

39. $\dfrac{\sin(x)}{(6x^8 + 2)e^x}$

40. $\dfrac{\ln(x)}{e\cos(x) + x^2}$

41. $\dfrac{e^x \cos(x)}{x + 2}$

42. $\dfrac{e^x}{x^3 - \sqrt{x}}$

43. $\dfrac{x^4 - 2^x}{7x\sin(x)}$

44. $\dfrac{4x^6 + 3x}{\ln(x)\cos(x)}$

In Exercises 45 and 46, consider the following model of resonance, where displacement is measured in millimeters and time in minutes (M):

$$\text{Res}(M) = 5\,M \cdot \sin(M)$$

Find and explain the meaning of each quantity.

45. (a) Res(7) (b) Res$'$(7)

46. (a) Res(29) (b) Res$'$(29)

In Exercises 47 and 48, consider the following model of a sound wave beat that outputs the amplitude of the wave measured in decibels as a function of the distance from its source measured in meters (M):

$$\text{Beat}(M) = 70\sin(0.09\,M)\sin(0.01\,M)$$

Find and explain the meaning of each quantity, using the extended differentiation rule for sine: $[\sin(mx + b)]' = m \cdot \sin(mx + b)$.

47. (a) Beat(150) (b) Beat$'$(150)

48. (a) Beat(240) (b) Beat$'$(240)

In Exercises 49 and 50, consider the following model of the number Use of monthly active Twitter users worldwide by quarter Q (i.e., for example, 10.25 represents April to June 2010):

$$\text{Use}(Q) = \frac{297.202473}{1 + 0.95939e^{-0.97214(Q - 12.25)}}$$

Find and explain the meaning of each quantity.

49. (a) Use(12) (b) Use$'$(12)

50. (a) Use(16.25) (b) Use$'$(16.25)

In Exercises 51 and 52, consider the following model of the population of Belgium in millions of people per year (Yr):

$$\text{Pop(Yr)} = \frac{7.8}{1 + 0.66e^{-0.026(\text{Yr}-1930)}} + 3.27$$

Find and explain the meaning of each quantity.

51. (a) Pop(1800)   (b) Pop$'$(1800)

52. (a) Pop(1975)   (b) Pop$'$(1975)

In Exercises 53 and 54 compare

$$\frac{d}{dx}\left[\frac{f(x)}{g(x)}\right]$$

with $f'(x)/g'(x)$ to show that the derivative of a quotient is NOT the quotient of the derivatives.

53. $f(x) = x^2$ and $g(x) = x^3$

54. $f(x) = x^5$ and $g(x) = x^2$

In Exercises 55 and 56, verify the differentiation rule using the quotient rule along with trigonometric definitions and identities.

55. $\dfrac{d}{dx}\left[\cot(x)\right] = -\csc^2(x)$

56. $\dfrac{d}{dx}\left[\sec(x)\right] = \sec(x)\tan(x)$

In Exercises 57–70, differentiate the function.

57. $\tan(x) + \csc(x)$ 64. $5\tan(x) + e^x$

58. $\cot(x) + \sec(x)$ 65. $\sec(x) + \ln(x)$

59. $x^2 \tan(x)$ 66. $4\sec(x) + x^3$

60. $e^x \csc(x)$ 67. $\cot(x) + 2^x$

61. $\ln(x)\cot(x)$ 68. $\cot(x)\cos(x)$

62. $(2^x + 3^x)\sec(x)$ 69. $\csc(x) + x^7$

63. $(5x+7)\tan(x)$ 70. $\csc(x)\sin(x)$

In Exercises 71–78, find $h'(a)$ for the function $h(x)$ when $x = a$ using the following input–output table:

x	1	2	4
$f(x)$	-2	3	1
$g(x)$	0	7	-4
$f'(x)$	5	6	8
$g'(x)$	2	8	-1

71. $h(x) = xf(x)$ for $x = 4$

72. $h(x) = (5x + 7)f(x)$ for $x = 0$

73. $h(x) = x^3 g(x)$ for $x = 1$

74. $h(x) = e^x g(x)$ for $x = 3$

75. $h(x) = f(x)g(x)$ for $x = 2$

76. $h(x) = f(x)g(x)$ for $x = 1$

77. $h(x) = \dfrac{f(x)}{g(x)}$ for $x = 4$

78. $h(x) = \dfrac{f(x)}{g(x)}$ for $x = 1$

In Exercises 79–86, find the value of the derivative of the function when $x = 1$ using the following input–output table:

x	1	2	5
$f(x)$	0	-1	7
$g(x)$	2	3	-1
$f'(x)$	5	5	5
$g'(x)$	1	4	8

79. $f(x) + g(x)$ 84. $f(x) + x^2 g(x)$

80. $f(x) - g(x)$

81. $f(x) + e^x$ 85. $\dfrac{f(x)}{g(x)}$

82. $f(x)g(x)$

83. $xg(x) + 2x$ 86. $\dfrac{f(x) + x}{g(x) + 4}$

In Exercises 87–94, find the linear approximation to the function $f(x)$ for the given input $x = a$.

87. $f(x) = x^2 e^x$ when $x = 1$

88. $f(x) = (x^2 + x)\ln(x - 2)$ when $x = 4$

89. $f(x) = x\tan(x)$ when $x = \frac{\pi}{4}$

90. $f(x) = e^x \cos(x)$ when $x = \pi$

91. $f(x) = \dfrac{x^3 - 2}{2 + \sin(x)}$ when $x = 0$

92. $f(x) = \dfrac{2x + 5}{3x + 4}$ when $x = 1$

93. $f(x) = \dfrac{e^x}{2x + 3}$ when $x = 2$

94. $f(x) = \dfrac{\sin(x)}{\cos(2x)}$ when $x = \dfrac{\pi}{2}$

RStudio. In Exercises $95 - 114$, use RStudio to differentiate each of the functions from Exercises $25 - 40$ and $47 - 50$.

95. Exercise 25

96. Exercise 26

97. Exercise 27

98. Exercise 28

99. Exercise 29

100. Exercise 30

101. Exercise 31

102. Exercise 32

103. Exercise 33

104. Exercise 34

105. Exercise 35

106. Exercise 36

107. Exercise 37

108. Exercise 38

109. Exercise 39

110. Exercise 40

111. Exercise 47

112. Exercise 48

113. Exercise 49

114. Exercise 50

In Your Own Words. In Exercises $115 - 120$, explain the following.

115. Product rule

116. Quotient rule

117. Exponent notation for trigonometric functions

118. Derivatives of trigonometric functions

119. Computing derivatives using tabular data

120. Interpreting $f(a)$ and $f'(a)$

4.5   The Chain Rule

Section 4.3 provided differentiation rules for the simplest forms of the common modeling functions and, together with Section 4.4, differentiation rules for all the basic arithmetic operations. This section continues the development of these calculational methods by studying the differentiation rule for the operation of function composition. This differentiation formula is called the *chain rule*. As with the operations of arithmetic, the derivative of a function created by composing two functions $f(x)$ and $g(x)$ depends on their derivatives $f'(x)$ and $g'(x)$.

A motivation for studying the chain rule includes much greater calculational power through the resulting significant expansion of the number of readily differentiable functions. Among other consequences of this increased computational power, many more models of our physical and social world can be analyzed via the derivative, because many phenomena exhibit subtle behaviors that can only be expressed via compositions of functions. Differentiating these models requires the chain rule.

Recall that the standard forms of all the common modeling functions introduced in Chapters 1 and 2 consist of compositions with linear functions. As a specific example, average maximum temperatures over the course of multiple years are often modeled by sine functions of the form

$$\text{Temp(Month)} \;=\; A \cdot \sin\left[\frac{2\pi}{P}\left(\text{Month} - h\right)\right] + v$$

Similarly, long-term population trends of humans and other living beings are often modeled by sigmoidal functions, which can be thought of as either quotients or compositions:

$$\text{Population(Year)} \;=\; \frac{L}{1 + Ce^{-k(\text{Year}-h)}} + v \;=\; L\left[1 + Ce^{-k(\text{Year}-h)}\right]^{-1} + v$$

In order to study these and other similarly more sophisticated models, a differentiation rule is needed for compositions of functions.

While learning any new differentiation rule, the reader will want to keep in mind the meaning and interpretation of the derivative. Recall that the derivative $f'(x)$ provides the instantaneous rate of change of a function $f(x)$ and enables analysis of the monotonic behavior of $f(x)$. The derivative is also used to obtain linear approximations of differentiable functions that provide a simple means to approximate functional values for inputs near (or local to) where the corresponding tangent line is based. Further applications of the derivative are explored in Chapter 5.

Composition of Functions

Recall from Section 1.1 that a **function** is a rule assigning every input to a unique output, which is often denoted by $y = f(x)$ or simply $f(x)$. For arithmetic operations, such as $f(x) + g(x)$, the output of each component function is computed and then these two intermediate outputs are combined together (by addition in this case) to produce the final output of this operation.

Composition offers an alternative approach to combining functions together, which is often denoted by $y = f[g(x)]$. Namely, composition proceeds in stages, first applying just the "inside" function $g(x)$ to the given input to obtain an intermediate output. Then, this intermediate output is used as an input to the "outside" function $f(x)$ to produce the final output of this operation. The order in which the two functions are applied is almost always vitally important. The formal definition of the composition of two functions is as follows:

Definition. If $f(x)$ and $g(x)$ are functions, then the **composition** of f with g is $f \circ g(x) = f[g(x)]$ when this expression is defined.

◆ **EXAMPLE 1** Let $f(x) = x^3$ and $g(x) = x + 1$. Compute both compositions and evaluate the resulting functions when $x = -2$ and $x = 0$:

(a) $f \circ g$ (b) $g \circ f$

Solution. An expression for each composition is found by substituting the innermost function and then the outermost, as follows:

(a) $f \circ g(x) = f[g(x)] = f[x+1] = (x+1)^3$

(b) $g \circ f(x) = g[f(x)] = g[x^3] = x^3 + 1$

The resulting functions are evaluated at the given inputs by substituting the numbers into these algebraic expressions:

When $x = -2$, then

(a) $f \circ g(-2) = (-2+1)^3 = (-1)^3 = -1$

(b) $g \circ f(-2) = (-2)^3 + 1 = -8 + 1 = -7$

When $x = 0$, then

(a) $f \circ g(0) = (0+1)^3 = (1)^3 = 1$

(b) $g \circ f(0) = (0)^3 + 1 = 0 + 1 = 1$

∎

As indicated by Example 1, the order in which two functions are composed plays a fundamental role in determining the exact resulting function. Almost always $f \circ g(x)$ is not equal to $g \circ f(x)$, although some exceptions do exist (such as when $f(x) = x$ and $g(x) = x$).

Example 1 also includes an illustration of how to show that two functions are different from one another. Namely, when two functions do not agree on some input (e.g., $x = -2$ in Example 1), then the two functions are different from one another, even if they happen to agree on other inputs (e.g., $x = 0$ in Example 1). As in other contexts, a single counterexample showing that two functions differ on even one input is sufficient to show the two functions are not identical.

▶ **QUESTION 1** Let $f(x) = 3x$ and $g(x) = x^2 - 4$. Compute each composition and evaluate the resulting functions when $x = -1$ and $x = 2$:

(a) $f \circ g$

(b) $g \circ f$

In order to differentiate a composition $f(g(x))$ of two functions, the first step is to identify its inside function $g(x)$ that is first applied to the given input and its outside function $f(x)$ that is then applied to the intermediate output from $g(x)$. After developing a facility with this identification process, the complete implementation of the chain rule is discussed.

◆ **EXAMPLE 2** For each composition $f(g(x))$, identify its inside function $g(x)$ and its outside function $f(x)$:

(a) $(x^2 - 3)^5$

(b) $e^{x^2 + 3x + 1}$

(c) $\sin(x^2) + 1$

Solution.

(a) The composition $(x^2 - 3)^5$ results from $g(x) = x^2 - 3$ and $f(x) = x^5$.

(b) The composition $e^{x^2 + 3x + 1}$ results from $g(x) = x^2 + 3x + 1$ and $f(x) = e^x$.

(c) The composition $\sin(x^2) + 1$ results from $g(x) = x^2$ and $f(x) = \sin(x) + 1$. Alternatively, one might consider $g(x) = \sin(x^2)$ and $f(x) = x + 1$, although this choice is less helpful for the chain rule as discussed shortly.

∎

➤ **QUESTION 2** For each composition $f(g(x))$, identify its inside function $g(x)$ and its outside function $f(x)$:

(a) $\ln(3x^2 + e^x)$ (b) $\cos[x - \ln(x)]$ (c) $e^{x^2} - 4$

➤ **QUESTION 3** For each composition $f(g(x))$, identify its inside function $g(x)$ and its outside function $f(x)$:

(a) $\dfrac{1}{\sin^3(x)}$ (b) $\left(\dfrac{e^x}{x^4 - x^3}\right)^7$ (c) $\dfrac{1}{\sqrt[3]{(x^2 + 27)^4}}$

Example 2 and Questions 2 and 3 highlight certain patterns that often hold when selecting the inside function $g(x)$, including the following:

- For trigonometric and logarithmic functions, $g(x)$ is often the expression that appears between the parentheses immediately after the function.

- For exponential functions, $g(x)$ is often the expression appearing in the exponent.

- $g(x)$ is often the expression appearing inside a power or square root, which may be part of the denominator of a quotient.

While these guidelines do not hold for every possible computation involving the chain rule, they apply in almost all common settings. With sufficient practice, selecting the inside and outside functions of a composition becomes more natural and automatic.

The Chain Rule

Many functions are the result of compositions and the chain rule dramatically increases the number of functions that can be readily differentiated. As with the operations of arithmetic, the derivative of a function created by composing two functions $f(x)$ and $g(x)$ depends on their derivatives $f'(x)$ and $g'(x)$. In terms of how these derivatives are combined to obtain a formula for the derivative of a composition, the chain rule is somewhat involved and, in this sense, is more similar to the product and quotient rules than to the addition and difference rules.

CHAIN RULE.
Let $f(x)$ and $g(x)$ be differentiable functions. When the compositions are defined, the following equality holds:

$$\frac{d}{dx}[f(g(x))] = f'[g(x)] \cdot g'(x)$$

In addition to this symbolic version of the chain rule, some people best remember this rule by learning the following phrase:

- "differentiate the outside, leave the inside alone, then multiply by the derivative of the inside," where "outside" refers to $f(x)$ and "inside" refers to $g(x)$.

A complete argument for the validity of the chain rule relies on the limit definition of the derivative discussed in Section 4.7, but lies beyond the scope of this book. Instead, this section focuses on developing a proficiency with using this rule.

◆ **EXAMPLE 3** Differentiate each function using the chain rule:

(a) $(x^4 + 2x)^3$ (b) $\sin(x^2 + 8)$ (c) $\ln(18 - x^3)$

Solution. First, identify the inside function $g(x)$ and the outside function $f(x)$ for the composition $f(g(x))$, and then differentiate using the chain rule $[f(g(x))]' = f'(g(x)) \cdot g'(x)$.

(a) The composition $f(g(x)) = (x^4 + 2x)^3$ results from $g(x) = x^4 + 2x$ with $g'(x) = 4x^3 + 2$ and $f(x) = x^3$ with $f'(x) = 3x^2$. Now apply the chain rule:

$$\frac{d}{dx}\left[(x^4 + 2x)^3\right] = 3(x^4 + 2x)^2 \cdot \frac{d}{dx}\left[x^4 + 2x\right] = 3(x^4 + 2x)^2 \cdot (4x^3 + 2)$$

(b) The composition $f(g(x)) = \sin(x^2 + 8)$ results from $g(x) = x^2 + 8$ with $g'(x) = 2x$ and $f(x) = \sin(x)$ with $f'(x) = \cos(x)$. Now apply the chain rule:

$$\frac{d}{dx}\left[\sin(x^2 + 8)\right] = \cos(x^2 + 8) \cdot \frac{d}{dx}\left[x^2 + 8\right] = \left[\cos(x^2 + 8)\right] \cdot (2x)$$

(c) The composition $f(g(x)) = \ln(8 - x^3)$ results from $g(x) = 18 - x^3$ with $g'(x) = -3x^2$ and $f(x) = \ln(x)$ with $f'(x) = \dfrac{1}{x}$. Now apply the chain rule:

$$\frac{d}{dx}\left[\ln(18 - x^3)\right] = \frac{1}{18 - x^3} \cdot \frac{d}{dx}\left[18 - x^3\right] = \frac{1}{18 - x^3} \cdot (-3x^2) = \frac{-3x^2}{18 - x^3} \quad ∎$$

For Example 3(c), recall that the domain of $\ln(x)$ is $(0, \infty)$, which means that the domain of the given function $\ln(18 - x^3)$ is $(-\infty, \sqrt[3]{18})$. In general, restrictions on the domain of a function extend to the domain of the function's derivative, although sometimes additional restrictions are needed. Therefore, in Example 3(c), the domain of the derivative $-3x^2/(18 - x^3)$ is also just $(-\infty, \sqrt[3]{18})$.

◆ **EXAMPLE 4** Differentiate each function using the chain rule:

(a) $4e^{x^2 - 9x}$ (b) $3^{7x + \cos(x)}$

Solution. First, identify the inside function $g(x)$ and the outside function $f(x)$ for the composition $f(g(x))$, and then differentiate using the chain rule $[f(g(x))]' = f'(g(x)) \cdot g'(x)$.

(a) The composition $f(g(x)) = 4e^{x^2 - 9x}$ results from $g(x) = x^2 - 9x$ with $g'(x) = 2x - 9$ and $f(x) = 4e^x$ with $f'(x) = 4e^x$. Now apply the chain rule:

$$\frac{d}{dx}\left[4e^{x^2 - 9x}\right] = 4e^{x^2 - 9x} \cdot \frac{d}{dx}\left[x^2 - 9x\right] = 4e^{x^2 - 9x} \cdot (2x - 9) = (8x - 36)e^{x^2 - 9x}$$

(b) The composition $f(g(x)) = 3^{7x+\cos(x)}$ results from $g(x) = 7x + \cos(x)$ with $g'(x) = 7 - \sin(x)$ and $f(x) = 3^x$ with $f'(x) = \ln(3) \cdot 3^x$. Now apply the chain rule:

$$\frac{d}{dx}\left[3^{7x+\cos(x)}\right] = \ln(3) \cdot 3^{7x+\cos(x)} \cdot \frac{d}{dx}\left[7x + \cos(x)\right]$$
$$= \ln(3) \cdot 3^{7x+\cos(x)} \cdot \left[7 - \sin(x)\right]$$

∎

➤ **QUESTION 4** Differentiate each function using the chain rule:

(a) $(3x^4 + 2x)^7$ (b) $5\cos(x^2 - e^x)$ (c) $e^{x^2 + 5\sin(x)}$

➤ **QUESTION 5** Differentiate each function using the chain rule:

(a) $\sin(e^x + \ln(x))$ (b) $2 \cdot 3^{x^5 - 8x}$ (c) $\dfrac{1}{\sqrt[3]{(x^2 + 1)^4}}$

The beginning of this section asserted that the chain rule enables the analysis of more sophisticated models of our physical and social world. The following example illustrates the first steps of such an analysis in an economics setting:

◆ **EXAMPLE 5** The average monthly stock price of Toyota from 1982 to 1998 is stored in `ToyotaMonthly`. Applying `fitModel` in RStudio produces the following model for this data set:

$$\text{Stock}(\text{Month}) = 0.18\text{Month} + 9\sin\left[\frac{2\pi}{100.56}(\text{Month} + 46.13)\right] + 11.18$$

Both the data and this model are graphed in Figure 1. Find and explain the meaning of Stock(96) and Stock'(96).

Solution. Stock(96) is found by substituting Month = 96 into the given function Stock(Month) to obtain Stock(96) = 33.11974. Thus, based on the model, Toyota stock was selling for \$33.12 per share in January 1990 (or 96 months after January 1982).

Stock'(96) is the rate of change in the price of Toyota stock in January 1990. In this setting, the variable "Month" is synonymous with the variable "x" in the differentation rules, and the sum, constant multiple, and chain rules are used to compute the derivative of the given function:

$$\text{Stock}'(\text{Month}) = 0.18 + 9\cos\left[\frac{2\pi}{100.56}(\text{Month} + 46.13)\right] \cdot \frac{2\pi}{100.56} + 0$$
$$= 0.18 + \frac{18\pi}{100.56}\cos\left[\frac{2\pi}{100.56}(\text{Month} + 46.13)\right]$$

Substituting Month = 96 into this derivative gives Stock'(96) = −0.3010981. Therefore, based on the model, the price of Toyota stock is decreasing at a rate of \$0.30 per share per month in January 1990 (or 96 months after January 1982).

∎

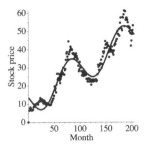

Figure 1: Price of Toyota stock from 1982 to 1998 for Example 5

◆ **EXAMPLE 6** Continue working with the model Stock(Month) for the average monthly price of Toyota stock from 1982 to 1998 given in Example 5. Find and explain the meaning of both Stock(168) and Stock$'$(168) using RStudio.

Solution. First define Stock(Month) in RStudio and evaluate this stored function at Month = 168. Then differentiate and evaluate this derivative at Month = 168.

```
Stock=makeFun(0.18*Month+9*sin((2*pi/100.56)*(Month+46.13))+11.18~Month)
Stock(Month=168)

[1] 47.9565

dStock=D(Stock(Month)~Month)
dStock(Month=168)

[1] 0.566552
```

Thus, based on the model, in January 1996 (or 168 months after January 1982), Toyota stock was selling for $47.96 per share, and this price was increasing at a rate of $0.57 per share per month. ■

➤ **QUESTION 6** The depth of the tide in feet relative to the MLLW (mean lower low water mark) in Pearl Harbor, Hawaii as a function of time measured in hours Hr is stored in `Hawaii`. Applying `fitModel` in RStudio produces the following model for this data set:

$$\text{Depth(Hr)} = 0.75 \sin\left[\frac{2\pi}{24.32}\left(\text{Hr} - 2.31\right)\right] - 0.55 \cos\left[\frac{2\pi}{12.31}\left(\text{Hr} - 1.17\right)\right] + 0.77$$

Both the data and this model are graphed in Figure 2. Find and explain the meaning of both Depth(40) and Depth$'$(40).

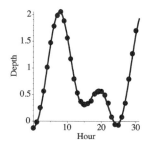

Figure 2: Tide depth relative to MLLW in Pearl Harbor for Question 6

➤ **QUESTION 7** Continue working with the model Depth(Hr) for the depth of the tide in feet relative to the MLLW (mean lower low water mark) in Pearl Harbor, Hawaii given in Question 6. Find and explain the meaning of both Depth(48) and Depth$'$(48) using RStudio.

Differentiating with Multiple Rules

As observed when discussing the product and quotient rules in Section 4.4, often more than one differentiation rule must be applied when computing the derivative of a given function. The key idea is to follow the standard order of operations *backwards*, applying the differentiation rules beginning with the outermost operation (which would apply to an input last) and working toward the innermost operation (which would apply to an input first). This process of differentiating in the reverse of the standard order of operations is illustrated in the next example.

◆ **EXAMPLE 7** Differentiate each function:

(a) $\left(5x^4 + 3\right)\ln(x^3 - 8)$ (b) $\cos\left[x^2 \ln(x)\right]$ (c) $\dfrac{e^x}{\sin(x^3)}$

Solution.

(a) The last, outermost operation is the multiplication of $\left(5x^4 + 3\right)$ with $\ln(x^3 - 8)$, which indicates that the product rule is used to differentiate this function. While doing so, the differentiation of $\ln(x^3 - 8)$ requires the chain rule, where the inside function is $g(x) = x^3 - 8$ with $g'(x) = 3x^2$ and the outside function is $f(x) = \ln(x)$ with $f'(x) = 1/x$. Combining this work together provides the derivative of the given function:

$$\frac{d}{dx}\left[\left(5x^4 + 3\right)\ln(x^3 - 8)\right] = \ln(x^3 - 8)\cdot\frac{d}{dx}\left[5x^4 + 3\right] + \left(5x^4 + 3\right)\cdot\frac{d}{dx}\left[\ln(x^3 - 8)\right]$$

$$= \ln(x^3 - 8)\cdot(20x^3 + 0) + \left(5x^4 + 3\right)\cdot\left[\frac{1}{x^3 - 8}\cdot 3x^2\right]$$

$$= 20x^3\ln(x^3 - 8) + \frac{15x^6 + 9x^2}{x^3 - 8}$$

(b) The chain rule is used to differentiate this composition of inside function $g(x) = x^2 \ln(x)$ and outside function $f(x) = \cos(x)$ with $f'(x) = -\sin(x)$. Because x^2 and $\ln(x)$ are multiplied together, the product rule is applied when differentiating $g(x) = x^2 \ln(x)$ as follows:

$$\frac{d}{dx}\left[\cos\left[x^2 \ln(x)\right]\right] = -\sin\left[x^2 \ln(x)\right] \cdot \frac{d}{dx}\left[x^2 \cdot \ln(x)\right]$$

$$= -\sin\left[x^2 \ln(x)\right] \cdot \left[\ln(x) \cdot \frac{d}{dx}\left[x^2\right] + x^2 \cdot \frac{d}{dx}\left[\ln(x)\right]\right]$$

$$= -\sin\left[x^2 \ln(x)\right] \cdot \left[\ln(x) \cdot (2x) + x^2 \cdot \frac{1}{x}\right]$$

$$= -\left[2x \ln(x) + x\right] \cdot \sin\left[x^2 \ln(x)\right]$$

(c) The last, outermost operation is division, which indicates that the quotient rule is used first when differentiating this function. For the numerator, recall that e^x is its own derivative. The denominator $\sin(x^3)$ is a composition of the inside function $g(x) = x^3$ with $g'(x) = 3x^2$ and the outside function $f(x) = \sin(x)$ with $f'(x) = \cos(x)$; thus, the chain rule must be applied to $\sin(x^3)$ to obtain its derivative of $\cos(x^3) \cdot (3x^2)$ for the appropriate portion of the quotient rule. Combining this work together provides the derivative of the given function:

$$\frac{d}{dx}\left[\frac{e^x}{\sin(x^3)}\right] = \frac{\sin(x^3) \cdot \frac{d}{dx}\left[e^x\right] - e^x \cdot \frac{d}{dx}\left[\sin(x^3)\right]}{\left[\sin(x^3)\right]^2}$$

$$= \frac{\sin(x^3) \cdot e^x - e^x \cdot \left[\cos(x^3) \cdot (3x^2)\right]}{\sin^2(x^3)}$$

$$= \frac{e^x \sin(x^3) - 3x^2 e^x \cos(x^3)}{\sin^2(x^3)}$$

■

Recall from Section 4.4 that for all "trig" functions (sin, cos, tan, csc, sec, and cot) and for all real numbers except $n = -1$, the notation $\text{trig}^n(x)$ denotes $\left[\text{trig}(x)\right]^n$. The denominator of the derivative in Example 7(c) is simplified from $\left[\sin(x^3)\right]^2$ to $\sin^2(x^3)$ using this notation.

➤ **QUESTION 8** Differentiate each function:

(a) $4x^6 e^{x^3 - 4x}$ (b) $\dfrac{\sin(x)}{\cos(e^x)}$ (c) $\ln\left(\dfrac{x}{3x + 2}\right)$ (d) $\dfrac{xe^x}{7x + 1}$

◆ **EXAMPLE 8** Find the linear approximation of $f(x) = \left(5x^4 + 3\right)\ln(x^3 - 8)$ when $x = 3.5$.

Solution. Recall from Section 4.3 that the tangent line $L(x) = f(a) + f'(a) \cdot (x - a)$ provides the linear approximation of $f(x)$ when $x = a$. Substituting $x = 3.5$ into the given expression for $f(x)$ and into the expression for $f'(x)$ from Example 7 provides the needed numeric values of $f(a)$ and $f'(a)$:

- $f(3.5) = \left[5 \cdot (3.5)^4 + 3\right] \cdot \ln[(3.5)^3 - 8] \approx 2675.593$

- $f'(3.5) = 20 \cdot (3.5)^3 \cdot \ln[(3.5)^3 - 8] + \dfrac{15 \cdot (3.5)^6 + 9 \cdot (3.5)^2}{(3.5)^3 - 8} \approx 3839.456$

Substituting these values provides the linear approximation of $f(x)$ when $x = 3.5$:

$$L(x) = f(3.5) + f'(3.5) \cdot (x - 3.5) \approx 2675.593 - 3839.456\,(x - 3.5)$$

■

▶ **QUESTION 9** Find the linear approximation of $g(x) = 4x^6 e^{x^3 - 4x}$ when $x = 1$.

Tabular Functions

Sometimes limited information is available about the values of functions and their derivatives, particularly when working with data sets. Even so, the differentiation rules can still be used to determine rates of change. In such settings, first compute the derivative with the standard differentiation rules and then substitute known functional values to find the instantaneous rate of change for the given inputs. This process is illustrated in the next example.

◆ **EXAMPLE 9** Find the value of $h'(a)$ for each function $h(x)$ and each input $x = a$ using the information in the following input–output table:

(a) $h(x) = f(2x) + g(4x^2)$ for $x = 2$

(b) $h(x) = x \cdot [g(x)]^2$ for $x = 4$

(c) $h(x) = f(4x) \cdot g(x^2)$ for $x = 4$

x	2	4	16
$f(x)$	3	1	-2
$g(x)$	-7	7	5
$f'(x)$	-3	6	2
$g'(x)$	8	9	3

Solution. First compute $h'(x)$ using the differentiation rules and then substitute $x = a$.

(a) For $h(x) = f(2x) + g(4x^2)$, apply the sum and chain rules to find $h'(x)$:

$$h'(x) = f'(2x) \cdot 2 + g'(4x^2) \cdot 8x = 2f'(2x) + 8xg'(4x^2)$$

Now substitute $x = 2$ to find $h'(2)$:

$$\begin{aligned} h'(2) &= 2 \cdot f'(2 \cdot 2) + 8 \cdot 2 \cdot g'(4 \cdot 2^2) \\ &= 2 \cdot f'(4) + 16 \cdot g'(16) = 2 \cdot 6 + 16 \cdot 3 = 60 \end{aligned}$$

(b) For $h(x) = x \cdot [g(x)]^2$, apply the product and chain rules to find $h'(x)$:

$$h'(x) = [g(x)]^2 \cdot 1 + x \cdot 2 \cdot [g(x)]^{2-1} \cdot g'(x) = [g(x)]^2 + (2x) \cdot g(x) \cdot g'(x)$$

Now substitute $x = 4$ to find $h'(4)$:

$$\begin{aligned} h'(4) &= [g(4)]^2 + (2 \cdot 4) \cdot g(4) \cdot g'(4) \\ &= 7^2 + 8 \cdot 7 \cdot 9 = 553 \end{aligned}$$

(c) For $h(x) = f(4x) \cdot g(x^2)$, apply the product and chain rules to find $h'(x)$:

$$h'(x) = g(x^2) \cdot [f'(4x) \cdot 4] + f(4x) \cdot [g'(x^2) \cdot 2x]$$

Now substitute $x = 4$ to find $h'(4)$:

$$\begin{aligned} h'(4) &= g(4^2) \cdot [f'(4 \cdot 4) \cdot 4] + f(4 \cdot 4) \cdot [g'(4^2) \cdot 2 \cdot 4] \\ &= g(16) \cdot f'(16) \cdot 4 + f(16) \cdot g'(16) \cdot 8 \\ &= 5 \cdot 2 \cdot 4 + (-2) \cdot 3 \cdot 8 = -8 \end{aligned}$$

$\blacksquare$

▶ **QUESTION 10** Find the value of $h'(a)$ for each function $h(x)$ and each input $x = a$ using the information in the following input–output table:

(a) $h(x) = [x^3 + g(x)]^2$ for $x = 2$

(b) $h(x) = \dfrac{f(5x)}{g(x^3)}$ for $x = 0$

(c) $h(x) = f[\sin(x - 1)]$ for $x = 1$

x	0	1	2
$f(x)$	4	5	-1
$g(x)$	2	9	6
$f'(x)$	-3	-1	4
$g'(x)$	3	7	1

Revisiting Extended Differentiation Rules

Sections 4.3 and 4.4 introduced various extended differentiation rules in order to enable the analysis of models of certain physical and social phenomena. These extensions all consisted of various functions composed with linear functions to include the following:

- $\dfrac{d}{dx}[(x - a)^n] = n(x - a)^{n-1}$

- $\dfrac{d}{dx}[e^{mx+b}] = me^{mx+b}$

- $\dfrac{d}{dx}[\sin(mx+b)] = m\cos(mx+b)$

- $\dfrac{d}{dx}[\cos(mx+b)] = -m\sin(mx+b)$

As might already be apparent from the other work in this section, these extended differentiation rules are valid as a direct result of the chain rule, as illustrated in the following example:

◆ **EXAMPLE 10** Use the chain rule to verify the extended exponential differentiation rule $\dfrac{d}{dx}[e^{mx+b}] = me^{mx+b}$.

Solution. The composition e^{mx+b} of the natural exponential function with an arbitrary linear function results from the inside function $g(x) = mx + b$ with derivative $g'(x) = m$ and the outside function $f(x) = e^x$ with derivative $f'(x) = e^x$. When applying the chain rule $[f(g(x))]' = f'(g(x)) \cdot g'(x)$ in this setting, observe that the factor of m in the differentiation rule results from the $g'(x)$ term as follows:

$$\frac{d}{dx}[e^{mx+b}] = e^{mx+b} \cdot \frac{d}{dx}[mx + b] = e^{mx+b} \cdot m = me^{mx+b}$$

$\blacksquare$

► **QUESTION 11** Use the chain rule to verify the extended cosine differentiation rule
$\frac{d}{dx}\left[\cos(mx+b)\right] = -m\sin(mx+b)$.

Evidence for the Chain Rule

A complete argument for the validity of the chain rule relies on the limit definition of the derivative discussed in Section 4.7, but lies beyond the scope of this book. Even so, some evidence can be developed in support of the chain rule.

First, recall that differentiable functions are locally linear, meaning that within a small enough region immediately around a point on its curve, every differentiable function is essentially a line. Thinking graphically, imagine zooming in very close on the graph of a differentiable function; zooming in close enough eventually results in even the most "curvy" of differentiable functions appearing to be a line. As such, a reasonable conjecture is that certain properties of linear functions extend to differentiable functions. While not all properties carry over for this reason, some do, including the chain rule.

Second, the chain rule holds for any pair of linear functions $f(x) = mx + b$ and $g(x) = nx + c$, as demonstrated by the following two sets of computations. Working directly with $f(g(x))$ yields the following:

$$f(g(x)) = f(nx+c) = m(nx+c)+b = mnx+mc+b = (mn)x+(mc+b)$$

Namely, the composition of two linear functions $f(x) = mx+b$ and $g(x) = nx+c$ results in a linear function with slope mn and vertical intercept $mc + b$. Differentiating using the line rule provides $\left[f(g(x))\right]' = mn$.

Now consider the formula part of the chain rule: $f'(g(x)) \cdot g'(x)$. Applying the line rule to $f(x) = mx+b$ yields $f'(x) = m$ and applying it to $g(x) = nx+c$ yields $g'(x) = n$. Notice that $f'(x) = m$ is a constant function (which graphically corresponds to a horizontal line) and produces the same output of m for every possible input. In other words, $f'(g(x)) = m$ for every possible input. Returning to the chain rule formula, substituting these values and multiplying provides $f'(g(x)) \cdot g'(x) = m \cdot n = mn$. Combining this work together yields

$$\left[f(g(x))\right]' = mn = f'(g(x)) \cdot g'(x)$$

Therefore, the chain rule holds for every pair of linear functions $f(x) = mx + b$ and $g(x) = nx + c$.

In summary, differentiable functions are locally linear and the chain rule holds for all linear functions, which suggests that the chain rule might hold for all differentiable functions. The reader is encouraged to keep in mind that this discussion only provides evidence for this differentiation rule. A complete proof of the chain rule is based on the definition of the limit and is discussed in detail in more advanced math books.

Summary

- If $f(x)$ and $g(x)$ are functions, then the *composition* of f with g is $f \circ g(x) = f[g(x)]$ when this expression is defined.

- *Chain rule*: For all differentiable functions $f(x)$ and $g(x)$ for which these expressions are defined, $\dfrac{d}{dx}[f(g(x))] = f'(g(x)) \cdot g'(x)$

Exercises

In Exercises 1–6, compute both compositions $f \circ g(x)$ and $g \circ f(x)$, and evaluate them when $x = 1$.

1. $f(x) = x^2 + 5$ and $g(x) = 2x$

2. $f(x) = x^3 + 1$ and $g(x) = \sqrt{x}$

3. $f(x) = 4x - 3$ and $g(x) = e^x$

4. $f(x) = 3x + 1$ and $g(x) = \ln(x)$

5. $f(x) = \cos(x)$ and $g(x) = \pi x$

6. $f(x) = \tan(\pi x)$ and $g(x) = 2^x$

In Exercises 7–14, identify the inside function $g(x)$ and the outside function $f(x)$ for each composite function $f[g(x)]$.

7. $(x^3 + 3x)^5$

8. $\sqrt{e^x - x}$

9. $\sqrt[3]{(x^6 - x)^2}$

10. $\sin(x^2 + 3x)$

11. $\cos(e^x + 2x)$

12. $\ln(\pi x + 4)$

13. $17e^{\sin(x)}$

14. $\tan(5x^6)$

In Exercises 15–18, differentiate the function using the chain rule.

15. $(x^3 + 3x)^5$

16. $(x^2 + 1)^{-4}$

17. $(x - e^x)^{-7}$

18. $[x^5 + \cos(x)]^9$

In Exercises 19–24, differentiate the function using the chain rule.

19. $\sin(2x^4 + 3)$

20. $\sin(ex - 6)$

21. $\sin(x^2 - e^x)$

22. $\cos(3x + 8)$

23. $\cos(6 - \pi x)$

24. $\cos(x^9 + 7x)$

In Exercises 25–30, differentiate the function using the chain rule.

25. $\ln(\pi x + 4)$

26. $\ln(5x - 25)$

27. $\ln(-4x - 8)$

28. $\ln(e^{2x})$

29. $\ln(5x^2 + 32)$

30. $\ln[8 - \sin(x)]$

In Exercises 31–36, differentiate the function using the chain rule.

31. $17e^{\sin(x)}$

32. $e^{6x + \cos(x)}$

33. $9^{x^2 + x}$

34. $5^{\sqrt{x} + \sin(x)}$

35. $16 \cdot 7^{3x^2 + 9}$

36. $[\cos(e^x)]^{99}$

In Exercises 37–60, differentiate the function.

37. $x^2 + \sin(x^3 + 6)$

38. $2x + \cos(e^{4x^3})$

39. $\cos(e^x) - 7x^4$

40. $\cos(x) \ln(x^2)$

41. $\sin(3x - e^{2x})$

42. $\sin(x + xe^x)$

43. $\ln(\sin(x) + x^5)$

44. $(x^3 + x)e^{x^2}$

45. $\sin[\cos(x) + x]$

46. $e^{x^2} \ln(5 - x^6)$

47. $e^{x^2+9x}\sin(x^4)$

48. $\cos(x)\ln(4-x)$

49. $6e^{3x-\cos(x)}$

50. $7^{x+\ln(x)} - \dfrac{1}{x}$

51. $3^{\ln(x)+x^2}$

52. $\dfrac{3^{x^2-9x}}{e^{x^4-2x}}$

53. $\dfrac{x+\sin(x^2)}{\ln(6x-9)}$

54. $\dfrac{x+5\cos(x)}{x^2-5x}$

55. $\dfrac{\sin(x^2)}{(6x^8+2)e^x}$

56. $\dfrac{\ln(x+2)}{e^{\cos(x)}}$

57. $\dfrac{e^{\cos(x)}}{x+2}$

58. $\dfrac{x^4-2^{3x}}{7x-6}$

59. $\dfrac{e^{x^4+5x}}{x^3-\sqrt{x}}$

60. $\dfrac{4x^6+3x}{e^x+5}$

In Exercises 61–66, find the rate of change of the functional model.

61. The probability ϕ of a real number x under the normal distribution with a mean of zero and a standard deviation of one is given by

$$\phi(x) = \frac{1}{\sqrt{2\pi}}e^{-x^2/2}$$

62. The probability N of a real number x under the normal distribution with a mean of 80 and a standard deviation of 10 is given by

$$N(x) = \frac{1}{\sqrt{200\pi}}e^{-(x-80)^2/200}$$

63. The number of monthly active Twitter users U worldwide by quarter Q (for example, 10.25 represents April to June 2010) is modeled by

$$U = \frac{297.202}{1+0.959e^{-0.972(Q-12.25)}}$$

64. The population P of Belgium in millions of people per year Y is reasonably modeled by

$$P = \frac{7.8}{1+0.66e^{-0.026(Y-1930)}} + 3.27$$

65. The number S of minutes after 3 p.m. until sunset in Greenwich, England during each month M is reasonably modeled by

$$S = 164.7\sin\left[\frac{\pi}{6}(M-2.6)\right] + 223.6$$

66. The electric bill B of a single-family home in Minnesota for each month M from 2000 through 2003 is reasonably modeled by

$$B = 23.2\sin\left[\frac{2\pi}{9.7}(M-3.8)\right] + 49.4$$

In Exercises 67–70, consider the following model of the average maximum temperature T in central Kentucky at the beginning of each month M since January 2006 (month 0):

$$T = 23.3\sin\left[\frac{2\pi}{11.98}(M-3.16)\right] + 65.4$$

Find and explain the meaning of each quantity.

67. (a) Temp(28)   (b) Temp$'$(28)

68. (a) Temp(38)   (b) Temp$'$(38)

69. (a) Temp(42)   (b) Temp$'$(42)

70. (a) Temp(58)   (b) Temp$'$(58)

In Exercises 71–74, use the chain rule to verify the extended differentiation rule.

71. $\dfrac{d}{dx}\left[(x-a)^n\right] = n(x-a)^{n-1}$

72. $\dfrac{d}{dx}\left[\sin(mx+b)\right] = m\cos(mx+b)$

73. $\dfrac{d}{dx}\left[e^{ax^2+bx}\right] = (2ax+b)e^{ax^2+bx}$

74. $\dfrac{d}{dx}\left[\cos(ax^2)\right] = -2ax\sin(ax^2)$

In Exercises 75–78, express the fraction as a product via $f(x)/g(x) = f(x) \cdot [g(x)]^{-1}$ and use the product and chain rules to verify the derivative rule.

75. Quotient rule

76. $\left[\tan(x)\right]' = \sec^2(x)$

77. $\left[\cot(x)\right]' = -\csc^2(x)$

78. $\left[\sec(x)\right]' = \sec(x)\tan(x)$

In Exercises 79–92, differentiate the function. Section 4.4 presents the derivatives of all the trigonometric functions.

79. $\tan(5x^6)$

80. $\csc(e^x)$

81. $\cot[\ln(x)]$

82. $\sec(2^x + 3^x)$

83. $\tan(5x+7)$

84. $\tan(x^2 + e^x)$

85. $\sec(\ln(x)+1)$

86. $\sec(2x+5)$

87. $\cot(e^x + 2^x)$

88. $\tan[\csc(x)]$

89. $\sec[\cot(x)]$

90. $\cot(e^x + 5x)$

91. $\csc(x^7 + 6x)$

92. $\csc(2^x + x^4)$

In Exercises 93–98, find $h'(a)$ for the function $h(x)$ when $x = a$ using the following input–output table:

x	1	2	4
$f(x)$	-2	3	1
$g(x)$	0	7	-4
$f'(x)$	5	6	8
$g'(x)$	2	8	-1

93. $h(x) = xf(x)$ for $x = 2$

94. $h(x) = (5x+7)f(4x+2)$ for $x = 0$

95. $h(x) = x^3 g(2x)$ for $x = 1$

96. $h(x) = e^x g(x+1)$ for $x = 3$

97. $h(x) = f(x^2)g(4x)$ for $x = 1$

98. $h(x) = \dfrac{f(2x)}{g(x^2)}$ for $x = 1$

In Exercises 99–106, find the value of the derivative of the function when $x = 1$ using the following input–output table:

x	1	2	5
$f(x)$	0	-1	7
$g(x)$	2	3	-1
$f'(x)$	5	5	5
$g'(x)$	1	4	8

99. $f(g(x))$

100. $f(g(2x))$

101. $f(g(x)+3)$

102. $g(f(x))$

103. $g(f(5x))+x$

104. $f(2x)g(x^2)$

105. $f(2x)g(5x)$

106. $\dfrac{f(x^2+x)}{g(4x+1)}$

In Exercises 107–110, find the linear approximation of $f(x)$ when $x = a$.

107. $f(x) = \dfrac{\sin(x)}{\cos(2x)}$ when $x = \dfrac{\pi}{2}$

108. $f(x) = (6x^2+5x)^{10}$ when $x = 0$

109. $f(x) = (\ln(2x)+x)^3$ when $x = e$

110. $f(x) = \sin(e^x + 2x^2)$ when $x = 2$

RStudio. In Exercises 111–122, use RStudio to differentiate the function from the earlier exercise.

111. Exercise 37

112. Exercise 39

113. Exercise 41

114. Exercise 43

115. Exercise 45

116. Exercise 47

117. Exercise 49

118. Exercise 51

In Your Own Words. In Exercises 123
and 124, explain the following.

123. Composition of functions

124. Chain rule

4.6   Partial Derivatives

This section extends the study of the derivative to the context of multivariable functions. Recall that the derivative provides the instantaneous rate of change of a function. Section 4.1 introduced an approach to approximating the derivative of a single-variable function $f(x)$ for an input $x = a$ that was based on finding the average rate of change of $f(x)$ on smaller and smaller intervals containing $x = a$. Sections 4.3 and 4.4 provided rules for exactly computing the derivatives of various functions presented analytically.

Transitioning these ideas to multivariable functions with their increased number of input variables raises some interesting questions. What might be meant by the derivative of a multivariable function? What input interval or intervals might be used to approximate a multivariable derivative? Which input variable or input variables should be used in derivative calculations?

Throughout this section, the central idea is to find the derivative of a multivariable function for each input variable separately, identifying the rate of change with respect to each variable one at a time. The initial study of multivariable functions in Section 1.2 addressed such questions, asking about functions increasing or decreasing in particular directions. Recall from Section 4.2 that a function is increasing when its derivative is positive, constant when its derivative is zero, and decreasing when its derivative is negative. These same relationships hold for multivariable functions and their *partial derivatives*, with the restriction that such behavior is always examined in a particular direction, or for one input variable at a time.

◆ **EXAMPLE 1**   Consider the contour plot of $f(x, y)$ in Figure 1. Starting at input $(x, y) = (2, 0.1)$, determine if $f(x, y)$ immediately increases or decreases and identify the sign of the corresponding derivative when

   (a) the x-inputs increase while the y-inputs are constant, and

   (b) the y-inputs increase while the x-inputs are constant.

Solution.

   (a) Letting the x-inputs increase while the y-inputs are constant corresponds to moving straight right on the countour map. At input $(x, y) = (2, 0.1)$, the value of $f(x, y)$ begins on the contour at level -8 and, moving straight right, increases toward the contour at level -6. Therefore, the derivative of $f(x, y)$ in the x-direction is positive.

   (b) Letting the y-inputs increase while the x-inputs are constant corresponds to moving straight up on the countour map. At input $(x, y) = (2, 0.1)$, the value of $f(x, y)$ begins on the contour at level -8 and, moving straight up, decreases toward the contour at level -10. Therefore, the derivative of $f(x,y)$ in the y-direction is negative. ■

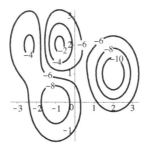

Figure 1: Contour plot of $f(x,y)$ for Example 1 and Question 1

➤ **QUESTION 1** Continue working with the contour plot of $f(x,y)$ in Figure 1. Starting at input $(x,y) = (1,1)$, determine if $f(x,y)$ immediately increases or decreases and identify the sign of the corresponding derivative when

(a) the x-inputs increase while the y-inputs are constant, and

(b) the y-inputs increase while the x-inputs are constant.

At first glance, identifying a function's monotonic behavior and its rate of change in just one direction at a time may appear quite limited in scope and usefulness. As it turns out, this approach is quite powerful, and researchers have found ways to ascertain significant information about a function based on these partial derivatives.

Working toward defining such partial derivatives, some relevant ideas are recalled from Section 4.1. First, the average rate of change of a function over an interval is computed using the following formula:

$$\text{average rate of change of } f(x) \text{ over } [a,b] \;=\; \frac{f(b) - f(a)}{b - a}$$

The instantaneous rate of change of a function $f(x)$ when $x = a$ (which is also called the derivative $f'(a)$ of $f(x)$ when $x = a$) is the number approached by the average rates of change of $f(x)$ as these rates are computed over smaller and smaller intervals containing $x = a$. Also, as illustrated in Figure 2, average rates of change provide slopes of secant lines (such as the lines through PQ, PR, and PS), and derivatives provide slopes of tangent lines (such as the horizontal line through point P).

This approach to defining derivatives of single-variable functions motivates the definition of partial derivatives for multivariable functions. Only now, because of having multiple input variables, a multivariable function has multiple rates of change based on allowing one input variable to change at a time while the other input variables are held constant. More specifically, a two-variable function $f(x,y)$ has two distinct average rates of change because of its two input variables: one in the x-direction and the other in the y-direction.

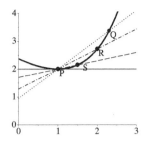

Figure 2: Geometry of single-variable derivatives

Definition. For a multivariable function $f(x, y)$ with input variables x and y,

- the **average rate of change with respect to** x of $f(x, y)$ over an x-interval $[a, c]$ when $y = b$ is given by

$$\frac{f(c, b) - f(a, b)}{c - a},$$

- the **average rate of change with respect to** y of $f(x, y)$ over an y-interval $[b, d]$ when $x = a$ is given by

$$\frac{f(a, d) - f(a, b)}{d - b}.$$

Some examples of rates of change are computed using these formulas shortly. For the moment, the key idea to keep in mind is that average rates of change are computed with respect to one input variable at a time. Instantaneous rates of change (and so derivatives) can be obtained for each input variable separately by computing average rates of change over smaller and smaller input intervals for that variable.

Definition. For a multivariable function $f(x, y)$ with input variables x and y, the **first partial derivatives** of $f(x, y)$ are the following:

- The **partial derivative of** $f(x, y)$ **with respect to** x when $(x, y) = (a, b)$ is the number approached by the average rates of change with respect to x of $f(x, y)$ when these rates are computed over smaller and smaller x-intervals containing $x = a$ while $y = b$ is held constant. This first partial with respect to x is denoted by $f_x(a, b)$.

> **Definition. (continued)**
>
> - The **partial derivative of** $f(x, y)$ **with respect to** y when $(x, y) = (a, b)$ is the number approached by the average rates of change with respect to y of $f(x, y)$ when these rates are computed over smaller and smaller y-intervals containing $y = b$ while $x = a$ is held constant. This first partial with respect to y is denoted by $f_y(a, b)$.

As with single-variable functions, these partial derivatives are sometimes referred to as instantaneous rates of change in the corresponding direction. In this way, $f_x(a, b)$ provides the instantaneous rate of change of $f(x, y)$ with respect to x when $(x, y) = (a, b)$, and $f_y(a, b)$ provides the instantaneous rate of change of $f(x, y)$ with respect to y when $(x, y) = (a, b)$.

A more precise definition of partial derivatives depents on the mathematical idea of a limit, which carefully describes the idea of "approaching a unique, fixed number." Section 4.7 discusses the relationships among limits, instantaneous rates of change, and derivatives for single-variable functions. The exploration of these ideas for multivariable functions lies beyond the scope of this book and is left for your studies in other courses.

The geometric interpretation of a single-variable derivative providing the slope of a tangent line extends to multivariable functions. In the same way, partial derivatives provide slopes of tangent lines, only now the lines are tangent to surfaces and they are oriented in the direction of the designated variable of differentiation.

In Figure 3(a), the numeric value of the partial derivative $f_x(a, b)$ is equal to the slope of the line that lies tangent to $f(x, y)$ in the x-direction. Similarly, the numeric value of the partial derivative $f_y(a, b)$ is equal to the slope of the line in Figure 3(b), which lies tangent to $f(x, y)$ in the y-direction. The end of this section discusses how these two partial derivatives can be combined to produce the equation of a plane tangent to the surface $f(x, y)$, and so provide a linear approximation of $f(x, y)$.

(a)   (b)

Figure 3: Geometry of partial derivatives

NOTATION. In parallel with the derivative notation for single-variable functions, the following Leibniz notation is sometimes used to identify the partial derivatives of a function $f(x, y)$:

- $\dfrac{\partial}{\partial x}\left[f(x, y)\right] = \dfrac{\partial f}{\partial x}(x, y)$ is the partial $f_x(x, y)$ of $f(x, y)$ with respect to x.

- $\dfrac{\partial}{\partial y}\left[f(x, y)\right] = \dfrac{\partial f}{\partial y}(x, y)$ is the partial $f_y(x, y)$ of $f(x, y)$ with respect to y.

While this discussion has focused on multivariable functions $f(x, y)$ with two input variables, these ideas extend to functions with three, four, and more inputs. Namely, a multivariable function has a distinct partial derivative for each of its input variables, which measure the function's rate of change with respect to that input with all others being held constant. Symbolically, the multivariable function $g(x, y, z)$ with three input variables of x, y, and z has three first partials of $g_x(x, y, z)$, $g_y(x, y, z)$, and $g_z(x, y, z)$. Similarly, the multivariable function $h(x, y, z, w)$ with four input variables of x, y, z, and w has the four first partials $h_x(x, y, z, w)$, $h_y(x, y, z, w)$, $h_z(x, y, z, w)$, and $h_w(x, y, z, w)$.

Approximating Partial Derivatives

Partial derivatives with respect to a particular input variable are defined in terms of average rates of change with respect to that variable. In this way, average rates of change enable the approximation of partial derivatives when limited information is available and a partial derivative cannot be computed exactly.

As with single-variable functions, many different approaches can be adopted to choose input intervals when computing average rates of change, and many of these approaches provide a reasonable approximation of the corresponding partial derivative. As might be expected, different choices for input intervals usually produce different (but still reasonable) approximations. The two guiding principles introduced in Section 4.1 for selecting input intervals when approximating derivatives by means of average rates of change are also relevant for approximating partial derivatives.

INTERVALS FOR AVERAGE RATES OF CHANGE.
When using average rates of change to approximate a partial derivative, input intervals should

- be as small as the available information allows, and

- contain the input for which the partial is being approximated.

Examples of average rates of change approximating partial derivatives are first developed for tabular presentations of functions, followed by functions presented analytically. The resulting numeric answers are also interpreted with particular attention to the monotonic behavior of the given function.

◆ **EXAMPLE 2** The National Weather Service Windchill Chart in Figure 4 provides windchill as a function of air temperature in degrees Fahrenheit and wind speed in miles per hour. Approximate and explain the meaning of each rate of change in the windchill when the temperature is $-5°$ Fahrenheit and the wind speed is 30 miles per hour:

(a) The rate of change in the windchill with respect to air temperature.

(b) The rate of change in the windchill with respect to wind speed.

Temperature (°F)

Calm	40	35	30	25	20	15	10	5	0	-5	-10	-15	-20	-25	-30	-35	-40	-45
5	36	31	25	19	13	7	1	-5	-11	-16	-22	-28	-34	-40	-46	-52	-57	-63
10	34	27	21	15	9	3	-4	-10	-16	-22	-28	-35	-41	-47	-53	-59	-66	-72
15	32	25	19	13	6	0	-7	-13	-19	-26	-32	-39	-45	-51	-58	-64	-71	-77
20	30	24	17	11	4	-2	-9	-15	-22	-29	-35	-42	-48	-55	-61	-68	-74	-81
25	29	23	16	9	3	-4	-11	-17	-24	-31	-37	-44	-51	-58	-64	-71	-78	-84
30	28	22	15	8	1	-5	-12	-19	-26	-33	-39	-46	-53	-60	-67	-73	-80	-87
35	28	21	14	7	0	-7	-14	-21	-27	-34	-41	-48	-55	-62	-69	-76	-82	-89
40	27	20	13	6	-1	-8	-15	-22	-29	-36	-43	-50	-57	-64	-71	-78	-84	-91
45	26	19	12	5	-2	-9	-16	-23	-30	-37	-44	-51	-58	-65	-72	-79	-86	-93
50	26	19	12	4	-3	-10	-17	-24	-31	-38	-45	-52	-60	-67	-74	-81	-88	-95
55	25	18	11	4	-3	-11	-18	-25	-32	-39	-46	-54	-61	-68	-75	-82	-89	-97
60	25	17	10	3	-4	-11	-19	-26	-33	-40	-48	-55	-62	-69	-76	-84	-91	-98

Wind (mph)

Frostbite Times   ▨ 30 minutes   ▨ 10 minutes   ▨ 5 minutes

Figure 4: Windchill chart for Example 2 and Question 2

Solution. Let WC(Temperature, Wind) denote the windchill.

(a) The rate of change in the windchill with respect to air temperature is approximated by computing its average rate of change on the smallest available temperature interval of $(-5, 0)$ with a left endpoint of $-5°$ Fahrenheit as follows:

$$\frac{\text{WC}(0, 30) - \text{WC}(-5, 30)}{0 - (-5)} = \frac{-26 - (-33)}{0 - (-5)} = \frac{7}{5} = 1.4$$

Therefore, when the temperature is $-5°$ Fahrenheit and the wind speed is 30 miles per hour, the wind chill is increasing at approximately $1.4°$ Fahrenheit per degree Fahrenheit (because the rate of change is measured with respect to air temperature). This result corresponds to the experience of feeling warmer as the air temperature increases and the wind speed remaining constant.

(b) The average rate of change of windchill with respect to wind is approximated on the smallest available wind speed interval of $(30, 35)$ with a left endpoint of 30 miles per hour as follows.

$$\frac{\text{WC}(-5, 35) - \text{WC}(-5, 30)}{35 - 30} = \frac{-34 - (-33)}{5} = -\frac{1}{5} = -0.2$$

Therefore, when the temperature is $-5°$ Fahrenheit and the wind speed is 30 miles per hour, the wind chill is decreasing at approximately $0.2°$ Fahrenheit per mile per hour (because the rate of change is measured with respect to wind speed). This result corresponds to the experience of feeling colder as the wind speed increases and the air temperature remains constant. ■

▶ **QUESTION 2** Use the National Weather Service Windchill Chart in Figure 4 to approximate each rate of change in the windchill when the temperature is 30° Fahrenheit and the wind speed is 50 miles per hour. Explain the meaning of these rates of change.

(a) The rate of change in the windchill with respect to air temperature.

(b) The rate of change in the windchill with respect to wind speed.

Average rates of change can also be used to approximate the partial derivatives of a function presented analytically. For such functions, recall that average rates of change are computed on smaller and smaller intervals containing the given input and that the desired partial derivative is equal to the number approached by these average rates of change. For analytic functions, the choice of input intervals is not constrained by limited data, so very small input intervals can be used to obtain a very accurate approximation.

◆ **EXAMPLE 3** Compute the average rate of change of the function $f(x, y) = xy^2 + x^3$ over each input interval of x-values for $y = 3$, and use these results to conjecture the partial derivative $f_x(2, 3)$ of $f(x, y)$ with respect to x when $(x, y) = (2, 3)$. Explain the meaning of the answers.

(a) $[2, 2.1]$ (b) $[2, 2.01]$ (c) $[2, 2.001]$ (d) $[1.99999, 2]$

Solution. Compute the average rate of change of $f(x, y)$ over each x-interval as follows:

(a) $\dfrac{f(2.1, 3) - f(2, 3)}{2.1 - 2} = \dfrac{[2.1 \cdot 3^2 + (2.1)^3] - [2 \cdot 3^2 + 2^3]}{0.1} \approx 21.61$

(b) $\dfrac{f(2.01, 3) - f(2, 3)}{2.01 - 2} = \dfrac{[2.01 \cdot 3^2 + (2.01)^3] - [2 \cdot 3^2 + 2^3]}{0.01} \approx 21.0601$

(c) $\dfrac{f(2.001, 3) - f(2, 3)}{2.001 - 2} = \dfrac{[2.001 \cdot 3^2 + (2.001)^3] - [2 \cdot 3^2 + 2^3]}{0.001} \approx 21.006$

(d) $\dfrac{f(2, 3) - f(1.99999, 3)}{2 - 1.99999} = \dfrac{[2 \cdot 3^2 + 2^3] - [1.99999 \cdot 3^2 + (1.99999)^3]}{0.00001} \approx 20.99994$

The average rate of change of $f(x, y)$ with respect to x becomes closer and closer to the number 21 as the intervals containing $x = 2$ become smaller and smaller. This observation indicates that $f_x(2, 3) = 21$, which means that $f(x, y)$ is increasing in the x direction at a rate of 21 when $(x, y) = (2, 3)$. ∎

▶ **QUESTION 3** Compute the average rate of change of the function $f(x, y) = xy + x^2$ over each input interval of y-values for $x = 2$, and use these results to conjecture the partial derivative $f_y(2, 3)$ of $f(x, y)$ with respect to y when $(x, y) = (2, 3)$. Explain the meaning of the answers.

(a) $[3, 3.1]$ (b) $[3, 3.01]$ (c) $[3, 3.001]$ (d) $[2.99999, 3]$

Differentiation Rules and Partial Derivatives

In some settings, derivatives must be approximated with average rates of change. However, for many functions presented analytically, known differentiation rules can be used to find the exact values of derivatives. Fortunately, the differentiation rules for single-variable functions extend to multivariable functions, enabling the computation of partial derivatives with comparable accuracy and ease.

When approximating a partial derivative with respect to an input value, the key idea is to allow one input value to change while holding all other inputs constant. This approach carries over to computing with the differentiation rules. Namely, when computing partial derivatives by applying the known differentiation rules, treat one designated input as the variable of differentiation and treat all the other inputs as constants.

COMPUTING PARTIAL DERIVATIVES.

Let $f(x, y)$ be a function whose first partials exist.

- Find $f_x(x, y)$ by differentiating $f(x, y)$ with respect to x while treating y as a constant.

- Find $f_y(x, y)$ by differentiating $f(x, y)$ with respect to y while treating x as a constant.

Perhaps the most challenging aspect of applying this strategy to compute partial derivatives is becoming used to treating some variables as variables and other variables as constants depending on the context. For years now, your study of mathematics mandated that variables either represent all numbers or some unknown number. But now, as part of answering a differentiation question, you must first identify which variable to treat as a variable quantity and which variables to treat as constants. Only after making this determination can the partial derivative be computed using the differentiation rules.

The following examples and questions use the differentiation rules discussed in Sections 4.3 – 4.5. Many of these rules are restated in the course of this discussion, but the reader is encouraged to revisit those sections if more details would be helpful.

◆ **EXAMPLE 4** Compute all first partial derivatives of each function:

(a) $f(x, y) = 2x + 7\cos(y)$ (b) $g(x, y) = x^2 y - \ln(y)$ (c) $h(x, y) = e^{xy} + \tan(x)$

Solution. These functions have two input variables x and y, which means that they have two first partial derivatives: one with respect to x and the other with respect to y.

(a) For $f_x(x, y)$, differentiate $f(x, y)$ with respect to the variable x while treating y as a constant. From this perspective, $f(x, y) = 2x + 3\cos(y) = mx + b$ is a linear function in x with $m = 2$ and $b = 7\cos(y)$. Applying the line rule $[mx + b]' = m$

yields $f_x(x, y) = 2$. For $f_y(x, y)$, differentiate $f(x, y)$ with respect to the variable y while treating x as a constant:

$$\frac{\partial}{\partial y}\left[\, 2x + 7\cos(y)\,\right] = \frac{\partial}{\partial y}\left[\, 2x\,\right] + \frac{\partial}{\partial y}\left[\, 7\cos(y)\,\right] \quad \text{Sum rule}$$

$$= 0 + 7 \cdot \frac{\partial}{\partial y}\left[\,\cos(y)\,\right] \qquad \text{Constant and constant multiple rules}$$

$$= 7 \cdot \left[-\sin(y)\right] \qquad \text{Cosine rule}$$

In summary, $f_x(x, y) = 2$ and $f_y(x, y) = -7\sin(y)$.

(b) For $g_x(x, y)$, differentiate $g(x, y)$ with respect to the variable x while treating y as a constant:

$$\frac{\partial}{\partial x}\left[\, x^2 y - \ln(y)\,\right] = \frac{\partial}{\partial x}\left[\, x^2 y\,\right] - \frac{\partial}{\partial x}\left[\,\ln(y)\,\right] \quad \text{Difference rule}$$

$$= y \cdot \frac{\partial}{\partial x}\left[\, x^2\,\right] - 0 \qquad \text{Constant multiple and constant rules}$$

$$= y(2x) = 2xy \qquad \text{Power rule and rearrange terms}$$

For $g_y(x, y)$, differentiate $g(x, y)$ with respect to the variable y while treating x as a constant:

$$\frac{\partial}{\partial y}\left[\, x^2 y - \ln(y)\,\right] = \frac{\partial}{\partial y}\left[\, x^2 y\,\right] - \frac{\partial}{\partial y}\left[\,\ln(y)\,\right] \qquad \text{Difference rule}$$

$$= x^2 - \frac{1}{y} \qquad \text{Line and logarithm rules}$$

In summary, $g_x(x, y) = 2xy$ and $g_y(x, y) = x^2 - \dfrac{1}{y}$.

(c) For $h_x(x, y)$, differentiate $h(x, y)$ with respect to the variable x while treating y as a constant. The e^{xy} term can be differentiated with respect to x using two different rules. Applying the extended exponential rule $\dfrac{d}{dx}\left[\, e^{mx+b}\,\right] = m e^{mx+b}$ to $e^{xy} = e^{yx} = e^{mx+b}$ with $m = y$ and $b = 0$ yields $\left[e^{xy}\right]_x = y e^{xy}$. Alternatively, the first term can be differentiated by applying the chain rule to e^{xy}, which has inside function $g(x) = xy$ with $g_x = y$ and outside function $f(x) = e^x$ with $f'(x) = e^x$; both inside and outside functions are identified as functions of x because the partial with respect to x is being computed. This chain rule approach also yields $\left[e^{xy}\right]_x = y e^{xy}$. The complete partial $h_x(x, y)$ is computed as follows:

$$\frac{\partial}{\partial x}\left[\, e^{xy} + \tan(x)\,\right] = \frac{\partial}{\partial x}\left[\, e^{xy}\,\right] - \frac{\partial}{\partial x}\left[\,\tan(x)\,\right] \qquad \text{Sum rule}$$

$$= y e^{xy} + \sec^2(x) \qquad \text{Chain and tangent rules}$$

For $h_y(x, y)$, differentiate $h(x, y)$ with respect to the variable y while treating x as a constant. As with $h_x(x, y)$, the partial of e^{xy} with respect to y can be found via the extended exponential rule or the chain rule:

$$\frac{\partial}{\partial y}\left[e^{xy} + \tan(x)\right] = \frac{\partial}{\partial y}\left[e^{xy}\right] - \frac{\partial}{\partial y}\left[\tan(x)\right] \qquad \text{Sum rule}$$

$$= xe^{xy} + 0 = xe^{xy} \qquad \text{Chain and constant rules}$$

In summary, $h_x(x, y) = ye^{xy} + \sec^2(x)$ and $h_y(x, y) = xe^{xy}$. ∎

➤ **QUESTION 4** Compute all first partials of each function:

(a) $f(x, y) = \sin(6x) + y^3$ (b) $g(x, y) = \ln(xy) - 8y$ (c) $h(x, y) = 7e^{x^2 + 9y}$

As with single-variable functions, the product rule, quotient rule, and chain rule dramatically increase the number of multivariable functions that can be differentiated. When applying these rules, continue to follow the computational approach of differentiating with respect to one input variable and treating the other input variables as constants.

◆ **EXAMPLE 5** Compute all first partial derivatives of each function:

(a) $a(x, y) = e^{4x}\cos(xy)$

(b) $b(x, y) = \dfrac{y\sin(x^2)}{5y + 3^{xy}}$

Solution. These functions have two input variables x and y, which means that they each have two first partial derivatives: one with respect to x and the other with respect to y.

(a) The product rule $[f(x) \cdot g(x)]' = g(x) \cdot f'(x) + f(x) \cdot g'(x)$ is used to compute the partial derivative $a_x(x, y)$, because both factors $f(x) = e^{4x}$ and $g(x) = \cos(xy)$ include the variable x. For this partial, $a(x, y)$ is differentiated with respect to variable x while y is treated as a constant:

$$\frac{\partial}{\partial x}\left[e^{4x}\cos(xy)\right] = \cos(xy) \cdot \frac{\partial}{\partial x}\left[e^{4x}\right] + e^{4x} \cdot \frac{\partial}{\partial x}\left[\cos(xy)\right] \quad \text{Product rule}$$

$$= \cos(xy) \cdot 4e^{4x} + e^{4x} \cdot [-y\sin(xy)] \qquad \text{Chain rule}$$

$$= 4e^{4x}\cos(xy) - ye^{4x}\sin(xy) \qquad \text{Rearrange terms}$$

For $a_y(x, y)$, differentiate $a(x, y)$ with respect to the variable y while x is treated as a constant, using the constant multiple and cosine rules because y only appears in $\cos(xy)$:

$$\frac{\partial}{\partial y}\left[e^{4x}\cos(xy)\right] = e^{4x} \cdot \frac{\partial}{\partial y}\left[\cos(xy)\right] = e^{4x} \cdot [-x\sin(xy)] = -xe^{4x}\sin(xy)$$

In summary, $a_x(x, y) = 4e^{4x}\cos(xy) - ye^{4x}\sin(xy)$ and $a_y(x, y) = -xe^{4x}\sin(xy)$.

(b) The quotient rule must be used to find both partials, because the numerator $f(x, y) = y\sin(x^2)$ and the denominator $g(x, y) = 5y + 3^{xy}$ are functions of both input variables. Recall that when $f(x)$ and $g(x)$ are differentiable and $g(x) \neq 0$, the quotient rule states

$$\frac{d}{dx}\left[\frac{f(x)}{g(x)}\right] = \frac{g(x) \cdot f'(x) - f(x) \cdot g'(x)}{[g(x)]^2}.$$

For $b_x(x, y)$, differentiate with respect to x while y is treated as a constant:

$$b_x(x, y) = \frac{(5y + 3^{xy}) \cdot \frac{\partial}{\partial x}\left[y\sin(x^2)\right] - \left[y\sin(x^2)\right] \cdot \frac{\partial}{\partial x}\left[5y + 3^{xy}\right]}{(5y + 3^{xy})^2}$$

$$= \frac{(5y + 3^{xy}) \cdot \left[2xy\cos(x^2)\right] - \left[y\sin(x^2)\right] \cdot \left[\ln(3)y3^{xy}\right]}{(5y + 3^{xy})^2}$$

For $b_y(x, y)$, differentiate with respect to y while x is treated as a constant:

$$b_y(x, y) = \frac{(5y + 3^{xy}) \cdot \frac{\partial}{\partial y}\left[y\sin(x^2)\right] - \left[y\sin(x^2)\right] \cdot \frac{\partial}{\partial y}\left[5y + 3^{xy}\right]}{(5y + 3^{xy})^2}$$

$$= \frac{(5y + 3^{xy}) \cdot \sin(x^2) - \left[y\sin(x^2)\right] \cdot \left[5 + \ln(3)x3^{xy}\right]}{(5y + 3^{xy})^2}$$

■

▶ **QUESTION 5** Compute all first-order partial derivatives of each function:

(a) $a(x, y) = yx\sin(x^3)$

(b) $b(x, y) = \dfrac{4x + 5y}{\tan(xy)}$

While this section has focused on finding partial derivatives of multivariable functions with two inputs, this same approach applies to functions of three, four, and more variables. Namely, compute the derivative of the given function with respect to one variable while treating all other variables as constants. For example, for a three-variable function $f(x, y, z)$, the first partial derivative with respect to x is found by treating both y and z as constants. Similarly, the first partial derivative of $f(x, y, z)$ with respect to y is found by treating both x and z as constants. Notice that the number of partials increases with the number of inputs, because a function has a first partial for each input variable.

◆ **EXAMPLE 6** Compute all first partials of $f(x, y, z) = xye^{2z} + y\sin(x) + xz$.

Solution. This function has three input variables x, y, and z, so three distinct first partial derivatives must be computed: a partial with respect to x, a partial with respect to y, and a partial with respect to z. For $f_x(x, y, z)$, differentiate with respect to x and treat y and z as constants, which yields $f_x(x, y, z) = ye^{2z} + y\cos(x) + z$. For $f_y(x, y, z)$, differentiate with respect to y and treat x and z as constants, which yields $f_y(x, y, z) = xe^{2z} + \sin(x) + 0 = xe^{2z} + \sin(x)$. For $f_z(x, y, z)$, differentiate with respect to z and treat x and y as constants, which yields $f_z(x, y, z) = xy \cdot 2e^{2z} + 0 + x = 2xye^{2z} + x$. In summary, $f(x, y, z)$ has the following three first partial derivatives:

$$f_x = ye^{2z} + y\cos(x) + z \qquad f_y = xe^{2z} + \sin(x) \qquad f_z = 2xye^{2z} + x$$

■

▶ **QUESTION 6** Compute all first partials of $g(x, y, z) = xy\ln(z) - y^2z^3 + x$.

▶ **QUESTION 7** Compute all first partials of $h(x, y, z) = \dfrac{z \sin(x)}{xy}$.

◆ **EXAMPLE 7** Using dimensional analysis in Section 2.7, the frequency f of the sound produced by a human's vocal chords was expressed as the following function of their length ℓ, tension s, and mass density d: $f(\ell, s, d) = k\, \ell^{-1}\, s^{\frac{1}{2}}\, d^{-\frac{1}{2}}$, where k is a constant. Compute all first partials of the frequency $f(\ell, s, d)$.

Solution. The frequency f is a function with three input variables ℓ, s, and d, and so has three first partial derivatives: a partial with respect to ℓ, a partial with respect to s, and a partial with respect to d. All three partials are computed using the constant multiple and power rules:

- $f_\ell(\ell, s, d) = \dfrac{\partial}{\partial \ell}\left[k\, \ell^{-1}\, s^{\frac{1}{2}}\, d^{-\frac{1}{2}}\right] = k\, s^{\frac{1}{2}}\, d^{-\frac{1}{2}} \cdot \dfrac{\partial}{\partial \ell}\left[\ell^{-1}\right] = k\, s^{\frac{1}{2}}\, d^{-\frac{1}{2}} \cdot \left[-\ell^{-2}\right]$

- $f_s(\ell, s, d) = \dfrac{\partial}{\partial \ell}\left[k\, \ell^{-1}\, s^{\frac{1}{2}}\, d^{-\frac{1}{2}}\right] = k\, \ell^{-1}\, d^{-\frac{1}{2}} \cdot \dfrac{\partial}{\partial \ell}\left[s^{\frac{1}{2}}\right] = k\, \ell^{-1}\, d^{-\frac{1}{2}} \cdot \left[\dfrac{s^{-\frac{1}{2}}}{2}\right]$

- $f_d(\ell, s, d) = \dfrac{\partial}{\partial \ell}\left[k\, \ell^{-1}\, s^{\frac{1}{2}}\, d^{-\frac{1}{2}}\right] = k\, \ell^{-1}\, s^{\frac{1}{2}} \cdot \dfrac{\partial}{\partial \ell}\left[d^{-\frac{1}{2}}\right] = k\, \ell^{-1}\, s^{\frac{1}{2}} \cdot \left[\dfrac{-d^{-\frac{3}{2}}}{2}\right]$

Simplifying provides the following first partials of the frequency $f(\ell, s, d) = k\, \ell^{-1}\, s^{\frac{1}{2}}\, d^{-\frac{1}{2}}$:

$$f_\ell = -k\, \ell^{-2}\, s^{\frac{1}{2}}\, d^{-\frac{1}{2}} \qquad f_s = \dfrac{k}{2}\, \ell^{-1}\, s^{-\frac{1}{2}}\, d^{-\frac{1}{2}} \qquad f_d = -\dfrac{k}{2}\, \ell^{-1}\, s^{-\frac{1}{2}}\, d^{-\frac{3}{2}}$$

■

▶ **QUESTION 8** In 1945, the United States detonated the first atomic bomb in the Trinity Test at White Sands, New Mexico. Using dimensional analysis in Section 2.7, the blast radius r was expressed as the following function of time t, air density d, and explosive energy n:

$$r(t, d, n) = k\sqrt[5]{\dfrac{t^2\, n}{d}} = \dfrac{k\sqrt[5]{t^2} \cdot \sqrt[5]{n}}{\sqrt[5]{d}} = k\, t^{2/5}\, d^{-1/5}\, n^{1/5}$$

where k is a constant. Compute all first partials of the blast radius $r(t, d, n)$.

Higher-Order Partial Derivatives

Section 4.2 introduced higher-order derivatives of single-variable functions. While the focus has primarily been on the first derivative $f'(x)$ of a function $f(x)$, the second derivative $f''(x) = [f'(x)]'$, third derivative $f'''(x) = f^{(3)}(x) = [f''(x)]'$, and higher-order derivatives were also discussed. In a similar fashion, one can compute second partials, third partials, and so on for multivariable functions.

As it turns out, the first partial derivatives of a multivariable function are themselves also multivariable functions with the same number of input variables as the original function. Therefore, each partial derivative has multiple partials, with the number of partials matching the number of input variables. The following table illustrates the resulting exponential growth in the number of higher-order partials for a two-variable function $f(x, y)$:

First partials	Second partials	Third partials
$f_x(x, y)$	$[f_x(x, y)]_x = f_{xx}(x, y)$	$[f_{xx}(x, y)]_x = f_{xxx}(x, y)$
		$[f_{xx}(x, y)]_y = f_{xxy}(x, y)$
	$[f_x(x, y)]_y = f_{xy}(x, y)$	$[f_{xy}(x, y)]_x = f_{xyx}(x, y)$
		$[f_{xy}(x, y)]_y = f_{xyy}(x, y)$
$f_y(x, y)$	$[f_y(x, y)]_x = f_{yx}(x, y)$	$[f_{yx}(x, y)]_x = f_{yxx}(x, y)$
		$[f_{yx}(x, y)]_y = f_{yxy}(x, y)$
	$[f_y(x, y)]_y = f_{yy}(x, y)$	$[f_{yy}(x, y)]_x = f_{yyx}(x, y)$
		$[f_{yy}(x, y)]_y = f_{yyy}(x, y)$

As suggested by this table, higher-order partial derivatives are computed by reading subscripts in the standard English order from left to right. This ordering becomes important when computing **mixed partials**, which involve differentiating with respect to different input variables. For example, $f_{xy}(x, y)$ is found by first differentiating with respect to x and then differentiating the result with respect to y. In contrast, $f_{yx}(x, y)$ is found by first differentiating with respect to y and then differentiating the result with respect to x.

While Leibniz's differential notation can be used to denote higher-order partial derivatives, this book mostly uses the subscript notation, because the partial derivatives in Leibniz notation are read in the reverse order from right to left.

◆ **EXAMPLE 8** Compute all second partial derivatives of $f(x, y) = x^2 y^3$.

Solution. Beginning with the first partials, $f_x(x, y)$ is found by differentiating f with respect to x while treating y as a constant, and $f_y(x, y)$ is found by differentiating f with respect to y while treating x as a constant. Both computations use the constant multiple and power rules:

- $f_x = \dfrac{\partial}{\partial x}\left[x^2\right] \cdot y^3 = 2x \cdot y^3 = 2xy^3$ • $f_y = x^2 \dfrac{\partial}{\partial y}\left[y^3\right] = x^2 \cdot 3y^2 = 3x^2 y^2$

Now find the four second partials of the given function by computing both partials of $f_x(x, y)$ and $f_y(x, y)$, again using the constant multiple and power rules.

- $f_{xx}(x, y) = \dfrac{\partial}{\partial x}\left[f_x(x, y)\right] = \dfrac{\partial}{\partial x}\left[2xy^3\right] = 2y^3 \cdot \dfrac{\partial}{\partial x}\left[x\right] = 2y^3 \cdot 1 = 2y^3$

- $f_{xy}(x, y) = \dfrac{\partial}{\partial y}\left[f_x(x, y)\right] = \dfrac{\partial}{\partial y}\left[2xy^3\right] = 2x \cdot \dfrac{\partial}{\partial y}\left[y^3\right] = 2x \cdot 3y^2 = 6xy^2$

- $f_{yx}(x,y) = \dfrac{\partial}{\partial x}\left[f_y(x,y)\right] = \dfrac{\partial}{\partial x}\left[3x^2y^2\right] = 3y^2\cdot\dfrac{\partial}{\partial x}\left[x^2\right] = 3y^2\cdot 2x = 6xy^2$

- $f_{yy}(x,y) = \dfrac{\partial}{\partial y}\left[f_y(x,y)\right] = \dfrac{\partial}{\partial y}\left[3x^2y^2\right] = 3x^2\cdot\dfrac{\partial}{\partial y}\left[y^2\right] = 3x^2\cdot 2y = 6x^2y$

■

▶ **QUESTION 9** Compute all second partials of $g(x,y) = x^4\sin(y)$.

As with single-variable functions, partial derivatives cannot always be computed for every function, which means that functions for which the first, second, third, and more partial derivatives can be found are distinctive and important. Among other things, the calculus results used to analyze multivariable mathematical models hold for such functions. These distinctive functions are called *smooth*, as stated more precisely in the following definition:

> **Definition.** A multivariable function $f(x,y)$ is **smooth** if and only if $f(x,y)$ has all partials of all orders.

Multivariable versions of the common modeling functions are all smooth on their domains as are the analytic functions that result from adding, subtracting, multiplying, dividing, and composing such functions. The name "smooth" was chosen for these and other such functions with partials of all orders because their graphs are indeed smooth in the common use of the word, describing surfaces of objects that are free of corners, cusps, holes, or other such irregularities.

Among many nice functional behaviors, the mixed partials of smooth functions are equal to each other independently of their order of differentiation. In Example 8, you may have observed that $f_{xy}(x,y) = 6xy^2$ and $f_{yx}(x,y) = 6xy^2$ and, similarly, in Question 9, both $g_{xy}(x,y)$ and $g_{yx}(x,y)$ are equal to $4x^3\cos(y)$. All smooth functions follow this pattern in which such mixed partials can be differentiated in either order and produce the same answer. The theorem asserting this property is named in honor of the French mathematician Alexis Clairaut, who affirmed and clarified results from Newton's *Principia* and helped ensure the dissemination of calculus in continental Europe in the early eighteenth century.

> **CLAIRAUT'S THEOREM.**
> If $f(x,y)$ is a smooth two-variable function, then $f_{xy}(x,y) = f_{yx}(x,y)$.

◆ **EXAMPLE 9** Verify that Clairaut's theorem holds for $f(x,y) = \sin(xy)$.

Solution. The two mixed second partials of $f(x,y)$ are computed to confirm they are equal. The first partials require the use of the chain rule:

- $f_x = \dfrac{\partial}{\partial x}\left[\sin(xy)\right] = y\cos(xy)$ • $f_y = \dfrac{\partial}{\partial x}\left[\sin(xy)\right] = x\cos(xy)$

Both the product and chain rules are needed to find the mixed second partials:

- $f_{xy} = \dfrac{\partial}{\partial y}\left[\dfrac{\partial}{\partial x}\left[f_x\right]\right] = \dfrac{\partial}{\partial y}\left[y\cos(xy)\right] = \cos(xy)\cdot 1 + y\cdot\left[-x\sin(xy)\right]$

- $f_{yx} = \dfrac{\partial}{\partial x}\left[\dfrac{\partial}{\partial y}\left[f_y\right]\right] = \dfrac{\partial}{\partial x}\left[x\cos(xy)\right] = \cos(xy)\cdot 1 + x\cdot\left[-y\sin(xy)\right]$

Simplifying these expressions to $\cos(xy) - xy\sin(xy)$ shows that $f_{xy}(x,y)$ and $f_{yx}(x,y)$ are equal and verifies that Clairaut's theorem holds for $f(x,y) = \sin(xy)$.   ■

▶ **QUESTION 10** Verify that Clairaut's theorem holds for $g(x,y) = e^{8x+9y}$.

Linear Approximation

Linear functions are the simplest, best understood nonconstant functions, which leads researchers to approximate other, more complicated functions with lines. For many single-variable functions, tangent lines provide good, local approximations. As illustrated in Figure 5(a), the tangent line to $f(x)$ when $x = a$ outputs values close to the function's outputs for inputs near $x = a$. Differentiable functions are said to be locally linear because of this very property.

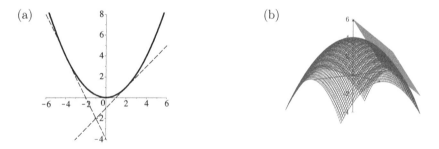

Figure 5: Linear approximations of single- and multivariable functions

The idea of using the tangent line as a linear approximation of a more complicated function extends to multivariable functions. Because of the increased number of input variables, two-dimensional lines are not usually used for linear approximations in such settings. Even so, the key goals remain the same: a linear approximation provides a simpler function and the output values of this simpler function are close to the output values of the given function for inputs near the point where the approximation is based. Figure 5(b) illustrates these properties for a two-variable function, where the linear approximation is a plane.

Recall from Section 4.3 that the equation for the linear approximation of a single-variable function $f(x)$ when $x = a$ is given by $L(x) = f(a) + f'(a)(x - a)$. The linear approximation for a two-variable function $f(x, y)$ is quite similar, except that now there are two partial derivatives $f_x(a, b)$ and $f_y(a, b)$, both of which appear in the equation. In this way, the seemingly limited scope of considering rates of change in one direction at a time is proven to be more expansive, providing more global information about the overall behavior of a function.

LINEAR APPROXIMATION.

Let $f(x, y)$ be a smooth function around the point (a, b). The **linear approximation** of $f(x, y)$ when $(x, y) = (a, b)$ is given by

$$L(x, y) = f(a, b) + f_x(a, b) \cdot (x - a) + f_y(a, b) \cdot (y - b)$$

While practicing with this linear approximation formula, the reader is encouraged to remember that the value of the linear approximation of $f(x, y)$ when $(x, y) = (a, b)$ provides a good estimate for the values of this function for inputs near the point $(x, y) = (a, b)$.

◆ **EXAMPLE 10** Find the linear approximation of $f(x, y) = xy + e^{xy}$ when $(x, y) = (1, 2)$.

Solution. The values of $f(1, 2)$, $f_x(1, 2)$, and $f_y(1, 2)$ are found and then substituted into the equation for the linear approximation. First, evaluate $f(x, y) = xy + e^{xy}$ at input $(x, y) = (1, 2)$ to obtain $f(1, 2) = (1 \cdot 2) + e^{(1 \cdot 2)} = 2 + e^2$. Applying the differentiation rules, the first partials of $f(x, y) = xy + e^{xy}$ are $f_x(x, y) = y + ye^{xy}$ and $f_y(x, y) = x + xe^{xy}$. Evaluating these partials when $(x, y) = (1, 2)$ yields $f_x(1, 2) = 2 + 2e^2$ and $f_y(1, 2) = 1 + e^2$. Substituting into the linear approximation equation produces the following:

$$L(x, y) = f(a, b) + f_x(a, b) \cdot (x - a) + f_y(a, b) \cdot (y - b)$$
$$= (2 + e^2) + (2 + 2e^2) \cdot (x - 1) + (1 + e^2) \cdot (y - 2)$$

∎

▶ **QUESTION 11** Find the linear approximation of $f(x, y) = x\sin(xy)$ when $(x, y) = (1, 0)$.

Working in RStudio

Recall from Section 4.3 that RStudio computes derivatives with the "**D**" command. For example, $f(x) = x^2 + e^{3x} - 4\sin(x)$ is differentiated by entering the command

```
D(x^2+exp(3*x)-4*sin(x)~x)
```

Essentially this same approach is followed when computing the partial derivatives of a multivariable function. In this setting, a function with multiple input variables is stated before the tilde "~" and the variable of differentiation is given immediately after the tilde. For example, xe^y is differentiated with respect to x by entering command (a) and xe^y is differentiated with respect to y by entering command (b):

(a) `D(x*exp(y)~x)`   (b) `D(x*exp(y)~y)`

Higher-order partial derivatives are computed by listing the variables of differentiation in order after the tilde "~" separated by the & symbol.

Examples of Commands:

- `D(6*x^3y-3*y^2~x)`

- `D(6*x^3y-3*y^2~y&x)`

◆ **EXAMPLE 11** Compute all first partials of $f(x, y) = xy^2 + e^{x+y}$ using RStudio.

Solution. The two partials $f_x(x, y)$ and $f_y(x, y)$ are found in RStudio as follows:

```
D(x*y^2+exp(x+y)~x)                  D(x*y^2+exp(x+y)~y)

function (x, y)                      function (y, x)
y^2 + exp(x + y)                     x * (2 * y) + exp(x + y)
```

Based on the final lines of output, $f_x(x, y) = y^2 + e^{x+y}$ and $f_y(x, y) = 2xy + e^{x+y}$. ∎

▶ **QUESTION 12** Compute all first partials of $g(x, y) = 5x^4y + 3\cos(y)$ using RStudio.

Summary

- For a multivariable function $f(x, y)$ with input variables x and y, the *average rate of change with respect to x* of $f(x, y)$ over the x-interval $[a, c]$ when $y = b$ and the *average rate of change with respect to y* of $f(x, y)$ over the y-interval $[b, d]$ when $x = a$ are given by

   ○ for x: $\dfrac{f(c, b) - f(a, b)}{c - a}$ ○ for y: $\dfrac{f(a, d) - f(a, b)}{d - b}$

Summary (continued)

- For a multivariable function $f(x, y)$, the *first partial derivatives* of $f(x, y)$ are as follows:

  - The *partial derivative $f_x(a, b)$ of $f(x, y)$ with respect to x* when $(x, y) = (a, b)$ is the number approached by the average rates of change with respect to x of $f(x, y)$ when these rates are computed over smaller and smaller x-intervals containing $x = a$ while $y = b$ is held constant.

  - The *partial derivative $f_y(a, b)$ of $f(x, y)$ with respect to y* when $(x, y) = (a, b)$ is the number approached by the average rates of change with respect to y of $f(x, y)$ when these rates are computed over smaller and smaller y-intervals containing $y = b$ while $x = a$ is held constant.

- $\dfrac{\partial}{\partial x}[f(x, y)] = \dfrac{\partial f}{\partial x}(x, y) = f_x(x, y)$ and $\dfrac{\partial}{\partial y}[f(x, y)] = \dfrac{\partial f}{\partial y}(x, y) = f_y(x, y)$.

- For approximations of a partial derivative by the average rate of change, the input intervals should

  - be as small as the available information allows, and

  - contain the input for which the partial is being approximated.

- $f_x(x, y)$ is computed by differentiating $f(x, y)$ with respect to x while treating y as a constant, and $f_y(x, y)$ by differentiating $f(x, y)$ with respect to y while treating x as a constant.

- A *smooth* multivariable function has all partials of all orders.

- *Mixed partials*, such as $f_{xy}(x, y)$ and $f_{yx}(x, y)$, involve differentiating with respect to different input variables.

- *Clairaut's theorem*: If $f(x, y)$ is a smooth function, then $f_{xy}(x, y) = f_{yx}(x, y)$.

- The *linear approximation* of a smooth function $f(x, y)$ when $(x, y) = (a, b)$ is given by $L(x, y) = f(a, b) + f_x(a, b) \cdot (x - a) + f_y(a, b) \cdot (y - b)$.

Exercises

In Exercises 1–8, determine if $f(x, y)$ immediately increases or decreases from the given point and identify the sign of the corresponding partial derivative using the following contour plot of $f(x, y)$:

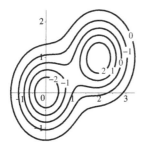

1. x-inputs from $(x, y) = (1, 0)$

2. y-inputs from $(x, y) = (1, 0)$

3. x-inputs from $(x, y) = (2, 1)$

4. y-inputs from $(x, y) = (2, 1)$

5. y-inputs from $(x, y) = (0, 0)$

6. x-inputs from $(x, y) = (0, 0)$

7. y-inputs from $(x, y) = (3, 1)$

8. x-inputs from $(x, y) = (3, 1)$

In Exercises 9–16, determine if $f(x, y)$ immediately increases or decreases from the given input and identify the sign of the corresponding partial derivative using the following contour plot of $f(x, y)$:

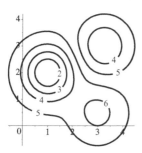

9. x-inputs from $(x, y) = (1, 1)$

10. y-inputs from $(x, y) = (1, 1)$

11. x-inputs from $(x, y) = (0, 0.75)$

12. y-inputs from $(x, y) = (0, 0.75)$

13. y-inputs from $(x, y) = (0.5, 2)$

14. x-inputs from $(x, y) = (0.5, 2$

15. y-inputs from $(x, y) = (3, 0)$

16. x-inputs from $(x, y) = (3, 0)$

In Exercises 17–24, find the average rate of change with respect to x of the function over the given x-interval for the given y-value.

17. $f(x, y) = x^2 + 4y$ for $[0, 2]$, $y = 2$

18. $f(x, y) = x^2 + 4y$ for $[0, 2]$, $y = 3$

19. $f(x, y) = xy + x$ for $[1, 3]$, $y = 1$

20. $f(x, y) = xy + x$ for $[1, 3]$, $y = 2$

21. $f(x, y) = e^x \ln(y)$ for $[-1, 1]$, $y = e$

22. $f(x, y) = e^x \ln(y)$ for $[0, 2]$, $y = e^3$

23. $f(x, y) = x^2 \sin(y)$ for $[-2, 3]$, $y = \frac{\pi}{2}$

24. $f(x, y) = \cos(xy)$ for $[0, \pi]$, $y = 5$

In Exercises 25–32, find the average rate of change with respect to y of the function over the given y-interval and for the given x-value.

25. $f(x, y) = x^2 + 4y$ for $[1, 5]$, $x = 1$

26. $f(x, y) = x^2 + 4y$ for $[3, 7]$, $x = 3$

27. $f(x, y) = xy + x$ for $[2, 4]$, $x = 0$

28. $f(x, y) = xy + x$ for $[-1, 2]$, $x = 4$

29. $f(x, y) = e^x \ln(y)$ for $[1, e^2]$, $x = 0$

30. $f(x, y) = e^x \ln(y)$ for $[e, e^2]$, $x = 2$

31. $f(x, y) = x^2 \sin(y)$ for $[-1, 1]$, $x = \pi$

32. $f(x,y) = \cos(xy)$ for $[0, \frac{\pi}{2}]$, $x = -4$

In Exercises 33–52, compute all first partials of the function.

33. $f(x,y) = 3x + 5$

34. $f(x,y) = x^8 + \sqrt{4x}$

35. $f(x,y) = y^3 + \pi\sqrt{y}$

36. $f(x,y) = ey - 6$

37. $f(x,y) = x^4 + y^2$

38. $f(x,y) = 8x^6 + 7y$

39. $f(x,y) = \sqrt{x^3} + 3x^4$

40. $f(x,y) = e^{4x} + \sin(2y)$

41. $f(x,y) = \cos(6x + 5) + \ln(y)$

42. $f(x,y) = \tan(7x) - \sec(4y)$

43. $f(x,y) = e^{3x+8} + xy$

44. $f(x,y) = e^x \cos(y) + 4$

45. $f(x,y) = x^3 y^5 + 7x$

46. $f(x,y) = ye^x + y^9$

47. $f(x,y) = \ln(x + y) + 4^{5x-3y}$

48. $f(x,y) = \sin(xy) + x^6 y$

49. $f(x,y) = 2^{xy} + \ln(x)\cos(y)$

50. $f(x,y) = 6^{2x+3y} + x^5 y^4$

51. $f(x,y) = x^2 y^3 + \sin(4x + 9y)$

52. $f(x,y) = \sqrt{x^5 y^4} + e^{x^2 y}$

In Exercises 53–62, compute all first partials of the function.

53. $f(x,y,z) = x^3 + y^3 + z^4$

54. $f(x,y,z) = x + \sqrt{3}y + 2z^5$

55. $f(x,y,z) = e^x + \sin(yz)$

56. $f(x,y,z) = \sin(4x + 5) + e^{yz}$

57. $f(x,y,z) = xy^2 + y^6 z^4$

58. $f(x,y,z) = x^3 yz^2 + z$

59. $f(x,y,z) = x\sin(z) + y^2 z^3$

60. $f(x,y,z) = ye^x + z2^x$

61. $f(x,y,z) = xy^2 \sin(z) + 5\pi$

62. $f(x,y,z) = xz\cos(y) + e^{xy}$

In Exercises 63–74, compute all first partials of the function.

63. $f(x,y) = xe^x + xy^2$

64. $f(x,y) = x^2 e^{4x} + y\sin(y)$

65. $f(x,y) = (3x + 7y)e^{xy}$

66. $f(x,y) = x^2 y^3 \sin(y)$

67. $f(x,y) = y^2 \cos(xy)$

68. $f(x.y) = 7^x \ln(6x + y^7)$

69. $f(x,y) = \dfrac{y}{x^4 + y^2}$

70. $f(x,y) = \dfrac{x^3 + y^5}{xe^y}$

71. $f(x,y) = \dfrac{xe^x}{\sin(y) + 4x}$

72. $f(x,y) = \dfrac{y}{x^4 \cos(y^3)}$

73. $f(x,y) = \dfrac{6x + 8y}{\tan(xy)}$

74. $f(x,y) = \dfrac{\sin(x)}{xy}$

In Exercises 75–78, compute all first partials of the function.

75. $f(x,y,z) = xze^y + y^2 z^3$

76. $f(x,y,z) = 3ze^{xy} + \ln(xz)$

77. $f(x,y,z) = \dfrac{2y}{3x + 5z}$

78. $f(x,y,z) = \dfrac{xyz}{y + e^{xz}}$

In Exercises 79–86, compute all second partials of the function.

79. $f(x, y) = 6x + y^5$

80. $f(x, y) = x^2 + 3y^4$

81. $f(x, y) = x^2 y$

82. $f(x, y) = xy^3$

83. $f(x, y) = x^4 e^y$

84. $f(x, y) = e^x \sin(2y)$

85. $f(x, y) = e^y \cos(x)$

86. $f(x, y) = y^7 \ln(x)$

In Exercises 87–92, find and interpret the rate of change in the interest $i = prt$ earned for a particular time period, where p is the principal (or initial amount of money invested), r is the interest rate, and t is the length of the time period.

87. $i_p = \frac{\partial i}{\partial p}$ when $(p, r, t) = (1000, 0.04, 4)$

88. $i_p = \frac{\partial i}{\partial p}$ when $(p, r, t) = (2000, 0.1, 5)$

89. $i_r = \frac{\partial i}{\partial r}$ when $(p, r, t) = (1000, 0.04, 4)$

90. $i_r = \frac{\partial i}{\partial r}$ when $(p, r, t) = (2000, 0.1, 5)$

91. $i_t = \frac{\partial i}{\partial t}$ when $(p, r, t) = (1000, 0.04, 4)$

92. $i_t = \frac{\partial i}{\partial t}$ when $(p, r, t) = (2000, 0.1, 5)$

In Exercises 93–102, find the rates of change of the variables in the Ideal Gas Law $pv = nrt$, where p is the absolute pressure of the gas, v is the volume of the gas, n is the amount of substance of gas (measured in moles), r is the ideal gas constant, and t is the absolute temperature of the gas. First, solve for the given output variable and then differentiate with respect to the given input variable.

93. $p_n = \frac{\partial p}{\partial n}$

94. $p_v = \frac{\partial p}{\partial v}$

95. $p_t = \frac{\partial p}{\partial t}$

96. $v_n = \frac{\partial v}{\partial n}$

97. $v_p = \frac{\partial v}{\partial p}$

98. $v_t = \frac{\partial v}{\partial t}$

99. $n_p = \frac{\partial n}{\partial p}$

100. $n_t = \frac{\partial n}{\partial t}$

101. $t_n = \frac{\partial t}{\partial n}$

102. $t_v = \frac{\partial t}{\partial v}$

103. Why is no partial computed with respect to r in Exercises 93–102?

In Exercises 104–109, find the linear approximation to the function at the input.

104. $f(x, y) = x + y$ when $(x, y) = (1, 1)$

105. $f(x, y) = x + y$ when $(x, y) = (5, 7)$

106. $f(x, y) = xy$ when $(x, y) = (2, 1)$

107. $f(x, y) = \frac{y}{x}$ when $(x, y) = (2, 5)$

108. $f(x, y) = x \sin(y) + xy$ when $(x, y) = (1, 0)$

109. $f(x, y) = e^{xy} + \cos(y)$ when $(x, y) = (1, 1)$

RStudio. In Exercises 110–119, use RStudio to compute all first partials of the function from the earlier exercise.

110. Exercise 33   115. Exercise 43

111. Exercise 35   116. Exercise 45

112. Exercise 37   117. Exercise 47

113. Exercise 39   118. Exercise 49

114. Exercise 41   119. Exercise 51

In Your Own Words. In Exercises 120–127, explain the following.

120. Average rate of change

121. Partial derivative

122. Differentiation rules for partials

4.7   Limits and the Derivative

Section 4.1 defined the derivative (and the instantaneous rate of change) of a given function as the number approached by the average rates of change of the function computed on smaller and smaller intervals. This definition is a specific example of what is known as a *limit*, which identifies a single value "approached" by certain numbers. Limiting processes are the fundamental theoretical building block of the calculus. In addition to the derivative being defined as the limit of average rates of change, the integral is defined as the limit of certain sums in Chapter 6.

This section defines the notion of limit, introduces how to compute certain limits, and discusses the limit definition of the derivative. The definition of the limit presented in this book captures the essence of this idea in a manner appropriate for the common modeling functions. This approach allow the exploration of certain subtleties inherent to the calculus, but does not address every possible functional setting. In fact, mathematicians have developed a more precise definition of the limit in response to questions that arose from deep studies of functions. This more technical definition lies beyond the scope of this book, and interested readers are encouraged to consult a traditional calculus or introductory real analysis textbook.

In order to find a limit, two pieces of information must be provided: a function $f(x)$ and an input value $x = a$. The result of the limit of $f(x)$ as x approaches a is a number that corresponds to the output values of $f(x)$ for inputs x very close to $x = a$. In many cases, these output values are close to $f(a)$ and, frequently, the limiting value is equal to the result of substituting $x = a$ into $f(x)$. But not always. In fact, the limit provides different information than the answer to the precalculus question of "what is the value of $f(x)$ when $x = a$?" As a graphical illustration, Figure 1(a) highlights the result of achieving the precalculus goal of finding a function value for a given input.

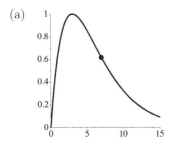
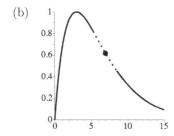

Figure 1: Precalculus versus calculus questions about $f(x)$ and $x = a$

In contrast, the limit answers the more dynamic question of "what is the value of $f(x)$ immediately around $x = a$?" As a graphical illustration, Figure 1(b) highlights the result of achieving the calculus goal of finding a limit, which focuses on a function's output values for inputs immediately both to the left and to the right of some given input. The following definition of the limit expresses the essence of this discussion:

Definition. The **limit** of a function $f(x)$ as the variable x approaches a number a is the number that the outputs of $f(x)$ approach as the inputs x get close (but not necessarily equal) to a from both the left and the right, and is denoted by $\lim\limits_{x \to a} f(x)$.

If $f(x)$ does not approach a unique number, then $\lim\limits_{x \to a} f(x)$ does not exist.

The phrase "the inputs x get close (but not equal) to a from both the left and the right" expresses the focus of a limit on the output values resulting from the inputs immmediately *around* given input $x = a$, as highlighted in Figure 1(b). Some people prefer to use the phrase "from either side" instead of "from both the left and the right." As will be demonstrated in Example 1, the value of $f(x)$ for exactly the input $x = a$ itself does not determine the value of $\lim\limits_{x \to a} f(x)$. In fact, $f(a)$ may be undefined even when $\lim\limits_{x \to a} f(x)$ has a specific value, or $f(a)$ may be defined but not equal to the value of $\lim\limits_{x \to a} f(x)$.

As it turns out, some functions have the distinctive property that $\lim\limits_{x \to a} f(x) = f(a)$, and such *continuous functions* are discussed in more detail later in this section. However, for the moment, keep in mind that such an equality of a limit and a function's value is special and that, in general, finding the number $\lim\limits_{x \to a} f(x)$ is different than finding the number $f(a)$.

Examples illustrating the process of evaluating limits are presented, beginning with limits of functions presented graphically and then numerically approximating limits of functions presented analytically.

◆ **EXAMPLE 1** Evaluate each limit using the graph of $f(x)$ given in Figure 2:

(a) $\lim\limits_{x \to 0} f(x)$ (b) $\lim\limits_{x \to 2} f(x)$ (c) $\lim\limits_{x \to 3} f(x)$

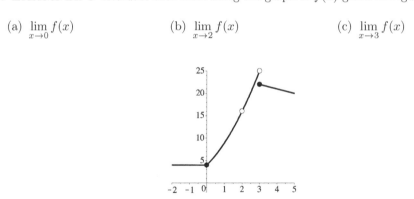

Figure 2: Graph of $f(x)$ for Example 1

Solution.

(a) As the input values get close to $a = 0$ from both the left and the right, the output values of $f(x)$ become close to the number 4. Thus, $\lim_{x \to 0} f(x) = 4$. In this case, $f(0)$ also happens to be equal to 4, but this one functional value of $f(0) = 4$ does not determine the value of the limit $\lim_{x \to 0} f(x)$. Rather, functional values resulting from inputs immediately around $x = 0$ must be examined.

(b) As the input values get close to $a = 2$ from both the left and the right, the output values of $f(x)$ become close to the number 15. Thus, $\lim_{x \to 2} f(x) = 15$. The hole in the graph indicates that $f(2)$ is undefined. However, the limit is based on the output values of $f(x)$ for inputs immediately around $a = 2$, rather for $a = 2$ itself; in other words, $f(2)$ being undefined does not influence $\lim_{x \to 2} f(x) = 15$.

(c) As the input values less than 3 (or to the left of 3) get close to $a = 3$, the output values of $f(x)$ become close to the number 25. In contrast, as the input values greater than 3 (or to the right of 3) get close to $a = 3$, the output values of $f(x)$ become close to the number 22. Therefore, $\lim_{x \to 3} f(x)$ does not exist, because the same output value is not approached when the inputs get close to $a = 3$ from both the left and the right.

■

Example 1(c) illustrates the distinction between left-hand and right-hand limits. For $f(x)$ in Figure 2, the left-hand limit $\lim_{x \to 3^-} f(x) = 25$ (note the $-$) and the right-hand limit $\lim_{x \to 3^+} f(x) = 22$ (note the $+$). A "whole" limit $\lim_{x \to a} f(x)$ exists if and only if the left-hand limit and the right-hand limit as x approaches a both exist and are equal.

➤ **QUESTION 1** Evaluate each limit using the graph of $g(x)$ given in Figure 3:

(a) $\lim_{x \to -1} g(x)$ (b) $\lim_{x \to 0} g(x)$ (c) $\lim_{x \to 1} g(x)$

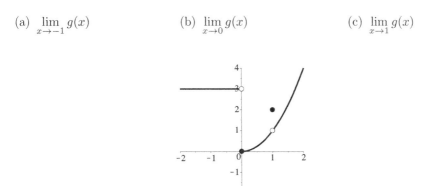

Figure 3: Graph of $g(x)$ for Question 1

For functions presented analytically, the value of a limit is approximated by substituting input values closer and closer to $x = a$ from both the left and the right. This approximation process parallels the approximations of a derivative by average rates of change using smaller and smaller input intervals.

◆ **EXAMPLE 2** Numerically approximate $\lim_{x \to 0} \dfrac{\sin(x)}{x}$ using a sequence of inputs.

Solution. The value of this limit is identified by substituting input values close to $a = 0$ into $f(x) = \dfrac{\sin(x)}{x}$ to determine whether they approach a single output value. As with graphical limits, inputs from both the left and the right must be used, both to the immediate left and to the immediate right of $a = 0$. Consider the following choices of inputs:

$$f(0.1) = \frac{\sin(0.1)}{0.1} \approx 0.99833 \qquad\qquad f(-0.1) = \frac{\sin(-0.1)}{-0.1} \approx 0.99833$$

$$f(0.01) = \frac{\sin(0.01)}{0.01} \approx 0.99998 \qquad\qquad f(-0.01) = \frac{\sin(-0.01)}{-0.01} \approx 0.99998$$

$$f(0.001) = \frac{\sin(0.001)}{0.001} \approx 1.00000 \qquad\qquad f(-0.001) = \frac{\sin(-0.001)}{-0.001} \approx 1.00000$$

The output values of $f(x)$ approach the number 1 as the inputs x approach $a = 0$ from both the left and right, which gives $\lim_{x \to 0} \dfrac{\sin(x)}{x} = 1$.   ■

In Example 2, observe that the function $\dfrac{\sin(x)}{x}$ is not defined when $x = 0$, because attempting to substitute $x = 0$ would result in a zero in the denominator. This example provides another setting in which a function $f(x)$ is not defined when $x = a$, but the limit $\lim_{x \to a} f(x)$ does exist.

▶ **QUESTION 2** Numerically approximate $\lim_{x \to 0} \dfrac{1 - e^x}{x}$ using a sequence of inputs.

Continuous Functions

One potential issue that arises with graphical and numerical determinations of limits is that they essentially amount to being guesses at the value of the limit. While these processes provide informed conjectures, reading values from a graph and substituting a sequence of input values can lead to incorrect results.

Fortunately, mathematicians have developed analytic methods that enable the exact evaluation of many limits. The most useful such method is by means of continuous functions, which also happen to provide the context for the most significant results about definite integrals discussed in Chapter 6. For now, this section focuses on the limit definition of continuous functions and their basic properties.

> **Definition.** A function $f(x)$ is **continuous** for $x = a$ if and only if
>
> $$\lim_{x \to a} f(x) = f(a)$$
>
> A function $f(x)$ is **continuous on its domain** exactly when $f(x)$ is continuous for every input in its domain.

Intuitively, a function is continuous for input $x = a$ if its limit as x approaches a is equal to its value by substitution. As discussed earlier in this section, the calculus question of "$\lim_{x \to a} f(x) = ?$" and the precalculus question of "$f(a) = ?$" are inherently different questions. However, for continuous functions, these questions are guaranteed to have exactly the same answer, as expressed symbolically in the definition by $\lim_{x \to a} f(x) = f(a)$. In this way, the process of finding a limit for a continuous function is greatly simplified, because one can substitute to evaluate the limit.

In addition, notice that a function $f(x)$ is continuous for $x = a$ exactly when the following three conditions must hold:

(1) $\lim_{x \to a} f(x)$ exists   (2) $f(a)$ is defined   (3) $\lim_{x \to a} f(x)$ equals $f(a)$

If even one of these conditions fails to hold, then $f(x)$ is not continuous for $x = a$. Some functions satisfy just one or two of these conditions for a given input, but not all three of them, resulting in the function not being continuous for that input.

◆ **EXAMPLE 3** Determine if $f(x)$ is continuous for each input $x = a$ using the graph of $f(x)$ given in Figure 4:

(a) $x = -2$ (b) $x = 0$ (c) $x = 2$

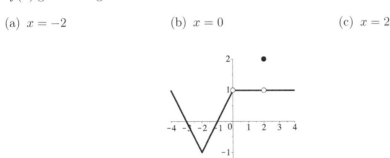

Figure 4: Graph of $f(x)$ for Example 3

Solution.

(a) As the input values x gets close to $a = -2$ from both the left and the right, the output values become close to the number -1 and so $\lim_{x \to -2} f(x) = -1$. Also, observe that $f(-2) = -1$. Thus, all three conditions for continuity hold: the limit exists, the function is defined, and these values are equal at $x = -2$, which means that $f(x)$ is continuous for $x = -2$.

(b) As the input values x gets close to $a = 0$ from both the left and the right, the output values become close to the number 1 and so $\lim\limits_{x \to 0} f(x) = 1$. However, the hole in the graph indicates that $f(0)$ is not defined, and so the second condition for continuous functions fails. Thus, $f(x)$ is not continuous for $x = 0$.

(c) As the input values x gets close to $a = 2$ from both the left and the right, the output values become close to the number 1 and so $\lim\limits_{x \to 2} f(x) = 1$. Also, $f(2) = 2$ as indicated by the height of the point on the graph of $f(x)$ directly above the input $x = 2$. The value of the limit and the value of the function are not equal to each other when $x = 2$, and so the third condition for continuous functions fails. Thus, $f(x)$ is not continuous for $x = 2$.

■

➤ **QUESTION 3** Determine if $g(x)$ is continuous for each input $x = a$ using the graph of $g(x)$ given in Figure 5:

(a) $x = -3$ 　　　　　　　　 (b) $x = 0$ 　　　　　　　　 (c) $x = 2$

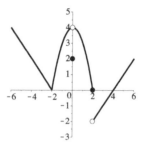

Figure 5: Graph of $g(x)$ for Question 3

An intuitive graphical perspective on continuous functions is that they can be hand-sketched without lifting the writing tool from the drawing surface. In Example 3 and Question 3, the points where the functions are not continuous correspond to places where you would have to lift your pen or pencil if you were tracing the graphs of the functions. In contrast, the functions are continuous over the intervals where their graphs can be traced without any interruptions.

Many analytically presented functions are continuous. Among others, all of the common modeling functions are continuous on their domains, as are all the standard combinations of these modeling functions. A function can be continuous on at most its domain because of the second condition requiring the function to be defined, and so these functions are as continuous as possible. Other analytically presented functions are also continuous, but this section focuses is on these familiar functions for now.

> **CONTINUOUS FUNCTIONS**.
>
> - Linear, exponential, power, trigonometric, and sigmoidal functions are continuous on their domains.
>
> - A function obtained as a sum, difference, product, quotient, or composition of the common modeling functions is continuous on its domain.

◆ **EXAMPLE 4** Identify where each function is continuous:

(a) $a(x) = e^{\sin(3x)} + \dfrac{1}{\sqrt{x}}$ (b) $b(x) = \dfrac{\cos(9x+5)}{4x-7}$ (c) $c(x) = \begin{cases} 2 & x < 4 \\ 8x & x \geq 4 \end{cases}$

Solution.

(a) The function $a(x)$ is continuous on its domain of all positive real numbers, or $(0, \infty)$. In more detail, $a(x)$ is the sum of a composite exponential $e^{\sin(3x)}$ and a power function $x^{-1/2}$. The composite $f(g(x)) = e^{\sin(3x)}$ results from an outside exponential function $f(x) = e^x$ and an inside trigonometric function $g(x) = \sin(3x)$, all of which are continuous on their domains of all real numbers. However, the power function $x^{-1/2}$ is continuous on only its domain of $(0, \infty)$, because one can neither take the square root of a negative number nor divide by zero.

(b) The function $b(x)$ is continuous on its domain consisting of all real numbers except $x = 7/4$. In more detail, $b(x)$ is the quotient of two continuous functions: the trigonometric function $\cos(9x + 5)$ and the linear function $4x - 7$, which are continuous on their domains of all real numbers. However, the input $x = 7/4$ would make the denominator $4x - 7$ equal to zero and the function undefined, and so must be excluded.

(c) The function $c(x)$ is continuous for every real number except $x = 4$, which happens to be just part of its domain of all real numbers. In more detail, both parts of the piecewise function $c(x)$ are linear and so continuous on all real numbers, which means that only the "boundary" between the two parts when $x = 4$ needs to be examined. As it turns out, $\lim\limits_{x \to 4} c(x)$ does not exist, and so the first condition for continuity fails. Namely, inputs to the immediate left of $x = 4$ all output the number 2, while inputs approaching $x = 4$ from the immediate right produce outputs becoming closer to the number $8 \cdot 4 = 32$. Therefore, $\lim\limits_{x \to 4} c(x)$ does not exist, because the same output value is not approached when the inputs get close to $a = 3$ from both the left and the right, which means that $c(x)$ is not continuous for $x = 4$. ■

➤ **QUESTION 4** Identify where each function is continuous:

(a) $a(x) = \dfrac{\sin(x-2)}{e^x - 1}$ (b) $b(x) = \dfrac{\sqrt[3]{x+5}}{x^2 - 1}$ (c) $c(x) = \begin{cases} x^2 & x < 0 \\ x & x \geq 0 \end{cases}$

Continuity is an incredibly useful tool for evaluating many limits. If a function $f(x)$ is continuous for $x = a$, then $\lim_{x \to a} f(x)$ is found by computing $f(a)$. While helpful for finding the limits of the common modeling functions, this approach proves even more useful when evaluating limits of more complicated functions.

◆ **EXAMPLE 5** Evaluate $\lim_{x \to 2} \dfrac{5x + e^{4x}}{3x - 1}$.

Solution. The function $\dfrac{5x + e^{4x}}{3x - 1}$ is the quotient of continuous functions, because the numerator is the sum of a linear and an exponential function, and the denominator is a linear function. Therefore, this function is continuous on its domain consisting of all real numbers except $x = 1/3$. Therefore, the given limit can be evaluated by substituting $x = 2$ as follows:

$$\lim_{x \to 2} \frac{5x + e^{4x}}{3x - 1} = \frac{5(2) + e^{4(2)}}{3(2) - 1} = \frac{10 + e^8}{5} \approx 598.192$$

■

▶ **QUESTION 5** Evaluate each limit using continuity:

(a) $\lim_{x \to 3} x^3 - 5$ (b) $\lim_{x \to \pi} \dfrac{x^2}{\cos(x) + 2}$ (c) $\lim_{x \to 0} \dfrac{x}{2^{9x+4}}$

Evaluating Limits at Discontinuities

Whenever a given function is continuous, substitution can be used to evaluate its limits. However, sometimes limits need to be evaluated for an input where a function is discontinuous. Recall that in Example 2 and Question 2, the following limits were numerically approximated:

$$\lim_{x \to 0} \frac{\sin(x)}{x} = 1 \qquad\qquad\qquad \lim_{x \to 0} \frac{1 - e^x}{x} = -1$$

Notice that both of these functions are not continuous for $x = 0$, because they are undefined when $x = 0$ and the second continuity condition fails. Therefore, the exact value of these limits cannot be found by simply substituting $x = 0$.

In addition to the graphical and numerical approximation techniques that have already been discussed, mathematicians have identified certain analytic methods for finding the exact values of such limits, including what this book calls *factor–cancel*, *expand–cancel*, and *multiply by the conjugate*, based on the key algebraic steps used in each. All three of these methods enable the evaluation of $\lim_{x \to a} f(x)$ by algebraically manipulating the given function $f(x)$ into an equivalent form with respect to limits that is both continuous for $x = a$ and matches $f(x)$ at every input except $x = a$. Then, $\lim_{x \to a} f(x)$ is evaluated by substituting $x = a$ into the transformed continuous function.

◆ **EXAMPLE 6** Evaluate $\lim_{x \to 2} \dfrac{x^2 + x - 6}{x - 2}$ using the factor–cancel method.

Solution. The function $(x^2 + x - 6)/(x - 2)$ is not continuous for $x = 2$, because substituing this input would make the denominator equal to zero, which means that this function is undefined when $x = 2$. Instead, the function is algebraically manipulated into an equivalent form for limits that is continuous when $x = 2$ and is equal to the given function everywhere except for $x = 2$.

$$\lim_{x \to 2} \frac{x^2 + x - 6}{x - 2} = \lim_{x \to 2} \frac{(x - 2)(x + 3)}{x - 2} \qquad \text{Factor the numerator}$$

$$= \lim_{x \to 2} x + 3 \qquad \text{Cancel } x - 2$$

$$= 2 + 3 = 5 \qquad \text{Substitute } x = 2$$

Substitution can be used to evaluate $\lim_{x \to 2} x + 3$ because the linear function $x + 3$ is continuous for all real numbers, including $x = 2$.   ∎

Factoring and canceling achieves the desired goal of enabling the computation of the given limit. The process of factoring is often the most challenging step in using this method, but, with persistent practice, you can strengthen your factoring abilities.

▶ **QUESTION 6** Evaluate each limit using the factor–cancel method. For (b), use the algebraic identity $x^3 - a^3 = (x - a) \cdot (x^2 + ax + a^2)$:

(a) $\displaystyle \lim_{x \to 3} \frac{x - 3}{x^2 - 9}$

(b) $\displaystyle \lim_{x \to 2} \frac{x^3 - 8}{x - 2}$

For the factor–cancel method, the first step is to "undo" some algebra, by writing an algebraic expression as a product. The expand–cancel method reverses this strategy by multiplying out, or expanding, some algebraic expression and then performing some additional algebra to obtain a limit that can be evaluated using continuity. The following example illustrates an application of this method:

◆ **EXAMPLE 7** Evaluate $\displaystyle \lim_{x \to 0} \frac{(2 + x)^2 - 4}{x}$ using the expand–cancel method.

Solution. The function $((2+x)^2 - 4)/x$ is not continuous for $x = 0$, because substituting this input would make the denominator equal to zero, which means that this function is undefined when $x = 0$. Instead, the function is algebraically manipulated into an equivalent form for limits that is continuous when $x = 0$ and is equal to the given function everywhere except for $x = 0$:

$$\lim_{x \to 0} \frac{(2 + x)^2 - 4}{x} = \lim_{x \to 0} \frac{(4 + 4x + x^2) - 4}{x} \qquad \text{Expand } (a + b)^2 = a^2 + 2ab + b^2$$

$$= \lim_{x \to 0} \frac{4x + x^2}{x} \qquad \text{Simplify } 4 - 4 = 0$$

$$= \lim_{x \to 0} \frac{x(4 + x)}{x} \qquad \text{Factor } x \text{ in the numerator}$$

$$= \lim_{x \to 0} 4 + x \qquad \text{Cancel } x$$

$$= 4 + 0 = 4 \qquad \text{Substitute } x = 0$$

Substitution can be used to evaluate $\lim\limits_{x \to 0} 4 + x$ because the linear function $4 + x$ is continuous for all real numbers, including $x = 0$. ■

As illustrated in Example 7, this analytic method usually involves more intermediate steps than factor–cancel. In this sense, the label "expand–cancel" is perhaps misleading. However, with practice, you will develop a facility with these intermediate steps, which usually involve some version of those illustrated in Example 7 of simplifying and factoring, followed by canceling and then evaluating the resulting limit.

▶ **QUESTION 7** Evaluate each limit using the expand–cancel method:

(a) $\lim\limits_{x \to 0} \dfrac{(x + 5)^2 - 25}{x}$

(b) $\lim\limits_{x \to 1} \dfrac{x - 1}{(x + 3)^2 - 16}$

Factoring, expanding, and canceling are familiar algebraic techniques from your previous studies of algebra. The third method, *multiply by the conjugate*, may be less familiar or indeed completely new. The **conjugate** of a sum or difference of square roots is essentially the same expression except that the sign of the second term is reversed as follows:

- The conjugate of $\sqrt{a} + \sqrt{b}$ is $\sqrt{a} - \sqrt{b}$.

- The conjugate of $\sqrt{a} - \sqrt{b}$ is $\sqrt{a} + \sqrt{b}$.

For the purposes of evaluating limits, the important feature of this algebraic object is that multiplying by the conjugate results in elimination of the square roots because the middle terms cancel:

$$(\sqrt{a} + \sqrt{b}) \cdot (\sqrt{a} - \sqrt{b}) = \sqrt{a}\sqrt{a} - \sqrt{a}\sqrt{b} + \sqrt{a}\sqrt{b} - \sqrt{b}\sqrt{b} = a - b$$

In essence, the conjugate relies on the familiar algebraic identify $(x+y)(x-y) = x^2 - y^2$, only now at least one of x or y is the square root of some expression.

The multiply by the conjugate method is used to evaluate a limit at a discontinuity when the function has a square root appearing in a sum or a difference in either the numerator or the denominator. Also, in order to preserve the value of the limit, this method requires multiplying by a clever form of "1" where the conjugate appears in both the numerator and the denominator, as illustrated in the next example.

◆ **EXAMPLE 8** Evaluate $\lim\limits_{x \to 0} \dfrac{\sqrt{4 + x} - 2}{x}$ using the multiply by the conjugate method.

Solution. The function $(\sqrt{4 + x} - 2)/x$ is not continuous for $x = 0$, because substituting this input would make the denominator equal to zero, which means that this function is undefined when $x = 0$. Instead, the function is algebraically manipulated into an equivalent form for limits that is continuous when $x = 0$ and is equal to the given function everywhere except for $x = 0$:

$$\lim_{x \to 0} \frac{\sqrt{4 + x} - 2}{x} = \lim_{x \to 0} \frac{\sqrt{4 + x} - 2}{x} \cdot \frac{\sqrt{4 + x} + 2}{\sqrt{4 + x} + 2} \qquad \text{Multiply by the conjugate}$$

$$= \lim_{x \to 0} \frac{(\sqrt{4+x})^2 - 2^2}{x(\sqrt{4+x} + 2)} \qquad \text{Expand } (a-b)(a+b) = a^2 - b^2$$

$$= \lim_{x \to 0} \frac{4 + x - 4}{x(\sqrt{4+x} + 2)} \qquad \text{Square terms in the numerator}$$

$$= \lim_{x \to 0} \frac{x}{x(\sqrt{4+x} + 2)} \qquad \text{Simplify } 4 - 4 = 0 \text{ in the numerator}$$

$$= \lim_{x \to 0} \frac{1}{\sqrt{4+x} + 2} \qquad \text{Cancel } x$$

$$= \frac{1}{\sqrt{4+0} + 2} = \frac{1}{4} \qquad \text{Substitute } x = 0$$

Substitution can be used to evaluate $\lim\limits_{x \to 0} \dfrac{1}{\sqrt{4+x}+2}$ because this function is continuous for all real numbers greater than or equal to -4, including $x = 0$. ∎

▶ **QUESTION 8** Evaluate each limit using the multiply by the conjugate method:

(a) $\displaystyle\lim_{x \to -2} \frac{\sqrt{x+6} - 2}{x + 2}$

(b) $\displaystyle\lim_{x \to 18} \frac{18 - x}{3 - \sqrt{x - 9}}$

Reprise: The Definition of the Derivative

Section 4.1 defined the **derivative** $f'(a)$ of a function $f(x)$ for an input $x = a$ as the number approached by the average rates of change of $f(x)$ on smaller and smaller intervals containing $x = a$. Considered from the perspective of the study of limits in this section, notice that this definition essentially describes a limit of average rates of change. Recall the formula for computing an average rate of change as illustrated in Figure 6(a):

$$\text{average rate of change of } f(x) \text{ over } [a, b] \; = \; \frac{f(b) - f(a)}{b - a}$$

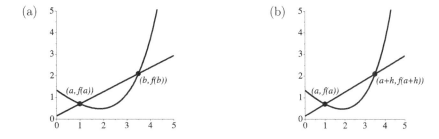

Figure 6: Finding the average rate of change

Finding the value of $f'(a)$ via average rates of change on smaller and smaller intervals containing $x = a$ amounts to computing the limit of average rates of change of $f(x)$ on intervals $[a, b]$ and $[b, a]$ as b approaches a. The definition of the limit requires the consideration of b's approaching a from both the left and the right, not just from the right as suggested by Figure 6(a), and so the need to work with intervals of both forms: $[a, b]$ and $[b, a]$.

Notice that in this context of computing slopes, the ordering of a and b in the quotient does not change its value. Namely, the slope m of the line through two points $(a, f(a))$ and $(b, f(b))$ can be calculated by using a consistent ordering of numerator and denominator as follows (when $a \neq b$):

$$m \; = \; \frac{f(b) - f(a)}{b - a} \; = \; \frac{f(a) - f(b)}{a - b}$$

In summary then, the definition of the limit considering b approaching a from both the left and the right enables a more precise definition of the derivative $f'(a)$ of a function $f(x)$ for an input $x = a$ as

$$f'(a) \; = \; \lim_{b \to a} \frac{f(b) - f(a)}{b - a}.$$

An alternative perspective on this same setting provides another, often more useful, approach to computing derivatives. As illustrated in Figure 6(b), one can think of the endpoint b of the given input interval as equal to $a + h$, where the real number h measures the width of the input interval. Adopting this point of view, the formula for the average rate of change can be expressed as follows:

$$\begin{aligned} \text{average rate of change of } f(x) \text{ over } [a, a + h] \;\; &= \;\; \frac{f(a + h) - f(a)}{(a + h) - a} \\ &= \;\; \frac{f(a + h) - f(a)}{h} \end{aligned}$$

Computing such average rates of change on smaller and smaller input intervals corresponds to letting the width of these intervals approach zero or, in terms of limits, taking the limit as h approaches 0 of this average rate of change difference quotient. The value h can approach 0 from both the left and the right, which means that h can be either positive for input intervals $[a, a + h]$ or negative for input intervals $[a + h, a]$. Either way of approaching 0 results in the same formula, which is used in the definition of the derivative.

Definition. The **derivative** of a function $f(x)$ for input $x = a$ is

$$f'(a) = \lim_{h \to 0} \frac{f(a + h) - f(a)}{h}$$

In order to develop a thorough understanding, this definition is used to numerically approximate some derivatives and then to exactly compute some derivatives.

◆ **EXAMPLE 9** The following table presents median home prices P in thousands of dollars each year Y from 2001 to 2010 according to the U.S. Census Bureau:

Year	2001	2002	2003	2004	2005	2006	2007	2008	2009	2010
Price	175.2	187.6	195.0	221.0	240.9	246.5	247.9	232.1	216.7	221.8

Applying `fitModel` in RStudio produces the following model for this data set:

$$P(Y) = 0.17(Y-2001)^4 - 3.2(Y-2001)^3 + 16.7(Y-2001)^2 - 10.9(Y-2001) + 177.7$$

Numerically approximate $P'(2005)$ using the definition of the derivative and explain the meaning of this value.

Solution. We numerically approximate the derivative by evaluating average rates of change of $P(Y)$ for $h = 1$, $h = 0.1$, and $h = 0.01$:

$$\frac{P(2005+1) - P(2005)}{1} = \frac{P(2006) - P(2005)}{1} \approx \frac{246.95 - 240.02}{1} \approx 6.93$$

$$\frac{P(2005+0.01) - P(2005)}{0.01} = \frac{P(2005.01) - P(2005)}{0.01} \approx \frac{241.14 - 240.02}{0.01} \approx 12.57$$

$$\frac{P(2005+0.0001) - P(2005)}{0.0001} = \frac{P(2005.0001) - P(2005)}{0.0001} \approx \frac{240.021 - 240.02}{0.0001} \approx 12.62$$

These values appear to approach the number 12.6 as h gets closer and closer to 0, which leads to the following conjecture:

$$P'(2005) = \lim_{h \to 0} \frac{P(2005+h) - P(2005)}{h} \approx 12.6$$

Therefore, based on the model, median home prices were increasing at a rate of $12,600 per year in 2005. ∎

In Example 9, the reader is encouraged to keep in mind that finding average rates of change for these values of h provides good evidence for conjecturing the value of $P'(2005) = 12.6$, but does not guarantee that 12.6 is, in fact, the exact value of $P'(2005)$. Analytic methods are needed to find this exact value.

➤ **QUESTION 9** Continue working with the model $P(Y)$ of median home prices in thousands of dollars given in Example 9. Numerically approximate $P'(2011)$ using $h = -1$, $h = -0.01$, and $h = -0.0001$ and explain the meaning of this value.

◆ **EXAMPLE 10** Let $f(x) = |x|$ be the absolute value function. Show that $f'(0)$ is undefined, using the definition of the derivative.

Solution. Evaluating the average rate of change of $f(x) = |x|$ for $x = 0$ and for $h = \pm 1$, $h = \pm 0.01$, and $h = \pm 0.001$ enables a numerical approximation of the left-hand and right-hand values of the limit in the definition of the derivative:

$$\frac{f(1)-f(0)}{1} = \frac{1-0}{1} = 1 \qquad\qquad \frac{f(-1)-f(0)}{-1} = \frac{1-0}{-1} = -1$$

$$\frac{f(0.01)-f(0)}{0.01} = \frac{0.01-0}{0.01} = 1 \qquad\qquad \frac{f(-0.01)-f(0)}{-0.01} = \frac{0.01-0}{-0.01} = -1$$

$$\frac{f(0.001)-f(0)}{0.001} = \frac{0.001-0}{0.001} = 1 \qquad\qquad \frac{f(-0.001)-f(0)}{-0.001} = \frac{0.001-0}{-0.001} = -1$$

The average rates of change of $f(x) = |x|$ when $x = 0$ are equal to 1 when h is positive, but they are equal to -1 when h is negative. A limit is defined when the output values approach the same number for inputs approaching a from both the left and the right. In this case, the outputs approach the two different numbers 1 and -1 as the inputs approach $a = 0$, which means that $\lim\limits_{h \to 0} \dfrac{f(0+h) - f(0)}{h}$ does not exist, and so $f'(0)$ is undefined.

◼

◆ **EXAMPLE 11** For $f(x) = x^2$, find $f'(1)$ using the definition of the derivative and verify the result using the power rule for differentiation.

Solution. Substituting $f(x) = x^2$ and $a = 1$ into the definition of the derivative yields the following limit:

$$f'(1) = \lim_{h \to 0} \frac{(1+h)^2 - 1}{h}$$

The function $((1+h)^2 - 1)/h$ is not continuous for $h = 0$, because substituting this input would make the denominator equal to zero, which means that this function is undefined when $h = 0$. Instead, the function is algebraically manipulated into an equivalent form for limits that is continuous when $h = 0$ and is equal to the given function everywhere except for $h = 0$. The expand–cancel method is used to evaluate this limit as follows:

$$\begin{aligned}
f'(1) &= \lim_{h \to 0} \frac{(1+h)^2 - 1}{h} && \text{Definition of derivative} \\
&= \lim_{h \to 0} \frac{(1+2h+h^2) - 1}{h} && \text{Expand } (a+b)^2 = a^2 + 2ab + b^2 \\
&= \lim_{h \to 0} \frac{2h+h^2}{h} && \text{Simplify } 1 - 1 = 0 \text{ in numerator} \\
&= \lim_{h \to 0} \frac{h(2+h)}{h} && \text{Factor } h \text{ in the numerator} \\
&= \lim_{h \to 0} 2 + h && \text{Cancel } h \\
&= 2 + 0 = 2 && \text{Substitute } h = 0
\end{aligned}$$

Substitution can be used to evaluate $\lim\limits_{h \to 0} 2 + h$ because the linear function $2 + h$ is continuous for all real numbers, including $h = 0$. From this computation, $f'(1) = 2$ when $f(x) = x^2$ and $a = 1$. This result matches the answer when using the power rule to compute $f'(x) = 2x$ and then substituting $a = 1$ to find $f'(1) = 2 \cdot 1 = 2$.

◼

▶ **QUESTION 10** For $f(x) = 3x + 7$, find $f'(4)$ using the definition of the derivative and verify the result using the line rule for differentiation.

The discussion in Section 4.1 of the derivative of a function $f(x)$ in terms of average rates of change transitioned from finding the derivative $f'(a)$ for a particular input $x = a$ to finding the derivative function $f'(x)$. This derivative function provides the instantaneous rate of change for all inputs on which the derivative was defined. This same transition can be made with the limit definition of the derivative.

Definition. The **derivative** of a function $f(x)$ is $f'(x) = \lim\limits_{h \to 0} \dfrac{f(x+h) - f(x)}{h}$.

The power of this definition is that it enables proofs of the differentiation rules. These analytic proofs are a significant improvement over the discussions of graphs in Section 4.3, ensuring that the differentiation rules introduced in this book hold in all possible settings.

◆ **EXAMPLE 12** Prove that $\dfrac{d}{dx}\left[\sqrt{2x + 1}\right] = \dfrac{1}{\sqrt{2x + 1}}$ using the definition of the derivative.

Solution. Substituting $f(x) = \sqrt{2x + 1}$ into the definition of the derivative yields the following limit:

$$f'(x) = \lim_{h \to 0} \frac{\sqrt{2(x+h) + 1} - \sqrt{2x + 1}}{h}$$

The function in this limit is not continuous for $h = 0$, because substituting this input would make the denominator equal to zero. Therefore, the multiply by the conjugate method is used to evaluate this limit:

$$
\begin{aligned}
f'(x) &= \lim_{h \to 0} \frac{\sqrt{2(x+h)+1} - \sqrt{2x+1}}{h} && \text{Definition of derivative} \\[2mm]
&= \lim_{h \to 0} \frac{\sqrt{2(x+h)+1} - \sqrt{2x+1}}{h} \cdot \frac{\sqrt{2(x+h)+1} + \sqrt{2x+1}}{\sqrt{2(x+h)+1} + \sqrt{2x+1}} && \text{Multiply by conjugate} \\[2mm]
&= \lim_{h \to 0} \frac{\left[\sqrt{2(x+h)+1}\right]^2 - \left[\sqrt{2x+1}\right]^2}{h\left(\sqrt{2(x+h)+1} + \sqrt{2x+1}\right)} && (a-b)(a+b) = a^2 - b^2 \\[2mm]
&= \lim_{h \to 0} \frac{(2x + 2h + 1) - (2x + 1)}{h\left(\sqrt{2(x+h)+1} + \sqrt{2x+1}\right)} && \text{Square in numerator} \\[2mm]
&= \lim_{h \to 0} \frac{2h}{h\left(\sqrt{2(x+h)+1} + \sqrt{2x+1}\right)} && \text{Simplify numerator} \\[2mm]
&= \lim_{h \to 0} \frac{2}{\sqrt{2(x+h)+1} + \sqrt{2x+1}} && \text{Cancel } h \\[2mm]
&= \frac{2}{\sqrt{2x+1} + \sqrt{2x+1}} && \text{Substitute } h = 0 \\[2mm]
&= \frac{1}{\sqrt{2x+1}} && \text{Simplify}
\end{aligned}
$$

Thus, based on this limit computation, $\dfrac{d}{dx}\left[\sqrt{2x+1}\right]=\dfrac{1}{\sqrt{2x+1}}$. $\blacksquare$

Notice that the derivative computed in Example 12 matches the result of applying the power rule and the chain rule to $\sqrt{2x+1}$:

$$
\begin{aligned}
\frac{d}{dx}\left[\sqrt{2x+1}\right] &= \frac{d}{dx}\left[(2x+1)^{\frac{1}{2}}\right] = \frac{1}{2}\cdot(2x+1)^{-\frac{1}{2}}\cdot\frac{d}{dx}\left[2x+1\right] \\
&= \frac{1}{2}\cdot\frac{1}{\sqrt{2x+1}}\cdot 2 = \frac{1}{\sqrt{2x+1}}
\end{aligned}
$$

➤ **QUESTION 11** Prove that $\dfrac{d}{dx}\left[5x-9\right]=5$ using the definition of the derivative.

Summary

- The **limit** of a function $f(x)$ as the variable x approaches a number a is the number that the outputs of $f(x)$ approach as the inputs x get close (but not necessarily equal) to a from both the left and the right, and is denoted by $\lim\limits_{x\to a} f(x)$. If $f(x)$ does not approach a unique number, then $\lim\limits_{x\to a} f(x)$ does not exist.

- A function $f(x)$ is *continuous* for an input $x=a$ if and only if $\lim\limits_{x\to a} f(x)=f(a)$. A function $f(x)$ is *continuous on its domain* exactly when $f(x)$ is continuous for every input in its domain.

- Linear, exponential, power, trigonometric, and sigmoidal functions are continuous on their domains. A function obtained as a sum, difference, product, quotient, or composition of the common modeling functions is continuous on its domain.

- We evaluate limits of functions at discontinuities using the analytic methods of *factor–cancel*, *expand–cancel*, and *multiply by the conjugate*.

- The *conjugate* of $\sqrt{a}+\sqrt{b}$ is $\sqrt{a}-\sqrt{b}$ and the conjugate of $\sqrt{a}-\sqrt{b}$ is $\sqrt{a}+\sqrt{b}$, where the sign of the second term is reversed in the sum or difference of square roots.

- The *derivative* of a function $f(x)$ for input $x=a$ is $f'(a)=\lim\limits_{h\to 0}\dfrac{f(a+h)-f(a)}{h}$.

- The *derivative* of a function $f(x)$ is given by $f'(x)=\lim\limits_{h\to 0}\dfrac{f(x+h)-f(x)}{h}$.

Exercises

In Exercises 1–8, evaluate the limit or explain why the limit does not exist using the following graph of $f(x)$:

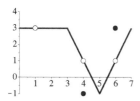

1. $\lim\limits_{x \to 0} f(x)$

2. $\lim\limits_{x \to 1} f(x)$

3. $\lim\limits_{x \to 2} f(x)$

4. $\lim\limits_{x \to 3} f(x)$

5. $\lim\limits_{x \to 4} f(x)$

6. $\lim\limits_{x \to 5} f(x)$

7. $\lim\limits_{x \to 6} f(x)$

8. $\lim\limits_{x \to 7} f(x)$

In Exercises 9–16, evaluate the limit or explain why the limit does not exist using the following graph of $g(x)$:

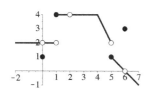

9. $\lim\limits_{x \to 0} g(x)$

10. $\lim\limits_{x \to 1} g(x)$

11. $\lim\limits_{x \to 2} g(x)$

12. $\lim\limits_{x \to 3} g(x)$

13. $\lim\limits_{x \to 4} g(x)$

14. $\lim\limits_{x \to 5} g(x)$

15. $\lim\limits_{x \to 6} g(x)$

16. $\lim\limits_{x \to 7} g(x)$

In Exercises 17–22, numerically approximate the limit using an appropriate sequence of inputs.

17. $\lim\limits_{x \to 0} \dfrac{\cos(x) - 1}{x}$

18. $\lim\limits_{x \to 2} \dfrac{2 - x}{\sin(x - 2)}$

19. $\lim\limits_{x \to 1} \dfrac{x - 1}{\ln(x)}$

20. $\lim\limits_{x \to 1} \dfrac{2^x - 2}{x - 1}$

21. $\lim\limits_{x \to 3} \dfrac{3 - x}{2^x - 8}$

22. $\lim\limits_{x \to 0} \dfrac{xe^{2x} - x}{x^2}$

In Exercises 23–28, explain why $f(x)$ is or is not continuous for the input $x = a$ using the graph of $f(x)$ given for Exercises 1–8.

23. $x = 0$

24. $x = 1$

25. $x = 2$

26. $x = 3$

27. $x = 4$

28. $x = 5$

In Exercises 29–34, explain why $g(x)$ is or is not continuous for the input $x = a$ using the graph of $g(x)$ given for Exercises 9–16.

29. $x = 0$

30. $x = 1$

31. $x = 2$

32. $x = 3$

33. $x = 4$

34. $x = 5$

In Exercises 35–48, identify where the function is continuous

35. $f(x) = (x + 1)^2$

36. $f(x) = (x + 1)e^x$

37. $f(x) = 2^x \tan(x)$

38. $f(x) = x \csc(x)$

39. $f(x) = e^x \cos(x)$

40. $f(x) = x^2 \ln(x)$

41. $f(x) = \dfrac{e^x}{x^2 - 1}$

42. $f(x) = \dfrac{\ln(x)}{x^2 - 4}$

43. $f(x) = \dfrac{x + 5}{\sin(x)}$

44. $f(x) = \dfrac{\cos(x)}{2x - 7}$

45. $f(x) = \begin{cases} x^2 + 4 & x < 3 \\ 2x & x \geq 3 \end{cases}$

46. $f(x) = \begin{cases} \sin(x) & x \leq 0 \\ x & x > 0 \end{cases}$

47. $f(x) = \begin{cases} \ln(x) & x \leq 1 \\ x - 1 & x > 1 \end{cases}$

48. $f(x) = \begin{cases} e^{x-2} & x < 2 \\ x - 2 & x > 2 \end{cases}$

In Exercises 49 – 58, evaluate the limit using continuity.

49. $\lim\limits_{x \to 4} 2$

50. $\lim\limits_{x \to \pi} e + 3$

51. $\lim\limits_{x \to 7} 3x - 2$

52. $\lim\limits_{x \to -4} -2x - 10$

53. $\lim\limits_{x \to -2} \pi x - \pi$

54. $\lim\limits_{x \to -1} x^2 + x$

55. $\lim\limits_{x \to 4} 2^x + x^3$

56. $\lim\limits_{x \to -\pi} \cos(2x)$

57. $\lim\limits_{x \to \pi} \sin(x) + 4$

58. $\lim\limits_{x \to e} \ln(x^2)$

In Exercises 59 – 64, evaluate the limit using the factor–cancel method.

59. $\lim\limits_{x \to 2} \dfrac{x^2 - 4}{x - 2}$

60. $\lim\limits_{x \to -3} \dfrac{x^2 - 9}{x + 3}$

61. $\lim\limits_{x \to -2} \dfrac{x^2 - 4}{x^2 + 2x}$

62. $\lim\limits_{x \to 3} \dfrac{x^2 - x - 6}{x^2 - 2x - 3}$

63. $\lim\limits_{x \to -1} \dfrac{x^2 + 4x - 5}{x^2 - 3x + 2}$

64. $\lim\limits_{x \to 0} \dfrac{x^2 + 2x}{x^3 + 4x}$

In Exercises 65 – 70, evaluate the limit using the expand–cancel method.

65. $\lim\limits_{x \to 4} \dfrac{(x - 2)^2 - 4}{x - 4}$

66. $\lim\limits_{x \to -2} \dfrac{x^2 + 2x}{(x + 3)^2 - 1}$

67. $\lim\limits_{x \to 0} \dfrac{x}{(x + 1)^2 - 1}$

68. $\lim\limits_{x \to 0} \dfrac{(2x + 3)^2 - 9}{x}$

69. $\lim\limits_{x \to 1} \dfrac{x(x + 5) - 6}{x - 1}$

70. $\lim\limits_{x \to -2} \dfrac{x^2 + x - 6}{x(x + 2) + 1}$

In Exercises 71 – 76, evaluate the limit using the multiply by the conjugate method.

71. $\lim\limits_{x \to 3} \dfrac{\sqrt{x + 6} - 3}{x - 3}$

72. $\lim\limits_{x \to -2} \dfrac{\sqrt{x + 6} - 2}{x + 2}$

73. $\lim\limits_{x \to -3} \dfrac{x + 3}{\sqrt{x + 12} - 3}$

74. $\lim\limits_{x \to 4} \dfrac{x - 4}{2 - \sqrt{x}}$

75. $\lim\limits_{x \to 1} \dfrac{\sqrt{x + 15} - 4}{x^2 - 1}$

76. $\lim\limits_{x \to 0} \dfrac{\sqrt{x + 9} - 3}{x^2 + x}$

In Exercises 77–94, evaluate the limit.

77. $\lim\limits_{x \to 0} \dfrac{e^x}{x+1}$

78. $\lim\limits_{x \to \pi} \dfrac{e^x}{\cos(x)}$

79. $\lim\limits_{x \to -2} \dfrac{3x^2 + 7x + 2}{4x^2 + 8x}$

80. $\lim\limits_{x \to 0} \dfrac{2x^3}{6x^4 + 5x}$

81. $\lim\limits_{x \to 0} \dfrac{2 - \sqrt{x+4}}{x}$

82. $\lim\limits_{x \to -1} \dfrac{x^2 + x}{x + \sqrt{x+2}}$

83. $\lim\limits_{x \to -1} \dfrac{x^2 - 1}{x(x+2) + 1}$

84. $\lim\limits_{x \to 2} \dfrac{(2x+1)^2 - 25}{x - 2}$

85. $\lim\limits_{x \to -2} \dfrac{x^2 - 4}{x + 3}$

86. $\lim\limits_{x \to -1} \dfrac{x+2}{\cos(\pi x)}$

87. $\lim\limits_{x \to 3} \dfrac{(x+1)^2 - 16}{3 - x}$

88. $\lim\limits_{x \to 3} \dfrac{3 - \sqrt{12 - x}}{x - 3}$

89. $\lim\limits_{x \to 3} \dfrac{x^3 - 27}{x^2 - 7x + 12}$

90. $\lim\limits_{x \to \pi} \dfrac{\sin(x) + 3}{2x}$

91. $\lim\limits_{x \to \pi} \dfrac{x - \pi}{x^2 - 2\pi x + \pi^2}$

92. $\lim\limits_{x \to -2} \dfrac{x+2}{4 - \sqrt{x+18}}$

93. $\lim\limits_{x \to -1} \dfrac{x(3x+2) - 1}{x^2 + 3x + 2}$

94. $\lim\limits_{x \to 3} \dfrac{3x + 5}{\cos(\pi x)}$

In Exercises 95–104, find $f'(a)$ using the definition of the derivative for the function $f(x)$ and input $x = a$.

95. $f(x) = 2$ when $x = 3$

96. $f(x) = 0$ when $x = 5$

97. $f(x) = x + 7$ when $x = -2$

98. $f(x) = -2x + 8$ when $x = e$

99. $f(x) = x^2$ when $x = 5$

100. $f(x) = -4x^2$ when $x = -1$

101. $f(x) = x^2 + x$ when $x = 0$

102. $f(x) = x^2 + x$ when $x = 3$

103. $f(x) = \sqrt{x}$ when $x = 4$

104. $f(x) = \sqrt{2x + 5}$ when $x = -1$

In Exercises 105–114, differentiate the function using the definition of the derivative.

105. $f(x) = 7$

106. $f(x) = \pi$

107. $f(x) = 2x + 7$

108. $f(x) = -x + 4$

109. $f(x) = x^2 + 4$

110. $f(x) = x^2 - x$

111. $f(x) = -2x^2$

112. $f(x) = \sqrt{x}$

113. $f(x) = \sqrt{x - 1}$

114. $f(x) = \sqrt{x + 7}$

In Your Own Words. In Exercises 115–120, explain the following.

115. Limit

116. Continuous function

117. Examples of continuous functions

118. Analytic methods for evaluating limits at discontinuities

119. Definition of $f'(a)$

120. Definition of $f'(x)$

Chapter 5

Optimization

Chapter 4 discussed Pierre de Fermat's resolution of the tangent line question for many functions by means of the derivative. One of Fermat's primary motivations for developing the derivative was an interest in identifying the maximum and minimum values of functions. The process of finding such *extreme values* of functions is known as *optimization*.

In many ways, optimization lies at the heart of everyday life as we seek the most efficient way to complete a task, an investment strategy that maximizes returns, or directions that minimize the amount of time it takes to travel to a particular destination. From the most mundane, small tasks perhaps worth pennies or seconds, to large-scale, life-transforming projects involving billions of dollars and years of effort, identifying such extreme values touches on many aspects of our personal, social, and economic lives.

Graphically, the extreme values of a function are located at the peaks and valleys of the function, which correspond to where the tangent line to the graph of the function is either horizontal or does not exist. In this way, the study of optimization relies on slopes of tangent lines and the derivative, and builds on the ideas and tools developed in Chapter 4 to identify the maximum and minimum values of functions.

This chapter uses the calculus to identify candidates for where the largest and smallest values of a function are located, and to classify these candidates as maximum or minimum values on global or local scales. Even more, these tasks are explored in both single-variable and multivariable settings, reflecting the breadth of their relevance to understanding real-world phenomena.

5.1   Global Extreme Values

The maximum and minimum values of a function can be considered on two different scales: from a global perspective asking for the extreme values over all numbers in a region of interest or from a local perspective focusing on relatively small regions near the extreme value. As you may recall, Section 1.2 introduced global extreme values for multivariable functions, and Section 5.5 continues this study of multivariable functions. This section focuses on global extreme values for single-variable functions, and the details for local extreme values are explored in Sections 5.2 and 5.3.

The intuitive idea of a global extreme value is that a global maximum is larger than all other outputs of the function and that a global minimum is smaller than all other outputs of the function. Figure 1 provides a graphical illustration of global extreme values for a particular function.

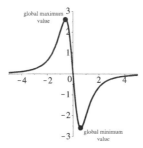

Figure 1: Global minimum and maximum values of a function

The precise mathematical definition of extreme values captures the intuitive idea of a global maximum value and a global minimum value not only for a domain of all real numbers, but also for settings over the smaller, more restricted input intervals that often arise in real-life settings. In such restricted cases, "global" refers to all inputs within a specified domain D of interest.

Definition. Let $f(x)$ be a function, let D be a set of real number inputs in the domain of $f(x)$, and let c be a real number in D.

- The **global maximum** of $f(x)$ on D is the value $f(c)$ such that $f(c) \geq f(x)$ for all x in D.

- The **global minimum** of $f(x)$ on D is the value $f(c)$ such that $f(c) \leq f(x)$ for all x in D.

- The **global extreme values** of $f(x)$ on D are the global maximum and global minimum values of $f(x)$ on D.

Real-life examples of global extreme values include the peaks and valleys of a mountain range. While a mountain range may contain many peaks, the global maximum value corresponds to the mountain peak that is the tallest (or peaks if more than one mountain has the exact same height), and the global minimum value corresponds to the lowest point in the deepest valley.

On an actual global scale for Earth, Mount Everest in the Himalaya Mountains of Asia provides the global maximum of 29,029 feet above sea level. Alternatively, restricting D to just the Appalachian Mountains of the eastern United States results in a global maximum at Mount Mitchell in North Carolina of 6648 feet above sea level. Keeping such a physical image in mind can be helpful when studying more abstract or subtle settings.

In the definition of global extreme values, notice that the output value $f(c)$ is identified as the extreme value. Sometimes the coordinates $(c, f(c))$ are requested, but, even then, the output values are the actual global maximum and minimum values, not the coordinates or the input values where these extreme values occur.

While the definition provides a clear statement articulating the key features of global extreme values, it does not provide an algorithm for locating them. Given the graph of a function or data that is modeled by a smooth function, locating the global maximum value and global minimum value of a given function is a reasonably straightforward task, as illustrated in the following examples.

◆ **EXAMPLE 1**   Identify the global maximum and global minimum values of the function $f(x)$ given in Figure 2 on the following two intervals D and their corresponding inputs $x = c$:

(a) $D = [-2, 7]$ (b) $D = [0, 5]$

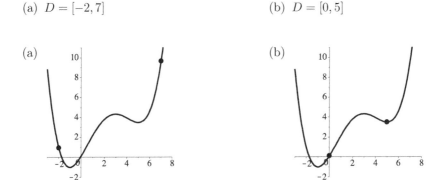

Figure 2: Graphs of $f(x)$ for Example 1

Solution.

(a) In Figure 2(a), the global maximum value is approximately $y = 9.5$ when $x = 7$, because this output is greater than any other output of $f(x)$ on the domain $[-2, 7]$. The global minimum is $y = -1$ when $x = -1$, because this output is less than any other output of $f(x)$ on the domain $[-2, 7]$.

(b) When the domain of the function is further restricted to $[0, 5]$ in Figure 2(b), the global minimum is $y = 0$ when $x = 0$, while the global maximum is approximately $y = 4.25$ when $x = 3$.

As illustrated in Example 1, the set D plays a vital role in determining the global maximum and minimum values, and these global extreme values may change depending on D. While a change in D does not necessarily produce different results, changing D often leads to different global extreme values. In addition, notice that all the extreme values in Example 1 occurred either for an input where $f(x)$ has a horizontal tangent line or at an endpoint of the interval D. This pattern will remain important and recurs throughout this study of global extreme values.

➤ **QUESTION 1** Identify the global maximum and global minimum values of the function $f(x)$ given in Figure 3 on the interval $D = [-2, 7]$ and their corresponding inputs $x = c$.

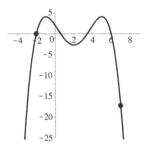

Figure 3: Graph of $f(x)$ for Question 1

Beyond studying smooth functions, as in Example 1 and Question 1, real-life data sets and their models are also important. In such settings, exact output values are not known for every possible input, but only for a relative handful of inputs. Global extreme values can still be identified, only now they must be thought of as approximate global extreme values that could change should additional data become available.

◆ **EXAMPLE 2** The table below provides the mean lower low water (MLLW) tidal depths (M) in feet in Pearl Harbor, Hawaii as a function of the hour (H). Approximate the global maximum and global minimum values of the output M on the input interval of $D = [0, 25]$ and their corresponding inputs.

H	0	1	2	3	4	5	6	7	8	9	10	11	12
M	−0.13	−0.02	0.25	0.56	1.01	1.47	1.78	1.98	2.05	1.88	1.56	1.17	0.79

H	13	14	15	16	17	18	19	20	21	22	23	24	25
M	0.53	0.37	0.32	0.33	0.38	0.51	0.55	0.56	0.49	0.34	0.08	−0.06	−0.07

Solution. Based on the data in the given table, the global maximum MLLW depth is approximately M = 2.05 feet when H = 8 hours, and the global minimum MLLW depth is approximately M = −0.13 feet when H = 0 hours. The graph of this data set in Figure 5 affirms and illustrates these conclusions.   ■

➤ **QUESTION 2** The table below provides the percent growth of the world population as a function of the year from 1970 to 2015 according to the U.S. Census Bureau, and Figure 5 provides a graph of this data. Approximate the global maximum and global minimum values of the percent growth on the interval of years $D = [1975, 2015]$ and their corresponding inputs.

Year	1975	1980	1985	1990	1995	2000	2005	2010	2015
Growth	1.739	1.866	1.732	1.562	1.408	1.208	1.203	1.132	1.079

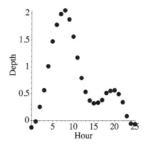

Figure 4: MLLW tidal depth each hour in Pearl Harbor for Example 2

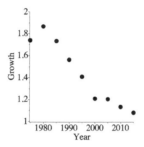

Figure 5: World population growth rate for Question 2

Extreme Value Theorem

Every function considered thus far has had both a global maximum and a global minimum value on the given set D. However, not every function on every set attains both a global maximum and a global minimum. In fact, some functions on some sets have neither global extreme value, and others have just one or the other of a global maximum or global minimum. Fortunately, mathematicians have identified certain conditions under which global extreme values are guaranteed to exist. The next theorem provides one such criteria that ensures global extreme values exist.

EXTREME VALUE THEOREM.
If $f(x)$ is a smooth function on a closed interval $D = [a, b]$, then $f(x)$ has both a global maximum value and a global minimum value on $D = [a, b]$.

The extreme value theorem is both incredibly useful and incredibly unhelpful. On the one hand, the theorem is quite powerful because it guarantees the existence of global extreme values for a large class of functions $f(x)$ on a large collection of sets D. Unfortunately, the extreme value theorem does not provide an algorithm for finding these global extreme values (although such an algorithm is detailed later in this section).

When the two assumptions of the extreme value theorem are satisfied for a given function $f(x)$ in a given set D, then $f(x)$ is guaranteed to have both a global maximum and a global minimum on D. The first assumption is that the given function must be smooth. Most functions considered in this book satisify this requirement, including the common modeling functions discussed in Chapter 2, which are smooth on their domains. However, if a given function is not smooth, then there is no guarantee that it has both global extreme values. In fact, such a function could have any of just a global maximum, just a global minimum, or neither.

The second assumption of the extreme value theorem is that the region of interest D must be a closed interval. If D is not a closed interval, then there is no guarantee that it has both global extreme values, just as when a function is not smooth. Similarly, a function on such a set D could have any of just a global maximum, just a global minimum, or neither.

The following graph provides a function that is neither smooth nor defined on a closed interval. The holes in the graph at $(-1, 4)$ and $(5, 4)$ make the function not smooth and also result in no global maximum value. The domain of $[-2, 7)$ is not a closed interval, and the hole in the graph at $(7, -17)$ results in the function not having a global minimum value.

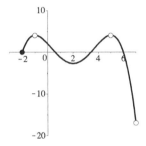

Figure 6: A nonsmooth function on an non-closed interval

Even when the assumptions of the extreme value theorem do not hold, global extreme values will sometimes still exist. For example, a drug administered to a patient produces a concentration in the bloodstream that can be modeled by $c(x) = \dfrac{e}{3} x e^{-x/3}$ (in milligrams per milliliter) x hours after being injected; see Figure 7 for a graph of $c(x)$. A natural domain of study for this function is $D = [0, \infty)$, which is not a closed interval. Therefore, the extreme value theorem does not guarantee that $c(x)$ has global extreme values on D. Even so, the graph in Figure 7 suggests a global maximum of approximately $y \approx 1$ when $x \approx 3$ and a global minimum of $y = 0$ when $x = 0$. A more precise analysis of this setting is given in Example 6 in Section 5.2.

Finally, while the statement of the extreme value theorem focuses on smooth functions, a more general version of this result is true for a broader class of functions. Namely, the extreme value theorem holds for *continuous* functions whose graphs can be hand-sketched without lifting the writing tool from the drawing surface. The precise definition of continuous functions was discussed in Section 4.7. Also, as the title "theorem" indicates, mathematicians have proven the extreme value theorem. However, a proof of this theorem lies beyond the scope of this book and is left for the reader's further studies in more advanced courses.

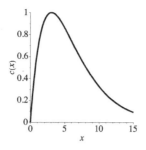

Figure 7: Bloodstream concentration of a drug

Critical Numbers

Approximating the global maximum and minimum values of a function $f(x)$ and the coordinates where they occur can be readily accomplished from the graph of $f(x)$. However, an analytic approach to locating global extreme values requires developing a method for finding them without consulting a graph. Among other benefits, such analytic methods transition from approximations of global extreme values to finding their exact values.

In Figure 1 at the beginning of this section, the global extreme values of the function occur where the function has a horizontal tangent line. Recall that the slope of a horizontal line is zero and that the derivative provides slopes of tangent lines. Therefore, at such extreme values, the derivative of the function is equal to zero.

Example 1 and Question 1 introduce another variation for locations of global extreme values. Namely, in addition to global extreme values occurring when the derivative is equal to zero, global extreme values can also occur at the endpoints of the interval D. Such patterns hold for most global extreme values, particularly for the common modeling functions.

Finally, consider Figure 8. For both of these functions, global extreme values occur when $x = 0$, only now the derivative is not zero. Rather, the derivative does not exist

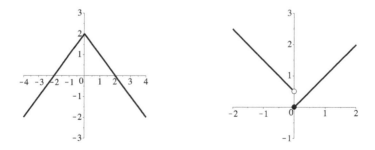

Figure 8: Global extrema can occur when $f'(c)$ does not exist

for both of these functions when $x = 0$. Example 3 in Section 4.2 provides a detailed discussion about derivatives not existing for particular inputs if the reader desires a review or more information about nondifferentiability.

In short, global extreme values can occur for inputs where the derivative of the function $f(x)$ is either zero or does not exist as well as at the endpoints of the input interval D. The inputs for which the derivative of a function is zero or does not exist are so significant that they are referred to as "critical."

Definition.   A **critical number** of a function $f(x)$ is an input c in the domain of $f(x)$ such that either $f'(c)$ is equal to zero or $f'(c)$ does not exist.

◆ **EXAMPLE 3** Identify the critical numbers of the function $f(x)$ given in Figure 9.

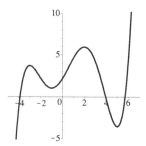

Figure 9: Graph of $f(x)$ for Example 3

Solution. The critical numbers of $f(x)$ given in Figure 9 are $x = -3$, $x = -1$, $x = 2$, and $x = 5$. In particular, each of these inputs corresponds to a horizontal tangent line, which means the derivative is zero for these inputs. Smooth functions, such as $f(x)$ in Figure 9, have a well-defined tangent line at every input. Therefore, the derivative exists at every input in the domain of $f(x)$, and this function does not have any critical numbers owing to the derivative not existing.   ■

◆ **EXAMPLE 4** Identify the critical numbers of the function $f(x)$ given in Figure 10.

Solution. The critical numbers of $f(x)$ given in Figure 9 are $x = 0$ and $x = 2$, because the derivative does not exist for these inputs. The derivative exists for all other inputs, but is never zero, because the function does not have any horizontal tangent lines. Therefore, there are no inputs that are critical numbers as a result of the derivative being zero.   ■

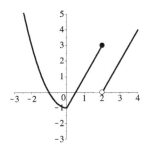

Figure 10: Graph of $f(x)$ for Example 4

▶ **QUESTION 3** Identify the critical numbers of the functions given in Figure 11.

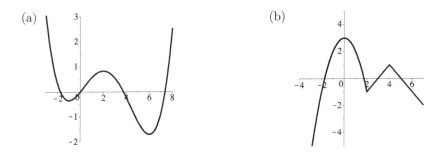

Figure 11: Graphs of functions for Question 3

◆ **EXAMPLE 5** The following table presents median home prices (P) in thousands of dollars each year (Y) from 2001 to 2010 according to the U.S. Census Bureau:

Year	2001	2002	2003	2004	2005	2006	2007	2008	2009	2010
Price	175.2	187.6	195.0	221.0	240.9	246.5	247.9	232.1	216.7	221.8

Applying `fitModel` in RStudio produces the following model for this data set:

$$P(Y) = 0.17(Y-2001)^4 - 3.2(Y-2001)^3 + 16.7(Y-2001)^2 - 10.9(Y-2001) + 177.7$$

Identify the critical numbers of the model P(Y) using the graph of this model and the data set in Figure 12.

Solution. Based on the graph in Figure 12, the critical numbers of the model are approximately Y = 201.5, Y = 2006, and Y = 2009.75, because the derivative of the model is zero for each of these inputs. The derivative $P'(Y)$ of the model is defined for every input, so no additional critical numbers result from the derivative not existing. ∎

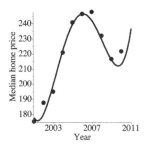

Figure 12: Median home prices from 2001 to 2010 for Example 5

Determining Critical Numbers from the Derivative

Although working directly with the graph of a function $f(x)$ to identify its critical numbers is helpful for understanding the definition of critical numbers, most often the critical numbers of $f(x)$ are found by working with its derivative $f'(x)$. Graphically, a function is zero when its graph intersects the x-axis and is undefined when a hole or gap appears in its graph. Thus, the critical numbers of a function $f(x)$ are the inputs for which the graph of its derivative $f'(x)$ intersects the x-axis or has holes or gaps.

◆ **EXAMPLE 6** Identify the critical numbers of a function $f(x)$ using the graph of its derivative $f'(x)$ given in Figure 13.

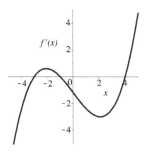

Figure 13: Graph of derivative function $f'(x)$ for Example 6

Solution. The derivative $f'(x)$ is equal to zero when its graph intersects the x-axis for inputs $x = -3$, $x = -1$, and $x = 4$. The derivative is defined for every input, because there are no holes or gaps in its graph. Therefore, $x = -3$, $x = -1$, and $x = 4$ are the critical numbers of $f(x)$. ∎

➤ **QUESTION 4** Identify the critical numbers of a function $f(x)$ using the graph of its derivative $f'(x)$ given in Figure 14.

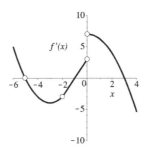

Figure 14: Graph of derivative function $f'(x)$ for Question 4

Beyond working with graphs of the derivative of a function, algebra is often used to identify the critical numbers of functions presented analytically. For such a function $f(x)$, its derivative $f'(x)$ is first computed using the differentiation rules developed in Chapter 4. The critical numbers of $f(x)$ are then found by solving the equation $f'(x) = 0$ to identify the locations of horizontal tangent lines and by looking for inputs that make $f'(x)$ undefined, such as zeros of the denominator or negatives in even roots.

◆ **EXAMPLE 7** Find the critical numbers of the function $f(x) = 3x^4 - 4x^3 - 12x^2 + 14$.

Solution. The derivative of the given function is $f'(x) = 12x^3 - 12x^2 - 24x$. This derivative is defined for all real number inputs, because the domain of every polynomial is the set of all real numbers. Therefore, the critical numbers of $f(x)$ are found by setting this derivative equal to zero and solving for x:

$$12x^3 - 12x^2 - 24x = 0 \qquad \text{Set } f'(x) \text{ equal to zero}$$
$$12x(x^2 - x - 2) = 0 \qquad \text{Factor } 12x \text{ from all terms}$$
$$12x(x + 1)(x - 2) = 0 \qquad \text{Factor } x^2 - x - 2$$
$$x = 0, -1, 2 \qquad \text{Find the zero of each factor}$$

Therefore, the critical numbers of $f(x)$ are $x = -1$, $x = 0$, and $x = 2$.   ∎

◆ **EXAMPLE 8** Find the critical numbers of the function $g(x) = \dfrac{x^2}{x + 1}$.

Solution. The derivative of $g(x)$ is computed using the quotient rule:

$$g'(x) = \frac{d}{dx}\left[\frac{x^2}{x + 1}\right] = \frac{(x + 1) \cdot 2x - x^2 \cdot 1}{(x + 1)^2} = \frac{2x^2 + 2x - x^2}{(x + 1)^2} = \frac{x^2 + 2x}{(x + 1)^2} = \frac{x(x + 2)}{(x + 1)^2}$$

This derivative $g'(x)$ is equal to zero when its numerator $x(x + 2)$ is equal to zero, which happens when $x = 0$ and $x = -2$. In addition, $g'(x)$ is undefined when its denominator

$(x+1)^2$ is equal to zero, which happens when $x = -1$. Therefore, the critical numbers of $g(x)$ are $x = -2$, $x = 0$, and $x = -1$.

■

▶ **QUESTION 5** Find the critical numbers of each function:

(a) $a(x) = 2x^3 - 3x^2 - 36x$

(b) $b(x) = \dfrac{4x + 5}{3x + 7}$

Locating Global Extreme Values

The various examples in this section illustrate the possible locations for global extreme values of functions. In more detail, in Example 1(a), the global minimum occurs at a critical number of the function, and the global maximum occurs at the right endpoint of the interval D. In both Examples 1(b) and 2, the global minimum occurs at the left endpoint of the interval D and the global maximum at a critical number.

Based on these and other such examples, one might reasonably conjecture that the global extreme values of a function $f(x)$ on a closed interval D occur for inputs that are critical numbers of $f(x)$ or for endpoints of D. In fact, this pattern does hold in general and leads to the following three-step process for finding the global extreme values of a smooth function on a closed interval D:

LOCATING GLOBAL EXTREME VALUES.

Let $f(x)$ be a smooth function on a closed interval $D = [a, b]$.

(1) Identify the critical numbers of the function $f(x)$.

(2) Evaluate the function $f(x)$ at the critical numbers that are inside the domain and at both endpoints $x = a$ and $x = b$ of D.

(3) The largest function value from step (2) is the global maximum, and the smallest function value from step (2) is the global minimum.

◆ **EXAMPLE 9** Find the global extreme values of the function $f(x) = 2x^3 - 3x^2 - 36x$ on the closed interval $D = [-3, 2]$.

Solution. The extreme value theorem guarantees that $f(x)$ has global maximum and global minimum values on D. Namely, the function is smooth, because $f(x)$ is a polynomial and all polynomials are smooth on their domains of all real numbers, which includes the given interval. Also, $D = [-3, 2]$ is a closed interval. The process given above is used to find these extreme values.

The first step in identifying the critical numbers of $f(x)$ was accomplished in Question 5(a). In more detail, the derivative is $f'(x) = 6x^2 - 6x - 36 = 6(x+2)(x-3)$, which is defined for every input. Therefore, the critical numbers of $f(x)$ result from $f'(x) = 0$, or when $x = -2$ and $x = 3$.

For the second step, the given interval $D = [-3, 2]$ contains only the critical number $x = -2$, so the other critical number $x = 3$ is discarded from consideration for this particular D. Evaluating $f(x)$ at the critical number $x = -2$ and at the endpoints of the given interval, $x = -3$ and $x = 2$, yields the output values in the following table:

x	-3	-2	2
$f(x)$	27	44	-68

The final step is to examine the output values and select the largest and smallest values as the global extreme values. By direct observation, the function $f(x) = 2x^3 - 3x^2 - 36x$ has a global maximum of $y = 44$ when $x = -2$ and a global minimum of $y = -68$ when $x = 2$ on the set $D = [-3, 2]$.   ■

◆ **EXAMPLE 10** Find the global extreme values of $f(x) = 3x^4 - 4x^3 - 12x^2 + 14$ on $D = [-1, 4]$.

Solution. The extreme value theorem guarantees that $f(x)$ has global maximum and global minimum values on D. Namely, $f(x)$ is a polynomial and so smooth on its domain of all real numbers, which includes the given interval. Also, $D = [-1, 4]$ is a closed interval. The process given above is used to find these extreme values.

Example 7 accomplished the first step in identifying the critical numbers of $f(x)$ as $x = -1$, $x = 0$, and $x = 2$. All of these critical numbers lie in the given interval $[-1, 4]$, including $x = -1$, which happens to be both a critical number and an endpoint. Now evaluate the function $f(x)$ at the critical numbers of $f(x)$ and at the endpoints of the given interval D to obtain the output values in the following table:

x	-1	0	2	4
$f(x)$	9	14	-18	334

Selecting the largest and smallest values as the global extreme values indicates that, on the set $D = [-1, 4]$, the function $f(x) = 3x^4 - 4x^3 - 12x^2 + 14$ has a global maximum of $y = 334$ when $x = 4$ and a global minimum of $y = -18$ when $x = 2$.   ■

◆ **EXAMPLE 11** Find the global extreme values of $f(x) = x^4 - 4x^2$ on $D = [-3, 3]$.

Solution. The extreme value theorem guarantees that $f(x)$ has global maximum and global minimum values on D, because $f(x)$ is a polynomial and so smooth on its domain of all real numbers, and $D = [-1, 4]$ is a closed interval.

The first step is to identify the critical numbers of the function using its derivative $f'(x) = 4x^3 - 8x$, which is defined for every input. Therefore, all critical numbers are found by setting $f'(x)$ equal to zero and solving for x:

$$4x^3 - 8x = 0 \qquad \text{Set } f'(x) \text{ equal to zero}$$
$$4x(x^2 - 2) = 0 \qquad \text{Factor } 4x \text{ from all terms}$$
$$4x(x + \sqrt{2})(x - \sqrt{2}) = 0 \qquad \text{Factor } x^2 - 2$$
$$x = 0, -\sqrt{2}, \sqrt{2} \qquad \text{Find the zero of each factor}$$

The critical numbers of $x = -\sqrt{2}$, $x = 0$, and $x = \sqrt{2}$ all lie within the given interval $D = [-3, 3]$. Evaluating $f(x)$ at these critical numbers and at the endpoints of D yields the output values in the following table:

x	-3	$-\sqrt{2}$	0	$\sqrt{2}$	3
$f(x)$	45	-4	0	-4	45

Thus, on the interval $D = [-3, 3]$, the function $f(x) = x^4 - 4x^2$ has a global maximum of $y = 45$ when $x = -3$ and $x = 3$. Similarly, $f(x)$ has a global minimum of $y = -4$ when $x = -\sqrt{2}$ and $x = \sqrt{2}$.

$\blacksquare$

▶ **QUESTION 6** Find the global extreme values of $f(x) = x^3 - 3x^2 - 9x + 4$ on each closed interval:

(a) $D = [-3, 4]$

(b) $D = [-2, 2]$

Working in RStudio

RStudio computes derivatives of functions previously defined via the `makeFun` command with the command "**D**", which is part of the `mosaicCalc` package. The derivative of a function `f` that is already defined in RStudio is computed and stored in RStudio by entering `df=D(f(x)~x)`. The independent, input variable `x` must be included with the function, as `f(x)`, in order for this command to execute properly.

The critical numbers where $f'(x) = 0$ are found using the `findZeros` command, provided the search is limited to the interval of interest. Similar to many of the commands already encountered, the syntax for using `findZeros` is the function, a tilde $\sim$, and then the input variable of the function. Additionally, the input interval to search for zeros is specified with the option `xlim=c(,)`, where the endpoints of the interval are listed on either side of the comma.

Evaluating $f(x)$ at the critical numbers and the endpoints of the given closed interval can be done one input at a time by entering `f(c)` or `f(x=c)`. Alternatively, RStudio can evaluate $f(x)$ for multiple inputs simultaneuously by means of the `rbind` command. Assuming the critical numbers are stored in the variable `critNum` and the endpoints are a and b, the RStudio command `evalNum=rbind(a,critNum,b)` combines the endpoint a, the critical numbers, and the endpoint b into a single variable named `evalNum`. The function is evaluated at all of these inputs simultaneously by entering `f(evalNum)`.

Examples of Commands:

- `f=makeFun(6*x^3-3*x^2~x)`   • `df=D(f(x)~x)`

- `critNum=findZeros(df(x)~x,xlim=c(-2,2))`

- `critNum`   • `evalNum=rbind(-2,critNum,2)`

- `f(evalNum)`

◆ **EXAMPLE 12** The following table presents median home prices P in thousands of dollars each year Y from 2001 to 2010 according to the U.S. Census Bureau:

Year	2001	2002	2003	2004	2005	2006	2007	2008	2009	2010
Price	175.2	187.6	195.0	221.0	240.9	246.5	247.9	232.1	216.7	221.8

Applying `fitModel` in RStudio produces the following model for this data set:

$$P(Y) = 0.17(Y-2001)^4 - 3.2(Y-2001)^3 + 16.7(Y-2001)^2 - 10.9(Y-2001) + 177.7$$

Find the global extreme median home prices on the interval of years $D = [2001, 2010]$ using RStudio.

Solution. First, define the function P(Y) in RStudio and evaluate its derivative:

```
P=makeFun(0.17*(Y-2001)^4-3.2*(Y-2001)^3+16.7*(Y-2001)^2-10.9*(Y-2001)
            +177.7~Y)
dP=D(P(Y)~Y)
```

Next, find the critical points, bind them together with the two endpoints $a = 2001$ and $b = 2010$, and then evaluate P(Y) on this collection of inputs:

```
critNum=findZeros(dP(Y)~Y,           evalNum=rbind(2001,critNum,2010)
        xlim=c(2001,2010))           P(evalNum)
critNum
                                            Y
        Y                            1 177.700
1 2001.36                            2 175.794
2 2006.10                            3 247.002
3 2009.66                            4 213.592
                                     5 214.870
```

Based on this RStudio output, for a domain of years $[2001, 2010]$, the model P(Y) for median home prices has an approximate global maximum of $247,002 in year $Y \approx$ 2006.10 and an approximate global minimum of $175,794 in year $Y \approx 2001.36$.   ∎

➤ **QUESTION 7** Find the global extreme values of $f(x) = e^{\cos(x)}$ on the interval $[-2, 5]$ using RStudio.

Summary

- Let $f(x)$ be a function, let D be a set of real number inputs in the domain of $f(x)$, and let c be a real number in D. The *global maximum* of $f(x)$ on D is the value $f(c)$ such that $f(c) \geq f(x)$ for all x in D. The *global minimum* of $f(x)$ on D is the value $f(c)$ such that $f(c) \leq f(x)$ for all x in D. The *global extreme values* of $f(x)$ on D are the local maximum and local minimum values of $f(x)$ on D.

- *Extreme value theorem*: If $f(x)$ is a smooth function on a closed interval $D = [a, b]$, then $f(x)$ has both a global maximum value and a global minimum value on $D = [a, b]$.

- A *critical number* of a function $f(x)$ is an input c in the domain of $f(x)$ such that either $f'(c)$ is equal to zero or $f'(c)$ does not exist.

- *Locating global extreme values*: Let $f(x)$ be a smooth function on a closed interval $D = [a, b]$.

  (1) Identify the critical numbers of the function $f(x)$.

  (2) Evaluate the function $f(x)$ at the critical numbers that are inside the domain and at both endpoints $x = a$ and $x = b$ of D.

  (3) The largest function value from step (2) is the global maximum, and the smallest function value from step (2) is the global minimum.

Exercises

In Exercises 1–6, identify the global extreme values (if they exist) of the function on the interval D and their corresponding inputs $x = c$.

1. $D = [-5, 5]$

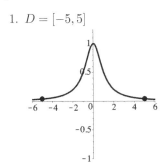

2. $D = [-2\pi, 2\pi]$

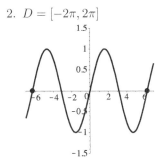

3. $D = [-2, 2]$

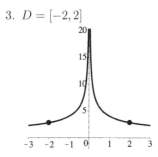

4. $D = [-1, 1]$

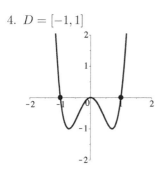

5. $D = [-2, 8]$

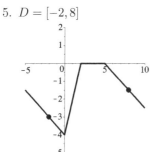

6. $D = [-2, 9]$

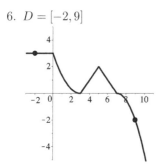

In Exercises 7 – 10, identify the global extreme values (if they exist) of the function graphed below on the interval D and their corresponding inputs $x = c$.

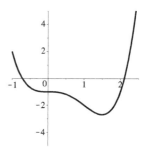

7. $D = [-1, 2.5]$ 9. $D = [0, 2]$

8. $D = [-1, 2]$ 10. $D = [0, 1]$

In Exercises 11 – 14, identify the global extreme values (if they exist) of the function graphed below on the interval D and their corresponding inputs $x = c$.

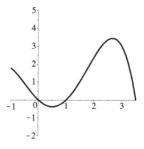

11. $D = [-1, 3.5]$ 13. $D = [0, 2]$

12. $D = [-1, 2]$ 14. $D = [1, 3]$

In Exercises 15 – 20, sketch the graph of a function with the stated features.

15. $D = [-4, 4]$; global maximum $y = 4$ when $x = 2$; global minimum $y = -3$ when $x = -1$.

16. $D = [-5, 1]$; global minimum $y = 2$ when $x = -5$; global maximum $y = 5$ when $x = 0$.

17. $D = [-4, 8]$; global minimum $y = 4$ when $x = 0$; global maximum $y = 7$ when $x = -3$ and $x = 6$.

18. $D = [-2, 2]$; global maximum and global minimum $y = 2$.

19. $D = [-4, 2]$; global minimum when $x = -4$ and $x = 1$; global maximum when $x = 0$.

20. $D = [-\infty, \infty]$ global maximum $y = 0$ when $x = 0$; no global minimum.

In Exercises 21–28, approximate the global extreme values of the tabular function and their corresponding inputs $x = c$.

21.
x	1.1	3.3	8.6	17.1	40.4
$f(x)$	−2.5	0	2.9	5.2	7.7

22.
x	1.1	2.2	2.9	3.6	4.5
$f(x)$	8.7	5.3	3.4	2.6	1.5

23.
x	15	18	24	30	31
$f(x)$	14	6	−7	−4	5

24.
x	1.5	2.6	3.4	5.3	8.7
$f(x)$	1.1	2.2	2.9	3.4	2.7

25.
x	15	18	24	30	31
$f(x)$	−7.4	65.2	91.1	−40.4	78

26.
x	0	1	2	3	4
$f(x)$	19.2	8.7	4.1	1.5	4.6

27.
x	1.1	2.2	2.9	3.6	4.5
$f(x)$	9.1	5.5	4.2	5.3	3.1

28.
x	0.4	2.1	3.3	5.2	6.7
$f(x)$	7.2	18	24.6	16.4	31

In Exercises 29–36, explain why the extreme value theorem does or does not apply to the function and set given here or in the earlier exercise.

29. Exercise 1

30. Exercise 3

31. Exercise 4

32. Exercise 6

33. $f(x) = x^2 - 17$ on $D = [0, 5]$

34. $f(x) = e^{2x+1}$ on $D = [-2, 4]$

35. $f(x) = \ln(2x + 1)$ on $D = [-1, 2]$

36. $f(x) = \sin(3x)$ on $D = (0, \pi]$

In Exercises 37–40, identify the critical numbers of the function from its graph in the earlier exercise.

37. Exercise 2

38. Exercise 4

39. Exercise 5

40. Exercise 6

In Exercises 41–44, identify the critical numbers of $f(x)$ using the graph of its derivative $f'(x)$.

41.
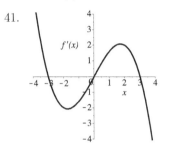

42.
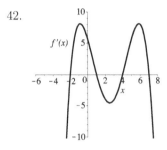

43.
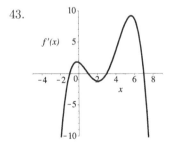

44.

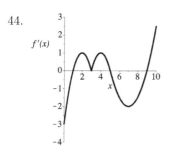

In Exercises 45–50, identify the critical numbers of $f(x)$.

45. $f(x) = x^2 - 2x + 6$

46. $f(x) = x^3 - 6x^2 + 7$

47. $f(x) = 3x + 8$

48. $f(x) = x^4 - 4x^3 + 15$

49. $f(x) = 4x^3 + 36x$

50. $f(x) = x^5 + 10x^3$

In Exercises 51–58, find the global extreme values of the function on the interval and the corresponding input $x = c$.

51. $f(x) = 2x - 4$ on $[3, 8]$

52. $f(x) = 2x - 4$ on $[-3, 2]$

53. $f(x) = 3 - 6x$ on $[3, 8]$

54. $f(x) = 3 - 6x$ on $[-3, 2]$

55. $f(x) = 3x^2 - 18x$ on $[0, 4]$

56. $f(x) = 3x^2 - 18x$ on $[4, 5]$

57. $f(x) = -4x^2 + 8x - 7$ on $[0, 3]$

58. $f(x) = -4x^2 + 8x - 7$ on $[-2, 4]$

In Exercises 59–70, find the global extreme values of the function on the interval and the corresponding input $x = c$.

59. $f(x) = x^3 - 6x^2 - 63x$ on $[-5, 10]$

60. $f(x) = x^3 - 12x$ on $[-1, 4]$

61. $f(x) = 2x^3 + 3x^2 - 12x$ on $[0, 2]$

62. $f(x) = x^3 - 9x^2 + 24x - 1$ on $[-2, 3]$

63. $f(x) = 2x^3 - 15x^2 + 24x$ on $[0, 5]$

64. $f(x) = x^3 - e^{x^3}$ on $[-2, 1]$

65. $f(x) = x^4 - 18x^2 + 40$ on $[-4, 4]$

66. $f(x) = -x^4 + 8x^2$ on $[-2, 4]$

67. $f(x) = x^4 - 2x^3$ on $[-2, 2]$

68. $f(x) = 6x^4 - 8x^3 - 24x^2$ on $[-3, 2]$

69. $f(x) = 3x^5 - 5x^4 + 1$ on $[-1, 1]$

70. $f(x) = 3x^5 + 15x^3$ on $[0, 2]$

In Exercises 71–78, find the global extreme values (if they exist) of the function on the interval and the corresponding input $x = c$.

71. $f(x) = x + \dfrac{1}{x}$ on $[\frac{1}{2}, 2]$

72. $f(x) = \dfrac{3}{1 + x^2}$ on $[-1, 1]$

73. $f(x) = \dfrac{x}{x^2 + 5}$ on $[1, 4]$

74. $f(x) = 1 - (x - 1)^{2/3}$ on $[0, 2]$

75. $f(x) = x^{5/3}$ on $[-2, 1]$

76. $f(x) = x - \cos(x)$ on $[0, 2\pi]$

77. $f(x) = x^2 e^{-x}$ on $[-1, 3]$

78. $f(x) = 2e^{-x^2}$ on $[-1, 2]$

RStudio. In Exercises 79–82, use RStudio to find the global extreme values of the function on the interval.

79. Cumulative number of Ebola cases in Sierra Leone from May 1, 2014 to December 16, 2015.

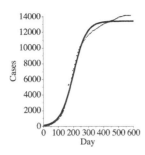

$$C = \frac{13420}{1 + 1.016e^{-0.023(D-200)}}$$

Interval: $[0, 500]$

80. Monthly stock prices for Toyota from 1982 to 1998.

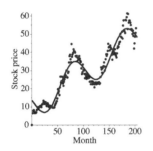

$$S = 11.18 + 0.18M$$
$$+ 9\sin\left(\frac{2\pi}{100.56}(M + 46.13)\right)$$

Interval: $[25, 140]$

81. Mean lower low water (MLLW) tidal depths in feet in Pearl Harbor, Hawaii as a function of the hour.

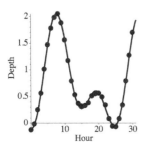

$$D = 0.75\sin\left(\frac{2\pi}{24.32}(H - 2.31)\right)$$
$$- 0.55\cos\left(\frac{2\pi}{12.31}(H - 1.17)\right)$$
$$+ 0.77$$

Interval: $[10, 20]$

82. U.S. carbon dioxide emissions in kT each year from 1960 to 2010.

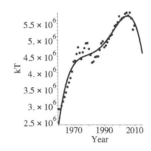

$$E = -5.22(Y - 1960)^4$$
$$+ 566.39(Y - 1960)^3$$
$$- 20771.19(Y - 1960)^2$$
$$+ 336677.76(Y - 1960)$$
$$+ 2433297.88$$

Interval: $[1970, 2010]$

RStudio. In Exercises 83–94, use RStudio to find the global extreme values of the function on the interval given here or in the earlier exercise.

83. Exercise 51   86. Exercise 59

84. Exercise 55   87. Exercise 65

85. Exercise 58   88. Exercise 70

89. $f(x) = e^{\sin(x)+3}$ on $[-1, 3]$

90. $f(x) = \ln(x) \cdot \sin(x)$ on $[1, 4]$

91. $f(x) = x^2 \sin(x)$ on $[-1, 4]$

92. $f(x) = \ln(\sin(x))$ on $\left[\frac{\pi}{6}, \frac{3\pi}{4}\right]$

93. $f(x) = e^{x\sin(x)}$ on $[-10, 10]$

94. $f(x) = e^{\ln(x)+\sin(x)}$ on $[1, 5]$

In Your Own Words. In Exercises 95 –
100, explain the following.

95. Global maximum of $f(x)$ on D

96. Global minimum of $f(x)$ on D

97. Extreme value theorem

98. Assumptions of the extreme value
theorem

99. Critical number

100. Locating global extreme values

5.2   Local Extreme Values

The maximum and minimum values of a function can be considered on two different
scales: from a global perspective asking for the extreme values over all numbers in a
region of interest or from a local perspective focusing on relatively small regions near
the extreme value. Section 5.1 discussed global extreme values in detail, particularly on
closed intervals. This section shifts to studying *local extreme values* of functions.

The intuitive idea of a local extreme value is that a local maximum is larger than all
the other nearby outputs of the function and that a local minimum is smaller than all
the other nearby outputs of the function. Figure 1 provides a graphical illustration of
local extreme values for a particular function.

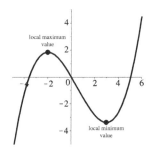

Figure 1: Local minimum and maximum values of a function

The following precise mathematical definition captures these intuitive ideas:

Definition. Let $f(x)$ be a function and let c be a real number in the domain
of $f(x)$.

- A **local maximum** of $f(x)$ is a value $f(c)$ such that $f(c) \geq f(x)$ for all inputs
  x near input c.

- A **local minimum** of $f(x)$ is a value $f(c)$ such that $f(c) \leq f(x)$ for all inputs
  x near input c.

- The **local extreme values** of $f(x)$ are the local maximum and local minimum
  values of the function.

Real-life examples of local extreme values include the peaks and valleys of a mountain range. Recall from Section 5.1 that a global maximum value corresponds to the mountain peak that is the tallest (or peaks if more than one mountain has the exact same height), and a global minimum value corresponds to the lowest point in the deepest valley. Switching perspective to local extreme values, every mountain peak corresponds to a local maximum value and every valley corresponds to a local minimum value. In the context of seeking local extreme values, the goal of optimization is to identify all of the peaks and all of the valleys without regard to which mountain peak is the tallest or which valley is the lowest. Such a physical image can be helpful to keep in mind while studying more abstract or subtle settings.

In drawing distinctions between global and local extreme values, notice that their definitions are almost identical, save for a subtle but very important distinction in the last few words of each definition. As just stated, the definition of a local maximum requires that $f(c) \geq f(x)$ for *all x near c*. In contrast, in Section 5.1, the definition of a global maximum requires that $f(c) \geq f(x)$ for *all x in the set D*. As a result, $f(c)$ can be a local maximum without being the global maximum; although sometimes an output value is both a local maximum and the global maximum. Parallel observations can be made about the definitions of global and local minimum values.

Note. Some books refer to "extreme values" in terms of the Latin origins and pluralization of these words, identifying a global extreme value as a global extremum (singular) and global extreme values as global extrema (plural). Similarly, local maxima (plural) and local minima (plural) identify multiple local maximum and local minimum values.

While the definition provides a clear statement articulating the key features of local extreme values, it does not provide an algorithm for finding them. Given the graph of a function or data that is modeled by a smooth function, locating the local maximum values and the local minimum values of a given function is a reasonably straightforward task, as illustrated in the following examples.

◆ **EXAMPLE 1** Identify the local maximum and minimum values of the function $f(x)$ given in Figure 2 along with their corresponding inputs $x = c$ and their coordinates.

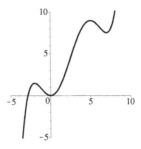

Figure 2: Graph of $f(x)$ for Example 1

Solution. Proceeding from left to right, the first local maximum value of the function $f(x)$ is approximately $y = 1.5$ when $x = -2$, because this output $y = 1.5$ is greater than all the other outputs of $f(x)$ when x *is near* -2. Likewise, $f(x)$ has another local maximum of $y = 9$ when $x = 5$. Thus, the local maximum values of $f(x)$ occur at approximately $(-2, 1.5)$ and at $(5, 9)$. For both of these maximum values, notice that $f(x)$ switches from increasing to decreasing at these maximum values, which correspond to $f'(x)$ switching from positive to negative.

Similarly, proceeding from left to right, the first local minimum value of the function $f(x)$ is $y = 0$ when $x = 0$, because this output $y = 0$ is less than all the other outputs of $f(x)$ when x *is near* 0. Likewise, $f(x)$ has another local minimum of approximately $y = 7$ when $x = 7.5$. Thus, the local minimum values of $f(x)$ occur at $(0, 0)$ and at approximately $(7, 7.5)$. For both of these minimum values, notice that $f(x)$ switches from decreasing to increasing at these minimum values, which correspond to $f'(x)$ switching from negative to positive. ∎

◆ **EXAMPLE 2** The table below presents the value (DJIA) of the Dow Jones Industrial Average at the end of each quarter (Qtr) from 2007 to 2011 as a function of quarters since 2000. For example, 11.25 represents the quarter April to June of the year 2011, when the value of the DJIA was 12,321. Approximate the coordinates of the local extreme values of the Dow Jones Industrial Average from 2007 to 2011.

Qtr	7	7.25	7.5	7.75	8	8.25	8.5	8.75	9	9.25
DJIA	12,460	12,355	13,410	13,896	13,262	12,267	11,345	10,847	8772	7606

Qtr	9.5	9.75	10	10.25	10.5	10.75	11	11.25	11.5	11.75
DJIA	8448	9712	10,431	10,857	9773	10,790	11,577	12,321	12,412	10,912

Solution. Based on the data in the table, the coordinates of the local minimum values appear to be $(7.25, 12, 355)$, $(9.25, 7606)$, and $(10.5, 9773)$. In this context, the two immediately adjacent data points provide all the inputs x near c for the local minimum. Namely, the output DJIA value of 12,355 for quarter 7.25 is less than the DJIA value of 12,460 for quarter 7 and 13,410 for quarter 7.5, resulting in DJIA = 12,355 being a local minimum when Qtr = 7. Similar observations justify the other two coordinates being local minimum values.

Switching perspectives to look for output DJIA values greater than the immediately adjacent output values provides the coordinates of the local maximum values: $(7.75, 13, 896)$, $(10.25, 10, 857)$, and $(11.5, 12, 412)$.

The graph of this data set presented in Figure 3 affirms and illustrates these conclusions. For some readers, such a graph might provide an easier, visual method for identifying extreme values, rather than directly comparing output values from the table. ∎

Example 2 suggests one limit inherent in using data to identify local extreme values. Data sets are discrete by their very nature, skipping over intermediate input–output pairs that might provide the actual local maximum or local minimum of the phenomenon being studied. In more detail, the actual local minimum value of the DJIA around quarter 7.25

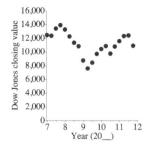

Figure 3: Graph of value of DJIA each quarter from 2007 to 2011 for Example 2

may occur for an input (say in months or days) before or after the point $(7.25, 12,355)$ identified in Example 2. However, without more data, a more precise value cannot be identified. And so, working with the available data leads to the best possible conclusion that the coordinates $(7.25, 12,355)$ provide a local minimum value of the DJIA.

➤ **QUESTION 1** Identify the coordinates of the local extreme values of the function $f(x)$ given in Figure 4.

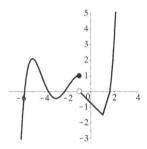

Figure 4: The graph $f(x)$ for Question 1

Critical Numbers

Approximating the local maximum and minimum values of a function $f(x)$ and the coordinates where they occur can be readily accomplished from the graph of $f(x)$. However, an analytic approach to identifying and classifying local extreme values requires developing a method for finding them without consulting a graph. Among other benefits, such analytic methods transition from approximations of local extreme values to finding their exact values.

As suggested by the discussion thus far, local maximum and minimum values occur when $x = c$ is a critical number. Recall the following definition from Section 5.1:

> **Definition.**   A **critical number** of a function $f(x)$ is an input c in the domain of $f(x)$ such that either $f'(c)$ is equal to zero or $f'(c)$ does not exist.

The graph of $f(x)$ in Figure 4 for Question 1 illustrates all the various combinations of behaviors: a local maximum value where $f'(x) = 0$ when $x = -5$ and where $f'(x)$ does not exist when $x = -1$, as well as a local minimum value where $f'(x) = 0$ when $x = -3$ and where $f'(x)$ does not exist when $x = 1$.

One natural question to consider is whether extreme values always correspond to critical numbers. Section 5.1 observed that global extreme values sometimes occur at the endpoints of a given closed interval D of interest, but otherwise the only other candidates for global extreme values were critical points from the interior of the interval D. As the following result states, critical numbers always provide the only other possible candidate inputs for extreme values:

> **FERMAT'S LOCAL EXTREME VALUE THEOREM.**
> If $f(c)$ is a local maximum value or a local minimum value of a function $f(x)$, then c is a critical number of $f(x)$.

Working toward understanding why Fermat's local extreme value theorem is true, consider a function $f(x)$ and inputs increasing from left to right. In order for $f(x)$ to have a local maximum, the function must switch from increasing to decreasing at the coordinates of the local maximum. Similarly, at the coordinates of a local minimum, $f(x)$ must switch from decreasing to increasing. Recall that when a function $f(x)$ is increasing, its derivative $f'(x)$ is positive and that when $f(x)$ is decreasing, $f'(x)$ is negative. The only inputs where the sign of the derivative $f'(x)$ can switch between positive and negative are where the derivative is zero or does not exist. In other words, local extreme values of $f(x)$ can only occur for inputs that are critical numbers of $f(x)$.

A full proof of Fermat's local extreme value theorem lies beyond the scope of this book, but is key to ensuring the validity of this section's approach to finding local extreme values. Namely, critical numbers provide all the possible candidates for local extreme values.

Notice that Fermat's local extreme value theorem is not an "exactly when" statement. While every local extreme value of a function corresponds to a critical number of the function, not every critical number provides a local extreme value, as illustrated in Figure 5. Although $x = -2$ and $x = 3$ are critical numbers, neither corresponds to a local extreme value of the function. Smooth functions, including the common modeling functions, do not exhibit the types of behaviors illustrated in Figure 5. However, the existence of such nonsmooth functions with such behaviors necessarily influences the theory and prevents Fermat's local extreme value theorem from being an "exactly when" result.

Moving forward then, processes are needed to classify each critical number as either a local maximum value, a local minimum value, or neither. This section uses monotonicity as a tool for making such distinctions, and the next section uses concavity.

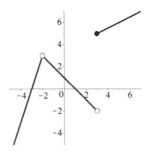

Figure 5: Sometimes critical points do not provide local extreme values

Critical Numbers and Monotonicity

Critical numbers provide the inputs where local extreme values may occur. The next step is to determine whether a critical number corresponds to a local maximum or local minimum value. The *first derivative test* provides a means for classifying critical numbers based on the monotonic behavior of the function. The following example illustrates the functional behaviors that enable this approach.

◆ **EXAMPLE 3** Identify the critical numbers c of the function $f(x)$ given in Figure 6 and graphically classify each $f(c)$ as a local maximum, local minimum, or neither. Also, discuss the monotonic behavior of $f(x)$ for inputs on both sides of each critical number.

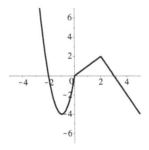

Figure 6: Graph of $f(x)$ for Example 3

Solution. The critical numbers of $f(x)$ given in Figure 6 are $x = -1$, $x = 0$, and $x = 2$. In particular, input $x = -1$ corresponds to a horizontal tangent line of $f(x)$, which means that $f'(-1) = 0$. Inputs $x = 0$, and $x = 2$ correspond to corners in the graph of $f(x)$, which means that $f'(0)$ and $f'(2)$ are undefined. The monotonicity of $f(x)$ and the corresponding signs of its derivative $f'(x)$ can be described as follows:

- When $x = -1$, the graph in Figure 6 shows that $f(-1) = -4$ is a local minimum value. For inputs to the left of $x = -1$, the function $f(x)$ is decreasing and its derivative $f'(x)$ is negative. For inputs to the immediate right of $x = -1$, $f(x)$ is increasing and $f'(x)$ is positive.

- When $x = 0$, the graph in Figure 6 shows that $f(0) = 0$ is neither a local maximum nor a local minimum value. For inputs to both the immediate left and the immediate right of $x = 0$, the function $f(x)$ is increasing and its derivative $f'(x)$ is positive.

- When $x = 2$, the graph in Figure 6 shows that $f(2) = 2$ is a local maximum value. For inputs to the immediate left of $x = 2$, the function $f(x)$ is increasing and its derivative $f'(x)$ is positive. For inputs to the right of $x = 2$, $f(x)$ is decreasing and $f'(x)$ is negative.

■

➤ **QUESTION 2** Identify the critical numbers c of the function $f(x)$ given in Figure 7 and graphically classify each $f(c)$ as a local maximum, local minimum, or neither. Also, discuss the monotonic behavior of $f(x)$ for inputs on both sides of each critical number.

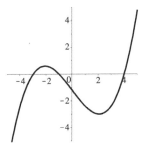

Figure 7: Graph of $f(x)$ for Question 2

First Derivative Test

Example 3 and Question 2 illustrate patterns that hold for many, diverse functions and that are summarized in the first derivative test. As the name suggests, this approach to classifying local extreme values focuses on the first derivative, with particular attention to the sign of the derivative for inputs to the immediate left and immediate right of the critical points of the function under consideration.

Focusing on the case of a local maximum for the moment, notice that as inputs increase from left to right, a function increases toward a local maximum and then decreases away from the local maximum. Consequently, the sign of the derivative switches from positive to negative at the local maximum. A parallel pattern occurs for local miniumum values, which enables the classification of extreme values as detailed in the following theorem:

> **FIRST DERIVATIVE TEST**.
>
> Let $f(x)$ be a smooth function with critical number c and consider the sign of $f'(x)$ as inputs increase from left to right along the x-axis.
>
> - If $f'(x)$ changes from positive to negative at input c, then $f(c)$ is a local maximum of $f(x)$.
>
> - If $f'(x)$ changes from negative to positive at input c, then $f(c)$ is a local minimum of $f(x)$.
>
> - If $f'(x)$ does not change sign at input c, then $f(c)$ is neither a local maximum nor a local minimum of $f(x)$.

The next examples and questions demonstrate the process of applying this test.

◆ **EXAMPLE 4** Find the critical numbers c of the function $f(x) = 2x^3 - 3x^2 - 12x + 2$ and use the first derivative test to classify each $f(c)$ as a local maximum, local minimum, or neither.

Solution. The derivative of $f(x)$ is $f'(x) = 6x^2 - 6x - 12$, which is a polynomial and defined for all real numbers. Therefore, the critical numbers of $f(x)$ are found by setting this derivative $f'(x)$ equal to zero and solving for x:

$$
\begin{array}{ll}
6x^2 - 6x - 12 = 0 & \text{Set } f'(x) \text{ equal to zero} \\
6(x^2 - x - 2) = 0 & \text{Factor 6 from all terms} \\
6(x + 1)(x - 2) = 0 & \text{Factor } x^2 - x - 2 \\
x = -1, 2 & \text{Find the zero of each factor}
\end{array}
$$

Thus, the critical numbers of $f(x)$ are $x = -1$ and $x = 2$.

Applying the first derivative test to these critical numbers involves determining the sign of $f'(x)$ to the immediate left and to the immediate right of each critical number. The derivative $f'(x)$ can only change sign at the critical numbers of $f(x)$, so the sign of $f'(x)$ remains the same throughout each interval determined by these critical numbers. Therefore, *test inputs* can be chosen from each such interval to determine the sign of the derivative. Many choices of test inputs are possible. This example considers test inputs of $x = -2$, $x = 0$, and $x = 3$, which lie on either side of the critical numbers $x = -1$ and $x = 2$. Substituting these test inputs into $f'(x)$ yields the following results:

$$
\begin{aligned}
f'(-2) &= 6(-2)^2 - 6(-2) - 12 = 24 \ > \ 0 \\
f'(0) &= 6(0)^2 - 6(0) - 12 = -12 \ < \ 0 \\
f'(3) &= 6(3)^2 - 6(3) - 12 = 24 \ > \ 0
\end{aligned}
$$

These computational results are summarized in the *sign chart*, given in Figure 8.

Figure 8: First derivative sign chart for Example 4

Because $f'(x)$ switches from positive to negative at $x = -1$, the first derivative test indicates that $f(-1) = 9$ is a local maximum value of $f(x)$. Similarly, because $f'(x)$ switches from negative to positive at $x = 2$, the first derivative test indicates that $f(2) = -18$ is a local minimum value of $f(x)$. ∎

The sign chart given in Figure 8 for Example 4 explicitly labels the alternating signs as being attributable to the derivative $f'(x)$ of $f(x)$. Such labels are essential for clearly communicating which function is being examined. Also, Example 4 mentioned the following important fact, which allows the use of test inputs when analyzing the sign of the derivative on whole intervals:

> ***Note.*** Because the derivative $f'(x)$ of a function $f(x)$ can only change sign at the critical numbers of $f(x)$, the sign of $f'(x)$ remains the same throughout each interval determined by these critical numbers, allowing the use of test inputs.

◆ **EXAMPLE 5** Find the critical numbers c of the function $f(x) = x^3 + 6$ and use the first derivative test to classify each $f(c)$ as a local maximum, local minimum, or neither.

Solution. The derivative of $f(x)$ is $f'(x) = 3x^2$, which is a polynomial and defined for all real numbers. Therefore, the critical numbers of $f(x)$ are found by setting $f'(x)$ equal to zero and solving for x. In this case, $f'(x) = 3x^2 = 0$ yields $x = 0$. Thus, the only critical number of $f(x)$ is $x = 0$.

Any positive number and any negative number can be used as test inputs to find the sign of $f'(x)$ to the immediate left and right of the sole critical number $x = 0$. This example considers $x = -1$ and $x = 1$. Substituting these test inputs into $f'(x)$ yields the following computational results and the sign chart in Figure 9.

$$f'(-1) = 3(-1)^2 = 3 > 0 \qquad\qquad f'(1) = 3(1)^2 = 3 > 0$$

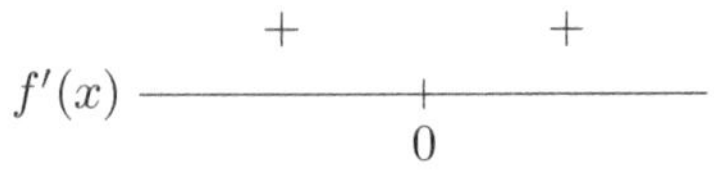

$$f'(x) \; \underline{\quad\; +\quad\;\; +\quad\;} \atop 0$$

Figure 9: First derivative sign chart for Example 5

Because $f'(x)$ does not change sign at $x = 0$, the first derivative test indicates that $f(0) = 6$ is neither a local maximum nor a local minimum. Therefore, $f(x) = x^3 + 6$ does not have any local extreme values. ∎

▶ **QUESTION 3** Find the critical numbers c of the function $f(x) = -x^3 - 3x^2 + 24x + 6$ and use the first derivative test to classify each $f(c)$ as a local maximum, local minimum, or neither.

▶ **QUESTION 4** Find the critical numbers c of the function $f(x) = \dfrac{-1}{x^2 + 1}$ and use the first derivative test to classify each $f(c)$ as a local maximum, local minimum, or neither.

Revisiting Global Extreme Values

In Section 5.1, the examples and questions about global extreme values provided a function $f(x)$ and set D satisfying the assumptions of the extreme value theorem: $f(x)$ is a smooth function on D and D is a closed interval. Therefore, these functions attained both a global maximum and global minimum value on the given set.

However, if $f(x)$ is not smooth on D or if D is not a closed interval, then the extreme value theorem does not provide any information about the existence of global extreme values, and the function may have both a global maximum and minimum, just a global maximum, just a global minimum, or neither on the given set. While the process in Section 5.1 for locating global extreme values does not apply in such cases, sometimes a successful analysis is still possible via the first derivative test. The next examples explore two such cases.

◆ **EXAMPLE 6** A drug administered to a patient produces a concentration in the bloodstream that is modeled by $c(x) = \dfrac{e}{3} x e^{-x/3}$ (in milligrams per milliliter) x hours after being injected. If possible, find the global maximum and global minimum concentration of the drug in the patient's bloodstream on the interval $D = [0, \infty)$.

Solution. This situation does not satisfy the assumptions of the extreme value theorem. Although $c(x)$ is a smooth function defined on all real number inputs, the interval $D = [0, \infty)$ is not a closed interval of the form $[a, b]$ specified in the extreme value theorem. (Note that D is technically a closed set, but D is not a closed interval, because it does not contain its right endpoint.) Consequently, the extreme value theorem does not guarantee that $c(x) = \dfrac{e}{3} x e^{-x/3}$ has a global extreme value on $D = [0, \infty)$.

Even so, $c(x)$ may still attain a global extreme value, and the critical numbers of $c(x)$ provide candidates for its extreme values. Working in this direction, $c'(x)$ is computed using the product, line, and chain rules:

$$
\begin{aligned}
c'(x) &= e^{-x/3} \cdot \frac{d}{dx}\left[\frac{e}{3}x\right] + \frac{e}{3}x \cdot \frac{d}{dx}\left[e^{-x/3}\right] && \text{Product rule} \\
&= e^{-x/3} \cdot \frac{e}{3} + \frac{e}{3}x \cdot \left(-\frac{1}{3}e^{-x/3}\right) && \text{Line and chain rules} \\
&= \frac{e}{3}e^{-x/3} - \frac{e}{9}xe^{-x/3} && \text{Simplify} \\
&= \frac{e}{3}e^{-x/3}\left(1 - \frac{1}{3}x\right). && \text{Factor } \frac{e}{3}e^{-x/3} \text{ from both terms}
\end{aligned}
$$

This derivative $c'(x)$ is defined for all real number inputs, because its exponential and polynomial factors are defined on all reals. Therefore, the critical numbers of $c(x)$ are found by setting $c'(x)$ equal to zero and solving for x. Exponentials are never equal to zero, so both $e/3$ and $e^{-x/3}$ are nonzero for any x. The linear factor $1 - \frac{1}{3}x$ provides the only zero $x = 3$ of $c'(x)$, which is the sole critical number of $c(x)$.

This unique critical number $x = 3$ of $c(x)$ lies in the given domain $D = [0, \infty)$. The next step in the standard process for finding global extreme values is to evaluate $c(x)$ at this critical number and at the endpoints of D. However, in this setting, ∞ cannot be substituted into $c(x)$, because ∞ is not a real number but rather indicates an unbounded increase of inputs x. Therefore, the first derivative test is needed to analyze how $c(x)$ behaves as the input values become larger.

With an endpoint $x = 0$ and a critical number $x = 3$, this example considers test inputs $x = 1$ and $x = 4$. Substituting these test inputs into $f'(x)$ yields the following computational results and the sign chart in Figure 10:

$$c'(1) = \frac{e}{3}e^1\left(1 - \frac{1}{3}(1)\right) = \frac{2e^2}{9} \approx 1.642 > 0$$

$$c'(4) = \frac{e}{3}e^{-4/3}\left(1 - \frac{1}{3}(4)\right) = -\frac{e^{-1/3}}{9} \approx -0.0796 < 0$$

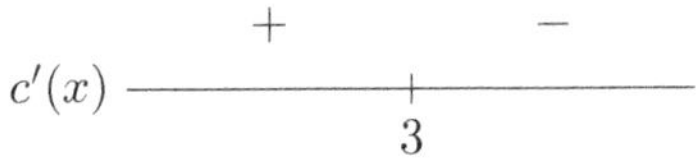

$$c'(x) \quad \overline{ \underset{3}{+} - }$$

Figure 10: First derivative sign chart for Example 6

The derivative $c'(x)$ switches from positive to negative at $x = 3$, and so the first derivative test indicates that $c(3)$ is a local maximum value of $c(x)$.

Because $c(x)$ has no other critical numbers, the monotonic behavior and sign of $c(x)$ must be examined to complete this analysis. Namely, $c'(x)$ is negative for all $x > 3$, which means that $c(x)$ is decreasing when $x > 3$. In addition, $c(x)$ is positive for all positive real numbers, including all $x > 3$. Therefore, $c(x)$ is decreasing from the local maximum of $c(3) = 1$ toward $y = 0$ as x becomes larger. Similarly, $c'(x)$ is positive for all inputs x in the interval $(0, 3)$, which means that $c(x)$ is increasing from $c(0) = 0$ toward $c(3) = 1$ throughout the interval $(0, 3)$.

Consequently, $c(3) = 1$ is the global maximum and $c(0) = 0$ is the global minimum of $c(x)$ on $D = [0, \infty)$. Interpreting this analysis in context, the patient's global maximum bloodstream concentration of the drug is $c(3) = 1$ milligrams per milliliter 3 hours after having the medication injected, and the patient's global minimum bloodstream concentration of the drug is $c(0) = 0$ milligrams per milliliter immediately prior to having the medication injected. The graph of $c(x)$ in Figure 11 affirms and illustrates these conclusions. ∎

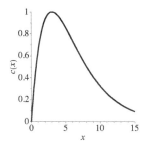

Figure 11: Bloodstream concentration of a drug for Example 6

◆ **EXAMPLE 7** If possible, find the global extreme values of the function $f(x) = 2x^3 + 3x^2 - 36x$ on the interval $D = (0, 4)$.

Solution. The set $D = (0, 4)$ is not a closed interval, because it does not contain its two endpoints $x = 0$ and $x = 4$, as indicated by parentheses (rather than brackets) defining the interval D. As a result, the extreme value theorem does not guarantee that $f(x)$ has either a global maximum or global minimum value on D. Even so, the function may still attain both, one, or no global extreme value on this interval.

The critical numbers of $f(x)$ are candidates for extreme values on D. The derivative of the function is $f'(x) = 6x^2 + 6x - 36$, which is a polynomial and defined for all real numbers. Therefore, the critical numbers of $f(x)$ are found by setting $f'(x)$ equal to zero and solving for x:

$$6x^2 + 6x - 36 = 0 \qquad \text{Set } f'(x) \text{ equal to zero}$$
$$6(x^2 + x - 6) = 0 \qquad \text{Factor } 6$$
$$6(x + 3)(x - 2) = 0 \qquad \text{Factor } x^2 + x - 6$$
$$x = -3, 2 \qquad \text{Find the zero of each term}$$

Thus, the critical numbers of $f(x)$ are $x = -3$ and $x = 2$, although only $x = 2$ lies in the open interval $D = (0, 4)$. With endpoints $x = 0$ and $x = 4$, and a critical number $x = 2$, this example considers test inputs $x = 1$ and $x = 3$. Substituting these test inputs into $f'(x)$ yields the following results and the sign chart in Figure 12:

$$f'(1) = 6(1)^2 + 6 \cdot 1 - 18 = 6 + 6 - 18 = -6 \; < \; 0$$
$$f'(3) = 6(3)^2 + 6 \cdot 3 - 18 = 54 + 18 - 18 = 54 \; > \; 0$$

$$f'(x) \; \frac{\quad - \qquad\qquad + \quad}{\underset{2}{\qquad\qquad + \qquad\qquad}}$$

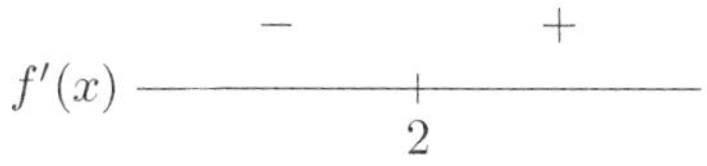

Figure 12: First derivative sign chart for Example 7

The derivative $f'(x)$ switches from negative to positive at $x = 2$, and so the first derivative test indicates that $f(2)$ is a local minimum value of $f(x)$. In addition, the signs of the derivative show that the function is decreasing toward this minimum on the interval $(0, 2)$ and increasing away from it on the interval $(2, 4)$. Therefore, $f(2) = -4$ when $x = 2$ is the global minimum value of $f(x)$ on D. In addition, because neither $x = 0$ nor $x = 4$ are included in $D = (0, 4)$ and there are no other critical points, $f(x)$ does not attain a global maximum value D.

■

◆ **EXAMPLE 8** Find and classify the local extreme values of $f(x) = x^3 - 3x + 6$ using RStudio.

Solution. First define $f(x) = x^3 - 3x + 6$ in RStudio and compute its derivative $f'(x)$. Polynomials are smooth functions, and their derivatives exist for all real number inputs, so the only possible critical numbers result from when the derivative is equal to zero. Therefore, **findZeros** is used to find the critical numbers of $f(x)$:

```
f = makeFun(x^3-3*x+6~x)          critNum
df = D(f(x)~x)
critNum=findZeros(df(x)~x)             x
                                  1 -1
                                  2  1
```

Thus, the critical numbers of $f(x)$ are $x = -1$ and $x = 1$. This example considers test inputs $x = -2$, $x = 0$, and $x = 2$ from either side of these critical numbers and substitutes these test inputs into the derivative:

```
df(-2)                            df(2)

[1] 9                             [1] 9

df(0)

[1] -3
```

Based on this output, $f'(x)$ switches sign at $x = -1$ from positive $(f'(-2) = 9 > 0)$ to negative $(f'(0) = -3 < 0)$, so the first derivative test indicates that $f(-1)$ is a local maximum. Similarly, $f'(x)$ switches sign at $x = 1$ from negative $(f'(0) = -3 < 0)$ to positive $(f'(2) = 9 > 0)$, so the first derivative test indicates that $f(1)$ is a local minimum. Finally, the actual outputs of the local extreme values are found by evaluating the function at the critical numbers:

```
f(critNum)

   x
1 8
2 4
```

In summary, the first derivative test shows that $f(x)$ has a local maximum of $f(-1) = 8$ when $x = -1$ and a local minimum of $f(1) = 4$ when $x = 1$.

■

▶ **QUESTION 5** Find and classify the local extreme values of $f(x) = x + e^{-x}$ using RStudio. *Hint:* Use a domain of $[-5, 5]$ to ensure that the **findZeros** command works for the transcendental function e^{-x}.

Summary

- Let $f(x)$ be a function and let c be a real number in the domain of $f(x)$. A *local maximum* of $f(x)$ is a value $f(c)$ such that $f(c) \geq f(x)$ for all inputs x near input c. A *local minimum* of $f(x)$ is a value $f(c)$ such that $f(c) \leq f(x)$ for all inputs x near input c. The *local extreme values* of $f(x)$ are the local maximum and local minimum values of the function.

- A *critical number* of a function $f(x)$ is an input c in the domain of $f(x)$ such that either $f'(c)$ is equal to zero or $f'(c)$ does not exist.

- *Fermat's local extreme value theorem*: If $f(c)$ is a local maximum value or a local minimum value of a function $f(x)$, then c is a critical number of $f(x)$.

- *First derivative test*: Let $f(x)$ be a smooth function with critical number c and consider the sign of $f'(x)$ as inputs increase from left to right along the x-axis.

  ○ If $f'(x)$ changes from positive to negative at input c, then $f(c)$ is a local maximum of $f(x)$.

  ○ If $f'(x)$ changes from negative to positive at input c, then $f(c)$ is a local minimum of $f(x)$.

  ○ If $f'(x)$ does not change sign at input c, then $f(c)$ is neither a local maximum nor a local minimum of $f(x)$.

- Because the derivative $f'(x)$ of a function $f(x)$ can change sign only at the critical numbers of $f(x)$, the sign of $f'(x)$ remains the same throughout each interval determined by these critical numbers, allowing the use of *test inputs*.

Exercises

In Exercises 1–6, identify the local extreme values of the function.

1.

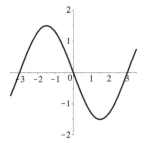

2.

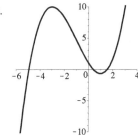

3.

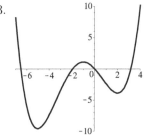

4.

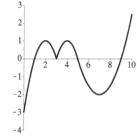

5.

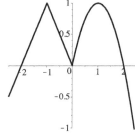

6.

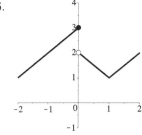

In Exercises 7–10, approximate the coordinates of the local extreme values of the data set.

7.

x	1	1.6	2.1	2.4	2.9	3.4
f	1.7	3.4	5.3	5.1	4.2	4.4

8.

x	1	1.6	2.1	2.4	2.9	3.4
f	4.7	1.4	−0.35	−2.0	−1.8	2.4

9.

x	2	2.4	3.2	3.9	4.2	5.0
f	1.2	1.6	1.4	1.1	0.9	1.4

10.

x	1	1.6	2.1	2.4	2.9	3.4
f	0.2	3.4	5.1	5.2	5.4	7.2

Your Turn. In Exercises 11–14, sketch the graph of two different functions with the local extreme values.

11. Local maximum when $x = 1$, $x = 5$
    Local minimum when $x = 4$

12. Local minimum when $x = 2$, $x = 7$
    Local maximum when $x = 6$

13. Local minimum when $x = 1$, $x = 5$

14. Local maximum when $x = 1$, $x = 2$

In Exercises 15–18, identify the critical numbers of the function.

15.
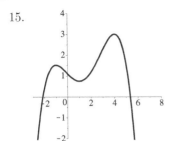

16.
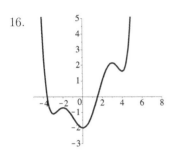

17.
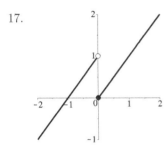

18.
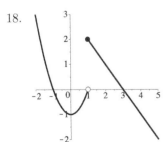

Your Turn. In Exercises 19–24, sketch the graph of two different functions with critical numbers at the specified inputs.

19. $x = -2$

20. $x = -3$, $x = 4$

21. $x = 0$, $x = 10$

22. $x = 1$, $x = 3$, $x = 6$

23. $x = -2$, $x = 0$, $x = 5$

24. $x = -10$, $x = -5$, $x = 3$, $x = 7$

In Exercises 25–34, find the critical numbers of the function.

25. $f(x) = 4x^2 - 2x + 3$

26. $f(x) = 2x^2 + 6x - 7$

27. $f(x) = -5x^2 + 2x - 4$

28. $f(x) = 2x^3 - 3x^2 + 7$

29. $f(x) = 2x^3 + 15x^2 + 24x + 4$

30. $f(x) = 2x^3 - 15x^2 + 36x - 8$

31. $f(x) = x^3 - 15x^2 + 72x - 96$

32. $f(x) = x^4 - 2x^2 + 1$

33. $f(x) = 3x^4 + 8x^3 + 6x^2 + 3$

34. $f(x) = 3x^4 - 16x^3 + 18x^2$

In Exercises 35–40, find the critical numbers of the function.

35. $f(x) = \cos(x) - x$

36. $f(x) = \cos(x) - \sin(x)$

37. $f(x) = e^{-x}$

38. $f(x) = xe^{-x}$

39. $f(x) = x\ln(x)$

40. $f(x) = 6\ln(x) - 2x$

In Exercises 41–48, classify the critical numbers of the function as a local maximum, a local minimum, or neither using the first derivative test.

41. $f(x) = 4x^2 - 2x + 3$

42. $f(x) = -2x^2 + 5x - 2$

43. $f(x) = x^3 - 3x^2$

44. $f(x) = 2x^3 - 3x^2 + 7$

45. $f(x) = x^3 - 15x^2 + 72x - 3$

46. $f(x) = x^4 - 2x^2 + 1$

47. $f(x) = 3x^4 - 16x^3 + 18x^2$

48. $f(x) = 2x^5 - 5x^2$

In Exercises 49–54, classify the critical numbers of the function as a local maximum, a local minimum, or neither using the first derivative test.

49. $f(x) = \cos(x) - x$

50. $f(x) = x\ln(x)$

51. $f(x) = 15\ln(x) - 60x$

52. $f(x) = 30xe^{-0.4x}$

53. $f(x) = \dfrac{20}{x} - x$

54. $f(x) = \sin(x) + \cos(x)$ on $[0, 2\pi]$

In Exercises 55–60, explain why the function does not have any local extreme values.

55. $f(x) = 2x + 4$

56. $f(x) = e^x$

57. $f(x) = 2^{-x}$

58. $f(x) = \ln(x)$

59. $f(x) = \tan(x)$

60. $f(x) = \cot(x)$

In Exercises 61–66, use the graph of the derivative $f'(x)$ to identify the critical numbers of $f(x)$.

61.

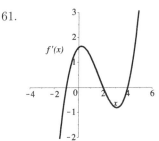

62.

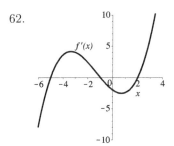

63.

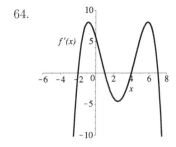

64.

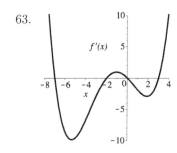

65.

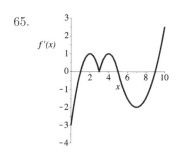

66.

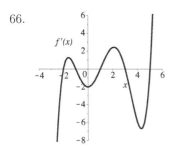

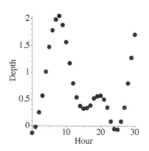

In Exercises 67–72 use the graph of the derivative $f'(x)$ from the earlier exercise to classify the critical numbers of $f(x)$ as local extreme values (or not) with the first derivative test.

67. Exercise 61   70. Exercise 64

68. Exercise 62   71. Exercise 65

69. Exercise 63   72. Exercise 66

In Exercises 73–76, use the first derivative test to identify the local extreme values of the function.

73. The data set stored in `ToyotaMonthly` contains the stock price of Toyota from 1982 to 1998. Applying `fitModel` in RStudio produces the best model for this data set shown in the following graph:

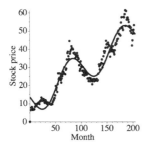

74. The following plot shows the mean lower low water (MLLW) tidal depths (in feet) in Pearl Harbor, Hawaii as a function of the hour:

75. The following table presents the population of Austin, Texas in thousands of people for each year:

Y	1950	1960	1970	1980	1990
P	132	187	252	346	466

The quadratic model $P = 0.11(Y - 1950)^2 + 3.73(Y - 1950) + 133.9$ fits this data well.

(a) Based on this model, determine the local extreme values of the population of Austin.

(b) Based on your answer to (a), does this model made reasonable predictions regarding the population of Austin for years before 1950 or after 1990?

76. The normal probability distribution function with mean zero and standard deviation 1 is given by $f(x) = \frac{1}{\sqrt{2\pi}}e^{-x^2/2}$.

In Exercises 77–80, consider $f(x) = k$, where k is a constant.

77. Use the definition to explain why $x = 4$ corresponds to a local maximum value of $f(x)$ for all values of k.

78. Use the definition to explain why $x = 4$ corresponds to a local minimum value of $f(x)$ for all values of k.

79. Carefully explain what the definition of local maximum tells us about every point on the graph of any constant function.

80. Carefully explain what the definition of local minimum tells us about every point on the graph of any constant function.

In Exercises 81–84, consider $f(x) = ax^2 + bx + c$, where $a \neq 0$, b, and c are real numbers.

81. Find the critical point of $f(x)$.

82. Assume $a > 0$. Use the first derivative test to classify the critical point of $f(x)$ from Exercise 81 as a local maximum or local minimum.

83. Assume $a < 0$. Use the first derivative test to classify the critical point of $f(x)$ from Exercise 81 as a local maximum or local minimum.

84. How does c effect the local extreme value of $f(x)$?

In Exercises 85–88, find the value of the parameter b such that $f(x)$ has a critical number when $x = 2$.

85. $f(x) = bx^2 + 4x + 2$

86. $f(x) = x^2 + bx + 6$

87. $f(x) = x^3 - bx$

88. $f(x) = 4(x - b)^2 + 5$

In Exercises 89–90, find values of the parameters a and b such that $f(1) = 2$ and $f(x)$ has a local minimum when $x = 1$.

89. $f(x) = ax^2 + x + b$

90. $f(x) = ae^x - bx$

RStudio. In Exercises 91–94, use RStudio to find and classify the local extreme values of the function.

91. $f(x) = x^2 - 3x - 4$

92. $f(x) = x^4 - 3x + 2$

93. $f(x) = x^7 - 6x^3 - 4$

94. $f(x) = x^8 - x^4 + x^3$

RStudio. In Exercises 95–98, use RStudio to find and classify the local extreme values of the function on the interval.

95. $f(x) = x^4 - 3x^2 + \sin(x)$ on $(-2, 2)$

96. $f(x) = \sin(x^2)$ on $(-\pi, 2\pi)$

97. $f(x) = x \sin(x)$ on $(-6, 6)$

98. $f(x) = e^x \sin(x)$ on $(2, 7)$

In Your Own Words. In Exercises 99–103, explain the following.

99. Local maximum

100. Local minimum

101. Critical number

102. Fermat's local extreme value theorem

103. First derivative test

5.3   Concavity and Extreme Values

Section 5.2 used the monotonic behavior of functions to classify their local extreme values. This method involving the first derivative test is extremely useful, but an alternative approach can be developed that relies on the concavity of functions. And, more than just serving as a useful tool for classifying local extreme values, the concavity of a function provides additional information about the behavior of a function that is interesting in its own right.

Section 1.4 provided an intuitive introduction to concavity that relied on graphical observations of how functions bend. This section develops such ideas more precisely in terms of the derivative, explores the information concavity provides regarding the behavior of a function, and introduces how concavity can be used to classify the local extreme values of a function via the *second derivative test*.

Concavity and Points of Inflection

The power functions $f(x) = x^2$ and $g(x) = -x^2$ graphed in Figure 1 provide classic examples of functions that are concave up and concave down.

The graph of $f(x) = x^2$ looks like a bowl that is rightside up and is said to be *concave up*. The graph of $g(x) = -x^2$ looks like a bowl that is upside down and is said to be *concave down*. Some people remember the names for such functional behavior by identifying concave up functions as looking like a "cup" (which rhymes with "up") and concave down functions as looking like a "frown" (which rhymes with "down"). By whatever means the reader chooses to remember these naming conventions, the respective orientation of these bowl shapes provide a means for visually recognizing where a function is concave up or concave down.

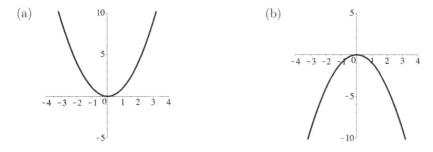

Figure 1: Graphs of (a) $f(x) = x^2$ and (b) $g(x) = -x^2$

Figure 1 also indicates certain recurring patterns in the behavior of the derivative of a function relative to the function's concavity. First, consider the graph of $f(x) = x^2$ as inputs are read from left to right along the x-axis. The slopes of its tangent lines are large negative numbers for inputs x toward $-\infty$ that become smaller negative numbers as x approaches 0, transition through zero where $f(x)$ has a horizonal tangent when $x = 0$, and become increasingly large positive numbers as inputs x increase toward ∞.

Alongside this graphical analysis, the derivative of $f(x) = x^2$ is $f'(x) = 2x$ by the power rule, which is an increasing linear function because its slope $m = 2$ is positive. In short, the derivative $f'(x) = 2x$ is an increasing function throughout its domain of all real numbers, which coincides with where $f(x) = x^2$ is concave up.

In parallel with this analysis of $f(x) = x^2$, the graph of $g(x) = -x^2$ in Figure 1 indicates that the slopes of its tangent lines are always decreasing as inputs are read from left to right along the x-axis. Computationally, the derivative of $f(x) = -x^2$ is $f'(x) = -2x$ by the power rule, which is a decreasing linear function because its slope

$m = -2$ is negative. In short, the derivative $g'(x) = -2x$ is a decreasing function throughout its domain of all real numbers, which coincides with where $g(x) = -x^2$ is concave down.

Similar observations can be made for all smooth functions and their derivatives, which leads to the following definitions:

Definition.   Let $f(x)$ be a smooth function and let (a, b) be an interval.

- $f(x)$ is **concave up** on (a, b) if and only if $f'(x)$ is increasing on (a, b).

- $f(x)$ is **concave down** on (a, b) if and only if $f'(x)$ is decreasing on (a, b).

The intervals on which functions are concave up and concave down are important features of functions, and so are the locations where functions transition between being concave up and being concave down. Such transition points have been given the name *points of inflection* as defined next.

Definition.   Let c be a real number in the domain of a function $f(x)$. The point $(c, f(c))$ is a **point of inflection** of $f(x)$ if and only $f(x)$ changes concavity at input c.

For the sake of completeness, note that when the derivative $f'(x)$ of a smooth function is neither increasing nor decreasing (nor a combination of these behaviors) on an interval (a, b), then the derivative is a constant function $f'(x) = m$ on (a, b). In such cases, $f(x)$ is a linear function, which concides with our intuition that a function that does not bend must be a straight line. This section focuses on functions that are a combination of concave up and concave down (rather than linear), because the second derivative test applies to such functions.

Determining Concavity Graphically

Given the graph of a function $f(x)$, identifying the input intervals where $f(x)$ is concave up and concave down as well as any points of inflection of $f(x)$ are reasonably straightforward tasks.

◆ **EXAMPLE 1** Identify the intervals on which the function $f(x)$ given in Figure 2 is concave up and concave down, and any points of inflection.

Solution. The function $f(x)$ is concave up on the input interval $(-2, \infty)$, because its derivative $f'(x)$ is increasing on this interval, as indicated by the increase in the slopes of its tangent lines as the inputs are read from -2 toward ∞. In addition, the curve in Figure 2 looks like a rightside up bowl (or a cup) on the interval $(-2, \infty)$.

Similarly, the function $f(x)$ is concave down on the input interval $(-\infty, -2)$, because its derivative $f'(x)$ is decreasing on this interval, as indicated by the decrease in the slopes of its tangent lines as the inputs are read from $-\infty$ toward -2. In addition, the curve in Figure 2 looks like an upside down bowl (or a frown) on $(-\infty, -2)$.

Based on this information, $(-2, f(-2)) \approx (-2, 1.5)$ is a point of inflection of $f(x)$, because the function changes from concave down to concave up at this point.

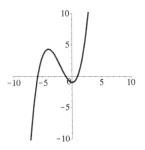

Figure 2: Graph of $f(x)$ for Example 1

▶ **QUESTION 1** Identify the intervals on which the functions given in Figure 3 are concave up and concave down, and any points of inflection.

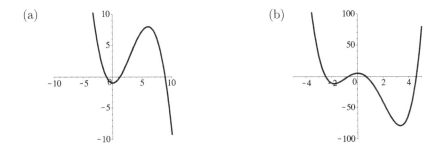

Figure 3: Graph of $f(x)$ for Example 1

◆ **EXAMPLE 2** The population of the Netherlands measured in millions of people at the beginning of each decade since 1700 is stored in `NetherlandsPopulation`. A plot of the data and the best possible model Pop(Year) found using `fitModel` are given in Figure 4. Estimate the intervals on which this model Pop(Year) of the population is concave up and concave down, and any points of inflection.

Solution. Based on the graph in Figure 4, Pop(Year) appears to be concave up on $(-\infty, 1950)$ and concave down on $(1950, \infty)$. Therefore, $(1950, \text{Pop}(1950)) \approx (1950, 10)$ is a point of inflection of Pop(Year), because this model changes from concave up to concave down at this point.

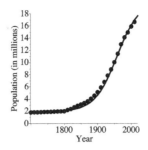

Figure 4: Population of the Netherlands for Example 2

▶ **QUESTION 2** The average monthly stock price of Toyota from 1982 to 1998 is stored in `ToyotaMonthly`. A plot of the data and the best possible model Stock(Month) found using `fitModel` are given in Figure 5. Estimate the intervals on which this model Stock(Month) of average monthly stock price of Toyota is concave up and concave down, and any points of inflection.

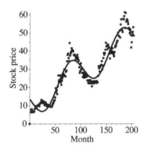

Figure 5: Price of Toyota stock from 1982 to 1998 for Questison 2

In addition to determining concavity and points of inflection directly from graphs of functions, the first and second derivatives are also important tools for carrying out such an analysis. In particular, the graph of a derivative $f'(x)$ provides concavity information about the original function $f(x)$ based directly on the definition of concavity: $f(x)$ is concave up exactly when $f'(x)$ is increasing, and $f(x)$ is concave down exactly when $f'(x)$ is decreasing. The next example illustrates such a derivative-based analysis of concavity.

◆ **EXAMPLE 3** Identify the intervals on which a function $f(x)$ is concave up and concave down, and any points of inflection using the graph of its derivative $f'(x)$ given in Figure 6.

Solution. The function $f(x)$ is concave up on the intervals $(-\infty, -2)$ and $(2, \infty)$, because $f'(x)$ is increasing on these intervals. The function $f(x)$ is concave down on the interval $(-2, 2)$, because $f'(x)$ is decreasing on this interval. Therefore, the points of inflection of $f(x)$ are $(-2, f(-2))$ and $(2, f(2))$, because the concavity of $f(x)$ changes at these two points.

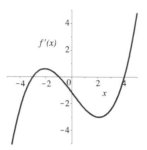

Figure 6: Graph of the derivative function $f'(x)$ for Example 3

In Example 3, notice that the points of inflection of $f(x)$ occur at the local maximum and local minimum values of $f'(x)$. These local extreme values are where $f'(x)$ changes between increasing and decreasing, and so $f(x)$ changes between concave up and concave down. Furthermore, because only $f'(x)$ is given and not $f(x)$, the y-coordinates of the points of inflection cannot be determined from the available information, only their x-coordinates of $x = -2$ and $x = 2$.

➤ **QUESTION 3** Identify the intervals on which a function $f(x)$ is concave up and concave down, and any points of inflection using the graph of its derivative $f'(x)$ given in Figure 7.

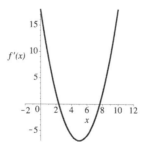

Figure 7: Graph of derivative function $f'(x)$ for Question 3

Determining Concavity from the Second Derivative

Approximating the input intervals where a function is concave up and concave down, and its points of inflection, can be readily accomplished from both a graph of the function and a graph of its derivative. An analytic approach to these questions removes the need to consider graphs of functions and derivatives, and also enables a transition from approximate solutions to exact answers.

The analytic approach to concavity hinges on the relationship between the monotonic behavior of a function and the sign of its derivative. Recall from Section 4.2 that a function is increasing when its derivative is positive, and that a function is decreasing when its derivative is negative. In the context of studying the concavity of a function $f(x)$, the monotonic behavior of $f'(x)$ is key, and this leads to examining the sign of the derivative of $f'(x)$ (i.e., the sign of the second derivative $f''(x)$ of the original function). The following relationships are fundamental to determining the concavity of smooth functions on intervals via derivatives:

$$f(x) \text{ is concave up} \quad \Leftrightarrow \quad f'(x) \text{ is increasing} \quad \Leftrightarrow \quad f''(x) \text{ is positive}$$

$$f(x) \text{ is concave down} \quad \Leftrightarrow \quad f'(x) \text{ is decreasing} \quad \Leftrightarrow \quad f''(x) \text{ is negative}$$

In particular, these relationships enable an algebraic analysis that determines the concavity of a function $f(x)$ based on its second derivative $f''(x)$.

CONCAVITY AND THE SECOND DERIVATIVE.

Let $f(x)$ be a smooth function and let (a, b) be an interval in the domain of $f(x)$.

- $f(x)$ is concave up on (a, b) if and only if $f''(x) > 0$ on (a, b).

- $f(x)$ is concave down on (a, b) if and only if $f''(x) < 0$ on (a, b).

These observations provide a standard algebraic approach to studying concavity: compute $f''(x)$, find where $f''(x)$ is equal to zero or does not exist, choose test inputs to determine the sign of $f''(x)$ on the corresponding intervals, and use this sign information to determine the concavity of $f(x)$ on each interval. The following examples illustrate this approach.

◆ **EXAMPLE 4** Find the intervals on which $f(x) = x^4 - 2x^3 - 12x^2 + 5$ is concave up and concave down, and any points of inflection.

Solution. The first derivative of $f(x)$ is $f'(x) = 4x^3 - 6x^2 - 24x$ and its second derivative is $f''(x) = 12x^2 - 12x - 24$, which is a polynomial and so defined on all real number inputs. Therefore, the zeros of $f''(x)$ providing the transitions in concavity for $f(x)$ are found by setting $f''(x)$ equal to zero and solving for x:

$$f''(x) = 12x^2 - 12x - 24 = 12(x^2 - x - 2) = 12(x + 1)(x - 2) = 0$$

Thus, $f''(x)$ is equal to zero when $x = -1$ and $x = 2$, which means that $f''(x)$ is exactly one of positive or negative on each of the intervals $(-\infty, -1)$, $(-1, 2)$, and $(2, \infty)$. The sign of $f''(x)$ is determined using test inputs from each interval; this example considers $x = -2$, $x = 0$, and $x = 3$:

$$f''(-2) = 48 > 0 \qquad f''(0) = -24 < 0 \qquad f''(3) = 48 > 0$$

Because $f''(x)$ is positive on $(-\infty, -1)$ and $(2, \infty)$, $f(x)$ is concave up on both of these intervals. Similarly, $f''(x)$ is negative on $(-1, 2)$, which means that $f(x)$ is concave down on this interval.

Based on this information, the points of inflection of $f(x)$ correspond to the transitions between these concavity intervals when $x = -1$ and $x = 2$. Substituting these inputs into the original function provides the points of inflection of $f(x)$ at $(-1, f(-1)) = (-1, -4)$ and $(2, f(2)) = (2, -43)$. The graph of $f(x)$ given in Figure 8 illustrates and affirms this analysis of the concavity of $f(x)$.

$\blacksquare$

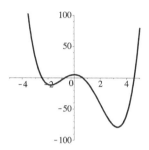

Figure 8: Graph of $f(x)$ for Example 4

This section's initial approach to concavity relied on graphical estimates. Comparing the answers to Question 1(b) and Example 4 illustrates the relative strength of the algebraic approach to studying concavity. Although determining concavity intervals graphically can be accomplished with relative ease when transitions occur for integer inputs, accurately identifying such transitions for noninteger inputs is much more challenging. Likewise, the coordinates of a point of inflection are often much more difficult to ascertain from a graph, in contrast with the precision that is possible with the algebraic approach.

In Examples 3 and 4, the points of inflection occured for inputs where the second derivative was equal to zero. Although this pattern holds for smooth functions, some functions are not smooth, and points of inflection can occur where the second derivative (or even the first derivative) does not exist.

In addition, a second derivative $f''(x)$ being equal to zero for some c does not guarantee that the point $(c, f(c))$ is a point of inflection of the function $f(x)$, because the second derivative might not change sign at this input. Namely, $f''(c)$ being equal to zero or not existing only indicates that $(c, f(c))$ is a candidate to examine as a possible point of inflection of $f(x)$. Only after verifying that $f(x)$ changes concavity at $(c, f(c))$ can this location be declared a point of inflection.

◆ **EXAMPLE 5** Find the intervals on which $f(x) = x^5 - 10x^3 + 20x^2 - 40$ is concave up and concave down, and any points of inflection.

Solution. The first derivative of $f(x)$ is $f'(x) = 5x^4 - 30x^2 + 40x$, and its second derivative is $f''(x) = 20x^3 - 60x + 40$, which is a polynomial and so defined on all real number inputs. Therefore, the zeros of $f''(x)$ providing the transitions in concavity for $f(x)$ are found by setting $f''(x)$ equal to zero and solving for x:

$$f''(x) = 20x^3 - 60x + 40 = 20(x^3 - 3x + 2) = 20(x + 2)(x - 1)^2 = 0$$

Because $f''(x)$ is equal to zero when $x = -2$ and $x = 1$, $f''(x)$ is exactly one of positive or negative on the intervals $(-\infty, -2)$, $(-2, 1)$, and $(1, \infty)$. The sign of $f''(x)$ is determined using test inputs from each interval; this example considers $x = -3$, $x = 0$, and $x = 2$:

$$f''(-3) = -320 < 0 \qquad f''(0) = 40 > 0 \qquad f''(2) = 80 > 0$$

Thus, $f''(x)$ is positive on $(-2, 1)$ and $(1, \infty)$, which means that that $f(x)$ is concave up on both of these intervals. Similarly, $f''(x)$ is negative on $(-\infty, -2)$, which means that $f(x)$ is concave down on this interval.

Based on this information, the only point of inflection of $f(x)$ occurs at the change in concavity when $x = -2$. Note that even though $f''(x)$ is equal to zero when $x = 1$, $f(x)$ does not change concavity at this input, and consequently $f(x)$ does not have a point of inflection when $x = 1$. Substituting $x = -2$ into the given function provides the point of inflection of $f(x)$ at $(-2, f(-2)) = (-2, 88)$. ∎

➤ **QUESTION 4** Find the intervals on which $f(x) = x^3 + 3x^2 + 12x - 18$ is concave up and concave down, and any points of inflection.

➤ **QUESTION 5** Find the intervals on which $f(x) = x^4 - 10x^3 + 24x^2 + 6$ is concave up and concave down, and any points of inflection.

Section 4.3 introduced and studied the law of diminishing returns in terms of the monotonic behavior of functions. This economic principle can also be considered from the perspective of concavity, enriching our understanding of this phenomenon.

◆ **EXAMPLE 6** Suppose the production function of some economic good is modeled by $p(x) = 100x^{0.35}$. Example 7 in Section 4.3 verified that $p(x)$ increases at a decreasing rate. Discuss the concavity of this production function.

Solution. In this setting, x represents the input to a production process, so the only situationally relevant values of x are positive real numbers. The function $p(x)$ is increasing, because its derivative $p'(x) = 100(0.35)x^{-0.65} = 35x^{-0.65}$ is positive for all positive inputs x. The production function's second derivative $p''(x)$ determines the monotonic behavior of the rate of change of $p(x)$. This rate $p'(x)$ is decreasing, because its derivative $p''(x) = 35(-0.65)x^{-1.65} = -22.75x^{-1.65}$ is negative for all positive inputs x. Therefore, $p(x)$ is concave down for all positive inputs, because its derivative $p'(x)$ is decreasing. Figure 9 illustrates and affirms this analysis of the concavity of $p(x)$. ∎

➤ **QUESTION 6** Discuss the concavity of a production function $p(x)$ that is increasing at an increasing rate.

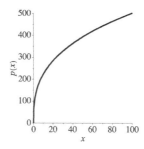

Figure 9: Graph of $p(x)$ for Example 6

Second Derivative Test

The concavity of a function is central to understanding the overall shape of a graph and to explaining how the derivative changes as inputs increase. Even more, the concavity of a function can help classify the critical numbers of many functions as either local maximum or local minimum values.

Recall the graphs of $f(x) = x^2$ and $g(x) = -x^2$ given in Figure 1 as classic examples of functions that are concave up and concave down. Note that $f(x) = x^2$ has a local minimum of $y = 0$ when $x = 0$. At this point, $f(x)$ is concave up and $f''(0)$ is positive. Similarly, $g(x) = -x^2$ has a local maximum of $y = 0$ when $x = 0$ and, at this point, $g(x)$ is concave down and $g''(0)$ is negative. This relationship between the concavity of a function at a critical number and its classification as a local maximum or as a local minimum holds for all smooth functions.

SECOND DERIVATIVE TEST.
Let $f(x)$ be a smooth function with a critical number at input c.

- If $f''(c) > 0$, then $f(c)$ is a local minimum of $f(x)$.

- If $f''(c) < 0$, then $f(c)$ is a local maximum of $f(x)$.

- If $f''(c) = 0$, then the first derivative test must be used to determine whether $f(c)$ is a local maximum, a local minimum, or neither for $f(x)$.

When using the second derivative test to find and classify the local extreme values of a function $f(x)$, the first step is to identify the critical numbers c by computing its first derivative $f'(x)$ and then determining where $f'(x)$ is equal to zero or is undefined. As with the first derivative test, these critical numbers provide candidate inputs for local extreme values.

The next step is to compute its second derivative $f''(x)$ and then check the sign of this second derivative at each critical number c by substituting $x = c$ and evaluating $f''(c)$. Applying the second derivative test to the resulting sign information distinguishes between whether $f(c)$ is a local maximum value or a local minimum value of $f(x)$, or whether the first derivative test must be used instead.

◆ **EXAMPLE 7** Find and classify each critical number of $f(x) = 2x^3 + 9x^2 - 60x + 30$ as a local maximum or local minimum value with the second derivative test, if possible.

Solution. The first derivative of $f(x)$ is $f'(x) = 6x^2 + 18x - 60$, which is a polynomial and defined for all real numbers. Therefore, the critical numbers of $f(x)$ are found by setting $f'(x)$ equal to zero and solving for x:

$$f'(x) = 6x^2 + 18x - 60 = 6(x^2 + 3x - 10) = 6(x + 5)(x - 2) = 0$$

Thus, the critical numbers of $f(x)$ are $x = -5$ and $x = 2$. Now evaluate the second derivative $f''(x) = 12x + 18$ at each critical number:

$$f''(-5) = 12(-5) + 18 = -42 < 0 \qquad f''(2) = 12(2) + 18 = 42 > 0$$

According to the second derivative test, $f(x)$ has a local maximum of $y = f(-5) = 305$ when $x = -5$, because $f''(-5)$ is negative, and a local minimum of $y = f(2) = -38$ when $x = 2$, because $f''(2)$ is positive ∎

For the sake of comparing the first derivative test and the second derivative test, a couple of examples from Section 5.2 are revisited and solved now with the second derivative test.

◆ **EXAMPLE 8** Find and classify each critical number of $f(x) = 2x^3 - 3x^2 - 12x + 2$ as a local maximum or local minimum value with the second derivative test, if possible.

Solution. Example 4 from Section 5.2 computed $f'(x) = 6x^2 - 6x - 12$ and found that the critical numbers of $f(x)$ are $x = -1$ and $x = 2$. The second derivative $f''(x) = 12x - 6$ is now evaluated at each critical number:

$$f''(-1) = 12(-1) - 6 = -18 < 0 \qquad f''(2) = 12(2) - 6 = 18 > 0$$

According to the second derivative test, $f(x)$ has a local maximum of $y = f(-1) = 9$ when $x = -1$, because $f''(-1)$ is negative, and a local minimum of $y = f(2) = -18$ when $x = 2$, because $f''(2)$ is positive. ∎

The result of applying the second derivative test in Example 8 matches the outcome of using the first derivative test in Example 4 from Section 5.2. Rather than working with test inputs on intervals between critical numbers, the second derivative test works exclusively with the critical numbers themselves, which some people find simpler and more direct. Even so, sometimes the second derivative test does not apply, which means that both tests are needed. Consider the following example of just such a case.

◆ **EXAMPLE 9** Find and classify each critical number of $f(x) = x^3 + 6$ as a local maximum or local minimum value with the second derivative test, if possible.

Solution. Example 5 from Section 5.2 computed $f'(x) = 3x^2$ and found that the only critical number of $f(x)$ is $x = 0$. The second derivative $f''(x) = 6x$ is evaluated at this critical number, producing $f''(0) = 6 \cdot 0 = 0$. Thus, the second derivative test is inconclusive, and the first derivative test must be applied to determine that $f(x)$ does not have any local extreme values.

∎

◆ **EXAMPLE 10** A drug administered to a patient produces a concentration in the bloodstream given by $c(x) = \dfrac{e}{3}xe^{-x/3}$ (in milligrams per milliliter) x hours after being injected. Identify the maximum concentration of the drug in the patient's bloodstream using the second derivative test, if possible.

Solution. Example 6 from Section 5.2 found that the only critical number of $c(x)$ is $x = 0$ from setting its first derivative $c'(x)$ as given below equal to zero and solving for x:

$$c'(x) = \frac{e}{3}e^{-x/3} - \frac{e}{9}xe^{-x/3} = \frac{e}{3}e^{-x/3}\left(1 - \frac{1}{3}x\right)$$

The second derivative $c''(x)$ is computed by applying the difference, constant multiple, chain, and product rules:

$$c''(x) = \frac{d}{dx}\left[\frac{e}{3}e^{-x/3}\right] - \frac{d}{dx}\left[\frac{e}{9}xe^{-x/3}\right] \qquad \text{Difference rule}$$

$$= \frac{e}{3}\frac{d}{dx}\left[e^{-x/3}\right] - \frac{e}{9}\frac{d}{dx}\left[xe^{-x/3}\right] \qquad \text{Constant multiple rule}$$

$$= \frac{e}{3}e^{-x/3}\frac{-1}{3} - \frac{e}{9}\frac{d}{dx}\left[xe^{-x/3}\right] \qquad \text{Chain rule}$$

$$= -\frac{e}{9}e^{-x/3} - \frac{e}{9}\left[e^{-x/3}\cdot 1 + x\cdot e^{-x/3}\frac{-1}{3}\right] \qquad \text{Product rule}$$

$$= -\frac{e}{9}e^{-x/3} - \frac{e}{9}e^{-x/3} + \frac{e}{27}x\cdot e^{-x/3} \qquad \text{Distribute and simplify}$$

$$= -\frac{e}{9}e^{-x/3}\left[2 - \frac{x}{3}\right] \qquad \text{Factor}$$

Substituting $x = 3$ into this second derivative yields

$$c''(3) = -\frac{e}{9}e^{-3/3}\left[2 - \frac{3}{3}\right] = -\frac{e}{9}\cdot e^{-1}\cdot 1 = -\frac{1}{9} < 0$$

According to the second derivative test, $c(x)$ has a local maximum of $y = c(3) = 1$ when $x = 3$, because $c''(3)$ is negative. Interpreting this result in context, the patient's global maximum bloodstream concentration of the drug is $c(3) = 1$ milligrams per milliliter three hours after having the medication injected.

∎

▶ **QUESTION 7** Find and classify the critical numbers of $f(x) = 2x^3 + 21x^2 + 50$ as local maximum or local minimum values with the second derivative test, if possible.

Working in RStudio

RStudio can be used to implement the second derivative test using the D and findZeros commands as introduced in detail in Sections 5.1 and 5.2. The second derivative is computed by applying the D command twice. As a reminder, the syntax for the relevant commands is demonstrated below.

Examples of Commands:

- f=makeFun(x^3-x^2+2~x)

- df=D(f(x)~x)

- critNum=findZeros(df(x)~x)

- critNum

- ddf=D(df(x)~x)

- ddf(critNum)

◆ **EXAMPLE 11** The following table presents median home prices P in thousands of dollars each year Y from 2001 to 2010 according to the U.S. Census Bureau:

Year	2001	2002	2003	2004	2005	2006	2007	2008	2009	2010
Price	175.2	187.6	195.0	221.0	240.9	246.5	247.9	232.1	216.7	221.8

Applying fitModel in RStudio produces the following model for this data set:

$$P(Y) = 0.17(Y-2001)^4 - 3.2(Y-2001)^3 + 16.7(Y-2001)^2 - 10.9(Y-2001) + 177.7$$

Find and classify the critical numbers of P(Y) as local maximum or local minimum values on $(2001, 2010)$ with the second derivative test, if possible, using RStudio.

Solution. First, define the function P(Y) in RStudio and compute its first derivative $P'(Y)$. This derivative is then used to find the critical numbers of P(Y) on the given input interval of $(2001, 2010)$.

```
P=makeFun(0.17*(Y-2001)^4-3.2*(Y-2001)^3+16.7*(Y-2001)^2-10.9*(Y-2001)
              +177.7~Y)
dP=D(P(Y)~Y)
critNum=findZeros(dP(Y)~Y,xlim=c(2001,2010))
critNum

        Y
1 2001.36
2 2006.10
3 2009.66
```

Next, compute the second derivative $P''(Y)$ and evaluate this second derivative at each critical number. The given model P(Y) is also evaluated at each critical number.

```
ddP=D(dP(Y)~Y)                            P(critNum)
ddP(critNum)
                                                Y
            Y                             1 175.794
1   26.6992                               2 247.002
2  -11.4676                               3 213.592
3   20.1029
```

This output provides the following sign information about $P''(Y)$:

$$P''(2001.4) \approx 26.699 > 0 \qquad P''(2006.1) \approx -11.468 < 0 \qquad P''(2009.7) \approx 20.103 > 0$$

According to the second derivative test, $P(Y)$ is concave up and has a local minimum of $P \approx 175.794$ when $Y \approx 2001.4$, because $P''(2001.4)$ is negative. Similarly, $P \approx 213.592$ is a local minimum of $P(Y)$ when $Y \approx 2009.7$. Finally, $P(Y)$ is concave down and has a local maximum of $P \approx 247.002$ when $Y \approx 2006.1$, because $P''(2006.1)$ is positive.

Interpreting these results in context, median home prices had local minimum values of \$175,794 in mid 2001 and \$213,592 in late 2009, and a local maximum value of \$247,002 in early 2006. ■

▶ **QUESTION 8** Find and classify the critical numbers of $f(x) = x\sin(x) - e^x$ as local maximum or local minimum values on the interval $(-15, 0)$ with the second derivative test, if possible, and using RStudio.

Summary

- If $f(x)$ is a smooth function and (a, b) is an interval, then $f(x)$ is *concave up* on (a, b) if and only if $f'(x)$ is increasing on (a, b), and $f(x)$ is *concave down* on (a, b) if and only if $f'(x)$ is decreasing on (a, b).

- Let c be a real number in the domain of a function $f(x)$. The point $(c, f(c))$ is a *point of inflection* of $f(x)$ if and only $f(x)$ changes concavity at input c.

- *Concavity and the second derivative*: If $f(x)$ is a smooth function and (a, b) is an interval in the domain of $f(x)$, then $f(x)$ is concave up on (a, b) if and only if $f''(x) > 0$ on (a, b), and $f(x)$ is concave down on (a, b) if and only if $f''(x) < 0$ on (a, b).

- *Second derivative test*: Let $f(x)$ be smooth with a critical number at c.

    ○ If $f''(c) > 0$, then $f(c)$ is a local minimum of $f(x)$.

    ○ If $f''(c) < 0$, then $f(c)$ is a local maximum of $f(x)$.

    ○ If $f''(c) = 0$, then the first derivative test must be used to determine whether $f(c)$ is a local maximum, a local minimum, or neither for $f(x)$.

Exercises

In Exercises 1–10, identify the intervals on which the function is concave up and concave down, and any points of inflection.

1.

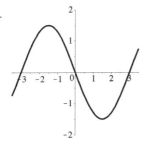

2.

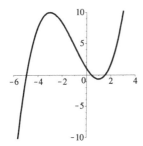

3.

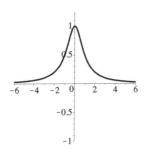

4.

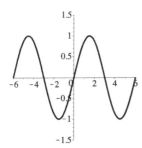

5.

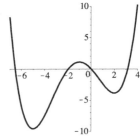

6.

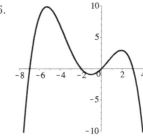

7.

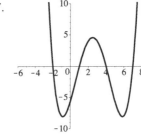

8.

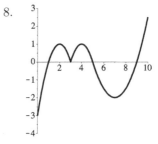

9.

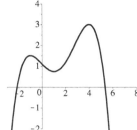

10.

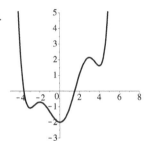

In Exercises 11 – 16, identify the intervals on which the model is concave up and concave down, and any points of inflection.

11. Mean lower low water (MLLW) tidal depths in feet in Pearl Harbor, Hawaii as a function of the hour.

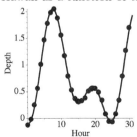

12. Closing stock market value of the Dow Jones Industrial Average at the end of each quarter from 2007 to 2011 as a function of quarters since 2000.

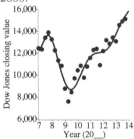

13. Closing stock market value of the Dow Jones Industrial Average at the end of each quarter from March 31, 1935 (quarter 1) to December 31, 2014 (quarter 320).

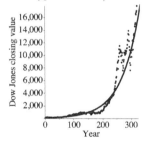

14. Estimated world population in billions of people every five years from 1950 to 2015.

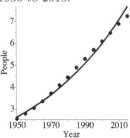

15. Cumulative number of Ebola cases
    in Sierra Leone from May 1, 2014 to
    December 16, 2015.

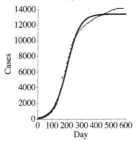

18.

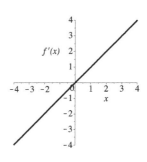

19.

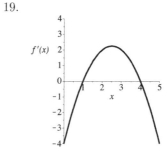

16. U.S. carbon dioxide emissions in kT
    each year from 1960 to 2010.

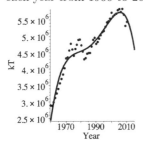

20.

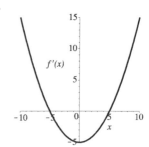

In Exercises 17–26, use the graph of the
derivative of $f'(x)$ to identify the intervals
on which the function $f(x)$ is concave up
and concave down, and any points of inflection.

17.

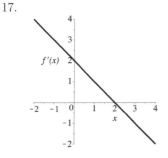

21.

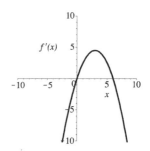

22.

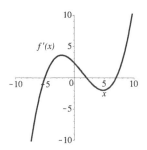

23.

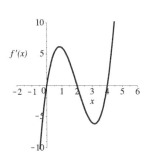

24.

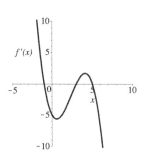

25.

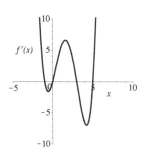

26.

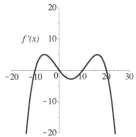

In Exercises 27–40, identify the intervals on which the function is concave up and concave down, and any points of inflection.

27. $f(x) = 3x^2 + 2x + 4$

28. $f(x) = x^2 - 2x + 3$

29. $f(x) = -2x^2 + 4x - 3$

30. $f(x) = x^3 - 3x^2$

31. $f(x) = x^3 - 3x^2 + 3x$

32. $f(x) = x^3 + x^2 - 5x + 7$

33. $f(x) = x^4 - 2x^2$

34. $f(x) = x^4 - 4x^3$

35. $f(x) = 3x^4 - 4x^3 - 6x^2 + 12x$

36. $f(x) = 3x^4 - 16x^3 + 18x^2$

37. $f(x) = 4x^4 - 7x^2 + 4x + 6$

38. $f(x) = x^5 - 5x^3$

39. $f(x) = x^5 - 5x^4$

40. $f(x) = 3x^5 - 5x^4 + 2$

In Exercises 41–54, find and classify each critical number of the function as a local maximum or a local minimum value with the second derivative test, if possible.

41. $f(x) = x^2 + x$

42. $f(x) = x^2 - 4x + 3$

43. $f(x) = x^3 - 1200x$

44. $f(x) = x^3 - 3x + 2$

45. $f(x) = x^3 - 3x^2$

46. $f(x) = x^3 - 9x^2 + 15x$

47. $f(x) = x^3 - 6x^2 + 9x + 2$

48. $f(x) = x^4 - 2x^2 - 12$

49. $f(x) = x^4 - 4x^3 + 16$

50. $f(x) = x^4 - 4x^3 + 12x + 3$

51. $f(x) = 3x^4 - 16x^3 + 18x^2$

52. $f(x) = 3x^5 - 5x^3 + 3$

53. $f(x) = 3x^5 - 5x^4 + 2$

54. $f(x) = x^5 - 5x^4$

In Exercises 55 – 66, find and classify each critical number of the function as a local maximum or local minimum value with the second derivative test, if possible.

55. $f(x) = x^{1/3}$

56. $f(x) = 1 - x^{1/3}$

57. $f(x) = x^{1/3}(8 - x)$

58. $f(x) = x^{3/5}(2 - x)$

59. $f(x) = x^{2/3}$

60. $f(x) = x^{2/3}(2 - x)$

61. $f(x) = x^{2/3}(6 - x)^{1/3}$

62. $f(x) = x^{2/3}(x^2 - 8)$

63. $f(x) = x^{5/3}$

64. $f(x) = x^{5/3} + 4$

65. $f(x) = \dfrac{x^2}{x^2 - 1}$

66. $f(x) = \dfrac{x^3}{x^2 - 1}$

In Exercises 67 – 70, find and classify each critical number of the function as a local maximum or local minimum value with the second derivative test, if possible.

67. $f(x) = e^{-x^2}$

68. $f(x) = e^{x^2 - 4x}$

69. $f(x) = e^{x^3 - x}$

70. $f(x) = e^{1/x}$

In Exercises 71 – 76, find and classify each critical number of the function as a local maximum or local minimum value with the second derivative test, if possible.

71. $f(x) = \sin^2(x)$ on $[-\pi, \pi]$

72. $f(x) = x - 2\sin(x)$ on $[0, \pi]$

73. $f(x) = x + \sin(x)$ on $[0, 2\pi]$

74. $f(x) = \cos^2(x) + \sin(x)$ on $[0, \pi]$

75. $f(x) = 2\sin(x) + \cos(2x)$ on $[0, 2\pi]$

76. $f(x) = \dfrac{x}{2} + \sin(x)$ on $[0, \pi]$

In Exercises 77 – 86, use the graph of the derivative of $f'(x)$ from the earlier exercise to identify the critical numbers of $f(x)$ and then classify each critical number as a local maximum or local minimum value with the second derivative test, if possible.

77. Exercise 17

78. Exercise 18

79. Exercise 19

80. Exercise 20

81. Exercise 21

82. Exercise 22

83. Exercise 23

84. Exercise 24

85. Exercise 25

86. Exercise 26

RStudio. In Exercises 87 – 98, use RStudio to find and classify each critical number of the function from the earlier exercise as a local maximum or local minimum value with the second derivative test, if possible.

In Your Own Words. In Exercises 99 – 103, explain the following.

99. Concave up

100. Concave down

101. Point of inflection

102. Identifying concavity algebraically

103. Second derivative test

5.4   Newton's Method and Optimization

This chapter has focused on studying algorithms for finding and classifying the global and local extreme values of a function. A key step in all of these processes has been determining the critical numbers of a given function, which are the inputs where the derivative is equal to zero or does not exist.

The examples and questions considered in previous sections involved critical numbers that are relatively easy to find. However, more often than not, determining where a derivative is equal to zero is algebraically difficult or even impossible. The following pairs of functions and their derivatives indicate the kinds of challenges that can arise:

$$f(x) = x^3 - \sin(x) \qquad g(x) = e^x + \cos(x) - 2x \qquad h(x) = \tfrac{1}{6}x^6 + x^3 + \tfrac{1}{2}x^2 + x$$
$$f'(x) = x^2 - \cos(x) \qquad g'(x) = e^x - \sin(x) - 2 \qquad h'(x) = x^5 + 3x^2 + x + 1$$

All three of these derivatives have zeros near the origin, as indicated by their graphs in Figure 1, but none of these zeros can be readily identified algebraically.

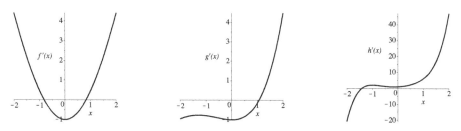

Figure 1: Graphs of $f'(x)$, $g'(x)$, and $h'(x)$

And yet, without knowing zeros of derivatives, none of the optimization tests studied in this chapter can be implemented for these functions.

This section introduces an algorithm called *Newton's method* that numerically approximates where a function is equal to zero. The advantages of using this process are at least threefold:

(1) Newton's method does not use algebra to find zeros of functions.

(2) Newton's method provides a straightforward formula.

(3) Newton's method quickly finds accurate approximations.

However, avoiding the algebra needed to obtain exact solutions does entail some sacrifice in that Newton's method only finds an approximate input x for which $f(x)$ is equal to zero. Even so, approximations are sufficient in most situations, making this sacrifice reasonable. In addition to introducing Newton's method, this section applies this algorithm in the context of optimization and then explores the reasons why this method successfully finds zeros of smooth functions.

Applying Newton's Method

Newton's method is an iterative process that approximates an input for which a given function is zero. The first step is to make an initial conjecture of an input where such a zero is located (some suggestions for choosing good first guesses are provided later in the section). This initial conjecture is called x_1 and is used to calculate a new, and hopefully better, second approximation x_2 for where the function is zero. This x_2 can then be used to calculate a third approximation x_3 that is hopefully even closer to the actual zero. This process can be repeated to obtain a fourth approximation x_4, a fifth approximation x_5, and so on, until a sufficiently accurate approximation has been produced. Newton's method uses a standard formula to calculate a new approximation x_{n+1} from the immediately previous approximation x_n.

NEWTON'S METHOD.

If $f(x)$ is a smooth function with a zero sufficiently near an input x_1, then more accurate approximations $x_2, x_3, \ldots, x_{n+1}$ of the zero are calculated as follows:

$$x_2 = x_1 - \frac{f(x_1)}{f'(x_1)} \qquad x_3 = x_2 - \frac{f(x_2)}{f'(x_2)} \qquad \cdots \qquad x_{n+1} = x_n - \frac{f(x_n)}{f'(x_n)}$$

The Newton's method formula is obtained by considering where a tangent line approximation of $f(x)$ intersects the x-axis. Further details about the derivation of this formula are discussed at the end of this section. For now, the focus is on developing a facility with using this algorithm.

◆ **EXAMPLE 1** Approximate a zero of $f(x) = xe^x - x - 1$ near $x_1 = 1$ with three iterations of Newton's method.

Solution. First compute the derivative $f'(x)$ that appears in the Newton's method formula. For $f(x) = xe^x - x - 1$, the first derivative is $f'(x) = e^x + xe^x - 1$, which yields the general formula for Newton's method in this setting:

$$x_{n+1} = x_n - \frac{f(x_n)}{f'(x_n)} = x_n - \frac{x_n e^{x_n} - x_n - 1}{e^{x_n} + x_n e^{x_n} - 1}$$

The first three iterations of Newton's method are computed as follows:

$$x_2 = x_1 - \frac{f(x_1)}{f'(x_1)} = 1 - \frac{f(1)}{f'(1)} = 1 - \frac{1e^1 - 1 - 1}{e^1 + 1e^1 - 1} \approx 0.838$$

$$x_3 = x_2 - \frac{f(x_2)}{f'(x_2)} = 0.838 - \frac{f(0.838)}{f'(0.838)} = 0.838 - \frac{0.838e^{0.838} - 0.838 - 1}{e^{0.838} + 0.838e^{0.838} - 1} \approx 0.807$$

$$x_4 = x_3 - \frac{f(x_3)}{f'(x_3)} = 0.807 - \frac{f(0.807)}{f'(0.807)} = 0.807 - \frac{0.807e^{0.807} - 0.807 - 1}{e^{0.807} + 0.807e^{0.807} - 1} \approx 0.806$$

Based on these numeric results, a reasonable conjecture for a zero of $f(x) = xe^x - x - 1$ near $x_1 = 1$ is $x_4 = 0.806$. One indicator that x_4 provides a good approximation of a zero is the agreement in the first two decimal places "0.80" of x_3 and x_4 and the relatively small change in the third decimal place. The graph of $f(x) = xe^x - x - 1$ in Figure 2 visually affirms that $x_4 = 0.806$ provides a good approximation of a zero of $f(x) = xe^x - x - 1$ near $x_1 = 1$. ■

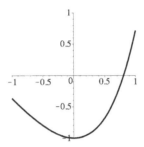

Figure 2: Graph of $f(x)$ for Example 1

Example 1 demonstrates that carrying out three iterations of Newton's method produces the fourth approximation x_4 of a zero of $f(x)$ because each iteration involves transitioning from approximation x_n to the next approximation x_{n+1}. In general, n iterations of Newton's method are needed to obtain the $(n+1)$th approximation x_{n+1}.

◆ **EXAMPLE 2** Approximate a zero of $f(x) = e^{\sin(x)} - 2$ near $x_1 = 1$ with three iterations of Newton's method.

Solution. The first derivative of $f(x) = e^{\sin(x)} - 2$ is $f'(x) = e^{\sin(x)} \cos(x)$, which yields the general formula for Newton's method in this setting:

$$x_{n+1} = x_n - \frac{f(x_n)}{f'(x_n)} = x_n - \frac{e^{\sin(x_n)} - 2}{e^{\sin(x_n)} \cos(x_n)}$$

The first three iterations of Newton's method are computed as follows:

$$x_2 = x_1 - \frac{f(x_1)}{f'(x_1)} = 1 - \frac{e^{\sin(1)} - 2}{e^{\sin(1)} \cos(1)} \approx 0.74487$$

$$x_3 = x_2 - \frac{f(x_2)}{f'(x_2)} = 0.7449 - \frac{e^{\sin(0.7449)} - 2}{e^{\sin(0.7449)} \cos(0.7449)} \approx 0.76580$$

$$x_4 = x_3 - \frac{f(x_3)}{f'(x_3)} = 0.7658 - \frac{e^{\sin(0.7658)} - 2}{e^{\sin(0.7658)} \cos(0.7658)} \approx 0.76585$$

Because x_3 and x_4 are equal up to four decimal places, a reasonable approximation of a zero of $f(x) = e^{\sin(x)} - 2$ near $x_1 = 1$ is $x = 0.7658$. Preserving even more decimal places would better clarify the slight difference between x_3 and x_4 in their fifth decimal places, and improve the accuracy of this conjectured zero of $f(x)$, as would computing additional iterations.

The graph of $f(x) = e^{\sin(x)} - 2$ in Figure 3 visually affirms that $x = 0.7658$ provides a good approximation of a zero of $f(x)$ near $x_1 = 1$. Even more, only three iterations of Newton's method were needed to obtain an approximation that is accurate to four decimal places. ∎

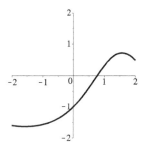

Figure 3: Graph of $f(x)$ for Example 2

Example 1 rounded intermediate answers to three decimal places and Example 2 to four decimal places. Different choices can be made depending on the setting or to achieve a higher degree of accuracy. In general, more decimal places should be saved from iteration to iteration than the level of accuracy needed for the final approximation.

➤ **QUESTION 1** Approximate a zero of $f(x) = x^5 + x + 1$ near $x_1 = -0.6667$ with three iterations of Newton's method.

Starting Values and Ending Criteria

One major strength of Newton's method is that this algorithm quickly produces a relatively accurate approximation of an input value where a given function is equal to zero. In Example 2, a zero of the given function was accurately approximated to four decimal places in only three iterations of Newton's method. While this algorithm requires more iterations in some settings to achieve such accuracy, Newton's method remains one of the quickest and most efficient methods for finding zeros of functions.

Implementing Newton's method requires two choices that are sometimes quite subtle: selecting an initial conjecture of a zero, and deciding when a sufficient number of iterations have been performed to produce a sufficiently accurate approximation. The factors involved in making these choices warrant further discussion.

Determining an appropriate initial conjecture can be challenging, and a poor choice for a starting value can cause Newton's method to fail. One approach to identifying an initial conjecture is to graph the function and visually estimate the desired root of the function. This approach oftens yields a good starting value from which Newton's method can quickly produce more accurate approximations. Alternatively, when given an input interval within which a zero is needed, the midpoint of the interval can provide a good initial conjecture.

Deciding how many iterations of Newton's method to perform is another important decision. In academic settings, this decision is often settled by the instructions. In real-life settings, this question can often be more subtle. One option is to always perform the same, fixed number of iterations of Newton's method (e.g., always compute x_2, x_3, and x_4). Because Newton's method provides accurate approximations very quickly, a small number of iterations will usually provide a sufficiently accurate approximation.

Another common approach is to perform Newton's method until successive iterations agree up to one decimal place more than the desired number. Successive iterations agreeing in this fashion indicates that the approximation is correct up to the desired number of decimal places of accuracy. Example 2 settling on $x = 0.7658$ illustrates this approach.

Determining Points of Intersection

In some circumstances, one needs to determine where a function is equal to a numeric value besides zero or, more generally, where two functions intersect. Newton's method can be applied in these settings.

As an illustrative example, suppose that some situation calls for determining an input for which a function $f(x)$ is equal to five, that is, $f(x) = 5$. Subtracting five from both sides of this equation yields $f(x) - 5 = 0$. Newton's method can then be applied to the new function $h(x) = f(x) - 5$ to approximate an input x for which $h(x) = 0$ or, equivalently, where $f(x) - 5 = 0$. In this way, the algorithm yields an approximation of input(s) for which $f(x) = 5$.

This same idea can be applied to determining where any two functions $f(x)$ and $g(x)$ intersect. Namely, such a point of intersection occurs for inputs where $f(x) = g(x)$ or, equivalently, where $h(x) = f(x) - g(x) = 0$.

FINDING POINTS OF INTERSECTION.
Let $f(x)$ and $g(x)$ be smooth functions. An input where $f(x) = g(x)$ can be approximated by applying Newton's method to $h(x) = f(x) - g(x)$. The approximate input x where $h(x) = 0$ is also an approximate input for where $f(x) = g(x)$.

◆ **EXAMPLE 3** Approximate the point of intersection of $f(x) = \cos(x)$ and $g(x) = x$ near $x_1 = 0.5$ accurate to three decimal places using Newton's method.

Solution. In order to approximate where $\cos(x) = x$ near $x_1 = -0.5$, Newton's method is applied to approximate where $h(x) = \cos(x) - x$ is equal to zero near x_1. Differentiating $h(x) = \cos(x) - x$ yields $h'(x) = -\sin(x) - 1$ and the general Newton's method formula:

$$x_{n+1} \;=\; x_n - \frac{h(x_n)}{h'(x_n)} \;=\; x_n - \frac{\cos(x_n) - x_n}{-\sin(x_n) - 1}$$

The first three iterations of Newton's method are computed as follows:

$$x_2 = x_1 - \frac{f(x_1)}{f'(x_1)} = 0.5 - \frac{\cos(0.5) - 0.5}{-\sin(0.5) - 1} \approx 0.75522$$

$$x_3 = x_2 - \frac{f(x_2)}{f'(x_2)} = 0.75522 - \frac{\cos(0.75522) - 0.75522}{-\sin(0.75522) - 1} \approx 0.73914$$

$$x_4 = x_3 - \frac{f(x_3)}{f'(x_3)} = 0.73914 - \frac{\cos(0.73914) - 0.73914}{-\sin(0.73914) - 1} \approx 0.73909$$

The repetition of the first three decimal places from x_3 to x_4 indicates that the approximations have stabilized at $x = 0.739$ and that subsequent iterations will produce the same result for these first three decimal places. Thus, $f(x) = \cos(x)$ and $g(x) = x$ intersect at approximately $(0.739, 0.739)$, as illustrated in the graph in Figure 4. ■

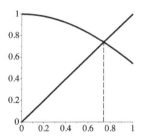

Figure 4: Graphs of $f(x)$ and $g(x)$ for Example 3

In Example 3, the y-coordinate of the point of intersection can be found by substituting the approximate x-coordinate into either of the two given functions, because their outputs are equal at this point of intersection. Namely, the following equalities hold:

$$(0.739, f(0.739)) = (0.739, \cos(0.739)) \approx (0.739, 0.739) = (0.739, g(0.739))$$

This pattern holds for every point of intersection of two functions.

▶ **QUESTION 2** Approximate the point of intersection of $f(x) = \sin(x)$ and $g(x) = x^2$ near $x_1 = 0.5$ accurate to three decimal places using Newton's method.

▶ **QUESTION 3** Approximate the input where $\cos(x) = 0.75$ near $x_1 = 1$ accurate to three decimal places using Newton's method.

Optimization Using Newton's Method

Newton's method is an excellent and interesting tool in its own right, but is particularly useful in the context of optimization. The first three sections of this chapter demonstrated that finding critical numbers is a key step in the process of optimization. In particular, local extreme values occur at critical numbers, which are inputs c where $f'(c)$ is equal to zero or does not exist. Thus, one of the first steps in identifying the extreme values of functions is to find the zeros of its first derivative $f'(x)$. This part of the optimization process can be carried out by means of Newton's method.

NEWTON'S METHOD FOR LOCATING CRITICAL NUMBERS.
If $f(x)$ is a smooth function with a critical number c near x_1, then more accurate approximations $x_2, x_3, \ldots, x_{n+1}$ of c are calculated as follows:

$$x_2 = x_1 - \frac{f'(x_1)}{f''(x_1)} \qquad x_3 = x_2 - \frac{f'(x_2)}{f''(x_2)} \qquad \cdots \qquad x_{n+1} = x_n - \frac{f'(x_n)}{f''(x_n)}$$

In more detail, the numerator of the Newton's method formula is the function for which zeros are being sought and the denominator is this function's derivative. Because Newton's method is being used to approximate the zeros of $f'(x)$ in this setting of seeking critical numbers of $f(x)$, the first derivative $f'(x)$ appears in the numerator and the second derivative $[f'(x)]' = f''(x)$ appears in the denominator. This new Newton's method formula approximates the critical numbers of $f(x)$ that result from when $f'(x)$ is equal to zero.

◆ **EXAMPLE 4** Approximate a critical number c of $f(x) = x^3 - 5x + 1$ near $x_1 = 2$ accurate to three decimal places using Newton's method.

Solution. The first derivative of $f(x)$ is $f'(x) = 3x^2 - 5$ and its second derivative is $f''(x) = 6x$. Substituting these derivatives yields the general formula for the approximation of critical numbers of $f(x)$ using Newton's method:

$$x_{n+1} \;=\; x_n - \frac{f'(x_n)}{f''(x_n)} \;=\; x_n - \frac{3(x_n)^2 - 5}{6x_n}$$

Performing five iterations of Newton's method yields an approximation of a critical number accurate to three decimal places near $x_1 = 2$:

$$x_2 = x_1 - \frac{f'(x_1)}{f''(x_1)} = 2 - \frac{3 \cdot (2)^2 - 5}{6 \cdot 2} \approx 1.41667$$

$$x_3 = x_2 - \frac{f'(x_2)}{f''(x_2)} = 1.41667 - \frac{3 \cdot (1.41667)^2 - 5}{6 \cdot 1.41667} \approx 1.29657$$

$$x_4 = x_3 - \frac{f'(x_3)}{f''(x_3)} = 1.29657 - \frac{3 \cdot (1.29657)^2 - 5}{6 \cdot 1.29657} \approx 1.29101$$

$$x_5 = x_4 - \frac{f'(x_4)}{f''(x_4)} = 1.29101 - \frac{3 \cdot (1.29101)^2 - 5}{6 \cdot 1.29101} \approx 1.29099$$

$$x_5 = x_4 - \frac{f'(x_4)}{f''(x_4)} = 1.29099 - \frac{3 \cdot (1.29099)^2 - 5}{6 \cdot 1.29099} \approx 1.29099$$

Thus, Newton's method approximates a critical number of $c \approx 1.291$ for $f(x) = x^3 - 5x + 1$ near $x_1 = 2$. ∎

   In Example 4, note the choice to preserve five decimal places when transitioning from one iteration to the next until the first three decimal places have stablized. This approach improves the speed and accuracy of Newton's method. In this setting, five iterations happen to be necessary to achieve the required level of accuracy.

➤ **QUESTION 4**   Approximate the critical number c of $f(x) = x^3 - 4x + 1$ near $x_1 = -2$ accurate to three decimal places using Newton's method.

◆ **EXAMPLE 5**   The following table presents median home prices P in thousands of dollars each year Y from 2001 to 2010 according to the U.S. Census Bureau:

Year	2001	2002	2003	2004	2005	2006	2007	2008	2009	2010
Price	175.2	187.6	195.0	221.0	240.9	246.5	247.9	232.1	216.7	221.8

Applying `fitModel` in RStudio produces the following model for this data set:

$$P(Y) = 0.17(Y-2001)^4 - 3.2(Y-2001)^3 + 16.7(Y-2001)^2 - 10.9(Y-2001) + 177.7$$

Approximate the maximum value of $P(Y)$ between 2001 and 2010 using Newton's method.

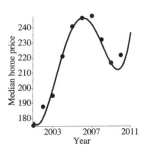

Figure 5: Median home prices from 2001 to 2010 for Example 5

Solution.   Compute the first and second derivatives of $P(Y)$ and state the general Newton's method formula for approximating the critical numbers of $P(Y)$:

$$P'(Y) = 0.68(Y-2001)^3 - 9.6(Y-2001)^2 + 33.4(Y-2001) - 10.9$$
$$P''(Y) = 2.04(Y-2001)^2 - 19.2(Y-2001) + 33.4$$
$$Y_{n+1} = Y_n - \frac{P'(Y)}{P''(Y)}$$
$$= Y_n - \frac{0.68(Y_n-2001)^3 - 9.6(Y_n-2001)^2 + 33.4(Y_n-2001) - 10.9}{2.04(Y_n-2001)^2 - 19.2(Y_n-2001) + 33.4}$$

From the graph of $P(Y)$ in Figure 5, $Y_1 = 2006$ is a reasonable initial conjecture for the critical point providing the designated maximum value. Now compute the first two iterations of Newton's method:

$$Y_2 = Y_1 - \frac{P'(Y_1)}{P''(Y_1)} = 2006 - \frac{P'(2006)}{P''(2006)} \approx 2006.095$$

$$Y_3 = Y_2 - \frac{P'(Y_2)}{P''(Y_2)} = 2006.095 - \frac{P'(2006.095)}{P''(2006.095)} \approx 2006.095$$

The identical results from these first two iterations indicate that $c \approx 2006.095$ is a critical number of P(Y) near 2006 and accurate to three decimal places. This input corresponds to the global maximum on $[2001, 2010]$, as indicated graphically in Figure 5 and demonstrated analytically using RStudio in Example 12 of Section 5.1. Substituting year $Y \approx 2006.095$ into the model P(Y) yields a maximum median home price of \$247,002.30 between 2001 and 2010.

◼

Understanding Newton's Method

Two fundamental ideas lie at the heart of Newton's method, one calculus-based and the other more algebraic in nature:

(1) The tangent line to a function provides a good, local approximation of the function near the point where the tangent line is based.

(2) The algebraic process of determining where a linear function is equal to zero is relatively straightforward.

For an input $x = x_1$, the tangent line to a function $f(x)$ has a slope $f'(x_1)$ and passes through the point $(x_1, f(x_1))$, which yields the following equation for this tangent line:

$$y - f(x_1) = f'(x_1)(x - x_1)$$

Because tangent lines provide good, local approximations of smooth functions, the input where the tangent line intersects the x-axis likewise provides a reasonable approximation to an input where the function is equal to zero, as illustrated in Figure 6.

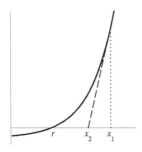

Figure 6: The zero of a tangent line approximates a zero of $f(x)$

The input where this tangent line intersects the x-axis is found by substituting $y = 0$ into the equation $y - f(x_1) = f'(x_1)(x - x_1)$ and then solving for x:

$$0 - f(x_1) = f'(x_1)(x - x_1) \qquad \text{Substitute } y = 0$$

$$-\frac{f(x_1)}{f'(x_1)} = x - x_1 \qquad \text{Divide both sides by } f'(x_1)$$

$$x_1 - \frac{f(x_1)}{f'(x_1)} = x \qquad \text{Add } x_1 \text{ to both sides}$$

This algebra yields the x-intercept of the equation of the tangent line that is labeled x_2 in Figure 6. This input x_2 provides a better approximation of the input for which the original function $f(x)$ is equal to zero (at least for many functions).

Even though x_2 is usually an improvement over x_1, more iterations are typically needed to obtain a sufficiently accurate approximation. The entire process discussed thus far can be repeated again, only now working with the equation of the tangent line to $f(x)$ for input x_2, which is given by $y - f(x_2) = f'(x_2)(x - x_2)$. The input where this second line intersects the x-axis is found by substituting $y = 0$ into the second tangent line equation and then solving for x:

$$0 - f(x_2) = f'(x_2)(x - x_2) \qquad \text{Substitute } y = 0$$

$$-\frac{f(x_2)}{f'(x_2)} = x - x_2 \qquad \text{Divide both sides by } f'(x_2)$$

$$x_2 - \frac{f(x_2)}{f'(x_2)} = x \qquad \text{Add } x_2 \text{ to both sides}$$

The x-intercept of this second tangent line is called x_3 and is usually even closer to the sought for zero of the original function $f(x)$, as illustrated in Figure 7.

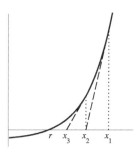

Figure 7: Approximations using Newton's method become more accurate

This graphical illustration of the zeros x_1, x_2, and x_3 approaching an actual zero of $f(x)$ indicates the improving accuracy of these approximations. If an even better approximation of the x-intercept of $f(x)$ is needed, this iterative process can be repeated multiple times to obtain a sufficient degree of accuracy. While this intuitive explanation is sufficient for this book, a more detailed analysis is undertaken in more advanced mathematics books.

Newton's Method and Dimensions

The Newton's method formula involving x, $f(x)$, and $f'(x)$ is the most natural and, from a certain perspective, the only dimensionally compatible way to arithmetically combine these three quantities in the context of seeking a zero of a function. For this discussion, recall from Section 2.7 that the square bracket notation [quantity] is used to denote the dimension of a quantity; other details about dimensions and dimensional quantities can be found in that same section.

The goal of Newton's method is to adjust a current approximation of a zero of $f(x)$ by some correction term to obtain a better approximation of the zero. Therefore, the Newton's method formula will be an expression of the form "$x \pm$ correction." Rule 1 of dimensional arithmetic mandates that dimensional quantities can only be added or subtracted if they have the same dimension, which yields the following information:

$$[\text{"correction"}] = [x] = [\text{input}]$$

From the perspective of dimensions, the only way to arithmetically combine the quantities $f(x)$ and $f'(x)$, which do not have the same dimensions, is to multiply or divide them, which makes the "correction" term have one of the following forms:

$$f(x) \cdot f'(x) \qquad \frac{f(x)}{f'(x)} \qquad \frac{f'(x)}{f(x)}$$

The dimensions of these three expressions can be written in terms of the dimensions [input] and [output]:

$$[f(x) \cdot f'(x)] = [f(x)] \cdot [f'(x)] = [\text{output}] \cdot [\text{output}][\text{input}]^{-1} = [\text{output}]^2[\text{input}]^{-1}$$

$$\left[\frac{f(x)}{f'(x)}\right] = \frac{[f(x)]}{[f'(x)]} = \frac{[\text{output}]}{[\text{output}][\text{input}]^{-1}} = \frac{1}{[\text{input}]^{-1}} = [\text{input}]$$

$$\left[\frac{f'(x)}{f(x)}\right] = \frac{[f'(x)]}{[f(x)]} = \frac{[\text{output}][\text{input}]^{-1}}{[\text{output}]} = [\text{input}]^{-1}$$

Therefore, because ["correction"] = [input], the only dimensionally compatible way to combine x, $f(x)$ and $f'(x)$ is one of the following two expressions:

$$x + \frac{f(x)}{f'(x)} \qquad x - \frac{f(x)}{f'(x)}$$

The Newton's method formula turns out to be a difference rather than a sum because of the relationship between the monotonicity of the function and the direction of the correction that needs to be made to the current approximation x_n provided by $f(x_n)/f'(x_n)$. This further detail is explored in the exercises.

Working in RStudio

RStudio can be used to compute numerous iterations of Newton's method by applying the `makeFun` and `D` commands. The key idea is to think of the Newton's method formula as providing a function that can be evaluated for multiple inputs. For each iteration, the current approximation x_n is substituted into a Newton's

method function defined by $x - f(x)/f'(x)$ in order to obtain the next approximation x_{n+1}. The example commands given below provide a standard approach to
implementing Newton's method using RStudio in the specific context of Example
1 for $f(x) = xe^x - x - 1$ with $x_1 = 1$.

Examples of Commands:

- f=makeFun(x*exp(x)-x-1~x)
- df=D(f(x)~x)
- NM=makeFun(x-f(x)/df(x)~~x)

- x2=NM(1)
- x2
- x2=NM(x3)

◆ **EXAMPLE 6** Approximate a zero of $f(x) = \dfrac{x^3 - 2x + 1}{x^2 + 3}$ near $x_1 = 0$ with six
iterations of Newton's method using RStudio.

Solution. Define the function $f(x)$, compute its derivative $f'(x)$, and store the Newton's
method formula in RStudio:

```
f=makeFun((x^3-2*x+1)/(x^2+3)~x)
df=D(f(x)~x)
NM=makeFun(x-f(x)/df(x)~x)
```

Now compute the first six iterations of Newton's method:

```
x2=NM(0)                    x4=NM(x3)                  x6=NM(x5)
x2                         x4                         x6

[1] 0.5                    [1] 0.61703                [1] 0.618034

x3=NM(x2)                  x5=NM(x4)                  x7=NM(x6)
x3                         x5                         x7

[1] 0.597015              [1] 0.618031               [1] 0.618034
```

Based on this output, Newton's method finds an approximate zero of $x = 0.618034$ for
$f(x) = \dfrac{x^3 - 2x + 1}{x^2 + 3}$ near $x_1 = 0$. This approximation is consistent with the zero of $f(x)$
nearest $x_1 = 0$ produced by the **findZeros** command:

```
findZeros(f(x)~x)

        x
1 -1.618
2  0.618
3  1.000
```

Different choices for x_1 may result in Newton's method approximating one of the other two zeros of this function.

■

▶ **QUESTION 5**   Approximate a zero of $f(x) = x^3 \ln(x) - x^2$ near $x_1 = 2$ with three iterations of Newton's method using RStudio.

Summary

- *Newton's method*: If $f(x)$ is a smooth function with a zero sufficiently near an input x_1, then more accurate approximations $x_2, x_3, \ldots, x_{n+1}$ of the zero are calculated as follows:

$$x_2 = x_1 - \frac{f(x_1)}{f'(x_1)} \qquad x_3 = x_2 - \frac{f(x_2)}{f'(x_2)} \quad \cdots \quad x_{n+1} = x_n - \frac{f(x_n)}{f'(x_n)}$$

- *Finding points of intersection*: Let $f(x)$ and $g(x)$ be two smooth functions. An input where $f(x) = g(x)$ can be approximated by applying Newton's method to approximate a zero of $h(x) = f(x) - g(x)$.

- *Locating critical numbers*: If $f(x)$ is a smooth function with a critical number c near x_1, then more accurate approximations $x_2, x_3, \ldots, x_{n+1}$ of c are calculated with Newton's method as follows:

$$x_2 = x_1 - \frac{f'(x_1)}{f''(x_1)} \qquad x_3 = x_2 - \frac{f'(x_2)}{f''(x_2)} \quad \cdots \quad x_{n+1} = x_n - \frac{f'(x_n)}{f''(x_n)}$$

Exercises

In Exercises 1–12, approximate a zero of the function near x_1 with three iterations of Newton's method.

1. $f(x) = x^2 - 8$ near $x_1 = -3$

2. $f(x) = x^2 - 8$ near $x_1 = 1$

3. $f(x) = x^2 - 8$ near $x_1 = 3$

4. $f(x) = x^2 - 17$ near $x_1 = -4$

5. $f(x) = x^2 - 17$ near $x_1 = 1$

6. $f(x) = x^2 - 17$ near $x_1 = 4$

7. $f(x) = x^3 - 5x + 1$ near $x_1 = 0$

8. $f(x) = x^3 - 5x + 1$ near $x_1 = -2$

9. $f(x) = x^4 - 3x - 1$ near $x_1 = 0$

10. $f(x) = x^6 - 3x^5 + 2$ near $x_1 = 0.5$

11. $f(x) = \sin(x) + \cos(x)$ near $x_1 = 0$

12. $f(x) = e^x - x^2$ near $x_1 = -1$

In Exercises 13–16, approximate a zero of $f(x) = x^2 - \sin(x)$ near x_1 with three iterations of Newton's method.

13. $x_1 = -0.5$ 15. $x_1 = 0.5$

14. $x_1 = 0.25$ 16. $x_1 = 1$

17. Based on the answers to Exercises 13–16, how each many different zeros do there appear to be in $[-1, 1]$? To which zero does each x_1 converge?

18. Comment on the number of decimal digits that appear consistent after three iterations for each x_1 in Exercises 13–16.

In Exercises 19–22, approximate a zero of $f(x) = x^2 \cos(x)$ near x_1 with three iterations of Newton's method.

19. $x_1 = -5$ 20. $x_1 = 1$

21. $x_1 = 5$ 22. $x_1 = 8$

23. Based on the answers to Exercises 19–22, how each many different zeros do there appear to be in $[-5, 8]$? To which zero does each x_1 converge?

24. Comment on the number of decimal digits that appear consistent after three iterations for each x_1 in Exercises 19–22.

In Exercises 25–32, approximate a zero of the function near x_1 accurate to three decimal places using Newton's method; that is, until successive approximations x_n and x_{n+1} agree to three decimal places.

25. $f(x) = x^2 - 26$ near $x_1 = 5$

26. $f(x) = x^2 - 31$ near $x_1 = -6$

27. $f(x) = 2x^3 - 4x^2$ near $x_1 = 2$

28. $f(x) = x^3 + 3x - 10$ near $x_1 = 1$

29. $f(x) = x^4 - 3x + 2$ near $x_1 = 0.6$

30. $f(x) = x^4 - 2x^2$ near $x_1 = 1.5$

31. $f(x) = x^5 - 3x$ near $x_1 = -1$

32. $f(x) = x^5 - 3x$ near $x_1 = 2$

In Exercises 33–38, approximate a zero of the function near x_1 accurate to three decimal places using Newton's method; that is, until successive approximations x_n and x_{n+1} agree to three decimal places.

33. $f(x) = x^3 - e^x + \sin(x)$ near $x_1 = 1.6$

34. $f(x) = x^3 - e^x + \sin(x)$ near $x_1 = 4.3$

35. $f(x) = x^2 \cos(x)$ near $x_1 = -4.5$

36. $f(x) = x^2 \cos(x)$ near $x_1 = 1.75$

37. $f(x) = x + e^{-x^2}$ near $x_1 = -0.5$

38. $f(x) = x^3 - e^x$ near $x_1 = $

In Exercises 39–46, approximate where $f(x) = 2$ accurate to three decimal places using Newton's method.

39. $f(x) = x^2$

40. $f(x) = x^2 - 9$

41. $f(x) = 4 - x^2$

42. $f(x) = x^3$

43. $f(x) = x^3 - 7$

44. $f(x) = e^x$

45. $f(x) = \ln(x)$

46. $f(x) = x \sin(x)$

In Exercises 47–54, approximate the point of intersection of $f(x)$ and $g(x)$ accurate to three decimal places using Newton's method.

47. $f(x) = x$ and $g(x) = x^2 - 3$

48. $f(x) = x - 1$ and $g(x) = x^4$

49. $f(x) = \sqrt{x}$ and $g(x) = x^2 - 1$

50. $f(x) = x^2$ and $g(x) = 1 - 3x$

51. $f(x) = \cos(x)$ and $g(x) = 4x$

52. $f(x) = 1$ and $g(x) = x \sin(x)$

53. $f(x) = 2x^3$ and $g(x) = x + 2$

54. $f(x) = \dfrac{1}{x}$ and $g(x) = x^2$

In Exercises 55–64, approximate an input corresponding to a local extreme value of the function on the interval using Newton's method.

55. $f(x) = x^5 - 4x^2 - 3$ on $[-1, 1]$

56. $f(x) = x^7 - x + 2$ on $[0, 1]$

57. $f(x) = x - e^x$ on $[-2, 2]$

58. $f(x) = x^2 \cos(x)$ on $[-4, -2]$

59. $f(x) = x^2 e^{-x}$ on $[0, 6]$

60. $f(x) = x \ln(x)$ on $[0.5, 1]$

61. $f(x) = \sin(x - 1)e^x$ on $[-1, 1]$

62. $f(x) = x \cos(e^x)$ on $[0, 1]$

63. $f(x) = x \cos(e^x)$ on $[0.5, 1.5]$

64. $f(x) = x^2 - e^x + \sin(x)$ on $[0.5, 1.5]$

RStudio. In Exercises 65–74, use RStudio to determine a reasonable starting value x_1 for Newton's method.

65. $f(x) = x^2 - 2$

66. $f(x) = 1 - 3x - x^2$

67. $f(x) = x^3 - 9$

68. $f(x) = x^3 - x - 1$

69. $f(x) = x^4 - 5x + 2$

70. $f(x) = 8 - x^4$

71. $f(x) = e^x - 7$

72. $f(x) = e^x - x$

73. $f(x) = e^{-x} - 22$

74. $f(x) = e^{-x} - 4x$

RStudio. In Exercises 75–86, use RStudio to approximate a zero of $f(x)$ near x_1 accurate to six decimal places using Newton's method.

75. $f(x) = e^{-x} \sin(x) + x$ near $x_1 = -14$

76. $f(x) = e^{\cos(x)} - x^2$ near $x_1 = 0.25$

77. $f(x) = \dfrac{x^3 - 2x + 1}{x^2 + 1}$ near $x_1 = 0$

78. $f(x) = \dfrac{\cos(e^x)}{x^2}$ near $x_1 = 1$

79. $f(x) = \dfrac{\sin(2x)}{e^x}$ near $x_1 = 2$

80. $f(x) = \dfrac{\sin(2x)}{e^x}$ near $x_1 = 4$

81. $f(x) = x^6 - 3x^5 + 2$ near $x_1 = 0.5$

82. $f(x) = x^4 - 3x - 1$ near $x_1 = 0$

83. $f(x) = \sin(x) + \cos(x)$ near $x_1 = 0$

84. $f(x) = \sin(x) + x - 2$ near $x_1 = 0.5$

85. $f(x) = e^x - x^2$ near $x_1 = -1$

86. $f(x) = \tan(x) - 2x$ near $x = 1.5$

RStudio. In Exercises 87–88, use RStudio to consider the application of Newton's method to $f(x) = x^3 - 6x^2 + 7x + 2$.

87. Compute the first six iterations of Newton's method for $f(x)$ near $x_1 = 1$.

88. Graph $f(x)$ and use this graph to discuss the results in Exercise 87.

In Exercises 89–92, use the monotonicity of $f(x)$, the sign of $f(x)$, and the relative location of x_n to verify that Newton's method is a difference rather than a sum.

89.

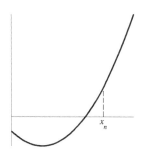

90.

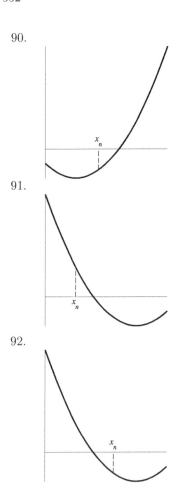

91.

92.

In Your Own Words. In Exercises 93 – 99, explain the following.

93. Goal of Newton's method

94. Formula of Newton's method

95. Choosing the initial conjecture x_1 for Newton's method

96. Deciding when to stop iterating for Newton's method

97. Finding points of intersection with Newton's method

98. Locating critical numbers with Newton's method

99. Newton's method and dimensions

5.5   Multivariable Optimization

Multivariable functions have maximum and minimum values, just like their single-variable counterparts as discussed in Sections 5.1 – 5.3. In many ways, the mathematics involved in finding and classifying extreme values for multivariable functions is similar to that for single-variable functions, although an additional layer of complexity arises from the existence of multiple first-order and second-order partial derivatives.

The intuitive idea of a local extreme value is that a local maximum is larger than all the other nearby outputs of the function and that a local minimum is smaller than all the other nearby outputs of the function. Visually identifying local extreme values from the surface plots of three-dimensional functions is relatively straightforward, as suggested by Figure 1.

(a)   (b)

Figure 1: Surface plots showing local extreme values

Namely, in Figure 1(a), input $(x, y) = (0, 0)$ produces a local maximum value and, in Figure 1(b), local maximum and minimum values occur for inputs in opposite quadrants of the xy-plane. Even more, all three of the extreme values presented in Figure 1 might be global extreme values, depending on the region $\mathcal{R}$ and the behavior of these functions beyond the part of the plane presented in these graphs. The following precise mathematical definition expresses these intuitive ideas for two-variable functions:

Definition. Let $f(x, y)$ be a two-variable function and let input (a, b) be in the domain of $f(x, y)$.

- A **local maximum** of $f(x, y)$ is a value $f(a, b)$ such that $f(a, b) \geq f(x, y)$ for all inputs (x, y) near (a, b).

- A **local minimum** of $f(x, y)$ is a value $f(a, b)$ such that $f(a, b) \leq f(x, y)$ for all inputs (x, y) near (a, b).

- The **local extreme values** of $f(x, y)$ are the local maximum and local minimum values of the function.

Real-life examples of local extreme values include the peaks and the valleys of a mountain range. Recall from Sections 5.2 and 5.3 that every mountain peak corresponds to a local maximum value and every valley to a local minimum value. The goal of optimization for local extreme values is to identify all of the peaks and all of the valleys in the mountain range without regard to which mountain peak is the tallest or which valley is the lowest.

This section expands previous work by seeking out extreme values in three-dimensional settings, rather than their analogues for two-dimensional functions on the plane. Keeping in mind a physical image, such as a mountain range, can be helpful while studying more abstract or subtle settings.

Contour Plots and Extreme Values

Although visually identifying extreme values from surface plots is worthwhile, identifying local extreme values from contour plots of functions is more important. The definition of local extreme values enables the approximation of inputs where local extreme values occur for functions presented via contour plots. The following examples illustrate this process as well as examining the corresponding monotonic behavior of functions in preparation for a pending discussion about critical points.

◆ **EXAMPLE 1** Approximate the value and coordinates of the local extreme values of a function $f(x, y)$ using its contour plot given in Figure 2. Also, discuss the monotonic behavior of $f(x, y)$ at each extreme value.

    (a) Local maximum of $f(x, y)$ (b) Local minimum of $f(x, y)$

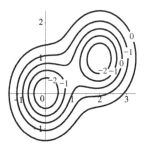

Figure 2: Contour plot of $f(x, y)$ for Example 1

Solution.

    (a) A local maximum of $f(x, y)$ is $z \approx 2.5$ when $(x, y) \approx (2, 1)$, because all of the function values for nearby inputs are smaller than this output. The exact value and location of this local maximum cannot be determined from the contour plot, but one can reasonably conjecture that this extreme value lies inside the highest contour $c = 2$ and so has a value greater than 5. Regarding monotonicity, the contours in Figure 2 indicate that $f(x, y)$ is decreasing from this maximum output in every direction that can be observed.

    (b) The local minimum of $f(x, y)$ is $z \approx -2.5$ when $(x, y) \approx (0, 0)$ because all of the function values for nearby inputs are greater than this output. Regarding monotonicity, the contours in Figure 2 indicate that $f(x, y)$ is increasing from this minimum output in every direction.

                                                                                 ■

◆ **EXAMPLE 2** Figure 3 presents a contour map of Mount St. Helens in the state of Washington with units of meters above sea level. Approximate the local maximum and its coordinates using this contour map and discuss the monotonic behavior of the mountain at this maximum.

Solution. The local maximum is approximately 2550 meters above sea level and its coordinates are approximately $(3.75, 3.75)$. Regarding monotonicity, the mountain is decreasing from this maximum peak in every direction.

                                                                                     ■

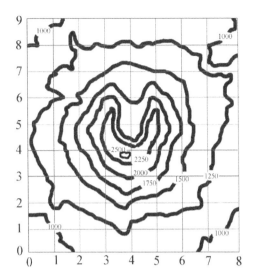

Figure 3: Contour map of Mount St. Helens for Example 2

➤ **QUESTION 1** Approximate the value and coordinates of the local extreme values of a function $f(x, y)$ using its contour plot given in Figure 4. Also, discuss the monotonic behavior of $f(x, y)$ at each extreme value.

(a) Local maximum of $f(x, y)$ (b) Local minimum of $f(x, y)$

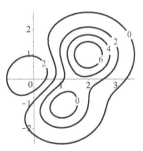

Figure 4: Contour plot of $f(x, y)$ for Question 1

Critical Points

Identifying extreme values graphically from surface plots and contour plots helps deepen an understanding of the definition of local extreme values and develops an intuition regarding their locations. However, many functions are presented via formulas, which necessitates the development of analytical methods for finding and classifying local extreme values.

In Figure 1 and in Examples 1 and 2, the local extreme values occur for inputs where the outputs of the function are either always increasing or always decreasing in every direction away from the extreme value. This functional behavior can only happen when both first-order partial derivatives of the function change sign for the corresponding input, which means that they must both be equal to zero at local extreme values. As with single-variable functions, such inputs are so significant for multivariable functions that they are referred to as *critical points*.

Definition.   A **critical point** of a two-variable function $f(x, y)$ is an input (a, b) in the domain of $f(x, y)$ such that both

$$\frac{\partial f}{\partial x}(a, b) = f_x(a, b) = 0 \qquad \text{and} \qquad \frac{\partial f}{\partial y}(a, b) = f_y(a, b) = 0$$

Both first-order partial derivatives of $f(x, y)$ must be equal to zero in order for an input (a, b) to be a critical point. Inputs with none or only one partial derivative equal to zero are not critical points. Also, for more complicated functions (which lie beyond the scope of this book), additional criteria are required for the definition of a critical point. The interested reader should consult a standard calculus textbook for more details.

The next examples develop our facility with finding critical points as a prelude to analytically finding and classifying local extreme values.

◆ **EXAMPLE 3** Find the critical points of $f(x, y) = 2 + e^{-(x^2 + y^2)}$.

Solution. The first-order partial derivatives of $f(x, y)$ are $f_x(x, y) = -2xe^{-(x^2+y^2)}$ and $f_y(x, y) = -2ye^{-(x^2+y^2)}$. Notice that the exponential function $e^{-(x^2+y^2)}$ appears as a factor in both partials and recall that exponentials are never equal to zero. Therefore, the other factors determine the zeros of these partials. Namely, $f_x(x, y) = 0$ when $-2x = 0$, and solving for x yields $x = 0$. Similarly, $f_y(x, y) = 0$ when $-2y = 0$, and solving for y yields $y = 0$. Thus, $(0, 0)$ is the sole critical point of $f(x, y)$.   ■

◆ **EXAMPLE 4** Find the critical points of $f(x, y) = x^2 + y$.

Solution. The first-order partials of $f(x, y)$ are $f_x(x, y) = 2x$ and $f_y(x, y) = 1$. While $f_x(x, y) = 0$ when $x = 0$, the other partial $f_y(x, y) = 1$ is never equal to zero. Therefore, $f(x, y)$ has no critical points.   ■

◆ **EXAMPLE 5** Find the critical points of $f(x, y) = x^3 + 3y^2 - 12x - 18y$.

Solution. The first-order partial derivatives of $f(x, y)$ are $f_x(x, y) = 3x^2 - 12$ and $f_y(x, y) = 6y - 18$. Setting $f_x(x, y) = 3x^2 - 12 = 3(x - 2)(x + 2)$ equal to zero and solving for x yields $x = 2$ or $x = -2$. Similarly, setting $f_y(x, y) = 6y - 18 = 6(y - 3)$

equal to zero and solving for y yields $y = 3$. Therefore, the two critical points of $f(x, y)$ are $(2, 3)$ and $(-2, 3)$.

◼

➤ **QUESTION 2** Find the critical points of $f(x, y) = x^2 + y^2 - 2x + 2y + 2$.

➤ **QUESTION 3** Find the critical points of $f(x, y) = x^3 + y^3 - 27x - 12y$.

The critical points of a function serve as the candidate inputs for where local extreme values can occur, but other functional behaviors can occur at these points as well. In addition to local maximum and local minimum values, sometimes a multivariable function has a *saddle point* for a critical point. Most saddle points resemble horse saddles or geographic saddles in a hilly region (thus the name), or alternatively Pringles® potato chips, as illustated in Figure 5.

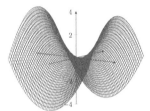

Figure 5: Example of a saddle point

In Figure 5, the function decreases away from the origin when considering inputs moving toward ∞ or toward $-\infty$ along the x-axis. Similarly, the function increases away from the origin when considering inputs moving toward ∞ or toward $-\infty$ along the y-axis. Variations of this type of functional behavior define saddle points.

Definition. A function $f(x, y)$ has a **saddle point** for input (a, b) if and only if $f(x, y)$ increases from $f(a, b)$ in some directions and decreases from $f(a, b)$ in other directions.

Multivariable Second Derivative Test

Critical points are classified as providing local maximum values, local minimum values, or saddle points of multivariable functions $f(x, y)$ by means of the multivariable second derivative test. As stated in Section 4.6, Clairaut's theorem ensures that the mixed partials of every smooth two-variable function $f(x, y)$ are equal; symbolically, this theorem asserts that $f_{xy} = f_{yx}$. As it turns out, all three of the distinct second-order partial derivatives f_{xx}, f_{xy}, and f_{yy} play a role in the second derivative test for multivariable functions.

MULTIVARIABLE SECOND DERIVATIVE TEST.

Let $f(x, y)$ be a smooth two-variable function, let (a, b) be a critical point of $f(x, y)$, and let $D(a, b) = f_{xx}(a, b) \cdot f_{yy}(a, b) - [f_{xy}(a, b)]^2$.

- If $D(a, b) > 0$ and $f_{xx}(a, b) > 0$, then $f(a, b)$ is a local minimum of $f(x, y)$.

- If $D(a, b) > 0$ and $f_{xx}(a, b) < 0$, then $f(a, b)$ is a local maximum of $f(x, y)$.

- If $D(a, b) < 0$, then $f(x, y)$ has a saddle point for input (a, b).

- If $D(a, b) = 0$, then the test is inconclusive and (a, b) may correspond to any or none of a local minimum, a local maximum, or a saddle point.

The expression $D(a, b) = f_{xx}(a, b) \cdot f_{yy}(a, b) - [f_{xy}(a, b)]^2$ is called the **discriminant** of $f(x, y)$ at (a, b), and it "discriminates" among these three graphical features of two-variable functions. In particular, the sign of $D(a, b)$ distinguishes the uniformly concave up and concave down bowl shapes of local extreme values from saddle points with their mix of concavities.

Once a positive $D(a, b)$ identifies a critical point as corresponding to a local extreme value, the sign of the second-order partial f_{xx} provides concavity information in the x-direction to decide whether the local extreme value is a maximum or minimum in the exact same fashion as the second derivative test for single-variable functions. This approach relies on the following relationships:

$$f(x, y) \text{ is concave up in the } x\text{-direction for } (a, b) \quad \Leftrightarrow \quad f_{xx}(a, b) \text{ is positive}$$

$$f(x, y) \text{ is concave down in the } x\text{-direction for } (a, b) \quad \Leftrightarrow \quad f_{xx}(a, b) \text{ is negative}$$

The second-order partial $f_{yy}(a, b)$ would provide exactly the same information and could have been used instead.

◆ **EXAMPLE 6** Classify each critical point of $f(x, y) = x^2 + y^2$ as a local maximum, a local minimum, or a saddle point using the multivariable second derivative test, or state that the test is inconclusive.

Solution. The critical points of $f(x, y)$ are determined by the zeros of its first-order partials $f_x(x, y) = 2x$ and $f_y(x, y) = 2y$. These partial derivatives are both equal to zero exactly when $x = 0$ and $y = 0$, which means that the sole critical point of $f(x, y)$ is $(0, 0)$. The second-order partials of $f(x, y)$ are needed to classify this critical point:

$$f_{xx}(x, y) = 2 \qquad\qquad f_{xy}(x, y) = 0 \qquad\qquad f_{yy}(x, y) = 2$$

Substituting these constant functions into the formula for the discriminant produces $D(0, 0) = 2 \cdot 2 - [0]^2 = 4$. Therefore, both $D(0, 0)$ and $f_{xx}(0, 0)$ are positive, which means that $f(x, y)$ has a local minimum value of $z = f(0, 0) = 0$ when $(x, y) = (0, 0)$ according to the multivariable second derivative test.   ∎

◆ **EXAMPLE 7** Classify each critical point of $f(x, y) = x^3 y$ as a local maximum, a local minimum, or a saddle point using the multivariable second derivative test, or state that the test is inconclusive.

Solution. The critical points of $f(x, y)$ are determined by the zeros of its first-order partials $f_x(x, y) = 3x^2 y$ and $f_y(x, y) = x^3$. Setting $f_y(x, y) = x^3$ equal to zero and solving for x yields $x = 0$. Substituting $x = 0$ into $f_x(x, y) = 3x^2 y$ yields $f_x(0, y) = 0$ for every possible value of y. Therefore, for every real number b, every point of the form $(0, b)$ is a critical point of $f(x, y)$; in other words, every point along the y-axis is a critical point of $f(x, y)$. The second-order partials of $f(x, y)$ are needed to classify this critical point:

$$f_{xx}(x, y) = 6xy \qquad\qquad f_{xy}(x, y) = 3x^2 \qquad\qquad f_{yy}(x, y) = 0$$

Substituting these partials and critical points into the formula for the discriminant produces $D(0, b) = (6 \cdot 0 \cdot b) \cdot 0 - [3 \cdot 0^2]^2 = 0 - 0 = 0$. Therefore, the multivariable second derivative test is inconclusive. ∎

The surface plot of $f(x, y) = x^3 y$ from Example 7 is given in Figure 6 and indicates that all points along the y-axis (namely, all the critical points) happen to be saddle points of $f(x, y)$. In particular, the function $f(x, y)$ is increasing in some directions and decreasing in other directions from each point along the y-axis. However, the discriminant $D(0, b) = 0$ for every b in the reals, which means that the multivariable second derivative test is not able to provide this information. Therefore, other methods must be used to reach these conclusions.

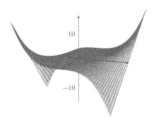

Figure 6: Graph of $f(x, y)$ for Example 7

► **QUESTION 4** Classify each critical point of $f(x, y) = 2x^2 + 6y^2 - 4x + 24y + 25$ as a local maximum, a local minimum, or a saddle point using the multivariable second derivative test, or state that the test is inconclusive.

► **QUESTION 5** Classify each critical point of $f(x, y) = x^2 - y^2$ as a local maximum, a local minimum, or a saddle point using the multivariable second derivative test, or state that the test is inconclusive.

Even though the preceding examples and questions only involved functions that are just polynomials, the multivariable second derivative test also applies to more complicated functions. Naturally, as the functions become more involved, the corresponding partial derivatives typically become more complicated as well. Even so, the multivariable second derivative test can still be applied, as demonstrated in the next two examples.

◆ **EXAMPLE 8** Suppose a heat source is placed under a flat piece of metal until the heat distribution of the metal sheet reaches an equilibrium state. A reasonable model for the intensity of the heat at any point on the metal sheet is given by $h(x, y) = e^{-(x+0.2)^2-(y-0.15)^2}$. Find the location of greatest heat intensity.

Solution. The critical points of $h(x, y)$ are determined by the zeros of the first-order partial derivatives of $h(x, y)$:

$$h_x(x, y) = -2(x + 0.2)e^{-(x+0.2)^2-(y-0.15)^2}$$

$$h_y(x, y) = -2(y - 0.15)e^{-(x+0.2)^2-(y-0.15)^2}$$

The exponential $e^{-(x+0.2)^2-(y-0.15)^2}$ appearing as a factor in both partials is never equal to zero, so the other factors determine the zeros of these partials. For $h_x(x, y)$, setting $x + 0.2$ equal to zero yields $x = -0.2$ and, for $h_y(x, y)$, setting $y - 0.15$ equal to zero yields $y = 0.15$. Thus, the sole critical point of $h(x, y)$ is $(-0.2, 0.15)$. The second-order partials of $h(x, y)$ are needed to classify this critical point:

$$h_{xx}(x, y) = -2e^{-(x+0.2)^2-(y-0.15)^2} + 4(x + 0.2)^2 e^{-(x+0.2)^2-(y-0.15)^2}$$

$$h_{xy}(x, y) = 4(x + 0.2)(y - 0.15)e^{-(x+0.2)^2-(y-0.15)^2}$$

$$h_{yy}(x, y) = -2e^{-(x+0.2)^2-(y-0.15)^2} + 4(y - 0.15)e^{-(x+0.2)^2-(y-0.15)^2}$$

Now substitute the critical point $(x, y) = (-0.2, 0.15)$ into these partials and compute the resulting discriminant:

$$h_{xx}(-0.2, 0.15) = -2e^{0^2-0^2} + 4(0)e^{0^2-0^2} = -2 \cdot e^0 + 0 = -2$$

$$h_{xy}(-0.2, 0.15) = 4(0)(0)e^{0^2-0^2} = 0$$

$$h_{yy}(-0.2, 0.15) = -2e^{0^2-0^2} + 4(0)e^{0^2-0^2} = -2 \cdot e^0 + 0 = -2$$

$$D(-0.2, 0.15) = (-2) \cdot (-2) - [0]^2 = 4$$

Therefore, $D(-0.2, 0.15)$ is positive and $h_{xx}(-0.2, 0.15)$ is negative, which means that $h(x, y)$ has a maximum value of $z = h(-0.2, 0.15) = 1$ when $(x, y) = (-0.2, 0.15)$ according to the multivariable second derivative test. Interpreting this result in context, the greatest heat intensity of the metal sheet occurs at the location determined by $(x, y) = (-0.2, 0.15)$. ■

◆ **EXAMPLE 9** Classify each critical point of $f(x, y) = xe^{-\frac{1}{2}x^2-\frac{1}{2}y^2}$ as a local maximum, a local minimum, or a saddle point using the multivariable second derivative test, or state that the test is inconclusive.

Solution. The critical points of $f(x, y)$ are determined by the zeros of its first-order partials, which are computed using the product and chain rules:

$$f_x(x, y) = e^{-\frac{1}{2}x^2-\frac{1}{2}y^2} \cdot 1 + x \cdot e^{-\frac{1}{2}x^2-\frac{1}{2}y^2}(-x) = (1 - x^2)e^{-\frac{1}{2}x^2-\frac{1}{2}y^2}$$

$$f_y(x, y) = xe^{-\frac{1}{2}x^2 - \frac{1}{2}y^2}(-y) = -xye^{-\frac{1}{2}x^2 - \frac{1}{2}y^2}$$

The exponential $e^{-\frac{1}{2}x^2 - \frac{1}{2}y^2}$ appearing as a factor in both partials is never equal to zero, so the other factors determine the zeros of these partials. For $f_x(x, y)$, setting $1 - x^2 = (1 + x)(1 - x)$ equal to zero yields $x = -1$ or $x = 1$. For $f_y(x, y)$, setting $-xy$ equal to zero when $x = \pm 1$ results in $y = 0$. Therefore, the two critical points of $f(x, y)$ are $(-1, 0)$ and $(1, 0)$. The second-order partials of $f(x, y)$ are needed to classify these critical points:

$$f_{xx}(x, y) = e^{-\frac{1}{2}x^2 - \frac{1}{2}y^2} \cdot (-2x) + (1 - x^2) \cdot e^{-\frac{1}{2}x^2 - \frac{1}{2}y^2}(-x) = (x^2 - 3x)e^{-\frac{1}{2}x^2 - \frac{1}{2}y^2}$$

$$f_{xy}(x, y) = (1 - x^2)e^{-\frac{1}{2}x^2 - \frac{1}{2}y^2}(-y) = -(1 - x^2)\, y\, e^{-\frac{1}{2}x^2 - \frac{1}{2}y^2}$$

$$f_{yy}(x, y) = e^{-\frac{1}{2}x^2 - \frac{1}{2}y^2} \cdot (-x) + (-xy)e^{-\frac{1}{2}x^2 - \frac{1}{2}y^2} \cdot (-y) = (xy^2 - x)e^{-\frac{1}{2}x^2 - \frac{1}{2}y^2}$$

Now substitute the critical points $(x, y) = (-1, 0)$ and $(x, y) = (1, 0)$ into these partials and compute the resulting discriminant:

$$f_{xx}(-1, 0) = (1 + 3)e^{-\frac{1}{2} - 0^2} = 4e^{-\frac{1}{2}}$$

$$f_{xy}(-1, 0) = -(1 - 1)(0)e^{-\frac{1}{2} - 0} = 0$$

$$f_{yy}(-1, 0) = (0 + 1)e^{-\frac{1}{2}} = e^{-\frac{1}{2}}$$

$$D(-1, 0) = 4e^{-\frac{1}{2}} \cdot e^{-\frac{1}{2}} - 0^2 = 4e^{-1}$$

$$f_{xx}(1, 0) = (1 - 3)e^{-\frac{1}{2} - 0^2} = -2e^{-\frac{1}{2}}$$

$$f_{xy}(1, 0) = -(1 - 1)(0)e^{-\frac{1}{2} - 0} = 0$$

$$f_{yy}(1, 0) = (0 - 1)e^{-\frac{1}{2}} = -e^{-\frac{1}{2}}$$

$$D(1, 0) = (-2e^{-\frac{1}{2}})(-e^{-\frac{1}{2}}) - 0^2 = 2e^{-1}$$

Therefore, both $D(-1, 0)$ and $f_{xx}(-1, 0)$ are positive, which means that $f(x, y)$ has a local minimum of $z = f(-1, 0) = e^{-\frac{1}{2}}$ when $(x, y) = (-1, 0)$ according to the multivariable second derivative test. Similarly, $D(1, 0)$ is positive and $f_{xx}(1, 0)$ is negative, which means that $f(x, y)$ has a local maximum of $z = f(1, 0) = e^{\frac{1}{2}}$ when $(x, y) = (1, 0)$. ■

▶ **QUESTION 6** The multivariable function $f(x, y) = 2\cos\left(\frac{\pi}{2}x\right)\sin\left(\frac{\pi}{2}y\right)$ has infinitely many critical points.

(a) Verify that $(0, 1)$ and $(0, -1)$ are critical points of $f(x, y)$.

(b) Classify $(0, 1)$ as a local maximum, a local minimum, or a saddle point using the multivariable second derivative test, or state that the test is inconclusive.

(c) Repeat (b) for the critical point $(0, -1)$.

Global Extreme Values

This section concludes with a limited exploration of global extreme values. Returning to the real-life example of mountainous terrain, global extreme values are found among the peaks and the valleys of a mountain range. While a mountain range may contain many peaks, the global maximum value corresponds to the mountain peak that is the tallest (or peaks if more than one mountain has the exact same height), and the global minimum value corresponds to the lowest point in the deepest valley.

The following mathematical definition of global extreme values not only captures the intuitive idea of a global maximum and a global minimum for a domain consisting of all points in the plane, but also for settings over the smaller, more restricted input intervals that often arise in real-life settings. In such restricted cases, "global" refers to all inputs within a specified region $\mathcal{R}$ of interest.

Definition. Let $f(x, y)$ be a two-variable function, let $\mathcal{R}$ be a set of points from the domain of $f(x, y)$, and let (a, b) be a point in $\mathcal{R}$.

- The **global maximum** of $f(x, y)$ on $\mathcal{R}$ is the value $f(a, b)$ such that $f(a, b) \geq f(x, y)$ for all (x, y) in $\mathcal{R}$.

- The **global minimum** of $f(x, y)$ on $\mathcal{R}$ is the value $f(a, b)$ such that $f(a, b) \leq f(x, y)$ for all (x, y) in $\mathcal{R}$.

- The **global extreme values** of $f(x, y)$ on $\mathcal{R}$ are the global maximum and global minimum values of $f(x, y)$ on $\mathcal{R}$.

Similar to smooth single-variable functions, smooth multivariable functions also attain both a global maximum and global minimum value on every closed region $\mathcal{R}$. For single-variable functions, such closed regions consist of closed intervals $D = [a, b]$ on the x-axis. For two-variable functions, closed regions consist of a variety of diverse regions on the plane of real numbers. This book just considers closed regions $\mathcal{R}$ formed by circles and rectangles, where these regions consist of both the interiors of these geometric shapes and their edges. Interested readers are encouraged to consult the multivariable chapters of a standard calculus textbook for information about more general regions.

Recall from Section 5.1 that the global extreme values of a single-variable function on a closed interval D occur for either critical points in D or endpoints of D. For multivariable functions, global extreme values on a closed region $\mathcal{R}$ occur for critical points inside $\mathcal{R}$ or on the edge of $\mathcal{R}$. In essence, the edge of $\mathcal{R}$ is similar to the endpoints of an interval D in the single-variable case. As just mentioned, this book only considers regions $\mathcal{R}$ formed by circles and rectangles. For a circular region $\mathcal{R}$, the outside edge is a circle, and for a rectangular region $\mathcal{R}$, the outside edge consists of four line segments that create the sides of the rectangle.

The process of finding and classifying global extreme values of a smooth multivariable function parallels that for single-variable functions.

LOCATING MULTIVARIABLE GLOBAL EXTREME VALUES.

Let $f(x, y)$ be a smooth two-variable function and let $\mathcal{R}$ be either a closed circular region or a closed rectangular region contained in the domain of $f(x, y)$.

(1) Identify the critical points of $f(x, y)$ in the region $\mathcal{R}$.

> **LOCATING MULTIVARIABLE GLOBAL EXTREMA VALUES. (CONTINUED)**
>
> (2) Substitute the equation(s) of the edge(s) of $\mathcal{R}$ into $f(x, y)$ to create a function(s) of one variable $g(x)$ or $g(y)$. For each such function, find its critical numbers on the closed interval defining the edge and the corresponding coordinate. Also, identify the endpoints of each closed interval.
>
> (3) Evaluate $f(x, y)$ at the collection of points identified in steps (1) and (2).
>
> (4) The largest output from step (3) is the global maximum, and the smallest output from step (3) is the global minimum of $f(x, y)$ in $\mathcal{R}$.

◆ **EXAMPLE 10** Find the global maximum and minimum values of $f(x, y) = xy^2 + x^2$ on the closed circular region $\mathcal{R}$ bounded by $x^2 + y^2 = 4$.

Solution.

Step (1): The critical points of $f(x, y)$ are determined by the zeros of its first-order partials $f_x(x, y) = y^2 + 2x$ and $f_y(x, y) = 2xy$. Setting the product $f_y(x, y)$ equal to zero and solving yields either $x = 0$ or $y = 0$. Now consider when the sum $f_x(x, y) = y^2 + 2x$ is equal to zero in each of these two cases. Namely, if $x = 0$, then $f_x(0, y) = y^2 = 0$ when $y = 0$. Similarly, if $y = 0$, then $f_x(x, 0) = 2x = 0$ when $x = 0$. Therefore, the sole critical point of $f(x, y)$ is $(x, y) = (0, 0)$, which happens to be the center of the circular region $\mathcal{R}$ and so is in $\mathcal{R}$.

Step (2): The equation $x^2 + y^2 = 4$ is substituted into $f(x, y)$ to produce a single-variable equation $g(x)$ or $g(y)$ for the bounding edge. This example follows the approach of expressing the equation for the edge as $y^2 = 4 - x^2$, which on substitution into $f(x, y)$ produces $g(x) = x(4 - x^2) + x^2 = -x^3 + x^2 + 4x$. In addition, observe that the bounding circle is centered at the origin and has a radius of $\sqrt{4} = 2$, which means that x is from the closed interval $[-2, 2]$.

The critical numbers of $g(x)$ inside the interval $[-2, 2]$ are found by setting its derivative $f'(x) = -3x^2 + 2x + 4$ equal to zero and solving for x. The quadratic formula yields the following two zeros, and the two corresponding y-coordinates are found by substituting these values into $y^2 = 4 - x^2$:

$$x = \frac{-2 + \sqrt{52}}{-6} \approx -0.869 \qquad\qquad x = \frac{-2 - \sqrt{52}}{-6} \approx 1.535$$
$$y \approx \pm 1.806 \qquad\qquad\qquad\qquad y \approx \pm 1.282$$

The endpoints of this closed interval $[-2, 2]$ are $x = -2$ with $y = 0$ and $x = 2$ with $y = 0$, and provide two additional candidates $(-2, 0)$ and $(2, 0)$ for global extreme values.

Step (3): The following table presents all seven inputs where the global extreme values of $f(x, y)$ can occur and their corresponding function values:

(x, y)	$f(x, y)$
$(0, 0)$	0
$(-0.869, 1.806)$	-2.079
$(-0.869, -1.806)$	-2.079
$(1.535, 1.282)$	4.879
$(1.535, -1.282)$	4.879
$(-2, 0)$	4
$(2, 0)$	4

Step (4): The largest output value from the table provides the global maximum of $z \approx 4.879$ for inputs $(x, y) \approx (-0.869, 1.806)$ and $(x, y) \approx (-0.869, -1.806)$ for $f(x, y)$ on the closed circular region $\mathcal{R}$ bounded by $x^2 + y^2 = 4$. The smallest output value from the table provides the global minimum of $z \approx -2.079$ for inputs $(x, y) \approx (1.535, 1.282)$ and $(x, y) \approx (1.535, -1.282)$. ∎

◆ **EXAMPLE 11** Find the global extreme values of $f(x, y) = x^2 + y^2 - 2x + 2y + 5$ on the closed rectangular region $\mathcal{R}$ bounded by the lines $y = -2$, $y = 2$, $x = 0$, and $x = 3$.

Solution.

Step (1): The critical points of $f(x, y)$ are determined by the zeros of its first-order partials $f_x(x, y) = 2x - 2$ and $f_y(x, y) = 2y + 2$. Setting $f_x = 2x - 2$ equal to zero yields $x = 1$, and setting $f_y = 2y + 2$ equal to zero yields $y = -1$. Thus, the sole critical point of $f(x, y)$ is $(x, y) = (1, -1)$.

Step (2): Additional candidates for maximum and minimum values of $f(x, y)$ might be found on the edges of the region $\mathcal{R}$. Because $\mathcal{R}$ is rectangular, the equation of the line defining each side must be separately substituted into $f(x, y)$ and the corresponding critical points identified, along with the endpoints of the interval associated with each side of the rectangle.

- The bottom side is given by $y = -2$ with interval of x-values $[0, 3]$. Substituting $y = -2$ into $f(x, y)$ gives $g(x) = f(x, -2) = x^2 + 4 - 2x - 4 + 5 = x^2 - 2x + 5$. The critical numbers of $g(x)$ are found by setting $g'(x) = 2x - 2$ equal to zero, which yields $x = 1$. Thus, along this edge, the input candidates for global extreme values of $f(x, y)$ are the critical point $(1, -2)$ and the interval endpoints $(0, -2)$ and $(3, -2)$.

- The top side is given by $y = 2$ with interval of x-values $[0, 3]$. Substituting $y = 2$ into $f(x, y)$ gives $g(x) = f(x, 2) = x^2 + 4 - 2x + 4 + 5 = x^2 - 2x + 13$. The critical numbers of $g(x)$ are found by setting $g'(x) = 2x - 2$ equal to zero, which yields $x = 1$. Thus, along this edge, the input candidates for global extreme values of $f(x, y)$ are the critical point $(1, 2)$ and the interval endpoints $(0, 2)$ and $(3, 2)$.

- The left side is given by $x = 0$ with interval of y-values $[-2, 2]$. Substituting $x = 0$ into $f(x, y)$ gives $g(y) = f(0, y) = 0 + y^2 - 0 + 2y + 5 = y^2 + 2y + 5$. The critical numbers of $g(y)$ are found by setting $g'(y) = 2y + 2$ equal to zero, which yields $y = -1$. Thus, along this edge, the input candidates for global extreme values of $f(x, y)$ are the critical point $(0, -1)$ and the interval endpoints $(0, -2)$ and $(0, 2)$.

- The right side is given by $x = 3$ with interval of y-values $[-2, 2]$. Substituting $x = 3$ into $f(x, y)$ gives $g(y) = f(3, y) = 9 + y^2 - 6 + 2y + 5 = y^2 + 2y + 8$. The critical numbers of $g(y)$ are found by setting $g'(y) = 2y + 2$ equal to zero, which yields $y = -1$. Thus, along this edge, the input candidates for global extreme values of $f(x, y)$ are the critical point $(3, -1)$ and the interval endpoints $(3, -2)$ and $(3, 2)$.

Step (3): The following table presents all nine inputs where the global extreme values of $f(x, y)$ can occur and their corresponding function values:

(x, y)	$f(x, y)$	(x, y)	$f(x, y)$	(x, y)	$f(x, y)$
$(1, -1)$	3	$(3, 2)$	16	$(1, -2)$	4
$(1, 2)$	12	$(3, -1)$	7	$(0, -2)$	5
$(0, 2)$	13	$(3, -2)$	8	$(0, -1)$	4

Step (4): The largest output value from the table provides the global maximum of $z = 16$ when $(x, y) = (3, 2)$ for $f(x, y)$ on the closed rectangular region $\mathcal{R}$. The smallest output value from the table provides the global minimum of $z = 3$ when $(x, y) = (1, -1)$.

∎

▶ **QUESTION 7** Find the global extreme values of $f(x, y) = x^2 - y^2$ on the closed circular region $\mathcal{R}$ bounded by $x^2 + (y - 1)^2 = 1$.

Summary

- Let $f(x, y)$ be a two-variable function and let input (a, b) be in the domain of $f(x, y)$. A *local maximum* of $f(x, y)$ is a value $f(a, b)$ such that $f(a, b) \geq f(x, y)$ for all inputs (x, y) near (a, b). A *local minimum* of $f(x, y)$ is a value $f(a, b)$ such that $f(a, b) \leq f(x, y)$ for all inputs (x, y) near (a, b). The *local extreme values* of $f(x, y)$ are the local maximum and local minimum values of the function.

- A *critical point* of a two-variable function $f(x, y)$ is an input (a, b) in the domain of $f(x, y)$ such that both $f_x(a, b) = 0$ and $f_y(a, b) = 0$.

- A function $f(x, y)$ has a *saddle point* for input (a, b) if and only if $f(x, y)$ increases from $f(a, b)$ in some directions and decreases from $f(a, b)$ in others.

- *Multivariable second derivative test*: Let $f(x, y)$ be a smooth function, (a, b) be a critical point of $f(x, y)$, and $D(a, b) = f_{xx}(a, b) \cdot f_{yy}(a, b) - [f_{xy}(a, b)]^2$.

  ○ If $D(a, b) > 0$ and $f_{xx}(a, b) > 0$, then $f(a, b)$ is a local minimum of $f(x, y)$.

  ○ If $D(a, b) > 0$ and $f_{xx}(a, b) < 0$, then $f(a, b)$ is a local maximum of $f(x, y)$.

Summary (continued)

○ If $D(a, b) < 0$, then $f(x, y)$ has a saddle point for input (a, b).

○ If $D(a, b) = 0$, then the test is inconclusive and (a, b) may correspond to any or none of a local minimum, a local maximum, or a saddle point.

- Let $f(x, y)$ be a two-variable function, let $\mathcal{R}$ be a set of points from the domain of $f(x, y)$, and let (a, b) be a real number in $\mathcal{R}$. The *global maximum* of $f(x, y)$ on $\mathcal{R}$ is the value $f(a, b)$ such that $f(a, b) \geq f(x, y)$ for all all (x, y) in $\mathcal{R}$. The *global minimum* of $f(x, y)$ on $\mathcal{R}$ is the value $f(a, b)$ such that $f(a, b) \leq f(x, y)$ for all (x, y) in $\mathcal{R}$. The *global extreme values* of $f(x, y)$ on $\mathcal{R}$ are the global maximum and global minimum values of $f(x, y)$ on $\mathcal{R}$.

- *Locating multivariable global extreme values*: Let $f(x, y)$ be a smooth two-variable function and let $\mathcal{R}$ be either a closed circular region or a closed rectangular region contained in the domain of $f(x, y)$.

  (1) Identify the critical points of $f(x, y)$ in the region $\mathcal{R}$.

  (2) Substitute the equation(s) of the edge(s) of $\mathcal{R}$ into $f(x, y)$ to create a function(s) of one variable $g(x)$ or $g(y)$. For each such function, find its critical numbers on the closed interval defining the edge and the corresponding coordinates, and the endpoints of each interval.

  (3) Evaluate $f(x, y)$ at the collection of points identified in steps (1) and (2).

  (4) The largest output from step (3) is the global maximum and the smallest output from step (3) is the global minimum of $f(x, y)$ in $\mathcal{R}$.

Exercises

In Exercises 1–12, find the critical points of the function $f(x, y)$, if any.

1. $f(x, y) = 2x^2 - y^2$

2. $f(x, y) = xy - 8x - y^2 + 12y$

3. $f(x, y) = x^2 - 2xy + 2y$

4. $f(x, y) = x^2 + y^2$

5. $f(x, y) = x^2 + 2y^2 - x$

6. $f(x, y) = 2x^2 + 3y^2 - 6x - 2y$

7. $f(x, y) = x^3 + 3y^2 - 12x + 18y$

8. $f(x, y) = x^3 + y^2$

9. $f(x, y) = x^3 - 8x + 2y^2$

10. $f(x, y) = x^3 y - x$

11. $f(x, y) = xy^3$

12. $f(x, y) = xy^3 - y^2$

In Exercises 13–18, find the critical points of the function $f(x, y)$, if any.

13. $f(x, y) = e^{-x-y}$

14. $f(x, y) = xe^{-x-y}$

15. $f(x, y) = xye^{-x-y}$

16. $f(x, y) = 3xe^y - x^3 - e^{3y}$

17. $f(x, y) = e^{x^2 y^2 - 2x}$

18. $f(x, y) = e^{x^2 y^2 - 2y}$

In Exercises 19–28, verify that the input (x, y) is a critical point of the function.

19. $(0, 0)$; $f(x, y) = 2 - xy^2$

20. $(0, 0)$; $f(x, y) = x^2 - xy$

21. $(0, 0)$; $f(x, y) = (x^2 + y^2)e^{-x}$

22. $(2, 0)$; $f(x, y) = (x^2 + y^2)e^{-x}$

23. $(1, 0)$; $f(x, y) = x^2 + y^2 - 2x$

24. $(2, -1)$; $f(x, y) = x^2 + y^2 - 4x + 2y$

25. $(-3, 1)$; $f(x, y) = x^2 + 2y^2 + 6x - 4y$

26. $(0, 0)$; $f(x, y) = x^3 + y^3 - 12xy$

27. $(4, 4)$; $f(x, y) = x^3 + y^3 - 12xy$

28. $(0, 0)$; $f(x, y) = 2x^4 + y^4$

In Exercises 29–36, approximate the value and coordinates of the local extreme values of the function.

29.

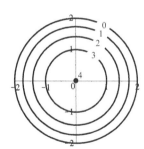

30.

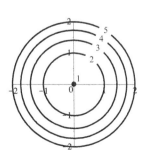

31.

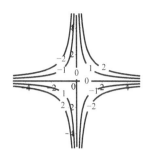

32.

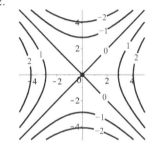

33.

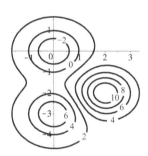

34.

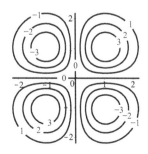

35.

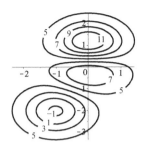

36.

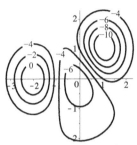

In Exercises 37–46, classify the critical point (x, y) as a local maximum, a local minimum, or a saddle point of the function from Exercises 19–28 using the multivariable second derivative test, or state that the test is inconclusive.

37. Exercise 19	42. Exercise 24
38. Exercise 20	43. Exercise 25
39. Exercise 21	44. Exercise 26
40. Exercise 22	45. Exercise 27
41. Exercise 23	46. Exercise 28

In Exercises 47–58, find and classify each critical point of the function $f(x, y)$ using the multivariable second derivative test, or state that the test is inconclusive.

47. $f(x, y) = x^2 + y^2$

48. $f(x, y) = x^2 - y^2$

49. $f(x, y) = x^2 - 2xy$

50. $f(x, y) = y^2 - xy$

51. $f(x, y) = x + y^2$

52. $f(x, y) = x + y$

53. $f(x, y) = x^2 - 6x - y^2 - 12y$

54. $f(x, y) = x^3 - y^3 + 3y$

55. $f(x, y) = x^3 + 3x$

56. $f(x, y) = xy^3 - y$

57. $f(x, y) = x^4 - 4xy - y^4$

58. $f(x, y) = x^5 - 5x + y^2$

In Exercises 59–64, find and classify each critical point of the function using the multivariable second derivative test, or state that the test is inconclusive.

59. $f(x, y) = e^{xy}$

60. $f(x, y) = xe^y$

61. $f(x, y) = (x + y)e^{xy}$

62. $f(x, y) = e^{x^2 y^2}$

63. $f(x, y) = e^{x^2 - y^2}$

64. $f(x, y) = e^{x^2 + y^2}$

In Exercises 65–68, use the multivariable second derivative test to show that the quantity is maximized.

65. A small drone has a remote control that is sensitive to both the temperature t (in degrees Celsius) and humidity h in the atmosphere. The range R in m over which the drone can be controlled from the ground is given by $R = 5000 - 2t^2 - 3ht - 1.5h^2 + 20t + 15h$. Find the atmospheric conditions that maximize the range R.

66. A cell phone manufactuerer produces two different kinds of cell phones. The daily profit, that is, the revenue minus the costs, is given by $P(x, y) = 500x + 400y - 0.04x^2 - 0.01y^2 - 0.01xy - 1,000,000$. Find the number of each type of cell

phone that should be produced to maximize profit and find this maximum profit.

67. A new shipping company has an exclusive deal with a massive online retailer to ship all of their small amd medium sized boxes. However, they can only handle boxes in which the sum of the length, width, and height is no more than 60 inches. The volume of such a box is given by $V = xy(60 - x - y)$. Find the dimensions of the box that maximize its volume.

68. Another new shipping company wants to seize the market for shipping larger boxes created by the company from Exercise 67. They are able to handle boxes in which the sum of the length, width, and height is no more than 120 inches. The volume of such a box is given by $V = xy(120 - x - y)$. Find the dimensions of the box that maximize its volume.

In Exercises 69–72, find the global extreme values of $f(x, y) = x^2 + y^2 - 2y$ on the closed region $\mathcal{R}$ bounded by the circle.

69. $x^2 + y^2 = 1$

70. $x^2 + y^2 = 4$

71. $x^2 + (y - 2)^2 = 9$

72. $(x - 2)^2 + (y - 1)^2 = 1$

In Exercises 73–76, find the global extreme values of $f(x, y) = x^2 + y^2 - 2x + 2$ on the closed region $\mathcal{R}$ bounded by the circle.

73. $x^2 + y^2 = 1$

74. $x^2 + y^2 = 4$

75. $x^2 + (y - 2)^2 = 4$

76. $(x - 1)^2 + (y - 1)^2 = 1$

In Exercises 77–80, find the global extreme values of $f(x, y) = 3x + y - 2xy$ on the closed rectangular region $\mathcal{R}$.

77. $0 \leq x \leq 1, 0 \leq y \leq 1$

78. $0 \leq x \leq 2, 0 \leq y \leq 1$

79. $0 \leq x \leq 1, 0 \leq y \leq 2$

80. $-1 \leq x \leq 1, 0 \leq y \leq 2$

In Exercises 81–84, find the global extreme values of $f(x, y) = x^3 - 12x + y^2$ on the closed rectangular region $\mathcal{R}$.

81. $0 \leq x \leq 1, 0 \leq y \leq 1$

82. $0 \leq x \leq 2, 0 \leq y \leq 1$

83. $-2 \leq x \leq 0, 0 \leq y \leq 2$

84. $-3 \leq x \leq 3, 0 \leq y \leq 1$

In Your Own Words. In Exercises 85–95, explain the following.

85. Local maximum

86. Local minimum

87. Local extreme value

88. Critical point

89. Saddle point

90. Multivariable second derivative test

91. Discriminant

92. Global maximum

93. Global minimum

94. Global extreme value

95. Locating multivariable global extreme values

5.6   Constrained Optimization

Section 5.5 explored the standard optimization questions for two-variable functions and developed methods for achieving goals such as

- finding critical points of multivariable functions,

- classifying critical points using the multivariable second derivative test, and

- determining global extreme values of multivariable functions on closed circular regions or closed rectangular regions.

This section shifts focus to considering *constrained optimization*, which seeks the extreme values of multivariable functions when the region $\mathcal{R}$ is restricted (or constrained) to a curve in the xy-plane. In more precise language, the local extreme values of a two-variable function $f(x, y)$ are sought subject to a constraint $g(x, y) = C$, which is a curve on the plane and corresponds to a single contour of the multivariable function $g(x, y)$.

   While such questions might appear esoteric at first blush, they arise in diverse real-life settings. As one example, consider a business that has inputs x and y (e.g., x might measure person hours and y might measure raw materials) and suppose they can produce a certain product as a function of these inputs as expressed by $f(x, y)$. However, often these inputs x and y are limited in some way (e.g., only so many person hours x are available or only so many raw materials y). A function $g(x, y)$ might express a budgetary relationship for the amount of money that can be spent on some combination of inputs x and y, where the total $g(x, y)$ must be no more than C. In other words, $g(x, y) \leq C$ because only a certain maximum amount of funds C are available.

   The goal in this optimizating context is to maximize the production of the product subject to this budgetary constraint. By identifying the best possible combination of inputs x and y, the company can thereby maximize their production of output $f(x, y)$. In addition, sometimes people are interested in knowing how much local extreme values might change with respect to the budget constraint value C changing. This section develops methods for addressing this question.

   As a prelude to this study, the gradient of a multivariable function is introduced as the most important tool for understanding and solving constrained optimization questions. The method of Lagrange multipliers is then developed as a solution method for identifying the extreme values of a function $f(x, y)$ subject to constraint $g(x, y) = C$.

The Gradient

Recall from Section 4.6 that partial derivatives provide the rate of change of multivariable functions in different directions. For a two-variable function $f(x, y)$, the partial derivatives $f_x(x, y)$ and $f_y(x, y)$ provide the rate of change of $f(x, y)$ in the x- and in the y-directions, respectively. The *gradient* is a vector-valued function that combines these first-order partials into a single vector that has these partials as its components.

> **Definition.** The **gradient** of a two-variable function $f(x, y)$ is a vector, denoted by $\nabla f(x, y)$, whose components are the partial derivatives of $f(x, y)$:
>
> $$\nabla f(x, y) \; = \; \begin{pmatrix} f_x(x, y) \\ f_y(x, y) \end{pmatrix}$$

As defined in Section 1.1, a vector-valued function has one or more inputs and two or more outputs. The gradient of $f(x, y)$ is a vector-valued function because each point (x, y), or pair of inputs, produces two outputs $f_x(x, y)$ and $f_y(x, y)$. Some examples of computing gradient vectors are presented, followed by a discussion of several important properties of the gradient.

◆ **EXAMPLE 1** Compute the gradient $\nabla f(x, y)$ of $f(x, y) = x^2 y + e^{xy}$.

Solution. The first-order partial derivatives of $f(x, y)$ are $f_x(x, y) = 2xy + ye^{xy}$ and $f_y(x, y) = x^2 + xe^{xy}$. Combining these first-order partials into a single vector produces the gradient of $f(x, y)$:

$$\nabla f(x, y) \; = \; \begin{pmatrix} 2xy + ye^{xy} \\ x^2 + xe^{xy} \end{pmatrix}$$

■

◆ **EXAMPLE 2** Laplace's equation is a partial differential equation $u_{xx} + u_{yy} = 0$ that models the steady-state heat distribution $u(x, y)$ of a surface over a region $\mathcal{R}$. For the square region $\mathcal{R} = \{(x, y) : 0 \le x \le 1, 0 \le y \le 1\}$, the solution of Laplace's equation that is equal to zero on the top, bottom, and left edges of the region and equal to $\sin(\pi y)$ on the right-hand edge of the region is given by the following function:

$$u(x, y) \; = \; \frac{e^{\pi x} - e^{-\pi x}}{e^{\pi} - e^{-\pi}} \sin(\pi y)$$

Compute the gradient $\nabla u(x, y)$ of $u(x, y)$.

Solution. First, compute the first-order partial derivatives of $u(x, y)$:

$$u_x(x, y) = \frac{\pi e^{\pi x} + \pi e^{-\pi x}}{e^{\pi} - e^{-\pi}} \sin(\pi y) = \pi \frac{e^{\pi x} + e^{-\pi x}}{e^{\pi} - e^{-\pi}} \sin(\pi y)$$

$$u_y(x, y) = \frac{e^{\pi x} - e^{-\pi x}}{e^{\pi} - e^{-\pi}} \cdot \pi \cos(\pi y) = \pi \frac{e^{\pi x} - e^{-\pi x}}{e^{\pi} - e^{-\pi}} \cos(\pi y)$$

Combining these first-order partials into a single vector produces the gradient of $u(x, y)$:

$$\nabla u(x, y) \; = \; \begin{pmatrix} \pi \dfrac{e^{\pi x} + e^{-\pi x}}{e^{\pi} - e^{-\pi}} \sin(\pi y) \\ \pi \dfrac{e^{\pi x} - e^{-\pi x}}{e^{\pi} - e^{-\pi}} \cos(\pi y) \end{pmatrix}$$

■

▶ **QUESTION 1** Compute the gradient of each function:

(a) $f(x,y) = x^2 + x\sin(y)$ (b) $g(x,y) = xe^{xy}$ (c) $h(x,y) = \ln(x+y)$

Properties of the Gradient

The gradient of a multivariable function has many interesting properties and is useful for the mathematical analysis of diverse scenarios. Three key properties of the gradient are needed to facilitate the discussion of constrained optimization and the method of Lagrange multipliers:

PROPERTIES OF THE GRADIENT. Let $f(x,y)$ be a smooth two-variable function, let the input (a,b) be a point on the plane, and let the vector $\nabla f(a,b)$ be graphed at (a,b) on the contour plot of $f(x,y)$.

  (1) Vector $\nabla f(a,b)$ points in the direction of greatest increase of $f(x,y)$ at (a,b), and $-\nabla f(a,b)$ points in the direction of greatest decrease of $f(x,y)$ at (a,b).

  (2) The length of $\nabla f(a,b)$ measures the steepness of $f(x,y)$ at (a,b).

  (3) At (a,b), $\nabla f(a,b)$ is perpendicular to the contour of $f(x,y)$ at level $f(a,b)$.

Figure 1 shows the contour plot of $f(x,y) = 4 - x^2 + y^2$ along with several gradient vectors to illustrate all three of these properties.

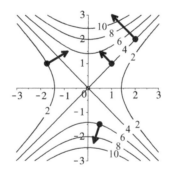

Figure 1: Contour plot of $f(x,y) = 4 - x^2 + y^2$ with $\nabla f(a,b)$ at several inputs

For property (1), $\nabla f(a,b)$ points in the direction of greatest increase at each input (a,b), although this direction is naturally quite different from point to point. While the function certainly increases in many other directions from each input (a,b), the gradient provides the direction in which the function increases the most. Similarly, the vector $-\nabla f(a,b)$ points in the direction of greatest decrease at each input (a,b). Although the vectors $-\nabla f(a,b)$ are not included in Figure 1, the reader is encouraged to picture them and consider this interpretation further. Namely, while the function certainly decreases in many other directions from each input (a,b), the negative of the gradient provides the direction in which the function decreases the most.

For property (2), the magnitude of $\nabla f(a, b)$ provides a measure of the steepness of $f(x, y)$ at each input (a, b). While both $\nabla f(1, 1)$ and $\nabla f(2, 2)$ point in the same direction in Figure 1, the magnitude of $\nabla f(2, 2)$ is greater than that of $\nabla f(1, 1)$, as indicated by $\nabla f(2, 2)$ being twice as long as $\nabla f(1, 1)$. This difference in magnitudes results from the function $f(x, y) = 4 - x^2 + y^2$ being relatively steeper at $(2, 2)$ and relatively flatter at $(1, 1)$.

Finally, Figure 1 also illustrates property (3): $\nabla f(a, b)$ is perpendicular to the contour of $f(x, y)$ level $f(a, b)$ for each input (a, b). Recall that two lines are identified as **perpendicular** when there is a $90°$ angle between them. In a setting involving curves, as in Figure 1, the tangent line to the curve at the point of interest is used for one of these two lines. From this perspective, the gradient $\nabla f(a, b)$ and the tangent line to the contour of $f(x, y)$ at level $f(a, b)$ are perpendicular at each point (a, b). While the gradient vector and tangent line change from point to point along the contour, they always remain perpendicular.

These three properties result from the study of *directional derivatives*, which provide a way to understand the rate of change of $f(x, y)$ in any direction (rather than just in the x- and y-directions). A more complete discussion of the many interesting properties of the gradient lies beyond the scope of this book. At the same time, while proofs of these properties are not completely discussed here, they are not beyond the scope of the reader, and several of them are explored in the exercises. For more details, the interested reader is encouraged to consult a standard multivariable calculus textbook.

Constrained Optimization

The goal of constrained optimization is to identify the local extreme values of a function $f(x, y)$ on some restricted (or constrained) domain identified using a curve $g(x, y) = C$. As a motivating example, consider the function $f(x, y) = 4 - x^2 + y^2$ with a restricted (or constrained) domain consisting of inputs on the curve $g(x, y) = x^2 + (y - 1)^2 = 1$ or, alternatively, the points on the plane lying on the circle of radius 1 centered at $(0, 1)$. Figure 2(a) shows a surface plot of $f(x, y) = 4 - x^2 + y^2$ illustrating this scenario, and its corresponding contour plot is shown in Figure 2(b).

(a)   (b)

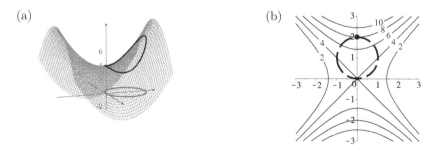

Figure 2: Function $f(x, y) = 4 - x^2 + y^2$ subject to the constraint $x^2 + (y - 1)^2 = 1$

On the surface plot of $f(x, y)$ in Figure 2(a), the constrained domain is presented as a gray circle on the xy-plane, and the solid curve on the surface plot provides the resulting curve in three-space for which the local extreme values are sought. The contour plot of $f(x, y)$ in Figure 2(b) parallels this presentation, with the dashed circle on the plane identifying the constrained domain $g(x, y) = x^2 + (y - 1)^2 = 1$.

Focusing on the global maximum value, the solid curve on the surface plot highlights the output values of $f(x, y) = 4 - x^2 + y^2$ for the constrained domain and makes it relatively easy to identify the global maximum value for this setting: the highest point on the solid curve in Figure 2(a). However, as with almost all such graphical investigations, the exact coordinates of the global maximum value are difficult to identify from the surface plot.

The contour plot of $f(x, y) = 4 - x^2 + y^2$ in Figure 2(b) makes identifying both the global maximum value and its input coordinates easier. The contours are shown as solid curves, with the domain constraint $x^2 + (y - 1)^2 = 1$ shown as a thick dashed circle (as on the surface plot). Considering only inputs on the constraint curve, the global maximum value occurs for input $(x, y) = (0, 2)$ on the contour for output $z = 8$.

While this discussion has focused on maximum values, optimization also often identifies minimum values. Again, the surface plot in Figure 2(a) does not provide definitive information. While the contour plot in Figure 2(b) is clearer, it too does not provide sufficient information to make an exact determination of the output value and the coordinates of the global minimum value of $f(x, y)$. Approximating from the contour plot, the gloal minimum value is between $z = 2$ and $z = 4$, because the dashed constraint curve passes between these two contours and the corresponding inputs appear to be $(x, y) \approx (0.75, 0.4)$ and $(x, y) \approx (-0.75, 0.4)$.

The precise mathematical definition of global extreme values subject to a constraint curve captures the intuitive ideas highlighted in this discussion. The following definition addresses the more general case of local extreme values:

Definition. Let $f(x, y)$ be a smooth two-variable function with constraint curve $g(x, y) = C$.

- $f(x, y)$ has a **local maximum** $f(a, b)$ **subject to the constraint** $g(x, y) = C$ if and only if $f(a, b) \geq f(x, y)$ for all points (x, y) near (a, b) and on the constraint curve $g(x, y) = C$.

- $f(x, y)$ has a **local minimum** $f(a, b)$ **subject to the constraint** $g(x, y) = C$ if and only if $f(a, b) \leq f(x, y)$ for all points (x, y) near (a, b) and on the constraint curve $g(x, y) = C$.

Examples involving the use of contour plots to locate extreme values subject to a constraint are presented, followed by the development of an analytical method for addressing these constrained optimization questions.

◆ **EXAMPLE 3** If possible, determine the local extreme values and their input coordinates for a function $f(x, y)$ subject to constraint $g(x, y) = C$ that is identified by the thick solid line on the contour plot of $f(x, y)$ in Figure 3.

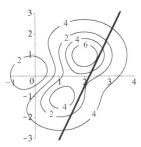

Figure 3: Contour plot of $f(x, y)$ and constraint $g(x, y) = C$ for Example 3

Solution. The extreme values of $f(x, y)$ subject to the constraint $g(x, y) = C$ correspond to the largest and smallest outputs of $f(x, y)$ resulting from inputs that lie on the thick solid line. The maximum value of $f(x, y)$ subject to the constraint is $z = 6$ when $(x, y) \approx (2.5, 1)$. The minimum value of $f(x, y)$ subject to the constraint is $z = 2$ and appears to occur when $(x, y) \approx (2, -0.5)$, $(1.4, -1.9)$, and $(2.8, 1.7)$. ∎

➤ **QUESTION 2** If possible, determine the local extreme values and their input coordinates for a function $f(x, y)$ subject to the constraint $g(x, y) = C$ identified by the thick solid line on the contour plot of $f(x, y)$ in Figure 4.

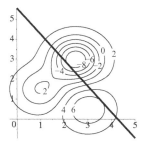

Figure 4: Contour plot of $f(x, y)$ and constraint $g(x, y) = C$ for Question 2

Understanding the Lagrange Multiplier

Recall the motivating example discussed at the beginning of this section. Suppose a company's output $f(x, y)$ of some product depends on inputs x and y, and that the constraint $g(x, y) = C$ expresses a budgetary restriction on how much money can be spent on some combination of inputs x and y. In such a context, the goal is to maximize the production $f(x, y)$ of the product subject to this budgetary constraint $g(x, y) = C$.

Beyond identifying such extreme values, businesses and economists also examine the question of how local extreme values change as the budgetrary constraint C is increased. In the study of economics, such changes are standardized to consider C increasing by one unit. This numeric value is approximately equal to the *Lagrange multiplier* and plays a prominent role in this book's approach to constrained optimization questions.

Definition. Let $f(x, y)$ be a smooth, two-variable function with a local extreme value subject to the constraint $g(x, y) = C$ for input (a, b).

- The **Lagrange multiplier**, denoted by λ, is the rate of change of the extreme values of $f(x, y)$ for (a, b) as the value of C increases.

- The value of λ is approximated by determining the change in the extreme values of $f(x, y)$ for (a, b) when C is increased by one unit.

◆ **EXAMPLE 4** The contour map of the function $f(x, y)$ given in Figure 5 presents constraint curves $g(x, y) = 1$ (thick solid line) and $g(x, y) = 2$ (thick dashed line).

(a) Approximate the local extreme values and their coordinates for $f(x, y)$ subject to the constraint $g(x, y) = 1$.

(b) Approximate the local extreme values and their coordinates for $f(x, y)$ subject to the constraint $g(x, y) = 2$.

(c) Approximate the value of the Lagrange multiplier λ for the minimum value of $f(x, y)$ on the constraint $g(x, y) = 1$ and explain the meaning of λ.

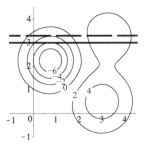

Figure 5: Contour plot of $f(x, y)$ and constraints $g(x, y) = C$ for Example 4

Solution.

(a) The maximum value of $f(x, y)$ subject to the constraint $g(x, y) = 1$ is between $z = 2$ and $z = 4$ when $(x, y) \approx (3.5, 3)$. The minimum value of $f(x, y)$ subject to the constraint $g(x, y) = 1$ is $z = -4$ when $(x, y) \approx (0.75, 3)$.

(b) The maximum value of $f(x, y)$ subject to the constraint $g(x, y) = 2$ is between $z = 2$ and $z = 4$ when $(x, y) \approx (3.5, 3.3)$. The minimum value of $f(x, y)$ subject to the constraint $g(x, y) = 2$ is $z = -2$ when $(x, y) \approx (0.75, 3.3)$.

(c) When the value of C increases from 1 to 2, the minimum value of $f(x, y)$ subject to the constraint $g(x, y) = C$ increases from $z = -4$ to $z = -2$, which means that $\lambda = (-2) - (-4) = 2$. This value of $\lambda = 2$ means that when C increases by one unit, the minimum values increase by two units when $(x, y) \approx (0.75, 3)$. ∎

In Example 4, the ambiguous maximum values of $f(x, y)$ when $C = 1$ and $C = 2$ mean that the value of the Lagrange multiplier λ for maxima cannot be determined in this context. The solution to Question 3 demonstrates that maximum and minimum values can have different Lagrange multipliers, which means that knowing λ for a minimum does not provide any information about λ for a maximum, and vice versa.

➤ **QUESTION 3** The contour map of the function $f(x, y)$ given in Figure 6 presents constraint curves $g(x, y) = 15$ (thick solid line) and $g(x, y) = 16$ (thick dashed line).

(a) Approximate the local extreme values and their coordinates for $f(x, y)$ subject to the constraint $g(x, y) = 15$.

(b) Approximate the local extreme values and their coordinates for $f(x, y)$ subject to the constraint $g(x, y) = 16$.

(c) Approximate the value of the Lagrange multiplier λ for both the maximum and minimum values of $f(x, y)$ on the constraint $g(x, y) = 15$ and explain the meaning of λ.

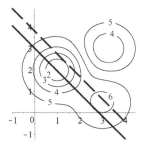

Figure 6: Contour plot of $f(x, y)$ and constraints $g(x, y) = C$ for Question 3

Method of Lagrange Multipliers

Contour plots enable the approximation of local extreme values subject to a constraint and illuminate the underlying ideas involved in constrained optimization. However, an analytical method must be used to find exact local extreme values and their coordinates. The method of Lagrange multipliers provides such an analytic approach to constrained optimization questions.

METHOD OF LAGRANGE MULTIPLIERS.
Let $f(x, y)$ and $g(x, y)$ be smooth two-variable functions. The local extreme values of $f(x, y)$ subject to the constraint $g(x, y) = C$ occur for inputs (x, y) that satisfy the equations $\nabla f(x, y) = \lambda \nabla g(x, y)$ and $g(x, y) = C$. Equivalently, such local extreme values occur for inputs (x, y) that satisfy the following system of equations:

$$f_x(x, y) = \lambda g_x(x, y) \qquad f_y(x, y) = \lambda g_y(x, y) \qquad g(x, y) = C$$

The vector equation $\nabla f(x,y) = \lambda \nabla g(x,y)$ yields the first two equations in this Lagrange multiplier system based on the fact that two vectors are equal exactly when their corresponding components are equal:

$$\nabla f = \lambda \nabla g \quad \Leftrightarrow \quad \begin{pmatrix} f_x \\ f_y \end{pmatrix} = \lambda \begin{pmatrix} g_x \\ g_y \end{pmatrix} = \begin{pmatrix} \lambda g_x \\ \lambda g_y \end{pmatrix} \quad \Leftrightarrow \quad \begin{array}{rcl} f_x(x,y) & = & \lambda g_x(x,y) \\ f_y(x,y) & = & \lambda g_y(x,y) \end{array}$$

Some examples utilizing the method of Lagrange multipliers are worked through to develop a facility with this approach, followed by a further discussion about why extreme values occur at the solutions to the Lagrange multiplier system of equations.

◆ **EXAMPLE 5** Let $f(x,y) = x^2 + y^2$ and $g(x,y) = x - y$. Find the local extreme values of $f(x,y)$ subject to the constraint $g(x,y) = 1$ using the method of Lagrange multipliers. Also, explain the meaning of λ in this context.

Solution. The first-order partial derivatives of $f(x,y) = x^2 + y^2$ are $f_x(x,y) = 2x$ and $f_y(x,y) = 2y$, and the first-order partials of $g(x,y) = x - y$ are $g_x(x,y) = 1$ and $g_y(x,y) = -1$, providing the following Lagrange multiplier system of equations:

$$2x = \lambda \qquad\qquad 2y = -\lambda \qquad\qquad x - y = 1$$

Solving the first two equations for x and y yields $x = \frac{1}{2}\lambda$ and $y = -\frac{1}{2}\lambda$. Substituting these expressions into the third equation and solving for λ produces the following:

$$x - y = \frac{1}{2}\lambda - \left(-\frac{1}{2}\lambda \right) = \left(\frac{1}{2} + \frac{1}{2} \right)\lambda = \lambda = 1$$

Substituting $\lambda = 1$ into the two equations for x and y yields their values:

$$x = \frac{1}{2}\lambda = \frac{1}{2} \cdot 1 = \frac{1}{2} \qquad\qquad y = -\frac{1}{2}\lambda = -\frac{1}{2} \cdot 1 = -\frac{1}{2}$$

Therefore, based on the method of Lagrange multipliers, the input $(x,y) = \left(\frac{1}{2}, -\frac{1}{2} \right)$ is the sole candidate for a local extreme value. Evaluating $f(x,y)$ for inputs immediately on either side of $\left(\frac{1}{2}, -\frac{1}{2} \right)$ that also lie on the constraint curve provides sufficient information to classify $f\left(\frac{1}{2}, -\frac{1}{2} \right)$ as a local maximum or local minimum. This example considers the x-intercept $(1,0)$ and y-intercept $(0,-1)$ of $g(x,y) = x - y = 1$, which are located on either side of $\left(\frac{1}{2}, -\frac{1}{2} \right)$:

$$f(1,0) = 1 \qquad\qquad f\left(\frac{1}{2}, -\frac{1}{2} \right) = \frac{1}{2} \qquad\qquad f(0,-1) = 1$$

Thus, the minimum value of $f(x,y) = x^2 + y^2$ subject to the constraint $x - y = 1$ is $z = f\left(\frac{1}{2}, -\frac{1}{2} \right) = \frac{1}{2}$ when $(x,y) = \left(\frac{1}{2}, -\frac{1}{2} \right)$.

The value $\lambda = 1$ means that the minimum value of $f(x,y)$ increases by approximately one unit if $C = 1$ is increased to $C = 2$. As just determined, the minimum value of $f(x,y) = x^2 + y^2$ subject to the constraint $g(x,y) = x - y = 1$ is $z = 0.5$. Applying this interpretation of the Lagrange multiplier, the minimum of $f(x,y) = x^2 + y^2$ subject to the constraint $g(x,y) = x - y = 2$ would be $z \approx 1.5$.   ■

In Example 5, $f(x, y) = x^2 + y^2$ increases when moving in either direction along the constraint $x - y = 1$ from $\left(\frac{1}{2}, -\frac{1}{2}\right)$, which means that the function $f(x, y) = x^2 + y^2$ has no maximum value on the constraint curve $g(x, y) = x - y = 1$. Keep in mind that this behavior is specific to this setting and that a different constraint curve could turn out to have local maximum values.

▶ **QUESTION 4** Let $f(x, y) = x^2 - y^2$ and $g(x, y) = 2x - y$. Find the maximum value of $f(x, y)$ subject to the constraint $g(x, y) = 6$ using the method of Lagrange multipliers. Also, explain the meaning of λ in this context.

Once the method of Lagrange multipliers has been learned, the consistently most challenging and most variable aspect of using this method is the relative difficulty of solving the resulting system of equations. Sometimes the Lagrange multiplier system of equations is relatively simple and straightforward, while at other times it is far more involved.

The next example illustrates such an increase in difficulty in the context of the graphical example used to introduce constrained optimization. While neither $f(x, y)$ nor $g(x, y)$ is an overly complicated looking function, solving the resulting Lagrange multiplier system of equations is more difficult than for the system in Example 5.

◆ **EXAMPLE 6** Let $f(x, y) = 4 - x^2 + y^2$ and $g(x, y) = x^2 + (y - 1)^2$. Find the local extreme values of $f(x, y)$ subject to the constraint $g(x, y) = 1$ using the method of Lagrange multipliers and explain the meaning of λ in this context.

Solution. The first-order partial derivatives of $f(x, y) = 4 - x^2 + y^2$ are $f_x(x, y) = -2x$ and $f_y(x, y) = 2y$, and the first-order partials of $g(x, y) = x^2 + (y - 1)^2$ are $g_x(x, y) = 2x$ and $g_y(x, y) = 2(y - 1)$, providing the following Lagrange multiplier system of equations:

$$-2x = 2x\lambda \qquad 2y = 2(y - 1)\lambda \qquad x^2 + (y - 1)^2 = 1$$

The first equation holds exactly when either $x = 0$ or $\lambda = -1$. Examining these possibilities one at a time provides the resulting solutions after further work with the other two equations.

> *Case (1) of $x = 0$:* Besides the first equation, x appears only in the third equation, and substituting $x = 0$ produces $(y - 1)^2 = 1$. Taking the square root of both sides gives $y - 1 = \pm 1$, and so $y = 0$ and $y = 2$. Substituting $y = 0$ into the second equation results in $\lambda = 0$. Substituting $y = 2$ into the second equation results in $4 = 2\lambda$, and so $\lambda = 2$. Thus, possible inputs producing extreme values of $f(x, y)$ subject to the constraint $g(x, y) = 1$ are $(x, y) = (0, 0)$ when $\lambda = 0$ and $(x, y) = (0, 2)$ when $\lambda = 2$.

> *Case (2) of $\lambda = -1$:* Besides the first equation, λ appears only in the second equation, and substituting $\lambda = -1$ produces $2y = -2(y - 1) = -2y + 2$. Solving for y yields $4y = 2$, and so $y = \frac{1}{2}$. Substituting $y = \frac{1}{2}$ into the third equation gives $x^2 + \left(\frac{1}{2} - 1\right)^2 = 1$, and so $x^2 + \frac{1}{4} = 1$. Solving for x yields $x^2 = \frac{3}{4}$ and taking the square root of both sides gives $x = \pm\frac{1}{2}\sqrt{3}$. Thus, possible inputs producing extreme values of $f(x, y)$ subject to the constraint $g(x, y) = 1$ are $(x, y) = \left(\frac{1}{2}\sqrt{3}, \frac{1}{2}\right)$ and $(x, y) = \left(-\frac{1}{2}\sqrt{3}, \frac{1}{2}\right)$, both when $\lambda = -1$.

The final step is to determine which of these points corresponds to the local maximum and minimum values of $f(x, y)$. Working in this direction, substitute the four possible candidates into $f(x, y)$ and evaluate:

$$f(0, 0) = 4 \qquad f(0, 2) = 8 \qquad f\left(\frac{\sqrt{3}}{2}, \frac{1}{2}\right) = 3.5 \qquad f\left(-\frac{\sqrt{3}}{2}, \frac{1}{2}\right) = 3.5$$

Observe that the largest output is $z = 8$ and the smallest output is 3.5. Therefore, $f(x, y) = 4 - x^2 + y^2$ subject to the constraint $g(x, y) = x^2 + (y - 1)^2 = 1$ has a local maximum value of $z = 8$ when $(x, y) = (0, 2)$ and a local minimum value of $z = 3.5$ when $(x, y) = \left(\frac{1}{2}\sqrt{3}, \frac{1}{2}\right)$ and $(x, y) = \left(-\frac{1}{2}\sqrt{3}, \frac{1}{2}\right)$.

The maximum and minimum values in this setting have two different values for λ. The local maximum value of $z = 8$ when $(x, y) = (0, 2)$ has a corresponding Lagrange multiplier of $\lambda = 2$. Thus, increasing the value of $C = 1$ by one unit to $C = 2$ would result in the maximum value of $z = 8$ increasing by two units to $z = 10$. The local minimum value of $z = 3.5$ when $(x, y) = \left(\pm\frac{1}{2}\sqrt{3}, \frac{1}{2}\right)$ has a corresponding Lagrange multiplier of $\lambda = -1$. Thus, increasing the value of $C = 1$ by one unit to $C = 2$ would result in the minimum value of $z = 3.5$ decreasing by one unit to $z = 2.5$. ∎

◆ **EXAMPLE 7** Suppose a new company decides to model their production using the Cobb–Douglas function $p(x, y) = 200x^{3/4}y^{1/4}$, where x represents units of labor and y represents units of material. Furthermore, each unit of labor costs \$150, each unit of material costs \$200, and the company has \$50,000 in startup funds available. Find the maximum number of items that they can produce at startup.

Solution. The function to be maximized is $p(x, y) = 200x^{3/4}y^{1/4}$. The constraint function results from the requirement that the costs of $150x + 200y$ must not exceed 50,000; that is, if the entire startup budget is spent, then $150x + 200y = 50,000$. In the symbolism of the method of Lagrange multipliers, the constraint is $g(x, y) = C$ with $g(x, y) = 150x + 200y$ and $C = 50,000$.

The first-order partials of $p(x, y) = 200x^{3/4}y^{1/4}$ are $p_x(x, y) = 150x^{-1/4}y^{1/4}$ and $p_y(x, y) = 50x^{3/4}y^{-3/4}$, and the first-order partials of $g(x, y) = 150x + 200y$ are $g_x(x, y) = 150$ and $g_y(x, y) = 200$, providing the following Lagrange multiplier system of equations:

$$150x^{-1/4}y^{1/4} = 150\lambda \qquad 50x^{3/4}y^{-3/4} = 200\lambda \qquad 150x + 200y = 50,000$$

Solving the first equation for λ yields $\lambda = x^{-1/4}y^{1/4}$, and solving the second equation for λ yields $\lambda = \frac{1}{4}x^{3/4}y^{-3/4}$. These two expressions are set equal to each other and simplified:

$$x^{-1/4}y^{1/4} = \frac{1}{4}x^{3/4}y^{-3/4} \qquad \text{Set } \lambda \text{ expressions equal}$$

$$\frac{y^{1/4}}{y^{-3/4}} = \frac{1}{4}\frac{x^{3/4}}{x^{-1/4}} \qquad \text{Divide by } x^{-1/4} \text{ and } y^{-3/4}$$

$$y = \frac{1}{4}x \qquad \text{Exponent rule } \frac{a^r}{a^s} = a^{r-s}$$

Next substitute this simplified expression into the third equation and solve for x:

$$150x + 200y = 50{,}000 \qquad \text{Third Lagrange multiplier equation}$$

$$150x + 200\left(\frac{1}{4}x\right) = 50{,}000 \qquad \text{Substitute } y = \frac{1}{4}x$$

$$150x + 50x = 50{,}000 \qquad \text{Simplify } 200 \cdot \frac{1}{4} = 50$$

$$200x = 50{,}000 \qquad \text{Combine like terms on left}$$

$$x = \frac{50{,}000}{200} = 250 \qquad \text{Divide by 200 and simplify}$$

Also, when $x = 250$, then $y = \frac{1}{4}(250) = 62.5$ and $\lambda = (250)^{-1/4}(62.5)^{1/4} \approx 0.707$. Therefore, based on the method of Lagrange multipliers, the input $(x, y) = (250, 62.5)$ when $\lambda \approx 0.707$ is the sole candidate for a local extreme value. Evaluating $p(x, y)$ for inputs immediately on either side of $(x, y) = (250, 62.5)$ that also lie on the constraint curve provides sufficient information to classify $p(250, 62.5)$ as a local maximum or local minimum. This example considers the x-intercept $(333.3, 0)$ and y-intercept $(0, 250)$ of $g(x, y) = 150x + 250y = 50{,}000$, which lie on either side of $(250, 62.5)$.

$$p(0, 250) = 0 \qquad p(250, 62.5) = 35{,}355.333 \qquad p(333.3, 0) = 0$$

Thus, the maximum value of $p(x, y) = 200x^{3/4}y^{1/4}$ subject to the constraint $g(x, y) = 150x + 250y = 50{,}000$ is $z = p(250, 62.5) = 35{,}355.333$ when $(x, y) = (250, 62.5)$.

The value $\lambda = 0.707$ means that the maximum value of $f(x, y)$ increases by approximately 0.707 units if C is increased by one unit. Interpreting this result in context, for every dollar the budget of \$50,000 is increased, the maximum number of items produced increases by approximately $\lambda = 0.707$ units; more specifically, a startup budget of \$50,001 increases production to 35,355.333+0.707 = 35,356.010 units. ∎

Understanding the Method of Lagrange Multipliers

Recall that the gradient $\nabla f(x, y)$ of a function $f(x, y)$ points in the direction of greatest increase of $f(x, y)$. In order to maximize $f(x, y)$ subject to the constraint $g(x, y) = C$, the ideal direction to travel is in the direction of its gradient. However, moving in the direction of the gradient $\nabla f(x, y)$ typically requires moving off the constraint curve $g(x, y) = C$. Therefore, some aspects of that directional information provided by $\nabla f(x, y)$ must be ignored in order to stay on the constraint curve $g(x, y) = C$.

The method of Lagrange multipliers provides a consistent, analytic method for resolving this tension and deciding how much directional information from the gradient $\nabla f(x, y)$ to preserve. A representative scenario is given in Figure 7, which presents the contour plot of a smooth two-variable function $f(x, y)$ along with a constraint curve $g(x, y) = C$ represented by the thick solid circle.

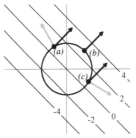

Figure 7: Contour plot of $f(x, y)$ and constraint $g(x, y) = C$

At points (a), (b), and (c), the gradient vector of $f(x, y)$ is indicated by the black arrow and the gradient of the constraint curve $g(x, y)$ by the gray arrow. At point (a), the gradient of $f(x, y)$ points up and to the right, while the gradient vector of $g(x, y)$ points up and to the left. This difference in directions indicates that the maximum output of $f(x, y)$ has not been attained at this input and that we need to move to a different point along the constraint curve $g(x, y) = C$ in an effort to find a candidate input (or inputs) for the maximum value of $f(x, y)$.

The gradient of $f(x, y)$ indicates that such a candidate lies up and to the right. However, the exact direction of $\nabla f(x, y)$ cannot be followed without moving off the constraint curve. Therefore, a compromise must be made. Namely, moving to the right along the constraint curve results in the output of $f(x, y)$ increasing and the two gradient vectors come more into alignment. At point (b), the gradients of $f(x, y)$ and $g(x, y)$ point in the same direction, thereby identifying a candidate input for the maximum value.

Similarly, note that at point (c), the two gradient vectors point in different directions: the gradient of $f(x, y)$ points up and to the right, while the gradient of $g(x, y)$ points down and to the left. Again, compromising in order to stay on the constraint curve, moving up and to the left along the constraint curve toward point (b) results in an increase in the output value of $f(x, y)$. And again, at point (b), the gradients of $f(x, y)$ and $g(x, y)$ point in the same direction, thereby identifying a candidate input for the maximum value.

These observations about this scenario based on Figure 7 hold in all such settings and are expressed by the Lagrange multiplier system of equations. The extreme values of $f(x, y)$ subject to the constraint $g(x, y) = C$ occur when the gradients of $f(x, y)$ and $g(x, y)$ are parallel and when the input is on the constraint curve $g(x, y) = C$. Recall from Section 3.1 that two nonzero vectors are **parallel** exactly when they are constant multiples of each other, which in this case happens when $\nabla f(x, y) = \lambda \nabla g(x, y)$.

Summary

- The *gradient* of a two-variable function $f(x, y)$ is a vector, denoted by $\nabla f(x, y)$, whose components are the partial derivatives of $f(x, y)$:

$$\nabla f(x, y) = \begin{pmatrix} f_x(x, y) \\ f_y(x, y) \end{pmatrix}$$

Summary (continued)

- *Properties of the gradient*: Let $f(x, y)$ be a smooth two-variable function, let the input (a, b) be a point on the plane, and let the vector $\nabla f(a, b)$ be graphed at (a, b) on the contour plot of $f(x, y)$.

  (1) The vector $\nabla f(a, b)$ points in the direction of greatest increase of $f(x, y)$ at (a, b) and $-\nabla f(a, b)$ points in the direction of greatest decrease of $f(x, y)$ at (a, b).

  (2) The length of $\nabla f(a, b)$ measures the steepness of $f(x, y)$ at (a, b).

  (3) At (a, b), $\nabla f(a, b)$ is perpendicular to the contour of $f(x, y)$ at level $f(a, b)$.

- Let $f(x, y)$ be a smooth two-variable function with constraint $g(x, y) = C$.

  ○ $f(x, y)$ has a *local maximum $f(a, b)$ subject to the constraint $g(x, y) = C$* if and only if $f(a, b) \geq f(x, y)$ for all points (x, y) near (a, b) and on the constraint curve $g(x, y) = C$.

  ○ $f(x, y)$ has a *local minimum $f(a, b)$ subject to the constraint $g(x, y) = C$* if and only if $f(a, b) \leq f(x, y)$ for all points (x, y) near (a, b) and on the constraint curve $g(x, y) = C$.

- Let $f(x, y)$ be a smooth, two-variable function with a local extreme value subject to the constraint $g(x, y) = C$ for input (a, b).

  ○ The *Lagrange multiplier*, denoted by λ, is the rate of change of the extreme values of $f(x, y)$ for (a, b) as the value of C increases.

  ○ The value of λ is approximated by determining the change in the extreme values of $f(x, y)$ for (a, b) when C is increased by one unit.

- *Method of Lagrange multipliers*: Let $f(x, y)$ and $g(x, y)$ be smooth two-variable functions. The local extreme values of $f(x, y)$ subject to the constraint $g(x, y) = C$ occur for inputs (x, y) that satisfy the equations $\nabla f(x, y) = \lambda \nabla g(x, y)$ and $g(x, y) = C$. Equivalently, such local extreme values occur for inputs (x, y) that satisfy the following system of equations:

$$f_x(x, y) = \lambda g_x(x, y) \qquad f_y(x, y) = \lambda g_y(x, y) \qquad g(x, y) = C$$

Exercises

In Exercises 1–14 compute the gradient of the function.

1. $f(x, y) = 2x + 7y$

2. $f(x, y) = 4x + 3y^2 - 2$

3. $f(x, y) = 5x^2 - y^5 + 8$

4. $f(x, y) = x^4 + y^3$

5. $f(x,y) = 3x^2 - 4y + 7$

6. $f(x,y) = x^3 + 3x^2$

7. $f(x,y) = \sin(x) + \cos(y)$

8. $f(x,y) = \cos(x+y)$

9. $f(x,y) = x\sin(y)$

10. $f(x,y) = \sin(xy) + 7y$

11. $f(x,y) = xe^y + y$

12. $f(x,y) = 12e^{xy} - 5x^2$

13. $f(x,y) = \ln(x+y) + xy$

14. $f(x,y) = 5\ln(x) - 2\ln(y)$

In Exercises 15–18, plot the gradient of $f(x,y) = x + y^2$ at the inputs.

15. $(1,0)$ and $(-1,0)$

16. $(0,1)$ and $(0,-1)$

17. $(1,1)$ and $(-1,-1)$

18. $(1,-1)$ and $(-1,1)$

In Exercises 19–22, plot the gradient of $f(x,y) = x^2 - y^3$ at the inputs.

19. $(1,0)$ and $(-1,0)$

20. $(0,1)$ and $(0,-1)$

21. $(1,1)$ and $(-1,-1)$

22. $(1,-1)$ and $(-1,1)$

In Exercises 23–30, for the contour plot, determine the local extreme values, if possible, subject to the constraint curve represented by the thick solid curve.

23.

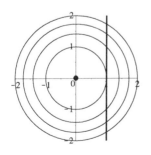

24.

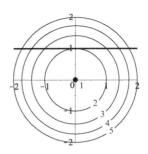

25.

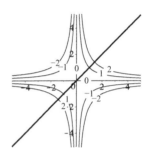

26.

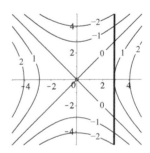

27.

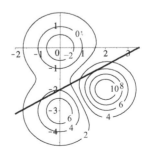

28.

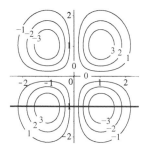

29.

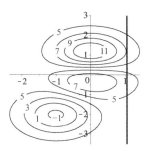

30.

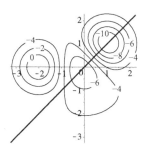

In Exercises 31–38, graphically estimate the value(s) of the Lagrange multiplier λ corresponding to the constraint
curves $g(x, y) = C$ (thick solid line) and
$g(x, y) = C + 1$ (thick dashed line).

31.

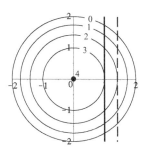

32.

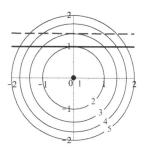

33.

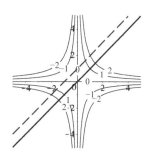

34.

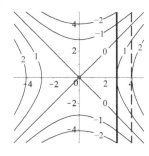

35.

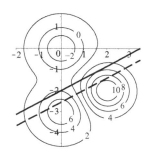

36.

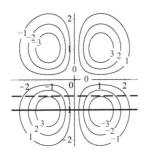

37.

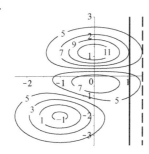

38.

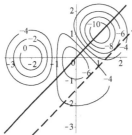

In Exercises 39–49, find the local extreme values of the function subject to the constraint using the method of Lagrange multipliers.

39. $f(x, y) = x^2 - y^2$
    constraint: $x - 2y = 0$

40. $f(x, y) = x^2 - y^2$
    constraint: $2x + y = 0$

41. $f(x, y) = 2x^2 - y$
    constraint: $x^2 + y^2 = 1$

42. $f(x, y) = 2x^2 - y$
    constraint: $x + y^2 = 1$

43. $f(x, y) = 2x^2 - y$
    constraint: $x^2 + y = 4$

44. $f(x, y) = x^2 + 2y^2 - x$
    constraint: $x^2 + y^2 = 1$

45. $f(x, y) = x^2 + 2y^2 - x$
    constraint: $x + 8y = 2$

46. $f(x, y) = x^2 + y^2$
    constraint: $x^2 - y = 1$

47. $f(x, y) = \frac{1}{3}x^3 + y$
    constraint: $x + y = 0$

48. $f(x, y) = 3x^2 - y^2$
    constraint: $x - y = 2$

49. $f(x, y) = 4x^2 + 2y^2$
    constraint: $x + y = 3$

In Exercises 50–57, find the local extreme values of the function subject to the constraint using the method of Lagrange multipliers.

50. $f(x, y) = 3x - y - 1$
    constraint $3x^2 + y^2 = 9$

51. $f(x, y) = 2x^2 + y^2 + 2$
    constraint $x^2 + 4y^2 = 4$

52. $f(x, y) = x^2 + y^2$
    constraint $2x^2 + 3xy + 2y^2 = 7$

53. $f(x, y) = xy$
    constraint $x^2 + y^2 - xy = 9$

54. $f(x, y) = x^{1/3}y^{2/3}$
    constraint $3x + 2y = 12$

55. $f(x, y) = 10x^{1/2}y^{1/2}$
    constraint $10x + 5y = 40$

56. $f(x, y) = 50x^{0.4}y^{0.6}$
    constraint $100x + 200y = 20000$

57. $f(x, y) = e^{2xy}$
    constraint: $x^2 + y^2 = 8$

In Exercises 58–67, explain the meaning of the Lagrange multiplier λ for the given constrained optimization scenario.

58. Maximum $f(3, 4) = 27$ with $\lambda = 4$ when $g(x, y) = 8$

59. Minimum $f(3, 4) = 15$ with $\lambda = 2$ when $g(x, y) = 8$

60. Maximum $f(-1, 1) = 4$ with $\lambda = -5$ when $g(x, y) = -6$

61. Minimum $f(-1, 1) = -7$ with $\lambda = 3$ when $g(x, y) = -6$

62. Maximum $f(8, 2) = 7$ with $\lambda = 9$ when $g(x, y) = 13$

63. Minimum $f(5, 8) = -11$ with $\lambda = -6$ when $g(x, y) = 13$

64. Maximum $f(2, 7) = 9$ with $\lambda = 3$ and minimum $f(1, 1) = 0$ with $\lambda = 2$ when $g(x, y) = 20$

65. Minimum $f(8, 9) = -13$ with $\lambda = 2$ and maximum $f(0, 0) = 3$ with $\lambda = -2$ when $g(x, y) = -10$

66. Maximum $f(7, 1) = 9$ with $\lambda = 8$ and minimum $f(4, 5) = 22$ with $\lambda = -1$ when $g(x, y) = -4$

67. Maximum $f(0, 1) = 16$ with $\lambda = 4$ and minimum $f(4, 2) = 5$ with $\lambda = 2$ when $g(x, y) = 15$

In Exercises 68–71 Let $f(x, y) = x + 2y$ and $g(x, y) = 5x^2 + y^2$.

68. Find the extreme values of $f(x, y)$ subject to the constraint $g(x, y) = 9$.

69. Find the extreme values of $f(x, y)$ subject to the constraint $g(x, y) = 10$.

70. Estimate the value of λ using the results of Exercises 68 and 69.

71. Compare the value of λ from Exercise 68 and your estimate of λ from Exercises 70.

In Exercises 72–75 Let $f(x, y) = x^2y$ and $g(x, y) = x^2 + y^2$.

72. Find the extreme values of $f(x, y)$ subject to the constraint $g(x, y) = 1$.

73. Find the extreme values of $f(x, y)$ subject to the constraint $g(x, y) = 2$.

74. Estimate the value of λ using the results of Exercises 72 and 73.

75. Compare the value of λ from Exercise 72 and your estimate of λ from Exercises 74.

The derivative of $f(x, y)$ at (a, b) in the direction of the vector of length one $\overline{u}$ is given by

$$D_{\overline{u}}f(a, b) = \nabla f(a, b) \cdot \overline{u}$$

and is called the directional derivative.
In Exercises 76–85, calculate the directional derivative of the given function at the point $(2, 1)$ in the given direction.

76. $f(x, y) = x^2y$; $\overline{u} = \begin{pmatrix} 1 \\ 0 \end{pmatrix}$

77. $f(x, y) = x^2y$; $\overline{u} = \begin{pmatrix} 1/\sqrt{2} \\ 1/\sqrt{2} \end{pmatrix}$

78. $f(x, y) = x^2 + 2xy + 4$; $\overline{u} = \begin{pmatrix} 0 \\ 1 \end{pmatrix}$

79. $f(x, y) = x^3y - 5xy$; $\overline{u} = \begin{pmatrix} -1/\sqrt{2} \\ 1\sqrt{2} \end{pmatrix}$

80. $f(x, y) = 3x^2 + xy$; $\overline{u} = \begin{pmatrix} \sqrt{3}/2 \\ 1/2 \end{pmatrix}$

81. $f(x, y) = xy^2 + 4x$; $\overline{u} = \begin{pmatrix} 5/13 \\ 12/13 \end{pmatrix}$

82. $f(x, y) = \sin(xy)$; $\overline{u} = \begin{pmatrix} 3/5 \\ 4/5 \end{pmatrix}$

83. $f(x, y) = x\cos(y)$; $\overline{u} = \begin{pmatrix} -1/2 \\ -\sqrt{3}/2 \end{pmatrix}$

84. $f(x, y) = e^{xy+1}$; $\overline{u} = \begin{pmatrix} -12/13 \\ -5/13 \end{pmatrix}$

85. $f(x, y) = xe^y$; $\overline{u} = \begin{pmatrix} -3/5 \\ 4/5 \end{pmatrix}$

In Exercises 86–89, let $f(x, y)$ be a smooth function of two variables.

86. One property of the dot product is that is can be written as $\overline{u} \cdot \overline{v} = \|\overline{u}\| \cdot \|\overline{v}\| \cos(\theta)$, where θ is the angle between $\overline{u}$ and $\overline{v}$. Rewrite the directional derivative of f in the direction of a unit vector $\overline{u}$ using this property of dot products.

87. Using the fact that $-1 \le \cos(\theta) \le 1$ and the result of Exercise 86, determine the largest value of $D_{\overline{u}}f$ and identify the angle θ that corresponds to this largest value.

88. Using the angle from Exercise 87, determine the direction $\overline{u}$ that corresponds to this maximum value.

89. Based on your answer to Exercise 88, what vector points in the direction of greatest increase of the function $f(x, y)$?

In Your Own Words. In Exercises 90–97, explain the following.

90. Gradient

91. Three properties of the gradient

92. Local maximum

93. Local minimum

94. Local extreme value

95. Finding Lagrange multiplier λ

96. Interpreting Lagrange multiplier λ

97. Method of Lagrange multipliers

Chapter 6

Accumulation and Integration

Chapters 4 and 5 focused on the derivative and questions about functional behavior that the derivative can address: monotonicity, concavity, finding and classifying extreme values as maxima or minima, both globally and locally, and for single-variable and multivariable functions. A diverse array of abstract and real-life functional behaviors can be studied, and with relative ease because of the computational tools developed for the derivative. And, as it turns out, the derivative is just one of the two major tools of calculus.

This chapter studies the second major tool of calculus: the definite integral. Rather than measuring rates of change, the integral provides a means for measuring the accumulation of a quantity over some interval of input values. Viewed from the right perspective, many different quantities can be thought of as accumulating, including money, populations, weight, area, volume, downloaded files, and more. In this way, the integral is a fundamental tool for analyzing diverse physical, social, and economic phenomenon, and helps enrich our understanding of our world.

As with derivatives, several important methods for calculating accumulation are developed. This study of the integral leads to two incredibly deep and significant results: the first and second fundamental theorems of calculus, which characterize the inverse relationship between the operations of differentiation and integration. This inverse relationship is one of the key factors for the success of calculus, enabling many of the successes of the Scientific Revolution, the Industrial Revolution, and our modern way of life. In this sense, calculus ranks "among our species' deepest, richest, farthest-reaching, and most beautiful intellectual achievements."

6.1   Accumulation

Recall that the derivative provides information about the rate at which the output values of a function change as the input values change. This section introduces a tool to measure the *accumulation* resulting from a functional rate of change. Namely, as a quantity changes over time, then the quantity accumulates as time passes. Various mathematical tools enable the approximation of this accumulation when the rate of change of the quantity is known. The following example illustrates this notion of accumulating a quantity in a familiar setting:

◆ **EXAMPLE 1** Suppose someone drives on the interstate with their cruise control set at exactly 73 miles per hour. How far does this person travel in 2 hours?

Solution. Because the person is driving at 73 miles per hour for 2 hours, a total of $73 \times 2 = 146$ miles is traveled during that time period. In other words, 146 miles have been accumulated in 2 hours. Figure 1 gives a plot of the velocity as a function of time. Notice that the total miles traveled is the product of the height of the velocity function and the width of the time interval.   ■

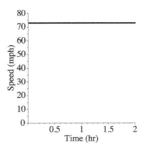

Figure 1: Graph of velocity versus time for Example 1

This scenario of traveling a given distance over a given time when the velocity is known provides an important illustrative example of accumulation. While working in such relatively simple settings or in more abstract settings, the reader is encouraged to keep in mind the diverse scenarios in which quantities accumulate, including those mentioned in the chapter introduction as well as appearing in the exercises.

Left and Right Approximations

In Example 1, the relative ease of computing the distance traveled by the car results from the fact that the velocity of the car is constant over the 2 hours. Even more, the observation that the distance traveled is a product of the height of the function and the width of the time interval provides the key insight for analyzing situations in which the rate of change varies from input to input. For example, the graph given in Figure 2 presents the velocity of a runner during the first 10 seconds of a sprint. Because the runner's velocity is not constant over the 10 second time interval, the distance accumulated cannot simply be calculated by multiplying the velocity of the runner by the time traveled. However, modifying this approach slightly enables us to find an approximation of the total distance traveled in this more nuanced setting.

Working with the graph of the velocity of the runner given in Figure 2, suppose that the total time interval of 10 seconds is split up into five equal-width subintervals that are each 2 seconds long. On each of these 2-second subintervals, the velocity does not change as much as it does over the span of the entire 10 second interval, and the velocity is therefore closer to being constant. Therefore, the total distance traveled can be approximated by multiplying the "constant" velocity on each subinterval by the length of the subinterval and then summing the resulting products together.

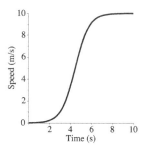

Figure 2: Velocity of a runner

The issue with this approach is that the runner's velocity is not actually constant on each subinterval of time. Consequently, some choice must be made when identifying the velocity on each subinterval for this calculation. Several different approaches can reasonably be followed to identify these velocities (or, more generally, the output values of the rate function) on each subinterval. The following definition articulates these ideas for two particular choices:

Definition.   Let $f(x)$ be the rate of change of a quantity $F(x)$ on an interval $[a, b]$, let $a = x_1 < x_2 < x_3 < \cdots < x_{n+1} = b$, let $[a, b]$ be subdivided into smaller subintervals $[x_1, x_2]$, $[x_2, x_3]$, ..., $[x_n, x_{n+1}]$, and let $\Delta x_k = x_{k+1} - x_k$ be the width of each subinterval.

- The **left approximation** to the accumulation of $F(x)$ on $[a, b]$ is

$$L_n = f(x_1)\Delta x_1 + f(x_2)\Delta x_2 + \cdots + f(x_n)\Delta x_n.$$

- The **right approximation** to the accumulation of $F(x)$ on $[a, b]$ is

$$R_n = f(x_2)\Delta x_1 + f(x_3)\Delta x_2 + \cdots + f(x_{n+1})\Delta x_n.$$

When possible, the subintervals $[x_k, x_{k+1}]$ are chosen with equal width, in which case $\Delta x = \dfrac{b - a}{n}$ for all n subintervals.

The only distinction between left and right approximations is whether the left or the right endpoint of each subinterval is used to determine the constant output value of the rate function over each subinterval. The products of these heights with the widths of the subintervals are then summed to obtain an approximation of the accumulation. The following examples enable the reader to develop a facility with calculating these left and right approximations:

◆ **EXAMPLE 2** Consider the graph given in Figure 2 of the velocity of a runner during the first 10 seconds of a sprint.

(a) Sketch the approximations L_5 and R_5 on $[0, 10]$ using equal-width subintervals.

(b) Calculate L_5 and R_5 on $[0, 10]$ using equal-width subintervals.

(c) Compare the approximations L_5 and R_5 with the actual distance traveled.

Solution.

(a) A sketch of L_5 is shown in Figure 3(a) and a sketch of R_5 in Figure 3(b).

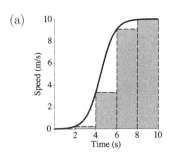 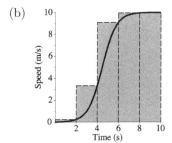

Figure 3: Graph of L_5 and R_5 for Examples 2 and 5

In more detail, the subscript "5" in L_5 and R_5 indicates that the interval $[0, 10]$ should be divided into five equal-width subintervals, which requires the following uniform width for all five subintervals:

$$\Delta x = \frac{b - a}{n} = \frac{10 - 0}{5} = \frac{10}{5} = 2$$

The resulting subintervals are $[0, 2]$, $[2, 4]$, $[4, 6]$, $[6, 8]$, and $[8, 10]$. For L_5, the heights are sketched by drawing a horizontal line from the height of the function at the left side of each subinterval across the entire subinterval. For example, on the subinterval $[4, 6]$, the horizontal line for the height is drawn from $f(4) \approx 3.25$ across all of $[4, 6]$. Similarly, for R_5, the heights are sketched by drawing a horizontal line from the height of the function at the right side of each subinterval across the entire subinterval. For example, on the subinterval $[4, 6]$, the horizontal line for the height is drawn from $f(6) = 9$ back across all of $[4, 6]$.

(b) The approximations L_5 and R_5 are calculated based on approximate output values of the rate function at the endpoints of each subinterval $[0, 2]$, $[2, 4]$, $[4, 6]$, $[6, 8]$, and $[8, 10]$. The graph of the velocity function given in Figure 2 (and in Figure 3) suggests the following output values:

$$f(0) = 0 \quad f(2) \approx 0.25 \quad f(4) \approx 3.25 \quad f(6) = 9 \quad f(8) \approx 9.9 \quad f(10) = 10$$

As calculated in part (a), every subinterval has width $\Delta x = 2$. Multiplying each function height by the width of its subinterval and adding the resulting products together gives the following:

$$L_5 = f(0)\Delta x_1 + f(2)\Delta x_2 + f(4)\Delta x_3 + f(6)\Delta x_4 + f(8)\Delta x_5$$
$$\approx 0 \cdot 2 + 0.25 \cdot 2 + 3.25 \cdot 2 + 9 \cdot 2 + 9.9 \cdot 2$$
$$= 0 + 0.5 + 6.5 + 18 + 19.8$$
$$= 44.8$$

$$R_5 = f(2)\Delta x_1 + f(4)\Delta x_2 + f(6)\Delta x_3 + f(8)\Delta x_4 + f(10)\Delta x_5$$
$$\approx 0.25 \cdot 2 + 3.25 \cdot 2 + 9 \cdot 2 + 9.9 \cdot 2 + 10 \cdot 2$$
$$= 0.5 + 6.5 + 18 + 19.8 + 20$$
$$= 64.8$$

Interpreting these results in context, the left approximation $L_5 = 44.8$ indicates that the runner traveled approximately 44.8 meters during the first 10 seconds of the sprint, and the right approximation of $R_5 = 64.8$ indicates an approximate traveled distance of 64.8 meters.

(c) The sketch of L_5 in Figure 3(a) indicates that L_5 is less than the total distance accumulated by the runner over the 10-second period, while Figure 3(b) indicates that R_5 is greater than the total distance accumulated by the runner over this time period. Combining this graphical analysis with the information from part (b) that $L_5 = 44.8$ and $R_5 = 64.8$ leads to the conclusion that the runner traveled between 44.8 and 64.8 meters during the first 10 seconds of the sprint ∎

➤ **QUESTION 1** Consider the graph given in Figure 4 of the growth rate of the mass of a male fetus during weeks 23 – 42 of pregnancy.

(a) Sketch the approximations L_4 and R_4 on $[24, 40]$ using equal-width subintervals.

(b) Calculate L_4 and R_4 on $[24, 40]$ using equal-width subintervals.

(c) Compare the approximations L_4 and R_4 with the actual growth of a male fetus.

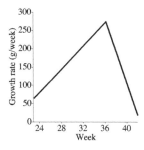

Figure 4: Graph of the growth rate of a male fetus for Question 1

Both Example 2 and Question 1 approximate accumulated quantities based on graphs of functional rates of change using equal-width subintervals. From a computational prespective, working with a uniform subinterval width Δx is almost always easier. However, the definition of left and right approximations does not mandate that the subintervals have equal width. In particular, when working with real-life data sets, the input values sometimes result in equal-width subintervals, but far more often do not. From this perspective, the flexibility to work with subintervals that are not equally wide is actually quite essential.

◆ **EXAMPLE 3** The following table presents the rate of change in the percent of high school graduates enrolled in a two-year or four-year college:

Year	2005	2006	2009	2010
Rate	−2.6	1.4	−2.0	−1.0

Calculate each approximation of the total change in the percent of high-school graduates enrolled in a two-year or four-year college from 2005 to 2010.

(a) L_3 (b) R_3

Solution.

(a) For computing L_3, the interval $[2005, 2010]$ must be divided into three subintervals. Even though the data points are not equally spaced, exactly three subintervals are formed by the given inputs: $[2005, 2006]$, $[2006, 2009]$, and $[2009, 2010]$. Thus, $\Delta x_1 = 2006 - 2005 = 1$, $\Delta x_2 = 2009 - 2006 = 3$, and $\Delta x_3 = 2010 - 2009 = 1$. Multiplying the height of the rate function at the left endpoint of each subinterval by the width of the corresponding subinterval and then adding these products together gives

$$L_3 = f(2005)\Delta x_1 + f(2006)\Delta x_2 + f(2009)\Delta x_3$$
$$= (-2.6) \cdot 1 + 1.4 \cdot 3 + (-2.0) \cdot 1$$
$$= -2.6 + 4.2 - 2.0 = -0.4$$

Because $L_3 = -0.4$, the left approximation estimates a change of -0.4% in the number of high-school graduates enrolled in a two-year or four-year college from 2005 to 2010. In other words, the percentage of high-school graduates enrolling in these forms of higher-education decreased by 0.4% from 2005 to 2010. This left approximation L_3 is depicted graphically in Figure 5(a). Notice that the rightmost data point $(2010, -1)$ is not used when calculating this left approximation.

(b) As when calculating L_3, the entire interval $[2005, 2010]$ is subdivided into three subintervals $[2005, 2006]$, $[2006, 2009]$, and $[2009, 2010]$, with the same resulting widths of $\Delta x_1 = 2006 - 3005 = 1$, $\Delta x_2 = 2009 - 2006 = 3$, and $\Delta x_3 = 2010 - 2009 = 1$. Multiplying the height of the rate function at the right endpoint of each subinterval by the width of the corresponding subinterval and then adding these products together gives

$$R_3 = f(2006)\Delta x_1 + f(2009)\Delta x_2 + f(2010)\Delta x_3$$
$$= 1.4 \cdot 1 + (-2.0) \cdot 3 + (-1.0) \cdot 1$$
$$= 1.4 - 6.0 - 1.0 = -5.6$$

Because $R_3 = -5.6$, the right approximation estimates that the percent of high-school graduates enrolled in a two-year or four-year college from 2005 to 2010 decreased by 5.6%. This right approximation R_3 is depicted graphically in Figure 5(b). Notice that the leftmost data point $(2010, -1)$ is not used when calculating this right approximation.

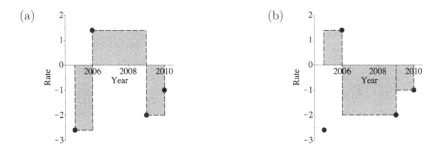

Figure 5: Graphs of L_3 and R_3 for Example 3

In Example 3, the data set has only four data points, which means that the entire input interval $[2005, 2010]$ can be divided into three subintervals in exactly one way. In settings where subintervals can be chosen in multiple ways, the most common approach is to work with subintervals that are as close as possible to having equal widths.

➤ **QUESTION 2** The following table presents the growth rate of the world population in billions of people per year as a function of the year:

Year	1995	2000	2005	2010	2015
Rate	1.41	1.26	1.20	1.13	1.06

Calculate each approximation of the increase in world population from 1995 to 2015:

(a) L_2 (b) R_4

The previous examples and questions explored left and right approximations of accumulation in the context of rate functions presented either by a graph or in a table. The process of calculating L_n or R_n is the same when a rate function is presented analytically, only now the rate function $f(x)$ must be computed at the appropriate endpoints x_1, x_2, $x_3, \ldots$ as part of evaluating the approximation formulas. The next example and question present calculations of these approximations for such analytic settings.

◆ **EXAMPLE 4** Let $f(x) = x^3 - 1$ be the rate of change of a quantity $F(x)$. Calculate each approximation of the accumulation of $F(x)$ on the interval $[0, 3]$ using equal-width subintervals.

(a) L_3 (b) R_6

Solution.

(a) Dividing the interval $[0, 3]$ into three equal-width subintervals requires the following uniform width for all three subintervals:

$$\Delta x = \frac{b - a}{n} = \frac{3 - 0}{3} = \frac{3}{3} = 1$$

The resulting subintervals are $[0, 1]$, $[1, 2]$, and $[2, 3]$. The left endpoints of these subintervals are $x = 0$, $x = 1$, and $x = 2$, and substituting these endpoints along with the uniform subinterval width $\Delta x = 1$ into the left approximation formula yields

$$L_3 = f(0)\Delta x + f(1)\Delta x + f(2)\Delta x$$
$$= (0^3 - 1) \cdot 1 + (1^3 - 1) \cdot 1 + (2^3 - 1) \cdot 1$$
$$= (-1) \cdot 1 + 0 \cdot 1 + 7 \cdot 1 = 6$$

This left approximation $L_3 = 6$ of the accumulation of $F(x)$ on $[0, 3]$ is depicted graphically in Figure 6(a).

(a) 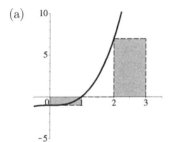   (b)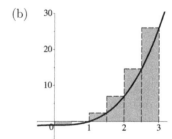

Figure 6: Graphs of L_3 and R_6 for Example 4

(b) Dividing the interval $[0, 3]$ into six equal-width subintervals requires the following uniform width for all six subintervals:

$$\Delta x = \frac{b - a}{n} = \frac{3 - 0}{6} = \frac{3}{6} = 0.5$$

The resulting subintervals are $[0, 0.5]$, $[0.5, 1]$, $[1, 1.5]$, $[1.5, 2]$, $[2, 2.5]$, and $[2.5, 3]$. The right endpoints of these subintervals are $x = 0.5$, $x = 1$, $x = 1.5$, $x = 2$, $x = 2.5$, and $x = 3$, and substituting these endpoints along with the uniform subinterval width $\Delta x = 0.5$ into the right approximation formula yields

$$R_6 = f(0.5)\Delta x + f(1)\Delta x + f(1.5)\Delta x + f(2)\Delta x + f(2.5)\Delta x + f(3)\Delta x$$
$$= (0.5^3 - 1) \cdot 0.5 + (1^3 - 1) \cdot 0.5 + (1.5^3 - 1) \cdot 0.5 + (2^3 - 1) \cdot 0.5$$
$$\quad + (2.5^3 - 1) \cdot 0.5 + (3^3 - 1) \cdot 0.5$$
$$= (-0.875) \cdot 0.5 + 0 \cdot 0.5 + 2.375 \cdot 0.5 + 7 \cdot 0.5 + 14.625 \cdot 0.5 + 26 \cdot 0.5$$
$$= 24.5625$$

This right approximation $R_6 = 24.5625$ of the accumulation of $F(x)$ on $[0,3]$ is depicted graphically in Figure 6(b).

■

▶ **QUESTION 3** Let $f(x) = 3x^2 + 5$ be the rate of change of a quantity $F(x)$. Calculate each approximation of the accumulation of $F(x)$ on $[0,4]$ using equal-width subintervals.

(a) L_2 (b) R_4

Midpoint Rule

The left approximation L_n and the right approximation R_n provide two different approaches to approximating the accumulation of $F(x)$ given its rate of change $f(x)$. However, other choices can be made that produce different approximations of this accumulation.

A consistent pattern in all such approximations is that they are sums of products of heights and widths. The most common modification is to determine the height in some other fashion than using the left and right endpoints of the subintervals (as when calculating L_n and R_n). The next most common choice is to use the midpoint of each subinterval to determine the height, and so calculate what is known as the midpoint approximation.

MIDPOINT APPROXIMATION.
Let $f(x)$ be the rate of change of a quantity $F(x)$ on an interval $[a, b]$, let $a = x_1 < x_2 < x_3 < \cdots < x_{n+1} = b$, let $[a, b]$ be subdivided into smaller, equal-width subintervals $[x_1, x_2]$, $[x_2, x_3]$, ..., $[x_n, x_{n+1}]$, and let $\Delta x = (b - a)/n$ be the uniform width of each subinterval. The **midpoint approximation** to the accumulation of $F(x)$ on the interval $[a, b]$ is

$$M_n = f\left(\frac{x_1 + x_2}{2}\right)\Delta x + f\left(\frac{x_2 + x_3}{2}\right)\Delta x + \cdots + f\left(\frac{x_n + x_{n+1}}{2}\right)\Delta x.$$

On first reading, the notation $f((x_1 + x_2)/2)$ might appear somewhat intimidating. The idea to keep in mind is that $(x_1 + x_2)/2$ provides the input halfway between x_1 and x_2, or the midpoint of the subinterval $[x_1, x_2]$. Similarly, $(x_2 + x_3)/2$ is the midpoint of the subinterval $[x_2, x_3]$, and so on.

This formulaic approach to finding the midpoint of each subinterval is one way to locate these inputs. In small-scale settings, another helpful approach to finding these inputs is to draw the x-axis, label the points x_1, x_2, x_3, ..., x_{n+1}, and then identify the midpoint of each subinterval on this graph of the number line.

◆ **EXAMPLE 5** Consider the graph given in Figure 2 (for Example 2) of the velocity of a runner during the first 10 seconds of a sprint.

(a) Calculate M_5 on $[0, 10]$ using equal-width subintervals.

(b) Compare M_5 with the approximations L_5 and R_5 from Example 2.

Solution.

(a) Dividing interval $[0, 10]$ into five equal-width subintervals requires the following uniform width for all five subintervals:

$$\Delta x = \frac{b - a}{n} = \frac{10 - 0}{5} = \frac{10}{5} = 2$$

The resulting subintervals are $[0, 2]$, $[2, 4]$, $[4, 6]$, $[6, 8]$, and $[8, 10]$, which have midpoints of $x = 1$, $x = 3$, $x = 5$, $x = 7$, and $x = 9$, respectively. The graph of the velocity function given in Figure 2 (and in Figure 3) suggests the following output values:

$$f(1) \approx 0.1 \quad f(3) = 1 \quad f(5) = 7 \quad f(7) \approx 9.8 \quad f(9) = 10$$

Substituting these output values and uniform subinterval widths into the midpoint approximation formula yields the following results:

$$\begin{aligned} M_5 &= f(1)\Delta x + f(3)\Delta x + f(5)\Delta x + f(7)\Delta x + f(9)\Delta x \\ &= 0.1 \cdot 2 + 1 \cdot 2 + 7 \cdot 2 + 9.8 \cdot 2 + 10 \cdot 2 \\ &= 0.2 + 2 + 14 + 19.8 + 20 \ = \ 55.8 \end{aligned}$$

This midpoint approximation $M_5 = 55.8$ of the accumulation of $F(x)$ is depicted graphically in Figure 7. Interpreting this result in context, the runner traveled approximately 55.8 meters during the first 10 seconds of the sprint.

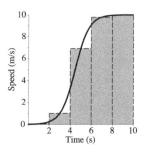

Figure 7: Sketch of M_5 for Example 5

(b) Recall from Example 2 that $L_5 = 44.8$ and $R_5 = 64.8$, and so $M_5 = 55.8$ falls between these two approximations. This pattern of relationships when M_n falling between L_n and R_n usually holds for most smooth functions, and, in general, M_n provides a more accurate approximation of accumulation than either L_n or R_n. ■

➤ **QUESTION 4** Let $f(x) = x^2 + 1$ be the rate of change of a quantity $F(x)$. Calculate the M_2 approximation of the accumulation of $F(x)$ on $[0, 8]$ using equal-width subintervals.

➤ **QUESTION 5** Let $f(x) = (x - 3)^2 - 1$ be the rate of change of a quantity $F(x)$. Calculate the M_4 approximation of the accumulation of $F(x)$ on $[1, 5]$ using equal-width subintervals.

Obtaining More Accurate Approximations

As apparent from the examples and questions in this section and even from their very names, the left approximation L_n, the right approximation R_n, and the midpoint approximation M_n are all *approximations*. They provide estimates of an accumulated quantity over some interval, rather than the exact accumulation. Important issues to consider for any approximation are whether a more accurate approximation can be produced and whether an exact value can be calculated.

Comparing the approaches discussed in this section, the midpoint approximation M_n usually provides a more accurate approximation of the accumulation of $F(x)$ on $[a, b]$ given its rate of change $f(x)$ than either the left approximation L_n or the right approximation R_n. Recall from Examples 2 and 5 the approximations of $L_5 = 44.8$, $R_5 = 64.8$, and $M_5 = 55.8$ of the distance traveled by a runner during the first 10 seconds of a sprint based on the velocity given in Figure 2. Among these three approximations, $M_5 = 55.8$ provides the best estimate of the total distance traveled by the runner.

Another way to make these approximations more accurate is to use a greater number of subintervals with decreasing widths. If equal-width subintervals are used, then increasing the number of subintervals results in a corresponding uniform decrease in the widths Δx. If variable-width subintervals are used, then in addition to increasing the number of subintervals, their widths must also be decreased.

Figure 8 presents the L_2, L_4, and L_8 left approximations of the accumulation of a quantity $F(x)$ with rate function $f(x)$ on the interval $[0, 4]$ using equal-width subintervals. As can be seen, increasing the number of subintervals decreases the amount that L_n overestimates the accumulation of $F(x)$ on the interval $[0, 4]$. As the number of subintervals increases, the amount that each rectangle overestimates the accumulation of $F(x)$ decreases. In this way, the increased number of subintervals results in a better approximation, at least when the rate of change is given by a smooth function $f(x)$.

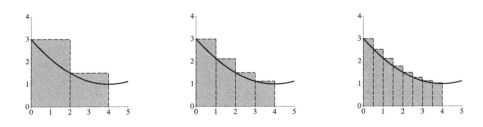

Figure 8: L_2, L_4, and L_8 approximating the accumulation of $F(x)$

▶ **QUESTION 6** Let $f(x) = x$ be the rate of change of a quantity $F(x)$. Calculate each approximation of the accumulation of $F(x)$ on $[0, 8]$ using equal-width subintervals. Compare the results with the actual accumulation of 32, which can be verified geometrically or using the methods in Section 6.4.

(a) R_2 (b) R_4 (c) R_8

Riemann Sums

The approximations studied in this section are specific instances of what is known as a *Riemann sum*. Two important aspects of a general Riemann sum are that (1) the interval can be subdivided into n subintervals of varying width and (2) the height of the corresponding rectangles can be the output value of the function resulting from any input in the subinterval (not just the left or right endpoint, or the midpoint). As such, there are infinitely many different approximations to the accumulation of $F(x)$ on the interval $[a, b]$. While this tool ultimately leads to the definition of the definite integral studied in Section 6.2, its full generality is particularly needed when approximating accumulations.

Riemann sums are named for the German mathematician Bernhard Riemann, who applied the idea of the limit developed in Section 4.7 to the study of accumulation in the mid 1800s. Further details of this powerful and sophisticated approach to understanding accumulation lie beyond the scope of this book; the interested reader is referred to standard textbooks for calculus and real analysis courses.

Summary

- Let $f(x)$ be the rate of change of a quantity $F(x)$ on an interval $[a, b]$, let $a = x_1 < x_2 < x_3 < \cdots < x_{n+1} = b$, let $[a, b]$ be subdivided into smaller subintervals $[x_1, x_2], [x_2, x_3], \ldots, [x_n, x_{n+1}]$, and let $\Delta x_k = x_{k+1} - x_k$ be the width of each subinterval.

Summary (continued)

       ◦ The *left approximation* to the accumulation of $F(x)$ on $[a, b]$ is

$$L_n = f(x_1)\Delta x_1 + f(x_2)\Delta x_2 + \cdots + f(x_n)\Delta x_n$$

       ◦ The *right approximation* to the accumulation of $F(x)$ on $[a, b]$ is

$$R_n = f(x_2)\Delta x_1 + f(x_3)\Delta x_2 + \cdots + f(x_{n+1})\Delta x_n$$

    When possible, the subintervals $[x_k, x_{k+1}]$ are chosen with equal width, in which case $\Delta x = \dfrac{b-a}{n}$ for all n subintervals.

- Let $f(x)$ be the rate of change of a quantity $F(x)$ on an interval $[a, b]$, let $a = x_1 < x_2 < x_3 < \cdots < x_{n+1} = b$, let $[a, b]$ be subdivided into smaller, equal-width subintervals $[x_1, x_2], [x_2, x_3], \ldots, [x_n, x_{n+1}]$, and let $\Delta x = (b-a)/n$ be the uniform width of each subinterval. The *midpoint approximation* to the accumulation of $F(x)$ on the interval $[a, b]$ is

$$M_n = f\left(\frac{x_1 + x_2}{2}\right)\Delta x + f\left(\frac{x_2 + x_3}{2}\right)\Delta x + \cdots + f\left(\frac{x_n + x_{n+1}}{2}\right)\Delta x$$

- If $f(x)$ is a smooth function providing the rate of change of a quantity $F(x)$, then the approximations L_n, R_n, and M_n calculated using $f(x)$ to estimate the accumulation of $F(x)$ become more accurate as the number of decreasing-width subintervals n is increased. Also, M_n generally provides a more accurate approximations of this accumulation than L_n and R_n.

Exercises

In Exercises 1–6, calculate the approximation of the accumulation of $F(x)$ on the interval using the graph of its rate of change.

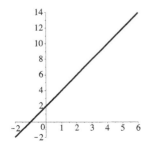

1. L_3; $[0, 6]$

2. R_3; $[0, 6]$

3. R_6; $[0, 6]$

4. L_6; $[0, 6]$

5. M_3; $[0, 6]$

6. M_6; $[0, 6]$

In Exercises 7–12, calculate the approximation of the accumulation of $F(x)$ on the interval using the graph of its rate of change.

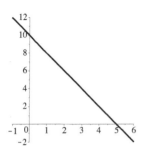

7. L_2; $[0, 5]$ 10. L_5; $[0, 5]$

8. R_2; $[0, 5]$ 11. M_2; $[0, 5]$

9. R_5; $[0, 5]$ 12. M_5; $[0, 5]$

In Exercises 13–18, calculate the approximation of the accumulation of $F(x)$ on the interval using the graph of its rate of change.

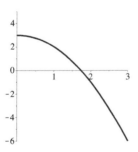

13. L_1; $[0, 3]$ 16. L_3; $[0, 3]$

14. R_1; $[0, 3]$ 17. M_1; $[0, 3]$

15. R_3; $[0, 3]$ 18. M_3; $[0, 3]$

In Exercises 19–26, calculate the approximation of the accumulation of $F(x)$ on the interval using the graph of its rate of change.

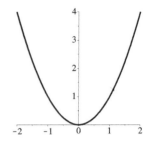

19. L_2; $[-2, 2]$ 23. M_2; $[-2, 2]$

20. R_2; $[-2, 2]$ 24. M_4; $[-2, 2]$

21. R_4; $[-2, 2]$ 25. R_2; $[0, 2]$

22. L_4; $[-2, 2]$ 26. L_2; $[0, 2]$

In Exercises 27–32, calculate the approximation of the accumulation of $F(x)$ on the interval using the graph of its rate of change.

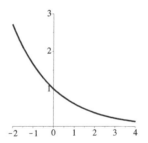

27. L_2; $[-2, 4]$ 30. R_3; $[-2, 4]$

28. R_2; $[-2, 4]$ 31. R_6; $[-2, 4]$

29. L_3; $[-2, 4]$ 32. L_6; $[-2, 4]$

In Exercises 33–38, calculate the approximation of the accumulation of $F(x)$ on the interval using the graph of its rate of change.

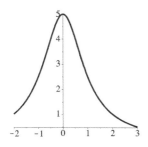

33. L_2; $[-2, 3]$ 36. R_5; $[-2, 3]$

34. R_2; $[-2, 3]$ 37. L_5; $[-2, 3]$

35. M_2; $[-2, 3]$ 38. M_5; $[-2, 3]$

In Exercises 39–44, calculate the approximations L_4 and R_4 of the accumulation of $F(x)$ using the input–output tables for its rate of change $f(x)$.

39.
x	15	18	21	24	27
$f(x)$	7.3	5.4	3	1.6	1.5

40.
x	1	2	3	4	5
$f(x)$	-7	65	92	-40	-78

41.
x	2	4	6	8	10
$f(x)$	14	6	-7	-21	-30

42.
x	-2	-1	0	1	2
$f(x)$	8.7	5.3	3.4	2.6	1.5

43.
x	5	8	11	14	17
$f(x)$	6.3	5.4	2	1.6	0.5

44.
x	4	8	12	16	20
$f(x)$	2	3	6	10	11

In Exercises 45–50, approximate the accumulation of $F(x)$ on the interval using the input–output table below for its rate of change $f(x)$. Subintervals will not always have equal width.

x	5	20	25	55	60
$f(x)$	900	625	400	400	225

45. R_2; $[5, 60]$ 48. R_2; $[5, 25]$

46. L_4; $[5, 60]$ 49. L_2; $[25, 60]$

47. R_4; $[5, 60]$ 50. R_2; $[20, 55]$

In Exercises 51–56, approximate the accumulation of $F(x)$ on the interval using the input–output table below for its rate of change $f(x)$. Subintervals will not always have equal width.

x	-5	0	10	15	30
$f(x)$	52	59	43	35	40

51. R_2; $[-5, 30]$ 54. R_2; $[5, 10]$

52. L_4; $[-5, 30]$ 55. L_3; $[-5, 15]$

53. R_4; $[-5, 30]$ 56. R_3; $[0, 30]$

In Exercises 57–62, approximate the accumulation of $F(x)$ on the interval using the input–output table below for its rate of change $f(x)$. Subintervals will not always have equal width.

x	-0.5	0.4	2.9	5.2	7.7
$f(x)$	0.2	1.5	18	181	2208

57. R_2; $[-0.5, 2.9]$

58. L_4; $[-0.5, 7.7]$

59. R_4; $[-0.5, 7.7]$

60. R_2; $[0.4, 5.2]$

61. L_2; $[2.9, 7.7]$

62. R_3; $[-0.5, 5.2]$

In Exercises 63–68, approximate the accumulation of $F(x)$ on the interval using the input–output table below for its rate of change $f(x)$. Subintervals will not always have equal width.

x	-0.5	0.4	2.9	5.2	7.7
$f(x)$	1.3	0.9	0.6	0.4	0.2

63. R_2; $[-0.5, 2.9]$

64. L_4; $[-0.5, 7.7]$

65. R_4; $[-0.5, 7.7]$

66. R_2; $[0.4, 5.2]$

67. L_2; $[2.9, 7.7]$

68. R_3; $[-0.5, 5.2]$

In Exercises 69–74, calculate the indicated approximation of the accumulation of $F(x)$ given its rate of change (R). Subintervals will not always have equal width.

69. Total change in the value of the S&P 500 stock market closing value each year (Y) from 1975 to 2005 with R_3.

Y	1975	1985	1995	2005
R	9.7	35.3	64.6	88.6

70. Total change in the percentage of unemployed Americans from 2007 to 2010 each year (Y) using L_3.

Y	2007	2008	2009	2010
R	4	5	7.8	9.7

71. Total change in annual sales (in thousands) of hybrid vehicles in the United States each year (Y) from 2010 to 2012 with R_2.

Y	2010	2011	2012
R	−8	169	61

72. U.S. field production of crude oil in billions of barrels from 1970 to 2005 in years since 1970 (Y) with L_4.

Y	0	10	20	30	35
R	3.5	3.1	2.7	2.1	1.9

73. Total change in the three-year average monthly pollen count in Brooklyn, New York City each month (M) from August through November with L_3.

M	8	9	10	11
R	−3.8	−2.9	−1.4	−0.2

74. Total change in the average monthly maximum temperature in degrees Fahrenheit in New York City each month (M) from April through December with R_3.

M	4	7	10	12
R	21.7	27.2	−20.5	−24.4

In Exercises 75–82, calculate the approximation of the accumulation of $F(x)$ using the rate of change $f(x)$ on the interval.

75. L_5; $f(x) = 2x - 5$; $[0, 5]$

76. R_6; $f(x) = 6 - 3.2x$; $[0, 12]$

77. R_4; $f(x) = x^2 - 3$; $[0, 8]$

78. M_4; $f(x) = 4 - x^2$; $[0, 4]$

79. L_6; $f(x) = e^{x-3}$; $[-2, 4]$

80. L_3; $f(x) = \sin(2x - 3)$; $[0, 3]$

81. M_5; $f(x) = \cos(x - 3)$; $[0, 5]$

82. R_5; $f(x) = xe^{-x}$; $[0, 10]$

In Exercises 83–88, calculate the approximation of the accumulation of $F(x)$ using its rate of change $f(x) = 3x + 1$ on $[0, 2]$.

83. R_2 86. L_6

84. R_4 87. L_8

85. R_6 88. L_{10}

89. Using the results of Exercises 83–88, conjecture the accumulation of $F(x)$ on $[0, 2]$.

In Exercises 90 – 95, calculate the approximation of the accumulation of $F(x)$ using its rate of change $f(x) = x^2 + 1$ on $[0, 2]$.

90. R_2

91. R_4

92. R_6

93. L_6

94. L_8

95. L_{10}

96. Using the results of Exercises 90 – 95, conjecture the accumulation of $F(x)$ on $[0, 2]$.

In Exercises 97 – 102, calculate the approximation of the accumulation of $F(x)$ given its rate of change $f(x) = (x-2)^2 - 1$ on $[0, 4]$.

97. R_2

98. R_4

99. R_8

100. L_2

101. L_4

102. L_{12}

103. Using the results of Exercises 97 – 102, conjecture the accumulation of $F(x)$ on $[0, 4]$.

In Your Own Words. In Exercises 104 – 107, explain the following.

104. Left approximation

105. Right approximation

106. Subinterval

107. Midpoint approximation

6.2 The Definite Integral

Section 6.1 introduced left, right, and midpoint approximations of the accumulation of a quantity given its rate of change, and these estimates of accumulation were obtained for graphical, tabular, and analytical presentations of functions. While approaches to improving the accuracy of these approximations were examined, they remain estimates that only sometimes provide the exact value of an accumulation. This section introduces the *definite integral* as a tool to calculate the exact value of such accumulations.

The definitions of left and right approximations motivate the definition of the definite integral. Recall that the goal of these methods is to determine the accumulation of a function $F(x)$ on an interval $[a, b]$ given its rate of change $f(x)$. When calculating left and right approximations, the interval $[a, b]$ is divided into shorter-width subintervals $[x_1, x_2]$, $[x_2, x_3]$, …, $[x_n, x_{n+1}]$, and then a constant output height is chosen for each subinterval. For a left approximation, this constant height is provided by the output value of the rate function $f(x)$ at each subinterval's left endpoint, while, for a right approximation, this constant height is provided by the output of $f(x)$ at each subinterval's right endpoint. The graph of L_5 in Figure 1(a) illustrates a left approximation, and the graph of R_5 in Figure 1(b) a right approximation, of the accumulation of $F(x)$ given its rate of change $f(x)$ on interval $[a, b] = [0, 4]$ with $n = 5$ equal-width subintervals.

The goal of these approximations is to estimate the exact value of the accumulation of $F(x)$ given its rate of change $f(x)$ on the interval $[a, b]$. Working toward finding the exact value of this accumulation, recall from the end of Section 6.1 that a greater number of subintervals with a corresponding smaller width Δx improves the accuracy of these approximations when the rate is a smooth function. Figure 2 illustrates the improved accuracy of left approximations resulting from the use of more but shorter subintervals

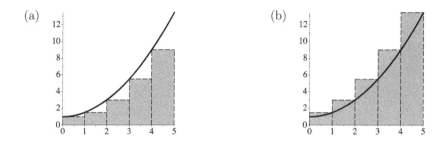

Figure 1: L_5 and R_5 approximating the accumulation of $F(x)$

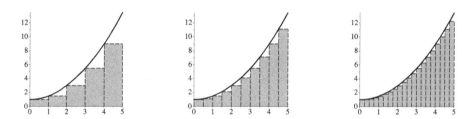

Figure 2: L_5, L_{10}, and L_{20} approximating the accumulation of $F(x)$

of equal width. As demontrated in Figure 2, the values of the left approximations L_n become closer to the actual accumulation of $F(x)$ as the number of equal-width subintervals increases from $n = 5$ to 10 to 20.

Similarly, the values of the right approximations R_n behave in exactly the same fashion and achieve greater accuracy as the number of equal-width subintervals increases from $n = 5$ to 10 to 20. Figure 3 illustrates this comparative improvement among these approximations.

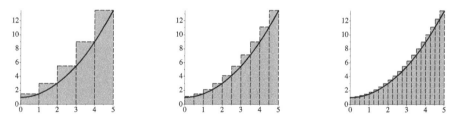

Figure 3: R_5, R_{10}, and R_{20} approximating the accumulation of $F(x)$

As suggested by Figures 2 and 3, the left and right approximations approach a single numeric value for a smooth rate function $f(x)$, which happens to be the actual accumlation of the quantity $F(x)$ with rate $f(x)$ on interval $[a, b]$. This common number approached by the left and right approximations is called the *definite integral* of $f(x)$ on $[a, b]$.

Definition.

- The **definite integral** $\displaystyle\int_a^b f(x)\,dx$ is the net accumulation of a quantity $F(x)$ with rate of change $f(x)$ on the interval $[a, b]$.

- When it exists, $\displaystyle\int_a^b f(x)\,dx$ is equal to the number approached by L_n and R_n when these approximations are computed for an increasing number of smaller and smaller width subintervals of $[a, b]$.

For a definite integral $\int_a^b f(x)\,dx$, the real numbers a and b are called the **limits of integration**, where a is the **lower limit** and b the **upper limit**. Also, $f(x)$ is called the **integrand** and dx the **differential**.

The left and right approximations L_n and R_n are depicted graphically in terms of rectangles, which means that this definition can be interpreted as referring to an increasing number of thinner and thinner rectangles, rather than smaller and smaller-width subintervals. In addition, while L_n and R_n are specifically identified in this definition, the midpoint approximation M_n can also be used to determine the value of definite integrals for smooth functions $f(x)$.

As observed in the definition, the definite integral $\int_a^b f(x)\,dx$ does not always exist, just as sometimes inverse functions do not exist (Section 1.5) and sometimes derivatives do not exist (Section 4.2). In contrast with the study of inverses and derivatives, this book does not explore the distinction between when definite integrals do and do not exist in much detail. However, Section 6.4 does state specific conditions under which $\int_a^b f(x)\,dx$ is guaranteed to exist.

NOTATION. The integral symbol $\int$ was first used by Gottfried Wilhelm von Leibniz, one of the co-founders of calculus, in the late seventeenth century. This symbolism was inspired by the latin word *summa*, which translates to "sum" in English, based on the addition of the various subinterval approximations. This symbol is an italicized version of the long "s" symbol from Latin cursive and was typically used when "s" appears as either the first letter or a middle letter in a word. The differential symbolism dx corresponds to the changing width Δx on smaller and smaller values in the approximations. Differentials are discussed in more detail in Section 6.5.

Finally, in Figure 2, notice that the left approximations L_n underestimate the actual accumulation, while, in Figure 3, the right approximations R_n overestimate it. These comparative relationships are specific to these settings. Depending on the rate-of-change function $f(x)$, the left and right approximations may reverse roles or adopt a more complicated mix of behaviors. Rather than overestimating and underestimating the accumulation, the key behavior to keep in mind is that of "approaching." Namely, as the approximations are refined via an increasing number of shorter and shorter subintervals, the values L_n and R_n approach the exact value of the accumulation.

Net Accumulation and Geometry

The connection between the net accumulation of $F(x)$ over $[a, b]$ and the notion of area from geometry has been in the background throughout this discussion. The net accumulation is approximated by summing together a product for each subinterval obtained by multiplying the output value of the rate function $f(x)$ for some input from the subinterval by the width of the subinterval (i.e., by Δx of the change in x). In other words, for each subinterval, the area of a rectangle is found where the length is an output from the function and the width is Δx.

In parallel with this analytic discussion, rectangles are used in Figures $1 - 3$ to graphically represent the quantities calculated to obtain L_n and R_n. Viewed from this perspective, the left and right approximations estimate the area between $f(x)$ and the x-axis, and increasing the number of smaller-width rectangles results in the values of these approximations approaching the exact value of this area. In other words, $\int_a^b f(x)\,dx$ is equal to the shaded area in Figure 4.

In light of the connection between net accumulation and its geometric representation as an area bounded by a curve and the x-axis on an interval, area formulas can sometimes be used to calculate definite integrals. In particular, the areas of certain geometric shapes are readily calculated, including rectangles, squares, triangles, circles, and more. This geometric approach enables the immediate calculation of certain definite integrals.

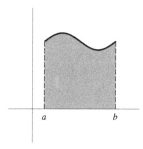

Figure 4: Area under the curve of $f(x)$

◆ **EXAMPLE 1** Evaluate each definite integral using the graph of the rate-of-change function $f(x)$ given in Figure 5:

(a) $\displaystyle\int_0^3 f(x)\,dx$ (b) $\displaystyle\int_3^7 f(x)\,dx$ (c) $\displaystyle\int_0^{10} f(x)\,dx$

Solution.

(a) The accumulation $\int_0^3 f(x)\,dx$ of $F(x)$ corresponds to the area of the triangle bounded by $f(x)$ and the x-axis on the interval $[0, 3]$, which is calculated using the formula $A = \frac{1}{2}bh$ for the area of a triangle:

$$\int_0^3 f(x)\,dx = \frac{1}{2}(3)(3) = \frac{9}{2} = 4.5$$

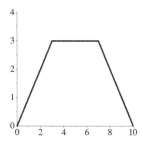

Figure 5: Graph of $f(x)$ for Example 1

(b) The accumulation $\int_3^7 f(x)\,dx$ of $F(x)$ corresponds to the area of the rectangle bounded by $f(x)$ and the x-axis on the interval $[3,7]$, which is calculated using the formula $A = bh$ for the area of a rectangle:

$$\int_3^7 f(x)\,dx \;=\; (4)(3) \;=\; 12$$

(c) The accumulation $\int_0^{10} f(x)\,dx$ of $F(x)$ corresponds to the area of the trapezoid bounded by $f(x)$ and the x-axis on $[0,10]$. Rather than using the area formula $A = \frac{1}{2}(b_1+b_2)h$, this example thinks of the bounded area as consisting of a triangle, a rectangle, and then another triangle. Using the area formulas for these geometric shapes, the accumulation is calculated as follows:

$$\int_0^{10} f(x)\,dx \;=\; \frac{1}{2}(3)(3) + (4)(3) + \frac{1}{2}(3)(3) \;=\; 4.5 + 12 + 4.5 \;=\; 21$$

An important subtlety of this geometric approach to evaluating definite integrals arises from the fact that many functions are negative on portions of their domains. Graphically, a function is negative where it appears below the x-axis. Recall from the definitions of left, right, and midpoint approximations of $\int_a^b f(x)\,dx$ that, on each subinterval, the output value of the function for some input in the subinterval is multiplied by the width of the subinterval and then all of these products are added together. If the function is negative (or below the x-axis), then the result of this product is negative.

Interpreting a negative product on some subinterval as the area of a rectangle does not make sense, because area cannot be negative. However, such a geometric interpretation of $\int_a^b f(x)\,dx$ as area is only one possible meaning of the definite integral. Among many possible options, $\int_a^b f(x)\,dx$ provides the total accumulation of such quantities as distance, population, money accrued, and more. In such contexts, a negative accumulation is reasonable, because people can move backward (negative distance), populations can decrease, and money can be lost.

In response to such realities, the definite integral $\int_a^b f(x)\,dx$ actually provides a measure of the *net* accumulation of $f(x)$ over $[a,b]$. If $f(x)$ is positive (or above the x-axis), then the area between $f(x)$ and the x-axis makes a positive contribution to the net accumulation. If $f(x)$ is negative (or below the x-axis), then the area between $f(x)$ and the x-axis makes a negative contribution to the net accumulation.

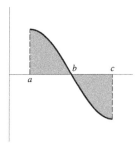

Figure 6: Directed, signed area of $f(x)$

Note. The numeric value of a definite integral $\int_a^b f(x)\,dx$ provides the **directed, signed area** of the region bounded by $f(x)$ and the x-axis on the interval $[a, b]$.

The adjective "signed" refers to the fact that the function may be above the x-axis and make a postive contribution, or may be below the x-axis and make a negative contribution to the accumulation as just discussed.

The adjective "directed" refers to the limits a and b of the definite integral $\int_a^b f(x)\,dx$. Thus far, the input a has been strictly less than b (i.e., $a < b$), which corresponds to the standard direction of traveling intervals from left to right.

However, sometimes a is strictly greater than b (i.e., $a > b$) and $\int_a^b f(x)\,dx$ has a lower limit of integration that is numerically greater than its upper limit. For example, consider the definite integral $\int_2^0 f(x)\,dx$, in which the interval forming the base begins at 2 and moves left toward 0, and so in the opposite direction to the standard approach when traveling intervals (i.e., moving from right to left). Computationally, the width of each subinterval $[x_k, x_{k+1}]$ is equal to $x_{k+1} - x_k$, which is found by subtracting the final x-value from the initial. Thus, if $a > b$, then $b - a$ results in a negative number. In this way, the definite integral provides a directed measure of area.

The idea of directed, signed area is explored further in the following examples, beginning with a focus on the signed aspect and then the directed aspect.

◆ **EXAMPLE 2** Evaluate each definite integral using the graph of the rate-of-change function $f(x)$ given in Figure 7:

(a) $\displaystyle\int_0^5 f(x)\,dx$ (b) $\displaystyle\int_7^9 f(x)\,dx$ (c) $\displaystyle\int_3^9 f(x)\,dx$

Solution. Each definite integral $\int_a^b f(x)\,dx$ is evaluated using the interpretation of the definite integral as equivalent to the directed, signed area bounded by $f(x)$ and the x-axis on $[a, b]$, as well as the area formulas for triangles $(A = \frac{1}{2}bh)$ and rectangles $(A = bh)$.

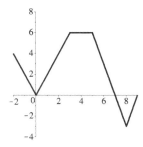

Figure 7: Graph of $f(x)$ for Example 2

(a) The accumulation $\int_0^5 f(x)\,dx$ corresponds to the area of the region bounded by $f(x)$ and the x-axis on $[0,5]$. This region consists of a triangle on the interval $[0,3]$ and a rectangle on the interval $[3,5]$, both of which lie above the x-axis. Now apply the area formulas for triangles and rectangles:

$$\int_0^5 f(x)\,dx \;=\; \frac{1}{2}(3)(6) + (2)(6) \;=\; 9 + 12 \;=\; 21$$

(b) The accumulation $\int_7^9 f(x)\,dx$ corresponds to the area of the triangle bounded by $f(x)$ and the x-axis on $[7,9]$. In addition, this definite integral is negative, because the region lies entirely below the x-axis. In the area formula, this observation is expressed by identifying the height of the triangle as negative:

$$\int_7^9 f(x)\,dx \;=\; \frac{1}{2}(2)(-3) \;=\; -3$$

(c) The accumulation $\int_3^9 f(x)\,dx$ corresponds to three distinct pieces of signed area bounded by $f(x)$ and the x-axis on $[3,9]$: a rectangle above the x-axis on $[3,5]$, a triangle above the x-axis on $[5,7]$, and a triangle below the x-axis on $[7,9]$. Calculating the area of each piece and incorporating the appropriate sign information provides the accumulation. For the third piece, the height of the triangle is negative, because that portion of $f(x)$ is below the x-axis. Thus,

$$\int_3^9 f(x)\,dx \;=\; (2)(6) + \frac{1}{2}(2)(6) + \frac{1}{2}(2)(-3) \;=\; 12 + 6 - 3 \;=\; 15$$

➤ **QUESTION 1** Evaluate each definite integral using the graph of the rate-of-change function $f(x)$ given in Figure 8:

(a) $\displaystyle\int_{-2}^{2} f(x)\,dx$ (b) $\displaystyle\int_{6}^{9} f(x)\,dx$ (c) $\displaystyle\int_{-2}^{5} f(x)\,dx$

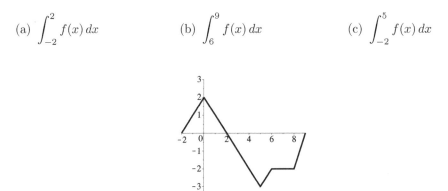

Figure 8: Graph of $f(x)$ for Question 1

With a facility in working with the signed aspect of definite integrals in hand, the next step is to incorporate the directed aspect. For a definite integral $\int_{a}^{b} f(x)\,dx$ with a less than b, the width of the base in the area formulas is positive (from $b - a$ being positive) as the input intervals are traveled in the standard direction from left to right. When b is less than a, the width of the base in the area formulas is negative (from $b - a$ being negative) as the input intervals are traveled in the opposite direction from right to left.

◆ **EXAMPLE 3** Evaluate each definite integral using the graph of the rate-of-change function $f(x)$ given in Figure 9:

(a) $\displaystyle\int_{-2}^{0} f(x)\,dx$ (b) $\displaystyle\int_{0}^{-2} f(x)\,dx$ (c) $\displaystyle\int_{9}^{7} f(x)\,dx$

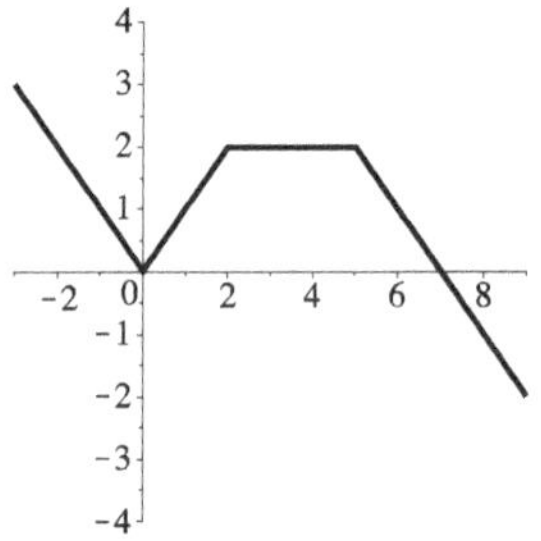

Figure 9: Graph of $f(x)$ for Example 3

Solution. Each definite integral $\int_{a}^{b} f(x)\,dx$ is evaluated using the interpretation of the definite integral as equivalent to the directed, signed area bounded by $f(x)$ and the x-axis on $[a, b]$, as well as using the area formulas for triangles and rectangles.

(a) The region bounded by $f(x)$ and the x-axis on $[-2, 0]$ consists of a triangle above the x-axis, which results in the height being positive. With lower limit $a = -2$ and upper limit $b = 0$, the interval is traversed from left to right, which results in the width being positive. Substituting into the area formula for the triangle yields

$$\int_{-2}^{0} f(x) \, dx \;=\; \frac{1}{2}(2)(2) \;=\; 2$$

(b) The region bounded by $f(x)$ and the x-axis on $[0, -2]$ consists of a triangle above the x-axis, which results in the height being positive. With lower limit $a = 0$ and upper limit $b = -2$, the interval is traversed from right to left, which results in the width being negative. Substituting into the area formula for the triangle yields

$$\int_{0}^{-2} f(x) \, dx \;=\; \frac{1}{2}(-2)(2) \;=\; -2$$

(c) The region bounded by $f(x)$ and the x-axis on $[9, 7]$ consists of a triangle below the x-axis, which results in the height being negative. With lower limit $a = 9$ and upper limit $b = 7$, the interval is traversed from right to left, which results in the width also being negative. Substituting into the area formula for the triangle yields

$$\int_{9}^{7} f(x) \, dx \;=\; \frac{1}{2}(-2)(-2) \;=\; 2$$

$\blacksquare$

➤ **QUESTION 2** Evaluate each definite integral using the graph of the rate-of-change function $f(x)$ given in Figure 10:

(a) $\displaystyle\int_{1}^{3} f(x) \, dx$ (b) $\displaystyle\int_{3}^{6} f(x) \, dx$ (c) $\displaystyle\int_{8}^{6} f(x) \, dx$

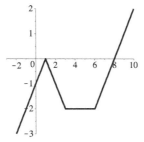

Figure 10: Graph of $f(x)$ for Example 2

Dimensions and Units of Definite Integrals

The connection between the net accumulation $\int_a^b f(x)\,dx$ and the directed, signed area of the region bounded by the function $f(x)$ and the x-axis on $[a, b]$ is incredibly useful when computing certain definite integrals. However, the reader must keep in mind that a definite integral actually provides a measure of area in only a relative handful of contexts. Rather, the definite integral measures the net accumulation of diverse quantities, including distance, volume, amount of money, number of people, size of a downloaded file, amount of medication in a patient's bloodstream, and more. Therefore, the dimension of a net accumulation is usually not equal to the dimension of area, which would be represented symbolically as L^2. Rather, the dimension of a definite integral depends on the given context.

In more detail, in order to determine the dimension and units of a definite integral, recall that the value of the integral is approached by left and right approximations consisting of sums of products. From Section 2.7, the dimension of a sum corresponds to the dimensions of the quantities being added, which all have to be identical. Therefore, the dimension of L_n and R_n is equal to the dimension of the products, or the dimension of the output times the dimension of the input:

$$\left[\int_a^b f(x)\,dx\right] = [L_n] = [R_n] = [f(x) \cdot x] = [f(x)] \cdot [x] = [\text{output}] \cdot [\text{input}]$$

Likewise, the units of left and right approximations, and so the definite integral, are equal to the product of the units of $f(x)$ and the units of x.

DIMENSION AND UNITS OF DEFINITE INTEGRALS.

- The dimension of $\int_a^b f(x)\,dx$ is the product of the dimension of $f(x)$ and the dimension of x, which is expressed symbolically as follows:

$$\left[\int_a^b f(x)\,dx\right] = [f(x)] \cdot [x] = [\text{output}] \cdot [\text{input}]$$

- The units of $\int_a^b f(x)\,dx$ are the product of the units of $f(x)$ and the units of x.

◆ **EXAMPLE 4** Supppose $w(t)$ models the rate of change in the weight of a female fetus in kilograms per week.

(a) State the meaning, units, and dimension of the definite integral $\int_{14}^{26} w(t)\,dt$.

(b) State a definite integral that provides the birth weight of a female baby, assuming an average gestation period of 40 weeks.

Solution.

(a) The definite integral $\int_{14}^{26} w(t)\,dt$ provides the total weight gain of a fetus from week 14 to week 26, or during the second trimester of pregnancy. The units are the product of the units of $w(t)$ and the units of t. Because the units of $w(t)$ are kilograms per week and the units of t are weeks, the units of $\int_{14}^{26} w(t)\,dt$ are kilograms. The corresponding dimension is mass, which is represented symbolically by M.

(b) Birth weight is equal to the total weight gain from week 0 to week 40, which is provided by the definite integral $\int_{0}^{40} w(t)\,dt$.

■

▶ **QUESTION 3** Suppose $b(t)$ models the rate of change of a student's debt in dollars per month.

(a) State the meaning, units, and dimension of the definite integral $\displaystyle\int_{12}^{23} b(t)\,dt$.

(b) State a definite integral that provides the total debt incurred during the four years the student is in college.

Properties of the Definite Integral

As with the derivative, the algebraic properties of the integral enable the successful application of this calculating tool to understanding our world. These properties are formally derived using a limit definition of the definite integral in parallel with how the limit is central to the definition of the derivative in Section 4.7. Alternatively, the geometric approach to definite integrals developed in this section motivates and illuminates algebraic properties of integrals, the most important of which include the following:

ALGEBRAIC PROPERTIES OF DEFINITE INTEGRALS.

(1) $\displaystyle\int_{a}^{a} f(x)\,dx = 0$

(2) $\displaystyle\int_{a}^{b} f(x)\,dx = -\int_{b}^{a} f(x)\,dx$

(3) $\displaystyle\int_{a}^{b} f(x) \pm g(x)\,dx = \int_{a}^{b} f(x)\,dx \pm \int_{a}^{b} g(x)\,dx$

ALGEBRAIC PROPERTIES OF DEFINITE INTEGRALS.
(CONTINUED)

(4) $\displaystyle\int_a^b k \cdot f(x)\,dx \;=\; \int_a^b f(x) \cdot k\,dx \;=\; k\int_a^b f(x)\,dx$

(5) $\displaystyle\int_a^b f(x)\,dx \;=\; \int_a^c f(x)\,dx + \int_c^b f(x)\,dx$ when $a \le c \le b$

Many of these algebraic properties of definite integrals directly parallel similar algebraic properties of the derivative. Several examples using these properties are studied to develop a proficiency in using them, followed by a geometry-based discussion of why these properties hold.

◆ **EXAMPLE 5** Given that

$$\int_0^2 2\,dx = 4 \qquad \int_0^2 4x\,dx = 8 \qquad \int_2^3 4x\,dx = 10 \qquad \int_0^2 3x^2\,dx = 8$$

use the algebraic properties of definite integrals to evaluate each integral:

(a) $\displaystyle\int_2^2 4x\,dx$

(c) $\displaystyle\int_0^2 12x\,dx$

(e) $\displaystyle\int_2^0 6x^2 - 12\,dx$

(b) $\displaystyle\int_0^2 4x - 2\,dx$

(d) $\displaystyle\int_0^2 6x^2 - 12\,dx$

(f) $\displaystyle\int_0^3 -12x\,dx$

Solution.

(a) Applying property (1) gives $\displaystyle\int_2^2 4x\,dx \;=\; 0.$

(b) Applying property (3) gives $\displaystyle\int_0^2 4x - 2\,dx = \int_0^2 4x\,dx - \int_0^2 2\,dx = 8 - 4 = 4.$

(c) Apply property (4) and use some arithmetic:

$$
\begin{aligned}
\int_0^2 12x\,dx \;&=\; \int_0^2 3 \cdot (4x)\,dx && \text{Factor } 12 = 3 \cdot 4\\[4pt]
&=\; 3\int_0^2 4x\,dx && \text{Property (4)}\\[4pt]
&=\; 3 \cdot 8 = 24 && \text{Simplify}
\end{aligned}
$$

(d) Apply properties (3) and (4), and use some arithmetic:

$$\int_0^2 6x^2 - 12\, dx \;=\; \int_0^2 6x^2\, dx - \int_0^2 12\, dx \qquad \text{Property (3)}$$

$$=\; \int_0^2 2 \cdot (3x^2)\, dx - \int_0^2 6 \cdot 2\, dx \qquad \text{Factor } 6 = 2 \cdot 3,\ 12 = 6 \cdot 2$$

$$=\; 2\int_0^2 3x^2\, dx - 6\int_0^2 2\, dx \qquad \text{Property (4)}$$

$$=\; 2 \cdot 8 - 6 \cdot 4 = -8 \qquad \text{Simplify}$$

(e) Use property (2) and the answer to part (d):

$$\int_2^0 6x^2 - 12\, dx \;=\; -\int_0^2 6x^2 - 12\, dx \;=\; -(-8) \;=\; 8$$

(f) Apply properties (4) and (5), and use some arithmetic:

$$\int_0^3 -12x\, dx \;=\; \int_0^3 (-3) \cdot (4x)\, dx \qquad \text{Factor } -12 = (-3) \cdot 4$$

$$=\; (-3)\left[\int_0^3 4x\, dx\right] \qquad \text{Property (4)}$$

$$=\; (-3)\left[\int_0^2 4x\, dx + \int_2^3 4x\, dx\right] \qquad \text{Property (5)}$$

$$=\; (-3)(8 + 10) = -54 \qquad \text{Simplify}$$

➤ **QUESTION 4** Given that

$$\int_0^3 4\, dx = 12 \qquad \int_0^3 2x\, dx = 9 \qquad \int_3^5 2x\, dx = 16 \qquad \int_0^3 x^2\, dx = 9$$

use the algebraic properties of definite integrals to evaluate each integral:

(a) $\displaystyle\int_3^3 4\, dx$ (c) $\displaystyle\int_0^3 8x\, dx$ (e) $\displaystyle\int_3^0 2x^2 - 12\, dx$

(b) $\displaystyle\int_0^3 2x - 4\, dx$ (d) $\displaystyle\int_0^3 2x^2 - 12\, dx$ (f) $\displaystyle\int_0^5 4x\, dx$

In Example 5 and Question 4, the values of some of the definite integrals can be found geometrically, while others rely on methods that will be developed in Section 6.4. Even when the integrand is not known, these algebraic properties can still be applied, which can be useful when studying real-life data sets for which models are not known.

◆ **EXAMPLE 6** Given that

$$\int_0^3 f(x)\,dx = 4 \qquad \int_2^3 f(x)\,dx = -2 \qquad \int_0^3 g(x)\,dx = 5 \qquad \int_0^5 g(x)\,dx = 3$$

use the algebraic properties of definite integrals to evaluate each integral:

(a) $\displaystyle\int_0^3 f(x) + g(x)\,dx$

(c) $\displaystyle\int_0^2 f(x)\,dx$

(b) $\displaystyle\int_0^3 3f(x) - 2g(x)\,dx$

(d) $\displaystyle\int_3^5 g(x)\,dx$

Solution.

(a) Applying property (3) yields the following:

$$\int_0^3 f(x) + g(x)\,dx \;=\; \int_0^3 f(x)\,dx + \int_0^3 g(x)\,dx \;=\; 4 + 5 \;=\; 9$$

(b) Apply properties (3) and (4) as follows:

$$\int_0^3 3f(x) - 2g(x)\,dx \;=\; \int_0^3 3f(x)\,dx - \int_0^3 2g(x)\,dx \qquad \text{Property (3)}$$

$$=\; 3\int_0^3 f(x)\,dx - 2\int_0^3 g(x)\,dx \qquad \text{Property (4)}$$

$$=\; 3(4) - 2(5) = 2 \qquad \text{Simplify}$$

(c) Apply property (5) and use some arithmetic:

$$\int_0^3 f(x)\,dx \;=\; \int_0^2 f(x)\,dx + \int_2^3 f(x)\,dx \qquad \text{Property (5)}$$

$$4 \;=\; \int_0^2 f(x)\,dx - 2 \qquad \text{Substitute given values}$$

$$6 \;=\; \int_0^2 f(x)\,dx \qquad \text{Solve and } 4 - (-2) = 6$$

(d) Apply property (5) and use some arithmetic:

$$\int_0^5 g(x)\,dx \;=\; \int_0^3 g(x)\,dx + \int_3^5 g(x)\,dx \qquad \text{Property (5)}$$

$$3 \;=\; 5 + \int_3^5 g(x)\,dx \qquad \text{Substitute given values}$$

$$-2 \;=\; \int_3^5 g(x)\,dx \qquad \text{Solve and } 3 - 5 = -2$$

■

➤ **QUESTION 5** Given that

$$\int_{-2}^{1} f(x)\, dx = -2 \qquad \int_{-2}^{4} f(x)\, dx = 7 \qquad \int_{-2}^{1} g(x)\, dx = 6 \qquad \int_{-2}^{0} g(x)\, dx = 1$$

use the algebraic properties of definite integrals to evaluate each integral:

(a) $\displaystyle\int_{-2}^{1} f(x) - g(x)\, dx$

(c) $\displaystyle\int_{1}^{4} f(x)\, dx$

(b) $\displaystyle\int_{-2}^{1} -f(x) + 3g(x)\, dx$

(d) $\displaystyle\int_{0}^{1} g(x)\, dx$

Understanding the Algebraic Properties of Integrals

Formal proofs of these algebraic properties of the definite integral lie beyond the scope of this book. Even so, an enriched understanding of many of these properties can be gained and an intuitive justification developed based on the geometric interpretation of the definite integral corresponding to the directed, signed area of the region bounded by a function and the x-axis.

Property (1) of $\int_{a}^{a} f(x)\, dx = 0$ states that if the input interval has no width (i.e., $\Delta x = 0$), then the net accumulation is equal to zero. Graphically, this claim corresponds with the observation that a vertical line segment has height, but no width for its base, and thus has an area of zero, as illustrated in Figure 11.

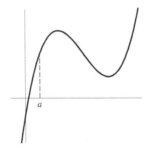

Figure 11: $\displaystyle\int_{a}^{a} f(x)\, dx$ yields zero net accumulation

Property (2) of $\int_{b}^{a} f(x)\, dx = -\int_{a}^{b} f(x)\, dx$ states that switching the upper and lower limit of integration results in the sign of the definite integral switching. Similar to the discussion in Example 3 about lower limit a being greater than upper limit b resulting in Δx being negative, this switching of limits changes the sign of Δx, which in turn changes the sign of the integral. As a parallel, recall that when computing slope, the change in output values is measured as x increases from left to right. If instead x decreases from right to left, then the sign of the slope switches. This same pattern holds for the definite integral.

Properties (3) and (4) indicate that integrals operate in exactly the same way as the derivative with respect to constants, sums, and differences. In particular, constants can be brought from the inside of a definite integral to the outside, and the definite integral distributes across both addition and subtraction. While the intuitive area interpretation of the definite integral as directed, signed area does not indicate why these properties hold, they do follow from working with the sums of products in the left and right approximations approaching $\int_a^b f(x)\,dx$. Further discussion lies beyond the scope of this book.

Finally, property (5) states that a definite integral can be split into two (or more) pieces by dividing the interval of integration into smaller subintervals as illustrated in Figure 12. Property (5) was used multiple times in the examples and questions in this section, when definite integrals were evaluated as a sum of areas of triangles and rectangles, rather than in terms of trapezoids.

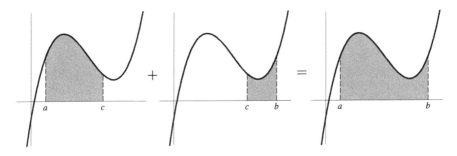

Figure 12: Subdividing the interval of integration

Summary

- The *definite integral* $\displaystyle\int_a^b f(x)\,dx$ is the net accumulation of a quantity $F(x)$ with rate of change $f(x)$ on the interval $[a, b]$.

- When it exists, $\int_a^b f(x)\,dx$ is equal to the number approached by L_n and R_n when these approximations are computed for an increasing number of smaller and smaller-width subintervals of $[a, b]$.

- For a definite integral $\int_a^b f(x)\,dx$, the real numbers a and b are called the *limits of integration*, where a is the *lower limit* and b is the *upper limit*. Also, $f(x)$ is called the *integrand* and dx is called the *differential*.

- The numeric value of a definite integral $\int_a^b f(x)\,dx$ is equal to the *directed, signed area* of the region bounded by $f(x)$ and the x-axis on the interval $[a, b]$.

Summary (continued)

- The dimension of $\int_a^b f(x)\,dx$ is the product of the dimension of $f(x)$ and the dimension of x, which is expressed symbolically as follows:

$$\left[\int_a^b f(x)\,dx\right] \;=\; [f(x)] \cdot [x] \;=\; [\text{output}] \cdot [\text{input}]$$

- The units of $\int_a^b f(x)\,dx$ are the product of the units of $f(x)$ and the units of x.

- Algebraic properties of the definite integral include the following:

$$(1) \quad \int_a^a f(x)\,dx \;=\; 0$$

$$(2) \quad \int_a^b f(x)\,dx \;=\; -\int_b^a f(x)\,dx$$

$$(3) \quad \int_a^b f(x) \pm g(x)\,dx \;=\; \int_a^b f(x)\,dx \pm \int_a^b g(x)\,dx$$

$$(4) \quad \int_a^b k \cdot f(x)\,dx \;=\; \int_a^b f(x) \cdot k\,dx \;=\; k\int_a^b f(x)\,dx$$

$$(5) \quad \int_a^b f(x)\,dx \;=\; \int_a^c f(x)\,dx + \int_c^b f(x)\,dx \quad \text{when } a \le c \le b$$

Exercises

In Exercises 1–8, state the meaning, units, and dimensions of the integral.

1. $\int_1^4 h(t)\,dt$, where $h(t)$ is the growth rate of the height of a persimmon tree in feet per year

2. $\int_0^6 p(r)\,dr$, where $p(r)$ is the rate of change in the population density of a city measured in miles from the center of the city

3. $\int_4^8 m(g)\,dg$, where $m(g)$ is the gas mileage of a car in miles per gallon

4. $\int_0^2 c(t)\,dt$, where $c(t)$ is the rate of change in the concentration (in mg/L) of a drug in the bloodstream per hour

5. $\int_8^{20} f(q)\,dq$, where $f(q)$ is the rate of change in Facebook stock value in dollars per quarter since 2010

6. $\int_1^4 g(s)\,ds$, where $g(s)$ is the rate of change in a college student's gpa per semester

7. $\int_{1980}^{2015} c(y)\, dy$, where $c(y)$ is the rate of change in atmospheric carbon dioxide in metric tons per year

8. $\int_{0}^{20} b(t)\, dt$, where $b(t)$ is the rate of change in a retirement account in dollars per year

In Exercises 9–16, state a definite integral that provides the requested information. Also, state the units and dimension of the integral.

9. Given the rate of change $\ell(w)$ in the length of a human fetus in centimeters per week, the length of the fetus at 28 weeks.

10. Given the pulse rate $p(t)$ in beats per minute, the total number of heartbeats in a 45-minute workout.

11. Given a download speed $s(t)$ in megabytes per second, the size of the file downloaded in 30 seconds.

12. Given a rainfall rate $r(m)$ in inches per minute, the amount of rain that fell between noon and 1 p.m.

13. Given the velocity $v(t)$ of a marathon runner in miles per hour, the total distance run between the first and second hours.

14. Given a rate of change $b(t)$ in a retirement account balance in dollars per quarter since 2000, the net amount of money earned during the past five years.

15. Given the infection rate $i(t)$ of a disease in persons per day, the number of people infected after three months.

16. Given the rate $w(c)$ of weight gain in grams per kilocalorie consumed per day, the weight gained from increasing caloric intake from 2000 to 2500 calories per day.

In Exercises 17–24, use the graph of $f(x) = 5$ to evaluate the integral.

17. $\int_{3}^{3} f(x)\, dx$

18. $\int_{3}^{4} f(x)\, dx$

19. $\int_{3}^{4} 3f(x)\, dx$

20. $\int_{4}^{3} f(x)\, dx$

21. $\int_{3}^{6} f(x)\, dx$

22. $\int_{0}^{2} f(x)\, dx$

23. $\int_{2}^{0} f(x)\, dx$

24. $\int_{3}^{0} f(x)\, dx$

In Exercises 25–32, use the graph of $g(x) = 2x - 2$ to evaluate the integral.

25. $\int_{1}^{2} g(x)\, dx$

26. $\int_{1}^{4} 5g(x)\, dx$

27. $\int_{4}^{1} g(x)\, dx$

28. $\int_{0}^{4} g(x)\, dx$

29. $\int_{-2}^{1} g(x)\, dx$

30. $\int_{0}^{2} 6g(x)\, dx$

31. $\int_{2}^{-3} g(x)\, dx$

32. $\int_{2}^{-2} 3g(x)\, dx$

In Exercises 33–40, use the graph of $h(x) = -\dfrac{1}{2}x + 4$ to evaluate the integral.

33. $\int_{2}^{2} h(x)\, dx$

34. $\int_{2}^{0} 3h(x)\, dx$

35. $\int_{2}^{-2} h(x)\, dx$

36. $\int_{4}^{10} h(x)\, dx$

37. $\int_{3}^{3} h(x)\, dx$

38. $\int_{4}^{0} 7h(x)\, dx$

39. $\int_{2}^{4} h(x)\, dx$

40. $\int_{3}^{0} -2h(x)\, dx$

In Exercises 41–48, use the following graph of $f(x)$ to evaluate the definite integral:

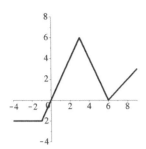

41. $\displaystyle\int_{-4}^{-2} f(x)\,dx$ 45. $\displaystyle\int_{-4}^{6} f(x)\,dx$

42. $\displaystyle\int_{0}^{4} f(x)\,dx$ 46. $\displaystyle\int_{0}^{8} f(x)\,dx$

43. $\displaystyle\int_{-1}^{4} f(x)\,dx$ 47. $\displaystyle\int_{3}^{3} f(x)\,dx$

44. $\displaystyle\int_{3}^{8} f(x)\,dx$ 48. $\displaystyle\int_{0}^{6} f(x)\,dx$

In Exercises 49–56, use the following graph of $f(x)$ to evaluate the definite integral:

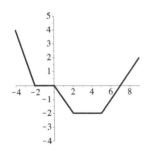

49. $\displaystyle\int_{-2}^{0} f(x)\,dx$ 51. $\displaystyle\int_{0}^{5} f(x)\,dx$

50. $\displaystyle\int_{-2}^{2} f(x)\,dx$ 52. $\displaystyle\int_{5}^{9} f(x)\,dx$

53. $\displaystyle\int_{0}^{9} f(x)\,dx$ 55. $\displaystyle\int_{5}^{2} 3f(x)\,dx$

54. $\displaystyle\int_{-4}^{9} f(x)\,dx$ 56. $\displaystyle\int_{2}^{-2} 5f(x)\,dx$

In Exercises 57–64, use the following graph of $f(x)$ to evaluate the definite integral:

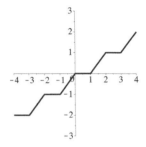

57. $\displaystyle\int_{-2}^{0} f(x)\,dx$ 61. $\displaystyle\int_{0}^{4} f(x)\,dx$

58. $\displaystyle\int_{-4}^{0} f(x)\,dx$ 62. $\displaystyle\int_{-4}^{4} 2f(x)\,dx$

59. $\displaystyle\int_{-2}^{2} f(x)\,dx$ 63. $\displaystyle\int_{1}^{3} 6f(x)\,dx$

60. $\displaystyle\int_{0}^{2} f(x)\,dx$ 64. $\displaystyle\int_{4}^{0} f(x)\,dx$

In Exercises 65–72, use the following graph of $f(x)$ to evaluate the definite integral:

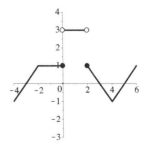

65. $\int_{-4}^{-2} f(x)\,dx$　　　69. $\int_{2}^{6} f(x)\,dx$　　　81. $\int_{6}^{1} 2s(t)\,dt$　　　83. $\int_{1}^{7} s(t)\,dt$

66. $\int_{0}^{2} f(x)\,dx$　　　70. $\int_{-2}^{6} 2f(x)\,dx$　　　82. $\int_{2}^{3} s(t)+1\,dt$　　　84. $\int_{2}^{7} s(t)\,dt$

67. $\int_{-4}^{2} f(x)\,dx$　　　71. $\int_{3}^{0} f(x)\,dx$

In Exercises 85–90, evaluate the definite integral using the values of the following integrals:

68. $\int_{0}^{4} -f(x)\,dx$　　　72. $\int_{-2}^{2} 5f(x)\,dx$

In Exercises 73–78, evaluate the definite integral using the values of the following integrals:

$$\int_{0}^{\pi/2} \sin(x)\,dx = 1 \qquad \int_{0}^{\pi} \sin(x)\,dx = 2$$

$$\int_{0}^{2\pi} \sin(x)\,dx = 0 \qquad \int_{0}^{2.5\pi} \sin(x)\,dx = 1$$

$$\int_{0}^{2} h(x)\,dx = 2 \qquad \int_{0}^{4} h(x)\,dx = 8$$

$$\int_{0}^{6} h(x)\,dx = 18 \qquad \int_{6}^{10} h(x)\,dx = 32$$

85. $\int_{0}^{\pi/2} 5\sin(x)\,dx$　　　88. $\int_{2.5\pi}^{2\pi} \sin(x)\,dx$

73. $\int_{0}^{2} 4h(x)\,dx$　　　76. $\int_{0}^{10} h(x)\,dx$

86. $\int_{0}^{\pi} \sin(x)+2\,dx$　　　89. $\int_{\pi}^{0} 4\sin(x)\,dx$

74. $\int_{2}^{4} h(x)+3\,dx$　　　77. $\int_{5}^{5} h(x)\,dx$

87. $\int_{\pi/2}^{\pi} \sin(x)\,dx$　　　90. $\int_{\pi}^{2\pi} 6\sin(x)\,dx$

75. $\int_{4}^{0} 4-3h(x)\,dx$　　　78. $\int_{6}^{4} h(x)\,dx$

In Your Own Words. In Exercises 91–99, explain the following.

91. Definite integral

92. Limits of integration

In Exercises 79–84, evaluate the definite integral using the values of the following integrals:

93. Integrand

94. Integral as signed area

95. Integral as directed area

$$\int_{1}^{3} s(t)\,dt = 4 \qquad \int_{1}^{6} s(t)\,dt = -2$$

96. Directed, signed area

$$\int_{1}^{2} s(t)\,dt = 3 \qquad \int_{3}^{7} s(t)\,dt = 7$$

97. Dimensions of definite integrals

98. Units of definite integrals

79. $\int_{1}^{3} -2s(t)\,dt$　　　80. $\int_{1}^{3} 2-s(t)\,dt$

99. Algebraic properties of integrals

6.3　First Fundamental Theorem

Our study of accumulations began with various approximation methods for quantities with known rates of change, even though their exact accumulations are often needed.

Then, we discussed how to evaluate the definite integrals of functions whose net accumulation corresponded to the directed, signed areas of such geometric figures as rectangles and triangles. However, most functions do not fall into this special category.

This section begins developing analytical methods for calculating definite integrals of functions that are presented analytically. As it turns out, integration and differentiation are intimately related to one another. Namely, they are inverse operations, which means that they undo each other in exactly the same way as $x+2$ and $x-2$ are inverses, as are e^x and $\ln(x)$. This inverse relationship enables the compututation of definite integrals by reframing integration questions in terms of differentiation.

This section begins by introducing the notion of an *antiderivative*, which reverses the process of differentiation by asking what function produces a given derivative. The first fundamental theorem of calculus is also studied, which among other things states a condition under which antiderivatives must exist as well as providing a particular form for computing them. First, the notion of an antiderivative of a function is defined:

Definition. An **antiderivative** of $f(x)$ is a function $F(x)$ such that $F'(x) = f(x)$.

Antiderivatives earn this name because each is a function that undoes the operation of differentiation. From this perspective, antiderivative computations amount to being given the derivative of some function and then being asked to find the original function. More symbolically, we might think about the following distinction, where "?" identifies the sought-for function.

- Derivative question: Given $f(x)$, $\dfrac{d}{dx}[f(x)] = f'(x) = \underline{\ ?\ }$.

- Antiderivative question: Given $f(x)$, $\dfrac{d}{dx}[\underline{\ ?\ }] = [\underline{\ ?\ }]' = f(x)$.

This study of antiderivatives begins by addressing how to verify that a correct antiderivative $F(x)$ has been identified for a given function $f(x)$. In particular, after computing $F'(x)$, if $F'(x) = f(x)$, then $F(x)$ is an antiderivative of $f(x)$ and, if not, then $F(x)$ is not an antiderivative of $f(x)$.

◆ **EXAMPLE 1** Determine if $F(x)$ is an antiderivative of $f(x)$:

(a) $F(x) = x^4 - 2$ for $f(x) = 4x^3$

(b) $F(x) = \ln(x) + 2x - 1$ for $f(x) = \dfrac{1}{x} + 2$

(c) $F(x) = \ln(x) + 2x + 7$ for $f(x) = \dfrac{1}{x} + 2$

(d) $F(x) = \cos(x) + e^x$ for $f(x) = \cos(x) + xe^{x-1}$

Solution. First compute $F'(x)$ and then compare the result with $f(x)$.

(a) The derivative of $F(x) = x^4 - 2$ is $F'(x) = 4x^3$. Thus, $F(x)$ is an antiderivative of $f(x)$, because $F'(x) = f(x)$.

(b) The derivative of $F(x) = \ln(x) + 2x - 1$ is $F'(x) = \dfrac{1}{x} + 2 = f(x)$. Thus, $F(x)$ is an antiderivative of $f(x)$, because $F'(x) = f(x)$.

(c) The derivative of $F(x) = \ln(x) + 2x + 7$ is $F'(x) = \dfrac{1}{x} + 2 = f(x)$. Thus, $F(x)$ is an antiderivative of $f(x)$, because $F'(x) = f(x)$.

(d) The derivative of $F(x) = \cos(x) + e^x$ is $F'(x) = -\sin(x) + e^x$. Thus, $F(x)$ is not an antiderivative of $f(x)$, because $F'(x) \neq f(x)$.

∎

Notice that $F(x)$ in part (b) and $F(x)$ in part (c) are both antiderivatives of $f(x) = \dfrac{1}{x} + 2$, even though the two functions $F(x)$ are different. This pattern is important and warrants further discussion.

➤ **QUESTION 1** Determine if $F(x)$ is an antiderivative of $f(x)$:

(a) $F(x) = x^2 + 3$ for $f(x) = 2x$ (c) $F(x) = x^3 - 7$ for $f(x) = 3x^2$

(b) $F(x) = x^3 - x$ for $f(x) = 3x^2$ (d) $F(x) = \sin(x)$ for $f(x) = \cos(x)$

From Example 1, recall that the function $f(x) = \dfrac{1}{x} + 2$ has two different antiderivatives: $F(x) = \ln(x) + 2x - 1$ from part (b) and $F(x) = \ln(x) + 2x + 7$ from part (c). Notice that these two antiderivatives are identical except for the constant term. In addition, recall that the derivative of a constant is zero, which means that "-1" and "$+7$" do not contribute to the resulting derivative. In fact, every function of the form $F(x) = \ln(x) + 2x + C$, where C is any real number, will have the same derivative $f(x) = \dfrac{1}{x} + 2 + 0 = \dfrac{1}{x} + 2$.

This pattern holds for any smooth function, meaning that an infinite number of different antiderivatives can be obtained simply by changing the constant term in the antiderivative. Even more, once one antiderivative $F(x)$ has been found, then every other antiderivative must be of the form $F(x) + C$, as summarized in the following theorem:

ANTIDERIVATIVES DIFFER BY A CONSTANT.

If $F(x)$ is an antiderivative of $f(x)$, then every antiderivative of $f(x)$ is of the form $F(x) + C$, where C is any real number.

This theorem is quite powerful, ensuring that once one antiderivative has been identified for a function, then essentially all antiderivatives have been found. The proof of this theorem relies on the mean value theorem, which states that for smooth functions, there exists an input on interval $[a, b]$ where the slope of the tangent line is equal to the average rate of change over $[a, b]$. A complete proof of this result lies beyond the scope of this book, and the interested reader is encouraged to consult a standard calculus textbook for more details.

The Net Accumulation Function

Formulas for the antiderivatives of most of the common modeling functions are presented soon. Before doing so, we consider the more general question of which functions have antiderivatives. A partial answer to this question is provided by means of net accumulation functions.

Definition. If $f(x)$ is a function and a is a real number in the domain of $f(x)$, then
$$A(t) = \int_a^t f(x)\, dx \text{ is the } \textbf{net accumulation function} \text{ of } f(x) \text{ with lower limit } a.$$

The net accumulation function is different from every other function encountered in this book. Throughout this discussion, the reader is encouraged to keep in mind that $A(t)$ is a function, where every input has exactly one output.

For this accumulation function $A(t)$, notice the location of the input value t as the upper limit in the integral. For each input t, the definite integral $\int_a^t f(x)\, dx$ is calculated with the input value as this upper limit in order to find the output value $A(t)$. For example, if $a = 1$, then the output associated with input $t = 2$ is determined by calculating the definite integral $A(2) = \int_1^2 f(x)\, dx$. And similarly for every other possible real number input.

A graphical illustration of varying inputs t_1, t_2, and t_3 providing varying outputs $A(t_1)$, $A(t_2)$, and $A(t_3)$ can be created based on the definite integral $\int_a^t f(x)\, dx$ corresponding to the directed, signed area of the region bounded by $f(x)$ and the x-axis on $[a, b]$. The areas of the shaded regions in Figure 1 are equal to the output values of $A(t_1)$, $A(t_2)$, and $A(t_3)$.

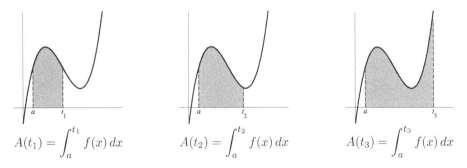

$$A(t_1) = \int_a^{t_1} f(x)\, dx \qquad A(t_2) = \int_a^{t_2} f(x)\, dx \qquad A(t_3) = \int_a^{t_3} f(x)\, dx$$

Figure 1: Net accumulation function $A(t)$ for various inputs t_1, t_2, and t_3

An important element of these net accumulation functions is the numeric value of the lower limit a. Indeed, one natural question to ask is: "Which value of a should be used as the lower limit of integration for a net accumulation function?" The only requirement for the value of this lower limit a is that it must be in the domain of the integrand function $f(x)$. Thus, the limited scale of real-life data sets might lead us to work with certain values of a.

In addition, observe that each value of a provides a different net accumulation function for $f(x)$, all of which are equally valid and will work equally well in the contexts discussed in this chapter. As it turns out, the different values of a correspond with the different values of C in the general form of an antiderivative and, again, particular contexts might lead us to work with a particular lower limit a.

The next examples will enable the reader to develop a better understanding of this net accumulation function $A(t)$ and a facility in calculating its various outputs.

◆ **EXAMPLE 2** For $a = -1$ and $f(x) = 6$, evaluate $A(t) = \displaystyle\int_{-1}^{t} 6\,dx$ for each input:

(a) $A(-1)$ (b) $A(1)$ (c) $A(5)$

Solution. Algebraic properties of definite integrals enable the evaluation of part (a), while for parts (b) and (c), the output values of $A(t)$ are determined by calculating the net area bounded by $f(x) = 6$ and the x-axis on $[-1, t]$, as illustrated in Figure 2.

(a) Algebraic property (1) of definite integrals gives $A(-1) = \int_{-1}^{-1} 6\,dx = 0$.

(b) Function $f(x) = 6$ is constant and its graph is a horizontal line. Therefore, the region bounded by $f(x)$ and the x-axis is a rectangle, and its area can be calculated using the formula $A = bh$. Thus, $A(1) = \int_{-1}^{1} 6\,dx = [1 - (-1)](6) = (2)(6) = 12$.

(c) As in part (b), the area of the corresponding bounded rectangle provides the exact value of the definite integral: $A(5) = \int_{-1}^{5} 6\,dx = [5 - (-1)](6) = (6)(6) = 36$. ■

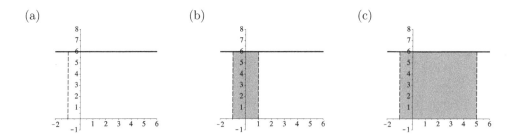

Figure 2: Graphs of (a) $A(-1)$, (b) $A(1)$, and (c) $A(5)$ for Example 2

◆ **EXAMPLE 3** For $a = 1$ and $f(x) = 2x + 1$, evaluate $A(t) = \displaystyle\int_{1}^{t} 2x + 1\,dx$:

(a) $A(2)$ (b) $A(3)$ (c) $A(-1)$

Solution. The net accumulation function is evaluated on these inputs by interpreting each definite integral as a directed, signed area bounded by $f(x)$ and the x-axis on $[1, t]$, as illustrated in Figure 3.

(a) For $A(2) = \int_{1}^{2} 2x + 1\,dx$, think of the trapezoidal region bounded by $f(x) = 2x + 1$ from $a = 1$ to $b = 2$ as consisting of a triangle stacked on top of a rectangle. Using the formulas $A = \frac{1}{2}bh$ for the area of a triangle and $A = bh$ for the area of a rectangle yields the following:

$$A(2) = \int_1^2 2x + 1\,dx = (1)(3) + \frac{1}{2}(1)(2) = 4$$

(b) For $A(3) = \int_1^3 2x + 1\,dx$, again think of the trapezoidal region as consisting of a triangle stacked on top of a rectangle and calculate the corresponding areas as follows:

$$A(3) = \int_1^3 2x + 1\,dx = (2)(3) + \frac{1}{2}(2)(4) = 10$$

(c) For $A(-1) = \int_1^{-1} 2x + 1\,dx$, apply algebraic property (2) of definite integrals $\int_a^b f(x)\,dx = -\int_b^a f(x)\,dx$ because $a = 1$ is greater than $b = -1$:

$$A(-1) = \int_1^{-1} 2x + 1\,dx = -\int_{-1}^1 2x + 1\,dx$$

The accumulation $\int_{-1}^1 2x + 1\,dx$ corresponds to two distinct pieces of signed area: a triangle below the x-axis on $[-1, 0]$ and a triangle above the x-axis on $[0, 1]$. Calculating the area of each piece and incorporating the appropriate sign information provides the value of $A(-1)$ as follows:

$$A(-1) = -\int_{-1}^1 2x + 1\,dx = -\left[\frac{1}{2}\left(\frac{1}{2}\right)(-1) + \frac{1}{2}\left(\frac{3}{2}\right)(3)\right] = -\left[-\frac{1}{4} + \frac{9}{4}\right] = -2$$

■

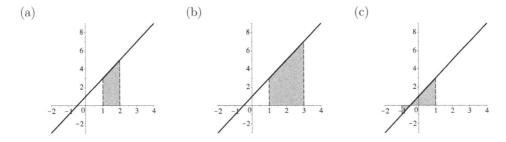

(a)   (b)   (c)

Figure 3: Graphs of (a) $A(2)$, (b) $A(3)$, and (c) $A(-1)$ for Example 3

➤ **QUESTION 2** For $a = 0$ and $f(x) = x + 1$, evaluate $A(t) = \int_0^t x + 1\,dx$:

(a) $A(0)$ (b) $A(2)$ (c) $A(3)$ (d) $A(-1)$

The First Fundamental Theorem of Calculus

Net accumulation functions turn out to be exceptionally important in the study of both integral and differential calculus. The first fundamental theorem of calculus provides a differentiation rule for these net accumulation functions. Even more, this theorem reveals that net accumulation functions provide a partial answer to the question of which functions have an antiderivative.

FIRST FUNDAMENTAL THEOREM OF CALCULUS.
If $f(x)$ is a smooth function and a is a real number in the domain of $f(x)$, then $A(t) = \int_a^t f(x)\, dx$ is an antiderivative of $f(x)$. Symbolically, the following holds:

$$\frac{d}{dt}\left[\int_a^t f(x)\, dx\right] = f(t)$$

The adjective "fundamental" in the title of this theorem reflects the deep significance of this result, which bridges and interconnects differentiation and integration. On a computational level, the first fundamental theorem of calculus provides a differentiation formula that can therefore be interpreted as providing antiderivatives for smooth functions.

Most importantly, the first fundamental theorem of calculus articulates the astonishing relationship between the derivative and the definite integral. While the idea of a slope of a tangent line (i.e., the derivative) and a net accumulation of a quantity (i.e., the definite integral) might appear completely unrelated, they are in fact deeply interconnected. In particular, this theorem asserts that the operation of differentiation undoes the definite integral operation, and is half of what must be shown to demonstrate that differentiation and integration are inverse operations. The next section introduces the second fundamental theorem of calculus, which provides the other half of this intimate relationship between derivatives and definite integrals.

The first fundamental theorem of calculus can be thought of as providing a differentiation rule in exactly the same way as the differentiation rules studied in Chapter 4. From an algorithmic perspective, when the lower limit is a constant and the upper limit is a variable, then the derivative of the integral with respect to the variable is obtained by substituting the variable into the integrand, as illustrated in the next example and question.

◆ **EXAMPLE 4** Evaluate each derivative:

(a) $\dfrac{d}{dt}\left[\displaystyle\int_2^t x^4 + 2x^2 + 1\, dx\right]$ (b) $\dfrac{d}{dt}\left[\displaystyle\int_t^1 e^{x^3}\, dx\right]$

Solution.

(a) The first fundamental theorem of calculus applies immediately as follows:

$$\frac{d}{dt}\left[\int_2^t x^4 + 2x^2 + 1\, dx\right] = t^4 + 2t^2 + 1$$

(b) The first fundamental theorem of calculus requires the lower limit to be a real number and the upper limit to be the input variable. Algebraic property (2) of definite integrals allows the limits to be swapped, provided the sign is also changed, and then the first fundamental theorem of calculus can be applied.

$$\frac{d}{dt}\left[\int_t^1 e^{x^3}\,dx\right] = \frac{d}{dt}\left[-\int_1^t e^{x^3}\,dx\right] \qquad \text{Algebraic property (2) of definite integrals}$$

$$= -\frac{d}{dt}\left[\int_1^t e^{x^3}\,dx\right] \qquad \text{Constant multiple differentiation rule}$$

$$= -e^{t^3} \qquad \text{First fundamental theorem of calculus}$$

■

➤ **QUESTION 3** Evaluate each derivative:

(a) $\dfrac{d}{dt}\left[\displaystyle\int_2^t x^2 + 2x\,dx\right]$ (b) $\dfrac{d}{dt}\left[\displaystyle\int_t^1 \ln(x^3)\,dx\right]$

The first fundamental theorem of calculus can also be thought of as providing antiderivatives for smooth functions. In particular, substituting a given function $f(x)$ as an integrand into a net accumulation function with a numeric lower limit from the domain of $f(x)$ and a variable upper limit produces an antiderivative of $f(x)$:

◆ **EXAMPLE 5**

(a) State an antiderivative of $f(x) = \ln(x^2 + 2x)$.

(b) State an antiderivative $A(t)$ of $f(x) = \sin(x^2)$ with $A(3) = 0$.

Solution.

(a) The first fundamental theorem of calculus states that a net accumulation function provides an antiderivative of a smooth function $f(x)$. Different choices can be made for the numeric lower limit a. This example considers $a = 1$, which lies in the domain of $f(x) = \ln(x^2+2x)$, to obtain the antiderivative $A(t) = \int_1^t \ln(x^2+2x)\,dx$. Different choices for the lower limit of integration produce different antiderivatives.

(b) From the first fundamental theorem of calculus $A(t) = \int_a^t \sin(x^2)\,dx$ is an antiderivative for every real number a in the domain of $f(x)$. Working toward ensuring that $A(3) = 0$, recall algebraic property (1) of definite integrals, which states that $\int_a^a f(x)\,dx$. Selecting $a = 3$ gives $A(3) = \int_3^3 \sin(x^2)\,dx = 0$, which means that $A(t) = \int_3^t \sin(x^2)\,dx$ is a requested antiderivative.

■

➤ **QUESTION 4**

(a) State an antiderivative of $f(x) = \cos(x^2)$.

(b) State an antiderivative $A(t)$ of $f(x) = e^{x^2}$ with $A(2) = 0$.

Antiderivatives of Modeling Functions

The first fundamental theorem of calculus reveals a deep connection between the definite integral and the notion of an antiderivative. Because of this connection a new notation is introduced for the antiderivative of a function.

NOTATION. The expression $\displaystyle\int f(x)\,dx$ denotes all antiderivatives of $f(x)$.

The first fundamental theorem provides one way to find the antiderivative of any smooth function. From this perspective, the first fundamental theorem of calculus provides a complete solution to the question of finding antiderivatives for a broad and important class of functions. However, on a practical level, this solution is only somewhat useful, because net accumulation functions require the calculation of directed, signed areas, which really only works in a relative handful of simple geometric settings.

Fortunately, the definition of the antiderivative of a function provides a simpler path forward for many functions that does not require the use of their net accumulation functions. Recall that antiderivatives are inverses of derivatives. Therefore, an antiderivative of a function that results from a known differentiation rule can be found by reversing the differentiation rule.

Functions resulting from known differentiation rules include many of the common modeling functions. Recall the following differentiation rules from Section 4.3 and the resulting antidifferentiation rules as examples of this approach:

Differentiation rule	Antidifferentiation rule
$\dfrac{d}{dx}\left[mx + C\right] = m$	$\displaystyle\int m\,dx = mx + C$
$\dfrac{d}{dx}\left[\dfrac{x^{n+1}}{n+1} + C\right] = x^n$	$\displaystyle\int x^n\,dx = \dfrac{x^{n+1}}{n+1} + C$
$\dfrac{d}{dx}\left[\dfrac{1}{m}\sin(mx)\right] = \cos(mx)$	$\displaystyle\int \cos(mx)\,dx = \dfrac{1}{m}\sin(mx) + C$

These antidifferentiation rules are only three of the many rules that can be stated in light of our extensive work with derivatives in Chapters 4 and 5. Some of the most important rules are summarized in the following theorem:

ANTIDERIVATIVES OF MODELING FUNCTIONS.

- $\displaystyle\int m\,dx = mx + C$ - $\displaystyle\int e^{mx+b}\,dx = \dfrac{1}{m}e^{mx+b} + C$

ANTIDERIVATIVES OF MODELING FUNCTIONS. (CONTINUED)

- $\displaystyle \int x^n \, dx \;=\; \frac{x^{n+1}}{n+1} + C, \, n \neq -1$ 　　　 • $\displaystyle \int \frac{1}{x} \, dx \;=\; \ln|x| + C$

- $\displaystyle \int \sin(mx + b) \, dx \;=\; -\frac{1}{m}\cos(mx + b) + C$

- $\displaystyle \int \cos(mx + b) \, dx \;=\; \frac{1}{m}\sin(mx + b) + C$

As discussed earlier in this section, every smooth function has infinitely many distinct antiderivatives. Furthermore, these antiderivatives differ from one another by some constant real number, because the derivative of a constant is zero. The infinite number of antiderivatives only differing by a constant is indicated by adding "$+C$" to the antiderivative, as included with the preceding formulas. This notation indicates the form of all possible antiderivatives with a relatively simple notation. In short, add "$+C$" when computing the antiderivative of a function.

In terms of basic understanding, these antidifferentiation formulas illuminate new relationships among known functions that would be hard to impossible to identify from just using the net accumulation approach to antiderivatives. On a practical level, they readily enable the determination of many antiderivatives. As with all such formulas, the best way to develop a proficiency in using them is practice.

The following examples and questions rely on the extension of algebraic properties (3) and (4) of definite integrals to this antiderivative, as just algebraic properties of integrals, which can be expressed as follows:

$$\text{Algebraic property (3):} \quad \int f(x) \pm g(x) \, dx \;=\; \int f(x) \, dx \pm \int g(x) \, dx$$

$$\text{Algebraic property (4):} \quad \int k \cdot f(x) \, dx \;=\; \int f(x) \cdot k \, dx \;=\; k \int f(x) \, dx$$

These antidifferentiation rules can be thought of as the reversals of the differentiation rules for addition, subtraction, and constant multiplication.

◆ **EXAMPLE 6** Evaluate each antiderivative:

(a) $\displaystyle \int 2x^3 + 4\sqrt{x} + 5 \, dx$ 　　　　　　　　 (b) $\displaystyle \int \frac{6}{x^4} - \frac{1}{x} \, dx$

Solution. Apply the algebraic properties of integrals and then the appropriate antidifferentiation rule, which requires expressing some integrands in an alternative form.

(a) The function $\sqrt{x}$ is expressed as $x^{1/2}$ to facilitate the application of the power rule for antiderivatives $\int x^n\, dx = x^{n+1} \cdot \dfrac{1}{n+1} + C$:

$$\int 2x^3 + 4\sqrt{x} + 5\, dx = 2\int x^3\, dx + 4\int x^{1/2}\, dx + \int 5\, dx$$

$$= 2\left(\frac{x^4}{4}\right) + 4\left(\frac{x^{3/2}}{3/2}\right) + 5x + C$$

$$= \frac{1}{2}x^4 + \frac{8}{3}x^{3/2} + 5x + C$$

(b) The function $1/x^4$ is expressed as x^{-4} using rules of exponents to enable the use of the power rule for antiderivatives. Also, the natural logarithm rule is used to antidifferentiate $1/x$ rather than the power rule:

$$\int \frac{6}{x^4} - \frac{1}{x}\, dx = 6\int x^{-4}\, dx - \int \frac{1}{x}\, dx = 6\frac{x^{-3}}{-3} - \ln|x| + C = \frac{-2}{x^3} - \ln|x| + C$$

∎

When first working with these antidifferentiation formulas, sometimes people attempt to use the power rule to antidifferentiate $1/x = x^{-1}$. However, following the power rule algorithm would result in the undefined, meaningless expression $x^{-1+1} \cdot \dfrac{1}{-1+1} = x^0 \cdot \dfrac{1}{0}$ with zero in the denominator. Fortunately, the rule resulting in the natural logarithm function provides a correct antiderivative.

◆ **EXAMPLE 7**   Evaluate each antiderivative:

(a) $\displaystyle\int 5e^{4x+5} + 9\sin(3x)\, dx$

(b) $\displaystyle\int e^{3-x} - 4\sin(4x+1)\, dx$

Solution. Apply the algebraic properties of integrals and then the antidifferentiation rules for the natural exponential and sine functions:

(a) $$\int 5e^{4x+5} + 9\sin(3x)\, dx = 5\int e^{4x+5}\, dx + 9\int \sin(3x)\, dx$$

$$= 5\cdot\frac{1}{4}e^{4x+5} + 9\left(-\frac{1}{3}\right)\cos(3x) + C$$

$$= \frac{5}{4}e^{4x+5} - 3\cos 3x + C$$

(b) $$\int e^{3-x} - 4\sin(4x+1)\, dx = \int e^{3-x}\, dx - 4\int \sin(4x+1)\, dx$$

$$= (-1)e^{3-x} - 4\left(-\frac{1}{4}\right)\cos(4x+1) + C$$

$$= -e^{3-x} + \cos(4x+1) + C$$

∎

The antidifferentiation rule for $\sin(x)$ includes a negative sign because $\dfrac{d}{dx}[\cos(x)] = -\sin(x)$. In contrast, the differentiation rule $\dfrac{d}{dx}[\sin(x)] = \cos(x)$ does not include a negative sign, so the antidifferentiation rule for $\cos(x)$ does not either. These sign properties result from and illustrate the inverse relationship between differentiation and antidifferentiation.

◆ **EXAMPLE 8** Evaluate each antiderivative:

(a) $\displaystyle\int 9x^2 + \cos(2x - 5)\,dx$

(b) $\displaystyle\int x^4 + \dfrac{1}{x^4} + \sqrt[4]{x} + \dfrac{1}{\sqrt[4]{x^7}}\,dx$

Solution. Apply the algebraic properties of integrals and then the appropriate antidifferentiation rules:

(a) $\displaystyle\int 9x^2 + \cos(2x - 5)\,dx = 9\int x^2\,dx + \int \cos(2x - 5)\,dx$

$$= 9\left(\frac{x^3}{3}\right) + \frac{1}{2}\sin(2x - 5) + C$$

$$= 3x^3 + \frac{1}{2}\sin(2x - 5) + C$$

(b) $\displaystyle\int \frac{1}{x^4} + x^4 + \sqrt[4]{x} + \frac{1}{\sqrt[4]{x^7}}\,dx = \int x^4\,dx + \int x^{-4}\,dx + \int x^{1/4} + \int x^{-7/4}\,dx$

$$= \frac{x^5}{5} + \frac{x^{-3}}{(-3)} + \frac{x^{5/4}}{5/4} + \frac{x^{-3/4}}{-3/4} + C$$

$$= \frac{x^5}{5} - \frac{1}{3x^3} + \frac{4\sqrt[4]{x^5}}{5} - \frac{4}{3\sqrt[4]{x^3}} + C$$

➤ **QUESTION 5** Evaluate each antiderivative:

(a) $\displaystyle\int x^4 - 5x^2 + 6\,dx$

(d) $\displaystyle\int 2x^4 - \frac{5}{x} + \frac{1}{7x^3}\,dx$

(b) $\displaystyle\int 6x + 4 + 3e^{4x-6}\,dx$

(e) $\displaystyle\int 3\cos(2x) + \sqrt{x} + 6\,dx$

(c) $\displaystyle\int 8\sin(7x + 4) + 5\cos(9x + 1)\,dx$

(f) $\displaystyle\int 6\sin(4x + 5) + 3e^{2x+5}\,dx$

Summary

- An *antiderivative* of $f(x)$ is a function $F(x)$ such that $F'(x) = f(x)$.

- *Antiderivatives differ by a constant*: If $F(x)$ is an antiderivative of $f(x)$, then every antiderivative of $f(x)$ is of the form $F(x) + C$, where C is any real number.

- If $f(x)$ is a function and a is a real number in the domain of $f(x)$, then $A(t) = \int_a^t f(x)\, dx$ is the *net accumulation function* of $f(x)$ with lower limit a.

- *First fundamental theorem of calculus*: If $f(x)$ is a smooth function and a is a real number in the domain of $f(x)$, then $A(t) = \int_a^t f(x)\, dx$ is an antiderivative of $f(x)$. Symbolically, the following holds.

$$\frac{d}{dt}\left[\int_a^t f(x)\, dx \right] = f(t)$$

- The expression $\int f(x)\, dx$ denotes all antiderivatives of $f(x)$.

- Antiderivatives of the common modeling functions:

$$\circ \int m\, dx = mx + C \qquad\qquad \circ \int x^n\, dx = \frac{x^{n+1}}{n+1} + C, n \neq -1$$

$$\circ \int e^{mx+b}\, dx = \frac{1}{m} e^{mx+b} + C \qquad\qquad \circ \int \frac{1}{x}\, dx = \ln|x| + C$$

$$\circ \int \sin(mx + b)\, dx = -\frac{1}{m} \cos(mx + b) + C$$

$$\circ \int \cos(mx + b)\, dx = \frac{1}{m} \sin(mx + b) + C$$

- Algebraic properties of integrals include the following:

$$\circ \text{ Property (3): } \int f(x) \pm g(x)\, dx = \int f(x)\, dx \pm \int g(x)\, dx$$

$$\circ \text{ Property (4): } \int k \cdot f(x)\, dx = \int f(x) \cdot k\, dx = k \int f(x)\, dx$$

Exercises

In Exercises 1–8, show that $F(x)$ is an antiderivative of $f(x)$.

1. $F(x) = x^4 - 2$; $f(x) = 4x^3$

2. $F(x) = x^3 - 2x + 1$; $f(x) = 3x^2 - 2$

3. $F(x) = e^{5x}$; $f(x) = 5e^{5x}$

4. $F(x) = 3e^{2x} - x^2$; $f(x) = 6e^{2x} - 2x$

5. $F(x) = xe^x$; $f(x) = e^x + xe^x$

6. $F(x) = \cos(4x + 3)$;
   $f(x) = -4\sin(4x + 3)$

7. $F(x) = \sin(3x - 2) + e^{2x}$;
   $f(x) = 3\cos(3x - 2) + 2e^{2x}$

8. $F(x) = \ln(x^2 + 1)$; $f(x) = \dfrac{2x}{x^2 + 1}$

In Exercises 9–12, evaluate the net accumulation function $A(t)$ for $f(x) = 2$ and lower limit $a = 1$ by using the graph of $f(x)$ and geometry.

9. $A(1)$

11. $A(6)$

10. $A(2)$

12. $A(-3)$

In Exercises 13–16, evaluate the net accumulation function $A(t)$ for $f(x) = -3$ and lower limit $a = -1$ by using the graph of $f(x)$ and geometry.

13. $A(-1)$

15. $A(10)$

14. $A(3)$

16. $A(-5)$

In Exercises 17–20, evaluate the net accumulation function $A(t)$ for $f(x) = 2x$ and lower limit $a = 0$ by using the graph of $f(x)$ and geometry.

17. $A(0)$

19. $A(6)$

18. $A(2)$

20. $A(-4)$

In Exercises 21–24, evaluate the net accumulation function $A(t)$ for $f(x) = 7x + 8$ and lower limit $a = 2$ by using the graph of $f(x)$ and geometry.

21. $A(2)$

23. $A(8)$

22. $A(6)$

24. $A(-4)$

In Exercises 25–28, evaluate the net accumulation function $A(t)$ for $f(x) = -5x + 4$ and lower limit $a = 3$ by using the graph of $f(x)$ and geometry.

25. $A(5)$

27. $A(3)$

26. $A(6)$

28. $A(0)$

In Exercises 29–32, evaluate the net accumulation function $A(t)$ for $f(x) = -x + 1$ and lower limit $a = -4$ by using the graph of $f(x)$ and geometry.

29. $A(-4)$

31. $A(2)$

30. $A(-3)$

32. $A(-6)$

In Exercises 33–38, evaluate the net accumulation function $A(t)$ for $f(x)$ graphed as follows and a lower limit $a = 0$:

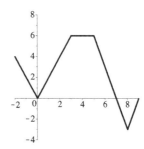

33. $A(2)$

36. $A(8)$

34. $A(4)$

37. $A(9)$

35. $A(7)$

38. $A(-2)$

In Exercises 39 – 46, evaluate the net accumulation function $A(t)$ for $f(x)$ graphed as follows and a lower limit $a = 2$:

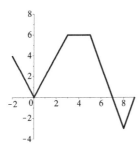

39. $A(0)$ 43. $A(7)$

40. $A(2)$ 44. $A(8)$

41. $A(3)$ 45. $A(9)$

42. $A(4)$ 46. $A(-2)$

In Exercises 47 – 52, evaluate the net accumulation function $A(t)$ for $f(x)$ graphed as follows and a lower limit $a = 0$:

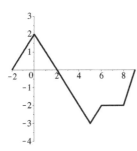

47. $A(-2)$ 50. $A(5)$

48. $A(0)$ 51. $A(6)$

49. $A(2)$ 52. $A(9)$

In Exercises 53 – 60, evaluate the net accumulation function $A(t)$ for $f(x)$ graphed as follows and a lower limit $a = 2$:

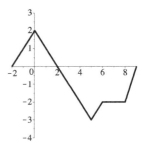

53. $A(-2)$ 57. $A(5)$

54. $A(-1)$ 58. $A(6)$

55. $A(0)$ 59. $A(8)$

56. $A(2)$ 60. $A(9)$

In Exercises 61 – 70, evaluate the derivative using the first fundamental theorem of calculus.

61. $\dfrac{d}{dt}\left[\displaystyle\int_2^t 2 - x^3 \, dx\right]$

62. $\dfrac{d}{dt}\left[\displaystyle\int_0^t xe^{-x} \, dx\right]$

63. $\dfrac{d}{dt}\left[\displaystyle\int_{-3}^t e^{-x^2} \, dx\right]$

64. $\dfrac{d}{dt}\left[\displaystyle\int_\pi^t \sin(x^2) \, dx\right]$

65. $\dfrac{d}{dt}\left[\displaystyle\int_0^t \tan(e^x) \, dx\right]$

66. $\dfrac{d}{dt}\left[\displaystyle\int_t^0 x^3 + 7x \, dx\right]$

67. $\dfrac{d}{dt}\left[\displaystyle\int_t^\pi x^2\sqrt{x} - x \, dx\right]$

68. $\dfrac{d}{dt}\left[\displaystyle\int_t^1 2e^x + 7x^3 \, dx\right]$

69. $\dfrac{d}{dt}\left[\displaystyle\int_t^1 \cos(2x - 3) \, dx\right]$

70. $\dfrac{d}{dt}\left[\displaystyle\int_t^3 xe^{-x^2} \, dx\right]$

In Exercises 71–78, find an antiderivative of $f(x)$ using the first fundamental theorem of calculus.

71. $f(x) = \sqrt{x^2 + x}$

72. $f(x) = e^x \cos(x)$

73. $f(x) = e^x \ln(x)$

74. $f(x) = 3\sin(x) + 4$

75. $f(x) = \dfrac{x^2 - \sin x}{e^x + 1}$

76. $f(x) = e^x + 1$

77. $f(x) = x^2 + \ln(x)$

78. $f(x) = \dfrac{x}{x^3 + 2x}$

In Exercises 79–94, antidifferentiate $f(x)$.

79. $f(x) = x^2 + 2x$

80. $f(x) = 3x^2 + 5x + 1$

81. $f(x) = x^3 - x^2 + x$

82. $f(x) = 3x^{12} - 4x^{10} + 1$

83. $f(x) = 5x^{20} + x^{-5} + 2$

84. $f(x) = \dfrac{32}{x^4} - \dfrac{3}{x^2} - 5$

85. $f(x) = \dfrac{1}{x} - \dfrac{1}{x^4}$

86. $f(x) = \dfrac{2}{x^5} + \dfrac{7}{x} - 9$

87. $f(x) = \sqrt{x} + \sqrt[3]{x} + 1$

88. $f(x) = \sqrt[4]{x} - 2\sqrt{x}$

89. $f(x) = \sqrt[3]{x^5} + \sqrt{x^7}$

90. $f(x) = 4\sqrt[5]{x^2} - 3\sqrt[7]{x^5}$

91. $f(x) = \dfrac{3}{\sqrt[5]{x}} + \dfrac{1}{x^2}$

92. $f(x) = \dfrac{5}{2\sqrt{x}} - \dfrac{1}{\sqrt{x^7}} - 9$

93. $f(x) = \dfrac{1}{\sqrt[3]{x^5}} - \dfrac{3}{\sqrt[7]{x^5}} + 2$

94. $f(x) = \dfrac{4}{\sqrt[4]{x^3}} + \dfrac{8}{x}$

In Exercises 95–100, antidifferentiate $f(x)$.

95. $f(x) = e^{2x+4}$

96. $f(x) = e^{3x-7}$

97. $f(x) = e^{-5x+8}$

98. $f(x) = e^{-9x-7}$

99. $f(x) = ex - 7$

100. $f(x) = \pi x + 8$

In Exercises 101–108, antidifferentiate $f(x)$.

101. $f(x) = \cos(5x + 2)$

102. $f(x) = \cos(-3x + 4)$

103. $f(x) = \cos(-2x - 9)$

104. $f(x) = \cos(4x - 8)$

105. $f(x) = \sin(5x + 2)$

106. $f(x) = \sin(-3x - 8)$

107. $f(x) = \sin(4x - 9)$

108. $f(x) = \sin(-2x + 7)$

In Exercises 109–122, antidifferentiate $f(x)$.

109. $f(x) = \ln(6) + \sqrt[4]{x} - e^{8x+4}$

110. $f(x) = \dfrac{1}{x^6} - \pi^3 + 4x + e$

111. $f(x) = \sqrt[4]{x^7} + e^{7x} - 4 - 3x^2$

112. $f(x) = \dfrac{5}{x} + 3\cos(\pi x + 2)$

113. $f(x) = ex + 7 - \dfrac{5}{x^7}$

114. $f(x) = 6e^{\pi x + 7} + \pi x + 7$

115. $f(x) = 35e^{6x+2} - 2\sin(8x+3) + 7$

116. $f(x) = \sqrt{e^{8x-4}} - \cos(3x-2) - \dfrac{3}{x}$

117. $f(x) = \sqrt{x^5} + 2\cos(3x+4) - 6$

118. $f(x) = 4x^7 - \dfrac{\pi}{ex} + e \cdot e^{13-ex}$

119. $f(x) = x^3 - \sqrt{x} - \sin(6 - \pi x)$

120. $f(x) = \dfrac{1}{\sqrt{x^8}} - \dfrac{1}{\sqrt[5]{x^3}} - 6e^{2x}$

121. $f(x) = \dfrac{x^4}{\sqrt[7]{x}} + 3\cos(9x+8) - e^{3x-4}$

122. $f(x) = e^{\pi x - 8} - 5\sin(x+8)$

In Your Own Words. In Exercises 123–128, explain the following.

123. Anti-derivative of $f(x)$

124. Antiderivatives differ by a constant

125. Net accumulation function $A(t)$

126. First fundamental theorem of calculus

127. $\displaystyle\int f(x)\,dx$

128. Antiderivatives of modeling functions

6.4   Second Fundamental Theorem

Section 6.3 introduced the antiderivative $F(x)$ of a function $f(x)$, which reverses the process of differentiation by finding $F(x)$ such that $F'(x) = f(x)$. The first fundamental theorem of calculus identifies a partial answer to the question "What functions have an antiderivative?" Namely, for every smooth function $f(x)$, the corresponding net accumulation function $A(t)$ for $f(x)$ with lower limit a from the domain of $f(x)$ is an antiderivative of $f(x)$. Symbolically, this result is expressed as follows:

$$\frac{d}{dt}\left[\int_a^t f(x)\,dx\right] = f(t)$$

The first fundamental theorem of calculus provides partial confirmation that the derivative and the integral are inverse operations.

This section studies the second fundamental theorem of calculus, which completes our understanding of the inverse relationship between the derivative and the definite integral of a function. Futhermore, this result answers a motivational question about why antiderivatives are worth studying by showing that they are integral to the process of calculating the net accumulation of a quantity given its rate of change $f(x)$ for smooth functions $f(x)$.

SECOND FUNDAMENTAL THEOREM OF CALCULUS.
If $f(x)$ is a smooth function on $[a,b]$ and $F(x)$ is any antiderivative of $f(x)$, then

$$\int_a^b f(x)\,dx = F(b) - F(a).$$

A more complete discussion about the meaning of this theorem and a partial proof of both fundamental theorems are provided at the end of this section. For now, keep in mind that this result again highlights an intimate connection between the seemingly disparate activities of finding slopes of tangent lines and net accumulation.

While many mathematicians contributed to the development of differential and integral calculus, and toward the recognition of these operations as inverses, Sir Isaac Newton and Gottfried Wilhelm von Leibniz from Saxony (now Germany) are recognized as the first people to articulate these relationships. During the seventeenth century, communication among people in different countries only took place in person, by personal letter, or through public news media or scientific journals. Consequently, Newton and Leibniz working in parallel to each other, at roughly the same time, independently articulated the first and second fundamental theorems of calculus. While they reached the same conclusions, their approaches differed from one another in various ways, and both of their perspectives and methods of "doing" calculus remain present in this area of mathematics to the present day.

As with the other results in this book, examples and questions working through the details of using this result are presented first. Then, we work on understanding why these fundamental theorems hold.

Using the Second Fundamental Theorem

Among other things, the second fundamental theorem of calculus provides a practical way to calculate the definite integral of a smooth function on an interval $[a, b]$. In effect, a definite integral can be calculated by first finding an antiderivative $F(x)$ of the integrand function $f(x)$, evaluating this antiderivative at the endpoints of the interval $[a, b]$, and then subtracting the values from one another. The evaluation and subtraction of the antiderivative are often linked together, providing the two-step process stated next.

A PROCESS FOR EVALUATING DEFINITE INTEGRALS.

If $f(x)$ is a smooth function on an interval $[a, b]$, then $\displaystyle\int_a^b f(x)\,dx$ can be evaluated using the following two steps:

(1) Find an antiderivative $F(x)$ of $f(x)$.

(2) Calculate $F(b) - F(a)$ by evaluating the antiderivative $F(x)$ at $x = b$ and $x = a$ and then subtracting these values.

Mathematicians have developed some standard notation that helps facilitate such calculations. In particular, step (1) involves finding an antiderivative of $f(x)$ and step (2) involves evaluating and subtracting this antiderivative at the limits of integration. Most often, these two steps are sufficiently intensive on their own that they are carried out as two distinct tasks rather than completing both at once. After completing step (1), the antiderivative is stated followed by a vertical bar with lower limit $x = a$ as a subscript and upper limit $x = b$ as a superscript. Then, step (2) of evaluating and subtracting is carried out. The intermediate vertical bar notation is often written as follows:

$$\int_a^b f(x)\,dx \;=\; F(x)\,\Big|_a^b \;=\; F(b) - F(a)$$

This notation mirrors the two-step process for calculating a definite integral. Also, this symbolism helps serve as a reminder that once an antiderivative has been found, the value of $F(b) - F(a)$ still needs to be calculated.

◆ **EXAMPLE 1** Evaluate each definite integral using the second fundamental theorem of calculus:

(a) $\displaystyle\int_{-1}^{3} 2x + 5\,dx$ (b) $\displaystyle\int_{1}^{e} \frac{2}{x} + \frac{1}{\sqrt[3]{x^2}}\,dx$ (c) $\displaystyle\int_{-\pi}^{\pi} 5e^{4x} - \sin(3x)\,dx$

Solution. The antiderivatives are found using the algebraic properties of integrals and the antidifferentiation formulas for basic modeling functions from Section 6.3:

(a) $\displaystyle\int_{-1}^{3} 2x + 5\,dx = 2\frac{x^2}{2} + 5x\,\Big|_{-1}^{3} = x^2 + 5x\,\Big|_{-1}^{3} = [3^2 + 5 \cdot 3] - [(-1)^2 + 5 \cdot (-1)] = 28$

(b) $\displaystyle\int_{1}^{e} \frac{2}{x} + \frac{1}{\sqrt[3]{x^2}}\,dx = \int_{1}^{e} 2 \cdot \frac{1}{x} + x^{-2/3}\,dx = 2\ln|x| + 3x^{1/3}\,\Big|_{1}^{e}$

$$= \left[2\ln(e) + 3e^{1/3}\right] - \left[2\ln(1) + 3 \cdot 1^{1/3}\right]$$

$$= 2 \cdot 1 + 3e^{1/3} - 2 \cdot 0 - 3 \cdot 1 \approx 3.187$$

(c) $\displaystyle\int_{-\pi}^{\pi} 5e^{4x} - \sin(3x)\,dx = 5 \cdot \frac{1}{4}e^{4x} - \left(\frac{-1}{3}\right)\cos(3x)\,\Big|_{-\pi}^{\pi}$

$$= \left[\frac{5}{4}e^{4\pi} + \frac{\cos(3\pi)}{3}\right] - \left[\frac{5}{4}e^{-4\pi} + \frac{\cos(-3\pi)}{3}\right]$$

$$= \frac{5}{4}e^{4\pi} + \frac{(-1)}{3} - \frac{5}{4}e^{-4\pi} - \frac{(-1)}{3} \approx 358,439.808$$

■

In Example 1, the antiderivatives of each integrand are chosen with constant term zero. However, from Section 6.3, smooth functions have infinitely many antiderivatives that all differ by a constant, which is why antiderivatives were expressed in the form $F(x) + C$. The second fundamental theorem states that *any* antiderivative of $f(x)$ may be used, and so the simplest antiderivative with $C = 0$ is typically used.

The freedom to use any antiderivative results from the subtraction $F(b) - F(a)$ in the second fundamental theorem formula. In particular, if $+C$ appears with an antiderivative then the following simplification can always be made in the final subtraction:

$$\begin{aligned} F(b) + C - [F(a) + C] &= F(b) + C - F(a) - C \\ &= F(b) - F(a) + C - C \\ &= F(b) - F(a) \end{aligned}$$

Because the $+C$'s always cancel, the choice of C does not impact the value of the definite integral and $C = 0$ is chosen to simplify the evaluation process.

➤ **QUESTION 1** Evaluate each definite integral using the second fundamental theorem of calculus:

(a) $\displaystyle\int_4^5 3x^2 + \sqrt{x}\,dx$ (b) $\displaystyle\int_0^{2\pi} 9x^3 + \cos(2x)\,dx$ (c) $\displaystyle\int_1^{2e} e^{3-x} - \frac{4}{x}\,dx$

◆ **EXAMPLE 2** Find the net accumulation of the quantity with rate of change $f(x)$ on the given domain $[a, b]$:

(a) $f(x) = x^3 + x$ on $[a, b] = [0, 1]$

(b) $f(x) = \dfrac{1}{x} - 2e^x$ on $[a, b] = [1, 3]$

(c) $f(x) = \cos(x) + \sin(x)$ on $[a, b] = [\pi, 2\pi]$

Solution. Net accumulation can be found by means of definite integrals, and the corresponding definite integrals are evaluated using the second fundamental theorem of calculus as follows:

(a) $\displaystyle\int_0^1 x^3 + x\,dx = \left.\frac{x^4}{4} + \frac{x^2}{2}\right|_0^1 = \left[\frac{1^4}{4} + \frac{1^2}{2}\right] - \left[\frac{0^4}{4} + \frac{0^2}{2}\right] = \frac{1}{4} + \frac{1}{2} - 0 - 0 = \frac{3}{4}$

(b) $\displaystyle\int_1^3 \frac{1}{x} - 2e^x\,dx = \left.\ln|x| - 2e^x\right|_1^3 = \left[\ln|3| - 2e^3\right] - \left[\ln|1| - 2e^1\right]$

$$= \ln(3) - 2e^3 - 0 - (-2e) = \ln(3) - 2e^3 + 2e \approx -33.636$$

(c) $\displaystyle\int_\pi^{2\pi} \cos(x) + \sin(x)\,dx = \left.\sin(x) - \cos(x)\right|_\pi^{2\pi}$

$$= [\sin(2\pi) - \cos(2\pi)] - [\sin(\pi) - \cos(\pi)]$$

$$= 0 - 1 - 0 + (-1) = -2$$

    ■

➤ **QUESTION 2** Find the net accumulation of the quantity with rate of change $f(x)$ on the given domain $[a, b]$:

(a) $f(x) = 2x^3 + 4\sqrt{x} + 5$ on $[a, b] = [0, 4]$

(b) $f(x) = 8 - \dfrac{3}{x} + \dfrac{3}{x^4}$ on $[a, b] = [1, 2]$

(c) $f(x) = 5e^{4x+5} - 9\sin(3x) - 1$ on $[a, b] = [0, \pi]$

◆ **EXAMPLE 3** The function $w(t)$ below provides the rate of change of the average weight of a male human fetus in pounds per week between weeks 23 and 40. At week 23, the average male fetus weighs 1.371 pounds. State and evaluate an expression involving definite integrals that provides the birth weight of an average male baby, assuming an average gestation of 40 weeks.

$$w(t) = \begin{cases} 0.334 & 23 \le t \le 34 \\ 0.525 & 35 \le t \le 40 \end{cases}$$

Solution. The following expression provides the birth weight of an average male baby. The average total weight at 23 weeks is added to integrals that provide the total weight gain from weeks 23 to 34, and from weeks 35 to 40:

$$1.371 + \int_{23}^{34} 0.334\,dt + \int_{35}^{40} 0.525\,dt$$

The second fundamental theorem of calculus is used to evaluate the integrals in this expression as follows.

$$1.371 + \int_{23}^{34} 0.334\,dt + \int_{35}^{40} 0.525\,dt = 1.371 + 0.334t \Big|_{23}^{34} + 0.525 \Big|_{35}^{40}$$
$$= 1.371 + 0.334(34 - 23) + 0.525(40 - 3535)$$
$$\approx 7.689$$

Interpreting this result in context, the average male baby weighs 7.689 pounds when born at 40 weeks.   ■

➤ **QUESTION 3** The function $w(t)$ below provides the rate of change of the average weight of a female human fetus in pounds per week between weeks 23 and 40. At week 23, the average female fetus weighs 1.294 pounds. State and evaluate an expression involving definite integrals that provides the birth weight of an average female baby, assuming an average gestation of 40 weeks.

$$w(t) = \begin{cases} 0.371 & 23 \leq t \leq 37 \\ 0.331 & 38 \leq t \leq 40 \end{cases}$$

Area Between Curves

The definite integral $\int_a^b f(x)\,dx$ can be interpreted as providing the directed, signed area of the region bounded by the integrand $f(x)$ and the x-axis on the interval $[a, b]$. As it turns out, the boundary of the x-axis on this region can be generalized from a smooth function $g(x) = 0$ to an arbitrary smooth function $g(x)$.

In this way, the definite integral can also be used to calculuate the area of the region bounded by any two smooth curves $f(x)$ and $g(x)$. Alternatively, such a definite integral provides the difference in the net accumulation of two functions. The following theorem states the formula for computations in such settings:

> **DIFFERENCE IN NET ACCUMULATION.**
> Let $f(x)$ and $g(x)$ be smooth functions with $f(x) \geq g(x)$ on an interval $[a, b]$ in the domain of both functions. The area of the region bounded by $f(x)$ and $g(x)$ on $[a, b]$, or equivalently the difference in the net accumulation of $f(x)$ and $g(x)$ on $[a, b]$ is given by the following definite integral:
>
> $$\int_a^b f(x) - g(x)\, dx$$

The information offered by Figure 1 together with the algebraic properties of integrals results in the following formula for the difference in net accumulation:

$$\text{Bounded area} = \int_a^b f(x)\, dx - \int_a^b g(x)\, dx = \int_a^b f(x) - g(x)\, dx$$

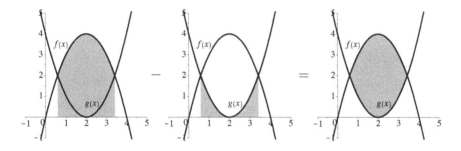

Figure 1: Difference in accumulation of two functions

◆ **EXAMPLE 4** Find the area of the region bounded by $y = x$ and $y = x^2$ on $[0, 1]$.

Solution. In order to use the formula for the difference in net accumulation, the upper bounding function $f(x)$ and the lower bounding function $g(x)$ must be identified. Proceeding graphically, Figure 2 indicates that $y = x$ is the upper function $f(x)$ and $y = x^2$ is the lower function $g(x)$. Therefore, the following integral provides the bounded area:

$$\int_0^1 f(x) - g(x)\, dx = \int_0^1 x - x^2\, dx = \frac{x^2}{2} - \frac{x^3}{3}\Big|_0^1 = \left[\frac{1^2}{2} - \frac{1^3}{3}\right] - \left[\frac{0^2}{2} - \frac{0^3}{3}\right]$$

$$= \frac{1}{2} - \frac{1}{3} - 0 + 0 = \frac{1}{6}$$

∎

▶ **QUESTION 4** Find the difference in the net accumulation of $y = 4 - (x - 2)^2 = 4x - x^2$ and $y = x$ on $[0, 3]$. The shaded region in Figure 3 identifies the corresponding bounded area.

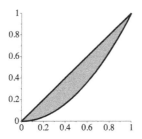

Figure 2: Graph of $y = x$ and $y = x^2$ on $[0, 1]$ for Example 4

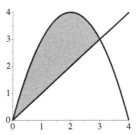

Figure 3: Graph of $y = 4x - x^2$ and $y = x$ on $[0, 3]$ for Question 4

Understanding the Fundamental Theorems

The power of the calculating tool stated by the second fundamental theorem of calculus is part of why this field of study is called calculus. Historically, many different algorithmic processes in diverse fields were referred to as calculus. Eventually, the power and effectiveness of this mathematical calculus resulted in the name of "calculus" exclusively referring to exactly the mathematical ideas and tools we have been studying.

In addition to facilitating calculations, the second fundamental theorem of calculus completes the argument that the derivative and the integral are inverse operations. This book has identified various, diverse inverse operations. For example, $x + 2$ and $x - 2$ are inverses because they undo each other. Namely, $(x + 2) - 2 = x$ and $(x - 2) + 2 = x$. Similarly, the natural exponential function e^x and the natural logarithmic funtion $\ln(x)$ are inverses because $e^{\ln(x)} = x$ and $\ln(e^x) = x$. In exactly the same way, the derivative and the integral are inverses so that applying one operation and then the other ultimately results in the original input.

The first fundamental theorem of calculus provides half of this inverse relationship. Recall that for a smooth function $f(x)$, its net accumulation function $A(t)$ is an antiderivative of $f(x)$, which is expressed symbolically by the following equation:

$$\frac{d}{dt}\left[\int_a^t f(x)\, dx\right] = f(t)$$

Thus, the derivative operation undoes the operation of the definite integral, resulting in a final output of the original integrand function.

The second fundamental theorem of calculus provides the other half of this inverse relationship between the derivative and the definite integral. Namely, the definite integral undoes the operation of differentation. This relationship may not be immediately apparent from this section's original statement of the second fundamental theorem. The inclusion of some more detail in this theorem will make this relationship more explicit. Recall that if $F(x)$ is an antiderivative of $f(x)$, then $\dfrac{d}{dx}[F(x)] = f(x)$. Substituting this expression into the conclusion of the second fundamental theorem yields the following:

$$\int_a^b f(x)\,dx \;=\; \int_a^b \frac{d}{dx}[F(x)]\,dx \;=\; F(b) - F(a)$$

From this perspective, the second fundamental theorem of calculus states that the definite integral operation undoes the operation of differentiation, completing the argument that the derivaive and definite integral are inverse operations.

A Partial Proof of the Fundamental Theorems

While complete proofs of the two fundamental theorems of calculus in their full generality lie beyond the scope of this book, an outline of the main ideas is presented here in a simplified context. This outline relies on the discussion of limits and their application to the definition of the derivative from Section 4.7.

The validity of the first fundamental theorem of calculus is established by showing that if $f(x)$ is a smooth function $f(x)$, then the following relationship holds:

$$\frac{d}{dt}[A(t)] \;=\; \frac{d}{dt}\left[\int_a^t f(x)\,dx\right] \;=\; f(t)$$

First, apply the definition of the derivative from Section 4.7 and algebraic property (5) for definite integrals:

$$
\begin{aligned}
\frac{d}{dt}[A(t)] \;&=\; \lim_{h\to 0} \frac{A(t+h) - A(t)}{h} && \text{Definition of the derivative}\\[2ex]
&=\; \lim_{h\to 0} \frac{\int_a^{t+h} f(x)\,dx - \int_a^t f(x)\,dx}{h} && \text{Substitute into } A(t)\\[2ex]
&=\; \lim_{h\to 0} \frac{\int_t^{t+h} f(x)\,dx}{h} && \text{Algebraic property (5)}
\end{aligned}
$$

The first fundamental theorem of calculus asserts that this limit is equal to $f(t)$. The rest of this partial proof assumes that $f(t)$ is an increasing function, as illustrated in Figure 4.

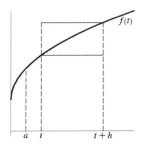

Figure 4: Increasing function $f(t)$ for partial proof of first fundamental theorem

Using the fact that $f(t)$ is increasing, applying the definitions of the left and right approximations L_1 and R_1 of $f(t)$ on $[t, t+h]$, and dividing through by h yields the following:

$$L_1 \leq \int_t^{t+h} f(x)\,dx \leq R_1 \qquad\qquad f(t) \text{ increasing}$$

$$hf(t) \leq \int_t^{t+h} f(x)\,dx \leq hf(t+h) \qquad\qquad \text{Definition of } L_1 \text{ and } R_1$$

$$\frac{hf(t)}{h} \leq \frac{\int_t^{t+h} f(x)\,dx}{h} \leq \frac{hf(t+h)}{h} \qquad\qquad \text{Divide all terms by } h$$

$$f(t) \leq \frac{\int_t^{t+h} f(x)\,dx}{h} \leq f(t+h) \qquad\qquad \text{Simplify}$$

Because the function $f(t)$ is smooth, when h becomes close to zero both $f(t)$ and $f(t+h)$ approach $f(t)$. In the language of limits, this pair of observations is expressed as follows:

$$\lim_{h \to 0} f(t) = f(t) \qquad \text{and} \qquad \lim_{h \to 0} f(t+h) = f(t)$$

In this setting, notice that the desired quantity is sandwiched between these values. Using a result called the *squeeze theorem*, the function in the middle must also have a limit of $f(t)$, because both the larger and smaller functions in the inequality have this same limit of $f(t)$. In summary, this discussion establishes the following limit for an increasing function $f(t)$:

$$\lim_{h \to 0} \frac{\int_t^{t+h} f(x)\,dx}{h} = f(t)$$

Pulling this collection equalities together yields the following:

$$\frac{d}{dt}\left[A(t)\right] = \frac{d}{dt}\left[\int_a^t f(x)\,dx\right] = \lim_{h \to 0} \frac{\int_t^{t+h} f(x)\,dx}{h} = f(t)$$

Thus, the first fundamental theorem of calculus holds for an increasing function. A very similar argument works for decreasing functions. The first fundamental theorem holds for all smooth functions, including those involving multiple monotonic behaviors, but the proof is more subtle and left for the reader's further studies in mathematics.

The validity of the second fundamental theorem of calculus relies on the fact from Section 6.3 that antiderivatives differ by a constant. The first fundamental theorem of calculus identifies $A(t) = \int_a^t f(x)\,dx$ as an antiderivative of $f(t)$. Therefore, if $F(t)$ is any antiderivative of $f(t)$, then $F(t) = A(t) + C$. Substituting a lower limit $t = a$ and an upper limit $t = b$ into this expression and subtracting yields the following:

$$
\begin{aligned}
F(b) - F(a) &= [A(b) + C] - [A(a) + C] && F(t) = A(t) + C \\
&= A(b) - A(a) && C - C = 0 \\
&= \int_a^b f(x)\,dx - \int_a^a f(x)\,dx && A(t) = \int_a^t f(x)\,dx \\
&= \int_a^b f(x)\,dx - 0 && \text{Algebraic property (1)} \\
&= \int_a^b f(x)\,dx && \text{Simplify}
\end{aligned}
$$

Thus, the second fundamental theorem of calculus holds for all smooth functions $f(x)$.

Working in RStudio

The "antiD" command in RStudio can be used to compute antiderivatives. The input syntax follows the standard RStudio approach used for makeFun, D, and many other commands. First, enter either the formula or a predefined name for the function to antidifferentiate, then the tilde "~", and finally the input variable of integration. For example, antiD(x^2~x).

The antiD command is capable of finding the symbolic antiderivative of some basic functions, but its capabilities in this regard are more limited. Therefore, this command is used primarily for computing the definite integral of a function. For such computations, the output of antiD is stored in RStudio with a name such as F = so that F(b)-F(a) can then be evaluated on the limits of integration a to b.

Examples of Commands:

- antiD(x^2+3*x-exp(x)~x)

- f=makeFun(x^2+3*x-exp(x)~x)

- F=antiD(f(x)~x)

- F(2)-F(0)

◆ **EXAMPLE 5** Evalute each definite integral using RStudio:

(a) $\displaystyle\int_1^2 x^2\, e^{x^3}\, dx$

(b) $\displaystyle\int_1^3 x^2\, (x^3 + 5)^4\, dx$

Solution. The following solutions illustrate the two possible approaches of directly entering a function into `antiD`, or using a predefined name for a function instead:

(a)   `F=antiD(x^2*exp(x^3)~x)`
      `F(2)-F(1)`

      `[1] 992.747`

(b)   `f = makeFun(x^2*(x^3+5)^4~x)`
      `F = antiD(f(x)~x)`
      `F(3)-F(1)`

      `[1] 2236444`

From this output, $\displaystyle\int_1^2 x^2\, e^{x^3}\, dx \approx 992.7466$ and $\displaystyle\int_1^3 x^2\, (x^3 + 5)^4\, dx \approx 2{,}236{,}444.$ ■

▶ **QUESTION 5** Evalute each definite integral using RStudio:

(a) $\displaystyle\int_2^4 e^{2x^3+4}\, x^2\, dx$

(b) $\displaystyle\int_0^1 \frac{3x^2\, \cos(x^3 + 1)}{\sin(x^3 + 1)}\, dx$

The antiderivatives of the integrands in the definite integrals from Example 5 and Question 5 may not be immediately apparent to the reader. And yet, without such antiderivatives, the second fundamental theorem of calculus cannot be used to evaluate these integrals. RStudio is certainly useful in such cases. In addition, Sections 6.5 and 6.6 introduce integration methods that will expand the collection of definite integrals that can be readily evaluated.

Summary

- *Second fundamental theorem of calculus*: If $f(x)$ is a smooth function on $[a, b]$ and $F(x)$ is any antiderivative of $f(x)$, then

$$\int_a^b f(x)\, dx \;=\; F(b) - F(a)$$

- *A process for calculating a definite integral*: If $f(x)$ is a smooth function on an interval $[a, b]$, then $\displaystyle\int_a^b f(x)\, dx$ can be evaluated using the following two steps:

(1) Find an antiderivative $F(x)$ of $f(x)$.

Summary (continued)

(2) Calculate $F(b) - F(a)$ by evaluating the antiderivative $F(x)$ at $x = b$ and $x = a$, and then subtracting these values.

- *Difference in net accumulation*: Let $f(x)$ and $g(x)$ be smooth functions with $f(x) \geq g(x)$ on an interval $[a, b]$ in the domain of both functions. The area of the region bounded by $f(x)$ and $g(x)$ on $[a, b]$, or equivalently the difference in the net accumulation of $f(x)$ and $g(x)$ on $[a, b]$, is given by the following definite integral:

$$\int_a^b f(x) - g(x)\, dx$$

- The fundamental theorems of calculus demonstrate that differentiation and integration are inverse operations.

Exercises

In Exercises 1–8, evaluate the definite integral using the second fundamental theorem of calculus.

1. $\int_0^3 2x + 1\, dx$

2. $\int_0^2 3x - 4\, dx$

3. $\int_1^4 -4x - 1\, dx$

4. $\int_{-1}^1 6x + 7\, dx$

5. $\int_0^4 4x^3 + 6x\, dx$

6. $\int_0^2 x^3 - 3x^2\, dx$

7. $\int_1^4 x^5 + ex\, dx$

8. $\int_1^5 x^6 - \pi x^2\, dx$

In Exercises 9–16, evaluate the definite integral using the second fundamental theorem of calculus.

9. $\int_1^2 \sqrt{x} + \frac{1}{\sqrt{x}}\, dx$

10. $\int_1^8 \sqrt[3]{x} + \frac{1}{\sqrt[3]{x}}\, dx$

11. $\int_0^1 3\sqrt{x} + \sqrt[5]{x^4}\, dx$

12. $\int_{-3}^{-1} \frac{1}{\sqrt[3]{x^2}} + x^4\, dx$

13. $\int_1^e \frac{3}{x} + ex\, dx$

14. $\int_e^{2e} \frac{5}{x} + \pi x\, dx$

15. $\int_{-2}^{-1} \frac{1}{x^2} + x^3\, dx$

16. $\int_3^7 \frac{1}{x^3} - \frac{1}{x^8}\, dx$

In Exercises 17–24, evaluate the definite integral using the second fundamental theorem of calculus.

17. $\int_0^2 e^x \, dx$

21. $\int_0^3 e^{-2x+4} \, dx$

18. $\int_{-2}^0 e^{-x} \, dx$

22. $\int_1^2 e^{2x-7} \, dx$

19. $\int_{-1}^1 e^{x+5} \, dx$

23. $\int_{-3}^{-2} e^{3x+9} \, dx$

20. $\int_{-2}^1 e^{-x-2} \, dx$

24. $\int_0^1 e^{-4x+1} \, dx$

In Exercises 25–32, evaluate the definite integral.

25. $\int_0^\pi \sin(x) \, dx$

26. $\int_0^\pi \cos(x) \, dx$

27. $\int_0^\pi \sin(x) - \cos(x) \, dx$

28. $\int_0^\pi \cos(x) - \sin(x) \, dx$

29. $\int_{-\pi}^\pi \sin(3x) - \sin(2x) \, dx$

30. $\int_{-\pi}^\pi \cos(3x) - \cos(4x) \, dx$

31. $\int_{-2\pi}^0 \cos(4x + 5) + 3x \, dx$

32. $\int_{-2\pi}^0 \sin(4x + 5) + 7x + \pi \, dx$

In Exercises 33–44, find the net accumulation of the function on the interval using the second fundamental theorem of calculus.

33. $f(x) = x^4 - 2x + 2$ on $[0, 2]$

34. $f(x) = x^3 + x^2 + x$ on $[-2, 1]$

35. $f(x) = 3 - 4x^3$ on $[-2, 2]$

36. $f(x) = \sqrt[5]{x} - 3x^2$ on $[0, 1]$

37. $f(x) = x - \sin(x)$ on $[0, \pi]$

38. $f(x) = x + \sin(x)$ on $[0, 2\pi]$

39. $f(x) = \cos(2x)$ on $[-2\pi, \pi]$

40. $f(x) = 2\sin(2x + 1)$ on $[1, 1.5]$

41. $f(x) = e^{2x} - \dfrac{2}{x}$ on $[2, 4]$

42. $f(x) = e^{-3x+1} + e^{-4x}$ on $[0, 1]$

43. $f(x) = \dfrac{2}{e^x} + x$ on $[-1, 2]$

44. $f(x) = \dfrac{4}{\sqrt[3]{x}} - \dfrac{1}{x^2}$ on $[1, 2]$

In Exercises 45–48, determine the average dimensions of male and female babies at birth using the piecewise linear models of growth rate.

45. The function $ML(t)$ provides the rate of change in the average length of a male human fetus in inches per week between weeks 23 and 40:

$$ML(t) = \begin{cases} 0.528 & 23 \le t \le 27 \\ 0.512 & 28 \le t \le 37 \\ 0.285 & 38 \le t \le 40 \end{cases}$$

At week 23, the average male fetus is 12.008 inches long. State an expression involving definite integrals that provides the length of an average male baby, assuming an average gestation of 40 weeks.

46. The function $FL(t)$ provides the rate of change in the average length of a female human fetus in inches per week between weeks 23 and 40:

$$FL(t) = \begin{cases} 0.546 & 23 \le t \le 30 \\ 0.419 & 31 \le t \le 40 \end{cases}$$

At week 23, the average female fetus is 11.772 inches long. State an expression involving definite integrals that provides the length of an average female baby, assuming an average gestation of 40 weeks.

47. The function $FH(t)$ provides the rate of change in the average head circumference of a female human fetus in inches per week between weeks 23 and 40:

$$FH(t) = \begin{cases} 0.367 & 23 \leq t \leq 34 \\ 0.191 & 35 \leq t \leq 40 \end{cases}$$

At week 23, the average female fetus has a head circumference of 8.189 inches. State an expression involving definite integrals that provides the head circumference of an average female baby, assuming an average gestation of 40 weeks.

48. The function $MH(t)$ provides the rate of change in the average head circumference of a male human fetus in inches per week between weeks 23 and 40:

$$MH(t) = \begin{cases} 0.370 & 23 \leq t \leq 33 \\ 0.276 & 34 \leq t \leq 38 \\ 0.079 & 39 \leq t \leq 40 \end{cases}$$

At week 23, the average male fetus has a head circumference of 8.386 inches. State an expression involving definite integrals that provides the head circumference of an average male baby, assuming an average gestation of 40 weeks.

In Exercises 49–60, find the difference in accumulation between the functions on the interval.

49. $f(x) = 9$; $g(x) = x$; $[0, 3]$

50. $f(x) = 1$; $g(x) = -x$; $[-1, 1]$

51. $f(x) = 2 - x$; $g(x) = x^2$; $[0, 1]$

52. $f(x) = 4 - x^2$; $g(x) = x^2 - 4$; $[-1, 2]$

53. $f(x) = 4$; $g(x) = x^2$; $[0, 2]$

54. $f(x) = x^3$; $g(x) = \sqrt[3]{x}$; $[-1, 1]$

55. $f(x) = e^x$; $g(x) = e^{-x}$; $[0, 1]$

56. $f(x) = e^{2x}$; $g(x) = e^{x+1}$; $[1, 2]$

57. $f(x) = \cos(x)$; $g(x) = \sin(x)$; $\left[0, \frac{\pi}{4}\right]$

58. $f(x) = \sin(x)$; $g(x) = \cos(x)$; $\left[\frac{\pi}{4}, \frac{5\pi}{4}\right]$

59. $f(x) = \cos(x)$; $g(x) = \cos(2x)$; $\left[0, \frac{\pi}{2}\right]$

60. $f(x) = \sin(2x)$; $g(x) = \sin(x)$; $\left[0, \frac{\pi}{4}\right]$

RStudio. In Exercises 61–76, use RStudio to evaluate the definite integrals given in the specified exercises.

61. Exercise 1

62. Exercise 3

63. Exercise 5

64. Exercise 7

65. Exercise 9

66. Exercise 11

67. Exercise 13

68. Exercise 15

69. Exercise 17

70. Exercise 19

71. Exercise 21

72. Exercise 23

73. Exercise 25

74. Exercise 27

75. Exercise 29

76. Exercise 31

In Your Own Words. In Exercises 77–82, explain the following.

77. Second fundamental theorem of calculus

78. Process for calculating a definite integral using the second fundamental theorem of calculus

79. Finding a difference in net accumulation

80. Finding the area of a region bounded by two functions

81. The relationship between differentiation and integration

82. Squeeze theorem

6.5   The Method of Substitution

This chapter has focused on the fundamental theorems of calculus and their identification of the inverse relationship between the derivative and the integral as means to understanding and calculating the accumulation of quantity with a known rate of change. While left, right, and midpoint approximations provide a successul way to estimate such accumulations, and geometric formulas enable the determination of exact accumulations in certain relatively simple contexts, the formula $\int F'(x)\,dx = F(b) - F(a)$ from the second fundamental theorem of calculus is the most broadly applicable and successful means for finding exact accumulations.

The process of antidifferentiating a rate function $f(x) = F'(x)$ is often the most challenging aspect of using the second fundamental theorem. The first fundamental theorem ensures that for any smooth function $f(x)$, the corresponding net accumulation function $A(t)$ of $f(x)$ is an antiderivative of $f(x)$. However, these antiderivatives are of the same limited usefulness as the initial L_n, R_n, and M_n approximations and as the geometric formulas.

Fortunately, the many known differentiation formulas can be reversed, or run backwards, to provide antidifferentiation formulas. Following this approach, Section 6.3 introduced such formulas as the following:

<u>Differentiation rule</u>   <u>Antidifferentiation rule</u>

$$\frac{d}{dx}\left[mx + C\right] = m \qquad\qquad \int m\,dx = mx + C$$

$$\frac{d}{dx}\left[\frac{x^{n+1}}{n+1} + C\right] = x^n \qquad \int x^n\,dx = \frac{x^{n+1}}{n+1} + C$$

$$\frac{d}{dx}\left[\frac{1}{m}\sin(mx)\right] = \cos(mx) \qquad \int \cos(mx)\,dx = \frac{1}{m}\sin(mx) + C$$

Similarly, the sum, difference, and constant multiple rules for differentiation have analogues that enable the antidifferentiation of many simple arithmetic combinations of the common modeling functions.

The next two sections focus on expanding the collection of functions that can be readily antidifferentiated by developing the reversals of two additional differentiation rules: the chain rule and the product rule. This section begins this expansion with the *method of substitution*, which runs the chain rule backwards. Just as the chain rule had the most significant impact in terms of increasing the collection of functions that can be readily differentiated, substitution is similarly the most broadly applicable method of integration. The next section considers a process for reversing the product rule called *integration by parts*. Moving forward in this direction, some features of the chain rule are recalled and then the method of substitution is discussed.

The Chain Rule

The chain rule provides a formula for differentiating the composition of two functions $f(x)$ and $g(x)$. Recall that for two functions, the composition of $f(x)$ with $g(x)$ is

$f \circ g(x) = f(g(x))$, when this expression is defined. For example, if $f(x) = \sqrt{x}$ and $g(x) = -x^2$, then $f(g(x)) = \sqrt{-x^2}$ can be expressed symbolically, but is only defined when $x = 0$. On the other hand, $g(f(x)) = -(\sqrt{x})^2$ is defined for all nonnegative reals numbers. Section 4.5 introduced the chain rule, which enables the differentiation of these and many other compositions of functions.

CHAIN RULE.

Let $f(x)$ and $g(x)$ be differentiable functions. When the compositions are defined, the following equality holds:

$$\frac{d}{dx}[\,f(g(x))\,] \;=\; f'[g(x)] \cdot g'(x)$$

A key step in using the chain rule, and so the method of substitution, is recognizing how to decompose a composition of two functions into its inside function $g(x)$ and its outside function $f(x)$. Developing a proficiency in using the chain rule involved acquiring some skill at this recognition process. As a refresher, some specific examples are considered here.

◆ **EXAMPLE 1** Identify the inside function $g(x)$ and the outside function $f(x)$ of each composition $f(g(x))$:

(a) $(2x^4 - 3)^6$ (b) $\ln(3x^2 + e^x)$ (c) $\left(\dfrac{e^x + \sin(x)}{x^3 + 2x^4 + 1} \right)^3$

Solution.

(a) The composition $(2x^4 - 3)^6$ results from the inside function $g(x) = 2x^4 - 3$ and the outside function $f(x) = x^6$.

(b) For $\ln(3x^2 + e^x)$, the inside function is $g(x) = 3x^2 + e^x$ and the outside function is $f(x) = \ln(x)$.

(c) The inside function is $g(x) = \dfrac{e^x + \sin(x)}{x^3 + 2x^4 + 1}$ and the outside function is $f(x) = x^3$.

                                                                             ■

▶ **QUESTION 1** Identify the inside function $g(x)$ and the outside function $f(x)$ of each composition $f(g(x))$:

(a) $e^{x^2 + 3x + 1}$ (b) $e^{x^2} + 1$ (c) $\dfrac{2}{\cos(x)}$

Antiderivatives via Substitution

The method of substitution provides an algorithmic process for finding the antiderivative of a product of two functions, where one term in the product is the composition of two functions and the other term is the derivative of the inside function of this composition.

An overview of this process is discussed, and then the method of substitution is applied to some specific examples.

Beginning with the big picture, suppose the antiderivative $\int f'(g(x))\, g'(x)\, dx$ is sought. The integrand consists of a product of a composition whose inside function is $g(x)$ with the derivative $g'(x)$ of this inside function. After recognizing this pattern, set $u = g(x)$ and compute its differential $du = g'(x)\, dx$. Differentials are explained at the end of this section; for now, think of $du = g'(x)\, dx$ as the derivative of $u = g(x)$. Substituting u and du into the original integral yields the following equation:

$$\int f'(g(x))\, g'(x)\, dx \;=\; \int f'(u)\, du$$

Ideally, the resulting integral $\int f'(u)\, du$ is simpler and more readily evaluated. Once this new integral has been antidifferentiated, the expression $u = g(x)$ is substituted for every u appearing in the antiderivative to obtain the final answer. This method of substitution is formally stated as follows:

INTEGRATION BY SUBSTITUTION.
Let $f(x)$ and $g(x)$ be smooth functions. When the compositions are defined, the following equality holds:

$$\int f'[\,g(x)\,]\, g'(x)\, dx \;=\; \int f'(u)\, du \;=\; f(g(x)) + C$$

As with all computational skills, practice is essential for developing a skilled proficiency with using them. The next few examples model specific applications of this method.

◆ **EXAMPLE 2** Find each antiderivative using the method of substitution:

(a) $\displaystyle\int 7(x^3 - 4)^6\, 3x^2\, dx$　　　　　　(b) $\displaystyle\int e^{x^4 + x}\, (4x^3 + 1)\, dx$

Solution.

(a) The integrand $7(x^3 - 4)^6\, 3x^2$ is the product of the composition of inside function $g(x) = x^3 - 4$ and outside function $f'(x) = 7x^6$, with the derivative of its inside function $g'(x) = 3x^2$. Set $u = x^3 - 4$ (the inside function), and compute its differential $du = 3x^2\, dx$. Applying the method of substitution yields the following solution:

$$\int 7(x^3 - 4)^6\, 3x^2\, dx \;=\; \int 7u^6\, du \qquad \text{Substitute } u \text{ and } du$$

$$= 7 \cdot \frac{u^7}{7} + C \qquad \text{Antidifferentiate } u^6$$

$$= u^7 + C \qquad \text{Simplify}$$

$$= (x^3 - 4)^7 + C \qquad \text{Back substitute } u = x^3 - 4$$

(b) The integrand $e^{x^4+x}(4x^3+1)$ is the product of the composition of inside function $g(x) = x^4 + x$ and outside function $f'(x) = e^x$, with the derivative of its inside function $g'(x) = 4x^3 + 1$. Set $u = x^4 + x$ (the inside function), and compute its differential $du = (4x^3 + 1)\,dx$. Applying the method of substitution yields the following solution:

$$\int e^{x^4+x}(4x^3+1)\,dx = \int e^u\,du \qquad \text{Substitute } u \text{ and } du$$
$$= e^u + C \qquad \text{Antidifferentiate } e^u$$
$$= e^{x^4+x} + C \qquad \text{Back substitute } u = x^4 + x$$

∎

While studying these examples, the reader will want to notice the various selections of inside functions $u = g(x)$ for each integrand. These representative examples reflect common choices when using the method of substitution. In Example 2(a), the inside function appeared inside a power function and, in Example 2(b), it was the exponent of a natural exponential expression.

◆ **EXAMPLE 3** Find each antiderivative using the method of substitution:

(a) $\displaystyle\int \frac{\cos(x)}{\sin^5(x)}\,dx$
(b) $\displaystyle\int \sin(x^2)\,2x\,dx$

Solution.

(a) The integrand is a product of a composition with the derivative of the inside function of this composition. The inside function is $u = \sin(x)$, which has differential $du = \cos(x)\,dx$, and the outside function is $f'(x) = \dfrac{1}{x^5}$ because in the denominator of the integrand $\sin^5(x) = [\sin(x)]^5$. Applying the method of substitution gives the following:

$$\int \frac{\cos(x)}{\sin^5(x)}\,dx = \int \frac{1}{u^5}\,du = \int u^{-5}\,du = \frac{u^{-4}}{-4} + C = -\frac{1}{4}\frac{1}{[\sin(x)]^4} + C$$

(b) The integrand is a product of a composition with the derivative of the inside function of this composition. The inside function is $u = x^2$, which has differential $du = 2x\,dx$, and the outside function is $f'(x) = \sin(x)$. Applying the method of substitution gives the following:

$$\int \sin(x^2)\,2x\,dx = \int \sin(u)\,du = -\cos(u) + C = -\cos(x^2) + C$$

∎

While a justification is not given in Example 3 for each equality as in Example 2, the reasons for each step are essentially the same. The reader is encouraged to think about why each equality holds in Example 3 and refer back to Example 2 as appropriate. Also, notice in Example 3 that $u = g(x)$ appeared inside a power function in part (a) and inside a trigonometric function in part (b).

➤ **QUESTION 2** Find each antiderivative using the method of substitution:

(a) $\displaystyle\int 5(x^2+1)^4\, 2x\, dx$

(c) $\displaystyle\int \frac{6x^5}{x^6+2}\, dx$

(b) $\displaystyle\int \cos(x^3+1)\, 3x^2\, dx$

(d) $\displaystyle\int e^{4x^3}\, 12x^2\, dx$

◆ **EXAMPLE 4** Find each antiderivative using the method of substitution:

(a) $\displaystyle\int \frac{\cos(x)}{e^{\sin(x)}}\, dx$

(b) $\displaystyle\int \sqrt{\ln(x)-x}\,\left(\frac{3}{x}-3\right)\, dx$

Solution.

(a) The integrand is a product of a composition with the derivative of the inside function of this composition. The inside function is $u=\sin(x)$, which has differential $du=\cos(x)\, dx$, and the outside function is $f'(x)=\dfrac{1}{e^x}=e^{-x}$. Applying the method of substitution gives the following:

$$\int \frac{\cos(x)}{e^{\sin(x)}}\, dx = \int \frac{1}{e^u}\, du = \int e^{-u}\, du = -e^{-u}+C = -e^{-\sin(x)}+C = -\frac{1}{e^{\sin(x)}}+C$$

(b) Let $u=\ln(x)-x$, which has differential $du=\dfrac{1}{x}-1\, dx$. After factoring a 3 from the final term in the integrand, the method of substitution yields the following:

$$\int \sqrt{\ln(x)-x}\cdot\left(\frac{3}{x}-3\right) dx = 3\int \sqrt{\ln(x)-x}\cdot\left(\frac{1}{x}-1\right) dx$$

$$= 3\int \sqrt{u}\, du$$

$$= 3\int u^{1/2}\, du$$

$$= 3\frac{u^{3/2}}{3/2}+C$$

$$= 3\frac{2}{3}[\ln(x)-x]^{3/2}+C$$

$$= 2\sqrt{[\ln(x)-x]^3}+C$$

■

➤ **QUESTION 3** Find each antiderivative using the method of substitution:

(a) $\displaystyle\int \frac{\cos(x)}{[\sin(x)+3]^2}\, dx$

(c) $\displaystyle\int \frac{4x-8}{\sqrt{x^2-4x}}\, dx$

(b) $\displaystyle\int \frac{1}{\ln(x)}\frac{1}{x}\, dx$

(d) $\displaystyle\int [2\ln(x)+1]^4\frac{2}{x}\, dx$

Substitution with Intermediate Algebra

Every application of the method of substitution considered thus far has involved the product of a composition $f'[g(x)]$ with the derivative of its inside function $g'(x)$. Sometimes the method of substitution can be applied even when the product does not present in exactly this fashion. In particular, even when the second term in the product is not exactly the derivative of the inside function, some intermediate algebra can still allow the use of this integration method.

The most common such circumstance arises when the derivative of the inside function and the second term in the product differ by a constant. For example, when finding the antiderivative $\int e^{x^2} x\,dx$, an appropriate choice of the inside function is $u = x^2$, which has differential $du = 2x\,dx$. However, the integrand contains $x\,dx$ rather than $2x\,dx$. In such a case, dividing by two yields $\frac{1}{2}\,du = x\,dx$, which allows the substitution to be made and the integral to be evaluated as follows:

$$\int e^{x^2} x\,dx \;=\; \int e^u \frac{1}{2}du \;=\; \frac{1}{2}\int e^u\,du \;=\; \frac{1}{2}e^u + C \;=\; \frac{1}{2}e^{x^2} + C$$

The second most common such circumstance arises when extra terms in the original variable x are not taken care of by either $u = g(x)$ or $du = g'(x)\,dx$ when substituting. In this situation, a sometimes successful strategy is to revisit the original relationship $u = g(x)$ between the variables u and x, solve for x in terms of u, and replace any remaining terms with this new expression for x in terms of u. Often the resulting integrand can be antidifferentiated, although sometimes another intergration method must be used. Example 5(b) presents a specific application of this process.

◆ **EXAMPLE 5** Find each antiderivative using the method of substitution:

(a) $\displaystyle\int x^2 \cos(x^3)\,dx$ (b) $\displaystyle\int x\sqrt{x+1}\,dx$

Solution.

(a) For this integrand, the inside function of $u = x^3$ yields the differential $du = 3x^2\,dx$. Notice that the outside function $\cos(x)$ is multiplied by x^2, rather than $3x^2$. Therefore, divide both sides of the differential expression $du = 3x^2\,dx$ by 3 to obtain $\frac{1}{3}du = x^2\,dx$. Now apply the method of substitution as follows:

$$\int x^2 \cos(x^3)\,dx \;=\; \frac{1}{3}\int \cos(u)\,du \;=\; \frac{1}{3}\sin(u) + C \;=\; \frac{1}{3}\sin(x^3) + C$$

(b) The most natural substitution choice for this integrand is the inside of the square root function $u = x + 1$, which yields $du = dx$. Substituting at this point leaves an x inside the integral:

$$\int x\sqrt{x+1}\,dx \;=\; \int x\sqrt{u}\,du$$

In this context, the variable x cannot be treated like a constant; for example, moving x to the front of the integral is not valid. Instead, the remaining, extra x must be replaced with an equivalent expression involving just u. Because $u = x+1$, rearranging terms yields $x = u - 1$, which can then be substituted into the integral to obtain an integrand expressed entirely in terms of the variable u. The resulting integral can then be antidifferentiated.

$$
\begin{aligned}
\int x\sqrt{x+1}\,dx &= \int (u-1)\sqrt{u}\,du && \text{Substitute } u,\ u-1,\ \text{and } du \\[2mm]
&= \int (u-1)u^{1/2}\,du && \text{Exponent rule } \sqrt[r]{a} = a^{1/r} \\[2mm]
&= \int u \cdot u^{1/2} - u^{1/2}\,du && \text{Distribute} \\[2mm]
&= \int u^{3/2} - u^{1/2}\,du && \text{Exponent rule } a^r a^s = a^{r+s} \\[2mm]
&= \frac{u^{5/2}}{5/2} - \frac{u^{3/2}}{3/2} + C && \text{Antidifferentiate} \\[2mm]
&= \frac{2}{5}(x+1)^{5/2} - \frac{2}{3}(x+1)^{3/2} + C && \text{Substitute } u = x+1
\end{aligned}
$$

$\blacksquare$

▶ **QUESTION 4** Find each antiderivative using the method of substitution:

(a) $\displaystyle \int x^3 \cos(x^4)\,dx$ (b) $\displaystyle \int x^2 \sqrt[3]{x+2}\,dx$

The Method of Substitution and Definite Integrals

The primary purpose of this integration method is to provide a process for antidifferentiating functions that are not among the common modeling functions or simple arithmetic combinations of such functions. In addition, the method of substitution also enables the evaluation of many more definite integrals, because the first step in calculating a definite integral is to identify the antiderivative of the integrand. The combination of the second fundamental theorem of calculus with the method of substitution is summarized below.

INTEGRATION BY SUBSTITUTION AND DEFINITE INTEGRALS.
Let $f(x)$ and $g(x)$ be smooth functions. When the compositions are defined, the following equalities hold:

$$
\int_a^b f'[\,g(x)\,]\,g'(x)\,dx \;=\; \int_{x=a}^{x=b} f'(u)\,du \;=\; f(u)\,\Big|_{x=a}^{x=b} \;=\; f(g(x))\,\Big|_a^b \;=\; f[\,g(b)\,] - f[\,g(a)\,]
$$

While the preceding long string of equalities may appear a bit overwhelming, the reader is encouraged to keep in mind the two steps for evaluating a definite integral remain the same. First, find an antiderivative of the integrand, and then calculate the difference of the antiderivative evaluated at the two endpoints. The appearance of greater complexity results from using substitution to find the antiderivative of the integrand. The next examples illustrate this process.

◆ **EXAMPLE 6** Evaluate the definite integral $\displaystyle\int_0^1 \frac{x^2}{(x^3+2)^2}\, dx$.

Solution. The integrand cannot be immediately antidifferentiated using a formula, so the method of substitution is applied. Set $u = x^3 + 2$, which yields $du = 3x^2\, dx$. Dividing both sides by 3 to make the differential match the integrand results in $\frac{1}{3} du = x^2\, dx$. The method of substitution and the second fundamental theorem of calculus give the following:

$$
\begin{aligned}
\int_0^1 \frac{x^2}{(x^3+2)^2}\, dx &= \int_{x=0}^{x=1} \frac{1}{u^2} \frac{1}{3}\, du && \text{Substitute } u \text{ and } du \\[2mm]
&= \frac{1}{3} \int_{x=0}^{x=1} u^{-2}\, du && \text{Exponent rule } \frac{1}{a^r} = a^{-r} \\[2mm]
&= \frac{1}{3} (-1)\, u^{-1} \Big|_{x=0}^{x=1} && \text{Antidifferentiate} \\[2mm]
&= -\frac{1}{3} \frac{1}{x^3+2} \Big|_0^1 && \text{Substitute } u = x^3 + 2 \\[2mm]
&= \left[-\frac{1}{3} \frac{1}{(1)^3+2} \right] - \left[-\frac{1}{3} \frac{1}{(0)^3+2} \right] && F(1) - F(0) \\[2mm]
&= -\frac{1}{9} + \frac{1}{6} = \frac{1}{18} && \text{Simplify}
\end{aligned}
$$

■

In Example 6, notice that after the substitution has been made and the integrand has been expressed in terms of u rather than x, the limits of integration are presented as $x = 0$ and $x = 1$, rather than as just 0 and 1. By convention, unlabeled limits are assumed to be values of the variable appearing in the integrand. Therefore, unchanged limits must be accurately labeled by the appropriate variable (in this case, $x = 0$ and $x = 1$) or the needed equality of integrals does not hold.

An alternative approach this is sometimes used is to evaluate $u = x + 1$ at $x = 0$ to find a new lower limit of $u = 1$, and $u = x + 1$ at $x = 1$ to find a new lower limit of $u = 2$. This approach is certainly correct and will work fine, but this book chooses to just label the limits to minimize the number of steps.

◆ **EXAMPLE 7** Evaluate the definite integral $\displaystyle\int_1^{e^\pi} \frac{\sin(\ln(x))}{x}\, dx$.

Solution. The integrand cannot be immediately antidifferentiated using a formula, so the method of substitution is applied. Set $u = \ln(x)$, which has differential $du = \frac{1}{x}\, dx$. The method of substitution and the second fundamental theorem of calculus give the following:

$$
\begin{aligned}
\int_1^{e^\pi} \frac{\sin(\ln(x))}{x}\, dx &= \int_{x=1}^{x=e^\pi} \sin(u)\, du = -\cos(u) \Big|_{x=1}^{x=e^\pi} = -\cos(\ln(x)) \Big|_1^{e^\pi} \\[2mm]
&= [-\cos(\ln(e^\pi))] - [-\cos(\ln(1))] = [-\cos(\pi)] - [-\cos(0)] \\[2mm]
&= -(-1) - (-1) = 2
\end{aligned}
$$

■

➤ **QUESTION 5** Evaluate each definite integral:

(a) $\displaystyle\int_{-1}^{0} x^2 \sqrt{x^3 + 1}\, dx$

(b) $\displaystyle\int_{0}^{1} \frac{x^3}{(5 - x^4)^6}\, dx$

Understanding Differentials

Throughout this section, substitutions of the form $u = g(x)$ have been made using its corresponding differential $du = g'(x)\, dx$. The dx accompanying the computations of the derivative of the inside function $u = g(x)$ may seem a bit curious. At some level, including the dx can be viewed as simply part of the notation, in parallel with dx being included as part of the notation of the definite integral. However, the reason that dx appears is a bit more subtle, and its meaning is briefly explored here.

Recall from Section 4.1 that the tangent line $y = f(a) + f'(a)(x - a)$ provides a good local, linear approximation to $f(x)$ for input $x = a$. In other words, the output $f(x) \approx f(a) + f'(a)(x - a)$ for inputs x near a, which can be expressed as $f(x) - f(a) \approx f'(x)(x - a)$. Now, let $\Delta y = f(x) - f(a)$ and let $\Delta x = x - a$ represent the changes in x and y, respectively, and substitute to obtain $\Delta y \approx f'(a)\Delta x$. Based on this equation, for x near a, the change in y is approximately equal to the product of the value of the derivative when $x = a$ multiplied by the change in x. Allowing the change in x to become infinitesimally small leads to the following definition:

Definition.   Let $f(x)$ be a smooth function and let dx be a small change in the input x. The **differential** $dy = f'(x)\, dx$ approximates the actual change in y provided by $\Delta y = f(x + dx) - f(x)$.

◆ **EXAMPLE 8** Consider $f(x) = e^{x^2}$.

(a) Compute an expression for the differential dy for a small change dx in the input.

(b) Compute and compare Δy and dy when x changes from 1 to 1.1

Solution.

(a) Because $f(x) = e^{x^2}$, the chain rule gives $f'(x) = 2xe^{x^2}$. Therefore, the differential is $dy = 2xe^{x^2}\, dx$.

(b) The input x changes from 1 to 1.1, which means $dx = 1.1 - 1 = 0.1$. The actual change in outputs is $\Delta y = f(x + dx) - f(x) = e^{(x+dx)^2} - e^{x^2}$ and, from part (a), $dy = 2xe^{x^2}\, dx$. Substituting $x = 1$ and $dx = 0.1$ into these expressions yields the following numeric results:

- $\Delta y = e^{(1+0.1)^2} - e^{1^2} = e^{(1.1)^2} - e \approx 0.635$
- $dy = 2(1)e^{1}(0.1) = 0.2e \approx 0.544$

The approximation dy is somewhat close to the actual change in output, Δy. In the context of an exponential function, $dx = 0.1$ is actually a relatively large change in input values. A smaller change dx would improve the accuracy of the approximation.   ■

➤ **QUESTION 6** Consider $f(x) = \cos(e^x)$.

(a) Compute an expression for the differential dy for a small change dx in the input.

(b) Compute and compare Δy and dy when x changes from 1 to 1.1

Recall that dx appears in a definite integral in connection with the width of the rectangles used to calculuate the left and right approximations of the definite integral. When using equal-width rectangles to compute R_n and L_n, an increase in the numeric value of n results in the width of the rectangles becoming smaller and smaller. As this width becomes infinitesimally small, the symbol dx is used instead of Δx.

As such, the definite integral $\int f'(g(x))g'(x)\,dx$ can be thought of as being computed by means of rectangles of width dx. Now apply this perspective to the method of substitution. When substituting $u = g(x)$, the width dx of the rectangles is not necessarily the exact same as the width du, but usually the width is either magnified or shrunk by a factor of $g'(x)$. Therefore, when substituting, dx is not replaced with just du, but with $du = g'(x)\,dx$ to reflect this change in widths and to preserve the equality of the integrals.

Summary

- *Chain rule*: Let $f(x)$ and $g(x)$ be differentiable functions. When the compositions are defined, the following equality holds:

$$\frac{d}{dx}[f(g(x))] = f'[g(x)] \cdot g'(x)$$

- The integration method of substitution results from running the chain rule backwards.

- *Integration by substitution*: Let $f(x)$ and $g(x)$ be smooth functions. When the compositions are defined, the following equality holds:

$$\int f'[g(x)]\,g'(x)\,dx = \int f'(u)\,du = f(g(x)) + C$$

- *Substitution and definite integrals*: Let $f(x)$ and $g(x)$ be smooth functions. When the compositions are defined, the following equalities hold:

$$\int_a^b f'[g(x)]\,g'(x)\,dx = \int_{x=a}^{x=b} f'(u)\,du = f(u)\Big|_{x=a}^{x=b}$$

$$= f(g(x))\Big|_a^b = f[g(b)] - f[g(a)]$$

Summary (continued)

- Let $f(x)$ be a smooth function and let dx be a small change in the input x. The *differential* $dy = f'(x)\,dx$ approximates the actual change in y provided by $\Delta y = f(x + dx) - f(x)$.

Exercises

In Exercises $1-12$, identify u and compute du when using the method of substitution to antidifferentiate the integral.

1. $\displaystyle \int \frac{2x}{x^2 + 2}\, dx$

2. $\displaystyle \int e^{x^3 - 5}\, 3x^2 \, dx$

3. $\displaystyle \int \frac{\sin(x)}{\cos^5(x)}\, dx$

4. $\displaystyle \int \frac{\cos(x)}{\sqrt{\sin(x)}}\, dx$

5. $\displaystyle \int (x^2 + x)^9 (2x + 1)\, dx$

6. $\displaystyle \int \ln(x^4)\, x^3 \, dx$

7. $\displaystyle \int x^3 \sqrt{x^4 + 5}\, dx$

8. $\displaystyle \int x^3 \cos(x^4)\, dx$

9. $\displaystyle \int \frac{x}{x^2 + 7}\, dx$

10. $\displaystyle \int x e^{x^2 + 3}\, dx$

11. $\displaystyle \int x\sqrt{3x + 4}\, dx$

12. $\displaystyle \int \frac{x}{x + 5}\, dx$

In Exercises $13-28$, antidifferentiate using the method of substitution.

13. $\displaystyle \int (x^2 + 1)^4 \cdot 2x \, dx$

14. $\displaystyle \int \frac{2x}{x^2 + 2}\, dx$

15. $\displaystyle \int 2x \cdot e^{x^2 + 3}\, dx$

16. $\displaystyle \int 6x^2 \cdot e^{2x^3 + 4}\, dx$

17. $\displaystyle \int \frac{\cos[\ln(x)]}{x}\, dx$

18. $\displaystyle \int \frac{2\cos(2x)}{\sin(2x)}\, dx$

19. $\displaystyle \int \frac{\cos(x)}{\sqrt{\sin(x)}}\, dx$

20. $\displaystyle \int \frac{6x^5 + 6x^2}{x^6 + 2x^3}\, dx$

21. $\displaystyle \int \frac{4e^{4x} + 4}{e^{4x} + 4x}\, dx$

22. $\displaystyle \int \cos(x^3 + 1)3x^2 \, dx$

23. $\displaystyle \int \frac{3x^2 \cos(x^3 + 1)}{\sin(x^3 + 1)}\, dx$

24. $\displaystyle \int \sin(x) e^{\cos(x) + 2}\, dx$

25. $\displaystyle \int \sin^2(x) \cos(x)\, dx$

26. $\int \cos^2(x)(-\sin(x))\,dx$

27. $\int \dfrac{\cos(x) + e^x}{(\sin(x) + e^x)^4}\,dx$

28. $\int (-\sin(x))\sqrt{4\cos(x) + 6}\,dx$

In Exercises 29–50, antidifferentiate using the method of substitution.

29. $\int x^3 \cos(x^4)\,dx$

30. $\int x^3 \sqrt{x^4 + 5}\,dx$

31. $\int x^4 \sqrt[3]{x^5 + 6}\,dx$

32. $\int \dfrac{x^2 + 1}{x^3 + 3x}\,dx$

33. $\int \dfrac{x}{x^2 + 7}\,dx$

34. $\int x \cdot e^{x^2}\,dx$

35. $\int x^2 \cdot e^{x^3 - 5}\,dx$

36. $\int \dfrac{x}{(x^2 + 1)^2}\,dx$

37. $\int \dfrac{\sin(x)}{\cos(x)}\,dx$

38. $\int \dfrac{\sin(x)}{\cos(x) + 2}\,dx$

39. $\int \dfrac{\sin(x)}{\cos^5(x)}\,dx$

40. $\int \dfrac{x^4 + 2x}{\sqrt{x^5 + 5x^2}}\,dx$

41. $\int x(x + 4)^{33}\,dx$

42. $\int x\sqrt{3x + 4}\,dx$

43. $\int \dfrac{x}{x + 5}\,dx$

44. $\int \dfrac{8x}{4x - 2}\,dx$

45. $\int \sqrt{x} \cdot \left(\sqrt{x^3} + 2\right)^7\,dx$

46. $\int \sin(x)e^{2\cos(x) + 4}\,dx$

47. $\int \dfrac{\sqrt{x} + 1}{\left(\sqrt{x^3} + \frac{3}{2}x\right)^2}\,dx$

48. $\int \dfrac{\cos(3x) + 1}{\sin(3x) + 3x}\,dx$

49. $\int (x^5 + x^2) \cdot e^{x^6 + 2x^3}\,dx$

50. $\int x^2 e^{x^3 + 6x} + 2e^{x^3 + 6x}\,dx$

In Exercises 51–60, evaluate the definite integral using the method of substitution and the second fundamental theorem of calculus.

51. $\int_0^1 \dfrac{-4x^3}{(5 - x^4)^6}\,dx$

52. $\int_0^2 2x\sqrt[3]{x^2 + 6}\,dx$

53. $\int_1^4 \dfrac{2x}{x^2 + 3}\,dx$

54. $\int_1^e \dfrac{\ln(x)}{x}\,dx$

55. $\int_{-4}^0 \dfrac{1}{x + 5}\,dx$

56. $\int_0^1 e^{x^3} 3x^2\,dx$

57. $\int_1^3 e^{-x^2}(-2x)\,dx$

58. $\int_0^{\pi/2} 2\cos(2x)\,dx$

59. $\int_0^{\pi/2} \sin^2(x) \cos(x)\, dx$

60. $\int_0^{\pi} \cos^2(x)(-\sin(x))\, dx$

In Exercises 61–74, evaluate the definite integral using the method of substitution and the second fundamental theorem of calculus.

61. $\int_{-2}^{2} x^3 \sqrt{x^4 + 5}\, dx$

62. $\int_{2}^{3} x^2 \sqrt{x^3 + 8}\, dx$

63. $\int_{-1}^{1} \sin(\pi x)\, dx$

64. $\int_{0}^{1} x \cdot e^{-x^2}\, dx$

65. $\int_{-1}^{0} x \cdot e^{-x^2}\, dx$

66. $\int_{0}^{e} \frac{x}{x^2 + 1}\, dx$

67. $\int_{0}^{\pi/3} \frac{\sin(x)}{\cos(x) + \frac{1}{2}}\, dx$

68. $\int_{\pi/2}^{3\pi/4} \frac{\cos(x)}{\sin^3(x)}\, dx$

69. $\int_{-1}^{-1/2} \frac{x^4 + 2x}{\sqrt{x^5 + 5x^2}}\, dx$

70. $\int_{1}^{\pi} \frac{x^5 + x^2}{x^6 + 2x^3}\, dx$

71. $\int_{0}^{1} x(3x + 4)^{33}\, dx$

72. $\int_{-4}^{0} \frac{x}{x + 5}\, dx$

73. $\int_{1}^{1.5} \frac{8x}{4x - 2}\, dx$

74. $\int_{-1}^{3} x\sqrt{3x + 4}\, dx$

In Exercises 75–82, compute the differential dy for the function.

75. $f(x) = x + 5$

76. $f(x) = x^2$

77. $f(x) = 5 - x^2 + 2$

78. $f(x) = e^x$

79. $f(x) = \sin(2x + 5)$

80. $f(x) = \ln(x) + e^x$

81. $f(x) = \cos(\ln(x))$

82. $f(x) = e^{x^2 + \sin(x)}$

In Exercises 83–88, approximate Δy using dy for the function, the input, and dx.

83. $f(x) = x^3 - 2x$; $x = 1$; $dx = 0.1$

84. $f(x) = \sin(x)$; $x = 0$; $dx = 0.2$

85. $f(x) = e^{x+3}$; $x = -2$; $dx = 0.01$

86. $f(x) = x\cos(x)$; $x = 1$; $dx = 0.1$

87. $f(x) = e^x \sin x$; $x = 0$; $dx = 0.15$

88. $f(x) = e^{\cos(x)+2}$; $x = 1$; $dx = -0.1$

In Exercises 89–92, verify each antidifferentiation formula using the given identity and the method of substitution.

89. $\int \tan(x)\, dx = \ln|\sec(x)| + C$

   using $\tan(x) = \dfrac{\sin(x)}{\cos(x)}$

90. $\int \cot(x)\, dx = \ln|\sin(x)| + C$

   using $\cot(x) = \dfrac{\cos(x)}{\sin(x)}$

91. $\int \csc(x)\, dx = \ln|\csc(x) + \cot(x)| + C$

   via $\csc(x) = \dfrac{\csc(x)[\csc(x) + \cot(x)]}{\csc(x) + \cot(x)}$

92. $\displaystyle\int \sec(x)\, dx = \ln|\sec(x)+\tan(x)|+C$

    via $\sec(x) = \dfrac{\sec(x)[\sec(x)+\tan(x)]}{\sec(x)+\tan(x)}$

In Your Own Words. In Exercises 93–98, explain the following.

93. Composition of functions

94. De-composing functions

95. Chain rule

96. Integration by substitution

97. Substitution and definite integrals

98. Differential

6.6   Integration by Parts

As introduced in Section 6.4 with the second fundamental theorem of calculus, antiderivatives play a central role in the evaluating of definite integrals used to find the net accumulation of a function over some interval. Section 6.5 developed the method of substitution as an algorithm for calculating the antiderivatives of certain functions by reversing the chain rule for derivatives. Namely, substitution works for products of functions when one term is a composition of functions $f'(g(x))$ and the other term is the derivative $g'(x)$ of the inside function of this composition.

This section develops another technique for finding antiderivatives known as integration by parts. This method results from running the product rule for differentiation backwards, as discussed in detail at the end of this section. For now, the reader should keep in mind that this method is also used when antidifferentiating products of functions, only without the stipulations that certain relationships hold between the two functions being multiplied together. While products consisting of a composition and the derivative of its inside function will continue to be evaluated using substitution, integration by parts is often used to antidifferentiate many other products of functions.

Method of Integration by Parts

The method of integration by parts provides a process for antidifferentiating integrands that can be interpreted as a product of two functions that will be labeled $f(x)$ and $g'(x)$ in the integration by parts formula stated next.

INTEGRATION BY PARTS.

Let $f(x)$ and $g(x)$ be smooth functions. Then the following equality holds:

$$\int f(x)\,g'(x)\, dx \;=\; f(x)\,g(x) - \int f'(x)\,g(x)\, dx$$

An initial inspection of the integration by parts formula might leave one wondering why the method is useful, because this formula expresses a given integral $\int f(x)\,g'(x)\, dx$ in a way that includes another integral $\int f'(x)\,g(x)\, dx$ that still must be calculated. Such an initial observation reveals a key insight into choosing the two functions $f(x)$

and $g'(x)$: they should ideally be chosen so that the antiderivative $\int f'(x)g(x)\,dx$ is easier to compute than $\int f(x)g'(x)\,dx$.

When using integration by substitution, the selection of the inside function $u = g(x)$ was a central, pivotal step in the process of applying the method. For integration by parts, the choices of $f(x)$ and $g'(x)$ from the given product are similarly central to the success of applying this method. Practice and some general guidelines will help develop a proficiency in making these choices. After considering some specific examples, some general guidelines for choosing $f(x)$ and $g'(x)$ are discussed.

◆ **EXAMPLE 1** Evaluate $\displaystyle\int x\cos(x)\,dx$ using integration by parts.

Solution. First, observe that this integral cannot be immediately evaluated by reversing a standard differentiation rule, nor does this product involve a composition indicating that the method of substitution would be appropriate. Integration by parts is the next option for evaluating this integral. The product in this integral suggests two natural choices for $f(x)$ and $g'(x)$ in the integration by parts formula: $f(x) = \cos(x)$ and $g'(x) = x$, or alternatively $f(x) = x$ and $g'(x) = \cos(x)$.

Both are discussed, beginning with $f(x) = \cos(x)$ and $g'(x) = x$. For this choice, $f'(x) = -\sin(x)$ and $g(x) = \frac{1}{2}x^2$, and substituting these expressions into the parts formula produces the following.

$$\int x\cos(x)\,dx \;=\; \cos(x)\,\frac{x^2}{2} - \int (-\sin(x))\,\frac{x^2}{2}\,dx$$

The resulting new integral $\displaystyle\int \frac{x^2}{2}\sin(x)\,dx$ appears even more difficult than the original given integral, which suggests that we might explore the other option.

If $f(x) = x$ and $g'(x) = \cos(x)$, then $f'(x) = 1$ and $g(x) = \sin(x)$, and substituting these expressions into the parts formula provides the following answer:

$$\begin{aligned}
\int x\cos(x)\,dx &= x\cdot\sin(x) - \int 1\cdot\sin(x)\,dx && \text{Integration by parts formula}\\[2mm]
&= x\sin(x) - (-\cos(x)) + C && \int \sin(x)\,dx = -\cos(x) + C\\[2mm]
&= x\sin(x) + \cos(x) + C && \text{Simplify}
\end{aligned}$$

∎

The validity of any antidifferentiation can be confirmed by taking the derivative of the answer. The original integrand should be produced if the antiderivative is correct. For Example 1, such a verification would mean that $\dfrac{d}{dx}[x\sin(x) + \cos(x) + C]$ is equal to $x\cos(x)$. Using the product rule and simplying produces the following results, and confirms the validity of the antiderivative from this example:

$$\begin{aligned}
\frac{d}{dx}[x\sin(x) + \cos(x) + C] &= \sin(x)\cdot 1 + x\cdot\cos(x) - \sin(x) + 0\\[2mm]
&= x\cos(x)
\end{aligned}$$

◆ **EXAMPLE 2** Evaluate $\int \frac{1}{2}x^2 e^x \, dx$ using integration by parts.

Solution. First, observe that this integral cannot be immediately evaluated by reversing a standard differentiation rule, nor does this product involve a composition that would indicate integration by substitution. Turning to integration by parts, the product in this integral suggests two natural choices for $f(x)$ and $g'(x)$ in the parts formula: $f(x) = e^x$ and $g'(x) = \frac{1}{2}x^2$, or alternatively $f(x) = \frac{1}{2}x^2$ and $g'(x) = e^x$.

Choosing $f(x) = e^x$ and $g'(x) = \frac{1}{2}x^2$ results in $g(x) = \frac{1}{6}x^3$, which would introduce an even higher power of x into the parts formula for $\int f'(x)\,g(x)\,dx$ and suggests that this choice is most likely not the best for $g'(x)$. Therefore, consider $f(x) = \frac{1}{2}x^2$ and $g'(x) = e^x$, which yields $f'(x) = x$ and $g(x) = e^x$. Substituting into the parts formula yields the following equation:

$$\int \frac{1}{2}x^2 e^x \, dx \;=\; \frac{1}{2}x^2 e^x - \int x e^x \, dx$$

While the antiderivative of xe^x still needs to be determined to obtain the final answer, this new integral involves a smaller power of x, which suggests that it will be easier to evaluate than the original integral.

As it turns out, the new integral also involves a product of two functions and is also an ideal candidate for integration by parts. Choosing $f(x) = x$ and $g'(x) = e^x$ results in $f'(x) = 1$ and $g(x) = e^x$. Applying integration by parts to this new integral as an extension of the initial work yields the following answer:

$$\begin{aligned}
\int \frac{1}{2}x^2 e^x \, dx \;&=\; \frac{1}{2}x^2 e^x - \int x e^x \, dx && \text{Parts with } f(x) = \frac{1}{2}x^2 \text{ and } g'(x) = e^x \\
&= \frac{1}{2}x^2 e^x - \left[x e^x - \int e^x \, dx \right] && \text{Parts with } f(x) = x \text{ and } g'(x) = e^x \\
&= \frac{1}{2}x^2 e^x - [x e^x - e^x] + C && \int e^x \, dx = e^x + C \\
&= \frac{1}{2}x^2 e^x - x e^x + e^x + C && \text{Simplify}
\end{aligned}$$

As with Example 1, the solution of Example 2 can be verified by differentiating the final answer.

▶ **QUESTION 1** Evaluate $\int 2x \ln(x) \, dx$ using integration by parts.

Choosing $f(x)$ and $g'(x)$

In the preceding integration by parts examples, both possible options for $f(x)$ and $g(x)$ were considered and the choice that resulted in an easier integral in the parts formula was used. While this exploratory process is certainly one valid approach to selecting $f(x)$ and $g'(x)$, mathematicians have articulated guidelines for choosing $f(x)$, and consequently $g'(x)$, that almost always work out quite well. These guidelines can be summarized as follows:

LPET GUIDELINES FOR INTEGRATION BY PARTS.
When using integration by parts to find an antiderivative, the function $f(x)$ should be chosen in the following order of preference:

(1) logarithmic function

(2) polynomial

(3) exponential function

(4) trigonometric function

This book follows these LPET guidelines when selecting $f(x)$ because this ordered list identifies a preference ranking that almost always yields the best path forward when applying integration by parts. If both terms of a product happen to appear on the list, then choose $f(x)$ to be the term that is ranked higher. For example, when evaluating $\int e^x \sin(x)\,dx$, choose $f(x) = e^x$ because (3) exponential function appears before (4) trigonometric function in the LPET guidelines.

◆ **EXAMPLE 3** Identify the best choice for $f(x)$, and so $g'(x)$, from each integral using the LPET guidelines for integration by parts:

(a) $\displaystyle\int x^2 \ln(x)\,dx$ \qquad\qquad (b) $\displaystyle\int e^x \cos(x)\,dx$

Solution.

(a) The integral contains the product of the polynomial x^2 with a logarithmic function $\ln(x)$. Because the LPET guidelines rank logarithmic functions first, choose $f(x) = \ln(x)$ and $g'(x) = \frac{1}{3}x^3$.

(b) The integral contains the product of the exponential e^x with a trigonometric function $\cos(x)$. Because the LPET guidelines rank exponential functions before trigonometric functions, choose $f(x) = e^x$ and $g'(x) = \cos(x)$.

■

▶ **QUESTION 2** Identify the best choice for $f(x)$, and so $g'(x)$, from each integral using the LPET guidelines for integration by parts:

(a) $\displaystyle\int x^2 \sin(x)\,dx$ \qquad\qquad (b) $\displaystyle\int 3x^2 \ln(2x)\,dx$

Integration by Parts and Definite Integrals

As with integration by method of substitution, integration by parts works equally well with definite integrals. This process is summarized next.

INTEGRATION BY PARTS WITH DEFINITE INTEGRALS.
Let $f(x)$ and $g(x)$ be smooth functions on $[a, b]$. Then the following equality holds:

$$\int_a^b f(x)g'(x)\,dx = f(x)g(x)\Big|_a^b - \int_a^b f'(x)g(x)\,dx$$

When combining the second fundamental theorem of calculus with integration by parts, be sure to evaluate the entire antiderivative on the limits of integration and then subtract. From a practical perspective, the initial product $f(x)g(x)$ is sometimes accidentally omitted or overlooked during the final evaluate-and-subtract step. Including the vertical bar with limits on the initial product $f(x)g(x)$ not only makes the equation correct, but also serves as a helpful reminder to include this expression in the final evaluation step.

◆ **EXAMPLE 4** Evaluate $\int_0^1 x^2 e^x\,dx$ using integration by parts.

Solution. First, observe that this integral cannot be immediately evaluated by reversing a standard differentiation rule, nor does this product involve a composition that would indicate integration by substitution.

Turning to integration by parts and following the LPET guidelines, choose polynomial $f(x) = x^2$ and exponential $g'(x) = e^x$. These choices result in $f'(x) = 2x$ and $g(x) = e^x$, and substituting into the parts formula yields the following:

$$\int_0^1 x^2 e^x\,dx = x^2 e^x\Big|_0^1 - \int_0^1 2x e^x\,dx$$

The new integral also contains a product, so integration by parts is applied a second time with polynomial $f(x) = 2x$ and exponential $g'(x) = e^x$, which results in $f'(x) = 2$ and $g(x) = e^x$. Applying integration by parts to this new integral as an extension of the initial work yields the following answer:

$$
\begin{aligned}
\int_0^1 x^2 e^x\,dx &= x^2 e^x\Big|_0^1 - \int_0^1 2x e^x\,dx && \text{Parts with } f(x) = x^2,\ g'(x) = e^x \\
&= x^2 e^x\Big|_0^1 - \left[2x e^x\Big|_0^1 - \int_0^1 2e^x\,dx \right] && \text{Parts with } f(x) = 2x,\ g'(x) = e^x \\
&= x^2 e^x - 2x e^x + 2e^x\Big|_0^1 && \int 2e^x\,dx = 2e^x + C \\
&= \left[1^2 e^1 - 2(1)e^1 + 2e^1 \right] - \left[0 - 0 + 2e^0 \right] && F(1) - F(0) \\
&= e - 2 \approx 0.718 && \text{Simplify}
\end{aligned}
$$

▪

▶ **QUESTION 3** Evaluate $\int_0^{\pi/2} x \cos(x)\,dx$ using integration by parts.

◆ **EXAMPLE 5** A drug administered to a patient produces a concentration in the bloodstream given by $c(x) = \frac{e}{3}xe^{-x/3}$ (in milligrams per milliliter) x hours after being injected. The bioavalability of a drug is related to how much and how fast the drug reaches its target inside the body and is partly determined by the area under the concentration curve. Focusing on the first six hours immediately after the drug is administered, calculate $\int_0^6 c(x)\,dx$.

Solution. The integrand $c(x) = \frac{e}{3}xe^{-x/3}$ cannot be immediately antidifferentiated by reversing a standard differentiation rule, nor does this product involve a composition that would indicate integration by substitution. Turning to integration by parts and following the LPET guidelines, choose polynomial $f(x) = \frac{e}{3}x$ and exponential $g'(x) = e^{-x/3}$, which result in $f'(x) = \frac{e}{3}$ and $g(x) = -3e^{-x/3}$. Applying integration by parts produces the following antiderivative of $c(x) = \frac{e}{3}xe^{-x/3}$:

$$\int \frac{e}{3}xe^{-x/3}\,dx = \frac{e}{3}x\left(-3e^{-x/3}\right) - \int \frac{e}{3}\left(-3e^{-x/3}\right)\,dx$$

$$= -(ex)e^{-x/3} + 3\frac{e}{3}\int e^{-x/3}\,dx$$

$$= -(ex)e^{-x/3} + e\left(-3e^{-x/3}\right) + C$$

$$= -(ex + 3e)e^{-x/3} + C$$

Next, evaluate this antiderivative on the limits of integration of 0 and 6, and subtract. Recall that when using the second fundamental theorem of calculus, the simplest constant of integration $C = 0$ is chosen, as in the following calculation:

$$\int_0^6 c(x)\,dx = \int_0^6 \frac{e}{3}xe^{-x/3}\,dx = -(ex + 3e)e^{-x/3}\Big|_0^6$$

$$= -(e \cdot 6 + 3e)e^{-6/3} - \left[-(e \cdot 0 + 3e)e^{-0/3}\right]$$

$$= -9e^{-2} + 3e \approx 6.937$$

Interpreting this answer in context, the area under the concentration curve $c(x) = \frac{e}{3}xe^{-x/3}$ over the first six hours after the drug is administered is equal to 6.937 hour milligrams per milliliter. ■

The units for the answer to Example 5 might seem a bit strange. However, they result immediately from taking the product of the units of the output (milligrams per milliliter) by the units of the input (hours) based on the discussion of dimensions and units in Section 6.2. Medical researchers are able to appropriately interpret this numeric result with its (strange) units in the context of analyzing drug bioavailability.

Understanding Integration by Parts

The method of substitution in Section 6.5 provides an algorithm for antidifferentiating functions that result from the chain rule, and so applies to a particular type of product of functions. Namely, the method of substitution is used for integrands when one term in the product is a composition of functions $f'(g(x))$ and the other term is the derivative $g'(x)$ of the inside function of this composition.

Integration by parts provides an algorithm for finding antiderivatives that undoes the product rule, and so applies to a different collection of products of functions. Recall that for smooth functions $f(x)$ and $g(x)$, the product rule for differentiation asserts the following equality:

$$\frac{d}{dx}\left[f(x)g(x)\right] \;=\; f(x)g'(x) + f'(x)g(x)$$

Integrating both sides of the product rule, using the fact that integration and differentiation are inverse operations, and distributing the integral across the sum, yields the following:

$$\int \frac{d}{dx}\left[f(x)g(x)\right]dx \;=\; \int f(x)g'(x) + f'(x)g(x)\,dx$$

$$f(x)g(x) \;=\; \int f(x)g'(x)\,dx \;+\; \int f'(x)g(x)\,dx$$

$$f(x)g(x) \;-\; \int f'(x)g(x)\,dx \;=\; \int f(x)g'(x)\,dx$$

Summary

- *Integration by parts*: Let $f(x)$ and $g(x)$ be smooth functions. Then the following equality holds:

$$\int f(x)\,g'(x)\,dx \;=\; f(x)\,g(x) - \int f'(x)\,g(x)\,dx$$

- *LPET guidelines for integration by parts*: When using parts to antidifferentiate, choose $f(x)$ in the following order of preference:

  (1) logarithmic function

  (2) polynomial

  (3) exponential function

  (4) trigonometric function

- *Integration by parts for definite integrals*: Let $f(x)$ and $g(x)$ be smooth functions on $[a, b]$. Then the following equality holds:

$$\int_a^b f(x)g'(x)\,dx \;=\; f(x)g(x)\,\Big|_a^b \;-\; \int_a^b f'(x)g(x)\,dx$$

- The integration by parts results from running the product rule backwards.

Exercises

In Exercises 1–8, identify the best choice of $f(x)$ and $g'(x)$ from the integral using the LPET guidelines for integration by parts.

1. $\int x^2 \ln(x)\, dx$

2. $\int x^3 e^{-x}\, dx$

3. $\int \ln(x) \cos(x)\, dx$

4. $\int (x^4 + x) \sin(x)\, dx$

5. $\int e^{2x} \cos(6x)\, dx$

6. $\int e^{3x} \ln(4x)\, dx$

7. $\int e^{-x} \sin(3x)\, dx$

8. $\int (x^5 - x) \ln(x)\, dx$

In Exercises 9–18, evaluate the integral using integration by parts.

9. $\int 3x^2 \ln(x)\, dx$

10. $\int x^4 \ln(x)\, dx$

11. $\int (8x^3 + 1) \ln(2x)\, dx$

12. $\int xe^x\, dx$

13. $\int (4x + 6)e^{2x+1}\, dx$

14. $\int 2xe^{-x}\, dx$

15. $\int x \sin(x)\, dx$

16. $\int (5x - 8) \sin(3x)\, dx$

17. $\int (3x - 4) \cos(x)\, dx$

18. $\int 6x \cos(9x + 2)\, dx$

In Exercises 19–26, evaluate the integral using integration by parts. The method may need to be applied more than once.

19. $\int x^2 \sin(x)\, dx$

20. $\int x^2 \sin(2x)\, dx$

21. $\int (9x^2 + 18) \cos(3x)\, dx$

22. $\int (4x^2 - x) \cos(x)\, dx$

23. $\int 9x^2 e^{3x}\, dx$

24. $\int (2x^2 - x + 1)e^x\, dx$

25. $\int x^3 e^{-x}\, dx$

26. $\int (x^3 - x)e^x\, dx$

In Exercises 27–36, evaluate the integral using integration by parts and the second fundamental theorem of calculus.

27. $\int_1^2 (2x + 1) \ln(x)\, dx$

28. $\int_2^4 3x^2 \ln(x)\, dx$

29. $\int_0^1 4xe^{-x}\, dx$

30. $\int_{-1}^1 8xe^x\, dx$

31. $\displaystyle\int_{-1/2}^{0} 8xe^{4x+2}\,dx$

32. $\displaystyle\int_{0}^{\pi/2} x\sin(x)\,dx$

33. $\displaystyle\int_{0}^{\pi} 2x\sin(x)\,dx$

34. $\displaystyle\int_{0}^{\pi} x^2\sin(2x)\,dx$

35. $\displaystyle\int_{-\pi}^{\pi} x^2\cos(x)\,dx$

36. $\displaystyle\int_{0}^{\pi} x\cos(4x)\,dx$

In Exercises 37–42, find the area of the region bounded by the two functions on the interval.

37. $f(x) = x$; $g(x) = x\sin(x)$; $[0, \frac{\pi}{2}]$

38. $f(x) = x$; $g(x) = x\cos(x)$; $[0, \frac{\pi}{2}]$

39. $f(x) = 1$; $g(x) = \ln(x)$; $[1, 2]$

40. $f(x) = x$; $g(x) = \ln(x)$; $[1, 3]$

41. $f(x) = xe^x$; $g(x) = x$; $[0, 2]$

42. $f(x) = x^2e^x$; $g(x) = x$; $[0, 1]$

RStudio. In Exercises 43–50, use RStudio to evaluate the definite integrals given in the specified exercises.

43. Exercise 27

44. Exercise 29

45. Exercise 31

46. Exercise 33

47. Exercise 35

48. Exercise 37

49. Exercise 39

50. Exercise 41

In Exercises 51–64, identify the method of integration used to evaluate the integral from among standard antidifferentiation formulas, substitution, and integration by parts.

51. $\displaystyle\int x^2 + x\,dx$

52. $\displaystyle\int x^2 \cdot x\,dx$

53. $\displaystyle\int x^2 e^x\,dx$

54. $\displaystyle\int x^2 e^{x^3}\,dx$

55. $\displaystyle\int e^{3x}\,dx$

56. $\displaystyle\int 5x^2\ln(x)\,dx$

57. $\displaystyle\int \ln(x)\,dx$

58. $\displaystyle\int \frac{\ln(x)}{x}\,dx$

59. $\displaystyle\int x\sin(4x^2)\,dx$

60. $\displaystyle\int x\sin(4x)\,dx$

61. $\displaystyle\int \sin(4x)\,dx$

62. $\displaystyle\int x + \cos(x)\,dx$

63. $\displaystyle\int x\cos(\pi x)\,dx$

64. $\displaystyle\int ex\cos(\pi x^2)\,dx$

In Exercises 65–78, evaluate the integrals using a standard antidifferentiation formula, substitution, or integration by parts.

65. Exercise 51

66. Exercise 52

67. Exercise 53

68. Exercise 54

69. Exercise 55

70. Exercise 56

71. Exercise 57

72. Exercise 58

73. Exercise 59

74. Exercise 60

75. Exercise 61

76. Exercise 62

77. Exercise 63

78. Exercise 64

In Your Own Words. In Exercises 79 –
83, explain the following.

79. Integration by parts

80. LPET guidelines for parts

81. Integration by parts and definite integrals

82. Product rule

83. Derivation of parts formula from the product rule

Appendix A

Answers to Questions

Section 1.1 Questions

1. (a) Nonfunction: a maps to both 2 and 16

    (b) Function

2. (a) Nonfunction; see $x = 0$

    (b) Function

    (c) Nonfunction; see $x = 1$

3. (a) Domain: $\{2008, 2009, 2010, 2012, 2013\}$
    Range: $\{5, 7.8, 9.7, 8.2, 7.9\}$

4. (a) $(-\infty, \infty)$

    (b) $(-\infty, 4]$

    (c) $(2, \infty)$

5. (a) $g(4) = 24$

    (b) $g(0) = 8$

    (c) $g(-3.2) = 18.24$

6. (a) $b(-0.1) = 0.1$

    (b) $b(0) = 0$

    (c) $b(30) = 30$

7. (a) `> 5*6-3/2`
       `[1] 28.5`

    (b) `> (5*6-3)/2`
       `[1] 13.5`

    (c) `> 5*(6-3)/2`
       `[1] 7.5`

    (d) `> abs((3*6^2-pi^3)/`
       `(4*5+7))`
       `[1] 2.851619`

    (e) `> abs((pi)^2-(4^3/2))`
       `[1] 22.1304`

    (f) `> (5^2-4^3)/2`
       `[1] -19.5`

8. `> f=makeFun(abs(4*x^3)+3*x-`
   `(1/(2*x))~x)`

    (a) `> f(x=10.34)`
       `[1] 4453.001`

    (b) `> f(x=pi)`
       `[1] 133.2907`

    (c) `> f(x=-2.42)`
       `[1] 49.6365`

    (d) `> f(x=-456)`
       `[1] 379273896`

9. `> plotFun(-3.546*x-9.128~x,`
   `xlim=range(-15,15))`

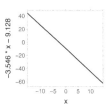

Section 1.2 Questions

1. (a) $g(1, 4, 2) = 5$

    (b) $g(2.2, 16.8, -4) = 2.4$

    (c) $g(10, -4, -16) = 30.25$

    (d) $g(3.4, 5.6, 0)$ is undefined

2. (a) $f(-1,4) = 30$

   (b) $f(0,2) = 28$

   (c) $f(2,-2) = 24$

3. (a) $f(0,1.25) = -6$

   (b) $f(-2.5,2) \approx -3.5$

   (c) $f(-1,2) \approx -1.5$

   (d) $f(2,1) \approx -10.5$

   (e) $f(x,y)$ immediately decreases

   (f) $f(x,y)$ immediately increases

4. For $C = -3$, $y = x^2 - 4$; for $C = -2$, $y = x^2 - 3$; for $C = -1$, $y = x^2 - 2$; for $C = 0$, $y = x^2 - 1$; and for $C = 1$, $y = x^2$.

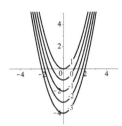

5. > g=makeFun(abs(x)-y^3+pi~x&y)

   (a) > g(x=1,y=2)
       [1] -3.858407

   (b) > g(x=-1,y=2)
       [1] -3.858407

   (c) > g(x=4.65,y=1.1)
       [1] 6.460593

   (d) > g(x=4.65,y=-1.1)
       [1] 9.122593

6. (a) > plotFun(x^2-y~x&y,
       surface=TRUE,xlim=range
       (-3,3),ylim=range(-3,3))

   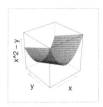

   (b) > plotFun(x^2-y~x&y,
       filled=FALSE,xlim=range

(-3,3),ylim=range(-3,3),
levels=c(0,2,4,6,8,10))

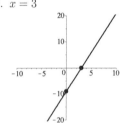

Section 1.3 Questions

1. $y = 1.55x - 1.95$

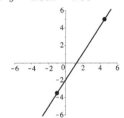

2. $x = 3$

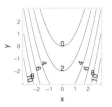

3. (a) $\left(\dfrac{-2}{7}, \dfrac{53}{7}\right)$

   (b) $(7,3)$

   (c) Parallel lines do not intersect

4. (a) $m = 0$

   (b) $m < 0$

   (c) m undefined

5. (a) Positive

   (b) Zero

   (c) Negative

6. (a) Constant on $(-\infty, \infty)$; zero slope

   (b) Increases on $(-1,1)$; decreases on $(-\infty, -1)$, $(2, \infty)$; constant on $(1,2)$

    (c) Increases on $(-\infty, \infty)$;
       positive slope

7. Decreases on $(-\infty, 0)$;
   increases on $(0, \infty)$

8. > plotFun(-x+5~x,xlim=
   range(-10,10))
   > plotFun(3*x-2~x,add=TRUE)

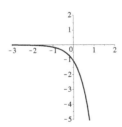

9. > findZeros(4*x-7~x,xlim=
   range(0,5))
       x
   1 1.75
   > -1.75+5
   [1] 3.25

Section 1.4 Questions

1. $C = \dfrac{-60}{e^4}$;   decreasing and concave down on $(-\infty, \infty)$;

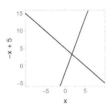

2. (a) Doubling time ≈ 2;
     $y = 3e^{0.347x}$

   (b) Halving time ≈ 4;
     $y = 100e^{-0.173x}$

3. y-intercept is C for each of $C = 1, 3, 7$ and shifts upward as C increases; increasing and concave up on $(-\infty, \infty)$; horizontal asymptote is $y = 0$

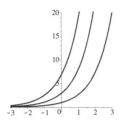

4. Vertical intercept is $(0, 1)$; horizontal asymptote is $y = 0$; decreasing and concave up on $(-\infty, \infty)$; as k becomes more negative from -1 to -2 to -3, $y = Ce^x$ decreases more rapidly.

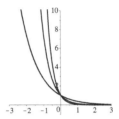

5. The exponentials $y = -e^{-x}$ and $y = -e^x$ are reflections across the y-axis. When $k < 0$ and $C < 0$, $y = Ce^x$ is increasing and concave down on $(-\infty, \infty)$. When $k > 0$ and $C < 0$, $y = Ce^x$ is decreasing and concave down on $(-\infty, \infty)$

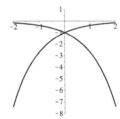

6. (a) Decreasing on $(-\infty, 1)$; increasing on $(1, 4)$; constant on $(4, \infty)$; concave up on $(-\infty, 4)$; linear on $(4, \infty)$.

   (b) Increasing on $(-\infty, -2)$ and $(2, \infty)$; decreasing on $(-2, 2)$; concave down on $(-\infty, 0)$; concave up on $(0, \infty)$.

7. Increasing and concave down on $(-\infty, 0)$; increasing and concave up on $(0, \infty)$.

8. (a) 11^{x-y} (b) $x^{2/y}$ (c) $\left(\dfrac{7}{y}\right)^x$ (d) x^{3y}

   (e) 11^{x+y} (f) $2^{1/x}$ (g) $(7y)^x$ (h) x^{-9}

9. ```
> plotFun(exp(-2*x)~x,xlim=
range(-3,3),ylim=c(-50,50))
> plotFun(-exp(2*x)~x,
add=TRUE)
```

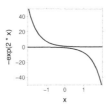

## Section 1.5 Questions

1. (a)

| $x$      | 24 | 34 | 37 | 38 | 39 | 42 |
|----------|----|----|----|----|----|----|
| $a^{-1}$ | 20 | 10 | 7  | 6  | 5  | 2  |

(b)

| $x$      | $-8$ | $-4$ | 1   | 2    | 6    |
|----------|------|------|-----|------|------|
| $b^{-1}$ | 8    | 4    | $-1$| $-2$ | $-6$ |

2. (a) How many million people used Twitter during a given quarter?

   (b) What quarter did a given number of people use Twitter? Alternatively, what quarter of the year did the number of Twitter users exceed a given number of people?

   (c)

   | Users | 68    | 85    | 101  |
   |-------|-------|-------|------|
   | Year  | 11    | 11.25 | 11.5 |
   | Users | 117   | 185   |      |
   | Year  | 11.75 | 12.75 |      |

3. (a)

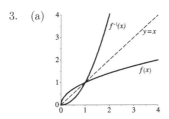

(b)

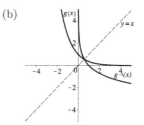

4. (a) $a(x)$ has an inverse

   | $x$      | 7    | 9 | 49   | 81 |
   |----------|------|---|------|----|
   | $a^{-1}$ | $-2$ | 2 | $-1$ | 1  |

   (b) $b(x)$ has no inverse because $b(-2) = 4$ and $b(2) = 4$.

5. The output 89 comes from the inputs of 1980 and 2001; or, the output 120 comes from the inputs of 1981 and 1990.

6. (a) Fails the horizontal line test. $y = 0$ intersects the curve twice.

   (b) Passes the horizontal line test.

7. (a) Switches from decreasing to increasing when $x = -1$ and does not have an inverse.

   (b) Decreases on $(-\infty, \infty)$ and has an inverse.

8. (a) ```
> plotFun((3*x-7)^(1/5)
~x,xlim=range(2,10))
```

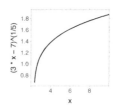

$$f^{-1}(x) = \frac{1}{3}x^5 + \frac{7}{3}$$

   (b) ```
> plotFun((5*x-3)/(7*x+4)
~x,xlim=range(0,10))
```

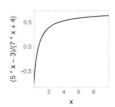
```

$$g^{-1}(x) = \frac{-4x-3}{7x-5}$$

Section 1.6 Questions

1. Figure 5(a) is $y = -2\ln(x+4)$.
   Figure 5(b) is $y = 4\ln(x-2)$.

2. (a) e (b) 0.25

3. (a) $\ln(3) + 2$
   (b) $2\ln(x) + \ln(y) - 3\ln(z)$

4. (a) $\ln(4x^2)$
   (b) $\ln\left[\dfrac{x+8}{x-1}\right]$

5. $y = -4e^{\ln(6)x/3}$ is decreasing and concave down on $(-\infty, \infty)$.

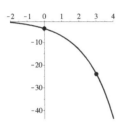

6. (a) \$2,060.91
   (b) 27.031 years

7. (a) > exp(log(exp(1)))
       [1] 2.718282
   (b) > log(exp(1/4))
       [1] 0.25

8. (a) > plotFun(log(exp(x))~x,
       xlim=range(0,10))

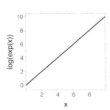

   (b) > plotFun(log(x^3)~x,
       xlim=range(0,10),
       ylim=range(-10,10))

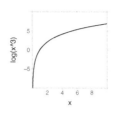

Section 1.7 Questions

1. (a) $300°$ (b) $210°$
   (c) π rad   (d) $\dfrac{3\pi}{4}$ rad

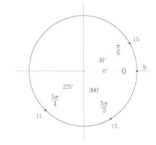

2. $\cos(x) = \dfrac{1}{6}$ $\sec(x) = 6$

   $\sin(x) = \dfrac{\sqrt{35}}{6}$ $\csc(x) = \dfrac{6}{\sqrt{35}}$

   $\tan(x) = \sqrt{35}$ $\cot(x) = \dfrac{1}{\sqrt{35}}$

3. $\sin(330°) = \dfrac{-1}{2}$ $\csc(330°) = -2$

   $\cos(330°) = \dfrac{\sqrt{3}}{2}$ $\sec(330°) = \dfrac{2}{\sqrt{3}}$

   $\tan(330°) = \dfrac{-1}{\sqrt{3}}$ $\cot(330°) = -\sqrt{3}$

4. (a) $\theta = \dfrac{\pi}{3}, \dfrac{4\pi}{3}$
   (b) $\theta = \dfrac{\pi}{6}, \dfrac{4\pi}{6}, \dfrac{7\pi}{6}, \dfrac{10\pi}{6}$

5. > sin(11*pi/6)
   [1] -0.5
   > cos(11*pi/6)
   [1] 0.866025
   > tan(11*pi/6)
   [1] -0.57735
   > 1/sin(11*pi/6)

```
[1] -2
> 1/cos(11*pi/6)
[1] 1.1547
> 1/tan(11*pi/6)
[1] -1.73205
```

6.
```
> plotFun(4*cos(x)-2~x,xlim=
    range(-2*pi,2*pi))
> plotFun(2*cos(2*x)~x,
add=TRUE)
```

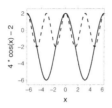

Section 2.1 Questions

1. (a) A linear model is not reasonable.

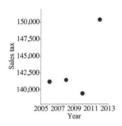

(b) A linear model appears reasonable.

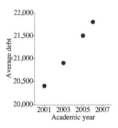

2. (a) A linear model is not reasonable.

Interval	Δx	Δy	Slope
2006–08	2	\$273	136.5
2008–10	2	\$–2008	–1004
2010–12	2	\$10,916	5458

(b) A linear model is reasonable.

Interval	Δx	Δy	Slope
2001–03	2	\$500	250
2003–05	2	\$600	300
2005–06	2	\$300	300

3. Question 2(b) average is
    $m = 283.33$.
    $b_1 = -546,550$, $b_2 = -546,626.67$,
    $b_3 = -546,583.33$, and
    $b_4 = -546,566.67$ average
    to $b = -546,581.67$.
    Debt $= 283.33$Year $- 546,581.67$

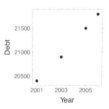

4. (a)
```
> Year=c(2001,2003,2005,
2006)
> Debt=c(20400,20900,
21500,21800)
> plotPoints(Debt~Year)
```

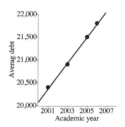

(b)
```
> bestModel=fitModel(Debt~
m*Year+b)
> coef(bestModel)
```

```
         m              b
   281.356 -542616.949
```

Debt $= 281.356$Year $- 542616.95$

```
> plotPoints(Debt~Year)
> plotFun(281.356*Year-
542616.949~Year,add=TRUE)
```

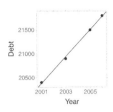

5. (a) A linear model appears reasonable.
   ```
   > names(WeightChange)
   ```

   ```
   [1] "Day"    "Weight"
   ```

   ```
   > plotPoints(Weight~Day,
   data=weightChange)
   ```

   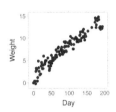

   (b) ```
   > weightModel=fitModel
   (Weight~m*Day+b,data=
   WeightChange)
   > coef(weightModel)
   ```

   ```
           m         b
   0.0632135 1.9033767
   ```

   Weight $= 0.0632$Day $+ 1.9033$
   ```
   > plotPoints(Weight~Day,
   data=WeightChange)
   > plotFun(0.0632135*Day+
   1.9033767~Day,add=TRUE)
   ```

   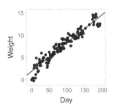

Section 2.2 Questions

1. An exponential model is reasonable, because the semi-log plot is nonconstant linear.

   - Standard plot

   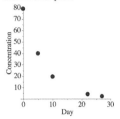

   - Semi-log plot

   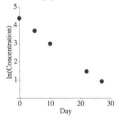

2. An exponential model is not reasonable, because the semi-log plot is constant linear.

   - Standard plot

   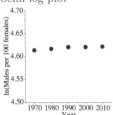

   - Semi-log plot

   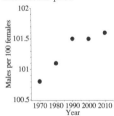

3. An exponential model is reasonable. $m_1 = 0.134$, $m_2 = 0.120$, $m_3 = 0.129$, and $m_4 = 0.116$ are approximately equal.
   ```

4. An exponential model is not reasonable. $m_1 = 0.0003$, $m_2 = 0.0004$, $m_3 = 0$, and $m_4 = 0.0001$ give $k \approx 0$, which means that a constant function is reasonable.

5. $m = 0.12475 = k$
$C_1 = 777$, $C_2 = 783.7$, $C_3 = 768.1$, $C_4 = 771.4$, and $C_5 = 757.1$
average to $C = 771.46$
Weight $= 771.46 \cdot e^{0.125(\text{Week}-25)}$.

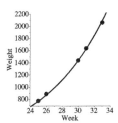

6.
```
> Day=c(0,5,10,22,27)
> Conc=c(79,40,19.6,4.3,2.5)
> bestConcModel=fitModel(
log(Conc)~m*Day+b)
> coef(bestConcModel)

        m           b
 -0.128379    4.325005
```

$\text{Conc} = e^{4.325005} \cdot e^{-0.128379\text{Day}}$
```
> plotPoints(Conc~Day)
> plotFun(exp(4.3250)*exp(
-0.1284*Day)~Day,add=TRUE)
```

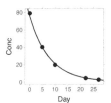

7.
```
> names(DJIACloseQuarterly)

[1] "Quarter" "Close"

> plotPoints(log(Close)~
Quarter,data=DJIACloseQuarterly)
```

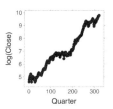

```
> bestDJIAModel=fitModel(log(
Close)~m*(Quarter-1)+b,
data=DJIACloseQuarterly)
> coef(bestDJIAModel)

        m           b
0.0160139   4.5161724
```

$\text{Close} = e^{4.51617} \cdot e^{0.01601(\text{Quarter}-1)}$
```
> plotPoints(Close~Quarter,
data=DJIACloseQuarterly)
> plotFun(exp(4.5161724)*
exp(0.0160139*(Quarter-1))~
Quarter,add=TRUE)
```

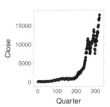

Section 2.3 Questions

1. A power function model is reasonable, because the log–log plot is nonconstant linear.

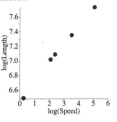

2. A power function model is reasonable. $m_1 = 0.289$, $m_2 = 0.225$, $m_3 = 0.232$, and $m_4 = 0.245$ are approximately equal and nonzero.

3. Question 2 average is $m = 0.25525 = k$.

$C_1 = 617.2$, $C_2 = 656.6$,
$C_3 = 650.7$, $C_4 = 634.2$, and
$C_5 = 624.2$ average to $C = 636.58$.
Speed $= 636.58 \cdot$ Length$^{0.25525}$

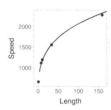

4. (a) A power function is reasonable, because the log–log plot is non-constant linear.
   ```
   > Length=c(1.3,8.1,11,34,
   160)
   > Speed=c(660,1120,1200,
   1560,2280)
   > plotPoints(log(Speed)~
   log(Length))
   ```

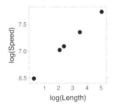

   (b)
   ```
   > bestModel=fitModel(log
   (Speed)~m*log(Length)+b)
   > coef(bestModel)

       m           b
   0.255245  6.455863
   ```

   Speed $= e^{6.455863}$Length$^{0.255245}$
   ```
   > plotPoints(Speed~Length)
   > plotFun(exp(6.455863)*
   Length^(0.255245)~Length,
   add=TRUE)
   ```

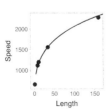

5. (a) The nonlinear log scale plot means the curve does not correspond to a power function.

   (b) The linear log scale plot means the curve corresponds to a power function.
   From $(1, 1)$ and $(3, 90)$,
   $k = \dfrac{\log_{10}(90)}{\log_{10}(3)} = 4.095903$ and
   $C = 1$, giving $y = x^{4.095903}$.

Section 2.4 Questions

1. (a) $A = 0.75$, $P = 1$, $h = 0.5$, $v = 2$

   (b) $A = 8$, $P = 4$, $h = 3$, $v = -10$

2. (a)
   ```
   > names(SunsetGreenwich)
   > plotPoints(Minutes~
   Month,data=SunsetGreenwich)


   [1] "Month"    "Minutes"
   ```

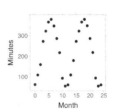

   (b) $A = 170$, $P = 12$, $h = 3$, $v = 210$

   $$\text{Min} = 170\sin\left[\frac{2\pi}{12}(\text{Month}-3)\right] + 210$$
   ```
   > plotFun(170*sin(2*pi/12*
   (Month-3))+210~Month,
   add=TRUE)
   ```

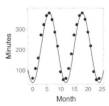

3.
   ```
   > bestSineModel=fitModel(
   Minutes~A*sin(2*pi/P*
   (Month-h))+v,data=SunsetGreenwich,
   ```

```
start=list(A=170,P=12,h=3,
v=210))
> coef(bestSineModel)
```

```
          A            P            h
   164.731336    12.013824    2.623384
```

```
          v
   223.613827
```

$$\text{Min} = 164.7 \sin\left[\frac{2\pi}{12.0}(\text{Month} - 2.6)\right] + 223.6$$

```
> plotPoints(Minutes~Month,
data=SunsetGreenwich)
> plotFun(bestSineModel(Month)~Month,
add=TRUE)
```

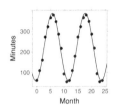

```
NetherlandsPopulation,
start=list(v=2,L=18,C=1.25,k=0.03054302))
> coef(bestPopModel)
```

```
          v            L            c
    1.82719      18.88981     1.25839
```

```
          k
    0.02608
```

$$P = \frac{18.8898}{1 + 1.258 e^{-0.026(Y - 1950)}} + 1.827$$

```
> plotPoints(Population~Year,
data=NetherlandsPopulation)
> plotFun(18/(1+1.25*
exp(-0.03054302*(Year-1950)))
+2~Year,add=TRUE)
> plotFun(bestPopModel(Year)~
Year,add=TRUE)
```

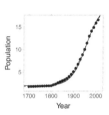

Section 2.5 Questions

1. (a) $v = 0$; $L = 40$;
   $(h, y(h)) = (-10, 18)$;
   $C = \dfrac{40}{18 - 0} - 1 = \dfrac{11}{9}$.

   (b) For $(0, 30)$, $k = 0.1299283$ and
   $y = \dfrac{40}{1 + (11/9)e^{-0.1299(x+10)}}$.

2. (a) The elongated "S" shape characteristic of sigmoidal data is apparent, including the asymptotic
   behavior for $y = v$ as inputs decrease.

   (b) $v = 2$; $v + L = 20$ gives $L = 18$;
   $(h, y(h)) = (1950, 10)$, $C = 1.25$;
   from $(1980, 14)$, $k = 0.030543$;
   $P = \dfrac{18}{1 + 1.25 e^{-0.031(\text{Yr} - 1950)}} + 2$

3. > names(NetherlandsPopulation)
   [1] "Year" "Population"

   ```
   > bestPopModel=fitModel
   (Population~L/(1+C*exp(-k*
   (Year-1950)))+v,data=
   ```

Section 2.6 Questions

1. (a) The log–log plot shows that a
   power function model is most reasonable.

   (b) > names(BlastData)

   ```
   [1] "X"        "time"      "radius"
   ```

   ```
   > radModel=fitModel(
   log(radius)~m*log(time)+b,
   data=BlastData)
   > coef(radModel)
   ```

   ```
           m          b
   0.386642  6.294689
   ```

   $\text{radius} = e^{6.2947} \cdot \text{time}^{0.3866}$

   (c) The model matches the data.
   ```
   > plotPoints(radius~time,
   data=BlastData)
   > plotFun(exp(6.2947)*time
   ^{0.38664}~time,add=TRUE)
   ```

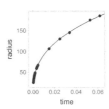

2. (a) The standard plot shows that a sine model is most reasonable.

(b) > names(WaterLevelsEastportMaine)

```
[1] "Hours"      "WaterLevel"
```

```
> WLModel=fitModel(
WaterLevel~A*sin(2*pi/P*
(Hours-h))+v,
data=WaterLevelsEastportMaine
start=list(A=16,
P=12,h=4,v=10))
> coef(WLModel)
```

```
        A          P          h
8.769145   12.429652   3.300070
```

```
        v
9.683791
```

$$W = 8.8 \sin\left[\frac{2\pi}{12.4}(H - 3.3)\right] + 9.7$$

(a) The model matches the data.

```
> plotPoints(WaterLevel~
Hours,
data=WaterLevelsEastportMaine)
> plotFun(8.769145*sin(
(2*pi/12.4297)*(Hours-3.3))
+9.683791~Hours,add=TRUE)
```

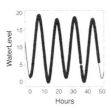

3. (a) The standard plot shows that a linear model is most reasonable.

(b) > names(NaturalGasConsumption)

```
[1] "Year"      "CubicFeet"
```

```
> natGasModel=fitModel(
CubicFeet~m*Year+b,
data=NaturalGasConsumption)
> coef(natGasModel)
```

```
      m              b
723130  -1404932269
```

$$CF = 723130 \cdot Yr - 1404932269$$

(b) The model matches the data, but not exactly because the data is not perfectly linear.

```
> plotPoints(CubicFeet~
Year,data=NaturalGasConsumption)
> plotFun(723130*Year-
1404932269~Year,add=TRUE)
```

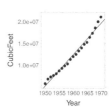

4. (a) > 723130*1949-1404932269
   ```
   [1] 4448101
   ```
   Reasonably close to the actual value of 4,971,152 because 1949 lies near the input interval used to create model.

(b) > 723130*1971-1404932269
   ```
   [1] 20356961
   ```
   Reasonably close to the actual value of 21,793,454, because 1971 lies near the input interval used to create the model.

(c) > 723130*1986-1404932269
   ```
   [1] 31203911
   ```
   Relatively far from the actual value of 16,221,296, because 1986 lies farther outside the input interval used to create the model. The linear trend does not seem to continue in future years.

Section 2.7 Questions

1. (a) Yard, nautical mile, furlong

(b) Volume: liter, cup, quart

2. (a) Invalid

    (b) Valid

    (c) Valid

    (d) Invalid

3. (a) $[a] = [b] = P$

    (b) $[a] = [b] = 1$

4. (a) $M \cdot LT^{-2} = MLT^{-2}$

    lb·miles per hr·sec

    (b) $M/L^3 = ML^{-3}$; g/m^3

5. (a) $[a] = L^{-2}$

    (b) $[a] = L^3 M^{-1}$

6. (a) Valid

    (b) Invalid

7. $[y] = [k] = MLT$; $[m] = 1$

8. (a) $[a] = PN^{-2}$; $[k] = P$

    (b) $[r] = 1$; $[q] = T^{-1}$

9. (a) $c = ka^x t^y$

    (b) $t = kp^x f^y c^z$

10. *Step (1):* Input variables length (ℓ), tension (s), and mass density (d). Output variable frequency (f).
    Step (2): $f = k\, \ell^x\, s^y\, d^z$
    Step (3): $[f] = T^{-1}$; $[k] = 1$;
    $[\ell] = L$; $[s] = MLT^{-2}$; $[d] = ML^{-1}$
    Step (4): $x = -1$; $y = 1/2$; $z = -1/2$
    Step (5): $f = k\, \ell^{-1}\, s^{1/2}\, d^{-1/2}$
    Step (6): Possible observations are: as tension increases, frequency increases; as length increases, frequency decreases; as mass density increases, frequency decreases; doubling length decreases frequency by a factor of two; doubling tension increases frequency by a factor of $\sqrt{2}$; doubling mass density decreases frequency by a factor of $\sqrt{2}$.

Section 3.1 Questions

1. (a) $\begin{pmatrix} 1 \\ 2 \end{pmatrix}$ (b) $\begin{pmatrix} -2 \\ 3 \end{pmatrix}$

    (c) $\begin{pmatrix} -4 \\ 3 \\ 1 \\ 5 \end{pmatrix}$ (d) $\begin{pmatrix} 12 \\ -4 \\ 1 \\ -1 \\ 2 \end{pmatrix}$

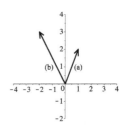

2. (a) $P = (3, 13)$

    (b) $Q = (16, 5, 4)$

3. $\sqrt{13} \approx 3.606$

4. (a) 5

    (b) $\sqrt{75} \approx 8.660$

5. (a) $\begin{pmatrix} 3 \\ 14 \\ 1 \\ 1 \end{pmatrix}$ (c) $\begin{pmatrix} 7 \\ 22 \\ 0 \\ 1 \end{pmatrix}$

    (b) $\begin{pmatrix} 5 \\ 2 \\ -3 \\ -1 \end{pmatrix}$ (d) $\begin{pmatrix} 5 \\ -14 \\ -6.5 \\ -3 \end{pmatrix}$

6. (a) $\begin{pmatrix} -4 \\ 8 \end{pmatrix}$ $\sqrt{80}$

    (b) $\begin{pmatrix} -1 \\ 2 \end{pmatrix}$ $\sqrt{5}$

    (c) $\begin{pmatrix} 3 \\ -6 \end{pmatrix}$ $\sqrt{45}$

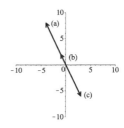

7. (a) $\begin{pmatrix} -2 \\ 6 \end{pmatrix}$ (b) $\begin{pmatrix} 1 \\ 5 \end{pmatrix}$ (c) $\begin{pmatrix} 1 \\ 5 \end{pmatrix}$

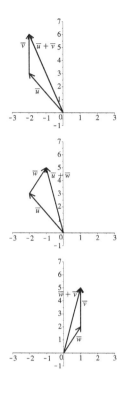

(b)

9.  ```
    > u=c(4,2,2)
    > v=c(1,1,-1)
    > 2*v
    [1]  2  2 -2
    > 3*u-4*v
    [1]  8  2 10
    > 6*u-1/2*v
    [1]  23.5 11.5 12.5
    ```

(a) $\begin{pmatrix} 2 \\ 2 \\ -2 \end{pmatrix}$ (b) $\begin{pmatrix} 8 \\ 2 \\ 10 \end{pmatrix}$ (c) $\begin{pmatrix} 23.5 \\ 11.5 \\ 12.5 \end{pmatrix}$

Section 3.2 Questions

1. (a) $20400 = 2001m + b$
    $20900 = 2003m + b$
    $21500 = 2005m + b$
    $21800 = 2006m + b$

    (b) $\begin{pmatrix} 20400 \\ 20900 \\ 21500 \\ 21800 \end{pmatrix} = m \begin{pmatrix} 2001 \\ 2003 \\ 2005 \\ 2006 \end{pmatrix} + b \begin{pmatrix} 1 \\ 1 \\ 1 \\ 1 \end{pmatrix}$

2. (a) $\begin{pmatrix} 8 \\ -2 \\ -8 \\ -10 \end{pmatrix}$ (b) $\begin{pmatrix} 4 \\ -15 \\ -4 \\ -3 \end{pmatrix}$

3. $x = 1,\ y = 2$

4. (a) 3×3; $\begin{pmatrix} 4 \\ -3 \\ 2 \end{pmatrix}$; $\begin{pmatrix} 1 \\ 16 \\ -1 \end{pmatrix}$; $\begin{pmatrix} 8 \\ -7 \\ 3 \end{pmatrix}$

    (b) 3×4; $\begin{pmatrix} -1 \\ 1 \\ -1 \end{pmatrix}$; $\begin{pmatrix} 2 \\ 0 \\ 5 \end{pmatrix}$; $\begin{pmatrix} -3 \\ 1 \\ -2 \end{pmatrix}$; $\begin{pmatrix} 4 \\ 0 \\ 4 \end{pmatrix}$

    (c) 4×2; $\begin{pmatrix} -1 \\ 1 \\ -1 \\ 5 \end{pmatrix}$; $\begin{pmatrix} 2 \\ 0 \\ 8 \\ 6 \end{pmatrix}$

8. (a) $\overline{G}(-1,-1) = \begin{pmatrix} \frac{-1}{\sqrt{6}} \\ \frac{1}{\sqrt{6}} \end{pmatrix}$

    $\overline{G}(0,-1) = \begin{pmatrix} 0 \\ \frac{1}{\sqrt{5}} \end{pmatrix}$

    $\overline{G}(1,-1) = \begin{pmatrix} \frac{1}{\sqrt{6}} \\ \frac{1}{\sqrt{6}} \end{pmatrix}$

    $\overline{G}(-1,0) = \begin{pmatrix} \frac{-1}{\sqrt{5}} \\ 0 \end{pmatrix}$

    $\overline{G}(0,0) = \begin{pmatrix} 0 \\ 0 \end{pmatrix}$

    $\overline{G}(1,0) = \begin{pmatrix} \frac{1}{\sqrt{5}} \\ 0 \end{pmatrix}$

    $\overline{G}(-1,1) = \begin{pmatrix} \frac{-1}{\sqrt{6}} \\ \frac{-1}{\sqrt{6}} \end{pmatrix}$

    $\overline{G}(0,1) = \begin{pmatrix} 0 \\ \frac{-1}{\sqrt{5}} \end{pmatrix}$

    $\overline{G}(1,1) = \begin{pmatrix} \frac{1}{\sqrt{6}} \\ \frac{-1}{\sqrt{6}} \end{pmatrix}$
    ```

5. $\begin{pmatrix} 1 \\ 4 \\ 5 \\ 7 \end{pmatrix} = \begin{pmatrix} -1 & 1 & -5 & 2 \\ 0 & 1 & 7 & -1 \\ 2 & 1 & 3 & -2 \\ 7 & 1 & 1 & 0 \end{pmatrix} \begin{pmatrix} x_1 \\ x_2 \\ x_3 \\ x_4 \end{pmatrix}$

6. (a) $1\begin{pmatrix} 7 \\ -1 \end{pmatrix} + 3\begin{pmatrix} 6 \\ 4 \end{pmatrix} = \begin{pmatrix} 25 \\ 11 \end{pmatrix}$

   (b) $2\begin{pmatrix} 1 \\ 0 \\ 1 \end{pmatrix} + 3\begin{pmatrix} 1 \\ 4 \\ 0 \end{pmatrix} + 4\begin{pmatrix} 1 \\ 1 \\ 2 \end{pmatrix} = \begin{pmatrix} 9 \\ 16 \\ 10 \end{pmatrix}$

7. (a) $\begin{pmatrix} -2 \\ 3 \end{pmatrix}$ (b) $\begin{pmatrix} -8 \\ -11 \end{pmatrix}$

8.
```
> u1=c(-1,0,2,7)
> u2=c(1,1,1,1)
> u3=c(-5,7,3,1)
> u4=c(2,-1,-2,0)
> U=matrix(c(u1,u2,u3,u4),
nrow=4,ncol=4)
> v=c(1,4,5,7)
> solve(U,v)
[1] 0.578 2.831 0.120 -0.325
```

$\bar{x} = \begin{pmatrix} 0.578 \\ 2.831 \\ 0.120 \\ -0.325 \end{pmatrix}$

Section 3.3 Questions

1. $\bar{v}$ is not a linear combination of $\bar{u}$

  (a)

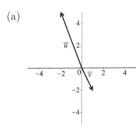

  (b) $\bar{v} \neq x\bar{u}$ for every scalar x

2. $\bar{v}$ is a linear combination of $\bar{u}$; $x = -\frac{1}{3}$

  (a)

  (b) $\bar{v} = (-3)\bar{u}$

3. $\bar{w}$ is a linear combination of $\bar{u}$ and $\bar{v}$; $x = -2$, $y = 0$

  (a) $\bar{w} = (-2)\bar{u} + 0\bar{v} = (-2)\bar{u}$

  (b)
```
> u = c(4,-1)
> v = c(-2,3)
> U = matrix(c(u,v),nrow=2,
ncol=2)
> w = c(-8,2)
> solve(U,w)
[1] -2 0
```

4.
```
> u1 = c(-2,5,2,1)
> u2 = c(0,7,3,-2)
> u3 = c(1,0,-8,3)
> u4 = c(-1,12,-3,2)
> U = matrix(c(u1,u2,u3,u4),
nrow=4,ncol=4)
> v = c(1,2,3,4)
> solve(U,v)
Error in solve.default(U,v):
  system is computationally
  singular
```
$\bar{v}$ is a not a linear combination of $\bar{u}_1$, $\bar{u}_2$, $\bar{u}_3$, and $\bar{u}_4$.

5. For $\begin{pmatrix} 20400 \\ 20900 \end{pmatrix} = \begin{pmatrix} 2001 & 1 \\ 2003 & 1 \end{pmatrix} \begin{pmatrix} m \\ b \end{pmatrix}$,

$m = 250$ and $b = -479{,}850$,

but $250\overline{\text{Year}} + (-479850)\overline{1}_4 \neq \overline{\text{Debt}}$.

6.
```
> u = c(0,1)
> v = c(1,1)
> U = matrix(c(u,v),nrow=2,
ncol=2)
> w = c(6.655,6.789)
> solve(U,w)
[1] 0.134 6.655
> WeekShift=c(0,1,5,6,8)
> Intercept=c(1,1,1,1,1)
> 0.134*WeekShift +
6.655*Intercept
[1] 6.655 6.789 7.325
7.459 7.727
```
No solution, because in the third components $7.325 \neq 7.269$.

Section 3.4 Questions

1. (a) $\bar{u} \cdot \bar{v} = 3$

   (b) $\bar{v} \cdot \bar{u} = 3$

(c) $(2\overline{u}) \cdot \overline{w} = -32$

(d) $\overline{u} \cdot (2\overline{w}) = -32$

2. (a) 22   (b) 0

3. (a) $\overline{u} \cdot \overline{v} = 45 \neq 0$; not orthogonal.

(b) Not orthogonal; the enclosed angle is not a right angle.

4. Examples are $\begin{pmatrix} 1 \\ -3 \end{pmatrix}$ and $\begin{pmatrix} -1.5 \\ 4.5 \end{pmatrix}$.

5. Only $\overline{u}$ and $\overline{v}$ are orthogonal.

6. $\overline{r} = \overline{v} - 1.5\overline{u} = \begin{pmatrix} -1.5 \\ 1.5 \\ 0.5 \\ 0 \end{pmatrix}$

7. (a) $\overline{r}_{\frac{1}{2}} = \begin{pmatrix} -1 \\ 4 \end{pmatrix}$;

$\overline{r}_{\frac{7}{8}} = \begin{pmatrix} -2.5 \\ 2.5 \end{pmatrix}$;

$\overline{r}_2 = \begin{pmatrix} -7 \\ -2 \end{pmatrix}$

(b) $\|\overline{r}_{\frac{1}{2}}\| = \sqrt{17} \approx 4.12$
$\|\overline{r}_{\frac{7}{8}}\| = \sqrt{12.5} \approx 3.54$
$\|\overline{r}_2\| = \sqrt{53} \approx 7.28$

(c)

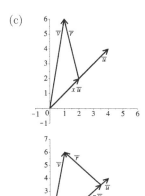

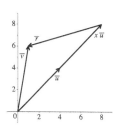

8. (a) $x = -\dfrac{41}{29}$

$x\overline{u} = \begin{pmatrix} -205/29 \\ 82/29 \end{pmatrix} \approx \begin{pmatrix} -7.069 \\ 2.828 \end{pmatrix}$

(b) $x = \dfrac{19}{91}$

$x\overline{u} = \begin{pmatrix} 19/91 \\ 95/91 \\ -133/91 \\ 76/91 \end{pmatrix} \approx \begin{pmatrix} 0.209 \\ 1.044 \\ -1.462 \\ 0.835 \end{pmatrix}$

9.
```
> u = c(1,8,2,5)
> v = c(1,0,9,11)
> dot(u,v)
[1] 74
> dot(u,3*v)
[1] 222
> dot(-1.1*v,4.3*u)
[1] -350.02
```

10.
```
> u = c(1,2,4)
> v = c(0,7,8)
> project(v~u)
       u
2.190476
> v-(2.190476)*u
[1] -2.1905  2.6191 -0.7619
```

$\overline{r} \approx \begin{pmatrix} -2.19 \\ 2.62 \\ -0.76 \end{pmatrix}$

Section 3.5 Questions

1. $D = 281.36Y - 542616.95$

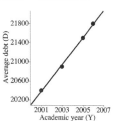

2. $\|r_1\| = 51.367 < \|r_2\| = 52.608$

3. > Y=c(2001,2003,2005,2006)
   > D=c(20400,20900,21500,21800)
   > project(D~Y+1)
     (Intercept)   Y
     -542616.9492   281.3559

   $D = 281.3559Y - 542616.9492$

   > plotPoints(D~Y)
   > plotFun(281.3559*Y-
   542616.9492~Y,add=TRUE)

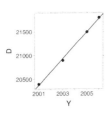

4. > names(WeightChange)
   [1] "Day"   "Weight"
   > project(Weight~Day+1,data=
   WeightChange)
     (Intercept)   Day
     1.9033767   0.0632135

   $\text{Weight} = 0.0632135 \cdot \text{Day} + 1.9033767$

   > plotPoints(Weight~Day,data=
   WeightChange)
   > plotFun(0.0632135*Day+
   1.9033767~Day,add=TRUE)

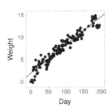

Section 4.1 Questions

1. Dimension is BT^{-1} with units of thousands of dollars per year.

   (a) $\dfrac{247.9 - 221}{2007 - 2004} = 8.967$

   (b) $\dfrac{216.7 - 221}{2009 - 2004} = -0.86$

   (c) $\dfrac{247.9 - 221}{2007 - 2004} = 5.178$

2. (a) $\dfrac{4-0}{4-0} = 1$

   (b) $\dfrac{2 - 16}{1 - (-2)} = \dfrac{14}{3}$

3. (a) $\dfrac{(3.1^2 + 3 \cdot 3.1) - (3^2 + 3 \cdot 3)}{3.1 - 3} = 9.1$

   (b) $\dfrac{(3.01^2 + 3 \cdot 3.01) - (3^2 + 3 \cdot 3)}{3.01 - 3}$
   $= 9.01$

   (c) $\dfrac{(3.001^2 + 3 \cdot 3.001) - (3^2 + 3 \cdot 3)}{3.001 - 3}$
   $= 9.001$

   (d) $\dfrac{(3^2 + 3 \cdot 3) - (2.9999^2 + 3 \cdot 2.9999)}{3 - 2.9999}$
   $= 8.9999$

   Conjecture $f'(3) = 9$.

4. (a) $\dfrac{P(2006) - P(2005)}{2006 - 2005} \approx 6.93$

   (b) $\dfrac{P(2005.01) - P(2005)}{2005.01 - 2005} \approx 12.57$

   (c) $\dfrac{P(2005.0001) - P(2005)}{2005.0001 - 2005} \approx 12.62$

   The instantaneous rate of change in median home prices is approximately $12,600.

5.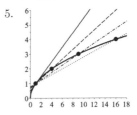

6. $f(3) = 12;\ f'(3) = 9;\ y = 12 + 9(x - 3)$

7. > f=makeFun(x^2+3*x~x)
   > AROC=makeFun((f(b)-f(a))/
   (b-a)~a&b)
   > AROC(a=3,b=3.1)
   [1] 9.1
   > AROC(a=3,b=3.01)
   [1] 9.01
   > AROC(a=3,b=3.0001)
   [1] 9.0001
   > AROC(a=2.99999,b=3)
   [1] 8.99999
   Conjecture $f'(3) = 9$.

8.
```
> P=makeFun(0.17*(Y-2001)^4
  -3.2*(Y-2001)^3+16.7*
  (Y-2001)^2-10.9*(Y-2001)
  +177.7~Y)
> avgP=makeFun((P(b)-P(a))/
  (b-a)~a&b)
> options(digits=10)
> avgP(a=2007,b=2007.1)
[1] -9.62903
> avgP(a=2007,b=2007.00001)
[1] -9.220041802
> avgP(a=2006.9,b=2007)
[1] -8.79337
> avgP(a=2006.99999,b=2007)
[1] -9.219958196
```
Conjecture $P'(2007) \approx -9.22$.

Section 4.2 Questions

1. • $f'(-2) \approx 10$

    • $f'(-1) \approx 0$

    • $f'(0) \approx -5$

    • $f'(2) \approx 0$

    • $f'(3.25) \approx 10$

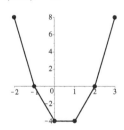

2. • $f'(1989) \approx -1.25$

    • $f'(1990) \approx -1.325$

    • $f'(1992) \approx -1.3025$

    • $f'(1996) \approx -0.8$

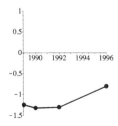

3. $x = 0$, $x = 1$, $x = 2$, $x = 4$, $x = 5$

4. (a) $s(8) = 493$ means the student drove 493 miles in eight hours.
    $[t] = T$ with units of hours.
    $[s(t)] = L$ with units of miles.

    (b) $s'(7.45) = 0$ means that 7 hours and 27 minutes into the trip, the rate of change is zero miles per hour.
    $[s'(t)] = LT^{-1}$ with units of miles per hour.

5. $f(x)$ is increasing on $(4,6)$
   $f(x)$ is constant on $(2,4) \cup (6,8)$
   $f(x)$ is decreasing on $(0,2) \cup (8,10)$

6. $f(x)$ is increasing on $(-\infty,-3.5)$, $(0,4)$, and $(4,\infty)$; $f(x)$ is constant on no intervals but $f'(x) = 0$ when $x = -3.5, 0, 4$; $f(x)$ is decreasing on $(-3.5,0)$.

7. Figure 16(b)

Section 4.3 Questions

1. $g'(-4) = -8$; $g'(-1) = -2$; $g'(3) = 6$; $g'(x) = 2x$

2. $h'(-3) = 3$; $h'(1) = 3$; $h'(8) = 3$; $h'(x) = 3$

3. (a) $f'(x) = 0$ (b) $g'(x) = \pi$

4. $\text{Gender}(2008) = 101.598$ is the number of males per 100 females in 2008.
   $\text{Gender}'(\text{Yr}) = 0.006$ is the rate of increase in the number of males per 100 females in 2008.

5. (a) $a'(x) = \dfrac{-1}{2}x^{-\frac{3}{2}}$

    (b) $b'(x) = \dfrac{-3}{4}x^{-\frac{7}{4}}$

6. (a) $a'(x) = 8x^7 + \dfrac{6}{x^2} + 12x^3$

    (b) $b'(x) = 6\sqrt{x} + 42x^5 - 16x$

7. $\text{Running}(30) = 768.723$ is the running speed in centimeters per second of a 30 centimeter long animal.
   $\text{Running}'(\text{Len}) = 22.892 \cdot \text{Len}^{0.046521}$ gives $\text{Running}'(30) = 26.816$ as the rate of increase in the running speed of a 30 centimeter long animal in centimeters per second per centimeter.

8.   (a) $a'(x) = \cos(x)$

     (b) $b'(x) = \ln(10) \cdot 10^x - e^x$

     (c) $c'(x) = \dfrac{5}{x}$ when $x > 0$

9. Nasdaq$(300) = 3890.814$ is the closing NASDAQ stock market value in U.S. dollars at the end of December 2012. Nasdaq$'(300)87.874$ is the rate of increase in the closing NASDAQ stock market value in U.S. dollars per quarter at the end of December 2012.

10. $f(x) = x^3$ has $n = 3$; $f'(x) = 3x^2$. $f(x)$ increases and $f'(x) > 0$ on $(-\infty, \infty)$.

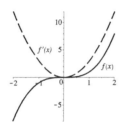

11. $g(3) = 18$; $g'(x) = 2x + 3$; $g'(3) = 9$; $y = 18 + 9(x - 3)$

12. $h(5) = \ln(5) + 5$; $h'(x) = \dfrac{1}{x} + 1$; $h'(5) = 1.2$; $y = \ln(5) + 5 + 1.2(x - 5)$

13.   (a) $L(\text{Yrs}) = 0.02663(\text{Yrs} - 60) + 0.1787$

     (b) Physicians$(62) = 0.24077$; $L(62) = 0.23196$

14. `> g=makeFun(3*x/pi+cos(x)~x)`
`> g(x=pi)`
`[1] 2`
`> dg=D(g(x)~x)`
`> dg(x=pi)`
`[1] 0.9549297`
$y = 2 + 0.9549297(x - \pi)$

Section 4.4 Questions

1. $[f(x) \cdot g(x)]' = 5x^4 \neq$ $f'(x) \cdot g'(x) = 4x^3$

2.   (a) $\sin(x) \cdot (3x^2 + 18x) + (x^3 + 9x^2) \cdot \cos(x)$

     (b) $\cos(x) \cdot e^x + e^x \cdot [-\sin(x)]$

     (c) $\ln(x) \cdot (2x) + x^2 \cdot \dfrac{1}{x}$

3. Amplitude$(1.5) = 0.935$ inches of displacement upward 1.5 seconds after hitting the pothole
Amp$'(\text{Sec}) =$
$\quad \cos(4\,\text{Sec}) \cdot \left(-2.25\,e^{-0.75\text{Sec}}\right)$
$\quad + 3\,e^{-0.75\text{Sec}} \cdot [-4 \cdot \sin(4\,\text{Sec})]$
Amplitude$'(1.5) = 0.387$ indicates that 1.5 seconds after hitting the pothole, the displacement is increasing at 0.387 inches per second.

4. `> A=makeFun(3*exp(-0.75*x)*`
`cos(4*x)~x)`
`> A(x=3.5)`
`[1] 0.02971563`
Amplitude$(3.5) = 0.03$ inches of displacement upward 3.5 seconds after hitting the pothole
`> dA=D(A(x)~x)`
`> dA(x=3.5)`
`[1] -0.883399`
Amplitude$'(3.5) = -0.883$ indicates that 3.5 seconds after hitting the pothole, the displacement is decreasing at 0.883 inches per second.

5.   (a) $\dfrac{(7x - 3) \cdot 4 - (4x + 5) \cdot 7}{(7x - 3)^2}$

     (b) $\dfrac{\ln(x) \cdot [-\sin(x)] - \cos(x) \cdot \frac{1}{x}}{[\ln(x)]^2}$

     (c) $\dfrac{\left[x^2 + \sin(x)\right] e^x - e^x [2x + \cos(x)]}{\left[x^2 + \sin(x)\right]^2}$

6. Cases$(60) = 487.658$ cases of Ebola on June 30, 2014 (60 days after May 1, 2014)
Cases$'(\text{Day}) =$
$\quad \dfrac{317.689 e^{-0.0233(\text{Day} - 200)}}{[1 + 1.016 e^{-0.0233(\text{Day} - 200)}]^2}$
Cases$'(60) = 10.950$ indicates an increase of approximately 11 cases of Ebola per day on June 30, 2014.

7. `> Cases=makeFun(13420/(1+1.016`
`*exp(-0.0233*(x-200)))~x)`
`> Cases(x=100)`
`[1] 1172.832`
Cases$(100) = 1173$ cases of Ebola on August 9, 2014 (90 days after May 1,

2014).

```
> dCases=D(Cases(x)~x)
> dCases(x=100)
[1] 24.93876
```

Cases$'(100) = 24.939$ indicates an increase of approximately 25 cases of Ebola per day on August 9, 2014.

8. $\dfrac{d}{dx}\left[\dfrac{1}{\sin(x)}\right] = \dfrac{\sin(x) \cdot 0 - 1 \cdot \cos(x)}{[\sin(x)]^2}$

$= \dfrac{-1}{\sin(x)} \cdot \dfrac{\cos(x)}{\sin(x)} = -\csc(x)\cot(x)$

9. (a) $\cot(x) \cdot \ln(2) \cdot 2^x + 2^x \cdot [-\csc^2(x)]$

(b) $\dfrac{3\sec(x) - (3x+7)\sec(x)\tan(x)}{\sec^2(x)}$

10. $\dfrac{(x+4)[\cos(x) - x\sin(x)] - [x\cos(x)]}{(x+4)^2}$

11. $g(0)=0$; $g'(0)=0.25$; $L(x)=0.25x$

12. (a) $h'(x) = e^x + g'(x)$;
$h'(0) = e^0 + g'(0) = 4$

(b) $h'(x) = [5g(x) + 1] \cdot f'(x) + f(x) \cdot 5g'(x)$
$h'(1) =$
$[5g(1) + 1]f'(1) + 5f(1)g'(1) =$
$[5 \cdot 9 + 1] \cdot (-1) + 5 \cdot 5 \cdot 7 = 129$

(c) $h'(x) =$
$\dfrac{[g(x) + 4]f'(x) - f(x)g'(x)}{[g(x) + 4]^2}$
$h'(2) =$
$\dfrac{[g(2) + 4]f'(2) - f(2)g'(2)}{[g(2) + 4]^2} =$
$\dfrac{[6 + 4] \cdot 4 - (-1) \cdot 1}{[6 + 4]^2} = 0.41$

Section 4.5 Questions

1. $f \circ g(x) = 3(x^2 - 4)$; $f[g(-1)] = -9$;
$f[g(2)] = 0$; $g \circ f(x) = (3x)^2 - 4$;
$f[g(-1)] = 5$; $f[g(2)] = 32$

2. (a) $g(x) = 3x^2 + e^x$; $f(x) = \ln(x)$

(b) $g(x) = x - \ln(x)$; $f(x) = \cos(x)$

(c) $g(x) = x^2$; $f(x) = e^x - 4$

3. (a) $g(x) = \sin(x)$; $f(x) = x^{-3}$

(b) $g(x) = \dfrac{e^x}{x^4 - x^3}$; $f(x) = x^7$

(c) $g(x) = x^2 + 27$; $f(x) = x^{-4/3}$

4. (a) $7(3x^4 + 2x)^6 \cdot (12x^3 + 2)$

(b) $5\left[-\sin(x^2 - e^x)\right] \cdot (2x - e^x)$

(c) $e^{x^2 + 5\sin(x)} \cdot [2x + 5\cos(x)]$

5. (a) $\cos\left[e^x + \ln(x)\right] \cdot \left[e^x + \dfrac{1}{x}\right]$

(b) $2 \cdot \ln(3) \cdot 3^{x^5 - 8x} \cdot (5x^4 - 8)$

(c) $(-4/3) \cdot (x^2 + 1)^{-7/3} \cdot (2x)$

6. Depth$(40) = 0.228$ is the tide depth in feet relative to MLLW after 40 hours.

Depth$'($Hr$) =$

$0.2127721\cos\left[\dfrac{2\pi}{24.51}\,(\text{Hr} - 2.13)\right]$

$+0.281413\sin\left[\dfrac{2\pi}{12.28}\,(\text{Hr} + 1.84)\right]$

Depth$'(40) = 0.030$ is the rate of increase in tide depth relative to MLLW in feet per hour after 40 hours.

7.
```
> Depth=makeFun(0.83*sin(
  (2*pi/24.51)*(Hr-2.13))
  -0.55*cos((2*pi/12.28)*
  (Hr+1.84))+0.76~Hr)
> Depth(Hr=48)
[1] -0.3528292
```
Depth$(48) = -0.353$ is the tide depth in feet relative to MLLW after 48 hours.
```
> dDepth=D(Depth(Hr)~Hr)
> dDepth(Hr=48)
[1] 0.2484316
```
Depth$'(48) = 0.248$ is the rate of increase in tide depth relative to MLLW in feet per hour after 48 hours.

8. (a) $24x^5 e^{x^3 - 4x} + 4x^6 e^{x^3 - 4x}(3x^2 - 4)$

(b) $\dfrac{\cos(e^x)\cos(x) - \sin(x)\cos(e^x)e^x}{\cos^2(e^x)}$

(c) $\dfrac{3x + 2}{x} \cdot \dfrac{(3x + 2) \cdot 1 - x \cdot 3}{(3x + 2)^2}$

(d) $\dfrac{(7x + 1) \cdot [e^x + xe^x] - xe^x \cdot 7}{(7x + 1)^2}$

9. $g(1) = 4e^{-3}$; $g'(1) = 20e^{-3}$;
$L(x) = 4e^{-3} + 20e^{-3}(x - 1)$

10. (a) $h'(x) = 2[x^3 + g(x)][3x^2 + g'(x)]$
$h'(2) = 2[8 + g(2)][12 + g'(2)]$
$= 364$

(b) $h'(x) =$
$$\frac{g(x^3)f'(5x)5 - f(5x)g'(x^3)3x^2}{[g(x^3)]^2}$$
$h'(0) =$
$$\frac{2 \cdot (-3) \cdot 5 - 4 \cdot 3 \cdot 0}{2^2} = -\frac{15}{2}$$

(c) $h'(x) = f'[\sin(x-1)]\cos(x-1)$
$h'(1) = f'[\sin(0)]\cos(0) =$
$f'(0) \cdot 1 = -3$

11. $g(x) = mx + b$ with $g'(x) = m$;
$f(x) = \cos(x)$ with $f'(x) = -\sin(x)$;

$$[f(g(x))]' = -\sin(mx + b) \cdot m$$
$$= -m\sin(mx + b)$$

Section 4.6 Questions

1. (a) $f(x, y)$ decreases; $f_x(1, 1) < 0$

   (b) $f(x, y)$ increases; $f_y(1, 1) > 0$

2. (a) $\dfrac{WC(35, 50) - WC(30, 50)}{35 - 30} =$
   $\dfrac{19 - 12}{5} = \dfrac{7}{5}$
   Wind chill is increasing at 1.4 with respect to air temperature.

   (b) $\dfrac{WC(30, 55) - WC(30, 50)}{55 - 50} =$
   $\dfrac{4 - 4}{5} = 0$
   Wind chill is constant with respect to wind speed.

3. (a) $\dfrac{f(2, 3.1) - f(2, 3)}{3.1 - 3} = 12.2$

   (b) $\dfrac{f(2, 3.01) - f(2, 3)}{3.01 - 3} = 12.02$

   (c) $\dfrac{f(2, 3.001) - f(2, 3)}{3.001 - 3} = 12.002$

   (d) $\dfrac{f(2, 3) - f(2, 2.99999)}{3 - 2.99999} =$
   11.99998

   $f_y(2, 3) \approx 12$; $f(x, y)$ is increasing in the y-direction at a rate of 12 when $(x, y) = (2, 3)$

4. (a) $f_x = 6\cos(6x)$; $f_y = 3y^2$

   (b) $g_x = \dfrac{y}{xy}$; $g_y = \dfrac{x}{xy} - 8$

   (c) $h_x = 14xe^{x^2+9y}$; $h_y = 63e^{x^2+9y}$

5. (a) $a_x = y\sin(x^3) + 3x^3 y\cos(x^3)$
   $a_y = x\sin(x^3)$

   (b) $b_x =$
   $$\frac{4\tan(xy) - (4x+5y)y\sec^2(xy)}{\tan^2(xy)}$$
   $b_y =$
   $$\frac{5\tan(xy) - (4x+5y)x\sec^2(xy)}{\tan^2(xy)}$$

6. $g_x = y\ln(z) + 1$; $g_y = x\ln(z) - 2yz^3$;
   $g_z = \dfrac{xy}{z} - 3y^2 z^2$

7. $h_x = \dfrac{xyz\cos(x) - yz\sin(x)}{(xy)^2}$;
   $h_y = \dfrac{-xz\sin(x)}{(xy)^2}$; $h_z = \dfrac{\sin(x)}{xy}$

8. $r_t = \frac{2}{5}kt^{-3/5}d^{-1/5}n^{1/5}$
   $r_d = -\frac{1}{5}kt^{2/5}d^{-6/5}n^{1/5}$
   $r_n = \frac{1}{5}kt^{2/5}d^{-1/5}n^{-4/5}$

9. $g_{xx} = 12x^2\sin(y)$; $g_{xy} = 4x^3\cos(y)$;
   $g_{yx} = 4x^3\cos(y)$; $g_{yy} = -x^4\sin(y)$

10. $g_x = 8e^{8x+9y}$; $g_y = 9e^{8x+9y}$;
    $g_{xy} = 8 \cdot 9e^{8x+9y} = 72e^{8x+9y}$
    $g_{yx} = 9 \cdot 8e^{8x+9y} = 72e^{8x+9y}$

11. $f_x = \sin(xy) + xy\cos(xy)$;
    $f_y = x^2\cos(xy)$
    $f(1, 0) = 0$; $f_x(1, 0) = 0$; $f_y(1, 0) = 1$
    $L(x, y) = 0 + 0 \cdot (x-1) + 1 \cdot (y-0) = y$

12. ```
    > D(3*x^4*y+3*cos(y)~x)
    function (x, y)
    3 * (4 * x^3) * y
    Warning message: ...
    > D(3*x^4*y+3*cos(y)~y)
    function (y, x)
    3 * x^4 - 3 * sin(y)
    Warning message: ...
    ```
    $g_x = 12x^3 y$; $g_y = 3x^4 - 3\sin(y)$

Section 4.7 Questions

1. (a) $\lim\limits_{x \to -1} g(x) = 3$

   (b) $\lim\limits_{x \to 0} g(x)$ does not exist;
   $\lim\limits_{x \to 0^-} g(x) = 3 \neq \lim\limits_{x \to 0^+} g(x) = 0$

   (c) $\lim\limits_{x \to 1} g(x) = 1$

2. $\dfrac{1 - e^{0.1}}{0.1} \approx -1.0517$;
    ```

$\dfrac{1-e^{0.001}}{0.001} \approx -1.0005;$

$\dfrac{1-e^{-0.00001}}{-0.00001} \approx -0.999995$

Thus, $\lim\limits_{x\to 0} \dfrac{1-e^x}{x} = -1.$

3. (a) $\lim\limits_{x\to -3} g(x) = 1 = g(-3) = 1,$ so $g(x)$ is continuous for $x = -3.$

   (b) $\lim\limits_{x\to 0} g(x) = 4 \neq g(0) = 2,$ so $g(x)$ is not continuous for $x = 0.$

   (c) $\lim\limits_{x\to 2} g(x)$ does not exist, because $\lim\limits_{x\to 2^-} g(x) = 0 \neq \lim\limits_{x\to 2^+} g(x) = -2,$ so $g(x)$ is not continuous for $x = 2$

4. (a) All reals except $x = 0$

   (b) All reals except $x = \pm 1$

   (c) All reals

5. (a) $3^3 - 5 = 22$

   (b) $\dfrac{\pi^2}{\cos(\pi)+2} = \pi^2$

   (c) $\dfrac{0}{2^4} = 0$

6. (a) $\lim\limits_{x\to 3} \dfrac{x-3}{(x+3)(x-3)} = \dfrac{1}{6}$

   (b) $\lim\limits_{x\to 2} \dfrac{(x-2)(x^2+2x+4)}{x-2} = 12$

7. (a) $\lim\limits_{x\to 0} \dfrac{x^2+10x}{x} = 10$

   (b) $\lim\limits_{x\to 1} \dfrac{x-1}{x^2+6x-7} = \dfrac{1}{8}$

8. (a) Conjugate for the numerator is $\sqrt{x+6}+2$ and yields limit $\dfrac{1}{4}.$

   (b) Conjugate for the denominator is $3+\sqrt{x-9}$ and yields limit 6.

9. $\dfrac{P(2011-1)-P(2011)}{-1} \approx -23.8$

   $\dfrac{P(2011-0.01)-P(2011)}{-0.01} \approx -42.9$

   $\dfrac{P(2011-0.0001)-P(2011)}{-0.0001} \approx -43.1$
   $P'(2011) \approx -43.2$ means that median home prices were decreasing at a rate of \$43,200 per year in 2011.

10. $f'(4) = \lim\limits_{h\to 0} \dfrac{3(4+h)+7-(3\cdot 4+7)}{h} = 3$

11. $f' = \lim\limits_{h\to 0} \dfrac{5(x+h)-9-(5\cdot x-9)}{h} = 5$

Section 5.1 Questions

1. Global maximum $y = 4$ when $x = -1$ and $x = 5$; global minimum $y = -17$ when $x = 7.$

2. Global maxiuumum percent growth is approximately 1.866 in 1980; global minimum percent growth is approximately 1.079 in 2015.

3. (a) From $f'(x) = 0$: $x = -1, 2, 6.$
      From $f'(x)$ undefined: none.

   (b) From $f'(x) = 0$: $x = 0.$
      From $f'(x)$ undefined: $x = 2, 4.$

4. From $f'(x) = 0$: $x = -1.$
   From $f'(x)$ undefined: $x = -5, -2, 0.$

5. (a) $a'(x) = 6x^2 - 3x^2 - 36x;$
      from $a'(x) = 0$: $x = -2, 3;$
      from $a'(x)$ undefined: none.

   (b) $b'(x) = \dfrac{-47}{(7x-3)^2};$
      from $b'(x) = 0$: none;
      from $b'(x)$ undefined: $x = \dfrac{3}{7}.$

6. (a) Global maximum $y = 9$ when $x = -1$; global minimum $y = -23$ when $x = -3$ and $x = 3.$

   (b) Global maximum $y = 9$ when $x = -1$; global minimum $y = -18$ when $x = 2.$

7. ```
> f=makeFun(exp(cos(x))~x)
> df=D(f(x)~x)
> critNum=findZeros(df(x)~x,
xlim=c(-2,5))
        x
1 -3.1416
2 0.0000
3 3.1416
4 6.2832
> evalPts=rbind(-2,critNum,5)
> f(evalNum)
        x
1 0.6595834
2 0.3678794
3 2.7182818
4 0.3678794
5 2.7182818
6 1.3279842
```
Global maximum $y = 2.7182818$ when $x = 0$ and $x = 2\pi$; global minimum $y = 0.3678794$ when $x = -\pi$ and $x = \pi.$
```

Section 5.2 Questions

1. Local maximum values at $(-5, 2)$ and $(-1, 1)$; local minimum values at $(-3, -0.5)$ and $(1, -1.5)$.

2. Critical numbers $x = -2$ and $x = 2$ from $f'(x) = 0$; none from $f'(x)$ does not exist.
   $f(-2) \approx 0.75$ is a local maximum; $f(x)$ switches from increasing to decreasing, so $f'(x)$ switches from positive to negative.
   $f(2) \approx -3$ is a local minimum; $f(x)$ switches from decreasing to increasing, so $f'(x)$ switches from negative to positive.

3. $f'(x) = -3x^2 - 6x + 24$ gives critical numbers $x = -4$ and $x = 2$ from $f'(x) = 0$; none from $f'(x)$ does not exist.
   $f'(-5) = -21 < 0$; $f'(0) = 24 > 0$; $f'(4) = -21 < 0$.
   $f(-4) = -74$ is a local minimum, because $f'(x)$ switches from negative to positive; and $f(2) = 34$ is a local maximum, because $f'(x)$ switches from positive to negative.

4. $f'(x) = \dfrac{2x}{(x^2 + 1)^2}$ gives critical number $x = 0$ from $f'(x) = 0$; none from $f'(x)$ does not exist.
   $f'(-1) = -0.5 < 0$; $f'(1) = 0.5 > 0$.
   $f(0) = -1$ is a local minimum, because $f'(x)$ switches from negative to positive.

5. ```
   > f=makeFun(x+exp(-x)~x)
   > df=D(f(x)~x)
   > findZeros(df(x)~x,
   xlim=c(-5,5))
     x
   1 0
   > df(-1)
   [1] -1.718282
   > df(1)
   [1] 0.6321206
   > f(0)
     x
   1 1
   ```
   $f'(x)$ switches from negative to positive: local minimum of $f(0) = 1$ when $x = 0$.

Section 5.3 Questions

1. (a) Concave up on $(-\infty, 3)$
       Concave down on $(3, \infty)$
       Point of inflection $\approx (3, 3.5)$

   (b) Concave up on $(-\infty, -1)$ and $(2, \infty)$
       Concave down on $(-1, 2)$
       Points of inflection $\approx (-1, -5)$ and $(2, -45)$

2. Concave up on $(0, 50)$ and $(100, 150)$
   Concave down on $(50, 100)$ and $(150, 200)$
   Points of inflection $\approx (50, 18)$, $(100, 32)$, and $(150, 35)$

3. Concave up on $(-\infty, 5)$
   Concave down on $(5, \infty)$
   Point of inflection $\approx (5, -7)$

4. $f'(x) = 3x^2 + 6x + 12$;
   $f''(x) = 6x + 6$; $f''(x) = 0$ when $x = -1$; $f''(x)$ never undefined
   Test inputs: $x = -2$ and $x = 0$
   $f''(-2) = -6 < 0$; $f''(0) = 6 > 0$
   Concave up on $(-1, \infty)$
   Concave down on $(-\infty, -1)$
   Point of inflection at $(-1, -28)$

5. $f'(x) = 4x^3 - 30x^2 + 48x$;
   $f''(x) = 12x^2 - 60x + 48$;
   $f''(x) = 0$ when $x = 1$ and $x = 4$;
   $f''(x)$ never undefined
   Test inputs: $x = 0$, $x = 2$, $x = 5$
   $f''(0) = 48 > 0$; $f''(2) = -24 < 0$;
   $f''(5) = 48 > 0$
   Concave up on $(-\infty, 1)$ and $(4, \infty)$
   Concave down on $(1, 4)$
   Points of inflection at $(1, 21)$ and $(4, 6)$

6. When $p'(x)$ is increasing, $p(x)$ is concave up.

7. $f'(x) = 6x^2 + 42x$; $f''(x) = 12x + 42$;
   $f'(x) = 0$ when $x = -7$ and $x = 0$;
   $f'(x)$ never undefined
   $f''(-7) = -42 < 0$; $f''(0) = 42 > 0$
   Local maxmimum $y = 393$ when $x = -7$; local minimum $y = 50$ when $x = 0$

8. ```
   > f=makeFun(x*sin(x)-exp(x)~x)
   > df=D(f(x)~x)
   > critNum=findZeros(df(x)~x,
   xlim=c(-15,0))
   > critNum          x
   1 -14.2074
   2 -11.0855
   ```

```
3  -7.9787
4  -4.9118
5  -2.0745
> ddf=D(df(x)~x)
> ddf(critNum)
            x
1 -14.312551
2  11.220358
3  -8.165690
4   5.203393
5  -2.907841
> f(critNum)
            x
1  14.172373
2 -11.040723
3   7.916385
4  -4.821824
5   1.691230
```
Local maximum $(-14.207, 14.172)$, $(-7.979, 7.916)$, and $(-2.075, 1.691)$
Local minimum $(-11.086, -11.041)$ and $(-4.912, -4.822)$

Section 5.4 Questions

1. $x_{n+1} = x_n - \dfrac{(x_n)^5 + x_n + 1}{5(x_n)^4 + 1}$

    $x_1 = -0.6667$; $x_2 = -0.76811$;
    $x_3 = -0.75516$; $x_4 = -0.75487$
    Zero ≈ -0.75

2. $h(x) = \sin(x) - x^2$

    $x_{n+1} = x_n - \dfrac{\sin(x_n) - (x_n)^2}{\cos(x_n) - 2(x_n)}$

    $x_1 = 0.5$; $x_2 = 2.374125$;
    $x_3 = 1.47028$; $x_4 = 1.059476$;
    $x_5 = 0.9058268$; $x_6 = 0.8777119$;
    $x_7 = 0.8767274$; $x_8 = 0.8767262$
    Intersection $\approx (0.8767, 0.7686)$

3. $h(x) = \cos(x) - 0.75$

    $x_{n+1} = x_n - \dfrac{\cos(x_n) - 0.75}{-\sin(x_n)}$

    $x_1 = 1$; $x_2 = 0.750796$;
    $x_3 = 0.723160$; $x_4 = 0.722734$;
    $x_5 = 0.722734$

4. $x_{n+1} = x_n - \dfrac{3(x_n)^2 - 4}{6x_n}$

    $x_1 = -2$; $x_2 = -1.33333$;
    $x_3 = -1.16667$; $x_4 = -1.15476$;
    $x_5 = -1.15470$
    Critical number $c \approx -1.155$

5.
```
> f=makeFun(x^3*log(x)-x^2~x)
> df=D(f(x)~x)
> NM=makeFun(x-f(x)/df(x)~x)
> x2=NM(2)
> x2
[1] 1.81423
> x3=NM(x2)
> x3
[1] 1.76635
> x4=NM(x3)
> x4
[1] 1.76324
```
An approximate zero of $f(x)$ is 1.76.

Section 5.5 Questions

1. Local maximum of $z \approx 6.5$ when $(x, y) \approx (2, 1)$; $f(x, y)$ decreases in every direction from its maximum; local maximum of $z \approx 2.5$ when $(x, y) \approx (-0.25, -0.25)$; local minimum of $z \approx -0.5$ when $(x, y) \approx (1, -1)$; $f(x, y)$ increases in every direction from its minimums.

2. $f_x = 2x - 2$; $f_x = 0$ when $x = 1$
    $f_y = 2y + 2$; $f_y = 0$ when $y = -1$
    Critical point $(x, y) = (1, -1)$

3. $f_x = 3x^2 - 27$; $f_x = 0$ when $x = \pm 3$
    $f_y = 3y^2 - 12$; $f_y = 0$ when $y = \pm 2$
    Critical points: $(3, 2)$, $(3, -2)$, $(-3, 2)$, and $(-3, -2)$

4. $f_x = 4x - 4$; $f_x = 0$ when $x = 1$
    $f_y = 12y + 24$; $f_y = 0$ when $y = -2$
    Critical point $(x, y) = (1, -2)$
    $f_{xx} = 4$; $f_{xy} = 0$; $f_{yy} = 12$; $D = 48$
    Local minimum of $z = -1$ when $(x, y) = (1, -2)$

5. $f_x = 2x$; $f_x = 0$ when $x = 0$
    $f_y = -2y$; $f_y = 0$ when $y = 0$
    Critical point $(x, y) = (0, 0)$
    $f_{xx} = 2$; $f_{xy} = 0$; $f_{yy} = -2$; $D = -4$
    Saddle point when $(x, y) = (0, 0)$

6. $f_x = -\pi \sin\left(\dfrac{\pi}{2}x\right) \sin\left(\dfrac{\pi}{2}y\right)$

    $f_y = \pi \cos\left(\dfrac{\pi}{2}x\right) \cos\left(\dfrac{\pi}{2}y\right)$

    (a) $f_x(0, 1) = 0 = f_y(0, 1)$
        $f_x(0, -1) = 0 = f_y(0, -1)$
    (b) $D(0, 1) = 0$, so inconclusive
    (c) $D(0, -1) = 0$, so inconclusive

7. $f_x = 2x$; $f_x = 0$ when $x = 0$
$f_y = -2y$; $f_y = 0$ when $y = 0$
Critical point $(x, y) = (0, 0)$ in $\mathcal{R}$
From $x^2 = 1 - (y-1)^2$, $g(y) = 2y - 2y^2$
on $[0, 2]$
$g'(y) = 2 - 4y = 0$ when $y = \frac{1}{2}$
Critical points: $\left(\pm\frac{\sqrt{3}}{2}, \frac{1}{2}\right)$
Endpoints: $(0, 0)$ and $(0, 2)$
Global maximum of $z = \frac{1}{2}$ when
$(x, y) = \left(\frac{\sqrt{3}}{2}, \frac{1}{2}\right)$ and $\left(-\frac{\sqrt{3}}{2}, \frac{1}{2}\right)$
Global minimum of $z = -4$ when
$(x, y) = (0, 2)$

Section 5.6 Questions

1. (a) $\nabla f = \begin{pmatrix} 2x + \sin(y) \\ x\cos(y) \end{pmatrix}$

(b) $\nabla g = \begin{pmatrix} xe^{xy} + xye^{xy} \\ x^2 e^{xy} \end{pmatrix}$

(c) $\nabla h = \begin{pmatrix} \dfrac{1}{x+y} \\ \dfrac{1}{x+y} \end{pmatrix}$

2. Local maximum value $z = 6$ when
$(x, y) \approx (3.6, 0.8)$; local minimum $z = -6$ when $(x, y) \approx (2.2, 2.8)$

3. (a) Local maximum $z = 6$ when
$(x, y) \approx (2.75, 0.5)$
Local minimum $z = 1$ when
$(x, y) \approx (1, 2)$

(b) Local maximum $z = 6.5$ when
$(x, y) \approx (3, 0.5)$
Local minimum $z = 2$ when
$(x, y) \approx (1.4, 2.25)$

(c) For maximum: $\lambda = 6.5 - 6 = 0.5$
For minimum: $\lambda = 2 - 1 = 1$

4. $\nabla f = \begin{pmatrix} 2x \\ -2y \end{pmatrix} = \lambda \begin{pmatrix} 2 \\ -1 \end{pmatrix} = \lambda \nabla g$
$g(x, y) = 2x - y = 6$
$x = 4$; $y = 2$; $\lambda = 4$
$f(3, 0) = 9$; $f(4, 2) = 12$;
$f(0, -6) = -36$
Local maximum $z = 12$ when
$(x, y) = (4, 2)$

Section 6.1 Questions

1. (a)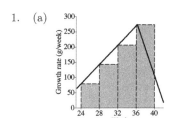

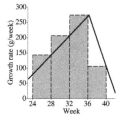

(b) $L_4 \approx 75 \cdot 4 + 140 \cdot 4 + 210 \cdot 4 + 265 \cdot 4 = 2760$
$R_4 \approx 140 \cdot 4 + 210 \cdot 4 + 265 \cdot 4 + 105 \cdot 4 = 2880$

(c) L_4 underestimates and R_4 overestimates the total growth of the fetus

2. (a) $L_2 = (1.41) \cdot 10 + (1.2) \cdot 10 = 26.1$

(b) $R_4 = (1.26) \cdot 5 + (1.2) \cdot 5 + (1.13) \cdot 5 + (1.06) \cdot 5 = 23.25$

3. $L_2 = (3 \cdot 0^2 + 5) \cdot 2 + (3 \cdot 2^2 + 5) \cdot 2 = 44$
$R_4 = (3 \cdot 1^2 + 5) \cdot 1 + (3 \cdot 2^2 + 5) \cdot 1 + (3 \cdot 3^2 + 5) \cdot 1 + (3 \cdot 4^2 + 5) \cdot 1 = 110$

4. $M_2 = (2^2 + 1) \cdot 4 + (6^2 + 1) \cdot 4 = 168$

5. $M_4 = ((1.5-3)^2 - 1) \cdot 1 + ((2.5-3)^2 - 1) \cdot 1 + ((3.5-3)^2 - 1) \cdot 1 + ((4.5-3)^2 - 1) \cdot 1 = 1$

6. $R_2 = 4 \cdot 4 + 8 \cdot 4 = 48$
$R_4 = 2 \cdot 2 + 4 \cdot 2 + 6 \cdot 2 + 8 \cdot 2 = 40$
$R_8 = 1 \cdot 1 + 2 \cdot 1 + 3 \cdot 1 + 4 \cdot 1 + 5 \cdot 1 + 6 \cdot 1 + 7 \cdot 1 + 8 \cdot 1 = 36$
All overestimates with more equal-width subintervals give a more accurate approximation of the exact accumulation of 32.

Section 6.2 Questions

1. (a) $\frac{1}{2} \cdot 4 \cdot 2 = 4$

   (b) $2 \cdot (-2) + \frac{1}{2} \cdot 1 \cdot (-2) = -5$

   (c) $\frac{1}{2} \cdot 4 \cdot 2 + \frac{1}{2} \cdot 3 \cdot (-3) = -0.5$

2. (a) $\frac{1}{2} \cdot 2 \cdot (-2) = -2$

   (b) $3 \cdot (-2) = -6$

   (c) $\frac{1}{2} \cdot (-2) \cdot (-2) = 2$

3. (a) $\int_{13}^{24} b(t)\, dt$ provides the debt accrued from months 13 to 24, or during the second year.
   Units of dollars and dimension of money with symbol B.

   (b) $\int_{1}^{48} b(t)\, dt$

4. (a) 0

   (b) $9 - 12 = -3$

   (c) $4 \cdot 9 = 36$

   (d) $2 \cdot 9 - 3 \cdot 12 = -18$

   (e) $-(2 \cdot 9 - 3 \cdot 12) = 18$

   (f) $2 \cdot 9 + 2 \cdot 16 = 50$

5. (a) $-2 - 6 = -8$

   (b) $-(-2) + 3 \cdot 6 = 20$

   (c) $7 - (-2) = 9$

   (d) $6 - 1 = 5$

Section 6.3 Questions

1. (a) $F'(x) = 2x = f(x)$, so yes.

   (b) $F'(x) = 3x^2 - 1 \neq f(x)$, so no.

   (c) $F'(x) = 3x^2 = f(x)$, so yes.

   (d) $F'(x) = \cos x = f(x)$, so yes.

2. (a) $A(0) = \int_0^0 x + 1\, dx = 0$

   (b) $A(2) = 2 \cdot 1 + \frac{1}{2} \cdot 2 \cdot 2 = 4$

   (c) $A(3) = 3 \cdot 1 + \frac{1}{2} \cdot 3 \cdot 3 = 7.5$

   (d) $A(-1) = \frac{1}{2} \cdot (-1) \cdot 1 = -0.5$

3. (a) $t^2 + 2t$

   (b) $-\ln(t^3)$

4. (a) $A(t) = \int_0^t \sin(x^2)\, dx$

   (b) $A(t) = \int_2^t e^{x^2}\, dx$

5. (a) $\dfrac{x^5}{5} - \dfrac{5x^3}{3} + 6x + C$

   (b) $6\left(\dfrac{x^2}{2}\right) + 4x + \dfrac{3}{4}e^{4x-6} + C$

   (c) $-\dfrac{8}{7}\cos(7x + 4) + \dfrac{5}{9}\sin(9x + 1) + C$

   (d) $\dfrac{2}{5}x^5 - 5\ln|x| - \dfrac{1}{7}\dfrac{x^{-2}}{2} + C$

   (e) $\dfrac{3}{2}\sin(2x) + \dfrac{x^{3/2}}{3/2} + 6x + C$

   (f) $-\dfrac{6}{4}\cos(4x + 5) + \dfrac{3}{2}e^{2x+5} + C$

Section 6.4 Questions

1. (a) $x^3 + \dfrac{2}{3}x^{3/2}\Big|_4^5 = \left[5^3 + \dfrac{2}{3}5^{3/2}\right] - \left[4^3 + \dfrac{2}{3}4^{3/2}\right] \approx 63.120$

   (b) $\dfrac{9}{4}x^4 + \dfrac{1}{2}\cos(2x - 5)\Big|_0^{2\pi} = 36\pi^4$

   (c) $-e^{3-x} - 4\ln|x|\Big|_1^{2e} = -e^{3-2e} - 4\ln(2e) + e^2 + 4\ln(1) \approx 0.529$

2. (a) $\int_0^4 2x^3 + 4\sqrt{x} + 5\, dx = x^3 + \dfrac{8}{3}x^{3/2} + 5x\Big|_0^4 = \dfrac{508}{3} \approx 169.33$

   (b) $\int_1^2 8 - \dfrac{3}{x} + \dfrac{3}{x^4}\, dx = 8x - 3\ln|x| - \dfrac{1}{x^3}\Big|_1^2 = \dfrac{71}{8} - 3\ln(2) \approx 6.796$

   (c) $\int_0^\pi 5e^{4x+5} - 9\sin(3x)\, dx = \dfrac{5}{4}e^{4\pi+5} - 6 - \pi - \dfrac{5}{4}e^5 \approx 5.32 \times 10^7$

3. $1.294 + \int_{23}^{37} 0.371\, dt + \int_{38}^{40} 0.331\, dt$

   $= 1.294 + 0.371\Big|_{23}^{37} + 0.331\Big|_{23}^{37}$

   ≈ 7.150

   Average female baby weighs 7.150 pounds when born at 40 weeks.

4. $\displaystyle\int_0^4 4x - x^2 - x\,dx = \int_0^4 3x - x^2\,dx$

$\displaystyle = \frac{3x^2}{2} - \frac{x^3}{3}\bigg|_0^4 = \frac{9}{2}$

5. (a)
```
> F=antiD(exp(2*x^3+4)*
  x^2~x)
> F(4)-F(2)
[1] 3.537695e+56
```
   (b)
```
> F=antiD(3*x^2*cos(x^3+
  1)/sin(x^3+1)~x)
> F(1)-F(0)
[1] 0.07752071
```

Section 6.5 Questions

1. (a) $g(x) = x^2 + 3x + 1$; $f(x) = e^x$

   (b) $g(x) = x^2$; $f(x) = e^x + 1$

   (c) $g(x) = \cos(x)$; $f(x) = \dfrac{2}{x}$

2. (a) $u = x^2 + 1$; $du = 2x\,dx$;
   $(x^2 + 1)^5 + C$

   (b) $u = x^3 + 1$; $du = 3x^2\,dx$;
   $\sin(x^3 + 1) + C$

   (c) $u = x^6 + 2$; $du = 6x^5\,dx$;
   $\ln(x^6 + 2) + C$

   (d) $u = 4x^3$; $du = 12x^2\,dx$;
   $e^{4x^3} + C$

3. (a) $u = \sin(x) + 3$; $du = \cos(x)\,dx$;
   $-\dfrac{1}{\sin x + 3} + C$

   (b) $u = \ln(x)$; $du = \frac{1}{x}\,dx$;
   $\ln|\ln(x)| + C$

   (c) $u = x^2 - 4x$; $du = (4x - 8)\,dx$;
   $4\sqrt{x^2 - 4x} + C$

   (d) $u = 2\ln(x) + 1$; $du = \frac{2}{x}\,dx$;
   $(2\ln(x) + 1)^5 + C$

4. (a) $u = x^4$; $\frac{1}{4}du = x^3\,dx$;
   $\dfrac{1}{4}\sin(x^4) + C$

   (b) $u = x + 2$; $du = dx$;
   $u - 2 = x$; $(u - 2)^2 = x^2$;
   $\dfrac{3}{10}(x+2)^{10/3} - \dfrac{12}{7}(x+2)^{7/3}$
   $+ 3(x+2)^{4/3} + C$

5. (a) $u = x^3 + 1$; $\frac{1}{3}du = x^2\,dx$;
   $\dfrac{2}{9}u^{3/2}\bigg|_{x=-1}^{0} = \dfrac{2}{9}$

   (b) $u = 5 - x^4$; $-\frac{1}{4}du = x^3\,dx$;
   $\dfrac{1}{24u^5}\bigg|_{x=0}^{1} = \dfrac{1}{24}\left[\dfrac{1}{5^5} - \dfrac{1}{4^5}\right]$
   $= \dfrac{2101}{64{,}000{,}000} \approx 3.28 \times 10^{-5}$

6. (a) $dy = -e^x \sin(e^x)$

   (b) $\Delta y = \cos(e^{0.1}) - \cos(e^0)$
   ≈ -0.0913
   $dy = -\sin(1)\cdot(0.1) \approx -0.0841$

Section 6.6 Questions

1. $f(x) = \ln(x)$; $f'(x) = \dfrac{1}{x}$
   $g'(x) = 2x$; $g(x) = x^2$
   $x^2 \ln(x) - \displaystyle\int \dfrac{1}{x}x^2\,dx$
   $= x^2 \ln(x) - \dfrac{1}{2}x^2 + C$

2. (a) $f(x) = x^2$; $g'(x) = \sin(x)$

   (b) $f(x) = \ln(2x)$; $g'(x) = 3x^2$

3. $f(x) = x$; $f'(x) = 1$
   $g'(x) = \cos(x)$; $g(x) = \sin(x)$
   $x\sin(x)\bigg|_0^{\pi/2} - \displaystyle\int_0^{\pi/2} \cos(x)\,dx$
   $= \dfrac{\pi}{2} - 1$

Appendix B

Answers to Odd-Numbered Exercises

Section 1.1 Exercises

1. Total profit made by a lemonade stand as a function of the number of cups of lemonade sold.

3. Cost of health care as a function of age, dependence, previous health conditions, and current health conditions.

5. An output of gas price from inputs of city and state.

7. Function; domain: $\{a, b, c, d\}$; range: $\{2, 3, 4, 5\}$

9. Function; domain: $\{a, b, d\}$; range: $\{6, 8, 9\}$

11. Function; domain: $\{2.5, 4.1, 8.7, 9\}$; range: $\{\pi, 4, \pi^2, 16\}$

13. Nonfunction; see $x = 4.1$

15. Function;
    domain: $\{2006, 2008, 2010, 2012\}$;
    range: $\{\$141.1, \$141.4, \$139.4, \$150.4\}$

17. Nonfunction; see $x = 6/11$

19.
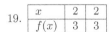

x	2	2
$f(x)$	3	3

21. Domain: $(-\infty, \infty)$; range: $[-4, \infty)$

23. Nonfunction; see $x = 0$

25. Nonfunction; see $x = 0$

27.
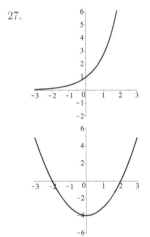

29. $(-\infty, \infty)$

31. $[-7, \infty)$

33. $(-\infty, -3] \cup [3, \infty)$

35. $(-\infty, -5) \cup (-5, 0) \cup (0, \infty)$

37. $(-\infty, -3) \cup (-3, 0) \cup (0, 3) \cup (3, \infty)$

39. $f(3) = 32$

41. $f(934.82) = 2.622 \times 10^6$

43. $f(-5.234) = 87.184$

45. $f(12) = 135$ million Twitter users

47. $f(12.5) = 168.4$ million Twitter users

49. $f(-10) = -26$

51. $f(0.5) = 1.75$

53. $f(-\pi) = 2 - \pi^2$

55. $f(-4.9) = -22.01$

57. $f(-10) = 203$

59. $f(2.1) = 0.3$

61. $f(\pi) = 3\pi - 6$

63. $f(-3.4) = 3$

Example function for Exercises 65 − 68:
$$f(x) = \begin{cases} x^2 - 7 & x \leq 0 \\ 2x + 8 & 0 < x \end{cases}$$

65. $f(4) = 16$

67. $f(0) = -7$

69. > (-4)*3-2
    [1] -14

71. > (3*pi+4*((2)^(1/2)))/2
    [1] 7.540816

73. > abs(pi^2-(5^3)*(3^2.1))
    [1] 1245.769

75. > ((2.34)^3+7)/(3.7*(4.3-3))
    [1] 4.119107

For Exercises 77 − 80:
> f=makeFun(4.3*(3*(x^5)-2.6)+2.5~x)

77. > f(x=-84.658)
    [1] -56095720663

79. > f(x=4.341)
    [1] 19876.92

For Exercises 81 − 84:
> f=makeFun(pi*(9.852*x^3-10.375*x)/
((9*pi)^2+27*x)~x)

81. > f(x=-10)
    [1] -57.8442

83. > f(x=13.987)
    [1] 71.56413

85. > plotFun(3*x+27~x,
        xlim=range(-10,10))

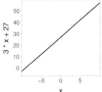

87. > plotFun(44.53*x-6.75~x,
        xlim=range(-50,150))

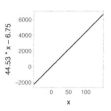

89. > plotFun(-26.79*x-145.62~x,
        xlim=range(-20,5))

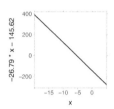

91. The modeling cycle:

    (1) Ask a question about reality.

    (2) Make some observations and collect the corresponding data.

    (3) Conjecture a model or modify a known model based on the data.

    (4) Test the model against known data (from Step (2)) and modify the model as needed.

    (5) Repeat Steps (2) – (4) to improve the model.

93. A rule assigning an input to more than one output.

95. A function with two or more inputs and one output.

97. The set of input values of a function.

99. For function $f(x)$ and input $x = a$, substitute a for every x appearing in $f(x)$ and perform the computations.

Section 1.2 Exercises

1. Total profit made by a lemonade stand as a function of the number of cups of lemonade sold.

3. Class taken as a function of professor, class time, and location.

5. 2

7. -36.545

9. -73

11. -0.31

13. 3

15. -6.790

17. -37

19. -175.84

21. For males 40–49 years old, the 50th percentile is 28.2.

23. For males 50–59 years old, the 25th percentile is 25.5.

25. For males 30–39 years old, the 10th percentile is 22.4.

27. At $(-68, 40)$, the sea surface is $77°$F.

29. At $(-64, 42)$, the sea surface is $60.8°$F.

31. At $(-60, 44)$, the sea surface is $48.2°$F.

33. 2.

35. Decrease

37. Increase

39. Approximate global minimum of -1 at $(1, -1)$.

41. 6

43. Decrease

45. Increase

47. Approximate global minimum of 1.5 at $(1, 2)$.

49. 0

51. Decrease

53. Increase

55. Approximate global minimum of -9 at $(2.5, 3)$.

57. D

59. F

61. A

63.

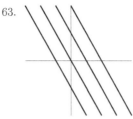

65.

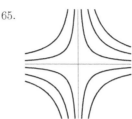

67.

For Exercises 69 – 72:
```
> f=makeFun(abs(x+5*y^3)~x&y)
```

69.
```
> f(-10,4)
  [1] 310
```

71.
```
> f(-1*pi,2*(pi^3))
  [1] 119236.832
```

For Exercises 73 – 76:
```
> f=makeFun(w-x*z^2+5*y~w&x&y&z)
```

73.
```
> f(1,2,3,4)
  [1] -16
```

75.
```
> f(9.623,1.579,-4.332,21.65)
  [1] -752.1498
```

77.
```
> plotFun(x+y+2~x&y,
    surface=TRUE)
```

79. > plotFun(x-y^2+2~x&y,
       surface=TRUE)

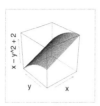

81. > plotFun(x^2-x*y+y^2~x&y,
       surface=TRUE)

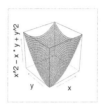

83. > plotFun(x+y+2~x&y,
       filled=FALSE,
       xlim=range(-5,5),
       ylim=range(-5,5),
       levels=c(0,2,4,6,8))

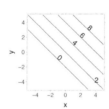

85. > plotFun(x-y^2+2~x&y,
       filled=FALSE,
       xlim=range(-8,8),
       ylim=range(-8,8),
       levels=c(0,2,4,6,8))

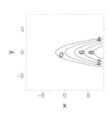

87. > plotFun(x^2-x*y+y^2~x&y,
       filled=FALSE,

       xlim=range(-5,5),
       ylim=range(-5,5),
       levels=c(0,2,4,6,8))

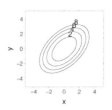

89. A rule that assigns two or more inputs to a unique output.

91. A three-dimensional graph where the value of each $f(x,y)$ is plotted as a height measured relative to the z-axis above or below the corresponding point (x,y) on the xy-plane.

93. A contour of $f(x,y)$ at level C is a curve in the xy-plane containing all points where $f(x,y) = C$.

Section 1.3 Exercises

1. $y = 3x - 8$

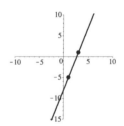

3. $y = \dfrac{5}{6}x - \dfrac{7}{2}$

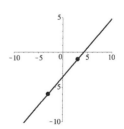

5. $y = -6$

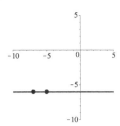

15. $y = 2x - 5$

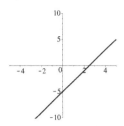

7. $y = -\dfrac{77}{20}x + \dfrac{1949}{200}$

17. $y = -3x - 11.5$

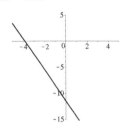

9. For $(-3, 1)$, $(3, 5)$: $y = \dfrac{2}{3}x + 3$

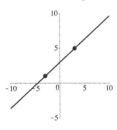

19. $y = 2.6x - 9$

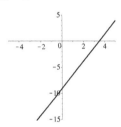

11. For $(2, 8)$, $(2, -2)$: $x = 2$

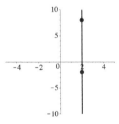

21. $y = 6.4$

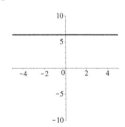

13. For $(-8, -2)$, $(-2, 8)$: $y = \dfrac{5}{3}x + \dfrac{34}{3}$

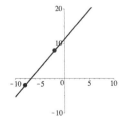

23. $x = -\pi$

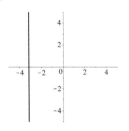

25. $y = 8$, $x = -2$

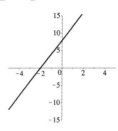

27. $y = 24$, $x = 3$

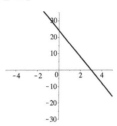

29. $y = 0$, $x = 0$

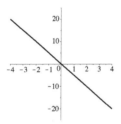

31. $y = -23.2$, no horizontal intercept

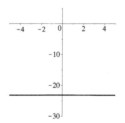

33. $y = 19.2$, $x = -16/3$

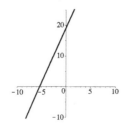

35. $(-7, -20)$

37. $(5, 14)$

39. Parallel lines do not intersect.

41. $(2.1, -2.23)$

43. Parallel lines do not intersect.

45. Positive ($m = 2$); positive ($b = 2$)

47. Negative ($m = -\dfrac{1}{2}$); positive ($b = 2$)

49. Positive ($m = \dfrac{1}{2}$); zero ($b = 0$)

51. $f(12) = 135$ million Twitter users in the first quarter of 2012.

53. The vertical intercept gives $y = -666.6$ million Twitter users in 2000 according to the model.

55. $f(2013) = 17.895$ is the model's projected spending on health care in the year 2013 as a percentage of total GDP.

57. The vertical intercept $y = -106.911$ is the healthcare spending in year 0 according to the model.

59. $f(2004) = 21.213$ means that the model predicts an average college debt load of $\$21,211$ in 2004.

61. The vertical intercept of $y = -546.5827$ means that the model predicts an average college debt load of $-\$546,583$ more than 2000 years ago in year 0.

63. Increasing on $(-\infty, \infty)$

65. Constant on $(-\infty, \infty)$

67. Increasing on $(-\infty, 5] \cup (5, \infty)$

69. Decreasing on $(-\infty, -2]$; increasing on $(-2, \infty)$

71. Constant on $(-\infty, 8]$; increasing on $(8, \infty)$

73. Increasing on $(-\infty, 0] \cup [5, \infty)$; decreasing on $(0, 5)$

75.
```
> plotFun(16*x+8~x,
     xlim=range(-3,5),
     ylim=range(-2,9))
```

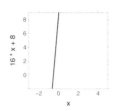

77. > plotFun(-8*x+4~x,
        xlim=range(-5,5),
        ylim=range(-5,5))

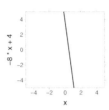

79. > plotFun(0.062*x-106.911~x,
        xlim=range(-100,2000),
        ylim=range(-200,100))

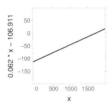

81. Among others, $y = 2x + 7$.

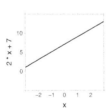

83. Among others, $y = -3x - 8$.

85. > findZeros(2*x+14~x,
        xlim=range(-20,20))
        x
      1 -7
      > 4*(-7)+8
      [1] -20
   Point of intersection is $(-7, -20)$.

87. > findZeros(15*x-25~x,
        xlim=range(-20,20))
        x
      1 1.7
      > -5*1.7-11
      [1] -19.5
   Point of intersection is $(1.7, -19.5)$.

89. > findZeros(5.6*x+17~x,
        xlim=range(-20,20))
        x
      1 -3
      > -3.6*(-3)+9.2
      [1] 20
   Point of intersection is $(-3, 20)$.

91. The change in y divided by the change in x; or $m = \dfrac{\Delta y}{\Delta x} = \dfrac{y_2 - y_1}{x_2 - x_1}$ when $x_1 \neq x_2$.

93. The x-coordinate a of the point $(a, 0)$ where the line intersects the x-axis, if such a point exists.

95. The sign of the slope of a line determines which direction the line "tilts."

97. Corresponds to whether and where the function is increasing, constant, or decreasing.

99. A function f is constant on (a, b) if and only if for all numbers c and d in (a, b), $f(c) = f(d)$.

Section 1.4 Exercises

1. $C = -\dfrac{2}{e^2}$

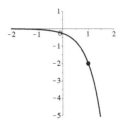

3. $C = 2e^3$

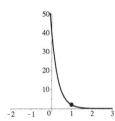

5. $C = 14e^{\frac{2}{7}}$

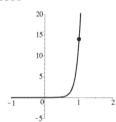

7. $C = 5/2^9$

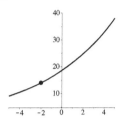

9. $C = 0.0014$

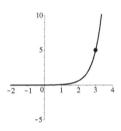

11. $y = 3e^{0.069x}$

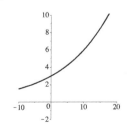

13. $y = -4e^{-0.231x}$

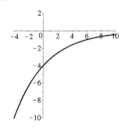

15. $y = 5.743e^{0.138x}$

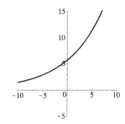

17. $y = 30.238e^{2.310x}$

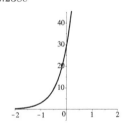

19. $y = 1.152e^{-0.110x}$

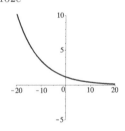

21. $(5, 10)$ and $(-2, 2)$; $y = 3.167e^{0.230x}$

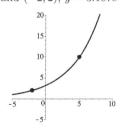

23. Not possible

25.

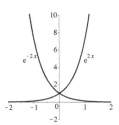

27.

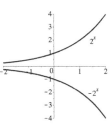

29.

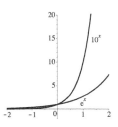

31.

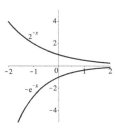

33.

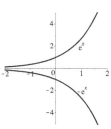

35. $k > 0$, $c < 0$, decreasing and concave down on $(-\infty, \infty)$

37. $k < 0$, $c > 0$, decreasing and concave up on $(-\infty, \infty)$

39. Yes, doubling time is 1

41. Yes, halving time is 1

43. No

45. This function is both increasing and decreasing and exponential functions exhibit only one monotonic behavior.

47. Exponentials do not cross the x-axis.

49. > plotFun(3*exp(2*x)~x,
        xlim=range(-4,4))

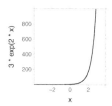

51. > plotFun(3*exp(-2*x)~x,
        xlim=range(-4,4))

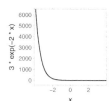

53. >   plotFun(4*10^x~x,
        xlim=range(-4,4))

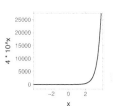

55. >   plotFun(4*10^(-x)~x,
        xlim=range(-4,4))

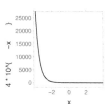

57. > plotFun(exp(4*x)~x,
       xlim=range(-2..2))
Doubling time is $\dfrac{\ln(2)}{4} \approx 0.173$.

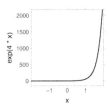

59. > plotFun(exp(-3*x)~x,
       xlim=range(-2,2))
Halving time is $\dfrac{-\ln(2)}{-3} \approx 0.231$.

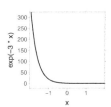

61. > plotFun(10^(2*x)~x,
       xlim=range(-2,2))
Doubling time is $\dfrac{\ln(2)}{4\ln(10)} \approx 0.151$.

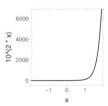

63. Decreasing on $(-\infty, 5)$; increasing on $[5, \infty)$

65. Decreasing on $(-\infty, 2]$; increasing on $(-2, \infty)$

67. $f(5) = 24.767$; In 1994 (5 years after 1989), the price of coal per short ton was $24.77.

69. The halving time is 14.441, so it takes 14.441 years for the price to decrease by half.

71. f is constantly decreasing.

73. $f(32) = 1855.107$; If born at 32 weeks, the average male baby weighs 1855.107 grams.

75. The doubling time is 5.582, so it takes 5.682 weeks for average male birth weight to double.

77. $f(15) = 11.079$; The plasma concentraion of Prozac is 11.079 ng/mL 15 days after the medication is stopped.

79. The halving time is 5.415, so it takes 5.415 days for the plasma concentration to decrease by half.

81. 4^{2x+8}

83. 6^{2x-1}

85. $x^{-\frac{13}{5}}$

87. 3^{2x+3}

89. 4^{2x+1}

91. A function of the form $f(x) = Ca^{kx}$ or $y = Ca^{kx}$, where a is a real number greater than one, $C \neq 0$ and $k \neq 0$ are constants, and x is the variable..

93. A horizontal line that a function gets closer and closer to as its inputs approach $\pm\infty$.

95. When the inside of a bend of a function opens upward, like a cup.

97. C provides the vertical intercept and its sign determines concavtiy. Together with the sign of k, the sign of C determines monotonicity.

99. If k is positive and $f(x_2) = 2f(x_1)$ with $x_2 > x_1$, then the doubling time is $\Delta x = x_2 - x_1$.

Section 1.5 Exercises

1.

x	7	9	6	8	0
$f^{-1}(x)$	1	2	3	4	5

3.

x	1	3	9	27	81
$f^{-1}(x)$	2	4	8	16	32

5.

x	d	b	a	e	c
$f^{-1}(x)$	r	s	t	u	v

7. What were the e-commerce sales in 2010?

9.

S	121	143	168	192
Y	2009	2010	2011	2012

11. When did the S&P 500 stock market close at 1191?

13. What was the percent growth in 2010?

15.

R	1.41	1.26	1.20	1.13	1.06
Y	1995	2000	2005	2010	2015

17. What year did the graduation rate reach 70.1 percent?

19. Was the temperature higher in month April or October?

21.

T	39.4	43.4	61.1	67.8	88.3
M	2	12	4	10	7

23. What month did the count fall below 1?

25. This function has an inverse.

x	−2	−1	1	2	3
$f^{-1}(x)$	4	1	2	3	5

27. No inverse: all inputs map to 1.

29. This function has an inverse.

x	−2	−1	0	1	2
$f^{-1}(x)$	−2	−1	0	1	2

31. This function has an inverse.

x	b	d	e	k	z
$f^{-1}(x)$	k	b	t	u	s

33. Does not have an inverse because $f(2006) = f(2008) = \$141$

35. Has an inverse.

G	17.66	17.68	17.71	17.91
Y	2010	2011	2009	2012

37. Does not have an inverse because $f(94) = f(98) = 2.7$

39. For example, the function

x	1	2	4	7	11
$f(x)$	16	22	29	37	46

has inverse function

x	16	22	29	37	46
$f^{-1}(x)$	1	2	4	7	11

41. Does not have an inverse; for example, $y = 1$ intersects the function twice.

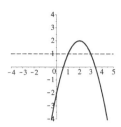

43. This function has an inverse.

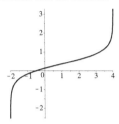

45. This function has an inverse.

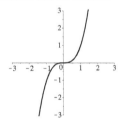

47. This function has an inverse.

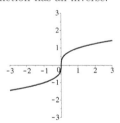

49. This function has an inverse.

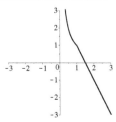

51. Changes from increasing to decreasing, so no inverse.

53. Constantly increasing, so has an inverse; see the answer to Exercise 43 for the graph.

55. Constantly increasing, so has an inverse; see the answer to Exercise 45 for the graph.

57. Constantly increasing, so has an inverse; see the answer to Exercise 47 for the graph.

59. Constantly decreasing, so has an iinverse; see the answer to Exercise 49 for the graph.

61.

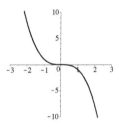

63.

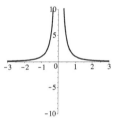

65.

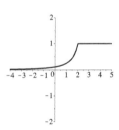

67. $y = \dfrac{1}{2}x + 2$

69. $y = -\dfrac{1}{3}x + 3$

71. $y = \sqrt[7]{x+1}$

73. $y = x^3 - 7$

75. $y = \dfrac{1}{x} + 2$

77. $y = \dfrac{2x + 3}{1 - x}$

79. $y = \dfrac{5x + 1}{2 + 3x}$

81. `> plotFun(x^7+x~x,`
    `    xlim=range(-10,10))`
    Constantly increasing, so has an inverse.

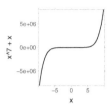

83. `> plotFun(x^4+x~x,`
    `    xlim=range(-10,10))`
    Changes from decreasing to increasing, so no inverse.

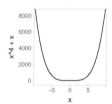

85. `> plotFun(abs(2*x-7)~x,`
    `    xlim=range(-10,10))`
    Changes from decreasing to increasing, so no inverse.

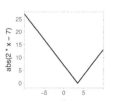

87. `> plotFun((x^2+5)^(1/4)~x,`
    `    xlim=range(-10,10))`
    Changes from decreasing to increasing, so no inverse.

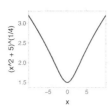

89. > plotFun(exp(x^2)~x,
       xlim=range(-2,2))

    Changes from decreasing to increasing, so no inverse.

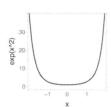

91. > plotFun(4*exp(x)-3~x,
       xlim=range(-10,10))

    Constantly increasing, so has an inverse.

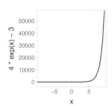

93. > plotFun(1/(1+exp(x))~x,
       xlim=range(-10,10))

    Constantly decreasing, so has an inverse.

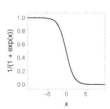

95. The inverse of a function $f(x)$ maps each output from $f(x)$ back to its corresponding input x.

97. A function mapping more than one input to some output.

99. A monotonic function increases or decreases on its entire domain and has an inverse.

Section 1.6 Exercises

1. 5

3. 8

5. -2

7. $\dfrac{3}{4}$

9. > exp(log(7))
   [1] 7

11. > exp(9*log(pi))
    [1] 29809.099

13. > log(exp(-16))
    [1] -16

15. > log((exp(22)^(1/17)))
    [1] 1.294118

17. $\ln(3) + \ln(x)$

19. $\ln(x - 2) + \ln(x + 2)$

21. $2\ln(x) + \ln(x - 9)$

23. $\ln(9) + \ln(x + 2)$

25. $\ln(x) + \ln(x + 1) - \ln(3) - 1$

27. $\ln(8x)$

29. $\ln(x^5)$

31. $\ln\left[\dfrac{(x + 1)^2}{(20)(x - 1)}\right]$

33. $\ln(e^3)$

35. $e^{\ln(7)} = 7$

37. $\ln(4 \cdot 3) = \ln(4) + \ln(3)$

39. $\ln(5^2) = 2\ln(5)$

41. $x = 0$

43. $x = -1$, $x = 4$

45. $x = 4$

47. $x = 9$

49. > plotFun(log(x-2)~x,
       xlim=range(2,10))

    Increasing and concave down.

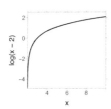

51. > plotFun(-log(x+4)~x,
    xlim=range(-4,10))
    Decreasing and concave up.

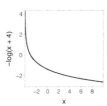

53. > plotFun(log(x)+2~x,
    xlim=range(0,10))
    Increasing and concave down.

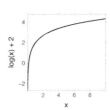

55. > plotFun(-3*log(x)+1~x,
    xlim=range(0,10))
    Decreasing and concave up.

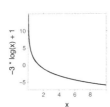

57. $y = e^{\frac{\ln(5)}{3}x}$; increasing and concave up.

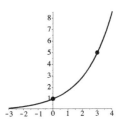

59. $y = 8e^{\frac{\ln(8)}{2}x}$; increasing and concave up.

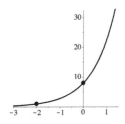

61. $y = -3e^{\frac{\ln(8/3)}{5}x}$; decreasing and concave down.

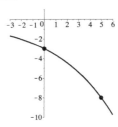

63. $y = -5e^{\frac{\ln(2/5)}{2}x}$; decreasing and concave down.

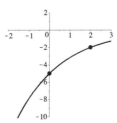

65. $B(2) = \$1020.20$

67. $t = 69.315$ years

69. $B(5) = \$2,699.72$

71. $t = 26.824$ years

73. $A(2865) = 7.071$ grams

75. $t = 19,034.648$ years

77. $A(29) = 12.300$ kilograms

79. $t = 137.870$ years

81. $P(2015) = 7.207$; The model predicts 7.207 billion people in the world in 2015.

83. $t = 2022.355$; The model predicts 8 billion people in the world during 2022.

85. $f(5) = 24.768$; In 1994, a short ton of coal costs $24.77.

87. $x = 4.806$; A short ton of coal costs $25 during 1993.

89. $f(32) = 1855.110$; After 32 weeks of gestation, an average male baby weighs 1855.110 grams.

91. $x = 26.935$; For an average male birth weight of 1000 grams, 26.935 weeks of gestation are needed.

93. $f(15) = 11.079$; After 15 days, the plasma concentration of Prozac is 11.079 ng/mL.

95. $x = 1.802$; A Prozac plasma concentration of 60 is reached 1.802 days after stopping the medication.

97. The natural logarithm $\ln(x)$ is the inverse of the natural exponential function e^x and has domain $(0, \infty)$. In general, natural logarithm functions are of the form $f(x) = C \ln(x - h) + v$ and have a domain of (h, ∞).

99. The parameter h corresponds to a horizontal shift; if $h > 0$, shift right and, if $h < 0$, shift left.

101. If (a, b) is a point from a given data set or curve, the semi-log plot includes the point $(a, \ln(b))$.

Section 1.7 Exercises

1. $\dfrac{\pi}{4}$

3. $\dfrac{2\pi}{3}$

5. $\dfrac{3\pi}{4}$

7. $\dfrac{\pi}{3}$

For Exercises 1, 3, 5, and 7

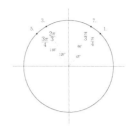

9. $0°$

11. $225°$

13. $300°$

15. $30°$

For Exercises 9, 11, 13, and 15

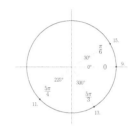

17. $\sin(45°) = \dfrac{\sqrt{2}}{2}$, $\cos(45°) = \dfrac{\sqrt{2}}{2}$, $\tan(45°) = 1$, $\csc(45°) = \sqrt{2}$, $\sec(45°) = \sqrt{2}$, $\cot(45°) = 1$

19. $\sin(135°) = \dfrac{\sqrt{2}}{2}$, $\cos(135°) = -\dfrac{\sqrt{2}}{2}$, $\tan(135°) = -1$, $\csc(135°) = \sqrt{2}$, $\sec(135°) = -\sqrt{2}$, $\cot(135°) = -1$

21. $\sin(0) = 0$, $\cos(0) = 1$, $\tan(0) = 0$, $\csc(0)$ is undefined, $\sec(0) = 1$, $\cot(0)$ is undefined

23. $\sin\left(\dfrac{5\pi}{3}\right) = -\dfrac{\sqrt{3}}{2}$, $\cos\left(\dfrac{5\pi}{3}\right) = \dfrac{1}{2}$, $\tan\left(\dfrac{5\pi}{3}\right) = -\sqrt{3}$, $\csc\left(\dfrac{5\pi}{3}\right) = -\dfrac{2}{\sqrt{3}}$, $\sec\left(\dfrac{5\pi}{3}\right) = 2$, $\cot\left(\dfrac{5\pi}{3}\right) = -\dfrac{1}{\sqrt{3}}$

25. $\cos(x) = \dfrac{\sqrt{56}}{9}$, $\tan(x) = \dfrac{5}{\sqrt{56}}$, $\csc(x) = \dfrac{9}{5}$, $\sec(x) = \dfrac{9}{\sqrt{56}}$, $\cot(x) = \dfrac{\sqrt{56}}{5}$

27. $\sin(x) = \dfrac{\sqrt{40}}{7}$, $\tan(x) = \dfrac{\sqrt{40}}{3}$,

$\csc(x) = \dfrac{7}{\sqrt{40}}$, $\sec(x) = \dfrac{7}{3}$, $\cot(x) = \dfrac{3}{\sqrt{40}}$

29. $\sin(x) = \dfrac{4}{\sqrt{137}}$, $\cos(x) = \dfrac{11}{\sqrt{137}}$,

$\csc(x) = \dfrac{\sqrt{137}}{4}$, $\sec(x) = \dfrac{\sqrt{137}}{11}$,

$\cot(x) = \dfrac{11}{4}$

31. $\sin(x) = \dfrac{\sqrt{165}}{13}$, $\cos(x) = \dfrac{2}{13}$,

$\tan(x) = \dfrac{\sqrt{165}}{2}$, $\csc(x) = \dfrac{13}{\sqrt{165}}$,

$\cot(x) = \dfrac{2}{\sqrt{165}}$

33. $x = 0, \pi, 2\pi$

35. $x = 0, \dfrac{\pi}{2}, \pi, \dfrac{3\pi}{2}, 2\pi$

37. $x = \dfrac{7\pi}{6}, \dfrac{11\pi}{6}$

39. $x = \dfrac{\pi}{2}, \dfrac{3\pi}{2}$

41. $x = \dfrac{\pi}{2}, \dfrac{3\pi}{2}$

43. $x = \dfrac{2\pi}{3}, \dfrac{4\pi}{3}$

45. $x = 0, \pi, 2\pi$

47. $x = 0, \dfrac{\pi}{2}, \pi, \dfrac{3\pi}{2}, 2\pi$

49. $x = \dfrac{2\pi}{3}, \dfrac{5\pi}{3}$

51. $x = \dfrac{\pi}{2}, \dfrac{3\pi}{2}$

53. $x = \dfrac{3\pi}{8}, \dfrac{7\pi}{8}, \dfrac{11\pi}{8}, \dfrac{15\pi}{8}$

55. $x = \dfrac{3\pi}{4}, \dfrac{7\pi}{4}$

57.
```
> 6*sin(pi)
[1] 7.347881e-16
```

59.
```
> 2*sin(3*pi/4)+5
[1] 6.414214
```

61.
```
> 4*cos(6)
[1] 3.840681
```

63.
```
> 2*tan(pi)+4
[1] 4
```

65.
```
> plotFun(3*sin(x)~x,
      xlim=range(0,2*pi))
```
Increasing on $\left[0, \frac{\pi}{2}\right] \cup \left(\frac{3\pi}{2}, 2\pi\right]$ and decreasing on $\left(\frac{\pi}{2}, \frac{3\pi}{2}\right)$. Concave down on $(0, \pi)$ and concave up on $(\pi, 2\pi)$.

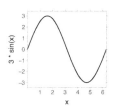

67.
```
> plotFun(-2*sin(x+pi)+4~x,
      xlim=range(0,2*pi))
```
Increasing on $\left[0, \frac{\pi}{2}\right] \cup \left(\frac{3\pi}{2}, 2\pi\right]$ and decreasing on $\left(\frac{\pi}{2}, \frac{3\pi}{2}\right)$. Concave down on $(0, \pi)$ and concave up on $(\pi, 2\pi)$.

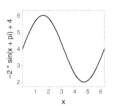

69.
```
> plotFun(3*cos(x)~x,
      xlim=range(0,2*pi))
```
Decreasing on $(0, \pi)$ and increasing on $(\pi, 2\pi)$. Concave down on $\left[0, \frac{\pi}{2}\right) \cup \left(\frac{3\pi}{2}, 2\pi\right]$ and concave up on $\left(\frac{\pi}{2}, \frac{3\pi}{2}\right)$.

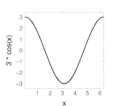

71.
```
> plotFun(2*tan(x+pi)+4~x,
      xlim=range(0,pi),
      ylim=range(-6,14))
```
Increasing on $\left[0, \frac{\pi}{2}\right) \cup \left(\frac{\pi}{2}, \pi\right]$. Concave up on $\left(0, \frac{\pi}{2}\right)$ and concave down on $\left(\frac{\pi}{2}, \pi\right)$.

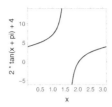

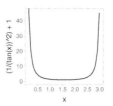

73. Min(5) = 380.239; In May, the sun sets at 9:20 p.m. (380.239 minutes after 3 p.m.) in Greenwich, England.

75. ```
> plotFun(164.7*sin(((pi/6)*
    (Month-2.6))+223.6~Month,
    xlim=range(0,12))
```

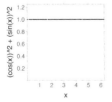

77. The amplitude of Min(Month) is 164.7.

79. Temp(10) = 67.082; In October 2013, the average maximum temperature was 67.082° Fahrenheit in New York City.

81. The period of Temp(Month) is 12.

83. ```
> plotFun((cos(x))^2+(sin(x))
    ^2~x,xlim=range(0,2*pi))
> f=makeFun(1~x)
> plotFun(f(x)~x,add=TRUE)
```

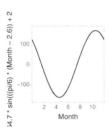

85. ```
> plotFun((1/(tan(x))^2)+1~x
    ,xlim=range(0,pi))
> plotFun(1/(sin(x))^2~x,
    add=TRUE)
```

87. ```
> plotFun(cos(2*x)~x,
    xlim=range(0,2*pi))
> plotFun((cos(x))^2-(sin(x))
    ^2~x,add=TRUE)
```

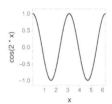

89. ```
> plotFun(cos(2*x)~x,
    xlim=range(0,2*pi))
> plotFun(1-2*((sin(x))^2)~x,
    add=TRUE)
```

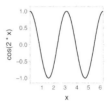

91. ```
> plotFun(cos(-x)~x,
    xlim=range(0,2*pi))
> plotFun(cos(x)~x,add=TRUE)
```

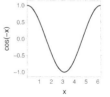

93. ```
> plotFun(sin(x+2*pi)~x,
    xlim=range(0,2*pi))
> plotFun(sin(x)~x,add=TRUE)
```
```

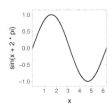

95.
```
> plotFun(sin(x+(3*pi/2))~x,
    xlim=range(0,2*pi))
> plotFun(-cos(x)~x,add=TRUE)
```

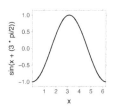

97. The unit circle is centered at the origin and has a radius of one.

99. $\sin(\theta)$ is the y-coordinate of the point θ radians around the unit circle.
$\cos(\theta)$ is the x-coordinate of the point θ radians around the unit circle.
$$\tan(\theta) = \frac{\sin(\theta)}{\cos(\theta)} \qquad \csc(\theta) = \frac{1}{\sin(\theta)}$$
$$\sec(\theta) = \frac{1}{\cos(\theta)} \qquad \cot(\theta) = \frac{\cos(\theta)}{\sin(\theta)}$$

101. $\sin(\theta)$ has amplitude 1, which is half the distance between its extreme values.

Section 2.1 Exercises

1. Linear; $y = 5x$

3. Not linear

5. Not linear

7. Linear; $y = 2x - 20$

9. Linear;

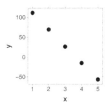

11. Not linear;

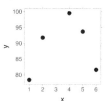

13. Linear;

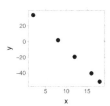

15. Linear;

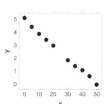

17. Linear;

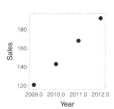

19. Linear;

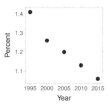

21. Not linear;

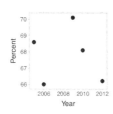

23. $m_1 = 2.1$, $m_2 = 4$, $m_3 = 2.9$, $m_4 = 2.1$;
    reasonable, $m = 2.78$

25. $m_1 = 3.3$, $m_2 = 2.5$, $m_3 = 2.5$, $m_4 = 3$;
    reasonable, $m = 2.83$

27. $m_1 = 7.3$, $m_2 = 4.5$, $m_3 = 7.8$, $m_4 = 5$, $m_5 = 6.9$, $m_6 = 6.8$, $m_7 = 6.2$, $m_8 = 5.2$, $m_9 = 5.6$;
    nonlinear

29. $m_1 = -42$, $m_2 = -43$, $m_3 = -41$, $m_4 = -42$;
    reasonable, $m = -42$

31. $m_1 = 13.4$, $m_2 = 3.85$, $m_3 = -5.8$, $m_4 = -12.1$;
    nonlinear

33. $m_1 = -5.3$, $m_2 = -5.3$, $m_3 = -5.3$, $m_4 = -5.3$;
    reasonable, $m = -5.3$

35. $m_1 = -0.14$, $m_2 = -0.11$, $m_3 = -0.09$, $m_4 = -0.09$, $m_5 = -0.11$, $m_6 = -0.09$, $m_7 = -0.06$, $m_8 = -0.09$, $m_9 = -0.13$;
    reasonable, $m = -0.10$

37. $b = 10$

39. $b = -102$

41.
```
x = c(1,2,3,4,5)
y = c(5.4,7.5,11.5,14.4,16.5)
bestModel = fitModel(y~x*m+b)
coef(bestModel)
      m      b
    2.91   2.33
```

43.
```
x = c(-5,-4,-3,-2,-1)
y = c(-14.2,-10.9,-8.4,-5.9,-2.9)
bestModel = fitModel(y~x*m+b)
coef(bestModel)
        m        b
     2.76   -0.18
```

45.
```
x = c(-5,-4,-3,-2,-1,0,1,2,3,4)
y = c(-28,-20.7,-16.2,-8.4,-3.4,3.5,
10.3,16.5,21.7,27.3)
bestModel = fitModel(y~x*m+b)
coef(bestModel)
         m         b
  6.187879  3.353939
```

47.

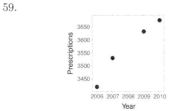

49. $y = 1.6x - 3060.5$

51. $f(2008) = 152.3$. The tornado season in the U.S. in 2008 was 152.3 days long.
    $f(2008) = 153.4$. The tornado season in the U.S. in 2008 was 153.4 days long.

53.

55. $y = 0.37x - 478.58$

57. From Exercise 55: $f(2008) = 264.38$. In the U.S. in 2008, 264.38 billion dollars were spent on retail prescription drugs.
    From Exercise 56: $f(2008) = 264.4$. In the U.S. in 2008, 264.4 billion dollars were spent on retail prescription drugs.

59.

61. $y = 68.5x - 133983.5$

63. From Exercise 61: $f(2008) = 3564.5$. In the U.S., there were 3564.5 million

prescriptions sold in 2008.
From Exercise 62: $f(2008) = 3564.5$.
In the U.S., there were 3564.5 million prescriptions sold in 2008.

65.

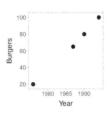

67. $y = 4.7x - 9266.8$

69. From Exercise 67: $f(1992) = 95.6$. In 1992, McDonald's will sell 95.6 billion hamburgers.
From Exercise 68: $f(1992) = 72.28$. In 1992, McDonald's will sell 72.28 billion hamburgers.

71.

73. $y = 0.0067x + 88.145$

75. From Exercise 73: $f(2000) = 101.545$. In the year 2000, for every 100 females, there will be 101.545 males.
From Exercise 74: $f(2000) = 101.55$. In the year 2000, for every 100 females, there will be 101.55 males.

77. `head(MonthlyUnemployment)`
`plotPoints(Rate~Months,`
`data=MonthlyUnemployment)`

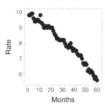

79. $f(36) = 7.61$. Meaning 36 months after January 2010, unemployment will be at a rate of 7.61 percent.

81. `head(WorldPopulation)`
`plotPoints(People~Year,`
`data=WorldPopulation)`

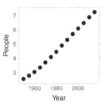

83. $f(1988) = 5.21$. Based on the model, in 1998, the world's population will be 5.21 billion people.

85. `head(Mortgage15YrAnnual)`
`plotPoints(Rate~Year,`
`data=Mortgage15YrAnnual)`

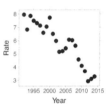

87. $f(2007) = 4.01$. Based on the model, in 2007, the 15-year mortgage rate will be 4.01 percent.

89. `head(FacebookUsers)`
`plotPoints(Users~Months,`
`data=FacebookUsers)`

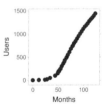

91. $f(13) = 386.41$. Based on the model, 13 months after the start of 2009, Facebook will have 386.41 million users.

93. `head(HSDropoutRate)`
`plotPoints(Rate~Year,`
`data=HSDropoutRate)`

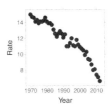

95. $f(2005) = 8.41$. Meaning that the percentage of highschoolers that will drop out in 2005 will be 8.41 percent.

97. ```
head(USCO2Emissions)
plotPoints(kT~Year,
data=USCO2Emissions)
```

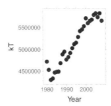

99. $f(2005) = 5,764,541.9$ Based on the model, the U.S. will emit 5,764,541.9 kT of $CO_2$ in 2005.

101. View a graph with the data plotted and determine if it fits a linear path

103. Compute the slope through successive data points and then average these slopes:

$$m = \frac{m_1 + m_2 + m_3 + \cdots + m_n}{n}$$

## Section 2.2 Exercises

1. Reasonable, because the semi-log plot is linear;
$C = e^3$, $k = -0.5$

3. Not reasonable, because the semi-log plot is not linear.

5. Not reasonable, because the semi-log plot is not linear.

7. Reasonable, because the semi-log plot is linear;
$C = e^{2.25}$, $k = 0.15$

9. ```
> input=c(15,18,24,30,31)
> output=c(7.3,5.4,3,1.6,1.5)
> plotPoints(output~input)
> plotPoints(log(output)~input)
```
Reasonable; $k = -0.1$; $C = e^2$

11. ```
> input=c(5,20,25,55,60)
> output=c(900,625,400,400,225)
> plotPoints(output~input)
> plotPoints(log(output)~input)
```
Not reasonable

13. ```
> input=c(-0.5,0.4,2.9,5.2,7.7)
> output=c(0.2,1.5,18,181,2208)
> plotPoints(output~input)
> plotPoints(log(output)~input)
```
Reasonable; $k = 1$, $b = e^{-1}$

15. ```
> input=c(-2,0,3,5,9)
> output=c(1,27,216,512,1728)
> plotPoints(output~input)
> plotPoints(log(output)~input)
```
Not reasonable

17. ```
> Y=c(0,10,20,30,40)
> V=c(95,192,545,1191,2077)
> plotPoints(V~Y)
> plotPoints(log(V)~Y)
```
Reasonable; $k = 0.08$, $C = e^{4.5}$

19. ```
> Y=c(2010,2011,2012,2013)—
> S=c(274,266,435,496)
> plotPoints(S~Y)
> plotPoints(log(S)~Y)
```
Not reasonable

21. ```
> M=c(8,9,10,11,12)
> C=c(8.4,4.6,1.7,0.3,0.1)
> plotPoints(C~M)
> plotPoints(log(C)~M)
```
Not reasonable

23. $m_1 = 0.223$, $m_2 = 0.2$, $m_3 = 0.198$, $m_4 = 0.202$; reasonable, $k = 0.205$

25. $m_1 = -1.809$, $m_2 = -1.799$, $m_3 = -1.838$, $m_4 = -1.763$; reasonable, $k = -1.802$

27. $m_1 = 0.916$, $m_2 = 0.885$, $m_3 = 0.931$, $m_4 = 0.878$, $m_5 = 0.901$, $m_6 = 0.9$, $m_7 = 0.899$; reasonable, $k = 0.9014$

29. $m_1 = -0.1005$, $m_2 = -0.098$, $m_3 = -0.105$, $m_4 = -0.065$; reasonable, $k = m = -0.0921$, $C = e^{16.378}$

31. $m_1 = -0.0243$, $m_2 = -0.0892$, $m_3 = 0$, $m_4 = -0.115$; not reasonable

33. $m_1 = 2.238$, $m_2 = 0.9938$, $m_3 = 1.003$, $m_4 = 1.0004$; not reasonable

35. $m_1 = 1.647$, $m_2 = 0.693$, $m_3 = 0.4315$, $m_4 = 0.304$; not reasonable

37. $C_1 = 14.78$, $C_2 = 14.86$, $C_3 = 14.74$, $C_4 = 16.14$; $C = 15.13$

39. $C_1 = 4.4$, $C_2 = 4.445$, $C_3 = 4.448$, $C_4 = 4.447$; $C = 4.435$

41. ```
> bestExpModel=fitModel(log(y)~
x*m+b)
> coef(bestExpModel)
      m   b
0.2044431 1.0858710
```

43. ```
> bestExpModel=fitModel(log(y)~
x*m+b)
> coef(bestExpModel)
      m         b
-1.80565 -0.02998
```

45. ```
> bestExpModel=fitModel(log(y)~
x*m+b)
> coef(bestExpModel)
      m   b
0.9007987 0.7648158
```

47.

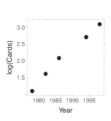

49. $C = 2.153 \cdot e^{0.1375(Y-1978)}$

51. #49 $f(2006) = 101.18$;
    #50 $f(2006) = 49.13$

53.

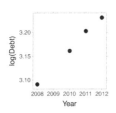

55. $D = 22.05 \cdot e^{0.035(Y-2008)}$

57. #55 $f(2009) = 22.84$;
    #56 $f(2009) = 22.81$

59.

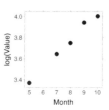

61. $V = 30.18 \cdot e^{0.1225(M-5)}$

63. #61 $f(6) = 34.1$;
    #62 $f(6) = 33.35$

65.

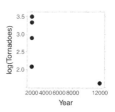

67. $T = 5.0925 \cdot e^{0.0625(Y-1975)}$

69. #67 $f(2006) = 35.3$;
    #68 $f(2006) = 37.3$

71.

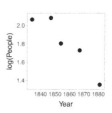

73. $P = 9.52 \cdot e^{-0.018(Y-1834)}$

75. #73 $f(1841) = 8.38$;
    #68 $f(1841) = 7.76$

77. ```
> plotPoints(log(AdjClose)~Month,
data=AAPLStockMonthly)
```
    $C = 0.19$, $k = 0.012$, $h = 1$

79. The closing price of Apple stock in August 2012 is 24.34

81. ```
> plotPoints(log(Close)~Quarter,
data=NASDAQQuarterly)
```
    $C = 4$, $k = 0.02$, $h = 1$

83. The closing NASDAQ stock market value on December 31, 1940 is 5.82.

85. ```
> plotPoints(log(Rate)~Year,
data=Mortgage30YrAnnual)
```
    $C = 2.09 \times 10^{33}$, $k = -0.03$, $h = 1981$

87. The average interest rate for convential 30-year mortgages in 1993 was 9.01%.

89.
```
> plotPoints(log(Population)~Year,
data=NetherlandsPopulation)
```
$C = 1$, $k = 0.008$, $h = 1700$

91. The population of the Netherlands in 2002 was 13.42 million.

93.
```
> plotPoints(log(Physicians)~
LifeExpectancy,
data=LifeExpectancyPhysicians)
```
$C = 2.34 \times 10^{-5}$, $k = 0.14$, $h = 44.83895$

95. The number of physicians per 1000 people for a country with an average life expectancy of 75 is 1.67.

97.
```
> plotPoints(log(Exams)~Year,
data=APCalculus)
```
$C = 4.35 \times 10^{-81}$, $k = 0.09$, $h = 1955$

99. The number of AP Calculus exams taken in 2000 was 165,581.2.

101. If the semi-log plot of the data set is nonconsant linear, then an exponential model is reasonable.

103. Let h equal the least (or leftmost) input value from the data set.

105. Substitute each data point from the origional data set into the original equation (using the conjectured value of k), and then average the intercepts;
$$C = \frac{C_1 + C_2 + C_3 + \cdots + C_n}{n}$$

Section 2.3 Exercises

1. Yes; $C = 1$, $k = 1$

3. No

5. Yes; $C = 1$, $k = -2$

7. Yes; $C = 3$, $k = \frac{1}{2}$

9. No

11. None of these

13. None of these

15. None of these

17. Linear

19. Yes; $C = 1$, $k = -5$

21. Nonlinear

23. Yes; $C = 1$, $k = -\frac{1}{3}$

25. Nonlinear

27. $y = x^{-5}$

29. $y = x^{-1/3}$

31. $y = 4x^{-1}$

33.
```
> x=c(1.1,2.2,2.9,3.6,4.5)
> y=c(8.7,5.3,3.4,2.6,1.5)
> plotPoints(log(y)~log(x))
```
Nonlinear log–log plot

35.
```
> x=c(2,2.5,3.1,4.5,6.2)
> y=c(4.02,6,9.73,19.93,38.12)
> plotPoints(log(y)~log(x))
```
$C \approx 1$, $k \approx 2$

37.
```
> x=c(0.5,1,3,4,8)
> y=c(1.43,1.01,0.59,0.49,0.35)
> plotPoints(log(y)~log(x))
```
$C \approx 1$, $k \approx -0.5$

39.
```
> Y=c(15,25,35,45)
> V=c(192,545,1191,2077)
> plotPoints(log(V)~log(Y))
```
$C \approx 1/2$, $k \approx 2$

41.
```
> Y=c(2010,2011,2012,2013)
> S=c(274,266,435,496)
> plotPoints(log(S)~log(Y))
```
Nonlinear log–log plot

43.
```
> Y=c(4,8,12,16)
> G=c(2.7,2.5,2.3,2.3)
> plotPoints(log(G)~log(Y))
```
Nonlinear log–log plot

45. $m_1 = 1.509$, $m_2 = 1.497$, $m_3 = 1.500$, $m_4 = 1.501$; power function reasonable, $k = m = 1.502$

47. $m_1 = 0.203$, $m_2 = 0.191$, $m_3 = 0.211$, $m_4 = 0.192$; power function reasonable, $k = m = 0.0.199$

49. $m_1 = 0.0653$, $m_2 = 0.0788$, $m_3 = 0.0883$, $m_4 = 0.0775$; power function reasonable, $k = m = 0.0775$

51. $m_1 = -0.715$, $m_2 = -1.607$, $m_3 = -1.241$, $m_4 = -2.465$; log–log data is not approximately linear

53. $m_1 = 1.598$, $m_2 = 2.247$, $m_3 = 1.924$, $m_4 = 2.024$; log–log data is approximately linear; $k = m = 1.948$; $C_1 = 1.042$, $C_2 = 1.007$, $C_3 = 1.074$, $C_4 = 1.064$, $C_5 = 1.090$; $C = 1.0554$

55. $m_1 = -0.502$, $m_2 = -0.489$, $m_3 = -0.646$, $m_4 = -0.485$, log-log data approximately linear $k = m = -0.531$ $C_1 = 0.990$, $C_2 = 1.01$, $C_3 = 1.057$, $C_4 = 1.023$, $C_5 = 1.056$; $C = 1.027$

57. $C_1 = 2.42$, $C_2 = 2.6$, $C_3 = 2.6$, $C_4 = 2.52$; $C = 2.535$

59. $C_1 = -0.8978$, $C_2 = -0.899$, $C_3 = -0.899$, $C_4 = -0.899$; $C = -0.899$

61.
```
> bestModel=fitModel(log(y)~
log(x)*m+b)
> coef(bestModel)
      m          b
1.5023751 -0.6966856
```

63.
```
> bestModel=fitModel(log(y)~
log(x)*m+b)
> coef(bestModel)
      m          b
0.1997668   4.9556026
```

65.
```
> bestModel=fitModel(log(y)~
log(x)*m+b)
> coef(bestModel)
      m            b
0.07340846   -1.12907950
```

67.
```
> plotPoints(log(Speed)~
log(Length),data=SwimmingSpeed)
```
$C \approx 8$, $k \approx 1$

69. $y = 195.5356$

71.
```
> plotPoints(log(Speed)~
log(Length),data=RunningSpeed)
```
$C \approx 22$, $k \approx 1$

73. 243.4739

75.
```
> plotPoints(log(Average)~
log(Year),data=MaunaLoaCO2
```
$C \approx 3 \times 10^{-25}$, $k \approx 8$

77. 312.922

79. 434.5187

81.
```
> bestModel=fitModel(log(Rate)~
log(Mass)*m+b,
data=BodyMassMetabolicRate)
> coef(bestModel)
```

```
      m          b
0.6528129  6.8147369
```

83. 15263.95

85. If (a, b) is a point from a data set, its log–log plot contains the point $(\ln(a), \ln(b))$.

87. Check if the log–log data is nonconstant linear by applying (1) to the transformed data $(\ln(a), \ln(b))$.

89. Substitute each data point from the original data set into the equation $y = Cx^k$, using the already conjectured value of k, and then average the intercepts:

$$C = \frac{C_1 + C_2 + C_3 + \cdots + C_n}{n}$$

Section 2.4 Exercises

1. Sinusoidal

3. None of these

5. Sinusoidal

7. Linear

9. $A = 5$; $v = 0$

11. $A = 1$; $v = 3$

13. $A = 5$; $v = -2$

15. $A = 0.5$; $v = 2$

17. $P = 5$; $h = 0$

19. $P = 3$; $h = 2$

21. $P = 6$; $h = 2$

23. $P = 0.5$; $h = 0.25$

25. $A = 4$; $v = 5$; $P = 2$; $h = 0$

27. $A = 4$; $v = 6$; $P = 8$; $h = 3$

29. $A = 3$; $v = 2$; $P = 4$; $h = 0$

31. $A = 6$; $v = 15$; $P = 10$; $h = 3$

33. $A = 6$; $v = 4$; $P = 3$; $h = 2$

35. $A = 6$; $v = 6$; $P = 8$; $h = 3$

37. $A = 4.5$; $v = -5$; $P = 5$; $h = 3$

39. $A = 1/2$; $v = 1$; $P = 7$; $h = 4$

41.

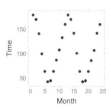

$A = 70$; $P = 12$; $v = 110$; $h = 10$

43.
```
sun = fitModel(Time~A*sin(2*
pi/P*(Month-h))+v,
   data=SunriseLA,
   start=list(A=70, P=12, v=110,
   h=10))
```
$$T = 65.44 \sin \left[\frac{2\pi}{12.01} (M - 9.86) \right] + 108.29$$

45. (41) The sunrise in June 2012 in Los Angeles, California was at 4:49 a.m. (43) The sunrise in June 2012 in Los Angeles was at 4:50 a.m..

47. $A = 75$; $P = 12$; $h = 3$; $v = 115$

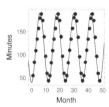

49. (46) The sunset in June 2014 in Los Angeles, California was at 7:10 p.m. (48) The sunset in June 2014 in Los Angeles was at 7:05 p.m.

51. $A = 29$; $P = 24$; $h = 8$; $v = 21$;

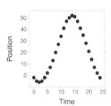

53.
```
SunMod=fitModel(Position~A*
sin(2*pi/P*(Time-h))+v,
   data=SunPositionAlaska,
```

```
   start=list(A=29,P=24,h=8,
   v=21))
```
$$P = 28.22 \sin \left[\frac{2\pi}{23.49} (T - 8.16) \right] + 22.13$$

55. (51) The altitude angle of the sun is 46.11 in Anchorage, Alaska at noon on July 1, 2014. (53) The altitude angle of the sun is 48.03 in Anchorage at noon on July 1, 2014.

57. $A = 7500$; $P = 4$; $h = 2006$; $v = 139,500$;

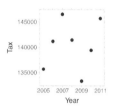

59.
```
TaxMod=fitModel(Tax~A*sin(2*
pi/P*(Year-h))+v,
   data=USRetailTax,
   start=list(A=7500,P=4,
h=2006,v=139500))
```
$$T = 5928.26 \sin \left[\frac{2\pi}{4.25} (Y - 2005.9) \right] + 140543.9$$

61. (57) The U.S. retail tax in 2020 is \$139,500 million dollars. (59) The U.S. retail tax in 2020 is \$145,975.2 million dollars.

63. $A = 25$; $P = 9$; $h = 4$; $v = 65$

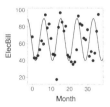

65. (62) The electric bill in January 2002 is \$89.62. (64) The electric bill in January 2002 is \$59.73.

67. $A = 1.2$; $P = 24$; $h = 5$; $v = 1$

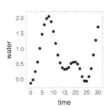

69. `tideMod=fitModel(water~A*sin(2`
    `*pi/P*(time-h))+v,`
    `    data=Hawaii,`
    `    start=list(A=1.2,P=24,h=5,v=1))`
    $$W = 0.804 \sin\left[\frac{2\pi}{24.66}\,(T - 2.52)\right]$$
    $$+0.732$$

71. (67) The water is at 0.4 when $t = 115$ in Pearl Harbor, Hawaii. (69) The water is at 0.434 when $t = 115$ in Pearl Harbor.

73. $y = A\sin\left[\frac{2\pi}{P}\,(x - h)\right] + v$

75. A gives half the height of the model; P tells how long it takes to repeat; h tells how far the starting point is shifted right of zero; v tells how far the midline is shifted up or down.

77. Find the midline of the sinusoidal model.

79. Find the distance between consecutive peaks or valleys.

Section 2.5 Exercises

1. Sinusoidal

3. Sigmoidal

5. Sigmoidal

7. Sine

9. Linear

11. $L = 45$

13. $L \approx 25.02$

15. $C = 1.5$

17. $C \approx -27.3$

19. $k = \ln(2) \approx 0.693$

21. $k = \ln(2)/2 \approx 0.347$

23. $v = 0;\ L = 60;\ (h, y(h)) = (1, 30);$ $C = 1$

25. $v = 60;\ L = 30;\ (h, y(h)) = (5.75, 75);$ $C = 1$

27. $v = -20;\ L = 6;\ (h, y(h)) = (3.5, -17);$ $C = 1;$

29. $v = 0;\ L = 2;\ (h, y(h)) = (2.75, 1);$ $C = 1$

31. $k = 0.5$

33. $k = 0.25$

35. $y = \dfrac{12}{1 + 3e^{-0.5(x-2)}} + 3$

37. $y = \dfrac{75}{1 + 25e^{-(x+4)}}$

39. $y = \dfrac{45}{1 + 15e^{-0.75(x-5)}} - 5$

41. $y = \dfrac{8}{1 + 2e^{-5x}}$

43. $y = \dfrac{100}{1 + 5e^{-2(x-4)}} - 20$

45. $y = \dfrac{5}{1 + 0.25e^{-(x-2)}}$

47. $v = 4\ L = 8,\ (h, y(h)) = (1930, 8);$ $C = 1;\ k = 0.034;$
    `head(PopulationBelgium)`
    `plotPoints(People~Year,`
    `data=PopulationBelgium)`

49. `BelModel=`
    `fitModel(People~L/(1+C*exp(`
    `-k*(Year-1930)))+v,`
    `data=PopulationBelgium,`
    `start=list(v=4, L=8, C=1,`
    ` k=0.034))`
    `coef(BelModel)`
    $$P = \frac{7.8348}{1 + 0.6599e^{-0.02555*(Y-1930)}}$$
    $$+3.2725$$

51. (47) The population of Belgium in 2020 is 11.64 million people. (49) The population of Belgium in 2020 is 10.62 million people.

53. $v = 0;\ L = 1600;\ (h, y(h)) = (100, 1100);\ C = 0.45;\ k = 0.063$

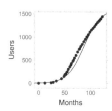

55. (52) The number of Facebook users in January 2010 is 14.67 million. (54) The number of Facebook users in January 2010 is -23.27 million.

57. $v = 490\ L = 35,\ (h, y(h)) = (1995, 507);\ C = 1.0588;\ k = 0.183;$
```
head(SATMathKentucky)
plotPoints(Score~Year,
data=SATMathKentucky)
```

59. ```
SATModel=
fitModel(Score~L/(1+C*exp(
-k*(Year-1995)))+v,
data=SATMathKentucky,
start=list(v=490, L=35, C=1.0588,
 k=0.183))
coef(SATModel)
```
$$S = \frac{22.74}{1 + 0.623e^{-0.248*(Y-1995)}} + 493.7$$

61. (57) The average SAT math score in Kentucky in 2020 is predicted to be 524.6. (59) The average SAT math score in Kentucky in 2020 is predicted to be 516.4.

63. $v = 40;\ L = 30;\ (h, y(h)) = (1990, 60);\ C = 0.5;\ k = 0.0998$

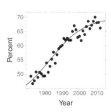

65. (62) The percentage of high school graduates in college in 1998 is predicted to be 64.5%. (64) The percentage of high school graduates in college in 1998 is predicted to be 64.8%.

67. $v = 0\ L = 14000,\ (h, y(h)) = (200, 7200);\ C = 0.944;\ k = 0.0367;$
```
head(EbolaSierraLeone)
plotPoints(Cases~Day,
data=EbolaSierraLeone)
```

69. ```
EbolaModel=
fitModel(Cases~L/(1+C*exp(
-k*(Day-200)))+v,
data=ElectronicMailOrderSales,
start=list(v=0, L=14000, C=0.944,
 k=0.0367))
coef(SalesModel)
```
$$S = \frac{14015.5}{1 + 0.952e^{-0.021*(Y-200)}} - 500.4$$

71. (72) The cummulative number of Ebola cases in Sierra Leone on day 700 is estimated to be 14,000. (74) The cummulative number of Ebola cases in Sierra Leone on day 700 is estimated to be 13,514.78.

73. $v = 50;\ L = 275;\ (h, y(h)) = (1992, 200);\ C = 0.83;\ k = 0.187$

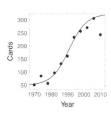

75. (72) The number of yellow cards in 1998 is predicted to be 266.5. (74) The number of yellow cards in 1998 is predicted to be 257.6.

77. $y = \dfrac{L}{1 + Ce^{-k(x-h)}} + v$

79. In the parametrized form of a sigmoidal function, v gives the vertical shift, $v+L$ gives the carrying capacity, $(h, y(h))$ are the coordinates of the point of inflection, C shifts the sigmoidal left and right, and k affects the steepness of the growth.

81. To conjecture L, first estimate the lower asymptote v and the upper asymptote $v + L$. Then, subtract to estimate L.

83. To conjecture C, first estimate, v, L, and $(h, y(h))$. Then compute

$$C = \frac{L}{y(h) - v} - 1$$

Section 2.6 Exercises

1. Sigmoidal

3. Linear

5. Sigmoidal

7. Power function; sine

9. Sine

11. Exponential

13. Linear

15. Power function

17. Linear

19. Exponential

21. Exponential

23. Sine

25. Exponential; sine

27. Linear; sigmoidal

29. Sine

31. Linear; `plotPoints(Sales~Year, data=ElectronicMailOrderSales)`

33. `SalesMod= fitModel(Sales~m*Year+b, data=ElectronicMailOrderSales) plotFun(SalesMod(Year)~Year, add=TRUE)`

35. `SalesMod(2001)`; 114.523

37. `SalesMod(2016)`; 368.790; this is an extrapolation.

39. Exponential; `plotPoints(log(Close)~ Quarter,data=NASDAQQuarterly)`

41. `NASDAQMod= fitModel(log(Close~m*Quarter+b, data=NASDAQQuarterly) plotFun(exp(1.49082)*exp(0.02258* Quarter)~Quarter, add=TRUE)`

43. `NASDAQMod(24)`; 7.64

45. `NASDAQMod(16)`; 6.37

47. Linear; `plotPoints(Rate~Year, data=Mortgage15YrAnnual)`

49. `RateMod= fitModel(Rate~m*Year+b, data=Mortgage15YrAnnual) plotFun(RateMod(Year)~Year, add=TRUE)`

51. `RateMod(2009)`; 4.52%

53. `RateMod(1981)`; 10.55%; this is an extrapolation.

55. Sine; `plotPoints(water~time, data=Hawaii)`

57. `TideMod= fitModel(water~A*sin(2*pi/P*(time -h))+v,data=Hawaii, start=list(A=1,P=24,h=4,v=1)) plotFun(TideMod(time))~time, add=TRUE)`

59. `TideMod(15)`; 0.702

61. `TideMod(115)`; 0.434; this is an extrapolation.

63. Power function; `plotPoints(log(Rate~log(Mass), data=BodyMassMetabolicRate)`

65. `BodyMassMod= fitModel(log(Rate)~m*log(Mass) +b,data=BodyMassMetabolicRate) plotFun(exp(6.81)*Mass^(0.652)~ Mass,add=TRUE)`

67. `exp(6.81)*(12)^(0.652)`; 4853.23

69. `exp(6.81)*(200)^(0.652)`; 28695.7

71. Linear; `plotPoints(kT~Year, data=USCO2Emissions)`

73. `EmissionsMod= fitModel(kT~m*Year+b, data=USCO2Emissions) plotFun(EmissionsMod(Year)~Year, add=TRUE)`

75. `EmissionsMod(2005)`; 5764625

77. `EmissionsMod(1972)`; 3870714; this is an extrapolation.

79. Sigmoidal; `plotPoints(Percent~Year, data=HSGradsinCollege)`

81. `GradMod=fitModel(Percent~L/(1+C*exp(-k*(Year-1990)))+v, data=HSGradsinCollege, start=list(L=30,v=40,C=0.5, k=0.0998))` `plotFun(GradMod(Year)~Year, add=TRUE)`

83. `GradMod(1982)`; 52.85%

85. `GradMod(2020)`; 67.67%; this is an extrapolation.

87. Power function; `plotPoints(log(Average)~log(Year), data=MaunaLoaCO2)`

89. `CO2Mod=fitModel(log(Average)~m* log(Year)+b,data=MaunaLoaCO2)` `plotFun(exp(-56.4615)* Year^(8.206)~Year,add=TRUE)`

91. `exp(-56.4615)*1960^(8.206)`; 312.85

93. `exp(-56.4615)*1900^(8.206)`; 242.40

95. Sine; `plotPoints(Position~Time, data=SunPositionAlaska)`

97. `SunMod= fitModel(Position~A*sin(2*pi/P*( Time-h))+v,data=SunPositionAlaska, start=list(A=25,P=24,h=9,v=20))` `plotFun(SunMod(Time)~Time, add=TRUE)`

99. `SunMod(3)`; -5.58

101. `SunMod(120)`; -6.03; this is an extrapolation.

103. Exponential; `plotPoints(log(Physicians)~ LifeExectancy, data=LifeExpectancyPhysicians)`

105. `LifeExpectMod= fitModel(log(Physicians)~m* LifeExpectancy+b, data=LifeExpectancyPhysicians)` `plotFun(exp(-3.98)*exp(0.1489*( Year-44.83895))~Year, add=TRUE)`

107. `exp(-3.98)*exp(0.149*(75-44.84))`; 1.67

109. `exp(-3.98)*exp(0.149*(35-44.84))`; 0.004

111. Linear; `plotPoints(Debt~Year, data=StudentDebt1)`

113. Continual linear growth is reasonable.

115. A power function model is more reasonable than the linear model in (114).

117. Linear; `plotPoints(Burgers~Year, data=McDBurgers1)`

119. Neither continued linear nor power function growth is reasonable.

121. A linear model is no longer reasonable; none of the other models are a reasonable fit.

123. Sine; `plotPoints(Volume~Date, data=FordMarketVolume1)`

125. The sinusoidal models make sense at a basic level, but do not necessarily capture all facets of the long-term behavior possible.

127. A linear model is no longer reasonable; maybe a sine model is better.

129. A power function or exponential model fit the data well; a sigmoidal model is more realistic in terms of expected long-term behavior of the data.

131. The five steps of the modeling cylce are as follows:

    (1) Ask a question about reality.

    (2) Make some observations and collect the corresponding data.

    (3) Conjecture a model or modify a known model based on the data.

    (4) Test the model against known data (from step (2)) and modify the model as needed.

    (5) Repeat steps (2) – (4) to improve the model.

133. The model must make sense in the context of the data and the expected long-term behavior of the physicial/social situation being modeled.

135. Extrapolation uses a mathematical model to attempt to predict an output value for an input that lies outside the input interval used to create the model.

Section 2.7 Exercises

1. 1 inch = 2.54 centimeters

3. 1 ft^2 = 0.092903 m^2

5. 10 miles per hour = 4.47 meters per second

7. 4.45 N = 1 lb(force)

9. 1000 g/m^3 = 1.8×10^{-5} slug/in^3

11. 24 hr = 1 day

13. $2\pi = 360°$

15. 1 person/mi^2 = 2.59 people/km^2

17. Valid

19. Valid

21. Valid

23. Valid

25. Valid

27. Invalid

29. Invalid

31. Valid: 3m + 6ft; invalid: 3m + 6g

33. Valid: 6m = 5ft; invalid: 6m = 5g

35. Valid: 6m × 5g; invalid: n/a

37. $[b] = [c] = L$

39. $[b] = [c] = T$

41. $[b] = M^{-1}; [c] = 1$

43. $b] = T$; $[c] = 1$

45. $[b] = LT^{-1}$; $[c] = L$

47. $[b] = LT^{-1}$; $[c] = L$

49. $[b] = M$; $[c] = L^{-1}T$

51. $[P] = [I] = B$; $[R] = T^{-1}; [T] = T$

53. $[N] = [L] = P$; $[c] = 1$; $[K] = T^{-1}$; $[t] = T$

55. $[a] = M$; $[b] = [c] = ML^{-2}$; $[d] = ML^{-3}$; $[h] = [r] = L$

57. $[dI] = PT^{-1}$; $[a] = T^{-1}P^{-1}$; $[b] = T^{-1}$; $[I] = [S] = P$

59. $[C] = ML^{-3}$; $[t] = T$; $[b] = T^{-1}$; $[a] = MT^{-1}L^{-3}$

61. $[m_1] = [m_2] = M$; $[r] = L$; $[F] = MLT^{-2}$; $[G] = M^{-1}T^{-2}L^2$

63. P

65. LTM

67. $P^3L^{-1}T^{-2}$

69. $M^3B^3P^{-2}$

71. $N^2T^{-3}M$

73. $w = ku^2v^4$

75. $c = ka^5b^2$

77. $s = kp^{-1}q^2r^3$

79. $s = kp^{-2}q^3r^2$

81. $d = ka^3b^4c^3$

83. $T = k\ell^{\frac{1}{2}}g^{-\frac{1}{2}}$

85. $S = k\ell^{\frac{1}{2}}g^{\frac{1}{2}}$

87. $C = kk_eq_1q_2r^{-2}$ (infinite number of answers)

89. No dimensionally compatible relationship exists.

91. Units are a way of assigning values to a dimension.

93. Dimensions derived from the powers of products of two or more fundamental dimensions.

95. If the operation under consideration is $-$, $+$, $=$, $>$, $<$, or any combination of those, then the two quantities being considered must have the same dimension to make the expression valid.

97. Consider a^r. r must be dimensionless. Also, either a must be dimensionless or the product of r with the exponent of the dimension of a is an integer.

99. The generalized product model sets the output of a function as the product of powers of the inputs and a scalar constant; $C = ka_1^{x_1}a_2^{x_2}\cdots a_n^{x_n}$

Section 3.1 Exercises

1. $\overline{PQ} = \begin{pmatrix} 6 \\ 2 \end{pmatrix}, \overline{QP} = \begin{pmatrix} -6 \\ -2 \end{pmatrix}$

3. $\overline{PQ} = \begin{pmatrix} -6 \\ -2 \end{pmatrix}, \overline{QP} = \begin{pmatrix} 6 \\ 2 \end{pmatrix}$

5. $\overline{PQ} = \begin{pmatrix} 4 \\ -9 \end{pmatrix}, \overline{QP} = \begin{pmatrix} -4 \\ 9 \end{pmatrix}$

7. $\overline{PQ} = \begin{pmatrix} 7 \\ -6 \\ 1 \end{pmatrix}, \overline{QP} = \begin{pmatrix} -7 \\ 6 \\ -1 \end{pmatrix}$

9. $\overline{PQ} = \begin{pmatrix} -18 \\ -12 \\ -12 \end{pmatrix}, \overline{QP} = \begin{pmatrix} 18 \\ 12 \\ 12 \end{pmatrix}$

11. $\overline{PQ} = \begin{pmatrix} 8 \\ -4 \\ 3 \\ -4 \end{pmatrix}, \overline{QP} = \begin{pmatrix} -8 \\ 4 \\ -3 \\ 4 \end{pmatrix}$

13. For example, $P = (10, 5)$, $Q = (1, 1)$, $\overline{PQ} = \begin{pmatrix} -9 \\ -4 \end{pmatrix}, \overline{QP} = \begin{pmatrix} 9 \\ 4 \end{pmatrix}$

15. For example, $P = (-1, 0, 1, 5)$, $Q = (2, 0, 5, -1)$, $\overline{PQ} = \begin{pmatrix} 3 \\ 0 \\ 4 \\ -6 \end{pmatrix}, \overline{QP} = \begin{pmatrix} -3 \\ 0 \\ -4 \\ 6 \end{pmatrix}$

17. $\begin{pmatrix} 6 \\ 1 \end{pmatrix}$

19. For example, $P = (3, 3)$, $Q = (9, 4)$

21. $P = (-3, 0)$

23. $P = (4, -7)$

25. $Q = (6, 4)$

27. $Q = (-3, 5)$

29. $\|\overline{v}\| = 5$

31. $\|\overline{v}\| = \sqrt{10} \approx 3.162$

33. $\|\overline{v}\| = \sqrt{26} \approx 5.100$

35. $\|\overline{v}\| = \sqrt{3} \approx 1.732$

37. For example, $\overline{v} = \begin{pmatrix} 2 \\ 4 \end{pmatrix}$, $\|\overline{v}\| = \sqrt{20} \approx 4.472$

39. For example, $\overline{v} = \begin{pmatrix} 1 \\ 2 \\ 3 \\ 4 \end{pmatrix}$, $\|\overline{v}\| = \sqrt{30} \approx 5.477$

41. $\begin{pmatrix} 12 \\ -9 \end{pmatrix}$

43. $\begin{pmatrix} -2 \\ -2 \end{pmatrix}$

45. $\begin{pmatrix} 6 \\ -1 \end{pmatrix}$

47. $\begin{pmatrix} 38 \\ -18 \end{pmatrix}$

49. u=c(4,-3)
   3*u
   [1] 12 -9
   $3\overline{u} = \begin{pmatrix} 12 \\ -9 \end{pmatrix}$

51. v=c(2,2)
   -v
   [1] -2 -2
   $-\overline{v} = \begin{pmatrix} -2 \\ -2 \end{pmatrix}$

53. u=c(4,-3)
   v=c(2,2)
   u+v
   [1] 6 -1
   $\overline{u} + \overline{v} = \begin{pmatrix} 6 \\ -1 \end{pmatrix}$

55. u=c(4,-3)
   v=c(2,2)
   8*u+3*v
   [1] 38 -18
   $8\overline{u} + 3\overline{v} = \begin{pmatrix} 38 \\ -18 \end{pmatrix}$

57. $\begin{pmatrix} 7 \\ -49 \\ 14 \end{pmatrix}$

59. $\begin{pmatrix} 0 \\ 9 \\ -27 \end{pmatrix}$

61. $\begin{pmatrix} 1 \\ -10 \\ 11 \end{pmatrix}$

63. $\begin{pmatrix} 7 \\ -64 \\ -59 \end{pmatrix}$

65. u=c(1,-7,2)
   7*u
   [1]   7 -49 14
   $7\overline{u} = \begin{pmatrix} 7 \\ -49 \\ 14 \end{pmatrix}$

67. v=c(0,-3,9)
   -3*v

[1]   0   9   -27

$$-3\overline{v} = \begin{pmatrix} 0 \\ 9 \\ -27 \end{pmatrix}$$

69. u=c(1,-7,2)
    v=c(0,-3,9)
    u+v
    [1]   1 -10   11

$$\overline{u} + \overline{v} = \begin{pmatrix} 1 \\ -10 \\ 11 \end{pmatrix}$$

71. u=c(1,-7,2)
    v=c(0,-3,9)
    7*u+5*v
    [1]   7 -64   59

$$7\overline{u} + 5\overline{v} = \begin{pmatrix} 7 \\ -64 \\ 59 \end{pmatrix}$$

73.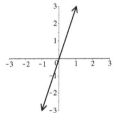

75. $\overline{v}$ and $-\overline{v}$ point in opposite directions.

77. $\overline{v}$ is three times shorter than $3\overline{v}$.

79. For example, $\overline{v} = \begin{pmatrix} -2 \\ -3 \end{pmatrix}$ and $m = -2$.

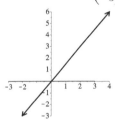

81. $\overline{v}$ and $m\overline{v}$ point in opposite directions.

83.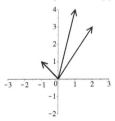

85. $\overline{u} - \overline{v}$; $-\overline{u} + \overline{v}$

87.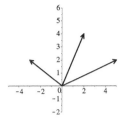

89. $\overline{u} - \overline{v}$; $-\overline{u} + \overline{v}$

91.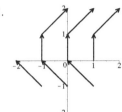

93.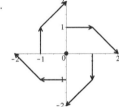

95.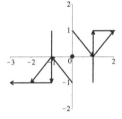

97. Each entry in the column of numbers making up a vector is called a component.

99. Vector addition is carried out by adding corresponding components.

101. Scalar multiplication $m\overline{v}$ is accomplished by multiplying each component of $\overline{v}$ by m.

103. Vector addition $\overline{u}+\overline{v}$ produces a vector corresponding to one of the diagonals of the parallelogram formed by $\overline{u}$ and $\overline{v}$.

105. A vector field is a vector-valued function that has two or more inputs

uniquely mapped to two or more outputs. A vector field $\overline{F}(x, y)$ assigns each input (x, y) to a unique vector output $\overline{F}(x, y) = \begin{pmatrix} f(x, y) \\ g(x, y) \end{pmatrix}$.

Section 3.2 Exercises

1. $\begin{pmatrix} 2 \\ -3 \end{pmatrix}$

3. $\begin{pmatrix} -3.4 \\ -7.7 \end{pmatrix}$

5. $\begin{pmatrix} 34 \\ -16 \end{pmatrix}$

7. $\begin{pmatrix} -1.0 \\ -7.6 \end{pmatrix}$

9. For example, let $\overline{u} = (1)$, let $\overline{v} = (3)$, and choose scalars $x = 2$ and $y = 4$. Then $x\overline{u} + y\overline{v} = 14$.

11. For example, let $\overline{u} = \begin{pmatrix} 2 \\ 3 \\ -1 \end{pmatrix}$,

    let $\overline{v} = \begin{pmatrix} -1 \\ 0 \\ 6 \end{pmatrix}$, and choose scalars $x = -2$ and $y = 3$. Then $x\overline{u} + y\overline{v} = \begin{pmatrix} -7 \\ -6 \\ 20 \end{pmatrix}$.

13. $\begin{pmatrix} 7 \\ 0 \end{pmatrix}$

15. The linear combination cannot be computed, because vectors 1 and 3 have two components, while vector 2 has three components.

17. 2×2; $\begin{pmatrix} -2 \\ 3 \end{pmatrix}, \begin{pmatrix} 7 \\ 2 \end{pmatrix}$

19. 4×3; $\begin{pmatrix} 7 \\ 0 \\ -1 \\ 2 \end{pmatrix}, \begin{pmatrix} 1 \\ 3 \\ -2 \\ 3 \end{pmatrix}, \begin{pmatrix} 2 \\ 4 \\ 1 \\ 4 \end{pmatrix}$

21. 3×4; $\begin{pmatrix} 7 \\ 0 \\ 1 \end{pmatrix}, \begin{pmatrix} 1 \\ 3 \\ 2 \end{pmatrix}, \begin{pmatrix} 2 \\ 4 \\ 1 \end{pmatrix}, \begin{pmatrix} 2 \\ 3 \\ 4 \end{pmatrix}$

23. $\begin{pmatrix} 15 \\ 6 \end{pmatrix}$

25. $\begin{pmatrix} 5 \\ 5 \end{pmatrix}$

27. $\begin{pmatrix} -7 \\ -3 \\ 5 \end{pmatrix}$

29. $\begin{pmatrix} 39 \\ 13 \\ 29 \\ 2 \end{pmatrix}$

31. $\begin{pmatrix} 5 \\ 7 \end{pmatrix} = x \begin{pmatrix} 3 \\ 1 \end{pmatrix} + y \begin{pmatrix} 2 \\ -4 \end{pmatrix}$;

    $\begin{pmatrix} 5 \\ 7 \end{pmatrix} = \begin{pmatrix} 3 & 2 \\ 1 & -4 \end{pmatrix} \begin{pmatrix} x \\ y \end{pmatrix}$

33. $\begin{pmatrix} 0 \\ -1 \end{pmatrix} = x \begin{pmatrix} 2 \\ 1 \end{pmatrix} + y \begin{pmatrix} 3 \\ 1 \end{pmatrix}$;

    $\begin{pmatrix} 0 \\ -1 \end{pmatrix} = \begin{pmatrix} 2 & 3 \\ 1 & 1 \end{pmatrix} \begin{pmatrix} x \\ y \end{pmatrix}$

35. $\begin{pmatrix} 4 \\ 5 \end{pmatrix} = \begin{pmatrix} 1 & -3 \\ 0 & 4 \end{pmatrix} \begin{pmatrix} x \\ y \end{pmatrix}$;

    $\begin{cases} 4 = x - 3y \\ 5 = 4y \end{cases}$

37. $\begin{pmatrix} -6 \\ 11 \end{pmatrix} = \begin{pmatrix} 7 & -2 \\ -9 & 5 \end{pmatrix} \begin{pmatrix} x \\ y \end{pmatrix}$;

    $\begin{cases} -6 = 7x - 2y \\ 11 = -9x + 5y \end{cases}$

39. $\begin{pmatrix} 4 \\ 1 \\ -3 \end{pmatrix} = \begin{pmatrix} 2 & 3 & -2 \\ 7 & 0 & 0 \\ -1 & 0 & 9 \end{pmatrix} \begin{pmatrix} x \\ y \\ z \end{pmatrix}$;

    $\begin{cases} 4 = 2x + 3y - 2z \\ 1 = 7x \\ -3 = -x + 9z \end{cases}$

41. $\begin{pmatrix} 1 \\ 1 \end{pmatrix} = x \begin{pmatrix} 3 \\ 2 \end{pmatrix} + y \begin{pmatrix} 8 \\ 1 \end{pmatrix}$;

    $\begin{cases} 1 = 3x + 8y \\ 1 = 2x + y \end{cases}$

43. $\begin{pmatrix} 0 \\ 1 \end{pmatrix} = x \begin{pmatrix} 7 \\ -2 \end{pmatrix} + y \begin{pmatrix} 3 \\ 5 \end{pmatrix}$;

    $\begin{cases} 0 = 7x + 3y \\ 1 = -2x + 5y \end{cases}$

45. For example, let $\overline{u} = \begin{pmatrix} 3 \\ 1 \end{pmatrix}$, $\overline{v} = \begin{pmatrix} -2 \\ 4 \end{pmatrix}$, and $\overline{w} = \begin{pmatrix} 0 \\ 1 \end{pmatrix}$. $\begin{cases} 3x - 2y = 0 \\ x + 4y = 1 \end{cases}$

47. $\begin{cases} 121 = m \cdot 2009 + b \\ 143 = m \cdot 2010 + b \\ 168 = m \cdot 2011 + b \\ 192 = m \cdot 2012 + b \end{cases}$

$$\begin{pmatrix} 121 \\ 143 \\ 168 \\ 192 \end{pmatrix} = m \begin{pmatrix} 2009 \\ 2010 \\ 2011 \\ 2012 \end{pmatrix} + b \begin{pmatrix} 1 \\ 1 \\ 1 \\ 1 \end{pmatrix}$$

49. $\begin{cases} 1.413 = m \cdot 1995 + b \\ 1.26 = m \cdot 2000 + b \\ 1.203 = m \cdot 2005 + b \\ 1.127 = m \cdot 2010 + b \end{cases}$

$$\begin{pmatrix} 1.413 \\ 1.26 \\ 1.203 \\ 1.127 \end{pmatrix} = m \begin{pmatrix} 1995 \\ 2000 \\ 2005 \\ 2010 \end{pmatrix}$$

51. $\begin{cases} 66.0 = m \cdot 2006 + b \\ 70.1 = m \cdot 2009 + b \\ 68.1 \\ m \cdot 2010 + b \\ 66.2 = m \cdot 2012 + b \end{cases}$

$$\begin{pmatrix} 66.0 \\ 70.1 \\ 68.1 \\ 66.2 \end{pmatrix} = m \begin{pmatrix} 2006 \\ 2009 \\ 2010 \\ 2012 \end{pmatrix} + b \begin{pmatrix} 1 \\ 1 \\ 1 \\ 1 \end{pmatrix}$$

53. $x = 5, \ y = 3$

55. $x = 4.3, \ y = -6$

57. $x = 1, \ y = 1$

59. $x = \dfrac{9}{7}, \ y = \dfrac{-13}{7}$

61. The vector equation does not have a solution, because $\begin{pmatrix} -1 \\ -1 \end{pmatrix}$ and $\begin{pmatrix} 3 \\ 3 \end{pmatrix}$ are scalar multiples of each other, but $\begin{pmatrix} 2 \\ 5 \end{pmatrix}$ is not a scalar multiple of either.

63. This vector equation does not have a solution, because the vectors do not have the same number of components.

65. $\begin{pmatrix} x \\ y \end{pmatrix} = \begin{pmatrix} -3 \\ -1 \end{pmatrix}$

67. $\begin{pmatrix} x \\ y \end{pmatrix} = \begin{pmatrix} -5 \\ 2 \end{pmatrix}$

69. $\begin{pmatrix} x \\ y \end{pmatrix} = \begin{pmatrix} -1 \\ 3 \end{pmatrix}$

71. $\begin{pmatrix} -4 \\ -1 \end{pmatrix} = \begin{pmatrix} 2 & -2 \\ 1 & 1 \end{pmatrix} \begin{pmatrix} x \\ y \end{pmatrix}$;

$\begin{pmatrix} x \\ y \end{pmatrix} = \begin{pmatrix} -2 \\ 1 \end{pmatrix}$

73. $\begin{pmatrix} 0 \\ -1 \end{pmatrix} = \begin{pmatrix} 2 & 3 \\ 1 & 1 \end{pmatrix} \begin{pmatrix} x \\ y \end{pmatrix}$;

$\begin{pmatrix} x \\ y \end{pmatrix} = \begin{pmatrix} -3 \\ 2 \end{pmatrix}$

75. $\begin{cases} 2 = x + y \\ 4 = y \\ 4 = y \end{cases}$

77.
```
u1 = c(-1,-1)
u2 = c(2,1)
U = matrix(c(u1,u2), nrow=2,
 ncol=2)
v = c(1,2)
solve(U,v)
[1] -3 -1
```

79.
```
u1 = c(1,0)  u2 = c(0,1)
U = matrix(c(u1,u2), nrow = 2,
 ncol = 2)
v = c(-5,2)
solve(U,v)
[1] -5  2
```

81.
```
u1 = c(4,1)  u2 = c(2,1)
U = matrix(c(u1,u2), nrow = 2,
 ncol = 2)
v = c(2,2)
solve(U,v)
[1] -1  3
```

83.
```
u1 = c(3,2,3)  u2 = c(8,1,2)
u3 = c(2,4,9)
U = matrix(c(u1,u2,u3),
nrow = 3, ncol = 3)
v = c(1,1,2)
solve(U,v)
[1] 0.16279  0.02326  0.16279
```

85.
```
u1 = c(7,-2,1)  u2 = c(3,5,2)
u3 = c(-5,-4,3)
U = matrix(c(u1,u2,u3),
nrow = 3, ncol = 3)
v = c(0,-1,3)
solve(U,v)
[1] 0.27358  0.41509  0.63208
```

87.
```
u1 = c(1,0)
u2 = c(-3,4)
U = matrix(c(u1,u2),
nrow=2, ncol=2)
```

```
v = c(4,5)
solve(U,v)
[1] 7.75 1.25
```
89.
```
u1 =c(7,-9)
u2 = c(-2,5)
U = matrix(c(u1,u2),
nrow=2, ncol=2)
v = c(-6,11)
solve(U,v)
[1]  -0.4705882 1.3529412
```
91.
```
u1 =c(2,7,-1)
u2 = c(3,0,0)
u3 = c(-2,0,9)
U = matrix(c(u1,u2,u3),
nrow=3, ncol=3)
v = c(4,1,-3)
solve(U,v)
[1]   0.1426   1.20646   -0.31746)
```
93. A system of linear equations consists of a collection of linear equations.

95. If $\overline{u}_1, \overline{u}_2, \ldots, \overline{u}_n$ are vectors and $x_1, x_2, \ldots, x_n$ are scalars, then $x_1\overline{u}_1 + x_2\overline{u}_2 + \cdots + x_n\overline{u}_n$ is a linear combination of $\overline{u}_1, \overline{u}_2, \ldots, \overline{u}_n$.

97. If $\overline{u}_1, \overline{u}_2, \ldots, \overline{u}_n$ and $\overline{v}$ are vectors, $x_1, x_2, \ldots, x_n$ are variables, and $a_1, a_2, \ldots, a_n$ are scalars, then $x_1 = a_1, x_2 = a_2, \ldots, x_n = a_n$ is a solution of vector equation $\overline{v} = x_1\overline{u}_1 + x_2\overline{u}_2 + \cdots + x_n\overline{u}_n$ when $\overline{v}$ is equal to $a_1\overline{u}_1 + a_2\overline{u}_2 + \cdots + a_n\overline{u}_n$.

99. A square matrix is an $n \times n$ matrix with the same number of rows and columns.

101. The solution $\begin{pmatrix} x \\ y \end{pmatrix}$ of the 2×2 matrix equation $\begin{pmatrix} r \\ s \end{pmatrix} = \begin{pmatrix} a & b \\ c & d \end{pmatrix} \begin{pmatrix} x \\ y \end{pmatrix}$ with $ad - bc \neq 0$ is

$$\begin{pmatrix} x \\ y \end{pmatrix} = \frac{1}{ad - bc} \begin{pmatrix} d & -b \\ -c & a \end{pmatrix} \begin{pmatrix} r \\ s \end{pmatrix}$$

If $ad - bc = 0$, then the matrix equation has no solution or infinitely many solutions.

Section 3.3 Exercises

1. Linear combination; $x = 5$

3. Linear combination; $x = 1$
5. Not a linear combination
7. Not a linear combination
9. Linear combination; $x = 2$
11. Not a linear combination
13. Linear Combination; $x = 5$
15. Linear Combination; $x = 1$
17. Not a linear combination
19. Not a linear Combination
21. Linear combination; $x = 2$
23. Linear combination; $x = 0.1$
25. Linear combination; $x = 5$
27. Not a linear combination
29. Linear combination; $x = 5$, $y = 0$
31. Linear combination; $x = 0$, $y = 0$
33. Linear combination; $x = 8$, $y = 1$ (infinitely many solutions)
35. Not a linear combination
37.
```
> u=c(1,2)
> v=c(2,1)
> U=matrix(c(u,v),nrow=2,ncol=2)
> w=c(5,10)
> solve(U,w)
[1] 5 0
```
Linear combination; $x = 5$, $y = 0$
39.
```
> u=c(1,2)
> v=c(2,1)
> U=matrix(c(u,v),nrow=2,ncol=2)
> w=c(-1.7,5.8)
> solve(U,w)
[1]   4.433333 -3.066667
```
Linear combination; $x = 4.433$, $y = -3.067$
41.
```
> u=c(1,2)
> v=c(-6,-3)
> U=matrix(c(u,v),nrow=2,ncol=2)
> w=c(5,10)
> solve(U,w)
[1] 5 0
```
Linear combination; $x = 5$, $y = 0$
43.
```
> u=c(1,2)
> v=c(-6,-3)
> U=matrix(c(u,v),nrow=2,ncol=2)
> w=c(-1.3,2.6)
> solve(U,w)
[1] 2.1666667 0.5777778
```
Linear combination; $x = 2.167$, $y = 0.578$

45. Linear combination; $x = 0.1$, $y = 0$, $z = 0$ (infinitely many solutions)

47. Not a linear combination

49. Linear combination; $x = 0.1$, $y = 0.2$, $z = 0.2$

51. Linear combination; $x = 5$, $y = -2$, $z = -2$

53. $\begin{pmatrix} 111 \\ 69 \\ 26 \\ -15 \\ -57 \end{pmatrix} = m \begin{pmatrix} 1 \\ 2 \\ 3 \\ 4 \\ 5 \end{pmatrix} + b \begin{pmatrix} 1 \\ 1 \\ 1 \\ 1 \\ 1 \end{pmatrix}$

No linear solution

55. $\begin{pmatrix} 77.4 \\ 81 \\ 84.1 \\ 87.5 \\ 92 \end{pmatrix} = m \begin{pmatrix} 1 \\ 2 \\ 4 \\ 5 \\ 6 \end{pmatrix} + b \begin{pmatrix} 1 \\ 1 \\ 1 \\ 1 \\ 1 \end{pmatrix}$

No linear solution

57. $\begin{pmatrix} 33.7 \\ 1.9 \\ -19.3 \\ -40.5 \\ -51.1 \end{pmatrix} = (-5.3) \begin{pmatrix} 2 \\ 8 \\ 12 \\ 16 \\ 18 \end{pmatrix} + (44.3) \begin{pmatrix} 1 \\ 1 \\ 1 \\ 1 \\ 1 \end{pmatrix}$

59. $\begin{pmatrix} 5 \\ 4.5 \\ 4 \\ 3.5 \\ 3 \\ 2 \\ 1.5 \\ 1 \\ .5 \\ 0 \end{pmatrix} = (-0.1) \begin{pmatrix} 0 \\ 5 \\ 10 \\ 15 \\ 20 \\ 30 \\ 35 \\ 40 \\ 45 \\ 50 \end{pmatrix} + (5) \begin{pmatrix} 1 \\ 1 \\ 1 \\ 1 \\ 1 \\ 1 \\ 1 \\ 1 \\ 1 \\ 1 \end{pmatrix}$

61. $\begin{pmatrix} 111 \\ 69 \\ 26 \\ -15 \\ -57 \end{pmatrix} = m \begin{pmatrix} 1 \\ 2 \\ 3 \\ 4 \\ 5 \end{pmatrix} + b \begin{pmatrix} 1 \\ 1 \\ 1 \\ 1 \\ 1 \end{pmatrix}$

```
> u1=c(1,2)
> u2=c(1,1)
> U=matrix(c(u1,u2),nrow=2,ncol=2)
> v=c(111,69)
> solve(U,v)
> [1] -42 153
> x=c(1,2,3,4,5)
> z=c(1,1,1,1,1)
> -42*x+153*z
[1] 111 69 27 -15 -57
```
No linear solution

63. $\begin{pmatrix} 77.4 \\ 81 \\ 84.1 \\ 87.5 \\ 92 \end{pmatrix} = m \begin{pmatrix} 1 \\ 2 \\ 4 \\ 5 \\ 6 \end{pmatrix} + b \begin{pmatrix} 1 \\ 1 \\ 1 \\ 1 \\ 1 \end{pmatrix}$

```
> u1=c(1,2)
> u2=c(1,1)
> U=matrix(c(u1,u2),nrow=2,ncol=2)
> v = c(77.4,81)
> solve(U,v)
[1] 3.6 73.8
> x=c(1,2,4,5,6)
> z=c(1,1,1,1,1)
> 3.6*x+73.8*z
[1] 77.4 81.0 88.2 91.8 95.4
```
No linear solution

65. $\begin{pmatrix} 5.0 \\ 4.5 \\ 4.0 \\ 3.5 \\ 3.0 \\ 2.0 \\ 1.5 \\ 1.0 \\ 0.5 \\ 0.0 \end{pmatrix} = m \begin{pmatrix} 0 \\ 5 \\ 10 \\ 15 \\ 20 \\ 30 \\ 35 \\ 40 \\ 45 \\ 50 \end{pmatrix} + b \begin{pmatrix} 1 \\ 1 \\ 1 \\ 1 \\ 1 \\ 1 \\ 1 \\ 1 \\ 1 \\ 1 \end{pmatrix}$

```
> u1=c(0,5)
> u2 = c(1,1)
> U=matrix(c(u1,u2),nrow=2,ncol=2)
> v=c(5,4.5)
> solve(U,v)
[1] -0.1 5.0
> x=c(0,5,10,15,20,30,35,40,45,50)
> z=c(1,1,1,1,1,1,1,1,1,1)
> -0.1*x+5.0*z
[1] 5.0 4.5 4.0 3.5 3.0 2.0 1.5
    1.0 0.5 0.0
```
Solution: $y = -0.1x + 5.0$

67. The first two rows produce $m = 22$ and $b = -4258$, but these give a third row of 153, which is not equal to 149.

69. The first two rows produce $m = -1.1$ and $b = 2469.3$, but these give a third row of 259.4, which is not equal to 266.8.

71. The first two rows produce $k = 0.1275$ and $b = 1.099$, but these give a third row of 2.119, which is not equal to 2.079.

73. The first two rows produce $k = 0.035$ and $b = 3.091$, but these give a third

row of 3.196, which is not equal to 3.203.

75. $\begin{pmatrix} 3419 \\ 3530 \\ 3633 \\ 3676 \end{pmatrix} = m \begin{pmatrix} 2006 \\ 2007 \\ 2009 \\ 2010 \end{pmatrix} + b \begin{pmatrix} 1 \\ 1 \\ 1 \\ 1 \end{pmatrix}$

77. > u1 = c(2006,2007)
     > u2 = c(1,1)
     > U=matrix(c(u1,u2),nrow=2,
         ncol=2)
     > v=c(3419,3530)
     > solve(U,v)
     [1] 111 -219247
     > x=c(2006,2007,2009,2010)
     > z=c(1,1,1,1)
     > 111*x-219247*z
     [1] 3419 3530 3752 3863
     No linear solution

79. The first two rows produce $m = 4.09$ and $b = -8063.24$, but these give a third row of 77.273, which is not equal to 80.

81. $\begin{pmatrix} 101.5 \\ 101.5 \\ 101.6 \\ 101.6 \end{pmatrix} = m \begin{pmatrix} 1990 \\ 1995 \\ 2005 \\ 2010 \end{pmatrix} + b \begin{pmatrix} 1 \\ 1 \\ 1 \\ 1 \end{pmatrix}$

83. > u1 = c(1990,1995)
     > u2 = c(1,1)
     > U=matrix(c(u1,u2),nrow=2,
         ncol=2)
     > v=c(101.5,101.5)
     > solve(U,v)
     [1] -3.205e-16 1.015e+02
     > x=c(1990,1995,2005,2010)
     > z=c(1,1,1,1)
     > (-3.205e-16)*x+(1.015e+02)*z
     [1] 101.5 101.5 101.5 101.5
     No linear solution

85. The first two rows produce $m = -0.306$ and $b = 62.46$, but these give a third row of 1.102, which is not equal to 1.203.

87. $\begin{pmatrix} 1.6094 \\ 2.0794 \\ 2.8903 \\ 3.3322 \\ 3.4965 \end{pmatrix} = m \begin{pmatrix} 0 \\ 10 \\ 20 \\ 25 \\ 30 \end{pmatrix} + b \begin{pmatrix} 1 \\ 1 \\ 1 \\ 1 \\ 1 \end{pmatrix}$

89. > u1 = c(1,10)
     > u2 = c(1,1)
     > U=matrix(c(u1,u2),nrow=2,
         ncol=2)
     > v=c(1.6094,2.0794)
     > solve(U,v)
     [1] 0.0470 1.6094
     > x=c(0,10,20,25,30)
     > z=c(1,1,1,1,1)
     > 0.0470*x+1.6094*z
     [1] 1.6094 2.0794 2.5494 2.7844
         3.0194
     No linear solution

91. The first two rows produce $k = 0.001215$ and $b = 2.0656$, but these give a third row of 2.089, which is not equal to 1.805.

93. $\begin{pmatrix} 3.3707 \\ 3.6454 \\ 3.7495 \\ 3.9435 \\ 4.0036 \end{pmatrix} = m \begin{pmatrix} 0 \\ 2 \\ 3 \\ 4 \\ 5 \end{pmatrix} + b \begin{pmatrix} 1 \\ 1 \\ 1 \\ 1 \\ 1 \end{pmatrix}$

95. > u1 = c(0,2)
     > u2 = c(1,1)
     > U=matrix(c(u1,u2),nrow=2,
         ncol=2)
     > v=c(3.3707,3.6454)
     > solve(U,v)
     [1] 0.13735 3.37070
     > x=c(0,2,3,4,5)
     > z=c(1,1,1,1,1)
     > 0.13735*x+3.37070*z
     [1] 3.37070 3.64540 3.78275
         3.92010 4.05745
     No linear solution

97. The first two rows produce $k = 0.0703$ and $b = 4.5538$, but these give a third row of 5.9611, which is not equal to 6.3007.

99. Graphically, a vector $\overline{v}$ is a linear combination of vector $\overline{u}$ if they lie on the same line.

101. If the vectors $\overline{u}_1$, $\overline{u}_2$, $\ldots$, $\overline{u}_n$ in the vector equation $\overline{v} = x_1\overline{u}_1 + x_2\overline{u}_2 + \cdots + x_n\overline{u}_n$ have more than n components, then determine if a solution exists via the following three steps:

     (1) Form a new vector equation using truncated versions of $\overline{u}_1$, $\overline{u}_2$, $\ldots, \overline{u}_n$ that contain just the first n components of the original vectors.

(2) Find the solution x_1, x_2, ..., x_n of the vector equation from step (1), if possible.

(3) Check whether the solution x_1, x_2, ..., x_n from step (2) satisfies the original vector equation by computing $x_1\overline{u}_1 + x_2\overline{u}_2 + \cdots + x_n\overline{u}_n$ and comparing the result with $\overline{v}$.

103. A model is linear in its parameters if the parameters are only multiplied by the variables or by functions of the variables in the modeling equation.

Section 3.4 Exercises

1. 0; orthogonal

3. 24; not orthogonal

5. 0; orthogonal

7. 0; orthogonal

9. -18; not orthogonal

11. 0; orthogonal

13. 5

15. $\sqrt{90}$

17. 4

19. $\sqrt{14}$

21. $\sqrt{42}$

23. $\sqrt{41}$

25. Orthogonal;

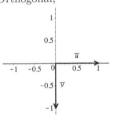

27. Not orthogonal;

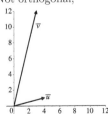

29. Orthogonal;

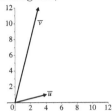

31. $\overline{r} = \begin{pmatrix} -2 \\ -1 \end{pmatrix}$

33. $\overline{r} = \begin{pmatrix} 0 \\ -6 \end{pmatrix}$

35. $\overline{r} = \begin{pmatrix} -0.5 \\ -0.5 \\ 1 \end{pmatrix}$

37. $\overline{r} = \begin{pmatrix} 0 \\ 0 \\ 0 \\ 0 \end{pmatrix}$

39. $x\overline{u} = \begin{pmatrix} 7 \\ 0 \end{pmatrix}$; $\overline{r} = \begin{pmatrix} 0 \\ 5 \end{pmatrix}$

41. $x\overline{u} = \begin{pmatrix} 6 \\ 6 \end{pmatrix}$; $\overline{r} = \begin{pmatrix} 1 \\ -1 \end{pmatrix}$

43. $x\overline{u} = \begin{pmatrix} 93/13 \\ 62/13 \end{pmatrix}$; $\overline{r} = \begin{pmatrix} -2/13 \\ 3/13 \end{pmatrix}$

45. $x\overline{u} = \begin{pmatrix} -3 \\ 0 \\ 0 \end{pmatrix}$, $\overline{r} = \begin{pmatrix} 0 \\ 5 \\ 4 \end{pmatrix}$

47. $x\overline{u} = \begin{pmatrix} 2 \\ 2 \\ 2 \end{pmatrix}$, $\overline{r} = \begin{pmatrix} -5 \\ 3 \\ 2 \end{pmatrix}$

49. $x\overline{u} = \begin{pmatrix} 32/35 \\ 96/35 \\ 160/35 \end{pmatrix}$, $\overline{r} = \begin{pmatrix} -137/35 \\ 79/35 \\ -20/35 \end{pmatrix}$

51. $x\overline{u} = \begin{pmatrix} 0 \\ -1.5 \\ -1.5 \end{pmatrix}$; $\overline{r} = \begin{pmatrix} 4 \\ -3.5 \\ 3.5 \end{pmatrix}$

53. $x\overline{u} = \begin{pmatrix} 69/15 \\ 23/15 \\ 46/15 \\ 23/15 \end{pmatrix}$; $\overline{r} = \begin{pmatrix} -24/15 \\ -8/15 \\ 29/15 \\ 22/23 \end{pmatrix}$

55. $x\overline{u} = \begin{pmatrix} 1.5 \\ .5 \end{pmatrix}, \overline{r} = \begin{pmatrix} -.5 \\ 1.5 \end{pmatrix}$

57. $x\overline{u} = \begin{pmatrix} -26/29 \\ 65/29 \end{pmatrix}, \overline{r} = \begin{pmatrix} 55/29 \\ 22/29 \end{pmatrix}$

59. For example, if $\overline{v} = \begin{pmatrix} 2 \\ 1 \end{pmatrix}, \overline{u} = \begin{pmatrix} 2 \\ 2 \end{pmatrix}$,

then $x\overline{u} = \begin{pmatrix} 1.5 \\ 1.5 \end{pmatrix}$, and $\overline{r} = \begin{pmatrix} .5 \\ -.5 \end{pmatrix}$

61. For example, if $\overline{v} = \begin{pmatrix} 6 \\ 8 \\ 1 \end{pmatrix}, \overline{u} = \begin{pmatrix} 2 \\ 2 \\ 5 \end{pmatrix}$,

then $x\overline{u} = \begin{pmatrix} 2 \\ 2 \\ 5 \end{pmatrix}$, and $\overline{r} = \begin{pmatrix} 4 \\ 6 \\ -4 \end{pmatrix}$

63. $\overline{r} = \begin{pmatrix} 3 \\ 5 \end{pmatrix}, ||\overline{r}|| = \sqrt{34} \approx 5.831$

65. $\overline{r} = \begin{pmatrix} 0 \\ 2 \end{pmatrix}, ||\overline{r}|| = 2$

67. $\overline{r} = \begin{pmatrix} -3 \\ -1 \end{pmatrix}, ||\overline{r}|| = \sqrt{10} \approx 3.162$

69. Neither $x = 0.5$ nor $x = 2$ can provide the projection $x\overline{u}$ of $\overline{v}$ onto $\overline{u}$, because the smallest vector $\overline{r}$ is needed, which is produced by $x = 1$.

71. $\overline{r} = \begin{pmatrix} 0.6 \\ -0.2 \end{pmatrix}$

73. $\overline{v} + 2\overline{u} = \begin{pmatrix} 5 \\ 13 \end{pmatrix};$

$||\overline{v} + 2\overline{u}|| = \sqrt{194} \approx 13.928$

75. $\overline{v} - 4\overline{u} = \begin{pmatrix} -7 \\ -23 \end{pmatrix};$

$||\overline{v} - 4\overline{u}|| = \sqrt{578} \approx 24.042$

77. $x\overline{u} = \begin{pmatrix} 0.9 \\ 1.2 \\ 1.5 \end{pmatrix}$

79. $||\overline{r}|| = \sqrt{1.5} \approx 1.225.$

81. $\overline{v} - \overline{u} = \begin{pmatrix} -1 \\ -3 \\ -4 \end{pmatrix};$

$||\overline{v} - \overline{u}|| = \sqrt{26} \approx 5.099$

83. The goal is to find the shortest residual vector, but $x = -2$, $x = 4$, and $x = 1$ all provide the projection $x\overline{u}$ of $\overline{v}$ onto $\overline{u}$ with residual lengths longer than $\sqrt{1.5}$.

85. $\overline{r} = \begin{pmatrix} -.4 \\ 3.2 \\ 10.8 \\ 3.4 \end{pmatrix}$

87. $\overline{v} + 2\overline{u} = \begin{pmatrix} -6 \\ 6 \\ 8 \\ 9 \end{pmatrix};$

$||\overline{v} + 2\overline{u}|| = \sqrt{217} \approx 14.731$

89. $\overline{v} - 4\overline{u} = \begin{pmatrix} 6 \\ 0 \\ 14 \\ -3 \end{pmatrix};$

$||\overline{v} - 4\overline{u}|| = \sqrt{211} \approx 14.526$

91. $y = 3x; \dfrac{\overline{y} \cdot \overline{x}}{\overline{x} \cdot \overline{x}} = 3$

93. $y = 5x; \dfrac{\overline{y} \cdot \overline{x}}{\overline{x} \cdot \overline{x}} = 5$

95. $y = 0.5x; \dfrac{\overline{y} \cdot \overline{x}}{\overline{x} \cdot \overline{x}} = \dfrac{1}{2}$

97.
```
u = c(7,6)
v = c(1,-4)
dot(u,v)
[1] -17
```

99.
```
u = c(2,3,4)
v = c(-2,2,-.5)
dot(u,v)
[1] 0
```

101.
```
u = c(1,2,3,4)
v = c(-2,0,1,5)
dot(u,v)
[1] 21
```

103.
```
u = c(3,4)
v = c(-2,5)
x = project(v~u)
x*u
[1] 1.68 2.24
v-x*u
[1] -3.68  2.76
```
$$x\overline{u} = \begin{pmatrix} 1.68 \\ 2.24 \end{pmatrix}, \ \overline{r} = \begin{pmatrix} -3.68 \\ 2.76 \end{pmatrix}$$

105.
```
u = c(0,3,-5)
v = c(12,1,3)
x = project(v~u)
x*u
[1]  0.000 -1.059  1.765
v-x*u
[1] 12.000  2.059  1.235
```
$$x\overline{u} \approx \begin{pmatrix} 0 \\ -1.059 \\ 1.765 \end{pmatrix}, \ \overline{r} \approx \begin{pmatrix} 12 \\ 2.059 \\ 1.235 \end{pmatrix}$$

107.
```
u = c(-1,5,-2,3)
v = c(5,1,1,-3)
x = project(v~u)
x*u
[1]  0.2820513 -1.4102564
0.5641026 -0.8461538
v-x*u
[1]  4.7179487  2.4102564
0.4358974 -2.1538462
```
$$x\overline{u} \approx \begin{pmatrix} 0.282 \\ -1.410 \\ 0.564 \\ -0.846 \end{pmatrix}, \ \overline{r} \approx \begin{pmatrix} 4.718 \\ 2.410 \\ 0.436 \\ -2.154 \end{pmatrix}$$

109. The dot product of two vectors $\overline{u}$ and $\overline{v}$ is $\overline{u} \cdot \overline{v} = u_1 v_1 + u_2 v_2 + \cdots + u_n v_n$.

111. Two vectors $\overline{u}$ and $\overline{v}$ are orthogonal when $\overline{u} \cdot \overline{v} = 0$.

113. The projection $x\overline{u}$ is the vector closest to $\overline{v}$ and occurs when $\overline{r}$ is orthogonal to $x\overline{u}$.

115. The key property of residual vectors in vector projection is that $\overline{r}$ is orthogonal to $x\overline{u}$.

Section 3.5 Exercises

1. $y = 0.357x + 0.2857$

3. $y = 0.4x + 0.9$

5. $y = 4.3x - 0.7$

7. $y = -2.11x - 2.03$

9. $y = 2.901x + 2.355$

11. $y = 2.757x - 0.167$

13. $y = 6.188x + 3.354$

15. $L = 1.77Y - 3403.57$

17. $S = 0.062Y - 120.931$

19. $S = 23.8Y - 47693.9$

21. $R = -0.0166Y + 1.378$

23. Answers will vary.

25. Answers will vary.

27. Answers will vary.

29. $\|\overline{r}\| = 6.873$

31. The length of the residual for the method of least squares is the shorter of the two residuals.

33. $\|\overline{r}\| = 3.54$

35. $\|\overline{r}\| = 0.0355$

37. The length of the residual for the method of least squares is the shorter of the two residuals.

39. $\|\overline{r}\| = 10.68$

41. $\|\overline{r}\| = 1.342$

43. The length of the residual for the method of least squares is the shorter of the two residuals.

45. $\|\overline{r}\| = 0.03496$

47.
```
x=c(2,5,7,8)
y=c(1,2,3,3)
project(y~x+1)
```

49.
```
x=c(0,1,2,3)
y=c(1,1,2,2)
project(y~x+1)
```

51.
```
x=c(2,3,5,6)
y=c(3,2,1,0)
project(y~x+1)
```

53.
```
x=c(-3,-2,-1,0,1)
y=c(4.2,2.4,-0.2,-1.7,-4.3)
project(y~x+1)
```

55.
```
x=c(1,2,3,4,5)
y=c(5.42,7.51,11.48,14.40,16.48)
project(y~x+1)
```

57.
```
x=c(-5,-4,-3,-2,-1)
y=c(-14.16,-10.88,-8.37,
     -5.93,-2.85)
project(y~x+1)
```

59. x=c(-5,-4,-3,-2,-1,0,1,2,3,4)
    y=c(-28.0,-20.7,-16.2,-8.4,
        -3.4,3.5,
    10.3,16.5,21.7,27.3)
    project(y~x+1)

61. Y=c(1990,2000,2005,2010)
    L=c(120,142,149,155)
    project(L~Y+1)

63. Y=c(2006,2007,2009,2010)
    S=c(3.42,3.53,3.63,3.68)
    project(S~Y+1)

65. Y=c(2009,2010,2011,2012)
    S=c(121,143,168,192)
    project(S~Y+1)

67. Y=c(0,5,10,15,20)
    P=c(1.41,1.26,1.20,1.13,1.06)
    project(Y~R+1)

69. project(Rate~Months+1,
    data=MonthlyUnemployment)
    Rate $= -0.0705$Months $+ 10.178$

71. -0.0705*36+10.178
    There is an unemployment rate of 7.64% in month 36.

73. project(People~Year+1,
    data=WorldPopulation)
    People $= 0.0749$Year $- 143.694$

75. 0.0749*1998-143.694
    The total midyear population for the world in 1988 is 5.9562 billion people.

77. project(Rate~Year+1,
    data=Mortgage15YrAnnual)
    Rate $= -0.216$Year $+ 437.526$

79. -0.216*2007+437.526
    The 15-year fixed-rate conventional mortgage was 4.014% in 2007.

81. project(Users~Months+1,
    data=FacebookUsers)
    Users $= 14.78$Months $- 424.72$

83. 14.78*30-424.72
    The number of Facebook users in month 30 is 18.68 million.

85. project(Rate~Year+1,
    data=HSDropoutRate)
    Rate $= -0.176$Year $+ 361.29$

87. -0.176*2005+361.29
    The high school dropout rate in 2005 is 8.41%.

89. project(kT~Year+1,
    data=USCO2Emissions)
    $kT = 57391.24$Year $- 109304814.11$

91. 57391.24*2000-109304814.11
    The U.S. carbon dioxide emissions in 2000 were 5,477,666 kT.

93. project(Rate~Year+1,
    data=Mortgage30YrAnnual.csv)
    Rate $= -0.318$year $+ 643.17$

95. -0.318*2000+643.17
    The convential 30-yr mortgage rate in 2000 was 7.17%.

97. Answers will vary

99. The best linear model $y = mx + b$ for a data set with n data points stored in an input vector $\overline{x}$ and an output vector $\overline{y}$ is given by the solution of the following matrix equation:

$$\begin{pmatrix} \overline{y} \cdot \overline{x} \\ \overline{y} \cdot \overline{1}_n \end{pmatrix} = \begin{pmatrix} \overline{x} \cdot \overline{x} & \overline{1}_n \cdot \overline{x} \\ \overline{x} \cdot \overline{1}_n & \overline{1}_n \cdot \overline{1}_n \end{pmatrix} \begin{pmatrix} m \\ b \end{pmatrix}$$

101. The linear combination $m\overline{x} + b\overline{1}_n$ closest to the target vector $\overline{y}$ is the one whose residual $\overline{r}$ is perpendicular to every linear combination of $\overline{x}$ and $\overline{1}_n$.

Section 4.1 Exercises

1. 1

3. -2

5. 5

7. 2

9. 4.5

11. 2

13. m

15. $a + b$

17. 0.15 billion dollars per year

19. 1.55 billion dollars per year

21. 2.25 billion dollars per year

23. -7.8 ng/mL per day

25. -4.08 ng/mL per day

27. -1.275 ng/mL per day

29. 3.8 million users per month

31. 25 million users per month

33. 22.73 million users per month

35. $(2-3)/(1-0) = -1$

37. $(6-2.25)/(2-0.5) = 2.5$

39. $(3-4)/(4-1) = -1/3$

41. $(3-4)/(2-1) = -1$

43. $(2-3)/(3-2) = -1$

45. f=makeFun(x+5~x)
    A=makeFun((f(b)-f(a))/(b-a)~a&b)
    A(1,3)

47. f=makeFun(3-2*x~x)
    A=makeFun((f(b)-f(a))/(b-a)~a&b)
    A(-4,-2)

49. f=makeFun(x^2+4~x)
    A=makeFun((f(b)-f(a))/(b-a)~a&b)
    A(0,5)

51. f=makeFun(x^2+4~x)
    A=makeFun((f(b)-f(a))/(b-a)~a&b)
    A(-1,3)

53. f=makeFun(-3*x^2+6*x~x)
    A=makeFun((f(b)-f(a))/(b-a)~a&b)
    A(-1,1)

55. f=makeFun(x^3-2*x+1~x)
    A=makeFun((f(b)-f(a))/(b-a)~a&b)
    A(0,2)

57. f=makeFun(exp(x)~x)
    A=makeFun((f(b)-f(a))/(b-a)~a&b)
    A(0,4)

59. f=makeFun(3*exp(-x)~x)
    A=makeFun((f(b)-f(a))/(b-a)~a&b)
    A(-1,1.5)

61. f=makeFun(exp(x^2)~x)
    A=makeFun((f(b)-f(a))/(b-a)~a&b)
    A(0,0.1)

63. f=makeFun(sin(x)~x)
    A=makeFun((f(b)-f(a))/(b-a)~a&b)
    A(0,0.5)

65. Answers will vary.

67. Answers will vary.

69. (a) 6.1; (b) 6.01; (c) 6.001; (d) $f'(3) \approx$ 6; (e) $y = 6x - 9$

71. (a) -3.9; (b) -3.99; (c) -3.999; (d) $f'(-2) \approx -4$; (e) $y = -4x - 4$

73. (a) $2c+0.01$; (b) $2c+0.001$; (c) $2c-0.01$; (d) $2c - 0.001$; (e) $f'(c) = 2c$

75. (a) 0.29978; (b) 0.29978; (c) 0.29978; (d) 0.29978; (e) $y = 0.29978x - 12.63416$

77. (a) 284.74; (b) 251.518; (c) 251.5163; (d) 251.5165; (e) $y - 2068.392 = 251.5165(x - 33)$

79. (a) -2.20504; (b) -2.209468; (c) -2.20967; (d) -2.209962; (e) $y - 1952.136 = -2.20967(x - 300)$

81.

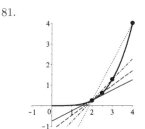

83.

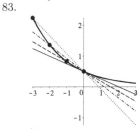

85. $f'(0) = 0$

87. $f'(-2) = -1$

89. $f'(5) = 2.5$

91. $f'(6) = \dfrac{1}{4}$

93. $\dfrac{f(b) - f(a)}{b - a}$; change of the output over change of the input; slope of the secant line joining the points $(a, f(a))$ and $(b, f(b))$.

95. The instantaneous rate of change of a function $f(x)$ when $x = a$ is the number approached by the average rates of change of $f(x)$ on smaller and smaller intervals containing $x = a$.

97. A secant line connects two points on the graph of a function; a tangent line $y = f(a) + f'(a)(x - a)$ locally touches the graph of a function $f(x)$ only at the point $(a, f(a))$.

99. Compute the slope of the tangent line at $x = a$ by finding the value approached by the average rates of change computed on smaller and smaller intervals containing $x = a$.

Section 4.2 Exercises

1. $f'(0) = f'(2) = -1/2$

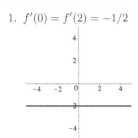

3. $f'(-2) = -2$; $f'(0) = 0$; $f'(2) = 2$

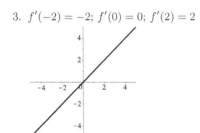

5.

Y	0	1	4	5
P'	-2	1.4	-2	-0.95

7.

Y	0	10	20	30
V'	9.7	35.3	64.6	88.6

9. (b)

11. (a)

13. Units of input: gigahertz; $[f] = T^{-1}$; units of output: gigahertz; $[c(f)] = T^{-1}$; units of derivative: gigahertz per second; $[c'(f)] = T^{-2}$

15. Units of input: units manufactured; $[u] = N$; units of output: dollars; $[p(u)] = B$; units of derivative: dollars per unit manufactured; $[p'(u)] = BN^{-1}$

17. Units of input: day; $[l] = T$; units of output: degress; $[T(l)] = \Theta$; units of derivative: degrees per day; $[T'(l)] = \Theta T^{-1}$

19. The concentration of salt disolved is 600 grams per liter when the bucket has 9 liters of water.

21. The concentration of salt is increasing at a rate of 50 grams per liter per liter when the bucket has 5 liters of water.

23. The concentration of the medicine in the bloodstream is 20 nanograms per mililiter 0 hours after the medicine was administered.

25. The concentration of the medicine in the bloodstream is 10 nanograms per mililiter 5 hours after the medicine was administered.

27. The concentration of the medicine in the bloodstream is decreasing at a rate of 3 nanograms per mililiter per hour 4 hours after the medicine was administered.

29. f is always increasing, and $f' > 0$ for all x.

31. f is increasing on the interval $(-\infty, 0)$ and decreasing on the interval $(0, \infty)$ $f' > 0$ on $(-\infty, 0)$, $f' < 0$ on $(0, \infty)$, and $f'(0) = 0$.

33. f is increasing on the interval $(-\infty, -2)$ and decreasing on the interval $(-2, \infty)$ $f' > 0$ on $(-\infty, -2)$, $f' < 0$ on $(-2, \infty)$, and $f'(2) = 0$.

35. f is always decreasing, and f' is always negative.

37. f is always increasing, and f' is always positive.

39.

Interval	f	f'
(0,2)	increasing	positive
(2,4)	decreasing	negative
(4,6)	increasing	positive

41.

Interval	f	f'
$(-\infty, -5)$	increasing	positive
$(-5, -2)$	decreasing	negative
$(-2, 0)$	increasing	positive
$(0, 3)$	decreasing	negative
$(3, \infty)$	increasing	positive

43.

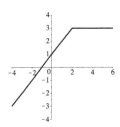

45.

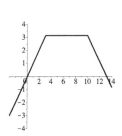

47.

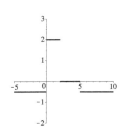

49.

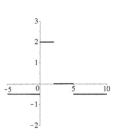

51.

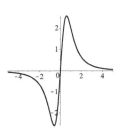

53.

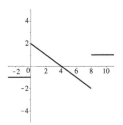

55. Differentiable everywhere

57. $x = 0$, $x = 8$

59. $x = -5$, $x = -2$, $x = 0$

61. Increasing on $(-\infty, -5)$, $(-1, 2)$;
    decreasing on $(-5, -1)$, $(2, \infty)$

63. Increasing on $(-5, -1)$, $(2, \infty)$;
    decreasing on $(-\infty, -5)$, $(-1, 2)$

65. Increasing on $(-1.5, 1.5)$;
    decreasing on $(\infty, -1.5)$, $(1.5, \infty)$

67. Increasing on $(-4, 0)$, $(0, 3)$;
    decreasing on $(-\infty, -4)$, $(3, \infty)$

69. Increasing on $(-7, -2)$, $(0, 3)$;
    decreasing on $(-\infty, -7)$, $(-2, 0)$,
    $(3, \infty)$

71. Increasing on $(-\infty, -4)$, $(1, 3)$;
    decreasing on $(-4, 1)$, $(3, \infty)$

73. Increasing on $(1, 3)$, $(3, 5)$, $(9, \infty)$;
    decreasing on $(-\infty, 1)$, $(5, 9)$

75.

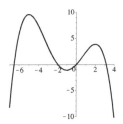

77.

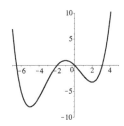

79.

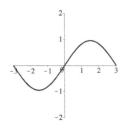

81.

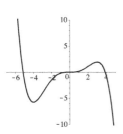

83.

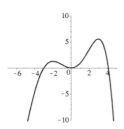

85.

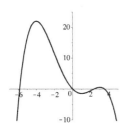

87.

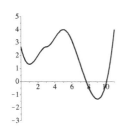

89.

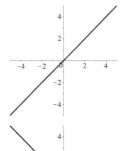

91.

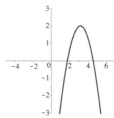

93.

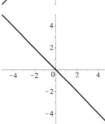

95.

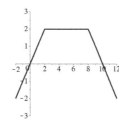

97. $g'(x) = f(x)$

99. $f'(x) = g(x)$

101. The derivative function of $f(x)$ equals the instantaneous rate of change of $f(x)$ at each input x. A function is called differentiable if and only its derivative exists.

103. Units of $f(x)$ divided by units of x.

105. $f(x)$ is increasing on (a, b) if and only if for all numbers c and d in (a, b), when $c < d$ then $f(c) > f(d)$; $f(x)$ is constant on (a, b) if and only if for all numbers c and d in (a, b), $f(c) = f(d)$; $f(x)$ is decreasing on (a, b) if and only if for all numbers c and d in (a, b), when $c < d$ then $f(c) < f(d)$.

Section 4.3 Exercises

1. 0

3. 0

5. 7

7. e

9. $8x^7$

11. $-6x^{-7}$

13. $\frac{1}{4}x^{-3/4}$

15. $\frac{3}{2}x^{1/2}$

17. $\frac{7}{3}x^{4/3}$

19. $\frac{7}{2}x^{5/2}$

21. $\frac{9}{2}x^{7/2}$

23. $\frac{11}{2}x^{9/2}$

25. $4e^{4x+8}$

27. $\pi(e^{\pi x+7})$

29. $2e^{2x+6}$

31. $-\pi(e^{-\pi x+8})$

33. $0 + 4x^3 - e^x$

35. $\frac{7}{4}x^{3/4} + 0 - 0 - 12x^3$

37. $e + 0 + \frac{14}{5}x^{-19/5}$

39. $3e^x - 0 - 2\cos(x)$

41. $\frac{1}{3}x^{-2/3} - \frac{10}{x} + 0$

43. $\frac{5}{2}x^{3/2} - \frac{\pi}{x} - 22e^{-x+13}$

45. $-5x^{-6} + \frac{3}{5}x^{-8/5} - \frac{6}{x}$

47. $\frac{27}{7}x^{20/7} - 3\sin(x) - 3e^x$

49. $\frac{4}{3x} - 0 + \ln(7)7^x$

51. $-\frac{1}{x} - \frac{1}{2x}$

53. (a) $HE(1995) = 12.887$

    (b) $HE'(1995) = 0.2998$

55. (a) $HE(2005) = 15.885$

    (b) $HE'(2005) = 0.2998$

57. (a) $RPM(0.4) = 18490.31$

    (b) $RPM'(0.4) = -15699.29$

59. (a) $RPM(200) = 2240.344$

    (b) $RPM'(200) = -3.80435$

61. (a) $Wt(20) = 425.6966$

    (b) $Wt'(20) = 51.76471$

63. (a) $Wt(30) = 1436.158$

    (b) $Wt'(30) = 174.6368$

65. $L(x) = 2 + 3(x - 2)$

67. $L(x) = 8 - 4(x + 1)$

69. $L(x) = e^2 + 2e^2(x - 1.5)$

71. $L(x) = 0 + 1(x - 0)$

73. $L(\text{Wk}) = 4.1154 + 0.0632(\text{Wk} - 35)$

75. $L(\text{Year}) = 376.9933 + 1.542952(\text{Year} - 2005)$

77. $f(x) = -x^3;\ f'(x) = -3x^2$

    $f(x)$ decreasing on $(-\infty, \infty)$

    $f'(x)$ negative on $(-\infty, \infty)$

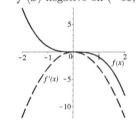

79. $f(x) = \ln(x);\ f'(x) = \frac{1}{x}$

    $f(x)$ increasing on $(0, \infty)$

    $f'(x)$ positive on $(0, \infty)$

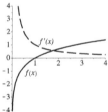

81. `> D(x^3-4*x^2+5*x-6~x)`

    $f'(x) = 3x^2 - 8x + 5$

83. `> D(3*2^(x+8)+exp(x)~x)`

    $f'(x) = 3\ln(2)2^{x+8} + e^x$

85. `> D((x^2+2*x)/(x^(1/2))`
    `+2^(3*(x-45)+13)~x)`

    $f'(x) = \dfrac{\sqrt{x}(2x+2) - (x^2+2x)\frac{1}{2}x^{-1/2}}{(\sqrt{x})^2}$

    $+ 3\ln(2)2^{3(x-45)+13}$

87. `> D(log((5*x)^4+9)`
    `-log((x)^(1/2))~x)`

    $f'(x) = \dfrac{20(5x)^3}{(5x)^4 + 9} - \dfrac{\frac{1}{2}(x)^{-1/2}}{\sqrt{x}}$

89.
```
> f=makeFun(x^2-3*x+4~x)
> f(2)
df=D(f(x)~x)
df(2)
```
$L(x) = 2 + 1(x - 2)$

91.
```
> f=makeFun(2^(3*x)-3^(4*x)~x)
> f(-1)
df=D(f(x)~x)
df(-1)
```
$L(x) = 0.1126543 + 0.2056777(x + 1)$

93.
```
> f=makeFun(5*cos(x+3)~x)
> f(1)
df=D(f(x)~x)
df(1)
```
$L(x) = -3.268218 + 3.784012(x - 1)$

95. $\dfrac{d}{dx}[mx + b] = m$

97. $\dfrac{d}{dx}[kf(x) \pm g(x)] = kf'(x) \pm g'(x)$

99. $\dfrac{d}{dx}[\sin(x)] = \cos(x)$

$\dfrac{d}{dx}[\cos(x)] = -\sin(x)$

Section 4.4 Exercises

1. $(45x^8) \cdot (x^3 + x^2) + (5x^9) \cdot (3x^2 + 2x)$

3. $(5x^4 - 1)\cos(x) + (x^5 - x)(-\sin(x))$

5. $(2x - 3x^{-4})e^x + (x^2 + x^{-3})e^x$

7. $3^x \ln(3)\cos(x) - 3^x \sin(x)$

9. $48x^5 \ln(x) + 8x^6 \cdot \dfrac{1}{x}$

11. $\dfrac{9x^8 \cdot (x^3 + 2) - (x^9 + 2) \cdot (3x^2)}{(x^3 + 2)^2}$

13. $(\cos(x) \cdot (x + \cos(x))$
$\qquad - \sin(x) \cdot (1 - \sin(x))) /$
$(x + \cos(x))^2$

15. $(-\sin(x) \cdot (3x^2 + 5x)$
$\qquad - \cos(x) \cdot (6x + 5)) /$
$(3x^2 + 5x)^2$

17. $\dfrac{\dfrac{1}{x} \cdot (x^4 + x^2) - \ln(x) \cdot (4x^3 + 2x)}{(x^4 + x^2)^2}$

19. $((2x - e^x)(2x^2 + \cos(x))$
$\qquad -(x^2 - e^x)(4x - \sin(x))) /$
$(2x^2 + \cos(x))^2$

21. $2x + \cos(x)$

23. $-\sin(x) - 28x^3$

25. $\cos(x) - 2e^{2x}$

27. $\dfrac{1}{x}\sin(x) + \ln(x)\cos(x) + 5x^4$

29. $(1 + \cos(x))\cos(x) +$
$\qquad (x + \sin(x))(-\sin(x))$

31. $e^x \sin(x) + e^x \cos(x)$

33. $6e^x - e6^x \ln(6)$

35. $7^x \ln(7) + \dfrac{x \cdot \dfrac{1}{x} - \ln(x) \cdot 1}{x^2}$

37. $\dfrac{(1 + \cos(x))\ln(x) - (x + \sin(x)) \cdot \dfrac{1}{x}}{(\ln(x))^2}$

39. $((6x^8 + 2)e^x \cdot \cos(x) -$
$\qquad \sin(x) \cdot ((48x^7)(e^x) + (6x^8 + 2)e^x)) /$
$((6x^8 + 2)e^x)^2$

41. $((e^x \cos(x) - e^x \sin(x))(x + 2) -$
$\qquad (e^x \cos(x))(1)) / ((x + 2)^2)$

43. $((4x^3 - 2^x \ln(2))(7x \sin(x)) -$
$\qquad (x^4 - 2^x)(7 \sin(x) + 7x \cos(x))) /$
$(7x \sin(x))^2$

45. (a) The displacement at 7 minutes is $35 \sin(7) \approx 22.99$ millimeters. (b) The displacement is changing at a rate of 29.67 millimeters per minute at 7 minutes.

47. (a) The amplitude of the wave is 56.12 decibels when measured 150 meters from its source. (b) The amplitude is changing at a rate of 3.78 decibels per meter when measured 150 meters from its source.

49. (a) The number of Twitter users worldwide in month 12 is 133.6744 million. (b) The number of Twitter users is increasing at a rate of 71.5 million users per month.

51. (a) The population of Belgium in 1800 is 3.65 million people. (b) The population of Belgium is increasing at a rate of 0.0095 million people per year.

53. $\dfrac{f'(x)}{g'(x)} = \dfrac{2}{3x}$; $\dfrac{d}{dx}\left[\dfrac{f(x)}{g(x)}\right] = -\dfrac{1}{x^2}$

55. $\dfrac{d}{dx}[\cot(x)] = \dfrac{d}{dx}\left[\dfrac{\cos(x)}{\sin(x)}\right]$

$\qquad = \dfrac{-\sin(x) \cdot \sin(x) - \cos(x) \cdot \cos(x)}{(\sin(x))^2}$

57. $\sec^2(x) - \csc(x)\cot(x)$

59. $2x\tan(x) + x^2\sec^2(x)$

61. $\dfrac{1}{x}\cdot\cot(x) + \ln(x)\cdot(-\csc^2(x))$

63. $5\tan(x) + (5x+7)\sec^2(x)$

65. $\sec(x)\tan(x) + \dfrac{1}{x}$

67. $-\csc^2(x) + 2^x\ln(2)$

69. $-\csc(x)\cot(x) + 7x^6$

71. $h'(4) = 33$

73. $h'(1) = 2$

75. $h'(2) = 66$

77. $h'(4) = -31/16$

79. 6

81. $5 + e$

83. 5

85. $5/2$

87. $y = e + 3e(x-1)$

89. $y = \pi/4 + (1 + \pi/8)(x - \pi/4)$

91. $y = -1 + 1\cdot(x - 0)$

93. $y = e^2/7 + 5e^2/49\cdot(x-2)$

95. D(sin(x)-exp(2*x)~x)

97. D(log(x)*sin(x)+x^5~x)

99. D((x+sin(x))*cos(x)~x)

101. D(exp(x)*sin(x)~x)

103. D(6*exp(x)-exp(1)*6^x~x)

105. D(7^x+log(x)/x~x)

107. D((x+sin(x))/log(x)~x)

109. D(sin(x)/((6*x^8+2)*exp(x))~x)

111. D(70*sin(0.09*M)*sin(0.01*M)~M)

113. D(297.202473/(1+0.95939*
     exp(-0.97214*(Q-12.25)))~Q)

115. $\dfrac{d}{dx}[f(x)g(x)] = g(x)\cdot f'(x) +$
     $\qquad f(x)\cdot g'(x)$

117. For all trigonometric functions and for
     all real numbers except $n = -1$, the
     notation $\text{trig}^n(x)$ denotes $[\text{trig}(x)]^n$

119. First, compute the derivative with the
     standard differentiation rules and then
     substitute known functional values to
     find the instantaneous rate of change
     for the given inputs.

Section 4.5 Exercises

1. $f \circ g(x) = (2x)^2 + 5$;
   $g \circ f(x) = 2(x^2 + 5)$

3. $f \circ g(x) = 4(e^x) - 3$;
   $g \circ f(x) = e^{4x-3}$

5. $f \circ g(x) = \cos(\pi x)$;
   $g \circ f(x) = \pi\cos(x)$

7. $g(x) = x^3 + x;\ f(x) = x^5$

9. $g(x) = (x^6 - x)^2;\ f(x) = \sqrt[3]{x}$

11. $g(x) = e^x + 2x;\ f(x) = \cos(x)$

13. $g(x) = \sin(x);\ f(x) = 17e^x$

15. $5(x^3 + 3x)^4 \cdot (3x^2 + 3)$

17. $(-7)(x - e^x)^{-8} \cdot (1 - e^x)$

19. $\cos(2x^4 + 3) \cdot (8x^3 + 0)$

21. $\cos(x^2 - e^x) \cdot (2x - e^x)$

23. $-\sin(6 - \pi x) \cdot (-\pi)$

25. $1/(\pi x + 4) \cdot \pi$

27. $1/(-4x - 8) \cdot (-4)$

29. $1/(5x^2 + 32) \cdot 10x$

31. $17e^{\sin(x)} \cdot \cos(x)$

33. $9^{x^2+x} \cdot \ln(9) \cdot (2x + 1)$

35. $16 \cdot 7^{3x^2+9} \cdot \ln(7) \cdot (6x + 0)$

37. $2x + \cos(x^3 + 6) \cdot (3x^2)$

39. $-\sin(e^x) \cdot e^x - 28x^3$

41. $\cos(3x - e^{2x}) \cdot (3 - 2e^{2x})$

43. $1/(\sin(x) + x^5) \cdot [\cos(x) + 5x^4]$

45. $\cos[\cos(x) + x] \cdot [-\sin(x) + 1]$

47. $\sin(x^4)e^{x^2+9x}(2x + 9) +$
    $\qquad e^{x^2+9x}(4x^3)\cos(x^4)$

49. $6e^{3x-\cos(x)}[3 + \sin(x)]$

51. $3^{\ln(x)+x^2} \cdot \ln(3) \cdot \left[\dfrac{1}{x} + 2x\right]$

53. $\Big(\ln(6x - 9)[1 + 2x\cos(x^2)]$
    $\qquad -[x + \sin(x^2)]\dfrac{6}{6x - 9}\Big)\Big/$
    $[\ln(6x - 9)]^2$

55. $((6x^8 + 2)e^x\cos(x^2)2x -$
    $\sin(x^2)[e^x(48x^7) + (6x^8 + 2)e^x])/$
    $[(6x^8 + 2)e^x]^2$

57. $\dfrac{(x+2)e^{\cos(x)}[-\sin(x)] - e^{\cos(x)} \cdot 1}{(x+2)^2}$

59. $((x^3 - \sqrt{x})e^{x^4+5x}(4x^3 + 5) -$
$e^{x^4+5x}(3x^2 - \frac{1}{2}x^{-1/2}))/$
$(x^3 - \sqrt{x})^2$

61. $\phi'(x) = \dfrac{1}{\sqrt{2\pi}}e^{-x^2/2} \cdot \dfrac{-2x}{2}$

63. $U'(Q) = (1+0.959e^{-0.972(Q-12.25)}(0) -$
$(297.202)(-0.932e^{-0.972(Q-12.25)}))/$
$(1+0.959e^{-0.972(Q-12.25)})^2$

65. $S'(M) = 164.7\sin\left[\dfrac{\pi}{6}(M - 2.6)\right] \cdot \dfrac{\pi}{6}$

67. (a) The average maximum temperature in month 28 is 75.8 degrees; (b) The average maximum temperature is increasing at a rate of 10.9 degrees per month in month 28

69. (a) The average maximum temperature in month 42 is 88.7 degrees; (b) The average maximum temperature is increasing at a rate of 0.69 degrees per month in month 42

71. $\dfrac{d}{dx}[(x-a)^n] = n(x-a)^{n-1} \cdot a$

73. $\dfrac{d}{dx}\left[e^{ax^2+bx}\right] = e^{ax^2+bx}\dfrac{d}{dx}[ax^2 + bx]$

75. $\dfrac{d}{dx}\left[\dfrac{f(x)}{g(x)}\right] = \dfrac{d}{dx}\left[f(x) \cdot (g(x))^{-1}\right]$
$= -(g(x))^{-2}g'(x)f(x) + f'(x) \cdot (g(x))^{-1}$

77. $\dfrac{d}{dx}[\cot(x)] = \dfrac{d}{dx}\left[\cos(x) \cdot (\sin(x))^{-1}\right]$
$= -(\sin(x))^{-2}\cos(x) \cdot \cos(x)$
$\quad +(-\sin(x))(\sin(x))^{-1}$

79. $\sec^2(5x^6) \cdot 30x^5$

81. $-\csc^2(\ln(x)) \cdot \dfrac{1}{x}$

83. $\sec^2(5x + 7) \cdot 5$

85. $\sec(\ln(x) + 1)\tan(\ln(x) + 1) \cdot \dfrac{1}{x}$

87. $-\csc^2(e^x + 2^x) \cdot (e^x + 2^x \ln(2))$

89. $\sec(\cot(x))\tan(\cot(x)) \cdot (-\csc^2(x))$

91. $-\csc(x^7 + 6x)\cot(x^7 + 6x) \cdot (7x^6 + 6)$

93. $h'(2) = 15$

95. $h'(1) = 37$

97. $h'(1) = -56$

99. 5

101. 5

103. 201

105. -50

107. $y = -1$

109. $y = (\ln(2e)+e)^3 + (3(\ln(2e)+e)^2(1/e + 1)) \cdot (x - e)$

111. `D(x^2+sin(x^3+6)~x)`

113. `D(sin(3*x-exp(2*x))~x)`

115. `D(sin(cos(x)+x)~x)`

117. `D(6*exp(3*x-\cos(x))~x)`

119. `D((x+sin(x^2))/(log(6*x-9))~x)`

121. `D(exp(cos(x))/(x+2)~x)`

123. If $f(x)$ and $g(x)$ are functions, then the composition of f with g is $f \circ g(x) = f[g(x)]$ when this expression is defined.

Section 4.6 Exercises

1. Decreases; $f_x < 0$

3. Decreases; $f_x < 0$

5. Increases; $f_y > 0$

7. Increases; $f_y > 0$

9. Increases; $f_x > 0$

11. Decreases; $f_x < 0$

13. Increases; $f_y > 0$

15. Increases; $f_y > 0$

17. 2

19. 2

21. 1.175

23. 1

25. 4

27. 0

29. $\dfrac{2}{e^2 - 1} \approx 0.313$

31. $\pi^2 \sin(1) \approx 8.305$

33. $f_x = 3$; $f_y = 0$

35. $f_x = 0$; $f_y = 3x^2 + \dfrac{\pi}{2}y^{-1/2}$

37. $f_x = 4x^3$; $f_y = 2y$

39. $f_x = 3/2x^{1/2} + 12x^3$

41. $f_x = -6\sin(6x+5)$; $f_y = 1/y$

43. $f_x = 3e^{3x+8} + y$; $f_y = x$

45. $f_x = 3x^2y^5 + 7$; $f_y = 5x^3y^4$

47. $f_x = 1/(x+y) + \ln(4) \cdot 5 \cdot 4^{5x-3y}$;
$f_y = 1/(x+y) + \ln(4) \cdot (-3) \cdot 4^{5x-3y}$

49. $f_x = \ln(2) \cdot y \cdot 2^{xy} + \dfrac{1}{x}\cos(y)$;
$f_y = \ln(2) \cdot x \cdot 2^{xy} + \ln(x)(-\sin(y))$

51. $f_x = 2xy^3 + 4\cos(4x+9y)$;
$f_y = 3x^2y^2 + 9\cos(4x+9y)$

53. $f_x = 3x^2$; $f_y = 3y^2$; $f_z = 4z^3$

55. $f_x = e^x$; $f_y = z\cos(yz)$;
$f_z = y\cos(yz)$

57. $f_x = y^2$; $f_y = 2xy + 6y^5z^4$;
$f_z = 4y^6z^3$

59. $f_x = \sin(z)$; $f_y = 2yz^3$;
$f_z = x\cos(z) + 3y^2z^2$

61. $f_x = y^2\sin(z)$; $f_y = 2xy\sin(z)$;
$f_z = xy^2\cos(z)$

63. $f_x = e^x + xe^x + y^2$; $f_y = 2xy$

65. $f_x = 3e^{xy} + (3x+7y)ye^{xy}$;
$f_y = 7e^{xy} + (3x+7y)xe^{xy}$

67. $f_x = y^2(-y\sin(xy))$;
$f_y = 2y\cos(xy) + y^2(-x\sin(xy))$

69. $f_x = ((x^4+y^2) \cdot 0 - y(4x^3))/(x^4 + y^2)^2$;
$f_y = ((x^4+y^2)(1) - y(2y))/(x^4+y^2)^2$

71. $f_x = ((\sin(y)+4x)(e^x+xe^x) - xe^x - 4)/(\sin(y)+4x)^2$;
$f_y = ((\sin(y)+4x) \cdot 0 - xe^x \cdot \cos(y))/(\sin(y)+4x)^2$

73. $f_x = (\tan(xy)\cdot 6 - (6x+8y)\sec^2(xy)\cdot y)/(\tan(xy))^2$;
$f_y = (\tan(xy)\cdot 8 - (6x+8y)\sec^2(xy)\cdot x)/(\tan(xy))^2$

75. $f_x = ze^y$; $f_y = xze^y + 2yz^3$;
$f_z = xe^y + 3y^2z^2$

77. $f_x = ((3x+5zz)\cdot 0 - 6y)/(3x+5z)^2$;
$f_y = ((3x+5z) \cdot 2 - (2y) \cdot 0)/(3x + 5z)^2$;
$f_z = ((3x+5z)\cdot 0 - 10y)/(3x+5z)^2$

79. $f_{xx} = f_{xy} = f_{yx} = f_{yy} = 0$

81. $f_{xx} = 2y$; $f_{xy} = 2x = f_{yx}$; $f_{yy} = 0$

83. $f_{xx} = 12x^2e^y$; $f_{xy} = 4x^3e^y = f_{yx}$;
$f_{yy} = x^4e^y$

85. $f_{xx} = -e^y\cos(x)$;
$f_{xy} = -e^y\sin(x) = f_{yx}$;
$f_{yy} = e^y\cos(x)$

87. $i_p = rt$; $i_p(1000, 0.04, 4) = 0.16$

89. $i_r = pt$; $i_r(1000, 0.04, 4) = 4000$

91. $i_t = pr$; $i_t(1000, 0.04, 4) = 40$

93. $p = (nrt)/v$; $p_n = (rt)/v$

95. $p = (nrt)/v$; $p_t = (nr)/v$

97. $v = (nrt)/p$; $v_p = nrtp^{-2}$

99. $n = (pv)/(rt)$; $n_p = v/(rt)$

101. $t = (pv)/(nr)$; $t_n = -(pv)/r \cdot n^{-2}$

103. r is a constant

105. $L(x, y) = 12 + 1(x-5) + 1(y-7)$

107. $L(x, y) = 5/2 - 5/4(x-2) + 1/2(y-5)$

109. $L(x, y) = (e + \cos(1)) + e(x-1) + (e - \sin(1))(y-1)$

111. D(y^3+pi*sqrt(y)~x);
D(y^3+pi*sqrt(y)~y)

113. D(x^(3/2)+sqrt(3)*x^4~x);
D(x^(3/2)+sqrt(3)*x^4~y)

115. `D(exp(3*x+8)+x*y~x);`
     `D(exp(3*x+8)+x*y~y)`

117. `D(\ln(x+y)+4^(5*x-3*y)~x);`
     `D(\ln(x+y)+4^(5*x-3*y)~y)`

119. `D(x^2*y^3+sin(4*x+9*y)~x);`
     `D(x^2*y^3+sin(4*x+9*y)~y)`

121. The partial derivative $f_x(x, y)$ of $f(x, y)$ with respect to x when $(x, y) = (a, b)$ is the number approached by the average rates of change with respect to x of $f(x, y)$ when these rates are computed over smaller and smaller x-intervals containing $x = a$ while $y = b$ is held constant. The partial derivative $f_y(x, y)$ of $f(x, y)$ with respect to y when $(x, y) = (a, b)$ is the number approached by the average rates of change with respect to y of $f(x, y)$ when these rates are computed over smaller and smaller y-intervals containing $y = b$ while $x = a$ is held constant.

123. The second-order partial derivatives are obtained by taking all partial derivatives of the multivariable functions f_x and f_y.

125. The results of taking higher-order derivatives with respect to different input variables.

127. The linear approximation of a smooth function $f(x, y)$ when $(x, y) = (a, b)$ is given by $L(x, y) = f(a, b) + f_x(a, b) \cdot (x - a) + f_y(a, b) \cdot (y - b)$.

Section 4.7 Exercises

1. 3

3. 3

5. 1

7. 1

9. 2

11. 4

13. 4

15. 0

17. $\lim\limits_{x \to 0} \dfrac{\cos(x) - 1}{x} = 0$

19. $\lim\limits_{x \to 1} \dfrac{x - 1}{\ln(x)} = 1$

21. $\lim\limits_{x \to 3} \dfrac{3 - x}{2^x - 8} = -\dfrac{1}{8 \ln(2)} \approx 0.18$

23. Continuous;
    $\lim\limits_{x \to 0^+} f(x) = \lim\limits_{x \to 0^-} f(x) = f(0) = 3$

25. Continuous;
    $\lim\limits_{x \to 2^+} f(x) = \lim\limits_{x \to 2^-} f(x) = f(2) = 3$

27. Not continuous; $\lim\limits_{x \to 4} f(x) = 1 \neq f(4)$

29. Not continuous; $\lim\limits_{x \to 0} g(x) = 2 \neq f(0)$

31. Not continuous; $f(2)$ not defined

33. Continuous;
    $\lim\limits_{x \to 4^+} g(x) = \lim\limits_{x \to 4^-} g(x) = 4 = g(4)$

35. $(-\infty, \infty)$

37. All real numbers except $x = k\dfrac{\pi}{2}$ where k is an integer

39. $(-\infty, \infty)$

41. $(-\infty, -1), (-1, 1), (1, \infty)$

43. All real numbers except $x = k\pi$, where k is an integer

45. $(-\infty, 3), (3, \infty)$

47. $(-\infty, 3)$

49. $(0, \infty)$

51. 2

53. 19

55. -3π

57. 80

59. 4

61. 4

63. 2

65. $-4/3$

67. 4

69. $1/2$

71. 7

73. $1/6$

75. 6

77. $1/2$

79. 1

81. $5/8$

83. $-1/4$

85. Does not exist

87. 0

89. -8

91. -27

93. Does not exist

95. -4

97. $\lim\limits_{h\to 0} \dfrac{f(3+h)-f(3)}{h} = 0$

99. $\lim\limits_{h\to 0} \dfrac{f(-2+h)-f(-2)}{h} = 1$

101. $\lim\limits_{h\to 0} \dfrac{f(5+h)-f(5)}{h} = 10$

103. $\lim\limits_{h\to 0} \dfrac{f(0+h)-f(0)}{h} = 1$

105. $\lim\limits_{h\to 0} \dfrac{f(4+h)-f(4)}{h} = \dfrac{1}{4}$

107. $\lim\limits_{h\to 0} \dfrac{f(x+h)-f(x)}{h} = 0$

109. $\lim\limits_{h\to 0} \dfrac{f(x+h)-f(x)}{h} = 2$

111. $\lim\limits_{h\to 0} \dfrac{f(x+h)-f(x)}{h} = 2x$

113. $\lim\limits_{h\to 0} \dfrac{f(x+h)-f(x)}{h} = -4x$

115. $\lim\limits_{h\to 0} \dfrac{f(x+h)-f(x)}{h} = \dfrac{1}{2}\dfrac{1}{\sqrt{x-1}}$

117. A function is continuous for an input $x = a$ if and only if $\lim\limits_{x\to a} f(x) = f(a)$. A function $f(x)$ is continuous on its domain exactly when $f(x)$ is continuous for every input in its domain.

119. Linear, exponential, power, trigonometric, and simoidal functions are continuous on their domains. A function obtained as a sum, difference, product, quotient, and composition of the common modeling functions is continuous on its domain.

121. $f'(a) = \lim\limits_{h\to 0} \dfrac{f(a+h)-f(a)}{h}$

Section 5.1 Exercises

1. Global maximum of 1 when $x = 0$; global minimum of 0.05 when $x = \pm 5$

3. No global maximum; global minimum of 3 when $x = \pm 2$

5. Global maximum of 0 when $2 \le x \le 5$; global minimum of -4 when $x = 0$

7. Global maximum of 5 when $x = 2.5$; global minimum of -2.75 when $x = 1.5$

9. Global maximum of -0.5 when $x = 2$; global minimum of -2.75 when $x = 1.5$

11. Global maximum of 3.5 when $x = 2.75$; global minimum of -0.5 when $x = 0.5$

13. Global maximum of 2.5 when $x = 2$; global minimum of -0.5 when $x = 0.5$

15.

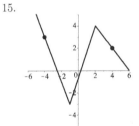

17.

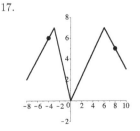

19.

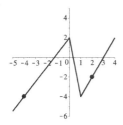

21. Global maximum of 7.7 when $x = 40.4$; global minimum of -2.5 when $x = 1.1$

23. Global maximum of 14 when $x = 15$; global minimum of -7 when $x = 24$

25. Global maximum of 91.1 when $x = 24$; global minimum of -40.4 when $x = 30$

27. Global maximum of 9.1 when $x = 1.1$; global minimum of 3.1 when $x = 4.5$

29. Applies because it is a smooth function on a closed interval.

31. Applies because it is a smooth function on a closed interval.

33. Applies because it is a smooth function on a closed interval.

35. Does not apply, because $\ln(2x + 1)$ is not defined on $[-1, -1/2]$.

37. $x \approx \pm 4.7$, $x \approx \pm 1.5$

39. $x = 0$, $x = 2$, $x = 5$

41. $x = -3, 0, 3$

43. $x = -1, 1, 3, 7$

45. $x = 1$

47. No critical numbers

49. No critical numbers

51. Global minimum of 2 when $x = 3$; global maximum of 12 when $x = 8$

53. Global minimum of -45 when $x = 8$; global maximum of -15 when $x = 3$

55. Global minimum of -27 when $x = 3$; global maximum of 0 when $x = 0$

57. Global minimum of -19 when $x = 3$; global maximum of -3 when $x = 1$

59. Global minimum of -392 when $x = 7$; global maximum of 108 when $x = -3$

61. Global minimum of -7 when $x = 1$; global maximum of 4 when $x = 2$

63. Global minimum of -16 when $x = 4$; global maximum of 11 when $x = 1$

65. Global minimum of -41 when $x = \pm 3$; global maximum of 40 when $x = 0$

67. Global minimum of -1.6875 when $x = 1.5$; global maximum of 32 when $x = -2$

69. Global minimum of -7 when $x = -1$; global maximum of 1 when $x = 0$

71. Global minimum of 2.0 when $x = 1$; global maximum of 2.5 when $x = 1/2$ and $x = 2$

73. Global minimum of 0.167 when $x = 1$; global maximum of 0.223 when $x = 2.2361$

75. Global minimum of $-\sqrt[3]{32}$ when $x = -2$; global maximum of 1 when $x = 1$

77. Global minimum of 0 when $x = 0$; global maximum of e when $x = -1$

79.
```
f=makeFun(13420/(1+1.016*
exp(-0.023*(x-200)))~x)
df=D(f(x)~x)
crit=findZeros(df(x)~x,
xlim=range(0,500))
eval=rbind(0,crit,500)
f(eval)
```

81.
```
f=makeFun(0.75*sin(2*pi/24.32*
(x-2.31))-0.55*cos(2*pi/12.31*
(x-1.17))+0.77~x)
df=D(f(x)~x)
crit=findZeros(df(x)~x,
xlim=range(10,20))
eval=rbind(10,crit,20)
f(eval)
```

83.
```
f=makeFun(2*x-4~x)
df=D(f(x)~x)
crit=findZeros(df(x)~x,
xlim=range(3,8))
eval=rbind(3,crit,8)
f(eval)
```

85.
```
f=makeFun(-4*x^2+8*x-7~x)
df=D(f(x)~x)
crit=findZeros(df(x)~x,
xlim=range(-2,4))
eval=rbind(-2,crit,4)
f(eval)
```

87. ```
f=makeFun(x^4-18*x^2+40~x)
df=D(f(x)~x)
crit=findZeros(df(x)~x,
xlim=range(-4,4))
eval=rbind(-4,crit,4)
f(eval)
```

89. ```
f=makeFun(exp(sin(x)+3)~x)
df=D(f(x)~x)
crit=findZeros(df(x)~x,
xlim=range(-1,3))
eval=rbind(-1,crit,3)
f(eval)
```

91. ```
f=makeFun(x^2*sin(x)~x)
df=D(f(x)~x)
crit=findZeros(df(x)~x,
xlim=range(-1,4))
eval=rbind(-1,crit,4)
f(eval)
```

93. ```
f=makeFun(exp(x*sin(x))~x)
df=D(f(x)~x)
crit=findZeros(df(x)~x,
xlim=range(-10,10))
eval=rbind(-10,crit,10)
f(eval)
```

95. The global minimum of $f(x)$ on D is the value $f(c)$ such that $f(c) \leq f(x)$ for all x in D.

97. If $f(x)$ is a smooth function on a closed interval $D = [a, b]$, then $f(x)$ has both a global maximum and a global minimum value on $D = [a, b]$.

99. A critical number of a function $f(x)$ is an input c in the domain of $f(x)$ such that either $f'(c)$ is equal to zero or $f'(c)$ does not exist.

Section 5.2 Exercises

1. Local maximum of 1.5 when $x = -1.5$; local minimum of -1.5 when $x = 1.5$

3. Local minimum of -10 when $x = -5$; local maximum of 1 when $x = -1$; local minimum of -4 when $x = 2$

5. Local maximum of 1 when $x = -1$ and $x = 1$; local minimum of 0 when $x = 0$

7. Local minimum of 1.7 when $x = 1$; local maximum of 5.3 when $x = 2.1$

9. Local minimum of 0.9 when $x = 4.2$; local maximum of 1.6 when $x = 2.4$

11.

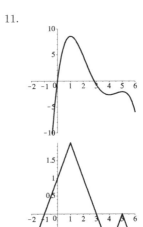

13.

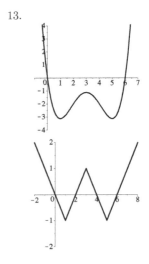

15. $x = -1, 1, 4$

17. $x = 0$

19.

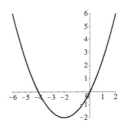

21.

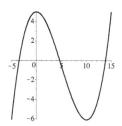

23.

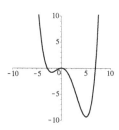

25. $x = 1/4$

27. $x = 1/5$

29. $x = -4, -1$

31. $x = 4, 6$

33. $x = -1, 0$

35. $x = \dfrac{3\pi}{2} + 2\pi k$, where k is any integer

37. No critical numbers

39. $x = e^{-1}$

41. Local minimum when $x = 1/4$

43. Local maximum when $x = 0$; local minimum when $x = 2$

45. Local maximum when $x = 4$; local minimum when $x = 6$

47. Local minimum when $x = 0$; local maximum when $x = 1$; local minimum when $x = 3$

49. All critical points neither maximum nor minimum

51. Local maximum at $x = 1/4$

53. No critical points

55. No critical points since strictly increasing

57. No critical points since strictly decreasing

59. No critical points since strictly increasing

61. $x = -1, 2, 4$

63. $x = -7, -2, 0, 3$

65. $x = 1, 3, 5, 9$

67. Local minimum when $x = -1, 4$; local maximum when $x = 2$

69. Local maximum when $x = -7, 0$; local minimum when $x = -2, 3$

71. Local minimum when $x = 1, 9$; local maximum when $x = 5$; neither local maximum nor local minimum when $x = 3$

73. Local maximum when $x \approx 100$ and $x \approx 190$; local minimum when $x \approx 25$ and $x \approx 125$

75. (a) local minimum of 102.3 thousand people in January 1933. (b) No, because the population would not grow indefinitely and would not be decreasing as it approaches 1950.

77. $f(4) \geq f(x)$ for all x near 4

79. According to the definition, every point $x = c$ on the graph of $f(x) = k$ is a local maximum, because $f(c) \geq f(x)$ for all x near c

81. $x = -b/(2a)$

83. If $a < 0$, there is a local maximum at $x = -\dfrac{b}{2a}$, because $f'(x) > 0$ when $x < -\dfrac{b}{2a}$ and $f'(x) < 0$ when $x > -\dfrac{b}{2a}$.

85. $b = -1$

87. $b = 12$

89. $a = -1/2$; $b = 3/4$

91.
```
f=makeFun(x^2-3*x-4~x)
df=D(f(x)~x)
crit=findZeros(df(x)~x)
f(1); f(3)
```

93.
```
f=makeFun(x^7-6*x^3-4~x)
df=D(f(x)~x)
crit=findZeros(df(x)~x)
f(-2); f(-1); f(1); f(2)
```

95.
```
f=makeFun(x^4-3*x^2+sin(x)~x)
df=D(f(x)~x)
crit=findZeros(df(x)~x)
f(-1.5); f(-0.5); f(0.5); f(1.5)
```

97.
```
f=makeFun(x*sin(x)~x)
df=D(f(x)~x)
crit=findZeros(df(x)~x)
f(-5); f(-3); f(-1); f(1); f(3); f(5)
```

99. A local maximum of $f(x)$ is the value $f(x)$ such that $f(c) \geq f(x)$ for all inputs x near input c.

101. A critical number of a function $f(x)$ is an input c in the domain of $f(x)$ such that either $f'(c)$ is equal to zero or $f'(c)$ does not exist.

103. Let $f(x)$ be a smooth function with critical number c and consider the sign of $f'(x)$ as inputs increase from left to right along the x-axis.

    ○ If $f'(x)$ changes from positive to negative at input c, then $f(c)$ is a local maximum of $f(x)$.

    ○ If $f'(x)$ changes from negative to positive at input c, then $f(c)$ is a local minimum of $f(x)$.

    ○ If $f'(x)$ does not change sign at input c, then $f(c)$ is neither a local maximum nor a local minimum of $f(x)$.

Section 5.3 Exercises

1. Concave up on $(0, \infty)$; concave down on $(-\infty, 0)$; $(0, 0)$ is a point of inflection

3. Concave up on $(-\infty, -1)$ and $(1, \infty)$; concave down on $(-1, 1)$; $(-1, 0.4)$ and $(1, 0.4)$ are points of inflection

5. Concave up on $(-\infty, -3)$ and $(1, \infty)$; concave down on $(-3, 1)$; $(-3, -4)$ and $(1, -2)$ are points of inflection

7. Concave up on $(-\infty, -0.5)$ and $(4.5, \infty)$; concave down on $(0.5, 4.5)$; $(0.5, -2)$ and $(4.5, -2)$ are points of inflection

9. Concave up on $(0, 3)$; concave down on $(-\infty, 0)$ and $(3, \infty)$; $(0, 1)$ and $(3, 2)$ are points of inflection

11. Concave up on $(-\infty, 4)$, $(12, 18)$, and $(22, \infty)$; concave down on $(4, 12)$ and $(18, 22)$; $(4, 1)$, $(12, 0.7)$, $(18, 0.5)$, and $(22, 0.4)$ are points of inflection

13. Concave up on $(-\infty, \infty)$; no points of inflection

15. Concave up on $(0, 200)$; concave down on $(200, \infty)$; $(200, 700)$ is a point of inflection

17. Concave down on $(-\infty, \infty)$; no points of inflection

19. Concave up on $(-\infty, 2.5)$; concave down on $(2.5, \infty)$; point of inflection at $x = 2.5$

21. Concave up on $(-\infty, 3)$; concave down on $(3, \infty)$; point of inflection at $x = 3$

23. Concave up on $(-\infty, 1)$ and $(3, \infty)$; concave down on $(1, 3)$; points of inflection at $x = 1$ and $x = 3$

25. Concave up on $(0.5, 1.5)$ and $(4, \infty)$; concave down on $(-\infty, -0.5)$ and $(1.5, 4)$; points of inflection at $x = -0.5$ and $x = 4$

27. Concave up on $(-\infty, \infty)$; no points of inflection

29. Concave down on $(-\infty, \infty)$; no points of inflection

31. Concave up on $(1, \infty)$; concave down on $(1, \infty)$; $(1, 1)$ is a point of inflection

33. Concave up on $(-\infty, -1/\sqrt{3})$ and $(1/\sqrt{3}, \infty)$; concave down on $(-1/\sqrt{3}, 1/\sqrt{3})$; $(\pm 1/\sqrt{3}, -5/9)$

35. Concave up on $(-\infty, -1/3)$ and $(1, \infty)$; concave down on $(-1/3, 1)$; $(-1/3, 285/81)$ and $(1, 5)$ are points of inflection

37. Concave up on $(-\infty, -\sqrt{7/24})$ and $(\sqrt{7/24}, \infty)$; concave down on $(-\sqrt{7/24}, \sqrt{7/24})$; $(-\sqrt{7/24}, 2.14)$ and $(\sqrt{7/24}, 6.46)$ are points of inflection

39. Concave up on $(-\infty, 0)$ and $(3, \infty)$; concave down on $(0, 3)$; $(0, 0)$ and $(3, -162)$ are points of inflection

41. Local minimum of $-1/4$ when $x = -1/2$

43. Local maximum of 16,000 when $x = -20$; local minimum of $-16,000$ when $x = 20$

45. Local maximum of 0 when $x = 0$; local minimum of -4 when $x = 2$

47. Local maximum of 6 when $x = 1$; local minimum of 2 when $x = 3$

49. Local minimum of -11 when $x = 3$

51. Local maximum of 5 when $x = 1$; local minimum of 0 at $x = 0$; local minimum of -27 when $x = 3$

53. Local minimum of -1.16 when $x = 4/3$

55. No local extreme values

57. Local maximum of $6\sqrt[3]{2}$ when $x = 2$

59. No local extreme values

61. Local maximum of $\sqrt[3]{32}$ when $x = 4$; local minimum of 0 when $x = 0$; local minimum of 0 when $x = 6$

63. No local extreme values

65. Local maximum of 0 when $x = 0$

67. Local maximum of 1 when $x = 0$

69. Local maximum of 1.469 when $x = -1/\sqrt{3}$; local minimum of 0.681 when $x = 1/\sqrt{3}$

71. Local maximum of 1 when $x = -\pi/2$; local minimum of 0 when $x = 0$; local maximum of 1 when $x = \pi/2$

73. No local extreme values

75. Local maximum of 1.5 at $x = \pi/6$; local minimum of 1 when $x = \pi/2$; local maximum of 1.5 when $x = 5\pi/6$; local minimum of -3 when $x = 3\pi/2$

77. Local maximum when $x = 2$

79. Local minimum when $x = 1$; local maximum when $x = 4$

81. Local minimum when $x = 0$; local maximum when $x = 6$

83. Local minimum when $x = 0$; local maximum when $x = 2$; local minimum when $x = 4$

85. Local maximum when $x = -1$; local minimum when $x = 0$; local maximum when $x = 3$; local minimum when $x = 5$

87.
```
f=makeFun(x^2+x~x)
df=D(f(x)~x)
crit=findZeros(df(x)~x)
ddf=D(df(x)~x)
ddf(crit)
```

89.
```
f=makeFun(x^3-3*x^2~x)
df=D(f(x)~x)
crit=findZeros(df(x)~x)
ddf=D(df(x)~x)
ddf(crit)
```

91.
```
f=makeFun(x^4-4*x^3+16~x)
df=D(f(x)~x)
crit=findZeros(df(x)~x)
ddf=D(df(x)~x)
ddf(crit)
```

93.
```
f=makeFun(3*x^5-5*x^4+2~x)
df=D(f(x)~x)
crit=findZeros(df(x)~x)
ddf=D(df(x)~x)
ddf(crit)
```

95.
```
f=makeFun(exp(x^3-x)~x)
df=D(f(x)~x)
crit=findZeros(df(x)~x)
ddf=D(df(x)~x)
ddf(crit)
```

97.
```
f=makeFun(x+sin(x)~x)
df=D(f(x)~x)
crit=findZeros(df(x)~x,
    xlim=range(-0,2*pi))
ddf=D(df(x)~x)
ddf(crit)
```

99. If $f(x)$ is a smooth function and (a, b) is an interval, then $f(x)$ is concave up on (a, b) if and only if $f'(x)$ is increasing on (a, b).

101. Let c be a real number in the domain of a function $f(x)$. The point $(c, f(c))$ is a point of inflection of $f(x)$ if and only if $f(x)$ changes concavity at input c.

103. Let $f(x)$ be smooth with a critical number at c.

   ○ If $f''(c) > 0$, then $f(c)$ is a local minimum of $f(x)$.

   ○ If $f''(c) < 0$, then $f(c)$ is a local maximum of $f(x)$.

   ○ If $f''(c) = 0$, then the first derivative test must be used to determine whether $f(c)$ is a local maximum, a local minimum, or neither for $f(x)$.

Section 5.4 Exercises

1. $x_4 \approx -2.828$

3. $x_4 \approx 2.828$

5. $x_4 \approx 4.283$

7. $x_4 \approx 0.202$

9. $x_4 \approx -0.329$

11. $x_4 \approx -0.785$

13. $x_4 \approx 0$

15. $x_4 \approx 1.059$

17. There appear to be three distinct zeros in $[-1, 1]$. $x_1 = -0.5$ and $x_1 = 0.25$ converge to 0, $x_1 = 0.5$ converges to approximately 1.06, and $x_1 = 1$ converges to approximately 0.876.

19. $x_4 \approx -4.712$

21. $x_4 \approx 4.712$

23. There appear to be four distinct zeros in $[-5, 8]$. $x_1 = \pm 5$ converge to approximately ± 4.712, $x_1 = 1$ converges to approximately -1.655, and $x_1 = 8$ converges to approximately 7.854.

25. $x_4 \approx 5.099$

27. $x_2 = 2$

29. $x_5 \approx 0.810$

31. $x_8 \approx -1.316$

33. $x_4 \approx 1.545$

35. $x_4 \approx -4.712$

37. $x_4 \approx -0.6529$

39. $x_5 \approx 1.414$

41. $x_5 \approx 1.414$

43. $x_4 \approx 2.0800$

45. $x_7 \approx 7.389$

47. $x_5 \approx 2.302776$

49. $x_5 \approx 1.490$

51. $x_4 \approx 0.2426$

53. $x_5 \approx 1.165$

55. $x = 0$

57. $x = 0$

59. $x = 0$ and $x = 2$

61. $x_5 \approx 0.2146$

63. $x_4 \approx 1.21785$

65. `plotFun(x^2-2~x)`
    $x_1 \approx \pm 1$

67. `plotFun(x^3-9~x)`
    $x_1 \approx 2$

69. `plotFun(x^4-5*x+2~x)`
    $x_1 \approx 1$; $x_1 \approx 1.5$

71. `plotFun(exp(x)-7~x)`
    $x_1 \approx 1$

73. `plotFun(exp(-x)-22~x)`
    $x_1 \approx -3$

75. `f=makeFun(exp(-x)*sin(x)+x~x)`
    `df=D(f(x)~x)`
    `nm=makeFun(x-f(x)/df(x)~x)`
    `x1=-14; x2=nm(x1) ...`

77. `f=makeFun((x^3-2*x+1)/(x^2+1)~x)`
    `df=D(f(x)~x)`
    `nm=makeFun(x-f(x)/df(x)~x)`
    `x1=0; x2=nm(x1) ...`

79. `f=makeFun(sin(2*x)/exp(x)~x)`
    `df=D(f(x)~x)`
    `nm=makeFun(x-f(x)/df(x)~x)`
    `x1=2; x2=nm(x1) ...`

81. `f=makeFun(x^6-3*x^5+2~x)`
    `df=D(f(x)~x)`
    `nm=makeFun(x-f(x)/df(x)~x)`
    `x1=0.5; x2=nm(x1) ...`

83. `f=makeFun(sin(x)+cos(x)~x)`
    `df=D(f(x)~x)`
    `nm=makeFun(x-f(x)/df(x)~x)`
    `x1=0; x2=nm(x1) ...`

85. `f=makeFun(exp(x)-x^2~x)`
    `df=D(f(x)~x)`
    `nm=makeFun(x-f(x)/df(x)~x)`
    `x1=-1; x2=nm(x1) ...`

87. `f=makeFun(x^3-6*x^2+7*x+2~x)`
    `df=D(f(x)~x)`
    `nm=makeFun(x-f(x)/df(x)~x)`
    `x1=1; x2=nm(x1) ... x7=nm(x6)`

89. $f'(x_n) > 0$; $f(x_n) > 0$; need to move left

91. $f'(x_n) < 0$; $f(x_n) > 0$; need to move right

93. To approximate the input where a smooth function $f(x)$ is zero.

95. One approach to identifying an initial conjecture is to graph the function and visually estimate the desired root of the function.

97. Let $f(x)$ and $g(x)$ be smooth functions. An input where $f(x) = g(x)$ can be approximated by applying Newton's

method to $h(x) = f(x) - g(x)$. The approximate input x where $h(x) = 0$ is also an approximate input where $f(x) = g(x)$.

99. Newton's method is the most natural and, from a certain perspective, the only dimensionally compatible way to use x, $f(x)$, and $f'(x)$ in the context of seeking a zero of a function.

Section 5.5 Exercises

1. $(0,0)$

3. $(1,1)$

5. $(1/2, 0)$

7. $(2, -3)$

9. $(\pm\sqrt{8/3}, 0)$

11. $(0,0)$

13. No critical points

15. $(0,0)$; $(1,1)$

17. No critical points

19. $f_x(0,0) = f_y(0,0) = 0$

21. $f_x(0,0) = f_y(0,0) = 0$

23. $f_x(1,0) = f_y(1,0) = 0$

25. $f_x(-3,1) = f_y(-3,1) = 0$

27. $f_x(4,4) = f_y(4,4) = 0$

29. Local maximum of 4 at $(0,0)$

31. Saddle point at $(0,0)$

33. Local minimum of approximately -3 at $(0,0)$; local maximum of approximately 11 at $(2,-2)$; local maximum of approximately 7 at $(0,-3)$

35. Local maximum of approximately 12 at $(0, 1.5)$; local maximum of approximately 8 at $(0, -0.5)$; local minimum of approximately -2 at $(-1, -2)$

37. Test is inconclusive

39. $(0,0)$ corresponds to a local minimum

41. $(1,0)$ corresponds to a local minimum

43. $(-3,1)$ corresponds to a local minimum

45. $(4,4)$ corresponds to a local minimum

47. $(0,0)$ corresponds to a local minimum

49. $(0,0)$ corresponds to a saddle point

51. No critical points

53. $(3, -6)$ corresponds to a saddle point

55. No critical points

57. $(0,0)$ corresponds to a saddle point

59. $(0,0)$ corresponds to a saddle point

61. No critical points

63. $(0,0)$ corresponds to a saddle point

65. $t = 5$, $h = 0$ maximize the range of R

67. $x = y = 20$ maximizes the volume of the box

69. Global maximum of $f(0,-1) = 3$; global minimum of $f(0,1) = -1$

71. Global maximum of $f(0,5) = 15$; global minimum of $f(0,1) = -1$

73. Global maximum of $f(-1,0) = 5$; global minimum of $f(1,0) = 1$

75. Global maximum of $f\left(\frac{-2}{\sqrt{5}}, 2 + \frac{4}{\sqrt{5}}\right) \approx 18.944$; global minimum of 1 at $(1,0)$

77. Global maximum of $f(0,0) = 0$; global minimum of $f(1,0) = 3$

79. Global maximum of $f(2,0) = 6$; global minimum of $f(0,0) = f(2,2) = 0$

81. Global maximum of $f(0,1) = 1$; global minimum of $f(1,0) = -11$

83. Global maximum of $f(-2,2) = 20$; global minimum of $f(0,0) = 0$

85. A local maximum of $f(x,y)$ is a value $f(a,b)$ such that $f(a,b) \geq f(x,y)$ for all inputs (x,y) near (a,b)

87. The local extreme values of $f(x,y)$ are the local maximum and local minimum values of the function.

89. A function $f(x,y)$ has a saddle point for input (a,b) if and only if $f(x,y)$ increases from $f(a,b)$ in some directions and decreases from $f(a,b)$ in others.

91. $D(a,b) = f_{xx}(a,b) \cdot f_{yy}(a,b) - [f_{xy}(a,b)]^2$

93. The global minimum of $f(x,y)$ on $\mathcal{R}$ is the value $f(a,b)$ such that $f(a,b) \leq f(x,y)$ for all (x,y) in $\mathcal{R}$.

95. Let $f(x,y)$ be a smooth two-variable function and let $\mathcal{R}$ be either a closed circular region or a closed rectangular region contained in the domain of $f(x,y)$.

(1) Identify the critical points of $f(x, y)$ in region $\mathcal{R}$.

(2) Substitute the equation(s) of the edge(s) of $\mathcal{R}$ into $f(x, y)$ to create a function(s) of one variable $g(x)$ or $g(y)$. For each such function, find its critical numbers on the closed interval defining the edge and the corresponding coordinates, and the endpoints of each interval.

(3) Evaluate $f(x, y)$ at the collection of points identified in steps (1) and (2).

(4) The largest output from step (3) is the global maximum and the smallest output from step (3) is the global minimum of $f(x, y)$ in $\mathcal{R}$.

Section 5.6 Exercises

1. $\nabla f = \begin{pmatrix} 2 \\ 7 \end{pmatrix}$

3. $\nabla f = \begin{pmatrix} 10x \\ -5y^4 \end{pmatrix}$

5. $\nabla f = \begin{pmatrix} 6x \\ -4 \end{pmatrix}$

7. $\nabla f = \begin{pmatrix} \cos(x) \\ -\sin(y) \end{pmatrix}$

9. $\nabla f = \begin{pmatrix} \sin(y) \\ x \cos(y) \end{pmatrix}$

11. $\nabla f = \begin{pmatrix} e^y \\ xe^y + 1 \end{pmatrix}$

13. $\nabla f = \begin{pmatrix} 1/(x+y) + y \\ 1/(x+y) + x \end{pmatrix}$

15.

17.

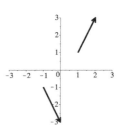

19.

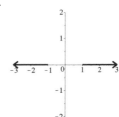

21.

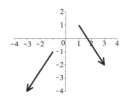

23. Local maximum of $f(1, 0) = 3$

25. Local minimum of $f(0, 0) = 0$

27. Local maximum of $f(1.5, -1.25) = 6$

29. Local maximum of $f(1, 1.25) = 7$

31. $\lambda \approx -1$

33. $\lambda \approx -0.5$

35. $\lambda \approx 4$

37. $\lambda \approx -2$

39. Local minimum of $f(0, 0) = 0$

41. Local maximum of $f(\pm\sqrt{(15/16)}, -1/4) = 2.125$; local minimum of $f(0, 1) = -1$; local minimum of $f(0, -1) = 1$

43. Local minimum of $f(0, 4) = -4$

45. Local minimum of $f(18/33, 6/33) \approx -0.182$

47. Local maximum of $f(-1, 1) = 2/3$; local minimum of $f(1, -1) = -2/3$

49. Local minimum of $f(1, 2) = 12$

51. Local maximum of $f(\pm 2, 0) = 10$; local minimum of $f(0, \pm 1) = 3$

53. Local maximum of $f(3,3) = f(-3,-3) = 9$; local minimum of $f(\sqrt{3}, -\sqrt{3}) = f(-\sqrt{3}, \sqrt{3}) = -3$

55. Local maximum of $f(2,4) \approx 28.28$

57. Local maximum of $f(2,2) = f(-2,-2) = e^8$; local minimum of $f(-2,2) = f(2,-2) = e^{-8}$

59. If the constraint is increased by one unit to $g(x,y) = 9$, the minimum value will increase to approximately 17.

61. If the constraint is increased by one unit to $g(x,y) = -5$, the minimum value will increase to approximately -4.

63. If the constraint is increased by one unit to $g(x,y) = 14$, the minimum value will decrease to approximately -17.

65. If the constraint is increased by one unit to $g(x,y) = -9$, the minimum value will increase to approximately -11 and the maximum will decrease to approximately 1.

67. If the constraint is increased by one unit to $g(x,y) = 16$, the minimum value will increase to approximately 7 and the maximum value will increase to approximately 20.

69. Local maximum of $f(\sqrt{10/105}, 10\sqrt{10/105}) = 21\sqrt{10/105}$; local minimum of $f(-\sqrt{10/105}, -10\sqrt{10/105}) = -21\sqrt{10/105}$

71. In Exercise 68, $\lambda = 0.3415$ for the maximum and $\lambda = -0.3415$ for the minimum. In Exercise 70, $\lambda \approx 0.3326$ for the maximum and $\lambda \approx -0.3326$ for the minimum.

73. Local maximum of $f(2/\sqrt{3}, \sqrt{2/3}) = f(-2/\sqrt{3}, \sqrt{2/3}) = 4/3\sqrt{2/3}$; local minimum of $f(2/\sqrt{3}, -\sqrt{2/3}) = f(-2/\sqrt{3}, -\sqrt{2/3}) = -4/3\sqrt{2/3}$

75. In Exercises 72, $\lambda = 1/\sqrt{3} \approx 0.577$ for the maximum and $\lambda = -1/\sqrt{3} \approx -0.577$ for the minimum. In Exercise 74, $\lambda \approx 0.7038$ for the maximum and $\lambda \approx -0.7038$.

77. $8/\sqrt{2}$

79. $-9/\sqrt{2}$

81. $73/13$

83. $-1/2 \cos(1) + \sqrt{3} \sin(1) \approx 1.187$

85. e

87. Maximum value of $D_{\bar{u}}f = \|\nabla f\|^2$; $\theta = 0$

89. ∇f points in the direction of greatest increase.

91. Let $f(x,y)$ be a smooth two-variable function, input (a,b) be a point on the plane, and vector $\nabla f(a,b)$ be graphed at (a,b) on the contour plot of $f(x,y)$.

    (1) The vector $\nabla f(a,b)$ points in the direction of greatest increase of $f(x,y)$ at (a,b), and $-\nabla f(a,b)$ points in the direction of greatest decrease of $f(x,y)$ at (a,b).

    (2) The length of $\nabla f(a,b)$ measures the steepness of $f(x,y)$ at (a,b).

    (3) At (a,b), $\nabla f(a,b)$ is perpendicular to the contour of $f(x,y)$ at level $f(a,b)$.

93. Let $f(x,y)$ be a smooth two-variable function with constraint $g(x,y) = C$. Then, $f(x,y)$ has a local minimum $f(a,b)$ subject to the constraint $g(x,y) = C$ if and only if $f(a,b) \leq f(x,y)$ for all points (x,y) near (a,b) and on the constraint curve $g(x,y) = C$.

95. The value of λ is approximated by determining the change in the extreme values of $f(x,y)$ for (a,b) when C is increased by one unit.

97. Let $f(x,y)$ and $g(x,y)$ be smooth two-variable functions. The local extreme values of $f(x,y)$ subject to constraint $g(x,y) = C$ occur for inputs (x,y) that satisfy the equations $\nabla f(x,y) = \lambda \nabla g(x,y)$ and $g(x,y) = C$. Equivalently, such local extreme values occur for inputs (x,y) that satisfy the following system of equations: $f_x(x,y) = \lambda g_x(x,y), f_y(x,y) = \lambda g_y(x,y)$, and $g(x,y) = C$.

Section 6.1 Exercises

1. 36

3. 54

5. 48

7. 37.5

9. 20

11. 25

13. 9

15. −5

17. 3

19. 8

21. 6

23. 4

25. 5

27. 9.9

29. 8.2

31. 6.5

33. 12.5

35. 10.25

37. 12

39. $L_4 = 51.9$; $R_4 = 34.5$

41. $L_4 = -16$; $R_4 = -104$

43. $L_4 = 45.9$; $R_4 = 28.5$

45. 15,875

47. 18,250

49. 15,125

51. 1445

53. 1500

55. 1520

57. 46.35

59. 5982.65

61. 493.9

63. 2.31

65. 3.73

67. 2.38

69. 1885

71. 230

73. −8.1

75. −5

77. −196.8

79. 1.57806

81. 1.095496

83. 11

85. 9

87. 7.25

89. 8

91. 5.75

93. 3.96296

95. 0.03424

97. 4

99. 1.5

101. 2

103. 1.333

105. Let $f(x)$ be the rate of change of a quantity $F(x)$ on an interval $[a, b]$, let $a = x_1 < x_2 < x_3 < \cdots < x_{n+1} = b$, let $[a, b]$ be subdivided into smaller subintervals $[x_1, x_2]$, $[x_2, x_3]$, ..., $[x_n, x_{n+1}]$, and let $\Delta x_k = x_{k+1} - x_k$ be the width of each subinterval. The right approximation to the accumulation of $F(x)$ on $[a, b]$ is $R_n = f(x_2)\Delta x_1 + f(x_3)\Delta x_2 + \cdots + f(x_{n+1})\Delta x_n$.

107. Let $f(x)$ be the rate of change of a quantity $F(x)$ on an interval $[a, b]$, let $a = x_1 < x_2 < x_3 < \cdots < x_{n+1} = b$, let $[a, b]$ be subdivided into smaller, equal-width subintervals $[x_1, x_2]$, $[x_2, x_3]$, ..., $[x_n, x_{n+1}]$, and let $\Delta x = (b - a)/n$ be the uniform width of each subinterval. The midpoint approximation to the accumulation of $F(x)$ on the interval $[a, b]$ is
$$M_n = f\left(\frac{x_1 + x_2}{2}\right)\Delta x$$
$$+ f\left(\frac{x_2 + x_3}{2}\right)\Delta x + \ldots$$
$$+ f\left(\frac{x_n + x_{n+1}}{2}\right)\Delta x$$

Section 6.2 Exercises

1. Total feet the tree grew from years 1 to 4; feet; [length] $= L$

3. Number of miles traveled when gallons 4 to 8 are used; miles; [length] $= L$

5. Price change in Facebook stock from last quarter of 2012 to last quarter of 2017; dollars; [money] $= B$

7. Amount of carbon dioxide added to the atmosphere from 1980 to 2015; metric tons; [mass] $=$ M

9. $\int_0^{28} \ell(w)\,dw$; centimeters; [length] $= L$

11. $\int_0^{30} s(t)\,dt$; megabytes; [amount] $= N$

13. $\int_1^2 v(t)\,dt$; miles; [length] $= L$

15. $\int_0^{90} i(t)\,dt$; people; [population] $= P$

17. 0

19. 15

21. 15

23. -50

25. 1

27. -9

29. -9

31. 15

33. 7

35. -16

37. 0

39. 0

41. -4

43. 13

45. 11

47. 0

49. 0

51. -8

53. -8

55. 18

57. $-\dfrac{3}{2}$

59. -1

61. 3

63. 9

65. 0

67. 8

69. 0

71. -6.5

73. 8

75. 0

77. 8

79. -8

81. 4

83. 11

85. 5

87. 1

89. -8

91. $\int_a^b f(x)\,dx$ is the net accumulation of a quantity $F(x)$ with rate of change $f(x)$ on the interval $[a, b]$.

93. For $\int_a^b f(x)\,dx$, $f(x)$ is called the integrand.

95. For $\int_a^b f(x)\,dx$, if $a < b$, then Δx is positive and $\int_a^b f(x)\,dx$ is the signed area. If $a > b$, then Δx is negative and $\int_a^b f(x)\,dx$ is the negative of the signed area; alternatively, $\int_a^b f(x)\,dx = -\int_b^a f(x)\,dx$.

97. $\left[\int_a^b f(x)\,dx\right] = [f(x)] \cdot [x]$ $= [\text{output}] \cdot [\text{input}]$

99. See the Summary at the end of Section 6.2.

Section 6.3 Exercises

1. $F'(x) = 4x^3 = f(x)$

3. $F'(x) = 5e^{5x} = f(x)$

5. $F'(x) = e^x + xe^x = f(x)$

7. $F'(x) = 3\cos(3x - 2) + 2e^{2x} = f(x)$

9. $A(1) = 0$

11. $A(6) = 10$

13. $A(-1) = 0$

15. $A(10) = -33$

17. $A(0) = 0$

19. $A(6) = 36$

21. $A(2) = 0$

23. $A(8) = 258$

25. $A(5) = -32$

27. $A(3) = 0$

29. $A(-4) = 0$

31. $A(2) = 12$

33. $A(2) = 4$

35. $A(7) = 27$

37. $A(9) = 24$

39. $A(0) = -4$

41. $A(3) = 5$

43. $A(7) = 23$

45. $A(9) = 20$

47. $A(-2) = -2$

49. $A(2) = 0$

51. $A(6) = -5$

53. $A(-2) = -4$

55. $A(0) = -2$

57. $A(5) = -\dfrac{9}{2}$

59. $A(8) = -11$

61. $2 - t^3$

63. e^{-t^2}

65. $\tan(e^t)$

67. $-t^2\sqrt{t} + t$

69. $-\cos(2t - 3)$

71. $A(t) = \displaystyle\int_0^t \sqrt{x^2 + x}\, dx$

73. $A(t) = \displaystyle\int_1^t e^x \ln(x)\, dx$

75. $A(t) = \displaystyle\int_1^t \dfrac{x^2 - \sin(x)}{e^x + 1}\, dx$

77. $A(t) = \displaystyle\int_2^t x^2 + \ln(x)\, dx$

79. $\frac{1}{3}x^3 + x^2 + C$

81. $\frac{1}{4}x^4 - \frac{1}{3}x^3 + \frac{1}{2}x^2 + C$

83. $\frac{5}{21}x^{21} - \frac{1}{4}x^{-4} + 2x + C$

85. $\ln(x) + \frac{1}{3}x^{-3} + C$

87. $\frac{2}{3}x^{3/2} + \frac{3}{4}x^{4/3} + x + C$

89. $\frac{3}{8}x^{8/3} + \frac{7}{8}x^{8/7} + C$

91. $\frac{15}{4}x^{4/5} - \dfrac{1}{x} + C$

93. $-\dfrac{3}{x^{2/3}} - \dfrac{21}{2}x^{2/7} + C$

95. $\frac{1}{2}e^{2x+4}$

97. $-\frac{1}{5}e^{-5x+8}$

99. e

101. $\frac{1}{5}\sin(5x + 2)$

103. $-\frac{1}{2}\sin(-2x - 9)$

105. $-\frac{1}{5}\cos(5x + 2)$

107. $-\frac{1}{4}\cos(4x + 9)$

109. $\ln(6)x + \frac{4}{5}x^{5/4} - \frac{1}{8}e^{8x+4} + C$

111. $\frac{4}{11}x^{11/4} + \frac{1}{7}e^{7x} - 4x - x^3 + C$

113. $\frac{1}{2}ex^2 + 7x + \frac{5}{6}x^{-6} + C$

115. $\frac{35}{6}e^{6x+2} + \frac{1}{4}\cos(8x + 3) + 7x + C$

117. $\frac{2}{7}x^{7/2} + \frac{2}{3}\sin(3x + 4) - 6x + C$

119. $\frac{1}{4}x^4 - \frac{2}{3}x^{3/2} - \frac{1}{\pi}\cos(6 - \pi x) + C$

121. $\frac{7}{34}x^{34/7} + \frac{1}{3}\sin(9x + 8) - \frac{1}{3}e^{3x-4} + C$

123. An antiderivative of $f(x)$ is a function $F(x)$ such that $F'(x) = f(x)$.

125. If $f(x)$ is a function and a is a real number in the domain of $f(x)$, then $A(t) = \displaystyle\int_a^t f(x)\, dx$ is the net accumulation function of $f(x)$ with lower limit a.

127. $\displaystyle\int f(x)\, dx$ denotes all antiderivatives of the function $f(x)$.

Section 6.4 Exercises

1. 12

3. -33

5. 304

7. $\dfrac{2016}{3} + 6e \approx 688.310$

9. $\dfrac{10\sqrt{2}}{3} - \dfrac{8}{3} \approx 2.047$

11. $\dfrac{23}{9} \approx 2.556$

13. $3 + \dfrac{e^2}{2} - \dfrac{e}{2} \approx 5.335$

15. -3.25

17. $e^2 - 1 \approx 6.389$

19. $e^6 - e^4 \approx 348.831$

21. $\dfrac{-1}{2e^2} + \dfrac{e^4}{2} \approx 27.231$

23. $\dfrac{e^3}{3} - \dfrac{1}{3} \approx 6.362$

25. 2

27. 2

29. 0

31. -59.218

33. 6.4

35. 12

37. $\dfrac{\pi^2}{2} - 2 \approx 2.935$

39. 0

41. $\dfrac{e^8 - e^4}{2} + 2\ln\left[\dfrac{1}{2}\right] \approx 1461.794$

43. $\dfrac{-2}{e^2} + 1.5 + 2e \approx 6.666$

45. The average male is 19.298 inches long at birth.

47. The average female head circumference at birth is 13.181 inches at birth.

49. 22.5

51. $\dfrac{7}{6} \approx 1.167$

53. 5.333

55. $e + \dfrac{1}{e} - 2 \approx 1.086$

57. $\sqrt{2} - 1 \approx 0.414$

59. 1

61. > F=antiD(2*x+1~x)
    > F(3)-F(0)
    [1] 12

63. > F=antiD(4*x^3+6*x~x)
    > F(4)-F(0)
    [1] 304

65. > F=antiD(x^(1/2)+1/(x^(1/2))~x)
    > F(2)-F(1)
    [1] 2.047

67. > F=antiD(3/x+exp(1)*x~x)
    > F(exp(1))-F(1)
    Error in stats::

69. > F=antiD(exp(x)~x)
    > F(2)-F(0)
    [1] 6.389056

71. > F=antiD(exp(-2*x+4)~x)
    > F(3)-F(0)
    [1] 27.231

73. > F=antiD(sin(x)~x)
    > F(pi)-F(0)
    [1] 2

75. > F=antiD(sin(3*x)-sin(2*x)~x)
    > F(pi)-F(-pi)
    [1] 0

77. Second fundamental theorem of calculus: If $f(x)$ is a smooth function on $[a, b]$ and $F(x)$ is any antiderivative of $f(x)$, then

$$\int_a^b f(x)\,dx = F(b) - F(a)$$

79. Difference in net accumulation: Let $f(x)$ and $g(x)$ be smooth functions with $f(x) \geq g(x)$ on an interval $[a, b]$ in the domain of both functions. The area of the region bounded by $f(x)$ and $g(x)$ on $[a, b]$, or equivalently the difference in the net accumulation of $f(x)$ and $g(x)$ on $[a, b]$, is given by the following definite integral:

$$\int_a^b f(x) - g(x)\,dx$$

81. The fundamental theorems of calculus demonstrate that differentiation and integration are inverse operations.

Section 6.5 Exercises

1. $u = x^2 + 2$, $du = 2x\,dx$

3. $u = \cos(x)$, $du = -\sin(x)\,dx$

5. $u = x^2 + x$, $du = (2x + 1)\,dx$

7. $u = x^4 + 5$, $du = 4x^3\,dx$

9. $u = x^2 + 7$, $du = 2x\,dx$

11. $u = 3x + 4$, $du = 3\,dx$

13. $\dfrac{1}{5}(x^2 + 1)^5 + C$

15. $e^{x^2+3} + C$

17. $\sin(\ln|x|) + C$

19. $2\sqrt{\sin(x)} + C$

21. $\ln|e^{4x} + 4x| + C$

23. $\ln|\sin(x^3 + 1)| + C$

25. $\dfrac{1}{3}\sin^3(x) + C$

27. $\dfrac{-1}{3}(\sin(x)+e^x)^{-3}+C$

29. $\dfrac{1}{4}\sin(x^4)+C$

31. $\dfrac{3}{20}(x^5+6)^{4/3}+C$

33. $\dfrac{1}{2}\ln|x^2+7|+C$

35. $\dfrac{1}{3}e^{x^3-5}+C$

37. $-\ln|\cos(x)|+C$

39. $\dfrac{-1}{4\cos^4(x)}+C$

41. $\dfrac{1}{35}(x+4)^{35}-\dfrac{4}{34}(x+4)^{34}+C$

43. $(x+5)-5(\ln|x+5|)+C$

45. $\dfrac{1}{12}(\sqrt{x^3}+2)^8+C$

47. $\dfrac{-2}{3(\sqrt{x^3}+\frac{3}{2}x)}+C$

49. $\dfrac{1}{6}e^{x^6+2x^3}+C$

51. $\dfrac{-1}{5\cdot4^5}+\dfrac{1}{5\cdot5^5}\approx-0.0001313$

53. $\ln(19)-\ln(4)\approx1.558$

55. $\ln(5)-\ln(1)=\ln(5)\approx1.609$

57. $e^{-9}-e^{-1}\approx-0.368$

59. $\dfrac{1}{3}\approx0.333$

61. $\dfrac{1}{6}(21)^{3/2}-\dfrac{1}{6}(21)^{3/2}=0$

63. $\dfrac{1}{\pi}-\dfrac{1}{\pi}=0$

65. $\dfrac{-1}{2}+\dfrac{1}{2e}\approx-0.316$

67. $\ln(1.5)\approx0.405$

69. $\dfrac{2}{5}\left[\dfrac{39}{32}\right]^{1/2}-\dfrac{2\sqrt{4}}{5}\approx-0.358$

71. $\dfrac{7^{35}}{105}-\dfrac{2\cdot7^{34}}{51}-\dfrac{4^{35}}{105}+\dfrac{2\cdot4^{34}}{51}$

73. $2+\dfrac{1}{2}\ln(4)-1-\dfrac{1}{2}\ln(2)\approx1.347$

75. $dy=dx$

77. $dy=-2x\,dx$

79. $dy=2\cos(2x+5)\,dx$

81. $dy=-\sin(\ln(x))\,\dfrac{1}{x}\,dx$

83. $\Delta y\approx dy=0.1$

85. $\Delta y\approx dy=e\cdot0.01\approx0.0272$

87. $\Delta y\approx dy=-0.15$

89. For $u=\cos(x)$, $du=-\sin(x)\,dx$

91. For $u=\sec(x)+\tan(x)$,
$du=\left[\sec(x)\tan(x)+\sec^2(x)\right]\,dx$

93. The composition of $f(x)$ with $g(x)$ is $f\circ g(x)=f(g(x))$, when this expression is defined.

95. Chain rule: Let $f(x)$ and $g(x)$ be differentiable functions. When the compositions are defined, then

$$\frac{d}{dx}\left[f(g(x))\right]=f'[g(x)]\cdot g'(x)$$

97. Substitution and definite integrals: Let $f(x)$ and $g(x)$ be smooth functions. When the compositions are defined, then

$$\int_a^b f'[g(x)]\,g'(x)\,dx=f[g(b)]-f[g(a)]$$

Section 6.6 Exercises

1. $f(x)=\ln(x);\ g'(x)=x^2$

3. $f(x)=\ln(x);\ g'(x)=\cos(x)$

5. $f(x)=e^{2x};\ g'(x)=\cos(6x)$

7. $f(x)=e^{-x};\ g'(x)=\sin(3x)$

9. $x^3\ln(x)-\dfrac{3}{2}x^2+C$

11. $(2x^4+x)\ln(2x)-x^4-2x+C$

13. $(4x+6)\dfrac{1}{2}e^{2x+1}-e^{2x+1}+C$

15. $-x\cos(x)+\sin(x)+C$

17. $(3x+4)\sin(x)+3\cos(x)+C$

19. $-x^2\sin(x)+2x\sin(x)+2\cos(x)+C$

21. $(9x^2+18x)\dfrac{1}{3}\sin(3x)$
$+(2x+2)\cos(3x)-\dfrac{2}{3}\sin(3x)+C$

23. $3x^2e^x-2xe^x+\dfrac{2}{3}e^{3x}+C$

25. $-x^3e^{-x}+3x^2e^{-x}-6xe^{-x}+6e^{-x}+C$

27. $6\ln(2)-2.5$

29. -4

31. $-\dfrac{1}{2}e^2+\dfrac{3}{2}$

33. 2π

35. -4π

37. $\dfrac{\pi^2}{8} - 1$

39. $2 - 2\ln(2)$

41. $e - 2.5$

43. ```
    > F=antiD((2*x+1)*log(x)~x)
    > F(2)-F(1)
    [1] 1.658883
    ```

45. ```
    > F=antiD(8*x*exp(4*x+2)~x)
    > F(0)-F(-1/2)
    [1] -2.194528
    ```

47. ```
    > F=antiD(x^2*cos(x)~x)
    > F(pi)-F(-pi)
    [1] -12.56637
    ```

49. ```
    > F=antiD(1-log(x)~x)
    > F(2)-F(1)
    [1] 0.6137056
    ```

51. Standard formula

53. Integration by arts

55. Standard formula

57. Integration by arts

59. Substitution

61. Standard formula

63. Integration by arts

65. $\dfrac{x^3}{3} + \dfrac{x^2}{2} + C$

67. $x^2 e^x - 2xe^x + 2e^x + C$

69. $\dfrac{e^{3x}}{3} + C$

71. $x\ln(x) - x + C$

73. $-\dfrac{1}{8}\cos(4x^2) + C$

75. $-\dfrac{1}{4}\cos(4x) + C$

77. $\dfrac{x}{\pi}\sin(\pi x) + \dfrac{1}{\pi^2}\cos(\pi x) + C$

79. Integration by parts: Let $f(x)$ and $g(x)$ be smooth functions. Then the following equality holds:
$$\int f(x)\, g'(x)\, dx =$$
$$f(x)\, g(x) - \int f'(x)\, g(x)\, dx$$

81. Integration by parts for definite integrals: Let $f(x)$ and $g(x)$ be smooth functions on $[a, b]$. Then the following equality holds:
$$\int_a^b f(x) g'(x)\, dx =$$
$$f(x)g(x)\Big|_a^b - \int_a^b f'(x)g(x)\, dx$$

83. The derivation of the integration by parts formula from the product rule appears just before the Summary for Section 6.6.

Appendix C

Getting Started with RStudio

R is the standard statistical software package used by academic and professional statisticians, and includes powerful tools for modeling and analyzing data sets. RStudio is an open source, freeware software package that provides an integrated interface for using R. Both of these tools are needed to take advantage of the RStudio instruction, examples, and exercises in this book.

The first step is to download the current version of R from https://www.r-project.org. Once R has been downloaded and installed, download the current version of RStudio from https://www.rstudio.com. As with all Internet addresses, sometimes specific links change, and the reader is encouraged to search for an updated web address if these addresses that were current at the time of publication do not work.

RStudio is freely available as both a desktop and a server application. The server version is an important asset that is highly recommended to anyone using this book in a classroom setting. Your institutional technology office can help set up an RStudio server with relative ease.

Once R and RStudio have been installed, R commands can be executed by entering them in the Console pane located on the bottom left of the standard RStudio environment. The size and position of the various panes can be adjusted to match user preference by dragging the bounding edges to the desired scale on the screen. Different working panes can be chosen via the Global Options option that can accessed via Tools from the main menu.

Installing Packages

Before using the commands described in this book, the appropriate packages must be installed with the `install.packages` command. The four packages used in this book are: `MMAC`, `mosaic`, `mosaicCalc`, and `manipulate`. The `MMAC` includes all of the data sets used in this book, and the other three packages provide the commands needed throughout the book. The following RStudio command installs the `MMAC` package in your local version of RStudio:

```
install.packages("MMAC")
```

The other three packages are installed similarly.

After a package has been installed, it must be loaded into RStudio's active, working memory in order to access and use its commands. This task is accomplished by means of the **require** command, as illustrated here for the MMAC package:

```
require(MMAC)
```

Once the **require(MMAC)**, **require(mosaic)**, etc. commands have been entered and executed, all of the commands and the datasets that are part of MMAC are available for use as detailed in this book. RStudio does not always keep installed packages in its active, working memory when shut down. Therefore, packages should loaded into RStudio using **require(MMAC)**, or the appropriate name for the package, every time RStudio is opened and its commands are needed.

Accessing Data outside of MMAC

A reader may want to work with a data set that is not included as part of MMAC. While a variety of file formats can be read by RStudio, comma-separated value (CSV) files are the recommended choice.

RStudio assumes that the first line of a CSV file is a header line containing the names of the variables, that the values of the variables are separated by commas, and that line breaks indicate a new set of variable values. Data stored in a CSV file can be uploaded into RStudio in two relatively simple ways:

(1) For files stored locally on a user's machine, the **read.csv** command provides an excellent option for uploading files. As an example, entering the following command will upload and store a CSV file **World-Population.csv** in RStudio:

```
Popdata=read.csv("World-Population.csv")
```

In this example command, notice the syntax: after the **read.csv** command and an opening parenthesis, the name of the file is given inside quotation marks followed by a closing parenthesis. In addition, the data set is stored under the name **Popdata** by means of the initial **Popdata=**. Data sets must be named in this way so that they can be referenced in subsequent commands, where the choice of an appropriate name is determined by the user, depending on the context.

The **read.csv** command assumes that the file is located in the working directory, which is identified in the Files pane on the bottom right of the standard RStudio environment. If the file is stored in a different location, then the entire pathname for the file needs to be included in the **read.csv** command as follows:

```
data=read.csv("~/Desktop/WorldPopulation.csv")
```

More details about the `read.csv` command can be obtained by searching for `read.csv` in the Help pane, which is also located on the bottom right of the standard RStudio environment, or via Help from the main menu.

(2) Alternatively, for files stored anywhere, the `fetchData` command from the `fetch` package can be used. As of the publication of this book, the `fetch` package is not available on CRAN and must be installed via GitHub. Therefore, three commands are needed to download the `fetch` package into your local or server version of RStudio. First, install the `devtools` package, load this package into RStudio's active, working memory with the `require` command, and then install `fetch` with the third command as follows:

```
install.packages("devtools")
require(devtools)
devtools::install_github("ProjectMOSAIC/fetch")
```

Once the `fetch` package has been installed, the command `fetchData` can be used to upload data into RStudio. This command provides a tool for obtaining data sets stored in CSV files from the Internet and storing them in the active, working memory of RStudio as demonstrated here:

```
require(fetch)
Popdata=fetchData("http://web.centre.edu/MMAC/WorldPopulation.csv")
```

As mentioned above, RStudio does not keep installed packages in its active, working memory when it is shut down. Therefore, in this setting, the `fetch` package must be loaded into RStudio using `require(fetch)` every time RStudio is opened. In addition, the entire pathname for the file must be given inside quotes when using the `fetchData` command. As with `read.csv`, data sets must be named via `Popdata=` only with some context appropriate name so that they can be referenced in subsequent commands.

The reader will want to decide for themselves the relative advantages of installing additional packages to use the `fetchData` command versus downloading data sets to their local machine and using `read.csv`.

Additional Help

The Help feature inside of RStudio is an excellent resource for learning how to use a specific command in RStudio. Additionally, there are a multitude of websites that demonstrate the many commands available in R, as well as many books detailing the commands and applications of this software package. All of the commands used in this book are detailed throughout on a just-in-time basis.

The interested reader is encouraged to further explore the possibilities using Help from the main menu, the Help pane located on the bottom right of the standard RStudio environment, the Internet, or consulting other books on R and RStudio.

Appendix D

Sources

Chapter 1

Section 1.1

- Figure 1 from "Dow Jones Industrial — U.S. — Stooq." Accessed on July 2, 2014. http://stooq.com/q/d/?s=%5Edji.
- Figure 2 from "Tropical Cyclone Report: Hurrican Ivan" by Stacy Steward, National Hurricane Center. Accessed on December 18, 2017. Public Domain. http://www.nhc.noaa.gov/data/tcr/AL092004_Ivan.pdf.
- Figure 3 from "Daily Weather Maps" by U.S. Department of Commerce, National Oceanic and Atmospheric Administration. Accessed on December 18, 2017. Public Domain. http://www.wpc.ncep.noaa.gov/dailywxmap/pdf/DWM4817.pdf.
- Franklin is the most common city name from "Fun Facts — Postal Facts." Accessed on August 13, 2014. https://about.usps.com/who-we-are/postal-facts/fun-facts.htm.
- Example 3 and Exercises 45 – 48. Millions of Twitter users per quarter from "Twitter: number of monthly active users 2010–2014 | Statistics." Accessed on July 10, 2014. http://www.statista.com/statistics/282087/number-of-monthly-active-twitter-users/.
- Question 3. Annual unemployment rate in the United States from "Bureau of Labor Statistics Data." Accessed on July 15, 2014. http://data.bls.gov/timeseries/LNS14000000.
- Exercise 15. Annual total retail sales taxes collected in the United States in each year from "Monthly & Annual Retail Trade, Main Page — U.S. Census Bureau." Accessed on July 14, 2014. http://www.census.gov/retail/.
- Exercise 16. Average debt in 2012 dollars of bachelor's degree recipients attending U.S. public colleges and universities who borrowed money to finance their education from "Average Debt Levels Public Sector Bachelor's Degree Recipients Over Time | Trends in Higher Eductation." Accessed on July 2, 2014. https://trends.collegeboard.org/student-aid/figures-tables/average-debt-levels-public-sector-bachelors-degree-recipients-over-time.
- Exercise 17. Daily gas prices in Los Angeles, California from "USA National Gas Price Heat Map - GasBuddy.com." Accessed on June 11 – 12, 2015. http://www.gasbuddy.com/GasPriceMap.

Section 1.2

- Figure 2 from NWS Winter Storm Safety Windchill Information and Chart. Accessed on December 18, 2017. http://www.nws.noaa.gov/om/cold/wind_chill.shtml.
- Example 7. The contour map is modified from Volcano Models. Accessed on June 18, 2015. http://volcano.oregonstate.edu/book/export/html/208. The image of Mount St. Helens is by Lyn Topinka (CVO Phot Archive) and is part of the Public Domain. Accessed on June 18, 2015. https://en.wikipedia.org/wiki/Mount_St_Helens.
- Example 9. For more details see "Cobb–Douglas function — Oxford Reference" at http://www.oxfordreference.com/search?q=Cobb%E2%80%93Douglas%20function or "Cobb–Douglas production function — Wikipedia" at https://en.wikipedia.org/wiki/Cobb%E2%80%93Douglas_production_function.

- Exercises 21 – 26. Percentile values of body mass index values for males aged 20 and over from "Anthropometric Reference Data for Children and Adults: United States, 2007–2010," Vital Health and Statistics, Series 11, Number 252, October 2012, U.S. Department of Health and Human Services, Centers for Disease Control and Prevention, National Center for Health Statistics.
- Exercises 27 – 32. NOAA/NESDIS Geo-polar blended 5 km sea surface temperature analysis for the North Atlantic in degrees Fahrenheit based on contour chart from "Sea Surface Temperature (SST) Contour Charts." Accessed on June 12, 2015. http://www.ospo.noaa.gov/Products/ocean/sst/contour/index.html.

Section 1.3

- Exercises 51 – 54. Millions of Twitter users per quarter from "Twitter: number of monthly active users 2010-2014 | Statistics." Accessed on July 10, 2014. http://www.statista.com/statistics/282087/number-of-monthly-active-twitter-users/.
- Exercises 55 – 58. The U.S. health expenditure total from the World Bank. Accessed on July 10, 2014. http://data.worldbank.org/country/united-states.
- Exercises 59 – 62. Average debt in 2012 dollars of bachelor's degree recipients attending U.S. public colleges and universities who borrowed money to finance their education from "Average Debt Levels Public Sector Bachelor's Degree Recipients Over Time | Trends in Higher Eductation." Accessed on July 2, 2014. https://trends.collegeboard.org/student-aid/figures-tables/average-debt-levels-public-sector-bachelors-degree-recipients-over-time.

Section 1.4

- Figure 1. Estimated world population data from 1000 to 1940 from "International Programs – Historical Estimates of World Population — U.S. Census Bureau." Accessed on June 16, 2014. http://www.census.gov/population/international/data/worldpop/table_history.php. Also, estimated world population data from 1950 to 2015 from "International Programs — Total Mid-Year Population for the World: 1950–2050 — U.S. Census Bureau." Accessed on June 16, 2014. http://www.census.gov/population/international/data/worldpop/table_population.php.
- Exercises 67 – 72. U.S. Energy Information Administration's data on U.S. coal prices in dollars per short ton in each year from "Coal — Data — U.S. Energy Information Administration (EIA)." Accessed on July 3, 2014. http://www.eia.gov/coal/data.cfm#prices.
- Exercises 73 – 76. Olsen, I. E., S. A. Groveman, M. L. Lawson, R. H. Clark, and B. S. Zemel. "New Intrauterine Growth Curves Based on United States Data." Pediatrics 125.2 (2010): E214–224. Accessed on August 14, 2014. http://pediatrics.aappublications.org/content/125/2/e214.full.pdf.
- Exercises 77 – 80. Plasma concentrations of Prozac based on "RxMed: Pharmaceutical Information — PROZAC." Accessed on August 20, 2014. http://www.rxmed.com/b.main/b2.pharmaceutical/b2.1.monographs/CPS-%20Monographs/CPS-%20(General%20Monographs-%20P)/PROZAC.html.

Section 1.5

- Figure 1. Estimated world population data from 1000 to 1940 from "International Programs — Historical Estimates of World Population — U.S. Census Bureau." Accessed on Jun 16, 2014. http://www.census.gov/population/international/data/worldpop/table_history.php. Also, estimated world population data from 1950 to 2015 from "International Programs — Total Mid-Year Population for the World: 1950–2050 — U.S. Census Bureau." Accessed on June 16, 2014. http://www.census.gov/population/international/data/worldpop/table_population.php.
- Example 2. Annual unemployment rate in the United States from "Bureau of Labor Statistics Data." Accessed on July 15, 2014. http://data.bls.gov/timeseries/LNS14000000.
- Question 2. Millions of Twitter users per quarter from "Twitter: number of monthly active users 2010–2014 | Statistics." Accessed on July 10, 2014. http://www.statista.com/statistics/282087/number-of-monthly-active-twitter-users/.
- Example 6. The global gender ratio based on the number of males per 100 females by year from "World Population Prospects, the 2012 Revision" by the United Nations Department of Economic and Social Affairs. Accessed on June 25, 2014. http://esa.un.org/unpd/wpp/Excel-Data/population.htm.
- Question 5. The length of the tornado season (number of days between the first and last tornado) each year in the 19 county warning area of the National Weather Service Office in Goodland, Kansas from "Tornado Graphs." Accessed on July 1, 2014. http://www.weather.gov/gld/tornado-tornadographs.

- Exercises 7 – 9. Annual e-commerce sales in the United States in billions of dollars from "Monthly & Annual Retail Trade, Main Page — US Census Bureau." Accessed on July 14, 2014. http://www.census.gov/retail/.

- Exercises 10 – 12. Standard & Poor's 500 stock market closing value on June 1 of each year in U.S. dollars from "GSPC Historical Prices | S&P 500 Stock — Yahoo! Finance." Accessed on July 1, 2014. http://finance.yahoo.com/q/hp?s=%5EGSPC.

- Exercises 13 – 15. World population growth rates from "International Programs — Total Midyear Population for the World 1950–2050." Accessed on June 16, 2014. http://www.census.gov/population/international/data/worldpop/table_ population.php.

- Exercises 16 – 18. Percent of high school graduates to enroll in a two-year or four-year college from National Center for Education Statistics. Accessed on July 2, 2014. http://nces.ed.gov/programs/digest/d13/tables/dt13_302.10.asp.

- Exercises 19 – 21. Average maximum temperature in degrees Fahrenheit each month in New York City in 2013 requested from National Oceanic and Atmospheric Administration. Accessed on June 25, 2014. http://www.noaa.gov.

- Exercises 22 – 24. Three-year average monthly pollen count in Brooklyn, New York City, New York from "Historic Allergy Index for 11203 | Pollen.com." Accessed on June 25, 2014. http://www.pollen.com/allergy-trends.asp?PostalCode=11203.

- Exercise 33. Annual total retail sales taxes collected in the United States in each year from "Monthly & Annual Retail Trade, Main Page — U.S. Census Bureau." Accessed on July 14, 2014. http://www.census.gov/retail/.

- Exercise 34. Average debt in 2012 dollars of bachelor's degree recipients attending U.S. public colleges and universities who borrowed money to finance their education from "Average Debt Levels Public Sector Bachelor's Degree Recipients Over Time | Trends in Higher Eductation." Accessed on July 2, 2014. https://trends.collegeboard.org/student-aid/figures-tables/average-debt-levels-public-sector-bachelors-degree-recipients-over-time.

- Exercise 35. The U.S. health expenditure total from the World Bank. Accessed on July 10, 2014. http://data.worldbank.org/country/united-states.

- Exercise 37. Average number of goals scored per game in World Cup tournaments from "FIFA World Cup Record — Organisation." Accessed on July 17, 2014. http://www.fifa.com/worldfootball/statisticsandrecords/tournaments/worldcup/organisation/index.html.

- Exercise 38. Total annual sales in thousands of hybrid vehicles in the United States from "Electric Drive Transportation Association." Accessed on June 26, 2014. http://electricdrive.org/index.php?ht=d/sp/i/20952/pid/20952.

Section 1.6

- A biography of John Napier is at http://www-groups.dcs.st-and.ac.uk/~history/Biographies/Napier.html. Accessed on July 8, 2016.

- Example 5. For more information about population growth, see Vandermeer, J. (2010) How Populations Grow: The Exponential and Logistic Equations. Nature Education Knowledge 3(10):15. Access on November 27, 2017. http://www.nature.com/scitable/knowledge/library/how-populations-grow-the-exponential-and-logistic-13240157.

- Exercises 73 – 76. The half-life of carbon-14 from "Periodic Chart of the Nuclides." Accessed on August 8, 2015. http://ie.lbl.gov/toi/perchart.htm.

- Exercises 77 – 80. The half-life of cesium-137 from "Periodic Chart of the Nuclides." Accessed on August 8, 2015. http://ie.lbl.gov/toi/perchart.htm.

- Exercises 85 – 88. U.S. Energy Information Administration's data on U.S. coal prices in dollars per short ton in each year from "Coal — Data — U.S. Energy Information Administration (EIA)." Accessed on July 3, 2014. http://www.eia.gov/coal/data.cfm#prices.

- Exercises 89 – 92. Olsen, I. E., S. A. Groveman, M. L. Lawson, R. H. Clark, and B. S. Zemel. "New Intrauterine Growth Curves Based on United States Data." Pediatrics 125.2 (2010): E214–224. Accessed on August 14, 2014. http://pediatrics.aappublications.org/content/125/2/e214.full.pdf.

- Exercises 93 – 96. Plasma concentrations of Prozac based on "RxMed: Pharmaceutical Information — PROZAC." Accessed on August 20, 2014. http://www.rxmed.com/b.main/b2.pharmaceutical/b2.1.monographs/CPS-%20Monographs/CPS-%20(General%20Monographs-%20P)/PROZAC.html.

Section 1.7

- Figure 1. Average maximum temperature in Danville, Kentucky at the beginning of each month since January 2006 from "noaa.gov." Accessed on June 25, 2014. http://www1.ncdc.noaa.gov/pub/orders/cdo/352625.pdf.

- An overview of the history of trigonometry is at http://www-history.mcs.st-andrews.ac.uk/HistTopics/Trigonometric_functions.html. Accessed on July 9, 2016.

- An overview of the history of Babylonian mathematics is at http://www-history.mcs.st-andrews.ac.uk/HistTopics/Babylonian_mathematics.html. Accessed on July 9, 2016.

- Figure 2 and 5. Modified from Example: Unit circle by Supreme Aryal. Accessed on December 21, 2017. http://www.texample.net/tikz/examples/unit-circle/. Permission granted by Creative Commons License 2.5.

- Exercises 73 – 77. Number of minutes after 3 p.m. until sunset in Greenwich, England since January 2010 from "Sunrise and sunset times in Greenwich Borough." Accessed on July 14, 2014. http://www.timeanddate.com/sun/uk/greenwich-city.

- Exercises 78 – 82. Average maximum temperature in degrees Fahrenheit each month in New York City in 2013 requested from National Oceanic and Atmospheric Administration. Accessed on June 25, 2014. http://www.noaa.gov.

Chapter 2

Section 2.1

- Examples 1(a), 2(a), 3, and 4. Millions of Twitter users per quarter from "Twitter: number of monthly active users 2010–2014 | Statistics." Accessed on July 10, 2014. http://www.statista.com/statistics/282087/number-of-monthly-active-twitter-users/.

- Examples 1(b) and 2(b). Olsen, I. E., S. A. Groveman, M. L. Lawson, R. H. Clark, and B. S. Zemel. "New Intrauterine Growth Curves Based on United States Data." Pediatrics 125.2 (2010): E214–224. Accessed on August 14, 2014. http://pediatrics.aappublications.org/content/125/2/e214.full.pdf.

- Questions 1(a) and 2(a). Annual total retail sales taxes collected in the United States in each year from "Monthly & Annual Retail Trade, Main Page — U.S. Census Bureau." Accessed on July 14, 2014. http://www.census.gov/retail/.

- Questions 1(b), 2(b), 3, and 4. Average debt load in thousands of 2012 dollars at the end of the spring term in each year for bachelor's degree recipients attending public four-year colleges and universities who borrowed money to finance their education. from "Average Debt Levels Public Sector Bachelor's Degree Recipients Over Time | Trends in Higher Eductation." Accessed on July 2, 2014. https://trends.collegeboard.org/student-aid/figures-tables/average-debt-levels-public-sector-bachelors-degree-recipients-over-time.

- Example 5. Total U.S. health expenditures as a percentage of GDP from the World Bank. Accessed on July 10, 2014. http://data.worldbank.org/country/united-states.

- Question 5. Personal data collection by one author's spouse. Used with permission.

- Exercise 17. Annual e-commerce sales in the United States in billions of dollars from "Monthly & Annual Retail Trade, Main Page — US Census Bureau." Accessed on July 14, 2014. http://www.census.gov/retail/.

- Exercise 18. Highest value of Facebook stock from "Facebook Historical Prices." Accessed on July 2, 2014. http://finance.yahoo.com/q/hp?s=FB&a=04&b=18&c=2012&d=05&e=27&f=2014&g=m.

- Exercise 19. World population growth rates from "International Programs — Total Midyear Population for the World 1950–2050." Accessed on June 16, 2014. http://www.census.gov/population/international/data/worldpop/table_population.php.

- Exercise 20. Average number of goals scored per game in World Cup tournaments from "FIFA World Cup Record — Organisation." Accessed on July 17, 2014. http://www.fifa.com/worldfootball/statisticsandrecords/tournaments/worldcup/organisation/index.html.

- Exercise 21. Percent of high school graduates to enroll in a two-year or four-year college from National Center for Education Statistics. Accessed on July 2, 2014. http://nces.ed.gov/programs/digest/d13/tables/dt13_302.10.asp.

- Exercise 22. Average maximum temperature in degrees Fahrenheit each month in New York City in 2013 requested from National Oceanic and Atmospheric Administration. Accessed on June 25, 2014. http://www.noaa.gov.

- Exercises 47–52. The length of the tornado season (number of days between the first and last tornado) each year in the 19 county warning area of the National Weather Service Office in Goodland, Kansas from "Tornado Graphs." Accessed on July 1, 2014. http://www.weather.gov/gld/tornado-tornadographs.

- Exercises 53–58. United States retail prescription drug sales in billions of dollars per year from the U.S. Census Bureau. Accessed on July 8, 2014. https://www.census.gov/compendia/statab/2012/tables/12s0159.xls.

- Exercises 59–64. The number of prescription drugs sold in the United States in millions per year from the U.S. Census Bureau. Accessed on July 8, 2014. https://www.census.gov/compendia/statab/2012/tables/12s0159.xls.

- Exercises 65–70. The total number of burgers sold by McDonald's in billions as of each year from "Over How Many Billion Served." Accessed on July 3, 2014. http://overhowmanybillionserved.blogspot.com/.

- Exercises 71–76. The global gender ratio based on the number of males per 100 females by year from "World Population Prospects, the 2012 Revision" by the United Nations Department of Economic and Social Affairs. Accessed on June 25, 2014. http://esa.un.org/unpd/wpp/Excel-Data/population.htm.

- Exercises 77-80. United States monthly unemployment rate from January 2010 to December 2014 from "Bureau of Labor Statistics Data." Accessed on June 22, 2015. http://data.bls.gov/timeseries/LNS14000000.

- Exercises 81-84. Total midyear population for the world from "International Programs — Total Midyear Population for the World: 1950–2050 — U.S. Census Bureau." Accessed on June 16, 2014. http://www.census.gov/population/international/data/worldpop/table_population.php.

- Exercises 85-88. Interest rates on 15-year, fixed-rate conventional home mortgages annually from 1992 to 2014 from "Mortgage Interest Rates History." Accessed on June 22, 2015. http://www.fedprimerate.com/mortgage_rates.htm.

- Exercises 89-92. Number of Facebook users in millions from 2009 through 2012 from "Number of active users at Facebook over the years — Yahoo News" and "Facebook: number of active users 2015 | Statistic." Accessed on June 22, 2015. http://news.yahoo.com/number-active-users-facebook-over-230449748.html and http://www.statista.com/statistics/264810/number-of-monthly-active-facebook-users-worldwide/.

- Exercises 93-96. The high school dropout rate in the United States from 1970 through 2012 from "Percentage of high school dropouts among persons 16 to 24 years old." Accessed on June 22, 2015. http://nces.ed.gov/programs/digest/d13/tables/dt13_219.70.asp.

- Exercises 97-100. United States carbon dioxide emissions in kT annually from 1960 to 2010 according to the World Bank at "Data | United States." Accessed on July 10, 2014. http://data.worldbank.org/country/united-states.

Section 2.2

- Examples 1 and 6, and Questions 3 and 5. Olsen, I. E., S. A. Groveman, M. L. Lawson, R. H. Clark, and B. S. Zemel. "New Intrauterine Growth Curves Based on United States Data." Pediatrics 125.2 (2010): E214–224. Accessed on August 14, 2014. http://pediatrics.aappublications.org/content/125/2/e214.full.pdf.

- Examples 2 and 4. Annual U.S. Federal Funds Interest Rate, which is the interest rate at which banks and credit unions lend money to each other overnight, in each year from "FRB: H.15 Release — Selected Interest Rates — Historical Data." Accessed on July 3, 2014. http://www.federalreserve.gov/releases/h15/data.htm.

- Examples 3 and 5. U.S. Energy Information Administration's data on United States coal prices in dollars per short ton in each year from "Coal — Data — U.S. Energy Information Administration (EIA)." Accessed on July 3, 2014. http://www.eia.gov/coal/data.cfm#prices.

- Questions 1 and 6. Plasma concentrations of Prozac based on "RxMed: Pharmaceutical Information — PROZAC." Accessed on August 20, 2014. http://www.rxmed.com/b.main/b2.pharmaceutical/b2.1.monographs/CPS-%20Monographs/CPS-%20(General%20Monographs-%20P)/PROZAC.html.

- Questions 2 and 4. United Nations global estimates of the number of males per 100 females at the beginning of the last five decades from "World Population Prospects, the 2012 revision." Accessed on June 25, 2014. http://esa.un.org/unpd/wpp/Excel-Data/population.htm.

- Example 7. Estimated world population data from 1950 to 2015 from "International Programs
  — Total Mid-Year Population for the World: 1950–2050 — U.S. Census Bureau." Accessed on
  June 16, 2014. http://www.census.gov/population/international/data/worldpop/table_popula
  tion.php. Estimated world population data from 1000 to 1940 from "International Programs —
  Historical Estimates of World Population — U.S. Census Bureau." Accessed on Jun 16, 2014.
  http://www.census.gov/population/international/data/worldpop/table_history.php.

- Question 7. Closing stock market value of the Dow Jones Industrial Average at the end of each
  quarter from March 31, 1930 through December 31, 2014 from "^DJI — Nasdaq Composite —
  U.S. — Stooq." Accessed on July 7, 2015. http://stooq.com/q/d/?s=^dji.

- Exercise 17. Standard & Poor's 500 stock market closing value on June 1 of each year in U.S.
  dollars from "GSPC Historical Prices | S&P 500 Stock — Yahoo! Finance." Accessed on July 1,
  2014. http://finance.yahoo.com/q/hp?s=%5EGSPC.

- Exercise 18. Annual unemployment rate in the United States from "Bureau of Labor Statistics
  Data." Accessed on July 15, 2014. http://data.bls.gov/timeseries/LNS14000000.

- Exercise 19. Total annual sales in thousands of hybrid vehicles in the United States from "Electric
  Drive Transportation Association." Accessed on June 26, 2014. http://electricdrive.org/index.
  php?ht=d/sp/i/20952/pid/20952.

- Exercise 20. U.S. field production of crude oil in billions of barrels from the U.S. Energy Information Administration from "U.S. Field Production of Crude Oil." Accessed on July 3, 2014.
  http://www.eia.gov/dnav/pet/hist/LeafHandler.ashx?n=PET&s=MCRFPUS1&f=M.

- Exercise 21. Three-year average monthly pollen count in Brooklyn, New York City from "Historic
  Allergy Index for 11203 | Pollen.com." Accessed on June 25, 2014. http://www.pollen.com/aller
  gy-trends.asp?PostalCode=11203.

- Exercise 22. Average maximum temperature in degrees Fahrenheit each month in New York City
  in 2013 requested from National Oceanic and Atmospheric Administration. Accessed on June
  25, 2014. http://www.noaa.gov.

- Exercises 47 – 52. Number of red cards and expulsions given per men's World Cup tournament
  from "Planet World Cup — Statistics — Discipline." Accessed on July 16, 2014. http://www.pla
  networldcup.com/STATS/stat_disc.html.

- Exercises 53 – 58. Average debt load in 2013 dollars of bachelor's degree recipients attending U.S.
  public colleges and universities who borrowed money to finance their education from "Average Cumulative Debt Load of Bachelor's Degree Recipients at Public Four-Year Institutions over Time —
  Trends in Higher Education." Accessed on July 5, 2015. https://trends.collegeboard.org/student-
  aid/figures-tables/average-debt-levels-public-sector-bachelors-degree-recipients-over-time.

- Exercises 59 – 64. Highest value of Facebook stock each month in 2013 from "FB Historical Prices
  | Facebook, Inc. Stock — Yahoo! Finance." Accessed on July 2, 2014. http://finance.yahoo.com/
  q/hp?s=FB.

- Exercises 65 – 70. Number of tornadoes each year in the 19-county warning area of the National
  Weather Service Office in Goodland, Kansas from "Tornago Graphs." Accessed on July 1, 2014.
  http://www.weather.gov/gld/tornado-tornadographs.

- Exercises 71 – 76. Population of Ireland in millions of people each year from "IRELAND: population growth of the whole country." Accessed on July 18,2014. http://www.populstat.info/Europe
  /irelandc.htm.

- Exercises 77 – 80. Closing price of Apple Inc. stock in U.S. dollars adjusted for dividends
  and splits at the beginning of each month from January 1981 through December 2014 from
  "AAPL Historical Prices | Apple Inc. Stock — Yahoo! Finance." Accessed on July 7, 2015.
  http://finance.yahoo.com/q/hp?s=AAPL.

- Exercises 81 – 84. Closing NASDAQ stock market value in U.S. dollars at the end of each quarter
  from March 1938 through December 2014 from "^NDQ — Nasdaq Composite — U.S. — Stooq."
  Accessed on July 7, 2015. http://stooq.com/q/d/?s=^ndq.

- Exercises 85 – 88. Average interest rate for conventional 30-year mortgages each year from 1981
  to 2012 from "Primary Mortgage Market Survey Archives — 30 Year Fixed Rate Mortgages —
  Freddie Mac." Accessed on July 7, 2015. http://www.freddiemac.com/pmms/pmms30.htm.

- Exercises 89 – 92. Population of the Netherlands each decade from 1700 through 2010 from "The
  NETHERLANDS : country population." Accessed on July 11, 2014. http://www.populstat.info/
  Europe/netherlc.htm.

- Exercises 93 – 96. Number of physicians per 1000 people as a function of average life expectancy
  in different countries in 2010 from "Physicians (per 1,000 people) | Data | Table." Accessed on
  July 10, 2014. http://data.worldbank.org/indicator/SH.MED.PHYS.ZS.

- Exercises 97 – 100. Total number of AP Calculus exams taken each year from 1955 to 2015 from personal correspondence with Stephen M. Kokoska on April 6, 2014.

Section 2.3

- Zipf's Law and Goetz's Law based on "Zipf's Law — from Wolfram Mathworld." Accessed on February 12, 2015. http://mathworld.wolfram.com/ZipfsLaw.html.
- Example 1 modified from http://en.wikipedia.org/wiki/Stevens'_power_law. Access on February 12, 2015.
- Examples 2 – 5. Revolutions per minute of engines as a function of engine mass from pages 60–61. McMahon, Thomas A., and John Tyler Bonner. *On Size and Life*. New York: Scientific American Library, 1983.
- Questions 1 – 4. Flying speed as length of animals from page 153. McMahon, Thomas A., and John Tyler Bonner. *On Size and Life*. New York: Scientific American Library, 1983.
- Exercise 39. Standard & Poor's 500 stock market closing value on June 1 of each year in U.S. dollars from "GSPC Historical Prices | S&P 500 Stock — Yahoo! Finance." Accessed on July 1, 2014. http://finance.yahoo.com/q/hp?s=%5EGSPC.
- Exercise 40. 1990 percentage of children with telephone service by parental income groups from Table 6.3. Mayer, Susan E. *What Money Can't Buy: Family Income and Children's Life Chances*. Cambridge, Massachusetts: Harvard University Press, 1997.
- Exercise 41. Total annual sales in thousands of hybrid vehicles in the United States from "Electric Drive Transportation Association." Accessed on June 26, 2014. http://electricdrive.org/index.php?ht=d/sp/i/20952/pid/20952.
- Exercise 42. United States field production of crude oil in billions of barrels from the U.S. Energy Information Administration from "U.S. Field Production of Crude Oil." Accessed on July 3, 2014. http://www.eia.gov/dnav/pet/hist/LeafHandler.ashx?n=PET&s=MCRFPUS1&f=M.
- Exercise 43. Average number of goals scored per game in World Cup tournaments from "FIFA World Cup Record — Organisation." Accessed on July 17, 2014. http://www.fifa.com/worldfootball/statisticsandrecords/tournaments/worldcup/organisation/index.html.
- Exercise 44. Average maximum temperature in degrees Fahrenheit each month in New York City in 2013 requested from National Oceanic and Atmospheric Administration. Accessed on June 25, 2014. http://www.noaa.gov.
- Exercises 67 – 70. Swimming speed and length of animals from page 152. McMahon, Thomas A., and John Tyler Bonner. *On Size and Life*. New York: Scientific American Library, 1983.
- Exercises 71 – 74. Running speed and length of animals from page 152. McMahon, Thomas A., and John Tyler Bonner. *On Size and Life*. New York: Scientific American Library, 1983.
- Exercises 75 – 79. Atmospheric carbon dioxide from Mauna Loa in ppmv (parts per million by volume) as a function of years from 1958 to 2008 from "Atmospheric Carbon Dioxide Record from Mauna Loa." Accessed on August 19, 2015. http://cdiac.ornl.gov/trends/co2/sio-mlo.html.
- Exercises 80 – 83. Hudson, L. N., Isaac, N. J. B., Reuman, D. C. (2013), The relationship between body mass and field metabolic rate among individual birds and mammals. Journal of Ecology, 82: 1009–1020. doi: 10.1111/1365-2656.12086. Accessed on August 19, 2015. http://onlinelibrary.wiley.com/doi/10.1111/1365-2656.12086/suppinfo.

Section 2.4

- Figure 1 and Examples 2 and 3. Average maximum temperature in Danville, Kentucky at the beginning of each month since January 2006 from "noaa.gov." Accessed on June 25, 2014. http://www1.ncdc.noaa.gov/pub/orders/cdo/352625.pdf.
- Questions 2 and 3. Number of minutes after 3 p.m. until sunset in Greenwich, England since January 2010 from "Sunrise and sunset times in Greenwich Borough." Accessed on July 14, 2014. http://www.timeanddate.com/sun/uk/greenwich-city.
- Exercises 41 – 45. Number of minutes after 4 a.m. until sunrise in Los Angeles, adjusted for Daylight Savings Time, from January 2010 (month 1) through December 2011 (month 24) from "Sunrise and sunset times in Los Angeles, December 2011." Accessed on July 14, 2014. http://www.timeanddate.com/sun/usa/los-angeles?month=12&year=2011.
- Exercises 46 – 50. Number of minutes after 4 p.m. until sunset in Los Angeles, California, adjusted for Daylight Savings Time, from January 2010 (month 1) through December 2013 (month 48) from "Sunrise and sunset times in Los Angeles, December 2011." Accessed on July 14, 2014. http://www.timeanddate.com/sun/usa/los-angeles?month=12&year=2011.

- Exercises 51–56. Altitude angle of the sun in Anchorage, Alaska, each hour from midnight on June 29, 2014 (hour 0) until midnight on June 30, 2014 (hour 24) from "Sun & moon times, Anchorage, Alaska, U.S.A." Accessed on July 1, 2014. http://www.timeanddate.com/astronomy/usa /anchorage.

- Exercise 57–61. Annual total retail sales taxes collected in the United States in each year from "Monthly & Annual Retail Trade, Main Page — U.S. Census Bureau." Accessed on July 14, 2014. http://www.census.gov/retail/.

- Exercises 62–66 and Exercises 67–71 are based on data sets from Project Mosaic. Accessed on August 19, 2015. http://www.mosaic-web.org.

Section 2.5

- Figure 1 and Question 2. Population of the Netherlands from "The NETHERLANDS : country populations." Accessed on July 11, 2014. http://www.populstat.info/Europe/netherlc.htm.

- Examples 2 and 3 from "Twitter : number of monthly active users 2010–2014 | Statistic." Accessed on July 2, 2014. http://www.statista.com/statistics/282087/number-of-monthly-active-twitter-users/.

- Exercises 47–51. Population of Belgium in millions of people by year from "Population of the Netherlands, Belgium, and Luxembourg." Accessed on June 26, 2014. http://www.tacitus.nu/ historical-atlas/population/benelux.htm.

- Exercises 52–56. Number of Facebook users in millions from 2009 through 2012 from "Number of active users at Facebook over the years — Yahoo News" and "Facebook: number of active users 2015 | Statistics." Accessed on June 22, 2015. http://news.yahoo.com/number-active-users-facebook-over-230449748.html and http://www.statista.com/statistics/264810/number-of-monthly-active-facebook-users-worldwide/.

- Exercises 57–61. Average SAT math score in Kentucky each year from 1980 to 2013 from College Board's 2013 SAT State Profile Report for Kentucky. Accessed on July 16, 2014. http://media.collegeboard.com/digitalServices/pdf/research/2013/KY_13_03_03_01.pdf.

- Exercises 62–66. Percent of high school completers to enroll in either a two-year or four-year college each year from 1972 to 2012 from "Recent high school completers and their enrollments in 2-year and 4-year colleges, by sex: 1960 through 2012." Accessed on July 1, 2014. http://nces.ed.gov/programs/digest/d13/tables/dt13_302.10.asp.

- Exercises 67–71. 2014 Ebola Outbreak in West Africa from Centers for Disease Control and Prevention. Accessed on June 2, 2016. http://www.cdc.gov/vhf/ebola/outbreaks/2014-west-africa/cumulative-cases-graphs.html.

- Exercises 72–76. Number of yellow cards given per men's World Cup tournament from "Planet World Cup." Accessed on July 16, 2014. http://www.planetworldcup.com.

Section 2.6

- Example 1. Total number of AP Calculus exams taken each year from 1955 to 2015 from personal correspondence with Stephen M. Kokoska on April 6, 2014.

- Example 2. 2014 Ebola Outbreak in West Africa from Centers for Disease Control and Prevention. Accessed on June 2, 2016. http://www.cdc.gov/vhf/ebola/outbreaks/2014-west-africa/cumula tive-cases-graphs.html.

- Question 1. Modified from: Taylor, G. "The Formation of a Blast Wave by a Very Intense Explosion. II. The Atomic Explosion of 1945." *Proceedings of the Royal Society A: Mathematical, Physical and Engineering Sciences*, 1950, 201: 175-86.

- Question 2. Observed water levels in Eastport, Maine from NOAA Tides and Currents. Accessed on June 2, 2016. http://tidesandcurrents.noaa.gov/waterlevels.html?id=8410140&units= standard&bdate=20160526&edate=20160527&timezone=GMT&datum=MLLW&interval=6& action=

- Question 3, Example 5, and Question 4. Natural Gas from U.S. Energy Information Administration. Accessed on June 9, 2016. http://www.eia.gov/dnav/ng/hist/n9140us2a.htm.

- Example 3. U.S. census data for 1950–2000 is from "Measuring America: The Decennial Censuses from 1790 to 2000." Accessed on June 9, 2016. https://www.census.gov/prod/2002pubs/pol02-ma.
  pdf. The 2010 data is from http://www.census.gov/2010census/popmap/.

- Example 4. Millions of Twitter users per quarter from "Twitter: number of monthly active users 2010-2016 | Statistics." Accessed on June 9, 2016. http://www.statista.com/statistics/282087/ number-of-monthly-active-twitter-users/.

- Example 6. World population from the U.S. Census Bureau. Accessed on June 3, 2016. http://www.census.gov/population/international/data/worldpop/table_population.php.

- Exercise 19. Annual unemployment rate in the United States from "Bureau of Labor Statistics Data." Accessed on July 15, 2014. http://data.bls.gov/timeseries/LNS14000000.

- Exercise 20. Annual e-commerce sales in the United States in billions of dollars from "Monthly & Annual Retail Trade, Main Page — US Census Bureau." Accessed on July 14, 2014. http://www.census.gov/retail/.

- Exercise 21. Standard & Poor's 500 stock market closing value on June 1 of each year in U.S. dollars from "GSPC Historical Prices | S&P 500 Stock — Yahoo! Finance." Accessed on July 1, 2014. http://finance.yahoo.com/q/hp?s=%5EGSPC.

- Exercise 22. 1990 percentage of children with telephone service by parental income groups from Table 6.3. Mayer, Susan E. *What Money Can't Buy: Family Income and Children's Life Chances.* Cambridge, Massachusetts: Harvard University Press, 1997.

- Exercise 23. Highest value of Facebook stock from "Facebook Historical Prices." Accessed on July 2, 2014. http://finance.yahoo.com/q/hp?s=FB&a=04&b=18&c=2012&d=05&e=27&f=2014&g=m.

- Exercise 24. Total annual sales in thousands of hybrid vehicles in the United States from "Electric Drive Transportation Association." Accessed on June 26, 2014. http://electricdrive.org/index.php?ht=d/sp/i/20952/pid/20952.

- Exercise 25. Three-year average monthly pollen count in Brooklyn, New York City from "Historic Allergy Index for 11203 | Pollen.com." Accessed on June 25, 2014. http://www.pollen.com/allergy-trends.asp?PostalCode=11203.

- Exercise 26. U.S. field production of crude oil in billions of barrels from the U.S. Energy Information Administration from "U.S. Field Production of Crude Oil." Accessed on July 3, 2014. http://www.eia.gov/dnav/pet/hist/LeafHandler.ashx?n=PET&s=MCRFPUS1&f=M.

- Exercise 27. Average number of goals scored per game in World Cup tournaments from "FIFA World Cup Record — Organisation." Accessed on July 17, 2014. http://www.fifa.com/worldfootball/statisticsandrecords/tournaments/worldcup/organisation/index.html.

- Exercise 28. Average maximum temperature in degrees Fahrenheit each month in New York City in 2013 requested from National Oceanic and Atmospheric Administration. Accessed on June 25, 2014. http://www.noaa.gov.

- Exercise 29. Percent of high school graduates to enroll in a two-year or four-year college from National Center for Education Statistics. Accessed on July 2, 2014. http://nces.ed.gov/programs/digest/d13/tables/dt13_302.10.asp.

- Exercise 30. World population growth rates from "International Programs — Total Midyear Population for the World 1950–2050." Accessed on June 16, 2014. http://www.census.gov/population/international/data/worldpop/table_population.php.

- Exercises 31 – 38. Total U.S. electronic and mail-order shopping sales in millions of dollars each year from 1999 to 2012 from "Monthly & Annual Retail Trade, Main Page — US Census Bureau." Accessed on July 16, 2014.

- Exercises 39 – 46. Closing stock market value of the Dow Jones Industrial Average at the end of each quarter from March 31, 1930 through December 31, 2014 from "^DJI — Nasdaq Composite — U.S. — Stooq." Accessed on July 7, 2015. http://stooq.com/q/d/?s=^dji.

- Exercises 47 – 54. Interest rates on 15-year, fixed-rate conventional home mortgages annually from 1992 to 2014 from "Mortgage Interest Rates History." Accessed on June 22, 2015. http://www.fedprimerate.com/mortgage_rates.htm.

- Exercises 55 – 62. Tidal measurements in Pearl Harbor, Hawaii based on data sets from Project Mosaic. Accessed on August 19, 2015. http://www.mosaic-web.org.

- Exercises 63 – 70. Hudson, L. N., Isaac, N. J. B., Reuman, D. C. (2013), The relationship between body mass and field metabolic rate among individual birds and mammals. Journal of Ecology, 82: 1009-1020. doi: 10.1111/1365-2656.12086. Accessed on August 19, 2015. http://onlinelibrary.wiley.com/doi/10.1111/1365-2656.12086/suppinfo.

- Exercises 71 – 78. U.S. carbon dioxide emissions in kT annually from 1960 to 2010 according to the World Bank at "Data | United States." Accessed on July 10, 2014. http://data.worldbank.org/country/united-states.

- Exercises 79 – 86. Percent of high school completers to enroll in either a two-year or four-year college each year from 1972 to 2012 from "Recent high school completers and their enrollments in 2-year and 4-year colleges, by sex: 1960 through 2012." Accessed on July 1, 2014. http://nces.ed.gov/programs/digest/d13/tables/dt13_302.10.asp.

- Exercises 87 – 94. Atmospheric carbon dioxide from Mauna Loa in ppmv (parts per million by volume) as a function of years from 1958 to 2008 from "Atmospheric Carbon Dioxide Record from Mauna Loa." Accessed on August 19, 2015. http://cdiac.ornl.gov/trends/co2/sio-mlo.html.
- Exercises 95 – 102. Altitude angle of the sun in Anchorage, Alaska, each hour from midnight on June 29, 2014 (hour 0) until midnight on June 30, 2014 (hour 24) from "Sun & moon times, Anchorage, Alaska, U.S.A." Accessed on July 1, 2014. http://www.timeanddate.com/astronomy/usa/anchorage.
- Exercises 103 – 110. Number of physicians per 1000 people as a function of average life expectancy in different countries in 2010 from "Physicians (per 1,000 people) | Data | Table." Accessed on July 10, 2014. http://data.worldbank.org/indicator/SH.MED.PHYS.ZS.
- Exercises 111 – 113. Average debt load in 2013 dollars of bachelor's degree recipients attending U.S. public colleges and universities who borrowed money to finance their education from "Average Cumulative Debt Load of Bachelor's Degree Recipients at Public Four-Year Institutions over Time — Trends in Higher Education." Accessed on July 5, 2015. https://trends.collegeboard.org/student-aid/figures-tables/average-debt-levels-public-sector-bachelors-degree-recipients-over-time.
- Exercises 114 – 116. The global gender ratio based on the number of males per 100 females by year from "World Population Prospects, the 2012 Revision" by the United Nations Department of Economic and Social Affairs. Accessed on June 25, 2014. http://esa.un.org/unpd/wpp/Excel-Data/population.htm.
- Exercises 117 – 119. The total number of burgers sold by McDonald's in billions as of each year from "Over How Many Billion Served." Accessed on July 3, 2014. http://overhowmanybillionserved.blogspot.com/.
- Exercises 120 – 122. Average interest rate for conventional 30-year mortgages each year from 1981 to 2012 from "Primary Mortgage Market Survey Archives — 30 Year Fixed Rate Mortgages — Freddie Mac." Accessed on July 7, 2015. http://www.freddiemac.com/pmms/pmms30.htm.
- Exercises 123 – 125. Ford Motor Company (F) stock market value quarterly in U.S. dollars. Accessed on July 7, 2014.http://stooq.com/q/d/?s=f.us.
- Exercises 126 – 128. U.S. field production of crude oil in billions of barrels from the U.S. Energy Information Administration from "U.S. Field Production of Crude Oil." Accessed on July 3, 2014. http://www.eia.gov/dnav/pet/hist/LeafHandler.ashx?n=PET&s=MCRFPUS1&f=M.
- Exercise 129. The U.S. Hispanic population has increased sixfold since 1970. Accessed on June 16, 2014. http://www.pewresearch.org/fact-tank/2014/02/26/the-u-s-hispanic-population-has-increased-sixfold-since-1970/.
- Exercise 130. Three-year average monthly pollen count in Los Angeles, California from "Historic Allergy Index for 90001 | Pollen.com." Accessed on June 25, 2014. http://www.pollen.com/allergy-trends.asp?PostalCode=90001.

Section 2.7

- Mars Climate Orbiter story based on "Metric mishap caused loss of NASA orbiter." Accessed on February 3, 2015. http://www.cnn.com/TECH/space/9909/30/mars.metric.02/.
- Example 3. Population and land area of Chicago from "Chicago Quick Facts from US Census Bureau." Accessed on February 3, 2015. http://quickfacts.census.gov/qfd/states/17/1714000.html.
- Question 4(b). Olympic swimming pool dimensions from "Olympic-size swimming pool." Accessed on June 16, 2015. https://en.wikipedia.org/wiki/Olympic-size_swimming_pool.
- Example 12 and 13. Modified from: Taylor, G. "The Formation of a Blast Wave by a Very Intense Explosion. II. The Atomic Explosion of 1945." *Proceedings of the Royal Society A: Mathematical, Physical and Engineering Sciences*, 1950, 201: 175–86.
- Figure 1. Trinity to Trinity Image Gallery. Accessed on December 28, 2017. Public domain courtesy of the U.S. Deparment of Energy. http://www.lanl.gov/about/history-innovation/trinity-to-trinity/gallery.php.
- Exercise 54. Based on Ledder, Glenn. "1.5 Optimization." In *Mathematics for the Life Sciences: Calculus, Modeling, Probability, and Dynamical Systems*, 42–43. Springer Verlag, 2013.
- Exercises 56 – 58. The SIR model was introduced by Kermack, W. O., and A. G. McKendrick. "A Contribution to the Mathematical Theory of Epidemics." *Proceedings of the Royal Society A: Mathematical, Physical and Engineering Sciences*, 1927, 115: 700–21. doi:10.1098/rspa.1927.0118.

Chapter 3

Section 3.1

- Figure 1: Vector field of arterial blood from Pedersen, M.M., et. al. *Arterial secondary blood flow patterns visualized with vector flow ultrasound.* Ultrasonics Sympsiums, 2011 IEEE International. Oct. 18–21, 2011, pp. 1242–1245.

Section 3.2

- Example 1. Millions of Twitter users per quarter from "Twitter: number of monthly active users 2010–2014 | Statistics." Accessed on July 10, 2014. http://www.statista.com/statistics/282087/number-of-monthly-active-twitter-users/.

- Question 1. Average debt load in 2012 dollars of bachelor's degree recipients attending U.S. public colleges and universities who borrowed money to finance their education from "Average Debt Levels Public Sector Bachelor's Degree Recipients Over Time | Trends in Higher Eductation." Accessed on July 2, 2014. https://trends.collegeboard.org/student-aid/figures-tables/average-debt-levels-public-sector-bachelors-degree-recipients-over-time.

- Exercise 47. Annual e-commerce sales in the United States in billions of dollars from "Monthly & Annual Retail Trade, Main Page — US Census Bureau." Accessed on July 14, 2014. http://www.census.gov/retail/.

- Exercise 48. Highest value of Facebook stock from "Facebook Historical Prices." Accessed on July 2, 2014. http://finance.yahoo.com/q/hp?s=FB&a=04&b=18&c=2012&d=05&e=27&f=2014&g=m.

- Exercise 49. World population growth rates from "International Programs — Total Midyear Population for the World 1950–2050." Accessed on June 16, 2014. http://www.census.gov/population/international/data/worldpop/table_ population.php.

- Exercise 50. Average number of goals scored per game in World Cup tournaments from "FIFA World Cup Record — Organisation." Accessed on July 17, 2014. http://www.fifa.com/worldfootball/statisticsandrecords/tournaments/worldcup/organisation/index.html

- Exercise 51. Percent of high school graduates to enroll in a two-year or four-year college from National Center for Education Statistics. Accessed on July 2, 2014. http://nces.ed.gov/programs/digest/d13/tables/dt13_302.10.asp

- Exercise 52. Average maximum temperature in degrees Fahrenheit each month in New York City in 2013 requested from National Oceanic and Atmospheric Administration. Accessed on June 25, 2014. http://www.noaa.gov.

Section 3.3

- Example 4. Millions of Twitter users per quarter from "Twitter: number of monthly active users 2010-2014 | Statistics." Accessed on July 10, 2014. http://www.statista.com/statis-tics/282087/number-of-monthly-active-twitter-users/.

- Question 5. Average total debt of bachelor's degree recipients attending public four-year colleges and universities in 2012 dollars from "Average Debt Levels Public Sector Bachelor's Degree Recipients Over Time | Trends in Higher Eductation." Accessed on July 14, 2014. https://trends.collegeboard.org/student-aid/figures-tables/average-debt-levels-public-sector-bachelors-degree-recipients-over-time.

- Example 5. U.S. Energy Information Administration's data on U.S. coal prices in dollars per short ton in each year from "Coal — Data — U.S. Energy Information Administration (EIA)." Accessed on July 3, 2014. http://www.eia.gov/coal/data.cfm#prices.

- Question 6. Olsen, I. E., S. A. Groveman, M. L. Lawson, R. H. Clark, and B. S. Zemel. "New Intrauterine Growth Curves Based on United States Data." Pediatrics 125.2 (2010): E214–224. Accessed on August 14, 2014. http://pediatrics.aappublications.org/content/125/2/e214.full.pdf.

- Example 6. City of Austin Population History. Accessed on July 17, 2015. http://www.austintexas.gov/sites/default/files/files/Planning/Demographics/population_history_pub.pdf.

- Exercises 67 and 68. The length of the tornado season (number of days between the first and last tornado) each year in the 19-county warning area of the National Weather Service Office in Goodland, Kansas from "Tornado Graphs." Accessed on July 1, 2014. http://www.weather.gov/gld/tornado-tornadographs.

- Exercises 69 and 70. U.S. retail prescription drug sales in billions of dollars per year from the U.S. Census Bureau. Accessed on July 8, 2014. https://www.census.gov/compendia/statab/2012/tables/12s0159.xls.

- Exercises 71 and 72. Number of red cards and expulsions given per men's World Cup tournament from "Planet World Cup — Statistics — Discipline." Accessed on July 16, 2014. http://www.planetworldcup.com/STATS/stat_disc.html.

- Exercises 73 and 74. Average debt load in 2013 dollars of bachelor's degree recipients attending U.S. public colleges and universities who borrowed money to finance their education from "Average Cumulative Debt Load of Bachelor's Degree Recipients at Public Four-Year Institutions over Time — Trends in Higher Education." Accessed on July 5, 2015. https://trends.collegeboard.org/student-aid/figures-tables/average-debt-levels-public-sector-bachelors-degree-recipients-over-time.

- Exercises 75 – 77. The number of prescription drugs sold in the United States in millions per year from the U.S. Census Bureau. Accessed on July 8, 2014. https://www.census.gov/compendia/statab/2012/tables/12s0159.xls.

- Exercises 78 – 80. The total number of burgers sold by McDonald's in billions as of each year from "Over How Many Billion Served." Accessed on July 3, 2014. http://overhowmanybillionserved.blogspot.com/.

- Exercises 81 – 83. The global gender ratio based on the number of males per 100 females by year from "World Population Prospects, the 2012 Revision" by the United Nations Department of Economic and Social Affairs. Accessed on June 25, 2014. http://esa.un.org/unpd/wpp/Excel-Data/population.htm.

- Exercises 84 – 86. World population growth rates from "International Programs — Total Midyear Population for the World 1950–2050." Accessed on June 16, 2014. http://www.census.gov/population/international/data/worldpop/table_population.php.

- Exercises 87 – 89. Number of tornadoes each year in the 19-county warning area of the National Weather Service Office in Goodland, Kansas from "Tornado Graphs." Accessed on July 1, 2014. http://www.weather.gov/gld/tornado-tornadographs.

- Exercises 90 – 92. Population of Ireland in millions of people each year from "IRELAND: population growth of the whole country." Accessed on July 18,2014. http://www.populstat.info/Europe/irelandc.htm.

- Exercises 93 – 95. Highest value of Facebook stock each month in 2013 from "FB Historical Prices | Facebook, Inc. Stock — Yahoo! Finance." Accessed on July 2, 2014. http://finance.yahoo.com/q/hp?s=FB.

- Exercises 96 – 98. Standard & Poor's 500 stock market closing value on June 1 of each year in U.S. dollars from "GSPC Historical Prices | S&P 500 Stock — Yahoo! Finance." Accessed on July 1, 2014. http://finance.yahoo.com/q/hp?s=%5EGSPC.

Section 3.5

- Examples 2 – 4. Millions of Twitter users per quarter from "Twitter: number of monthly active users 2010–2014 | Statista." Accessed on July 10, 2014. http://www.statista.com/statistics/282087/number-of-monthly-active-twitter-users/.

- Questions 1 – 3. Average debt load in thousands of 2012 dollars at the end of the spring term in each year for bachelor's degree recipients attending public four-year colleges and universities who borrowed money to finance their education. from "Average Debt Levels Public Sector Bachelor's Degree Recipients Over Time | Trends in Higher Eductation." Accessed on July 2, 2014. https://trends.collegeboard.org/student-aid/figures-tables/average-debt-levels-public-sector-bachelors-degree-recipients-over-time.

- Example 5. The U.S. health expenditure total as a percengate of GDP from the World Bank. Accessed on July 10, 2014. http://data.worldbank.org/country/united-states.

- Question 4. Personal data collection by one author's spouse. Used with permission.

- Exercises 15, 29 – 31, and 61. The length of the tornado season (number of days between the first and last tornado) each year in the 19-county warning area of the National Weather Service Office in Goodland, Kansas from "Tornado Graphs." Accessed on July 1, 2014. http://www.weather.gov/gld/tornado-tornadographs.

- Exercises 16, 32 – 34, and 62. U.S. retail prescription drug sales in billions of dollars per year from the U.S. Census Bureau. Accessed on July 8, 2014. https://www.census.gov/compendia/statab/2012/tables/12s0159.xls.

- Exercises 17, 35 – 37, and 63. The number of prescription drugs sold in the United States in millions per year from the U.S. Census Bureau. Accessed on July 8, 2014. https://www.census.gov/compendia/statab/2012/tables/12s0159.xls.

- Exercises 18, 38 – 40, and 64. The total number of burgers sold by McDonald's in billions as of each year from "Over How Many Billion Served." Accessed on July 3, 2014. http://overhowmany billionserved.blogspot.com/.

- Exercises 19, 41–43, and 65. Annual e-commerce sales in the United States in billions of dollars from "Monthly & Annual Retail Trade, Main Page — US Census Bureau." Accessed on July 14, 2014. http://www.census.gov/retail/.

- Exercises 20, 44–46, and 66. The global gender ratio based on the number of males per 100 females by year from "World Population Prospects, the 2012 Revision" by the United Nations Department of Economic and Social Affairs. Accessed on June 25, 2014. http://esa.un.org/unpd/wpp/Excel-Data/population.htm.

- Exercises 21 and 67. World population growth rates from "International Programs — Total Midyear Population for the World 1950–2050." Accessed on June 16, 2014. http://www.census.gov/population/international/data/worldpop/table_population.php.

- Exercises 22 and 68. Average number of goals scored per game in World Cup tournaments from "FIFA World Cup Record — Organisation." Accessed on July 17, 2014. http://www.fifa.com/worldfootball/statisticsandrecords/tournaments/worldcup/organisation/index.html.

- Exercises 69-72. U.S. monthly unemployment rate from January 2010 to December 2014 from "Bureau of Labor Statistics Data." Accessed on June 22, 2015. http://data.bls.gov/timeseries/LNS14000000.

- Exercises 73-76. Total midyear population for the world from "International Programs — Total Midyear Population for the World: 1950–2050 — U.S. Census Bureau." Accessed on June 16, 2014. http://www.census.gov/population/international/data/worldpop/table_population.php.

- Exercises 77-80. Interest rates on 15-year, fixed-rate conventional home mortgages annually from 1992 to 2014 from "Mortgage Interest Rates History." Accessed on June 22, 2015. http://www.fedprimerate.com/mortgage_rates.htm.

- Exercises 81-84. Number of Facebook users in millions from 2009 through 2012 from "Number of active users at Facebook over the years — Yahoo News" and "Facebook: number of active users 2015 | Statistics." Accessed on June 22, 2015. http://news.yahoo.com/number-active-users-facebook-over-230449748.html and http://www.statista.com/statistics/264810/number-of-monthly-active-facebook-users-worldwide/.

- Exercises 85-88. The high school dropout rate in the United States from 1970 through 2012 from "Percentage of high school dropouts among persons 16 to 24 years old." Accessed on June 22, 2015. http://nces.ed.gov/programs/digest/d13/tables/dt13_219.70.asp.

- Exercises 89-92. U.S. carbon dioxide emissions in kT annually from 1960 to 2010 according to the World Bank at "Data | United States." Accessed on July 10, 2014. http://data.worldbank.org/country/united-states.

- Exercises 93–96. Average interest rate for conventional 30-year mortgages each year from 1981 to 2012 from "Primary Mortgage Market Survey Archives — 30 Year Fixed Rate Mortgages — Freddie Mac." Accessed on July 7, 2015. http://www.freddiemac.com/pmms/pmms30.htm.

Chapter 4

Section 4.1

- Questions 1, 4, and 8. Median home prices in thousands of dollars from 2001 to 2010 according to the U.S. Census Bureau from "census.gov." Accessed on October 8, 2014. http://www.census.gov/const/uspriceann.pdf.

- A biography of Pierre de Fermat is at http://www-groups.dcs.st-and.ac.uk/~history/Biographies/Fermat.html. Accessed on July 8, 2016.

- Exercises 17–22. Annual total retail sales taxes collected in the United States in each year from "Monthly & Annual Retail Trade, Main Page — U.S. Census Bureau." Accessed on July 14, 2014. http://www.census.gov/retail/.

- Exercises 23–28. Plasma concentrations of Prozac based on "RxMed: Pharmaceutical Information — PROZAC." Accessed on August 20, 2014. http://www.rxmed.com/b.main/b2.pharmaceutical/b2.1.monographs/CPS-%20Monographs/CPS-%20(General%20Monographs-%20P)/PROZAC.html.

- Exercises 29–34. Number of Facebook users in millions from 2004 through 2009 from "Number of active users at Facebook over the years — Yahoo News" and "Facebook: number of active users 2015 | Statistics." Accessed on June 22, 2015. http://news.yahoo.com/number-active-users-facebook-over-230449748.html and http://www.statista.com/statistics/264810/number-of-monthly-active-facebook-users-worldwide/.

- Exercises 74 and 75. The U.S. health expenditure total as a percengage of GDP from the World Bank. Accessed on July 10, 2014. http://data.worldbank.org/country/united-states.

- Exercises 76 and 77. Olsen, I. E., S. A. Groveman, M. L. Lawson, R. H. Clark, and B. S. Zemel. "New Intrauterine Growth Curves Based on United States Data." Pediatrics 125.2 (2010): E214–224. Accessed on August 14, 2014. http://pediatrics.aappublications.org/content/125/2/e214.full.pdf.

- Exercises 78 and 79. Revolutions per minute of engines as a function of engine mass from pages 60–61. McMahon, Thomas A., and John Tyler Bonner. *On Size and Life*. New York: Scientific American Library, 1983.

Section 4.2

- Example 2. Average SAT math score in Kentucky each year from 1980 to 2013 from College Board's 2013 SAT State Profile Report for Kentucky. Accessed on July 16, 2014. http://media.collegeboard.com/digitalServices/pdf/research/2013/KY_13_03_03_01.pdf.

- Question 2. U.S. Energy Information Administration's data on United States coal prices in dollars per short ton in each year from "Coal — Data — U.S. Energy Information Administration (EIA)." Accessed on July 3, 2014. http://www.eia.gov/coal/data.cfm#prices.

- Example 6. Number of monthly active Twitter users worldwide from 1st quarter 2010 to 1st quarter 2016 (in millions) from "Twitter: number of monthly active users 2010–2016 | Statista." Accessed on June 9, 2016. http://www.statista.com/statistics/282087/number-of-monthly-active-twitter-users/.

- Question 7. Modified from Exercise 28 in Section 2.3. Burden, Richard L., J. Douglas Faires, and Annette M. Burden. *Numerical Analysis*. 10th ed. Boston, MA: Cengage, 2016.

- Exercise 5. Percent of high school graduates to enroll in a two-year or four-year college from National Center for Education Statistics. Accessed on July 2, 2014. http://nces.ed.gov/programs/digest/d13/tables/dt13_302.10.asp.

- Exercise 6. Percent of Americans with incomes below the poverty line in each year from "Income and Poverty in the United States: 2014," Current Population Reports, U.S. Department of Commerce, Economics and Statistics Administration, U.S. Census Bureau. Accessed on July 6, 2016. http://www.census.gov/content/dam/Census/library/publications/2015/demo/p60-252.pdf.

- Exercise 7. Standard & Poor's 500 stock market closing value on June 1 of each year in U.S. dollars from "GSPC Historical Prices | S&P 500 Stock — Yahoo! Finance." Accessed on July 1, 2014. http://finance.yahoo.com/q/hp?s=%5EGSPC.

- Exercise 8. Average maximum temperature in degrees Fahrenheit each month in New York City in 2013 requested from National Oceanic and Atmospheric Administration. Accessed on June 25, 2014. http://www.noaa.gov.

Section 4.3

- Example 3. U.S. monthly unemployment rate from January 2010 to December 2014 from "Bureau of Labor Statistics Data." Accessed on June 22, 2015. http://data.bls.gov/timeseries/LNS14000000.

- Question 4. Global gender ratio based on the number of males per 100 females by year from "World Population Prospects, the 2012 Revision" by the United Nations Department of Economic and Social Affairs. Accessed on June 25, 2014. http://esa.un.org/unpd/wpp/Excel-Data/population.htm.

- Example 6. Hudson, L. N., Isaac, N. J. B., Reuman, D. C. (2013), The relationship between body mass and field metabolic rate among individual birds and mammals. Journal of Ecology, 82: 1009–1020. doi: 10.1111/1365-2656.12086. Accessed on August 19, 2015. http://onlinelibrary.wiley.com/doi/10.1111/1365-2656.12086/suppinfo.

- Question 7. Running speed and length of animals from page 152. McMahon, Thomas A., and John Tyler Bonner. *On Size and Life*. New York: Scientific American Library, 1983.

- Example 9 and Question 13. Number of physicians per 1000 people as a function of average life expectancy in different countries in 2010 from "Physicians (per 1,000 people) | Data | Table." Accessed on July 10, 2014. http://data.worldbank.org/indicator/SH.MED.PHYS.ZS.

- Question 9. Closing NASDAQ stock market value in U.S. dollars at the end of each quarter from March 1938 through December 2014 from "^NDQ — Nasdaq Composite — U.S. — Stooq." Accessed on July 7, 2015. http://stooq.com/q/d/?s=^ndq.

- Example 12. Median home prices in thousands of dollars from 2001 to 2010 according to the U.S. Census Bureau from "census.gov." Accessed on October 8, 2014. http://www.census.gov/const/uspriceann.pdf.

- Exercises 53–56. Total U.S. health expenditures as a percentage of GDP from the World Bank. Accessed on July 10, 2014. http://data.worldbank.org/country/united-states.

- Exercises 57–60. Revolutions per minute of engines as a function of engine mass from pages 60-61. McMahon, Thomas A., and John Tyler Bonner. *On Size and Life*. New York: Scientific American Library, 1983.
- Exercises 61–64. Olsen, I. E., S. A. Groveman, M. L. Lawson, R. H. Clark, and B. S. Zemel. "New Intrauterine Growth Curves Based on United States Data." Pediatrics 125.2 (2010): E214–224. Accessed on August 14, 2014. http://pediatrics.aappublications.org/content/125/2/e214.full.pdf.
- Exercise 73. Personal data collection by one author's spouse. Used with permission.
- Exercise 74. Flying speed as length of animals from page 153. McMahon, Thomas A., and John Tyler Bonner. *On Size and Life*. New York: Scientific American Library, 1983.
- Exercise 75. Atmospheric carbon dioxide from Mauna Loa in ppmv (parts per million by volume) as a function of years from 1958 to 2008 from "Atmospheric Carbon Dioxide Record from Mauna Loa." Accessed on August 19, 2015. http://cdiac.ornl.gov/trends/co2/sio-mlo.html.
- Exercise 76. Estimated world population data from 1950 to 2015 from "International Programs — Total Mid-Year Population for the World: 1950–2050 — U.S. Census Bureau." Accessed on June 16, 2014. http://www.census.gov/population/international/data/worldpop/table_population.php.

Section 4.4

- Example 4. Modified from Exercise 28 in Section 2.3. Burden, Richard L., J. Douglas Faires, and Annette M. Burden. *Numerical Analysis*. 10th ed. Boston, MA: Cengage, 2016.
- Example 7. Population of the Netherlands from "The NETHERLANDS : country populations." Accessed on July 11, 2014. http://www.populstat.info/Europe/netherlc.htm.
- Exercises 45 and 46. For more information about resonance, visit "Resonance — Wikipedia." Accessed on December 29, 2017. https://en.wikipedia.org/wiki/Resonance.
- Exercises 47 and 48. For more information about acoustic beats, visit "Beats (acoustic) — Wikipedia." Accessed on December 29, 2017. https://en.wikipedia.org/wiki/Beat_(acoustics).
- Exercise 49 and 50. Number of monthly active Twitter users worldwide from 1st quarter 2010 to 1st quarter 2016 (in millions) from "Twitter : number of monthly active users 2010–2016 | Statista." Accessed on June 9, 2016. http://www.statista.com/statistics/282087/number-of-monthly-active-twitter-users/.
- Exercise 51 and 52. Population of Belgium in millions of people by year from "Population of the Netherlands, Belgium, and Luxembourg." Accessed on June 26, 2014. http://www.tacitus.nu/historical-atlas/population/benelux.htm.

Section 4.5

- Examples 5 and 6. Toyota Motors Corporation (TM) stock market value in U.S. dollars from "Yahoo Finance | TM Historical Prices." Accessed on August 13, 2016. http://finance.yahoo.com/quote/TM/history?p=TM.
- Questions 6 and 7. Sinusoidal model for tidal measurements in Pearl Harbor, Hawaii based on data sets from Project Mosaic. Accessed on August 19, 2015. http://www.mosaic-web.org.
- Exercises 61 and 62. For more about probability distributions, consult an introductory statistics textbook.
- Exercise 63. Number of monthly active Twitter users worldwide from 1st quarter 2010 to 1st quarter 2016 (in millions) from "Twitter : number of monthly active users 2010–2016 | Statista." Accessed on June 9, 2016. http://www.statista.com/statistics/282087/number-of-monthly-active-twitter-users/.
- Exercise 64. Population of Belgium in millions of people by year from "Population of the Netherlands, Belgium, and Luxembourg." Accessed on June 26, 2014. http://www.tacitus.nu/historical-atlas/population/benelux.htm.
- Exercise 65. Number of minutes after 3 p.m. until sunset in Greenwich Borough, England since January 2010 from "Sunrise and sunset times in Greenwich Borough." Accessed on July 14, 2014. http://www.timeanddate.com/sun/uk/greenwich-city.
- Exercise 66. Data set providing the electric bill of a single-family home in Minnesota for each month from 2000 through 2003 from Project Mosaic. Accessed on August 19, 2015. http://www.mosaic-web.org.
- Exercises 67–70. Average maximum temperature in Danville, Kentucky at the beginning of each month since January 2006 from "noaa.gov." Accessed on June 25, 2014. http://www1.ncdc.noaa.gov/pub/orders/cdo/352625.pdf.

Section 4.6

- Example 2 and Question 2. Figure 4 from NWS Winter Storm Safety Windchill Information and Chart. Accessed on January 30, 2015. http://www.nws.noaa.gov/om/winter/windchill.shtml.
- Question 8. Modified from: Taylor, G. "The Formation of a Blast Wave by a Very Intense Explosion. II. The Atomic Explosion of 1945." *Proceedings of the Royal Society A: Mathematical, Physical and Engineering Sciences*, 1950, 201: 175-86.
- A biography of Alexis Clairaut is at http://www-history.mcs.st-and.ac.uk/Biographies/Clairaut.html. Accessed on August 1, 2016.

Section 4.7

- Example 9 and Question 9. Median home prices in thousands of dollars from 2001 to 2010 according to the U.S. Census Bureau from "census.gov." Accessed on October 8, 2014. http://www.census.gov/const/uspriceann.pdf.

Chapter 5

Section 5.1

- For more information about Mount Everest, see https://en.wikipedia.org/wiki/Mount_Everest. Accessed on December 2, 2017.
- For more information about Mount Mitchell, see https://en.wikipedia.org/wiki/Mount_Mitchell. Accessed on December 2, 2017.
- Exercise 2. Tidal measurements in Pearl Harbor, Hawaii based on data sets from Project Mosaic. Accessed on August 19, 2015. http://www.mosaic-web.org.
- Question 2. World population from the U.S. Census Bureau. Accessed on June 3, 2016. http://www.census.gov/population/international/data/worldpop/table_population.php.
- Examples 5 and 12. Median home prices in thousands of dollars from 2001 to 2010 according to the U.S. Census Bureau from "census.gov." Accessed on October 8, 2014. http://www.census.gov/const/uspriceann.pdf.
- Exercise 79. 2014 Ebola Outbreak in West Africa from Centers for Disease Control and Prevention. Accessed on June 2, 2016. http://www.cdc.gov/vhf/ebola/outbreaks/2014-west-africa/cumulative-cases-graphs.html.
- Exercise 80. Toyota Motors Corporation (TM) stock market value in U.S. dollars from "Yahoo Finance | TM Historical Prices." Accessed on August 13, 2016. http://finance.yahoo.com/quote/TM/history?p=TM.
- Exercise 81. Tidal measurements in Pearl Harbor, Hawaii based on data sets from Project Mosaic. Accessed on August 19, 2015. http://www.mosaic-web.org.
- Exercise 82. U.S. carbon dioxide emissions in kT annually from 1960 to 2010 according to the World Bank at "Data | United States." Accessed on July 10, 2014. http://data.worldbank.org/country/united-states.

Section 5.2

- Example 2 from "Dow Jones Industrial — U.S. — Stooq." Accessed on July 2, 2014. http://stooq.com/q/d/?s=%5Edji.
- Example 6. Modified from Exercise 28 in Section 2.3. Burden, Richard L., J. Douglas Faires, and Annette M. Burden. *Numerical Analysis*. 10th ed. Boston, MA: Cengage, 2016.
- Exercise 73. Toyota Motors Corporation (TM) stock market value in U.S. dollars from "Yahoo Finance | TM Historical Prices." Accessed on August 13, 2016. http://finance.yahoo.com/quote/TM/history?p=TM.
- Exercise 74. Tidal measurements in Pearl Harbor, Hawaii based on data sets from Project Mosaic. Accessed on August 19, 2015. http://www.mosaic-web.org.
- Exercise 75. City of Austin Population History. Accessed on July 17, 2015. http://www.austintexas.gov/sites/default/files/files/Planning/Demographics/population_history_pub.pdf.
- Exercise 76. For more about probability distributions, consult an introductory statistics textbook.

Section 5.3

- Example 2. Population of the Netherlands from "The NETHERLANDS : country populations." Accessed on July 11, 2014. http://www.populstat.info/Europe/netherlc.htm.
- Example 2. Toyota Motors Corporation (TM) stock market value in U.S. dollars from "Yahoo Finance | TM Historical Prices." Accessed on August 13, 2016. http://finance.yahoo.com/quote/TM/history?p=TM.

- Example 10. Modified from Exercise 28 in Section 2.3. Burden, Richard L., J. Douglas Faires, and Annette M. Burden. *Numerical Analysis*. 10th ed. Boston, MA: Cengage, 2016.

- Example 11. Median home prices in thousands of dollars from 2001 to 2010 according to the U.S. Census Bureau from "census.gov." Accessed on October 8, 2014. http://www.census.gov/const/uspriceann.pdf.

- Exercise 11. Tidal measurements in Pearl Harbor, Hawaii based on data sets from Project Mosaic. Accessed on August 19, 2015. http://www.mosaic-web.org.

- Exercises 12 and 13. Closing stock market value of the Dow Jones Industrial Average at the end of each quarter from March 31, 1930 through December 31, 2014 from "^DJI — Nasdaq Composite — U.S. — Stooq." Accessed on July 7, 2015. http://stooq.com/q/d/?s=^dji.

- Exercise 14. Estimated world population data from 1950 to 2015 from "International Programs — Total Mid-Year Population for the World: 1950–2050 — U.S. Census Bureau." Accessed on June 16, 2014. http://www.census.gov/population/international/data/worldpop/table_population.php.

- Exercise 15. 2014 Ebola Outbreak in West Africa from Centers for Disease Control and Prevention. Accessed on June 2, 2016. http://www.cdc.gov/vhf/ebola/outbreaks/2014-west-africa/cumulative-cases-graphs.html.

- Exercise 16. U.S. carbon dioxide emissions in kT annually from 1960 to 2010 according to the World Bank at "Data | United States." Accessed on July 10, 2014. http://data.worldbank.org/country/united-states.

Section 5.4

- Example 5. Median home prices in thousands of dollars from 2001 to 2010 according to the U.S. Census Bureau from "census.gov." Accessed on October 8, 2014. http://www.census.gov/const/uspriceann.pdf.

Section 5.5

- Example 2. The contour map is modified from Volcano Models at http://volcano.oregonstate.edu/book/export/html/208. Accessed on June 18, 2015.

- Exercise 65. Modified from Exercise 18 in Section 8.5 of Hughes-Hallett, Gleason, Lock, Flath, et.al. *Applied Calculus*. 5th ed. Wiley, 2014.

- Exercise 66. Modified from Example 2 in Section 13.1 of Sydsæter, K. and Hammond, P. *Essential Mathematics for Economic Analysis*. 3rd ed. Prentice Hall, 2008.

- Exercises 67 and 68. Modified from Example 4 in Section 12.8 of Briggs, Cochran, and Gillett. *Calculus: Early Transcendentals*. 2nd ed. Pearson, 2015.

Section 5.6

- Example 7. For more details about the Cobb–Douglas function visit "Cobb–Douglas function — Oxford Reference" at http://www.oxfordreference.com/search?q=Cobb%E2%80%93Douglas%20function or "Cobb–Douglas production function — Wikipedia" at https://en.wikipedia.org/wiki/Cobb%E2%80%93Douglas_production_function. Accessed on December 29, 2017.

Chapter 6

Section 6.1

- Question 1. Olsen, I. E., S. A. Groveman, M. L. Lawson, R. H. Clark, and B. S. Zemel. "New Intrauterine Growth Curves Based on United States Data." Pediatrics 125.2 (2010): E214–224. Accessed on August 14, 2014. http://pediatrics.aappublications.org/content/125/2/e214.full.pdf.

- Example 3. Percent of high school graduates to enroll in a two-year or four-year college from National Center for Education Statistics. Accessed on July 2, 2014. http://nces.ed.gov/programs/digest/d13/tables/dt13_302.10.asp.

- Question 2. World population growth rates from "International Programs — Total Midyear Population for the World 1950–2050." Accessed on June 16, 2014. http://www.census.gov/population/international/data/worldpop/table_population.php.

- A biography of Bernhard Riemann is at http://www-history.mcs.st-and.ac.uk/Biographies/Riemann.html. Accessed on December 29, 2017.

- Exercise 69. Standard & Poor's 500 stock market closing value on June 1 of each year in U.S. dollars from "GSPC Historical Prices | S&P 500 Stock — Yahoo! Finance." Accessed on July 1, 2014. http://finance.yahoo.com/q/hp?s=%5EGSPC.

- Exercise 70. Annual unemployment rate in the United States from "Bureau of Labor Statistics Data." Accessed on July 15, 2014. http://data.bls.gov/timeseries/LNS14000000.

- Exercise 71. Total annual sales in thousands of hybrid vehicles in the United States from "Electric Drive Transportation Association." Accessed on June 26, 2014. http://electricdrive.org/index.php?ht=d/sp/i/20952/pid/20952.
- Exercise 72. U.S. field production of crude oil in billions of barrels from the U.S. Energy Information Administration from "U.S. Field Production of Crude Oil." Accessed on July 3, 2014. http://www.eia.gov/dnav/pet/hist/LeafHandler.ashx?n=PET&s=MCRFPUS1&f=M.
- Exercise 73. Three-year average monthly pollen count in Brooklyn, New York City from "Historic Allergy Index for 11203 | Pollen.com." Accessed on June 25, 2014. http://www.pollen.com/allergy-trends.asp?PostalCode=11203.
- Exercise 74. Average maximum temperature in degrees Fahrenheit each month in New York City in 2013 requested from National Oceanic and Atmospheric Administration. Accessed on June 25, 2014. http://www.noaa.gov.

Section 6.2

- An overview of the historical development of calculus is at http://www-history.mcs.st-andrews.ac.uk/HistTopics/The_rise_of_calculus.html. Accessed on December 29, 2017.

Section 6.4

- A biography of Sir Isaac Newton is at http://www-history.mcs.st-and.ac.uk/Biographies/Newton.html. Accessed on December 29, 2017.
- A biography of Gottfried Wilhelm von Leibniz is at http://www-history.mcs.st-andrews.ac.uk/Biographies/Leibniz.html. Accessed on December 29, 2017.
- Example 3, Question 3, and Exercises 45 – 48. Data used to determine models of average fetal growth rates Olsen, I. E., S. A. Groveman, M. L. Lawson, R. H. Clark, and B. S. Zemel. "New Intrauterine Growth Curves Based on United States Data." Pediatrics 125.2 (2010): E214–224. Accessed on August 14, 2014. http://pediatrics.aappublications.org/content/125/2/e214.full.pdf.

Section 6.6

- Example 5. Modified from Exercise 28 in Section 2.3. Burden, Richard L., J. Douglas Faires, and Annette M. Burden. *Numerical Analysis*. 10th ed. Boston, MA: Cengage, 2016.

Index

Printed in the USA/Agawam, MA
January 19, 2024

859702.131